TRACE ELEMENTS AS CONTAMINANTS AND NUTRIENTS

TRACE ELEMENTS AS CONTAMINANTS AND NUTRIENTS

Consequences in Ecosystems and Human Health

Edited by

M. N. V. Prasad

A JOHN WILEY & SONS, INC., PUBLICATION

Published by John Wiley & Sons, Inc., Hoboken, New Jersey
Published simultaneously in Canada

For general information on our other products and services or for technical support, please contact our Customer Care Department within the United States at (800) 762-2974, outside the United States at (317) 572-3993 or fax (317) 572-4002.

Wiley also publishes its books in variety of electronic formats. Some content that appears in print may not be available in electronic formats. For more information about Wiley products, visit our web site at www.wiley.com.

Library of Congress Cataloging-in-Publication Data:

Prasad, M. N. V. (Majeti Narasimha Vara), 1953–
  Trace Elements as Contaminants and Nutrients: Consequences in Ecosystems and Human Health / M.N.V. Prasad.
    p. cm.
  Includes index.
  ISBN 978-0-470-18095-2 (cloth)
1. Trace elements–Environmental aspects. I. Title.
  QH545.T7P73 2008
  613.2′85–dc22

                                                                        2007050456

Printed in the United States of America

10 9 8 7 6 5 4 3 2 1

CONTENTS

Foreword   **xix**

Preface   **xxiii**

Acknowledgments   **xxv**

Contributors   **xxvii**

**1 The Biological System of Elements: Trace Element Concentration and
   Abundance in Plants Give Hints on Biochemical Reasons of
   Sequestration and Essentiality**   **1**
   Stefan Fränzle, Bernd Markert, Otto Fränzle and Helmut Lieth

  1. Introduction   1
     1.1 Analytical Data and Biochemical Functions   1
  2. Materials and Methods   6
     2.1 Data Sets of Element Distribution Obtained in Freeland
         Ecological Studies: Environmental Analyses   6
     2.2 Conversion of Data Using Sets of Elements with Identical
         BCF Values   8
     2.3 Definition and Derivation of the Electrochemical Ligand
         Parameters   10
  3. Results   11
     3.1 Abundance Correlations Among Essential and
         Nonessential Elements   11
     3.2 (Lack of) Correlation and Differences in Biochemistry   14
     3.3 Implication for Biomonitoring: Corrections by Use of
         Electrochemical Ligand Parameters and BCF-Defined
         Element Clusters   14
  4. Discussion   15
  5. Conclusion   18
  References   19

**2 Health Implications of Trace Elements in the Environment
   and the Food Chain**   **23**
   Nelson Marmiroli and Elena Maestri

  1. Trace Elements Important in Human Nutrition   24
  2. The Main Trace Elements: Their Roles and Effects   25

2.1   Arsenic   25
2.2   Cadmium   29
2.3   Chromium   30
2.4   Cobalt   30
2.5   Copper   30
2.6   Fluorine   30
2.7   Iodine   31
2.8   Iron   31
2.9   Lead   31
2.10 Manganese   32
2.11 Mercury   32
2.12 Molybdenum   32
2.13 Nickel   32
2.14 Selenium   33
2.15 Silicon   33
2.16 Tin   33
2.17 Vanadium   34
2.18 Zinc   34
2.19 Hypersensitivity Issues   34

3.   Issues of Environmental Contamination of the Food Chain   37
4.   Legislation Concerning Trace Elements   38
     4.1   Elements in Soils and the Environment   38
     4.2   Elements in Foods   39
     4.3   Supplementation of Minerals to Foods   41
5.   Food Chain Safety   42
     5.1   Soil and Plants   42
     5.2   Animal Products   43
     5.3   Geological Correlates   44
     5.4   Intentional Contamination   45
     5.5   Availability of Minerals   46
6.   Biofortification   47
7.   Concluding Remarks   48
     Acknowledgments   49
     References   49

3 Trace Elements in Agro-ecosystems   55
Shuhe Wei and Qixing Zhou

1.   Introduction   55
2.   Biogeochemistry of Trace Elements in Agro-ecosystems   56
     2.1   Input and Contamination   56
     2.2   Translation, Translocation, Fate, and Their
           Implication to Phytoremediation   60
3.   Benefit, Harmfulness, and Healthy Implication
     of Trace Elements   65
     3.1   Benefit to Plant/Crop   65
     3.2   Harmfulness to Plant/Crop Physiology   65

3.3   Soil Environmental Quality Standards and Background
of Trace Elements   66
4.   Phytoremediation of Trace Element Contamination   68
4.1   Basic Mechanisms of Phytoremediation   68
4.2   Research Progress of Phytoextraction   72
4.3   Discussion on Agro-Strengthen Measurements   73
Acknowledgments   76
References   76

4 Metal Accumulation in Crops—Human Health Issues   81
Abdul R. Memon, Yasemin Yildizhan and Eda Kaplan

1.   Introduction   81
2.   The Concept of Ionomics and Nutriomics in the Plant Cell   83
3.   The Trace Element Deficiencies in the Developing World   84
4.   Improvement of Trace Metal Content in Plants Through
Genetic Engineering   85
5.   Genetic Engineering Approaches to Improve the Bioavailability
of Iron and Zinc in Cereals   88
6.   Decreasing the Content of Inhibitors of Trace Element Absorption   91
7.   Increasing the Synthesis of Promoter Compounds   92
8.   Conclusions   93
Acknowledgments   93
References   93

**5 Trace Elements and Plant Secondary Metabolism: Quality
and Efficacy of Herbal Products   99**
*Charlotte Poschenrieder, Josep Allué, Roser Tolrà,
Mercè Llugany and Juan Barceló*

1.   Coevolutionary Aspects   99
2.   Environmental Factors and Active Principles   102
3.   Influence of Macronutrients   102
4.   Influence of Micronutrients   104
5.   Trace Elements as Elicitors of Active Principles   106
6.   Trace Elements as Active Components of Herbal Drugs   107
7.   Trace Elements in Herbal Drugs: Regulatory Aspects   111
Acknowledgments   112
References   112

6 Trace Elements and Radionuclides in Edible Plants   121
Maria Greger

1.   Introduction   121
2.   Plant Uptake and Translocation of Trace Elements   122
3.   Distribution and Accumulation of Trace Elements in Plants   124

	4.	Vegetables, Fruit, and Berries	125
	5.	Cereals and Grains	128
		5.1 Cadmium in Wheat	128
		5.2 Arsenic in Rice	129
	6.	Aquatic Plants	129
	7.	Fungi	130
	8.	How to Cope with Low or High Levels of Trace Elements	131
		References	132

7 Trace Elements in Traditional Healing Plants—Remedies or Risks   **137**
M. N. V. Prasad

	1.	Introduction	137
	2.	The Indigenous System of Medicine	138
	3.	Herbal Drug Industry	139
	4.	Notable Medicinal and Aromatic Plants that have the Inherent Ability of Accumulating Toxic Trace Elements	141
	5.	Cleanup of Toxic Metals from Herbal Extracts	149
	6.	Polyherbal Preparation and Traditional Medicine Pharmacology	150
	7.	Conclusions	152
		References	155

8 Biofortification: Nutritional Security and Relevance to Human Health   **161**
M. N. V. Prasad

	1.	Introduction	161
	2.	Bioavailablity of Micronutrients	168
	3.	Social Acceptability of Biofortified Crops	169
	4.	Development and Distribution of the New Varieties	169
	5.	Selected Examples of Biofortified Crops Targeted by Harvestplus in Collaboration with a Consortium of International Partners	169
		5.1 Rice	170
		5.2 Wheat	171
		5.3 Maize	172
		5.4 Beans	173
		5.5 *Brassica juncea* (Indian Mustard)	174
	6.	Selenium-Fortified Phytoproducts	175
	7.	Sources of Selenium in Human Diet	175
	8.	Selenium (Se) and Silica (Si) Management in Soils by Fly Ash Amendment	175
	9.	Chromium for Fortification Diabetes Management	176
	10.	Silica Management in Rice—Beneficial Functions	177
	11.	Conclusions	178
		Acknowledgments and Disclaimer	179
		References	179

9 **Essentiality of Zinc for Human Health and Sustainable Development**   **183**
   M. N. V. Prasad

   1. Biogeochemical Cycling of Zinc   185
   2. Distribution of Zinc Deficiency in Soils on a Global Level   186
   3. Zinc Intervention Programs   188
   4. Zinc-Transporting Genes in Plants   191
   5. Addressing Zinc Deficiency Without Zinc Fortification   204
   6. Zinc Deficiency is a Limitation to Plant Productivity   204
   Acknowledgments and Disclaimer   205
   References   205

10 **Zinc Effect on the Phytoestrogen Content of Pomegranate Fruit Tree**   **217**
   Fatemeh Alaei Yazdi and Farhad Khorsandi

   1. Introduction   217
   2. Materials and Methods   220
   3. Results and Discussions   222
       3.1 Pomegranate Yield   222
       3.2 Pomegranate Zinc Content   223
       3.3 Phytoestrogen Content   225
   4. Summary and Conclusions   227
   Acknowledgments   227
   References   228

11 **Iron Bioavailability, Homeostasis through Phytoferritins and
   Fortification Strategies: Implications for Human
   Health and Nutrition**   **233**
   N. Nirupa and M. N. V. Prasad

   1. Introduction   233
   2. Iron Importance   234
   3. Iron Toxicity   235
   4. Interactions with Other Metals   235
   5. Iron Acquisition by Plants   238
   6. Translocation of Iron in Plants   238
   7. Iron Deficiency in Humans   239
   8. Amelioration of Iron Deficiencies   241
   9. Ferritin   242
   10. Ferritin Structure   243
   11. Mineral Core Formation   247
   12. Ferritin Gene Family and Regulation   248
   13. Developmental Regulation   249
   14. Role of Ferritin   251
   15. Metal Sequestration by Ferritin: Health Implications   254

16. Overexpression of Ferritin   254
Acknowledgments   257
References   257

**12 Iodine and Human Health: Bhutan's Iodine
      Fortification Program   267**
Karma Lhendup

1. Role of Iodine   267
2. Iodine Deficiency Disorders (IDD)   268
3. Sources of Iodine   269
4. Recommended Intake of Iodine   270
5. Indicators for Assessment of Iodine Status and Exposure   270
6. Control of IDD   271
7. IDD Scenario in Bhutan: Past and Present   272
8. Toward IDD Elimination in Bhutan: Highlights of the IDD
    Control Program   273
    8.1 IDD Survey   273
9. 1996 Onward: Internal Evaluation of the IDDCP through
    Cyclic Monitoring   277
10. Conclusion   278
References   278

**13 Floristic Composition at Kazakhstan's Semipalatinsk Nuclear
      Test Site: Relevance to the Containment of Radionuclides to
      Safeguard Ecosystems and Human Health   281**
K. S. Sagyndyk, S. S. Aidossova and M. N. V. Prasad

1. Introduction   281
2. Kazakhstan: Semipalatinsk Nuclear Test Site   283
3. Flora of Nuclear Test Site   286
4. Fodder Plants   292
5. Conclusions   293
Acknowledgments and Disclaimer   293
References   293

**14 Uranium and Thorium Accumulation in
      Cultivated Plants   295**
Irina Shtangeeva

1. Introduction: Uranium and Thorium in the Environment   295
2. Uranium and Thorium in Soil   296
    2.1 Soil Characteristics Affecting Uranium and Thorium
         Plant Uptake   297
    2.2 Effects of Soil Amendments   300
3. Radionuclides in Plants   301

3.1 Accumulation of Uranium and Thorium
      in Plant Roots   302
3.2 Differences in U and Th Uptake by Different Plant Species (in the
      example of wheat *Triticum aestivum* and Rye *Secale cereale*)   303
3.3 Effects of U and Th Bioaccumulation on
      Distribution of Other Elements in Rye and Wheat   311
3.4 Relationships Between U and Th in Soils and in
      Different Plant Parts   312
3.5 Phytotoxicity of U and Th   314
3.6 Effects of U and Th on Leaf Chlorophyll Content
      and the Rhizosphere Microorganisms   321
3.7 Temporal Variations of U and Th in Plants   325
3.8 Effects of Thorium on a Plant During Initial Stages
      of the Plant Growth   328
4.   Potential Health Effects of Exposure to U and Th   333
References   336

**15 Exposure to Mercury: A Critical Assessment of Adverse Ecological
and Human Health Effects   343**
Sergi Díez, Carlos Barata and Demetrio Raldúa

1.   Human Health Effects   343
1.1 Introduction   343
1.2 Sources and Cycling of Mercury to the Global Environment   344
1.3 Methylmercury   346
2.   Adverse Ecological Effects   349
2.1 Laboratory Toxicity Studies   349
2.2 Biochemical Approaches to Study Bioavailability and Effects   351
2.3 Methods   353
2.4 Results and Discussion   354
3.   Case Study: Mercury-Cell Chlor-Alkali Plants as a Major Point
      Sources of Mercury in Aquatic Environments—The Case of
      Cinca River, Spain   357
3.1 Introduction   357
3.2 The Case of Mercury Pollution in Cinca River, Spain   358
References   364

**16 Cadmium as an Environmental Contaminant: Consequences to
Plant and Human Health   373**
Saritha V. Kuriakose and M. N. V. Prasad

1.   Introduction   373
2.   Cadmium is Natural   374
3.   Past and Present Status   375
3.1 Natural Sources   376
3.2 Technogenic Sources   376
3.3 In Agricultural Soils: Cadmium from
      Phosphate Fertilizers   378

3.4	Induction of Oxidative Stress as a Fall-Out of Cadmium Toxicity	378
3.5	Oxidative Damage to Membranes	378
3.6	Oxidative Damage to Chloroplasts	379
3.7	Protein Oxidation	379
3.8	Oxidative Damage to DNA	380
3.9	Antioxidant Defense Mechanisms in Response to Cadmium Toxicity	382
3.10	Cadmium Availability and Toxicity in Plants	384
3.11	Metal–Metal Interactions	387
3.12	Uptake and Transport of Cadmium by Plants	388
3.13	Consequences to Human Health	389
3.14	Options for Cadmium Minimization	392
3.15	Molecular and Biochemical Approaches	392
3.16	Breeding Strategies	394
3.17	Soil Cadmium Regulation	394
4.	Conclusions	396
	References	397

17 Trace Element Transport in Plants **413**
Danuta Maria Antosiewicz, Agnieszka Sirko and Paweł Sowiński

1.	Introduction	413
2.	Short-Distance Transport	416
	2.1 Metal Uptake Proteins	416
	2.2 Metal Efflux Proteins	423
	2.3 Alternative Plant Metal Transporter	433
3.	Intercellular and Long-Distance Transport	433
4.	The Importance of Plant Mineral Status for Human Health	438
	Acknowledgments	438
	References	439

18 Cadmium Detoxification in Plants: Involvement of ABC Transporters **449**
Sonia Plaza and Lucien Bovet

1.	Cadmium in Plants	449
	1.1 Cadmium Effects in Plants	449
	1.2 Genes Regulated by Cd Stress	450
2.	ABC Transporters	451
	2.1 Functions of ABC Transporters in Plants	451
	2.2 Characteristics of ATP-Binding Cassette Transporters	451
	2.3 Subfamilies of ATP-Binding Cassette Proteins	452
	2.4 Involvement of ABC Transporters in Cadmium Detoxification in Plants	452
3.	Conclusion	462
	Acknowledgments	463
	References	463

19 Iron: A Major Disease Modifier in Thalassemia   **471**
Sujata Sinha

1. Introduction   471
    1.1 Hemoglobin: The Tetramer Molecule   472
    1.2 Erythropoiesis and Erythroid Differentiation   472
    1.3 Pathophysiology of Thalassemia   474
2. Iron Metabolism: Current Concepts and
    Alterations in Thalassemia   474
    2.1 Iron Absorption and Uptake   476
    2.2 Regulation of Expression of Transferrin Receptors   477
    2.3 Alterations in Iron Absorption and Uptake in Thalassemia   479
3. Heme Synthesis and Its Role in Regulation of Erythropoiesis   480
    3.1 Role of Heme in Globin Regulation and
         Erythroid Differentiation   481
    3.2 Pivotal Role of HRI in Microcytic Hypochromic Anemia   481
    3.3 Role of HRI in Beta Thalassemia Intermedia   482
    3.4 Iron and Pathobiology of Thalassemia   482
    3.5 Iron Storage and Its Effects on Parenchymal Tissues
         and Organs   483
4. Effect of Transfusional Iron Overload on Iron Homeostasis and
    Morbidity and Mortality   484
    4.1 Iron Homeostasis in Transfusional Iron Overload   484
    4.2 Transfusion Iron Overload-Associated Morbidity
         and Mortality   485
    4.3 Endocrinopathy in Thalassemia   485
    4.4 Liver Disease   485
    4.5 Heart Disease   486
5. Evaluation and Management of Iron Overload   486
    5.1 Evaluation of Iron Overload   486
    5.2 Basis of Iron Chelation Therapy and Iron Chelator Drugs   487
    5.3 Potential Role of Iron Chelation Therapy in Improving
         Basic Pathophysiology of Beta Thalassemia   488
6. Summary   488
References   489

20 Health Implications: Trace Elements in Cancer   **495**
Rafael Borrás Aviñó, José Rafael López-Moya and Juan Pedro Navarro-Aviño

1. Introduction   495
    1.1 General Nutritional and Medical Benefits   496
2. Toxic Heavy Metals   496
    2.1 Mercury   497
    2.2 Arsenic   500
    2.3 Chromium   508
    2.4 Cadmium   511
    2.5 Lead   515
    2.6 Benefits in Cancer   517

3. General Conclusions   519
References   519

21 Mode of Action and Toxicity of Trace Elements   523
Arun K. Shanker

1. Introduction   523
2. Mode of Action and Toxicity of Trace Elements in General   525
3. Specific Mode of Action of Major Trace Elements   528
    3.1 Arsenic   528
    3.2 Cadmium   532
    3.3 Chromium   537
4. Specific Mode of Action of Other Metals   542
    4.1 Nickel   542
    4.2 Lead   544
    4.3 Mercury   545
5. Mode of Action: What is the Future?   549
References   550

**22 Input and Transfer of Trace Metals from Food via
Mothermilk to the Child: Bioindicative Aspects to Human Health   555**
*Simone Wuenschmann, Stefan Fränzle, Bernd Markert and
Harald Zechmeister*

1. Introduction   555
2. Aims and Scopes   556
3. Principles   558
    3.1 Transfer of Chemical Elements   558
    3.2 Physiology of Lactation   559
    3.3 Transfer of Chemical Elements into Human Milk   560
4. Materials and Methods   561
    4.1 A Comparison of the Two Experimental Regions Euroregion
         Neisse and Woivodship Małopolska with Respect to
         Factors that Cause Environmental Burdens   561
    4.2 Origins and Sampling of Food and Milk Samples   564
    4.3 Analytical Methods   567
    4.4 Quality Control Measures for Analytic Data   569
    4.5 Calculation of Transfer Factors in the System
         Food/Mother's Milk   570
5. Results   570
    5.1 A Comparison of Element Concentrations Detected
         in Colostrum and Mature Milk Sampled
         in Different Countries   570
    5.2 Transfer Factors for All the Investigated Elements
         (Specific Ones) in the Food/Milk System and Extent
         of Partition of Elements into Mother's Milk   574

6. Discussion 577
   6.1 Physiological and Dynamic Features of Chemical
      Elements in the Food/Milk System 577
   6.2 Lack of an Effect of Regional Pollution on Chemical
      Element Composition in Mother's Milk 582
7. Conclusion: Is There a Role for Human Milk in Metal Bioindication? 584
References 588

23 Selenium: A Versatile Trace Element in Life and Environment 593
Simona Di Gregorio

1. What is Selenium? 593
   1.1 Selenium Industrial Applications 593
   1.2 Selenium in the Environment 594
2. Biological Reactions in Selenium Cycling 596
   2.1 Microbial Assimilatory Reduction 597
   2.2 Microbial Dissimilatory Reduction 597
   2.3 Detoxification of Se Oxyanions by Reduction Reactions
      in Aerobiosis 599
   2.4 Regulation of Reducing Equivalents 601
   2.5 Oxidation of Reduced Se Forms 602
   2.6 Selenium Volatilization, Se Methylation
      and Demethylation 602
3. Selenium in Humans and Animals 603
4. Selenium in Plants 605
5. Selenium of Environmental Concern: Exploitation of
   Biological Processes for Treatment of Selenium Polluted Matrices 607
   5.1 Microbe-Induced Bioremediation 608
   5.2 Selenium Plant-Assisted Bioremediation
      (Phytoremediation) 609
   5.3 Plant–Microbe Interaction: Selenium
      Phytoremediation Processes 611
References 612

**24 Environmental Contamination Control of Water
Drainage from Uranium Mines by Aquatic Plants 623**
Carlos Paulo and João Pratas

1. Introduction 623
2. Uranium Mining: Environmental and Health 624
   2.1 Uranium Toxicity 627
   2.2 Uranium Mining History in Portugal 629
3. Phytoremediation of Metals with Aquatic Plants as Strategies
   for Mine Water Remediation 631
   3.1 Uranium Accumulation in Aquatic Plants and
      Phytoremediation Studies 632

4. Case Study: Water Drainage from Uranium Mines Control by
    Aquatic Plants in Central Portugal   634
    4.1 Selection of Aquatic Macrophytes: Field Studies   634
    4.2 Laboratory Experiments: Uranium Accumulation
         by *C. stagnalis*   640
    4.3 Phytoremediation Laboratory Prototype   644
5. Future Prospects of Water Phytoremediation   646
Acknowledgments   647
References   647

25 **Copper as an Environmental Contaminant: Phytotoxicity
    and Human Health Implications**   **653**
    Myriam Kanoun-Boulé, Manoel Bandeira De Albuquerque,
    Cristina Nabais and Helena Freitas

1. Copper and Humans: A Relation of 10,000 Years   653
2. Copper: Identity Card, Main Sources, and
    Environmental Pollution   654
    2.1 Copper in the Atmosphere   654
    2.2 Copper in the Hydrosphere   654
    2.3 Copper in the Lithosphere and Pedosphere   655
3. Copper in Plants   656
    3.1 Metabolic Functions of Copper   656
    3.2 Toxicity of Copper   657
    3.3 Copper and Human Health   663
4. Further Research Topics   670
References   671

26 **Forms of Copper, Manganese, Zinc, and Iron in Soils
    of Slovakia: System of Fertilizer Recommendation
    and Soil Monitoring**   **679**
    Bohdan Jurani and Pavel Dlapa

1. Forms of Trace Elements in Heterogeneous Soil Materials   679
2. Concept of Micronutrients Used in Agriculture of
    Former Czechoslovakia   682
3. Determination of Available Forms of Some Micronutrients in Soil Based
    on the Rinkis Method   682
4. Results of Modified Rinkis Method of Available Copper, Manganese,
    and Zinc in Soils of Slovakia   685
5. More Suitable Method for Determination of Plant Available
    Forms of Copper, Manganese, Zinc, and Iron in Soils   686
6. Limits to Lindsay—Norvell Method   687
7. Some Results Concerning Using Lindsay—Norvell Method   690
8. System of Micronutrients Application: Copper, Manganese, Zinc, and
    Iron for Agricultural Crops, Recommended in Slovakia   692

9. Remarks to the System used for Copper, Manganese, Zinc, and
    Iron Available Forms Determination and Fertilizers Recommendation   694
10. New Priorities in Research of Trace Elements in Soils of
    Slovakia—Soil Monitoring   695
References   697

27 Role of Minerals in Halophyte Feeding to Ruminants   **701**
Salah A. Attia-Ismail

1. Introduction   701
2. Ash and Mineral Contents of Halophytes   702
3. Factors Affecting Mineral Contents of Halophytes   702
4. Salt-Affected Soils   706
5. Irrigation with Saline Water   706
6. Salinity Level   706
7. Plant Species   708
8. Mineral Role in Ruminant Nutrition   708
9. Recommended Mineral Allowances   708
10. Minerals Deficiency in Halophyte Included Diets   710
11. Excessive Minerals in Livestock Rations in Dry Areas   713
12. Effect of Halophytes Feeding on Mineral Utilization   713
13. Effect of Minerals on Rumen Function   714
14. Effect of Minerals on Feed Intake   715
15. Effect of Minerals on Water Intake and Nutrient Utilization   716
16. Effect of Minerals on Microbial Community in the Rumen   717
References   717

28 Plants as Biomonitors of Trace Elements Pollution in Soil   **721**
Munir Ozturk, Ersin Yucel, Salih Gucel,
Serdal Sakçali and Ahmet Aksoy

1. Introduction   721
2. Soils and Trace Elements   722
3. Plants as Biomonitors of Trace Elements   725
4. Conclusions   735
References   735

29 Bioindication and Biomonitoring as Innovative
Biotechniques for Controlling Trace Metal
Influence to the Environment   **743**
Bernd Markert

1. Introduction   743
2. Definitions   745

3. Comparision of Instrumental Measurements and the Use of
    Bioindicators with Respect to Harmonization and Quality Control   746
4. Examples for Biomonitoring   748
    4.1 Mosses for Atmospheric Pollution Measurements   748
    4.2 Is There a Relation Between Moss Data and
         Human Health?   750
5. What do Bioaccumulation Data Really Tell Us?   752
6. Future Outlook: Breaking "Mental" Barriers Between
    Ecotoxicologists and Medical Scientists   754
    References   757

Biodiversity Index   **761**

Subject Index   **769**

FOREWORD

From the very beginning, metals such as gold, silver, copper, and iron have played a major role in the development and history of human societies and civilizations. Metals are dispersed on and in the Earth's crust, and methods for obtaining them from natural deposits have evolved over time. The distribution of metals is not uniform, and localized deposits serve as ores for metals, usually found as compounds, combined with other minerals and inorganic anions. If the concentration of the desired metal is high enough in the deposit for an economical extraction, then the ore can be exploited for a short or long period, depending on the state of the art and technology of mining. Most metals have to be purified or refined and then reduced to the metallic state before use. For example, the production of steel from iron requires the elimination of impurities present in the rocks, followed by the addition of other metals to obtain steel with the desired properties, such as hardness and resistance to corrosion. The science and technology of metals is precisely called "metallurgy." Our post-modern society is still based on the use of metals, and some major applications are briefly mentioned below:

- Potassium chloride is used as a fertilizer, and potash (K_2CO_3) is used in making soft soaps, pottery, and glass. Potassium hydroxide is an electrolyte in alkaline batteries, and NaOH is the most important base for industry. Soda ash (Na_2CO_3) is mainly used to make glass, but is also required to prepare chemicals, paper, and detergents. $NaHCO_3$ is an additive to control water pH in swimming pools, as well as to provide the fizz and neutralize excess stomach acid in analgesic drugs.

- Magnesium and calcium are good heat and electricity conductors. Alloyed with aluminum, Mg produces a strong structural metal. Another use of Mg is in fireworks. Epsom salt ($MgSO_4$) is useful in the tanning of leather and to treat fabrics. Milk of magnesia ($Mg(OH)_2$) has antacid and laxative properties. $CaCl_2$ is used to remove moisture from very humid places; CaO is a major ingredient in Portland cement, and partially dehydrated $CaSO_4$ (gypsum) produces plaster of Paris.

- Chromium is resistant to corrosion and is excellent as a protective coating over brass, bronze, and steel. Chromium is also needed to produce alloys such as stainless steel or nichrome; the latter is often used as the wire heating element in various devices such as toasters. Compounds of Cr have many practical applications, such as for pigments production and leather tanning. The main use of manganese is as an additive to steel and in the preparation of different alloys.

- Iron and its alloys have such physical properties that they have been put to more uses than any other metal. Nickel is one of our most useful metals; in its pure state, it resists corrosion, and it is thus frequently layered on iron and steel as a protective coating by electrolysis. When alloyed with iron or with copper, Ni makes the metal more ductile and resistant to corrosion and to impact.

- Copper has a very high electrical and thermal conductivity and is thus used in electrical wiring. It is also resistant to corrosion and thus appropriate to carry hot and cold water in buildings. Cu does oxidize slowly in air; and when CO_2 is also present, its surface becomes coated with a green film.

- Zinc provides a protective coating on steel, in a process called galvanizing. It is also used in various alloys, like brass (Cu and Zn) and bronze (Cu, Sn, and Zn). Zinc is important in the manufacture of zinc–carbon dry cells and other batteries. Zinc oxide is used in sunscreens and to make quick-setting dental cements. Zinc sulfide is suitable to prepare phosphors that glow when submitted to UV light or high-energy electrons of cathode rays, like the inner surface of TV picture tubes and the displays of computer monitors. Cadmium is useful as a protective coating on other metals and for making Ni–Cd batteries.

- In the past, lead was used for pipes and as an additive to gasoline. Nowadays, Wood's metal consists of an alloy of Bi, Pb, Sn, and Cd, melting at 70°C only, used to seal the heads of overhead sprinkler systems: A fire triggers the system automatically by melting the alloy. Different lead oxides are also needed in making pottery glazes and fine lead crystal; in corrosion-inhibiting coatings applied to structural steel; and as the cathode in lead storage batteries.

However, metals not only play an essential role in our daily life, but also are released into the environment in an uncontrolled way and become contaminants, or even pollutants. A contaminant is present where it would not normally occur, or at concentrations above natural background, whereas a pollutant is a contaminant that cause adverse biological effects to ecosystems and/or human health. In such a context, green plants play a key role in the availability and mobility of metals. Plants can remove metals from contaminated soils and water for cleanup purposes. Several plant species, hyperaccumulating elements like nickel, gold, or thallium, can be used for phytomining. On the other hand, crops with a reduced capacity to accumulate toxic metals in edible parts should be valuable to improve food safety. In contrast, crop plants with an enhanced capacity to accumulate essential minerals in an easily assimilated form can help to feed the rapidly increasing world population and improve human health through balanced mineral nutrition. Because many metals hyperaccumulated by plants are also essential nutrients, food fortification and phytoremediation are thus two sides of the same coin. The different chapters of this book do address the dual role of trace elements as nutrients and contaminants and review the consequences for ecosystems and health.

Dr. Jean-Paul Schwitzguébel

Chairman of COST Action 859
Laboratory for Environmental Biotechnology (LBE)
Swiss Federal Institute of Technology Lausanne (EPFL),
Station 6, CH 1015, Lausanne, Switzerland

PREFACE

It is a general belief that the fruits and vegetables that our parents ate when they were growing up were more nutritious and enriched with essential mineral nutrients and were less contaminated with toxic trace elements than the ones that are being consumed by us currently. A study of the mineral content of fruits and vegetables grown in Great Britain between 1930 and 1980 has added weight to that belief with findings of such decreases in nutrient density. The study, conducted by scientists in Great Britain, found significantly lower levels of calcium, magnesium, copper, and sodium in vegetables, as well as significantly lower levels of magnesium, iron, copper and potassium in fruits. Research studies are showing that the reducing nutritional value and the problem of contamination associated with food quality is increasing at an alarming rate. The decline in quality of agricultural produce has corresponded to the period of increased industrialization of our farming systems, where emphasis has been on cash crop cultivation that demands high doses of agrochemicals—that is, fertilizers and pesticides.

Several of the trace elements are essential for human as well as animal health. However, nutritionally important trace elements are deficient in soils in many regions of the world and the health problems associated with an excess, deficiency, or uneven distribution of these essential trace elements in soils are now a major public health issue in many developing countries. Therefore, the development of "foods and animal feeds" fortified with essential nutrients is now one of the most attractive research fields globally. In order to achieve this, knowledge of the traditional forms of agriculture, along with conservation, greater use of native bio-geo-diversity, and genetic diversity analysis of the cultivable crops, is a must.

A number of trace elements serve as cofactors for various enzymes and in a variety of metabolic functions. Trace elements accumulated in medicinal plants have the healing power for numerous ailments and disorders. Trace elements are implicated in healing function and neurochemical transmission (Zn on synaptic transmission); Cr and Mn can be correlated with therapeutic properties against diabetic and cardio-vascular diseases. Certain transition group elements regulate hepatic synthesis of cholesterol. Nutrinogenomics, pharmacogenomics, and metallomics are now emerging as new areas of research with challenging tasks ahead.

Soil, sediment, and urban dust, which originate primarily from the Earth's crust, is the most pervasive and important factor affecting human health and well-being. Trace element contamination is a major concern because of toxicity and the threat to human life and the environment. A variety of elements commonly found in the urban environment originate technogenically. In an urban environment, exposure of human beings to trace elements takes place from multiple sources, namely, water transported material from surrounding soils and slopes, dry and wet atmospheric deposition, biological inputs, road surface wear, road paint degradation, vehicle wear (tyres, body, brake lining, etc.), and vehicular fluid and particulate emissions. Lead and cadmium are the two elements that are frequently studied in street dust, but very little attention has been given to other trace elements such as Cr, Cu, Zn, and Ni, which are frequently encountered in the urban environment.

Street dusts often contain elevated concentrations of a range of toxic elements, and concerns have been expressed about the consequences for both environmental quality and human health, especially of young children because of their greater susceptibility to a given dose of toxin and the likelihood to ingest inadvertently significant quantities of dust. Sediment and dust transported and stored in the urban environment have the potential to provide considerable loadings of heavy metals to receiving water and water bodies, particularly with changing environmental conditions. On land, vegetables and fruits may be contaminated with surficial deposits of dusts. Environmental and health effects of trace metal contaminants in dust are dependent, at least initially, on the mobility and availability of the elements, and mobility and availability is a function of their chemical speciation and partitioning within or on dust matrices. The identification of the main binding sites and phase associations of trace metals in soils and sediments help in understanding geochemical processes and would be helpful to assess the potential for remobilization with changes in surrounding chemistry (especially pH and Eh). Sophisticated analytical and speciation techniques and synchrotron research are being applied to this field of research in developed nations.

This book covers both the benefits of trace elements and potential toxicity and impact of trace elements in the environment in the chosen topics by leaders of the world in this area.

M. N. V. Prasad

University of Hyderabad
Hyderabad, India

ACKNOWLEDGMENTS

I am thankful to Padmasri Professor Seyed Ehtesham Hasnain, Vice-Chancellor, University of Hyderabad for inspiring me to focus research in the area of health and nutritional science which gained considerable momentum under his dynamic leadership. I am grateful to all authors for cogent reviews which culminated in the present form.

Thanks are due to Anita Lekhwani, Senior Acquisitions Editor, Chemistry and Biotechnology for laying the foundation for this fascinating subject in 2005. I wish to place on record my appreciation for Rebekah Amos, Senior Editorial Assistant; Kellsee Chu, Senior Production Editor at John Wiley and Sons for superb and skillful technical assistance in production of this work punctually.

Dr K. Jayaram and Mr. H. Lalhruaitluanga helped in the preparation of the Index and their assistance is greatly appreciated. Last, but not least, I must acknowledge the excellent cooperation of my wife, Savithri.

CONTRIBUTORS

S. S. AIDOSSOVA,   Botany Department, Biology Faculty, Kazakh National al-Farabi University, Almaty 050040, Republic of Kazakhstan

AHMET AKSOY,   Biology Department, Faculty of Science & Arts, Erciyes University, 38039 Kayseri, Turkey

JOSEP ALLUÉ,   Department of Plant Physiology, Bioscience Faculty, Autonomous University of Barcelona, E-08193 Bellaterra, Spain

DANUTA MARIA ANTOSIEWICZ,   Department of Ecotoxicology, Faculty of Biology, The University of Warsaw, 02-096 Warsaw, Poland

SALAH A. ATTIA-ISMAIL,   Desert Research Center, Matareya, 11753 Cairo, Egypt

RAFAEL BORRÁS AVIÑÓ,   ABBA Chlorobia S.L., Citriculture Department, School of Agronomists, Polytechnic University of Valencia, 46022 Valencia, Spain

MANOEL BANDEIRA DE ALBUQUERQUE,   Center for Functional Ecology, Department of Botany, University of Coimbra, 3001-455 Coimbra, Portugal

JUAN BARCELÓ,   Department of Plant Physiology, Bioscience Faculty, Autonomous University of Barcelona, E-08193 Bellaterra, Spain

CARLOS BARATA,   Environmental Chemistry Department, IIQAB-CSIC, 08034 Barcelona, Spain

LUCIEN BOVET,   Philip Morris International R & D, Philip Morris Products SA, 2000 Neuchâtel, Switzerland

SIMONA DI GREGORIO,   Department of Biology, University of Pisa, 56126 Pisa, Italy

SERGI DÍEZ,   Environmental Geology Department, ICTJA-CSIC, 08028 Barcelona, Spain; and Environmental Chemistry Department, IIQAB-CSIC, 08034 Barcelona, Spain

PAVEL DLAPA,   Department of Soil Science, Faculty of Natural Sciences, Comenius University, 842 15 Bratislava, Slovak Republic

OTTO FRÄNZLE,   Christian-Albrechts-University Kiel, Ecology Centre, Olshausenstr. 40, D-24089 Kiel, Germany

STEFAN FRÄNZLE,   International Graduate School (IHI) Zittau, Department of Environmental High Technology, D-02763 Zittau, Germany

HELENA FREITAS, Center for Functional Ecology, Department of Botany, University of Coimbra, 3001-455 Coimbra, Portugal

MARIA GREGER, Department of Botany, Stockholm University, 106 91 Stockholm, Sweden

SALIH GUCEL, Centre for Environmental Studies, Near East University, Nicosia, 33010 North Cyprus

BOHDAN JURANI, Department of Soil Science, Faculty of Natural Science, Comenius University, 842 15 Bratislava, Slovak Republic

MYRIAM KANOUN-BOULÉ, Center for Functional Ecology, Department of Botany, University of Coimbra, 3001-455 Coimbra, Portugal

EDA KAPLAN, Department of Biology, Istanbul University, 34134 Eminou, Istanbul, Turkey

FARHAD KHORSANDI, Department of Agronomy, Islamic Azad University—Darab Branch, Darab, Fars Province, I.R. of Iran

SARITHA V. KURIAKOSE, Department of Plant Sciences, University of Hyderabad, Hyderabad 500 046, India

KARMA LHENDUP, Faculty of Agriculture, College of Natural Resources, Lobesa, PO Box Wangduephodrang, Bhutan

HELMUT LIETH, Wipperfürther Strasse 147, D-51515 Kürten, Germany

MERCÈ LLUGANY, Department of Plant Physiology, Bioscience Faculty, Autonomous University of Barcelona, E-08193 Bellaterra, Spain

JOSÉ RAFAEL LÓPEZ-MOYA, ABBA Chlorobia S.L., Citriculture Department, School of Agronomists, Polytechnic University of Valencia, 46022 Valencia, Spain

ELENA MAESTRI, Division of Genetics and Environmental Biotechnologies, Department of Environmental Sciences, University of Parma, Parma 43100, Italy

BERND MARKERT, International Graduate School (IHI) Zittau, Department of Environmental High Technology, D-02763 Zittau, Germany

NELSON MARMIROLI, Division of Genetics and Environmental Biotechnologies, Department of Environmental Sciences, University of Parma, Parma 43100, Italy

ABDUL R. MEMON, Institute of Genetic Engineering and Biotechnology, 41470 Gebze, Kocaeli, Turkey

CRISTINA NABAIS, Center for Functional Ecology, Department of Botany, University of Coimbra, 3001-455 Coimbra, Portugal

JUAN PEDRO NAVARRO-AVIÑO, ABBA Chlorobia S.L., Citriculture Department, School of Agronomists, Polytechnic University of Valencia, 46022 Valencia, Spain; and Department of Agrarian Sciences and of the Natural Environment,

School of Technology and Experimental Sciences, University "Jaume I," 12071 Castellón, Spain

N. NIRUPA, Department of Plant Sciences, University of Hyderabad, Hyderabad 500 046, India

MUNIR OZTURK, Botany Department, Science Faculty, Ege University, 35100 Bornova, Izmir, Turkey

CARLOS PAULO, Earth Sciences Department, Faculty of Sciences and Technology of the University of Coimbra, 3000-272 Coimbra, Portugal

SONIA PLAZA, Plant Biology, University of Fribourg, 1700 Fribourg, Switzerland

CHARLOTTE POSCHENRIEDER, Department of Plant Physiology, Bioscience Faculty, Autonomous University of Barcelona, E-08193 Bellaterra, Spain

M. N. V. PRASAD, Department of Plant Sciences, University of Hyderabad, Hyderabad 500 046, India

JOÃO PRATAS, Earth Sciences Department, Faculty of Sciences and Technology of the University of Coimbra, 3000-272 Coimbra, Portugal

DEMETRIO RALDÚA, Laboratory of Environmental Toxicology (UPC), 08220 Terrassa, Spain

K. S. SAGYNDYK, Botany Department, Biology Faculty, Kazakh National al-Farabi Universiy, Almaty 050040, Republic of Kazakhstan

SERDAL SAKÇALI, Biology Department, Faculty of Science & Arts, Fatih University, 34500 Hadimkoy, Istanbul, Turkey

ARUN K. SHANKER, Central Research Institute for Dryland Agriculture (CRIDA), Indian Council of Agricultural Research (ICAR), Santoshnagar, Hyderabad, 500 059, India

IRINA SHTANGEEVA, Chemical Department, St. Petersburg University, St. Petersburg 199034, Russia

SUJATA SINHA, BPS LAB-Centre for Diagnostic Hematology, Sankatmochan, Varanasi-221005 (UP), India

AGNIESZKA SIRKO, Institute of Biochemistry and Biophysics, Polish Academy of Sciences, 02-106 Warsaw, Poland

PAWEŁ SOWIŃSKI, Department of Plant Physiology, Institute of Botany, Faculty of Biology, University of Warsaw, 02-096 Warsaw, Poland; and Plant Biochemistry and Physiology Department, Plant Breeding and Acclimatization Institute, 05-870 Błonie, Radzików, Poland

ROSER TOLRÀ, Department of Plant Physiology, Bioscience Faculty, Autonomous University of Barcelona, E-08193 Bellaterra, Spain

SHUHE WEI, Key Laboratory of Terrestrial Ecological Process, Institute of Applied Ecology, Chinese Academy of Sciences, Shenyang 110016, People's Republic of China

SIMONE WUENSCHMANN, Fliederweg 17, D-49733 Haren, Germany

FATEMEH ALAEI YAZDI, Department of Agronomy, Yadz Agricultural and Natural Resources Research Center, Yazd, Yazd Province, I.R. of Iran

YASEMIN YILDIZHAN, Institute of Genetic Engineering and Biotechnology, 41470 Gebze, Kocaeli, Turkey

ERSIN YUCEL, Biology Department, Science Faculty, Anadoulu University, 26470 Eskisehir, Turkey

HARALD ZECHMEISTER, University of Vienna, Faculty of Life Sciences, Department of Conservation Biology, Vegetation, and Landscape Ecology, A-1090, Vienna, Austria

QIXING ZHOU, Key Laboratory of Terrestrial Ecological Process, Institute of Applied Ecology, Chinese Academy of Sciences, Shenyang 110016, People's Republic of China; and College of Environmental Science and Engineering, Nankai University, Tianjin 300071, People's Republic of China

1 The Biological System of Elements: Trace Element Concentration and Abundance in Plants Give Hints on Biochemical Reasons of Sequestration and Essentiality

STEFAN FRÄNZLE and BERND MARKERT*

Department of Environmental High Technology, International Graduate School (IHI) Zittau, D-02763 Zittau, Germany

OTTO FRÄNZLE

Christian-Albrechts-University Kiel, Ecology Centre, Olshausenstr. 40, D-24089 Kiel, Germany

HELMUT LIETH

Wipperfürther Strasse 147, D-51515 Kürten, Osnabrueck, Lower Saxony, Germany

1 INTRODUCTION

1.1 Analytical Data and Biochemical Functions

With ongoing improvement of analytic gear, it has already become commonplace to detect the vast majority of (stable) elements in biological samples [Garten, 1976; Markert, 1996, Lieth and Markert, 1990] as well as in soils [Kabata-Pendias and Pendias, 1984; Fränzle, 1990] or seawater [Nozaki, 1997]. The concentrations there may be considerably lower than in environmental compartments, down to the pico- or even femtomolar levels, because metal ions may not undergo bioaccumulation in plants or fungi [Lepp et al., 1987; Fränzle, 1993; Markert et al., 2003] in

Bernd Markert's present address: Fliederweg 17, D-49733 Haren-Erika, Germany.

Trace Elements as Contaminants and Nutrients: Consequences in Ecosystems and Human Health, Edited by M. N. V. Prasad

the same manner as unpolar organics. As a rule, there is "genuine" soil/plant bioconcentration (i.e., BCF [Biological Concentration Factor] > 1) for only a few metals (Mg, Zn, K) in green plants, with that of others being rare.

The presence of some chemical element in biomass, be it in substantial amounts, does by no means imply that it exerts some biochemical function. Several elements that are very abundant in the environment (particularly in soils) are not known to be essential for any kind of organism (e.g., Al, Ti) [Fränzle and Markert, 2007a,b]. Yet, besides the principal nonmetals C, H, O, N, S, and P, metals and other copious or trace elements were involved in biology quite apparently from the very beginnings of biological evolution [Beck and Ling, 1977; Kobayashi and Ponnamperuma, 1985a,b; Williams and da Silva, 1996; Frausto and Williams, 2001]—that is, from biogenesis itself—whereas during chemical evolution metal ions are (were) rarely required to afford crucial intermediates or catalyze transformations providing important structural features [Fränzle, 2007; Fränzle and Markert, 2002]. All living beings, even those that appear least advanced with respect to biological or biochemical complexity, share the requirement for at least seven metals (namely, K, Mg, Mn, Fe, Cu, Zn, Mo [or W in hyperthermophilic creatures]). These metals obviously differ considerably in their chemical properties; the latter is to be anticipated because metals in biology serve to effect or catalyze rather different transformations, causing them or just increasing selectivities of transformations. Nevertheless, the exact function is not yet established for all of these $(7 + n)$ metals (neither for humans nor for any other species, irrespective of full genetic sequencing such as with *Arabidopsis thaliana*).

Moreover, there are fairly many cases of metals acting in bioinorganic chemistry which differ from the optimum catalysts as defined by all the experiences in inorganic and metal–organic catalytic chemistry [Fränzle, 2007]. In addition, there are biochemical processes transforming substantial amounts of matter in the biosphere which rely upon combinations of various metals. The most prominent example of this is *photosynthesis*: Mn ($+$ Ca) ions are located in the center of photosystem II, affording oxidation of water, with the electrons thus liberated being shuttled to chlorophyll and Rubisco (both containing Mg) in order to bind and reduce CO_2 or to restore chlorophyll as a neutral molecule, respectively. Hence, photosynthesis will occur in an efficient manner only if some stoichiometric relationship between Mg and Mn is kept within a green plant; the empirical value for green leaves Mg/Mn is close to 5 (stoichiometric, not mass, ratio), with conifer needles differing somewhat from this value.

The same reasoning on mandatory metal ratios holds for animal metabolism, too, here concerning, for example, the Mo/Mg and Cu/Mg ratios for combinations of oxidizing substrates (Mo, Cu in redox enzymes) and for storing energy from this process (Mg in kinases, NTPases). For the sake of efficient metabolism, living organisms must keep these ratios in their bodies rather constant throughout lifetimes. In those parts of their corresponding environments from which they retrieve metals (soil, ambient water, food organisms), the respective interelemental ratios most likely will differ from the demands of the organism under consideration and possibly even vary with time. Hence the organism need not just obtain several different metals by complexation but also has to achieve some fractionation among them. Because metal–biomass interactions, starting with sequestration from food or environment, depend on *complexation* of the metal ions to biomass or some carrier within or (with root exudate) outside the organism, this fractionation will be accomplished due to either unequal complex formation equilibria or selective transport across membranes. Moreover, the number of different sequestration agents in/around roots, fungal mycelia, or the guts of some animals is considerably smaller than that of different metal ions; accordingly, different metals are transported by the same carriers and compete for their binding sites [Duffield and Taylor, 1987].

Now, correlations among element abundances were produced for a number of plant species some time ago [Markert, 1996; see Section 2.1]. Here, abundances of element pairs were compared in 13 different plant species and correlated to each other. Conspicuously, abundances of chemically similar elements like P and As or Ca and Ba are not correlated, whereas the REE (rare earth elements) abundances are closely correlated once again. From the correlation analysis a so-called BSE has been established (Fig. 1, see also Section 2.1).

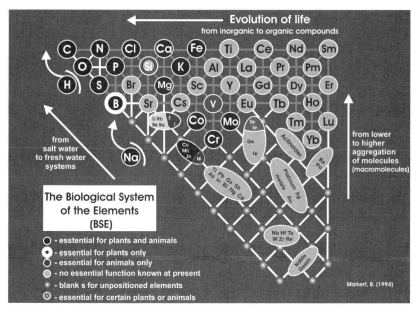

Figure 1. The Biological System of Elements as introduced by Markert (1994). Links between adjacent dots refer to abundance correlations among elements investigated in earlier works. The respective biological functions of the elements are given in different shadowings. Although after chemical evolution, its biological counterpart and successor acted to introduce a considerable number of metals into biomass where they behave e.g. as biocatalyst components, generally speaking the concentrations of the latter metals (even Ca, Mg or Fe, Zn) are low, far lower than in the Earth's crust which gives rise to a more prominent role for non-metals in biology/biochemistry which are gathered in the very left tip of the triangle. This latter fact corresponds to the fact that non-metals form the back-bones of biological materials, both in bulk and membranes.

Accordingly, abundances of essential elements may but need not be positively correlated. In addition, there is a need for definition of binding properties of metal ions toward (different kinds of) biological material which can account for enrichments in certain samples. For this purpose, a general relationship that links concentrations or, more precisely, bioconcentration factors (BCF [values]) to complex stabilities of metals taken up by some organism must be constructed (and expressed by some quotient k', see below). This approach also can be extended to trophic chains. There are several reasons for this particular approach in biochemistry:

- Metals interact with biological material by coordinating to it.
- Not all metals cause pronounced biochemical effects (i.e., many ones are neither essential nor considerably toxic), hence can be expected just to follow chemical equilibria by speciation into biological material rather than being selectively enriched or expelled/retained/linked to certain "controlling" sites/molecules.
- Complex ligands using the same functional groups produce complexes of closely similar stabilities.
- Complex formation usually occurs close to chemical equilibrium under physiological conditions and can be described by perturbation theory because there are only small effects on metal ions brought about by electronic properties and energies.

Any living being thus must cope with the endeavor to

- obtain the essential elements in a mixture that meets the demands of the corresponding organism while being constantly connected to an environment (e.g., soil, surrounding water) that usually does not match the respective demands directly (cf. Liebiǵ minimum principle),
- keep away toxic elements, in some cases even, and
- maintain a certain ratio between different essential elements involved in catalyzing the same biochemical transformation (e.g., Mn and Mg in photosynthesis).

There are several matters that render this a complicated endeavor: The relative stabilities of (chelate) complexes of (divalent) metal ions usually change according to a certain sequence, referred to as Irving–Williams sequence [Irving and Williams, 1953], which itself is not related to the specific amounts required by an organism, nor to the abundances of metals in soil or fresh water (or the inverse behavior, thereby permitting organisms to retrieve similar amounts of Mn, Cu, Zn, and so on in spite of their highly different complex stabilities). Yet, the Irving–Williams sequence rather is something like a rule-of-thumb. With ligands other than dicarboxylates, amino acids, or phenol(-ic) carboxylates, there are (often several) "inversions" of the stability series. Accordingly, living beings might select appropriate metals by producing and delivering suitable ligands, with the above ones not being the only ones that could be produced in substantial amounts and given away, for example, by roots or mycelia. On the other hand, soil organic matter (SOM) or aquatic organics

(DOM) contain certain ligand functions capable of retaining metals from transfer into living beings. Thus there is some competition for the metals between plant or fungus and soil. The data that are derived from analyses of (plant) biomass (e.g., Markert [1996]) thus correspond to some superposition of effects.

Therefore, organisms have to change concentrations of the metal ions encountered in their environment/food actively[1] in a typical way; for example, wood-degrading (basidiomycete) fungi need much more Fe and Cu, sometimes also V, to accomplish oxidative degradation of lignin than green plants require for sustaining their unlike biochemistry. The latter in turn have larger demands for Mg and Mn owing to photosynthesis. Plants and fungi might—and do—deliver different ligands to cope with this: citric, malic, and oxalic acids in green plants and peptides; hydroxamates and sometimes amino acids in fungi (and also in soil bacteria) [Kaim and Schwederski, 1993; Farago, 1986; Haas and Purvis, 2006]. How, then, do these ligands compare with respect to metal binding affinities and selectivities to the former ones?

Complex stabilities depend on the extent of metal ion–ligand interactions; at the same time, due to orbital interactions, the more the energy levels of the central metal ions are changed, the stronger the interactions become. In metal ions that are susceptible to redox reactions, the shift of orbital energies can be detected directly by change of redox potential of the altered complex. Hence there should be a relationship between complex stability and the potential shift caused by some ligand in a standard system (the so-called electrochemical ligand parameter $E_L(L)$ [Lever, 1990]), with the latter providing a measure for binding capability [Fränzle, 2007]. This argument from perturbation theory can be represented in the following equation, with complex stabilities at a given metal ion taken from experiment or literature (aqueous medium, 25°C, I about 0.2 M/kg) correlated to the above electrochemical ligand parameter $E_L(L)$ by linear regression analysis:

$$-\log k_{\text{diss}} = x \times E_L(L) + c, \tag{1}$$

Here, x is the slope parameter of the correlation between (logarithmic) complex formation constant (taken, e.g., from Furia [1972] and Moeller et al. [1965]) and the electrochemical ligand parameter [Lever, 1990; Fränzle, 2007; Fränzle and Markert, 2007a] for a given metal ion—say, in a series of Zn(II) complexes with bidentate ligands (e.g., glycinate, oxalate, lactate, ethylene diamine, and aminomethanephosphonate); then $c = 5.15$ and $x = +8.69$ for Zn^{2+} (tables for >50 different metal ions were published elsewhere, e.g., Fränzle and Markert [2007a]) while c gives the axis intercept at $E_L(L) = 0$. After rearrangement of the above Eq. (1), one obtains

$$E_L(L)_{\text{eff}} = [+c - \log k_{\text{diss}}]/x \tag{2}$$

to define fractionation behavior via an effective electrochemical ligand parameter. Thus a large, if not comprehensive, set of parameters were produced by the first

[1]This is not to suggest involvement of active transport in any case but might also refer to processes that rely upon metabolic energy—for example, associated with biosynthesis and eventual oxidative destruction of sequestrants connected with the Krebs tricarboxylate cycle, with destruction of these primary ligands being necessary to "hand over" the metal ions to other (cytosol) ligands in the root itself.

author (e.g., Fränzle and Markert [2007a]) which permit to estimate hydrolytic complex formation stabilities, with $E_L(L)$ being typical of a given ligand (donor) moiety. When different ligands are present, speciation or distribution/partition may also be inferred from the above equation. In soil, kinds and properties of ligands (SOM) change upon humification, and so do the ligand affinities of the metal ions. Then soil composition (C/N ratios, etc.) and speciation of N-free versus nitrogeneous ligands control which metals will be passed into green plants or fungi, respectively.

2 MATERIALS AND METHODS

2.1 Data Sets of Element Distribution Obtained in Freeland Ecological Studies: Environmental Analyses

The first data on element (abundance) correlations among green plant species later on to be used in this study were obtained at Grasmoor (literally, "grassy bog") natural reserve near Osnabrück, Lower Saxony, Germany around 1990 (Figs. 2a and 2b).

   The original abundance correlations were derived from these data for a total of 13 plant species [Markert, 1996] and called the Biological System of Elements (Figs. 1 and 3), with the name "Biological System of Elements" alluding to the chemical periodic system of elements (cf. Railsback [2003]). From these data already the clear-cut limits of that analogy become apparent: There is nothing like (some) "biological/ biochemical group[s] of elements" which could directly be related/compared to the chemical groupings of the PSE. While abundances of most REEs among each other and with Al correlate strongly, there are almost no relationships, for example, for distributions of P and As or of Ca and Ba. Recently, when converting these data from mere correlations of abundances into a kind of parameter which describes both (a) the general binding features of some kind of bioorganic ligand system/tissue to retain and accumulate metals and (b) their capacity of fractionation among the latter, comparative data for the same species at other sites (*Betula pendula*) and for quite different (including aquatic) plants were calculated (Table 2). For essential elements, this is expected to correspond to the demands that relatively differ among different plant species, with distributions/BCF values of certain nonessential elements to be used as a benchmark for this. The focus of interest thus shifted from mere comparison to producing a scale for the underlying biochemistry; for the latter purpose, bioconcentration factors soil/photosynthetic organs[2] (neglecting the extent of bioavailability of metals in certain soils) are used to get an idea on bioinorganic

---

[2]The question for binding stabilities of metals or some fractions of the latter to soil [Tyler, 2004] need not be considered here; it suffices to use the overall soil concentrations (produced, e.g., by *aqua regia* extraction or nitric acid/HF digestion of some soil sample) for this purpose: As plants partly use the same ligands in retrieval of heavy metals (e.g., citric acid) as in this procedure, corresponding barriers—which will, moreover, hold to most of chemically similar metals like the REEs and Al—blocking some part of soil heavy metal content from the plants will just provide a constant correction term. As comparisons and analysis of data are based on clusters of elements displaying equal BCF likewise, the effects from but partial bioavailability cancel out in this approach.

Figure 2. (**a**, **b**) The "primary study site": Grasmoor bog area near Osnabrück/FGR [www.nlwkn.niedersachsen.de (Niedersächsischer Landesbetrieb für Wasserwirtschaft, Küsten- und Naturschutz)].

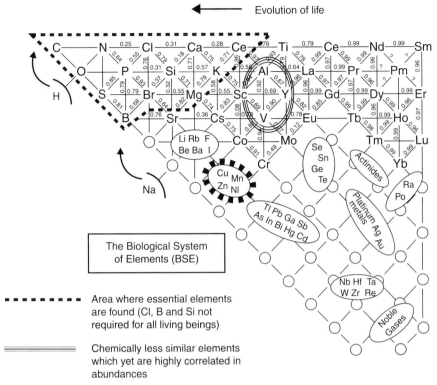

Figure 3. Depiction of the correlations that comprise the Biological System of Elements with correlation coefficients partly reported (numbers on lines connecting elements noted next to each other). There are both clusters of essential elements and highly correlated metals which do not resemble each other very mcuh in chemical terms (AI, V. Y) besides REE intercorrelations.

(metal ion) concentration processes and eventually characterize fractionation inside plants and their corresponding rhizospheres (see below; k' index).

2.2   Conversion of Data Using Sets of Elements with Identical BCF Values

The distribution of the essential metal Mg among the 13 species is highly correlated with those of (other essential elements *italicized*) Al ($r = +0.60$), Ca (0.71), Cs (0.74), Cu (0.61), Eu (0.74), K (0.79), N (0.70), Pb (0.66), Si[3] (0.71), Sc (0.66), and Sr (0.63), but not *Mn* (-0.13; in spite of the coupling between Mn [PS II] and Mg [chlorophyll, rubisco] in photosynthesis), Zn (-0.09), V (-0.73), or Ba ($+0.27$) [Markert, 1996]. There are positive abundance correlations between Mg and the REEs La, Ce, and Dy, an even more strongly positive one ($r = +0.738$)

---

[3]Si is essential for certain (fairly many) terrestrial plants, algae.

with Eu, and negative correlations of Mg abundances with those of Y, Gd, Tb, and Ho through Yb, whereas occurrences of Pr, Nd, and Lu (which latter, strictly speaking, is not a REE anymore as the filled f^{14} orbital set is no longer activated by oxidation or certain coordination effects) in the 13 plant species investigated by Markert are not at all correlated to the abundances of Mg.

So, most of these data suggest that there are some "high-concentration-biochemistry" plants in which essential elements (almost) altogether are present in elevated levels besides others which contain appreciably less of all these elements—metals and nonmetals alike. Yet, Mn and Zn deviate from this pattern. Accordingly, water photooxidation, providing electrons for reduction of CO_2 (based on Mn + Ca) and carboanhydrase and hydrolytic activities (both based on Zn usually), must be decoupled from the general high-concentration/low-concentration "antipodal" relationship. However, if either essential or highly toxic elements are considered, it may well be that these are either enriched (essential trace elements) or rejected by specific chemical means, such as *chaperons* [Rosenzweig, 2001], some of which also control metal distribution in plants, supporting rejection of As, Cd, and Pb there while controlling transport and allocation of Cu or Zn [Tottey et al., 2005]. Though also some major essential elements make their way through biological materials/tissues without any such contributions from chaperons (Mg, Mn, mostly also Ni), it is advisable for understanding the consecutive chemical steps in element transport (Fig. 4, Fränzle [2007]) to focus on such elements (i.e., metal ions) which are neither essential nor prominently toxic.

A large group of such elements which fulfill this condition are rare earth elements (lanthanoids, REE), including yttrium (cf. Jakubowski et al. [1999]). Unlike conventional chemical "wisdom" has it, REE chemical properties differ sufficiently from each other—especially concerning complex formation [Moeller et al., 1965]—also to undergo substantial fractionation in biomass [Emsley, 2001] and also bring about distinct differences in toxicities. As the data set [Markert, 1996] includes all Y and the two La–Nd and Sm–Lu series, REEs and some other elements (Sr, Al) can be used for benchmarking metal fractionation in a plant. For all these elements, c and x values for mono- and bidentate binding are available [Fränzle, 2007; Fränzle and Markert, 2007a,b]. Thus Eq. (2) can be applied for calculating effective electrochemical ligand parameters for biomass (assuming bidentate coordination as corresponding to average interaction constants of numerous metal ions with biological material [data from Williams and Silva, 1996]) from REE distributions, with their large number of 15 elements (omitting promethium) allowing for construction of meaningful clusters of identical BCF. These in turn can be used to determine k', a term describing metal transport and propensity for accumulation of weakly binding elements like the heavier alkaline earths beyond Mg within a plant (Section 3.3).

2.3 Definition and Derivation of the Electrochemical Ligand Parameters

The electrochemical ligand parameters as introduced in this chapter are now known for hundreds of donor compounds or ions [Lever, 1990], mainly from direct

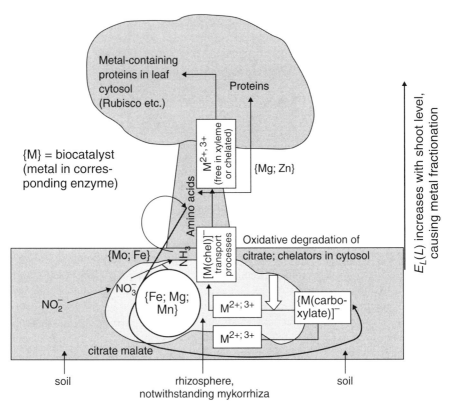

Figure 4. Schematic picture of metal transport through a plant. Most of root exudate components are taken from the tricarboxylate cycle (citrate, malate; a comprehensive list of exuded ligands can be found in Dakora and Phillips [2002]), with the original ligands removed by oxidation after resorption and replaced by carrier proteins in xylem and histosol, with the ions eventually ending up in leaf proteins. Swift brackets denote catalysts (matalloenzymes containing the corresponding metal). Owing to oxidative cleavage of the original ligand, metals may be transported and accumulated efficiently further on even when forming weak complexes (Ba, REEs in certain nutforming trees) but likewise reside in the roots without any further vertical transport (Cd in corn).

electrochemical studies and an additivity assumption; others are derived from correlations with other systems (Os or Re complexes, Mo(dppe)$_2$ or Cr polycarbonyl species). More recently, data were added by Fränzle using Eq. (1) to calculate $E_L(L)$ from known stability constants; in addition, aqueous measurements of Ru(II/III) redox ligands involving biorelevant ligands were done, producing a correction factor for solvation differences versus data obtained in CH$_3$CN [Fränzle, 2007]. The list includes including many ligands that are involved in biological processes, be it amino acids, their side chains (e.g., imidazole [histidine], dialkyl sulfide [methionine]), heterocycles (purines, flavines), salicylic acid (some kind of prototype of tannin derivatives), or phosphorylated species. For orientation, a couple of data for species involved in soil chemistry and general biochemistry is given in (Table 1).

TABLE 1. Ligands Involved in Soil and Biochemistry

Ligand	$E_L(L)$ [V]	Denticity n	Biologically Relevant Ligand (Example)
α-Aminocarboxylate	−0.05	2	Glycinate
Porphyrinate	0.00	4	Haemin, chlorophyll
Sugar	[+0.11]	2 (?)	**anion** of fructose ($pK_a = 12.3$)
Nucleic acids	ca. −0.30 V	2–3	
Pyridine	+0.24	1	Vitamin B_6
Hydroxamate	+0.02	2	Rhodotorulic acid
Hydrogen carbonate	[−0.37]	2	
Carboxylate	ca. −0.21	1	Acetate, aspartate, or glutamate residues (deprotonated)
Oxalate	−0.17	2	
Hydroxipolycarboxylate	[−0.26... −0.13]	2 or 3	Malate, citrate
Phenolate	[+0.23]	1	Tyrosine residue, gallic and caffeic acids, humic acids
Chelating 2-acylphenolates	ca. −0.07 V	2	Salicylaldehyde anion, 2-acetylphenolate
Salicylate	[ca. +0.1]	2	
SO_4^{2-}	[−0.33]	1	
NO_3^-	−0.11	1 or 2	
Cl^-	−0.22	1	
Carboxamide	+0.03	1	Peptide, asparagine residue, urea
Carboxamidate	−0.38	1	

$E_L(L)$ denotes the electrochemical ligand parameter, the denticity is the number of binding sites which link metal in centers and the ligand given in column 1. $E_L(L)$ values in brackets [0] were calculated from complex stability constants with various metal ions in aqueous media rather than determined by electrochemical methods. **Bold:** ligands which are involved in metal transport in plants.

3 RESULTS

3.1 Abundance Correlations Among Essential and Nonessential Elements

There is an apparent maximum in the distribution of effective $E_L(L)$ at −0.17 V, corresponding to $E_L(L)$ of oxalate ion. As yet, it must be left to forthcoming investigations whether this is a result of an abundant biochemical setting in terms of transport and metal retention in photosynthetic organs merely or can be interpreted as an adaption to the plants (frequent) requirement to maintain certain ratios of essential elements for proper biochemical interconnections among reactions promoted by unlike elements. Yet there is some hint concerning this: The CEP (competitive exclusion principle) theorem states that different biological species must "tap" resources in unlike ways in order to be able to coexist in the same habitat (see below). This

TABLE 2. Data for Effective Electrochemical Ligand Parameters of Several Different Plant Species

Plant Species	Tested Organ	Groups of Metal Ions with Equal $BCF_{soil/plant}$ Values	Effective Electrochemical Ligand Parameter for the Metal Ion Group	Average Value (weighted)	Biochemical Remarks
Lolium perenne	"Green" (above ground)	(1) Eu, Gd, Tb, Er; (2) Y, La, Nd; (3) Ce, Dy, Yb		−0.19	
Taraxacum officinale	Leaves	Cd ≈ Cu; Sr ≈ Zn and Sr ≈ Cu ≈ Zn (other sampling site)	(1) −0.29, (2) −0.26, (3) −0.27	−0.27	
Vaccinium vitis-idaea	Leaves			−0.04	The two *Vaccinium* species strongly differ in $E_L(L)_{eff}$ and usually do not coexist (competitive exclusion principle)
Vaccinium myrtillus	Leaves			−0.25	
Deschampsia flexuosa	"Green" (above ground)			−0.165	
Molinia coerulea				−0.135	

Betula pendula	Leaves	(a) Grasmoor; (b) Grown from seedlings on sewage sludge in Lithuania: Mn, Cu, Ni, Pb (BCF ca. 0.45 each); (c) upper Lusatia, including soil samples	(a) -0.17 V (b) -0.19 V (c) -0.20 V	Data from several, also polluted, sites (Lithuania) agree
Spruce	Needles (1 year c d)		-0.18	Mg/Mn ratio different from deciduous plants
Spruce	Needles (2 years old)		-0.24	
Pinus sylvestris	Needles	(1) Y, La, Pr, Nd, Sm; (2) Gd, Tb, Er, Lu	(1) -0.164 (2) -0.13	
Lemna trisulca	Total	La, Ce, Pr	-0.17	
Azolla filiculoides		La, Sm, Eu	-0.06	
Ceratophyllum demersum		Ce, Sm	-0.17	

Values were calculated using Eq. (2).

apparently is the case with Ericaceae *Vaccinium myrtillus* and *Vaccinium vitis-idaea*, respectively, the effective $E_L(L)$ of which differ by more than 0.2 volts (see Table 2).

3.2 (Lack of) Correlation and Differences in Biochemistry

As noted before (Section 2.2), the abundances of the elements essential for plants do not always correlate positively, which means there are some deviations from any possible general pattern of identical concentration relationships at either higher or lower levels. Apart from necessities of efficient coupling of different metal-catalyzed biochemical pathways (e.g., in photosynthesis), there is highly positive correlation among most of these elements yet, holding for metals and nonmetals alike (also cf. Fig. 3). Apart from this, biochemically relevant ligands will control resorption, retention, and transport of metals according to Eq. (1) all the way from rhizosphere upward to leaf tip or fruit, implying that—like in Fig. 6—complex formation constants must be calculated using the electrochemical ligand parameters. In addition, metal fractionation is described by an effective electrochemical ligand parameter [Eq. (2)] derived from the fact that entire plant organs tend to behave as if consisting of a single, homogeneous kind of ligand. The latter values from sets of identical BCF values and Eq. (2) are given in Table 2.

Data for birch and spruce trees and dandelion (*Taraxacum officinale* L.) directly correspond to soil analyses, whereas the Ln^{3+} patterns of the data set provided by Markert are related to average REE contents of Central European soils. If the molar ratio Mg/Mn is about five already in soil—which frequently is the case—$E_L(L)_{eff} = -0.07$ V keeps this relationship also in the plant (photosynthetic organ); plants can produce this situation by delivering simple amino acids (rather than hydroxicarboxylates [citrate, malate]) to the soil.

3.3 Implication for Biomonitoring: Corrections by Use of Electrochemical Ligand Parameters and BCF-Defined Element Clusters

Since the original data show that, before a prediction of BCF values of yet other elements in a given soil-dwelling organism can be made, the relationship between

- empirical (sets of identical) element BCF values and
- the corresponding calculated (once again, identical) complex formation constants for bidentate binding

must be derived. The various BCF clusters for one species differ with respect to BCF and to $-\log k_{diss}$ calculated from this for each set of equal-BCF-metal ions, with the quotient

$$k' = \Delta \log k / \Delta \log \mathrm{BCF} \tag{3}$$

that links both depending on the corresponding species even if $E_L(L)_{eff}$ is similar in a couple of plant species—for example, about -0.17 V. The size of this quotient k' in addition gives some measure of whether the plant might be able to hyperaccumulate certain metals that form particularly stable complexes with their biomass(-es): With $E_L(L)_{\mathrm{eff}}$ being negative in general and $k' \approx \pm 0$, no hyperaccumulation will take place even concerning metals with either "exceptionally" positive (U, Cr(III), V(IV), Cu, Al) or negative (Nd, Ti, Zr) x values. Conversely, different values of k' $\ll 0$ or $k' \gg 0$ may result in hyperaccumulation which, however, can only be realized if soil chemistry is appropriate. The value of k' in addition gives a piece of information on the consecutive steps of metal (ion) sequestration and transport in some plant: A cascade of ligands (sequestrants, protein carriers) with steadily increasing or decreasing $E_L(L)$ (possibly plus active transport through membranes) will produce corresponding bioinorganic amplication, notable by $|k'| \gg 0$, which renders (hyper-)accumulation even of metals forming only weak complexes possible (see below). Other series or sequestrants and carriers that do not produce a steady change of $E_L(L)$ from rhizosphere up to leaf tip or fruit, but "arbitrarily" change differences of metal ion attraction from step to step of transport, will exclude amplification, thus producing $k' \approx 0$.

Highly negative values of k' in addition imply the possibility to retain or even enrich elements that will form rather labile complexes given the effective electrochemical ligand parameter of the plant species and Eq. (1). These include Sr, Ba, or Mn and the REEs (except Sm and Tb) if $E_L(L)_{\mathrm{eff}}$ is close to zero in the latter case, all of which are known to be hyperaccumulated in some plants—for example, Ba and Mn in Brazil nuts [Emsley, 2001]; also, among our test set of plant species, Mn gets substantially enriched in blueberries (both leaves [to which data reported here [Markert, 1996] pertain] and fruits). Thus k' thus is a kind of measure for amplification of differences in the sequence of transport within some plant, from sequestration in/by root exudates to deposition in the tips of leaves, in needles,[4] or in fruits, respectively. In addition, a negative k' enables the plant to cope with high environmental levels of metal ion toxicants which form rather stable complexes—in particular, Al and Cu—whereas Ni and Zn are accumulated in specially adapted (serpentine) flora [Lee et al., 1975] rather than efficiently repelled. The values are given in Table 3.

In both *Vaccinium* species, there is potential for thorough amplification of small differences in complex stabilities within the plant and thus for hyperaccumulation; this mainly refers to *Vaccinium myrtillus* because its effective electrochemical ligand parameter is fairly low (-0.25 V). k' values for grasses and trees (both deciduous ones and coniferes) vary around $k' \pm 0$, except for *Deschampsia flexuosa*.

4 DISCUSSION

In any biocoenosis (= community of different populations), *by definition* different organisms do coexist. When they derive nutrients from a common source—for example, soil (as, say, soil bacteria, green plants, fungi, and earthworms do)—they will compete for the same resources such as essential metal ions. One condition of

---

[4]Note that there is a difference in effective $E_L(L)$ of different ages in spruce needles.

TABLE 3. Sensitivity of BCF Toward Changes in Complex Formation Constant According to Eqs. (1) and (3): $k' = \Delta \log k / \Delta \log BCF$

Species	BCF for Element Cluster 1	$-\log k_{\mathrm{diss}}$ for Element Cluster 1	BCF for Element Cluster 2	$-\log k_{\mathrm{diss}}$ for Element Cluster 2	$\Delta \log k$	$\Delta \log$ BCF	k'
Lolium perenne	0.016	6.82	0.007	6.30	−0.52	−0.36	+1.4
Betula pendula	0.004	6.82	0.011	5.99	−0.83	+0.44	−1.9
Vaccinium myrtillus	0.004	6.54	0.008	9.59	3.05	+0.30	+10.1
Vaccinium vitis-idaea	0.0045	6.82	0.007	8.36	1.54	+0.19	+8.1
Pinus sylvestris	0.004	6.10	0.008	6.64	0.54	+0.30	+1.8
Molinia coerulea	0.008	6.35	0.013	6.00	−0.35	+0.21	−1.7
Deschampsia flexuosa	0.0055	6.82	0.0105	5.97	−0.85	+0.28	−3.0

k' (positive or negative deviation from zero) denotes the propensity for hyperaccumulation of some element. As usual, data on REEs were used for scaling.

long-term coexistence will be that all these organisms do not "tap" resources in like manners, regardless of the chemical kind of resources (metal ions, anions like sulfate, nitrate, vitamines, etc.). This theorem is called the competitive exclusion principle (CEP) [Rastetter and Ågren, 2002], essentially a chemical description of the idea of ecological niche.

Obviously the metabolic pathways of the above organisms differ, even between autotrophy (green plants) and heterotrophy (the others mentioned above except for some soil bacteria and possibly archaea); accordingly, they are likely to differ in their essential element demands also, even though 13 of them will be required by all the organisms coexisting on or in soil (like with all others). Yet, as they dwell on the same soil at a given spot, they must extract the required set of elements in corresponding amounts by unlike means. The challenge of obtaining the corresponding metal mixture can then be tackled by applying different ligands, such as shown in the following picture (Fig. 5).

So metal uptake is regulated yet also nonessential and even toxic metals will be admitted by way of the same carrier molecules to these organisms to some extent, a prominent example being the role of *Amanita muscaria* (fly agaric toadstool) sporophores in short-time circulation of Cd and V close to the soil−air interface [Lepp et al., 1987].

Using Eq. (1), one obtains numerous lines that show complex stabilities, one for each metal ion/oxidation state, both essential ones the various organisms are after and compete for (Fig. 5) and concomitant bioconcentration of toxic ions (broken lines in Fig. 6).

But this effect holds for fungi in general, allowing them to accumulate sufficient Fe, Cu, and V, each components of fungal oxidation biocatalysts, to decompose wood by corresponding enzymes [Machuca et al., 2001]. Likewise, it can be shown that lichens will get their share of metals of a "typical" support if, and only if, they use

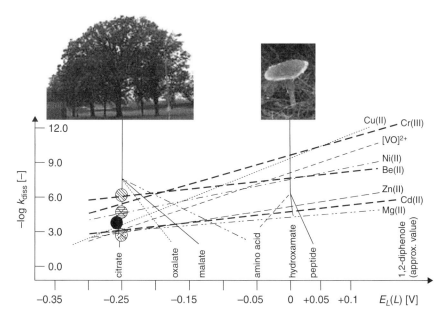

Figure 5. Different organisms deliver different ligands to upper soil layers, mobilizing metal ions from there.

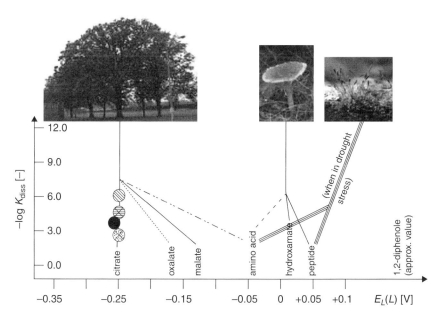

Figure 6. Complex formation constants of some essential versus toxic (Cr, Be, Cd, bold broken lines) as depending on electrochemical ligand parameters given at the lower axis. The values correspond to bidentate binding of ligands, except for citrate for citrate (the stability constants of some corresponding citratocomplexes [from above: cu, Zn, V, Mg] are displayed as dots in the left part of the diagram).

some ligand distinguished by $E_L(L) \approx -0.18$ V for metal sequestration. Accordingly, they deliver oxalic acid to their environment $[E_L(L) = -0.17$ V$]$ [Haas and Purvis, 2006], rather than amino acids [Sharma and Dietz, 2006] or malic and citric acids in "pure" (nonsymbiotic, not being a member of some lichen) fungi and green plants, respectively.

For any species that exists in some terrestrial ecosystem of substantial size (there are horizontal, vertical, and seasonal variations in the elemental composition of seawater [Nozaki, 1997] also, but these differences are far less pronounced than with either soil, food organisms, or limnetic waters), there is the challenge of maintaining internal element ratios. This implies that these ratios be conserved in order to keep various metals involved in the same kind of biochemical transformation in that way of balance necessary for optimum yields and efficiencies: Best coupling of water oxidation and CO_2 sequestering/reduction in photosynthesis demands Mg/Mn $\approx$ 5:1. This ratio is commonplace in leaves, also approximately in soil, but not in aquatic environments. So, while conditions improved for plants in terms of more efficient photosynthesis when going ashore in evolution in the late Ordivician or early Silurian [Sillen, 1967; Rothe, 2000] in several ways, including metal supply for photosynthesis, it just in turn necessitated formation of tougher internal structures and some mechanisms of drought protection, whereas, even using chemically active roots, other elements might not as easily be obtained from soil as from (fresh or ocean) waters.

There are few couples of closely related organisms in Markert's original set of data or among the ones that now provided additional data, though conclusion on the following topic must be considered tentative when going beyond general statements corroborated by the CEP. The only closely related plants are two *Vaccinium* species, with grasses *Deschampsia flexuosa* and *Molinia coerulea* sometimes coexisting in forests of intermediate soil humidity. Yet it is obvious that both couples of plants avoid "counter-CEP" competition due to differences in both $E_L(L)_{eff}$ and k' values. A simple estimate using the effects of some replacement of a donor site in some carrier protein by mutation suggests a change of $\approx \pm 0.02$ V in $E_L(L)_{eff}$ owing to a single mutation in a carrier. This might already alter k' substantially as well, permitting a counterintuitive enrichment pattern in e.g., the (Ca; Sr) versus (Cu; Al) couples (i.e., a negative k'). Likewise, negative k' values might also result from pronounced retention of strongly coordinating ions by soil ligands (cf. Tyler [2004] and soil/[beech] microroot BCF values reported there).

5   CONCLUSION

The advanced description of the modes of behavior of metal ions in biomass derived from the BSE will link extracorporal biochemistry, soil biogeochemistry, and strategies used by different organisms both to get their share of required elements and fulfill the CEP upon placing within an ecological niche. The term k' gives a concise description of the reaction sequence in metal transport within living beings which can also be extended to heterotrophs and to entire trophic chains.

REFERENCES

Beck MT, Ling J. 1977. Transition-metal complexes in the prebiotic soup. *Naturwissenschaften* **64**:91.

Dakora FD, Phillips DA. 2002. Root exudates as mediators of mineral acquisition in low-nutrient environments. *Plant and Soil* **245**:35–47.

Duffield JR, Taylor DM. 1987. A spectroscopic study of the binding of plutonium(IV) and its chemical analogues to transferrin. *Inorg Chimi Acta* **140**:365–367.

Emsley J. 2001. *Naturés Building Blocks. An A–Z Guide to the Elements*. Oxford: Oxford University Press.

Farago ME. 1986. Metal ions and plants. In: Xavier AV, editor. *Frontiers in Bioinorganic Chemistry*. Weinheim: VCH, pp. 106–122.

Fränzle O. 1990. Representative sampling of soils in the Federal Republic of Germany and the EC countries. In: Lieth H, Markert B, editors. *Element Concentration Cadasters in Ecosystems—Methods of Assessment and Evaluation*. Weinheim: VCH, pp. 41–76.

Fränzle O. 1993. *Contaminants in Terrestrial Environments*. Berlin: Springer.

Fränzle S. 2007. Prinzipien und Mechanismen der Verteilung und Essentialität von chemischen Elementen in pflanzlicher Biomasse—Ableitungen aus dem Biologischen System der Elemente. *Habilitation* thesis, 164 pp.

Fränzle S, Markert B. 2002. Three ways ashore. Where did terrestrial plants, arthropods and vertebrates come from? Are essentiality patterns of chemical elements living bio(-geo)chemical"fossils"? Inferences from essentiality patterns and the Biological System of Elements. *Chem Inzynieria Eco* **9**:949–967.

Fränzle S, Markert B. 2007a. What does bioaccumulation really tell us—Analytic data in their natural environment. *Chem Inzyn Ecol* **14**:7–23.

Fränzle S, Markert B. 2007b. Implications of bioconcentration of metals for evaluation of biomonitoring data: A method for correction of selective uptake from soil and water allows for "realistic" biomonitoring by means of plants. Conference proceedings der Tagung EcOpole 2006 in Duzhniki Zdroj/Polen, pp. 25–31.

Frausto Da Silva JJR, Williams RJP. 2001. *The Biological Chemistry of the Elements. The Inorganic Chemistry of Life*. Oxford: Oxford University Press.

Furia T. 1972. *First Stability Constants of Various Metal Chelates. CRC Handbook of Food Additives*. Baton Rouge, LA: CRC Press.

Garten CT. 1976. Correlations between concentrations of elements in plants. *Nature* **261**:686–688.

Haas JR, Purvis OW. 2006. Lichen biogeochemistry. In: Gadd GM, editor. *Fungi in Biogeochemical Cycles*. Cambridge: Cambridge University Press, pp. 106–119.

Irving H, Williams RJP. 1953. The stability series for complexes, of divalent ions. *J Chem Soc* 3192–3205.

Jakubowski N, Brandt R, Stuewer D, Eschnauer HR, Görtges S. 1999. Analysis of wines by ICP-MS: is the pattern of the rare earth elements a reliable fingerprint for the provenance? *Fresenius J Anal Chem* **364**:424–428.

Kabata-Pendias A, Pendias H. 1984. *Trace Elements in Soils and Plants.* Boca Raton, FL: CRC Press.

Kaim W, Schwederski B. 1993. *Bioanorganische Chemie.* Stuttgart: Teubner.

Kobayashi K, Ponnamperuma C. 1985a. Trace elements in chemical evolution, part I. *Origins of Life* **16**:41–55.

Kobayashi K, Ponnamperuma C. 1985b. Trace elements in chemical evolution, part II: Experimental results. *Origins of Life* **16**:57–67.

Lee J, Brooks RR, Reeves RD, Boswell CR. 1975. Soil factors controlling a New Zealand serpentine flora. *Plant and Soil* **42**:153–160.

Lepp NW, Harrison SCS, Morrell BG. 1987. A role for *Amanita muscaria* L. in the circulation of cadmium and vanadium in a non-polluted woodland. *Environ Geochem Health* **9**:61–64.

Lever ABP. 1990. Electrochemical parametrization of metal complex redox potentials, using the ruthenium(III)/ruthenium(II) couple to generate a ligand electrochemical series. *Inorg Chem* **29**:1271–1285.

Lieth H, Markert B. 1988. *Aufstellung und Auswertung ökosystemarer Element-Konzentrations-Kataster—Eine Einführung.* Berlin: Springer.

Lieth H, Markert B, editors. 1990. *Element Concentration Cadasters in Ecosystems—Methods of Assessment and Evaluation.* Weinheim: VCH.

Machuca A, Napoleão D, Milagres AMF. 2001. Detection of metal-chelating compounds from wood-rotting fungi *Trametes versicolor* and *Wolfiporia cocos. World J Microbiol Biotech* **17**:687–690.

Markert B. 1994. The Biological System of the Elements (BSE) for terrestrial plants (glycophytes). *Sci Total Environ* **155**:221–228.

Markert B. 1996. *Instrumental Element and Multi-Element Analysis of Plant Samples— Methods and Applications.* New York: John Wiley & Sons.

Markert B, Breure A, Zechmeister H, editors. 2003. *Bioindicators and Biomonitors: Principles, Concepts and Applications.* Amsterdam: Elsevier.

Moeller T, Martin DF, Thompson LC, Ferrus R, Feistel GR, Randall WJ. 1965. The coordination chemistry of yttrium and the rare earth metal ions. *Chem Rev* **65**:1–50.

Nozaki Y. 1997. A fresh look at element distribution in the North Pacific. *Eos* **78**:207–211.

Railsback LB. 2003. An earth scientist's periodic table of the elements and their ions. Geology **31**:737–740.

Rastetter EB, Ågren GI. 2002. Changes in individual allometry can lead to species coexistence without niche separation. *Ecosystems* **5**:789–801.

Rosenzweig AC. 2001. Copper delivery by metallochaperone proteins. *Accts Chem Res* **34**:119–128.

Rothe P. 2000. Erdgeschichte. *Spurensuche im Gestein.* Darmstadt: Wissenschaftliche Buchgesellschaft.

Sharma SS, Dietz K-J. 2006. The significance of amino acids and and amino acid-derived molecules in plant responses and adaptation to heavy metal stress. *J Exp Bot* **57**:711–726.

Sillen LG. 1967. How have seawater and air got their present compositions? *Chem Britain* **1**:291–297.

Tottey S, Harvie DR, Robinson NJ. 2005. Understanding how cells allocate metals using metal-sensors and metallochaperones. *Accts Chem Res* **38**:775–783.

Tyler G. 2004. Ionic charge, radius, and potential control root/soil concentration ratios of fifty cationic elements in the organic horizon of a beech (*Fagus sylvatica*) forest podzol. *Sci Total Environ* **329**:231–239.

Williams RJP, Frausto da Silva JJR. 1996. *The Natural Selection of the Chemical Elements.* Oxford: Clarendon Press.

2 Health Implications of Trace Elements in the Environment and the Food Chain

NELSON MARMIROLI and ELENA MAESTRI

Division of Genetics and Environmental Biotechnologies, Department of Environmental Sciences, University of Parma, Parma 43100, Italy

The importance of trace elements in nutrition and health cannot be underestimated. Literature reports several examples of diseases and symptoms which can be correlated to mineral deficiency. One example is the Keshan disease, widespread and endemic in some areas of China, known for more than 100 years, leading to myocardial fibrosis and necrosis. In the 1970s, deficiency of selenium was discovered to be one of the main determinant factors, through two main observations: Crops in the area had extremely low selenium content, and administration of selenium before establishment of the disease could improve the conditions [Tan et al., 2002]. Similarly, occurrence of goiter in the "goiter belt" of the United States was associated with insufficient iodine in soils and plants in the 1930s [White and Zasoski, 1999].

"Medical geochemistry" studies examples of endemic diseases linked to mineral distribution and availability in the diet, such as: dental fluorosis in China caused by excess fluoride in waters; iodine deficiency disorders in tropical developing countries, including goiter and cretinism; chronic arsenic poisoning in different countries [Dissayanake and Chandrajith, 1999].

Environmental conditions and agricultural practices can affect in a dramatic way the presence of minerals in the diet; the effects are more relevant in countries where food is still grown locally and populations have close contacts with their natural environment, whereas in Northern–Western countries the manifestations are not so evident. In this social environment, rather, the increased use of dietary supplements, foods with special formulations, and fortified foods prompts a discussion on risks linked to high intakes and definition of safety limits.

Trace Elements as Contaminants and Nutrients: Consequences in Ecosystems and Human Health, Edited by M. N. V. Prasad

The present review deals with (a) issues concerning the presence of trace elements in the food supply chain and (b) effects on human health.

1   TRACE ELEMENTS IMPORTANT IN HUMAN NUTRITION

The generally accepted definition of trace elements describes them as elements which occur in the organism in very small quantities, less than 0.01% [Oliver, 1997]. It encompasses both essential elements with physiological relevance (microelements or micronutrients), elements which can become toxic at high concentrations, and elements that are altogether toxic [Reinhold, 1975]. Different organisms have different nutritional requirements, so some elements can be essential for one organism and toxic to another. In animals, and therefore also in humans, a possible classification concerns:

- Elements essential in nutrition: calcium, cobalt, chromium (III), copper, fluorine, iodine, iron, magnesium, manganese, molybdenum, potassium, selenium, sodium, zinc.
- Elements with possible beneficial effects: boron, nickel, silicon, vanadium.

Currently, the trace elements considered essential for nutrition by the WHO are iron, zinc, copper, chromium, iodine, cobalt, molybdenum, and selenium [WHO, 2002b]. The elements silicon, manganese, nickel, boron, and vanadium are considered to be probably essential.

- Elements without any beneficial effects: aluminium, antimony, arsenic, barium, beryllium, cadmium, lead, mercury, silver, strontium, thallium, tin. This list encompasses elements that can be found in the environment and for which human exposure should be limited.

The criteria for essentiality is that absence or deficiency of an element brings abnormalities that can be connected to specific biochemical changes reversed by supplying the element. Most of these elements act primarily as catalysts for enzymes (see Table 1), and all of them can be toxic at high concentrations. In tissues and fluids, metals are mostly present as complexes with organic compounds: amino acids, proteins and peptides, organic acids, glutathione. Transporters of metals and transporters of their complexes are major players in the homoeostasis and in mediating effects of toxic metals, because some of them are not highly specific and interact with multiple metals [Ballatori, 2002]. One very important feature when considering health effects of trace elements is their slow accumulation in tissues even at low doses. Hence, very rarely are acute effects reported, whereas chronic exposure can lead to buildup of higher concentrations and onset of disease. Trace element toxicity can manifest with nonspecific symptoms, and often epidemiology is the only possible approach to ascertain their role.

TABLE 1. Biochemical Role, Ascertained or Putative, of Main Trace Elements in Humans

Element	Biochemical Role
Arsenic	Unknown, involved in methionine metabolism
Chromium	Potentiates insulin action; maintenance of glucose metabolism
Cobalt	Integral part of vitamin B_{12}
Copper	Involved in the function of several enzymes, including cytochrome c oxidase, superoxide dismutase
Fluoride	Unknown; protective for skeleton and enamel
Iodine	Part of the thyroid hormones T4 and T3 (controlling growth, development and metabolic processes)
Iron	Present in heme proteins, enzymes (electron carrier)
Manganese	Component of many enzymes and activator of enzymes, especially glycosyl transferases
Molybdenum	Component of molibdoenzymes, oxidoreductases, as part of the molybdenum cofactor (e.g., xanthine oxidase)
Nickel	Probably cofactor for some enzymes (e.g. urease); role in folate metabolism
Selenium	Component of selenocysteine, incorporated into proteins, including glutathione peroxidases, iodothyronine deiodinases, thioredoxin reductases
Silicon	Probably component of glycosaminoglycans complexes in connective tissue; role in cartilage composition
Vanadium	Unknown; regulatory effects
Zinc	Constituent of over 300 metalloenzymes, contributes to membrane structure, regulates gene expression, has a central role in the immune system

2 THE MAIN TRACE ELEMENTS: THEIR ROLES AND EFFECTS

Table 1 reports a brief summary of the biochemical roles demonstrated or presumed for the main trace elements. Table 2 reports data gathered from several sources about nutritional requirements and tolerable levels for most elements mentioned, including a summary on main food sources, deficiency, and toxicity symptoms. The list also includes the main trace elements of no physiological significance constituting an environmental concern due to their occurrence in pollution: cadmium, lead, mercury, tin. Figure 1a shows how, for essential elements, the safe range of intake is comprised between the levels leading to deficiency and toxicity.

2.1 Arsenic

Arsenic is a poison and a carcinogen, but there are evidences of its involvement in the metabolism of methionine [Uthus and Seaborn, 1996] suggesting that it may be essential for man [WHO, 2001b]. It can also be used for therapeutic purposes,

TABLE 2. Summary of Element Properties: Occurrence in Foods, Effects of Deficiency and Toxicity, Dietary Limits

Element	Source in Food Chain	Effects of Deficiency in Animals	Toxicity Symptoms	RDA-US[a]	RDA-EU[a]	RDI[b]	UL[c]	RNI[e]	Normative Requirement According to WHO[f]	RLV[g]
Arsenic	Dairy, meat, poultry, fish, grains, cereals	Reproductive effects, decreased growth	Inorganic arsenic is a poison, known carcinogen	nd			nd. PTWI[d] 0.015 mg/kg bw for inorganic			
Chromium	Cereals, meat, poultry, fish, beer, yeast	Diabete-like state, elevated bloodlipids, decreased fertility	Renal failure	35 µg		120 µg	nd			
Cobalt	Fish, nuts, leafy vegetables, cereals	Wasting disease in animals	Cardiomyopathy, protein deficiency in animals			(in vitamin B_{12})				
Copper	Shellfish, nuts, seeds, liver, kidneys, legumes	High blood pressure, fatigue, anemia, fragile bones	Nausea, muscle pain	900 µg	1.15	1.4 mg (70 kg)	10 mg (60 kg)		M 1.35 mg/day, F 1.15 mg/day	To be established
Fluoride	Water, seafood, tea	Tooth decay	Ulcers, nausea, vomiting, fluorosis	4 mg		3.5 mg	10 mg			
Iodine	Shellfish, seaweed, some fish types	Goiter, mental retardation	Impaired thyroid function	150 µg	130 µg	140 µg/day	1100 µg/day	130 µg/day	100–150 µg/day	150 µg
Iron	Beef, poultry, fish, oysters, egg yolks	Ulcers, fatigue, anemia	Gastrointestinal diseases	8 mg	14 mg	15 mg	45 mg	9–27 mg/day		14 mg

Element	Food sources	Deficiency effects	Toxicity effects							
Manganese	Grains, nuts, bran, fruits, green vegetables, tea	Low SOD activity, cancer susceptibility, skin anomalies	Kidney and liver damage, neurotoxicity	2.3 mg	5 mg		11 mg			
Molybdenum	Legumes, grains, nuts	Reduced fertility (in animals)	Gastrointestinal problems, anemia, elevated uric acid, reproductive effects	45 µg	75 µg		2 mg			0.1–0.3 mg
Nickel	Nuts, legumes, cereals, chocolate	Depressed growth, reproductive performance, altered distribution of elements	Decreased body weight gain	nd	100–300 µg		1 mg/day			
Selenium	Beef, poultry, brown rice, whole grains, offal, brazil nuts	Muscle weakness and fatigue	Fatigue, hair loss, nausea	55 µg	55 µg	35 µg	400 µg/day	34 µg/day	M 27.3 µg/day, F 20.2 µg/day	To be established
Silicon	Plant-based foods, additives	Anomalous composition of cartilage and bone	Decrease in antioxidant enzymes, calculi	nd	20–50 mg		1500 mg/day (60 kg)			
Vanadium	Mushrooms, shellfish, parsley	Iodine metabolism, abortion	Gastrointestinal problems, renal lesions	nd	10–20 µg		1.8 mg/day			
Zinc	Beef, pork, poultry, seafood, whole grains, bran	Growth retardation, reproductive problems, immune disorders, dermatitis	Reduced copper status	11 mg	15 mg	15 mg	40–45 mg	4.2–14 mg/day	M 1.4 mg/day, 15 mg F 1 mg/day	

(Continued)

TABLE 2. *Continued*

Element	Source in Food Chain	Effects of Deficiency in Animals	Toxicity Symptoms	RDA-US[a]	RDA-EU[a]	RDI[b]	UL[c]	RNI[e]	Normative Requirement According to WHO[f]	RLV[g]
Cadmium	Cereals, fruits and vegetables, meat and fish, liver, kidney, crustaceans, molluscs, cephalopods	Depressed growth					PTWI[d] 0.49 mg/week (70 kg)			
Lead	Potatoes, wine, game, fish, meat	Depressed growth, altered iron metabolism					1.75 mg/week			
Mercury	Fish, shellfish (methylmercury), fruits, vegetables, mushrooms (inorganic)						0.35 mg/week: for MeHg 112 µg/week			
Tin	Stored foods in cans	Depressed growth, altered mineral composition	Gastrointestinal problems				980 mg/week			

[a]Recommended Dietary Allowance, the average daily intake sufficient to meet the requirements of nearly all healthy individuals in a life-stage and gender group. Values are reported for a male of 25 years.

[b]Reference Daily Intake, replaced RDA.

[c]Upper limits (ULs) of nutrient intake are defined as the maximum intake from food, water, and supplements that is unlikely to pose risk of adverse health effects from excess in almost all (97.5%) apparently healthy individuals in an age- and sex-specific population group.

[d]Provisional Tolerable Weekly Intakes (PTWI) is an estimate of the amount that can be ingested weekly over a lifetime without appreciable risks to health.

[e]Recommended nutrient intake (RDI) is the daily intake, which meets the nutrient requirements of almost all (97.5%) apparently healthy individuals in an age- and sex-specific population group. Values are reported for a male of 19–65 years, weight 65 kg. The definition of RDI used is equivalent to that of the recommended dietary allowance (RDA) as used by the Food and Nutrition Board of the United States National Academy of Sciences.

[f]The quantity sufficient to maintain tissue stores judged to be desirable, as defined by FAO and WHO.

[g]Reference Labelling Values provided by Codex Guidelines on Nutrition Labelling [1985] for some vitamins and minerals, to be used for labeling purposes.

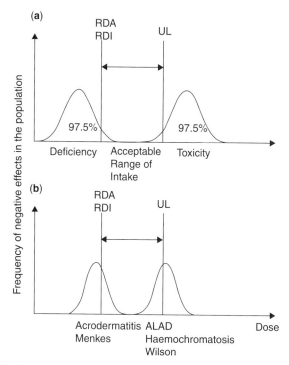

Figure 1. Distribution of negative effects due to trace elements in normal (**a**) and susceptible (**b**) populations. RDA, recommended daily allowance; RDI, recommended nutrient intake; UL, upper limit. (a) RDA and RDI are sufficient for avoiding deficiency symptoms in 97.5% of healthy individuals, whereas UL are sufficient to avoid toxicity symptoms in 97.5% of healthy individuals. (b) Susceptible populations may manifest deficiency symptoms at levels above RDA-RDI and may show toxicity symptoms below the UL. Examples of susceptibility syndromes are reported.

even against some types of cancer [Florea and Büsselberg, 2006]. Arsenate, As(V), is similar to phosphate and is transported with the same mechanisms.

Deficiency of arsenic in animals affects heart muscle and mitochondria; it also leads to altered methionine metabolism. In man it is related to nervous system disorders and vascular diseases.

Arsenic occurs in many organic and inorganic forms in food. In the marine environment it is often organic and can reach 50 mg/kg of wet weight in some seaweed and animals. The most toxic forms are As(III) and As(V), whereas the methylated organic forms are considered non toxic. Toxicity involves dermatosis, liver damage, neurological problems, and cancers.

2.2 Cadmium

Cadmium enters the environment because of human activities, mainly concerning its industrial use and waste disposal [WHO, 1992]. The main source of human exposure is food, but tobacco smoke is important also.

The kidney is the target organ for toxicity, and chronic accumulation in the kidney cortex leads to disfunctions and loss of proteins, amino acids, and glucose in urine. Cadmium toxicity also brings to reproduction problems, cancer, cardiovascular diseases and hypertension. Cadmium is classified as human carcinogen.

A well-known endemic disease associated to cadmium is the Itai-itai disease in Japan, with bone abnormalities and defects in calcium metabolism. In the area, food contains high quantities of cadmium, lead, and zinc. The disease affects almost exclusively females [WHO, 1992].

2.3   Chromium

Chromium exists in soils as Cr(III), whereas Cr(VI) is produced by human activities [WHO, 1988a]. Chromium is an essential element, required for carbohydrate and lipid metabolism, and its deficiency may be associated to cardiovascular disease.

Excess of chromium is very unlikely. However, it interacts with iron binding to transferrin, so it can alter the metabolism and storage of iron. Chronic exposure to Cr(VI) induces renal and liver failure. Cr(VI) is more toxic than Cr(III), genotoxic and probably carcinogenic.

2.4   Cobalt

Cobalt is an essential element as an integral part of vitamin B_{12} and is therefore essential for folate and fatty acid metabolism.

Deficiency in humans is not reported, but deficiency in animals causes anemia.

Toxicity from excess cobalt leads to cardiomyopathy, with damage to the heart muscle due to anoxia [Expert Group, 2003].

2.5   Copper

Copper is essential for the functioning of many metalloproteins and enzymes, and it also plays a role in regulation of gene expression. It is required for growth, defense, bone strength, blood cells production, iron transport, and metabolism. Copper uptake and excretion are efficiently regulated in humans, so deficiency and toxicity are quite rare.

Copper deficiency leads to anemia, decrease in white blood cells, neurological diseases, osteoporosis, and disorders of the connective tissues. It can derive from an X-linked inherited disease, Menkes' disease, characterized by reduced intestinal absorption.

Toxicity from excess copper is very rare, mainly from contaminated water: It causes gastrointestinal problems. Also in this case, it may occur in specific diseases: Wilson's disease and idiopathic copper toxicosis. In Wilson's disease, excretion in bile is reduced and copper accumulates in liver, brain, and kidney [Sadhra et al., 2007].

Copper interacts with other elements, and especially with zinc. High zinc intake inhibits intake of copper by competition for transporters.

2.6   Fluorine

Fluorine comes especially from water, whereas plants are a very poor source in human diet [WHO, 2002a]. It stimulates bone growth and improves dental health, but in excess it causes fluorosis with calcification of ligaments, increased bone density, and bone formation in cartilages, and it interferes with calcium metabolism. The correlation between fluorides in groundwater and dental health of communities is one famous example showing how element distribution affects human health [Dissanayake and Chandrajith, 1999].

2.7 Iodine

Iodine is required for the synthesis of thyroid hormones T_4 and T_3, and it accumulates in the thyroid gland. The hormone is involved in growth and development, and in controlling the metabolic processes.

Deficiency of iodine during the early development of brain and nervous system leads to cretinism, which is irreversible. At a later age it leads to hypothyroidism and goiter, the first endemic disease that was attributed to environmental factors. It seems now that selenium also plays a role in goiter and that other trace elements may be involved. Most locations affected by iodine deficiency are in tropical environments [Dissanayake and Chandrajith, 1999]. Several substances, the so-called goitrogens, interfere with iodine absorption by blocking thyroidal uptake.

Iodine excess can be harmful too, because it inhibits the synthesis of thyroid hormones, especially in children. Excess in plasma concentrations inhibits uptake in the thyroid, and iodine is transformed from inorganic to organic [EC SCF, 2002].

2.8 Iron

Iron is the most abundant trace element in human body, a structural component in heme proteins: hemoglobin, myoglobin, and cytochrome-dependent proteins. A large quantity is also stored in proteins like ferritin and hemosiderin. Its role is in oxidoreduction, by conversion of the Fe(II) and Fe(III) forms. The free ion would cause oxidative stress.

Iron deficiency leads to severe anemia.

Excess iron can give gastrointestinal problems, vomiting, and diarrhea, ending with cirrhosis for chronic exposure. It may also be carcinogenic. Hemochromatosis is a genetic disease in which iron is absorbed in excess from the diet and accumulates in different tissues with altered distribution [Griffiths, 2007].

Iron interacts with other metals: copper, manganese, zinc, and chromium. Iron-binding proteins can bind other cations, such as manganese, zinc, and vanadium, thereby contributing to their transport [Ballatori, 2002]. Its absorption is inhibited by calcium.

2.9 Lead

Lead has been known to be toxic since ancient times [WHO, 1995]. It is emitted by several human activities and reaches humans through the food chain and in drinking water. It causes reproduction failures, encephalopathy, neurophysical defects, anemia, renal damage, hypertension, and poisoning. One of the most important effects concerns inhibition of heme synthesis. In children, impairment of neurological development is the worst implication. The similarity in chemical properties between lead and calcium causes the accumulation of lead in bones, where it is stored.

2.10   Manganese

Manganese is a constituent in many enzymes, and it also acts as activator [WHO, 1980b]. It is necessary for connective tissues and bones, for general metabolism and reproductive functions. Food is the major source.

Deficiency leads to alterations in hair, nails, and skin. In animals it is associated with bone malformations, impaired growth.

Toxicity in polluted working environments leads to manganism, a neurological disease.

Manganese competes with iron for the same absorption sites.

2.11   Mercury

Methylmercury is the most toxic form of mercury, and it is present in fish and seafood products [EFSA, 2004]. For physiological reasons, some fish species concentrate mercury more easily: Methylmercury accumulates along the food chain, and therefore predatory animals, like swordfish and tuna, have higher contents. Methylmercury could be found also in animal meat after feeding with fish products. In other foods, mercury is inorganic and therefore not toxic [WHO, 1991b].

The developing brain is the most important target for methylmercury toxicity, and therefore the susceptible life stage is the developing fetus during pregnancy. In addition, it may increase the risk for cardiovascular disease.

2.12   Molybdenum

Molybdenum is present in molybdoenzymes, mainly oxidoreductases in which the element changes the valency state.

Deficiency has not been described in humans. In animals it is associated with reduced fertility and increased mortality. A lethal inherited disease involves molybdenum deficiency, with neurological disorders and other symptoms.

Molybdenum absorption is affected by the presence of copper and sulfates.

2.13   Nickel

Nickel is a cofactor for some enzymes, urease and others [WHO, 1991a]. It is involved in the metabolism of methionine, vitamin B_{12}, and folate, and therefore it is suggested that it may be essential [Uthus and Seaborn, 1996].

Deficiency leads to reduced growth, reproductive problems, and altered distribution of other elements.

Toxicity also leads to anemia and decreased growth. Sensitive individuals may react to low doses.

2.14   Selenium

Selenium is quite unique as trace element, because it is a component of an amino acid, selenocysteine, typical of selenoproteins and therefore involved in very specific biological roles. Enzymes depending on selenium perform very important roles in the cells, as shown in a recent review [Rayman, 2000]. The main roles are in protection against oxidative damages, defenses against infection, and modulation of growth and development. The main exposure to selenium occurs through food, and its distribution in the natural environment has a marked effect on its content in soils, crops, and the human body [WHO, 1987].

In most instances, a reduced intake of selenium can be connected to several effects: loss of immunocompetence, increased virulence of viral diseases including HIV, early pregnancy loss, depression and other negative mood states, hypothyroidism, cardiovascular diseases, inflammation states, and cancer incidence. Many of these effects can be linked to roles of the selenoproteins, and especially of glutathione peroxidases active as antioxidants and cell protecting agents. Deficiency of selenium in the diet causes diseases which often are endemic to specific regions. Keshan disease is a cardiomyopathy distributed in regions in China, attributed to a viral myocarditis enhanced by selenium deficiency in tissues. Kashin–Beck disease is a deforming arthritis (enlarged joints), endemic to China and Siberia. In both cases, the levels of glutathione peroxidase are reduced.

However, selenium is a toxic mineral with a narrow window for beneficial effects, and there are sensitive individuals who have to limit their intake [Rayman, 2000]. Overexposure leads to hair loss and gastrointestinal disturbance. Endemic diseases, chronic selenosis, are linked to selenium excess in China.

Selenium interacts with iodine because the enzyme deiodinase, which converts T_4 to T_3, contains selenium. Effects of iodine deficiency are worsened by selenium deficiency.

2.15   Silicon

Silicon influences bone formation and in particular the cartilage composition.

In animals, deficiency symptoms concern the metabolism of connective tissue and bone.

Toxicity symptoms concern the reduction of antioxidant enzymes and the formation of renal calculi [Uthus and Seaborn, 1996]. It has been classified as known human carcinogen by inhalation.

2.16   Tin

Inorganic tin can be present at oxidation states $+2$ and $+4$. No physiological function has been described for tin in humans [WHO, 1980]. Food is the main source of tin for man. Most importantly, tin can be leached in food and beverage packaged in steel cans coated with tin. However, it seems that health effects of inorganic tin found as leachate in canned food are negligible [Blunden and Wallace, 2003]. Organic compounds (organotin) are applied in agriculture as fungicides, herbicides, antifouling agents, and so on. Organic compounds present in food have higher absorption rates than inorganic tin. Bivalve molluscs are the main source of organotin compounds, together with marine and fresh water fish [EC, 2003c].

Deficiency has never been demonstrated, so its role as essential element is disputed. In rats, however, it is essential.

Toxicity from acute poisoning with inorganic tin gives gastrointestinal problems.

2.17 Vanadium

It has regulatory effects on enzymes as vanadate; in vivo, vanadyl forms would not have the same effects. It is considered an essential element for small animals, but for man there is no daily dietary requirement [WHO, 1988b].

In animals, deficiency affects the iodine metabolism and the lipid metabolism, and it causes high abortion rate and higher death rate. Evidence for vanadium deficiency in man is not available.

The main source for human exposure is food. Vanadate is structurally similar to phosphate and can compete with it for transport. The element can cross the blood–brain barrier. Toxicity is manifest especially after inhalation; symptoms include green tongue, and diarrhea, cramps [Uthus and Seaborn, 1996; Mukherjee et al., 2004]. Toxicity from food is limited because of the very low absorption in the gastrointestinal tract.

2.18 Zinc

Zinc is an essential nutrient, present in all tissues of the human body; it is a structural components of over 300 enzymes, important for metabolism of all macromolecules, in the metabolism of nucleic acids and in metabolism of other minerals [EC SCF, 2003]. It also has an important role for gene expression as a constituent of transcription factors (zinc finger domains). In the human genome, about 10% of proteins have the potential for binding zinc [Andreini et al., 2006].

Zinc deficiency is rare, but suboptimal intake of zinc is frequent. Deficiency therefore impacts many aspects of health, growth, reproduction, susceptibility to infections, and behavior. Acrodermatitis enteropathica is a genetic disease that leads to zinc deficiency.

Excess zinc derives from pollution [WHO, 2001a]. Symptoms of acute poisoning are nausea, vomiting, diarrhea, lethargy, and fever. Chronic exposure should interfere with copper status and immune response interfering with reproduction.

Zinc and copper are mutual antagonists, interfering for absorption in the intestine. Zinc also interferes with iron absorption.

2.19 Hypersensitivity Issues

Current guidelines and dietary recommendations (see Table 2 and Fig. 1a) are formulated for healthy individuals representing the average population. In the population, however, there may be individuals who show hypersensitivity or hypersusceptibility to some elements. For these persons, a normal intake according to recommendations could be detrimental or toxic: toxicity may ensue at levels below the Upper Tolerable Limit (UL), or deficiency symptoms may appear even at the Recommended Daily Allowance (RDA) (Fig. 1b). Susceptibility may be caused by factors such as nutrition, disease, physiological status, gender, and life habits, but it often depends on genetic factors, polymorphisms in genes which affect the reaction of the individual to a contaminant or a trace element. Polymorphisms for susceptibility are sometimes not manifested in visible phenotypes, and therefore recognition of carriers requires identification of the genetic variant through molecular analyses. In recent times, the Environmental Genome Project has appeared as an important resource that collects polymorphism data on genes involved in susceptibility to environmental contaminants [Tiffany-Castiglioni et al., 2005].

One well-known example concerns susceptibility to lead toxicity. One of the main targets in lead toxicity is the enzyme delta-aminolevulinic acid dehydratase (ALAD), involved in the synthesis of heme. Lead interaction with this enzyme displaces zinc and inhibits its function, leading to anemia and accumulation of aminolevulinic acid in blood and urine. It has been demonstrated that the ALAD gene is polymorphic in humans, with two alleles differing for one base substitution leading to one amino acid change in residue 59 (lysine to asparagine). Even if the two enzyme forms do not differ in enzyme activity, the allele ALAD-2 seems to increase sensitivity to lead toxicity in the carriers, due to alterations in lead metabolism and delivery to target organs [Kelada et al., 2001; Montenegro et al., 2006]. Studies on exposed individuals also demonstrated a relationship between ALAD-2 genotype and lead-associated renal effects: It is hypothesized that ALAD-2 individuals apportion less lead to bones, thereby increasing the availability to kidneys and other organs [Smith et al., 1995]. In particular, it has been demonstrated that the enzyme encoded by ALAD-2 binds lead more strongly than the enzyme encoded by ALAD-1. This polymorphism could also protect some organs from harmful effects, because ALAD-2 may sequester lead making it less available. This genetic variant is present in 11–20% of Caucasians, who have the highest frequency [Kelada et al., 2001].

It has also been reported that all polymorphisms affecting red blood cells metabolism can increase lead susceptibility: hemoglobin anomalies, thalassaemia, glucose-6-phosphate dehydrogenase variants [Goyer and Mahaffey, 1972].

Menkes' disease and Wilson disease are extreme forms of hypersensitivity, because they result in manifest disease, allowing recognition of hypersensitive individuals. Menkes' disease is a defect in the copper transporter gene ATP7A; copper becomes trapped in intestinal cells, does not reach the liver, and is not distributed to the body. Only intravenous administration of copper from the early stages of life can avoid the development of symptoms. In Wilson disease, excretion of copper to the bile is reduced, leading to accumulation in the liver [Cox, 1999]. In both cases, P-type ATPases are involved, and they are localized to intracellular vesicles [Ballatori, 2002].

Acrodermatitis enteropathica is a genetic disease with autosomal recessive inheritance which causes zinc deficiency [Eide, 2006]. The defect concerns the gene ZIP4/SLC39A4 on chromosome 8, encoding for a zinc transporter of the ZIP family (Zrt-, Irt-like protein), expressed in the plasma membrane of the apical surface of enterocytes and affecting zinc uptake in the intestine. The symptoms of the disease overlap with those of zinc deficiency, mainly dermathological lesions (hence the name), immune defects, reproductive problems, defects in growth, and digestive problems. ZIP transporters have been studied in all organisms, even fungi and plants, and they are typically up-regulated by zinc deprivation; the human genome contains about 12 ZIP genes (mammalian ZIP are correctly designated as SLC39).

Hemochromatosis is a genetic disorder leading to excess in iron absorption [Griffiths, 2007]. Normally, only about 10% of the iron in diet is absorbed, whereas affected people absorb up to 30%, highly in excess with respect to the physiological needs. Elimination of iron absorbed in excess is not possible, and so it accumulates in the liver, heart, skin, and pancreas, eventually leading to arthritis, heart failure, cirrhosis, diabetes, and tumors, culminating in death in some instances. The gene involved is HFE, on chromosome 6, with recessive mutations C282Y and H63D: The gene product interacts with the transferrin receptor. Homozygotes for C282Y have a frequency of 0.5% in US Caucasians, whereas carriers are 10%; symptoms are developed only by some homozygote individuals, or in compound heterozygotes C282Y/H63D, and therefore penetrance is incomplete. This disease is also known as the Celtic curse, due to its high occurrence in persons with this genetic background (Irish, Welsh, Scots). It has been suggested that these mutations were subject to positive selection in a period in human history when iron in the diet was scarce. Other types of hemochromatosis exist, which are caused by different genes. Symptoms of the disease are quite subtle, and diagnosis is difficult; usually diagnosis is earlier in men than in women: Loss of iron during menstruation helps alleviate the symptoms until menopause ensues. Pain in joints is the most common symptom. Affected people, but also carriers, have to avoid taking iron supplements and vitamin C supplements which enhance iron absorption, avoid cooking in iron utensils, and reduce consumption of red meat and ethanol. Due to the difficult diagnosis and frequency of occurrence, it has been suggested that genetic screening could be beneficial in early identification of affected people, improving their treatment. Tests are, however, still too expensive to be widely applied: iron load in blood, transferrin saturation test, ferritin level, or genetic analyses for the mutations.

As is always the case when genetic screening is involved, several issues have to be considered: the frequency of affected patients and carriers; the risk for negative effects in affected people with no apparent symptoms; the usefulness of the available tests in accurate diagnosis; the effectiveness of early treatments in preventing full developed symptoms; the risks associated to the screening tests and to treatment. At the current state of knowledge, screening is considered beneficial only for relatives of affected patients.

In view of the increasing levels of environmental contamination, which includes pollution by metals and trace elements, the issue of protection of hypersensitive individuals becomes more and more important. Regulators establishing risk levels and limits for concentrations of pollutants should be made aware of these issues.

3 ISSUES OF ENVIRONMENTAL CONTAMINATION OF THE FOOD CHAIN

Trace elements are present in the environment because they derive from parent material of soil. In this case, their concentration depends on the content in the underlying rock. This leads to regional variations in soil concentrations.

Increasingly, in recent times, their natural cycling has been perturbed by anthropogenic causes, since human activities disseminate them into the environment: Because soil is the main reservoir of elements, their input to soil derives from atmospheric deposition, water, and direct input. When elements exceed their natural background and derive from anthropogenic source, contamination arises. A thorough examination of pollution with toxic metals and trace elements is beyond the scope of this review, and therefore we will describe only the issues more tightly linked to contamination of the food supply chain.

In particular, direct input to soils comes from (i) commercial mineral fertilizers, (ii) agricultural chemicals, (iii) compost, sewage sludge, and manure, and (iv) waste disposal.

Mineral fertilizers contain several trace elements, depending on the parent material. Phosphatic and NPK fertilizers, especially, can contain cadmium [AROMIS, 2005]. It has to be considered that fertilizers can modify availability of trace elements in the soils to plants, therefore interfering with plant nutrition.

Many agrochemicals like fungicides and pesticides can be organometallic compounds containing specific metals in high concentrations—for example, copper in vineyards sprays, arsenic in orchards, mercury for seed dressing [Senesi et al., 1999].

Sewage sludge contains metals deriving from the input wastes, less from industrial sources and more from domestic sources [AROMIS, 2005]. Composts are not commonly used in agriculture, but their usage is due to increase in future. The variability in the raw material leads to extreme variability in their content of elements [AROMIS, 2005]. Manure from animal farming can contribute metals supplied as dietary supplements (see below). Other types of wastes, even from industrial sources, are applied to agricultural soils for conditioning or for nutrients, and these may contain specific elements.

Water used in irrigation and effluents spread on the soil can also contain trace elements [Senesi et al., 1999]. Other sources of elements in soils are flooding events, erosion from mining sites, runoffs from roads, and corrosion of structures covered with metals. In this case, elements in the soil are taken up by plants and microorganisms, but they are also subjected to percolation phenomena.

Atmospheric deposition derives from traffic, combustion emissions, and manufacturing and mining processes, and it is a long-distance source of contamination. The particles containing the elements fall to earth by dry or wet deposition, on the surface of plants or directly on the soil. Transport of particles depend on the source and on meteorological conditions. In Europe, zinc is the element deposited in largest amounts from atmosphere, followed by lead and copper [AROMIS, 2005]. For several years, lead was a major pollutant coming from traffic emissions, due to leaded gasoline, contaminating plants through deposition and root uptake [WHO, 1995].

As far as waters are concerned in contamination of the food chain, it must be considered that different aquatic animals act as concentrators of polluting elements and therefore their consumption is risky: Filter-feeding molluscs can be accumulators, and also crustaceans can concentrate metals in the hepatopancreas. Oysters are, for instance, accumulators of cadmium [WHO, 1992]. Aquatic organisms are known to concentrate organic arsenic and methylmercury species.

4   LEGISLATION CONCERNING TRACE ELEMENTS

Several acts of legislation deal with trace elements.

First, there is a group of norms concerning the presence in the environment of elements, mainly heavy metals and toxic elements. Second, there is a group of norms dealing with elements in foods. In this case, there are laws about the content of toxic elements (contaminants) and norms regulating the supplementation of elements and the labelling for nutritional claims.

4.1   Elements in Soils and the Environment

Most countries, and all the countries in the EU, have a legislation controlling the distribution of metals in the environment by means of sewage sludge [Hillman et al., 2003; AROMIS, 2005]. Directive 86/278/EEC [EC, 1986] gives to Member States the task of defining limit concentrations in soils, and in certain conditions spreading of sludge is forbidden: Limit values concern cadmium, chromium, copper, mercury, nickel, lead, and zinc in soil, in sludge, and as annual input (Table 3). The conditions may vary in the different countries.

National legislations also can limit the amount of metals in compost or sludge. Since there is no direct correlation between toxicity and amount of metals in the soil, the legislation takes into account that pH can affect bioavailability of metals, in particular for zinc, copper, and nickel; at lower pHs, the permitted concentration

TABLE 3. Limits for Heavy Metal Content in Agricultural Soils, and Allowed Metal Distribution Through Sludges According to Directive 86/278/EEC

Element	Limit in Soils (if pH is Between 6 and 7), mg/kg Dry Weight	Limit in Sludge, mg/kg Dry Weight	Maximum Amount Allowed to be Distributed, kg/ha year
Cadmium	1–3	20–40	0.15
Copper	50–140	1000–1750	12
Nickel	30–75	300–400	3
Lead	50–300	750–1200	15
Zinc	150–300	2500–4000	30
Mercury	1–1.5	16–25	0.1
Chromium	nd[a]	nd	nd

[a]nd, not determined.

of metals is lower, because their availability is greater. Analysis of soil pH and of sludge composition is required before ascertaining applicability of sludges. Application to arable crops is allowed only in certain periods, whereas pasture can be supplemented with sludge in all periods [Hillman et al., 2003]. In fact, application is timed with planting, grazing, and harvesting. Application to grassland is mainly on the surface, and it can adhere to the surface of plants. Livestock can therefore be exposed to metals by eating the plants, but also by ingestion of soils.

Metals in manure are controlled by Directive 96/61/EC [EC, 1996]. Composition of fertilizers is also controlled, especially in the case of organic agriculture, by Regulation (EC) 2003/2003 [EC, 2003b].

Indirectly, also legislation concerning supply of elements to livestock concerns environmental protection. Directive 2002/32/EC, amended by Directive 2003/100/EC [EC, 2002, 2003a], states that arsenic, lead, fluorine, mercury, and cadmium are undesirable substances. Maximum limits in different feeds are established.

Finally, water protection according to Directive 2006/11/EC limits input of toxic elements to the environment and the food chain [EC, 2006b].

4.2 Elements in Foods

In many countries there are regulations on food contaminants, expressing concentration limits for specific classes of compounds. The Codex Alimentarius Commission—in particular the Codex Committee on Food Additives and Contaminants (CCFAC)—has prepared several documents with standards for different foods. At the same time, the European Commission has issued legislation concerning contaminants in food. National requirements may also exist.

In general, the concentrations of contaminants reported in the legislation, maximum limits (MLs), are based on the ALARA principle: as low as reasonably achievable. There should be scientific justification for these limits, but this is sometimes unavailable; the precautionary principle can be applied in these cases. The species of the elements may be in some cases be very important: Mercury as methylmercury rather than inorganic; organotin compounds rather than inorganic tin; arsenic species with different toxicity; Cr(III) or Cr(VI). Iron is one element for which speciation issues are important, since its form affects the bioavailability: Heme iron from meat is more readily available than non-heme iron from vegetables. Reliable procedures are needed to assess speciation of the elements, in addition to determination of the total element content. Legislation concerning elements in food usually refers to the total element content, at the moment [Cornelis et al., 2001].

In view of the problems linked to international trade, the choice is to move towards international standards, with lower MLs, to protect the consumers [Berg and Licht, 2002]. In this direction, also trace elements that may give allergy problems, such as nickel, will have to be considered. In the General Standard CODEX STAN 193–1995 [Codex Alimentarius Commission, 1995] the MLs for some metals are reported (Table 4).

The principles of EU legislation on food contaminants are expressed in Council Regulation 315/93/EEC of 8 February 1993 [EC, 1993]. A contaminant is defined

TABLE 4. Maximum Limits of Heavy Metals in Food Products, as Established by Commission Regulation (EC) 1881/2006, and Comparison with the CODEX Standard STAN 193 of 1995

Element	Concentration Limit	Commission Regulation	CODEX Standard
Arsenic	0.1 mg/liter		Mineral waters
	0.1 mg/kg wet weight		Fats and oils
	0.5 mg/kg wet weight		Salt
Lead	0.01 mg/liter		Mineral waters
	0.02 mg/kg wet weight	Milk, infant formulae	Milk, infant formulae, wine
	0.05 mg/kg wet weight	Fruit juices	Fruit juices
	0.1 mg/kg wet weight	Meat (bovine, sheep, pig, poultry), some vegetables, fruits, fats and oils	Fruits, bulb, root and fruiting vegetables, meat, fats and oils
	0.2 mg/kg wet weight	Cereals, legumes and pulses, berries, wine	Berries, cereal grains
	0.3 mg/kg wet weight	Fish muscle, brassica, leafy vegetables, fungi	Brassicas and leafy vegetables, fish
	0.5 mg/kg wet weight	Edible offal, crustaceans	Offal
	1.0 mg/kg wet weight	Cephalopods	Canned fruits and vegetables, jams
	1.5 mg/kg wet weight	Bivalve molluscs	Tomato concentrates
	2.0 mg/kg wet weight		Salt
Cadmium	0.003 mg/liter		Mineral waters
	0.05 mg/kg wet weight	Meat (bovine, sheep, pig, poultry), some fish muscle, some vegetables, fruits	Brassicas, bulbs and fruiting vegetables
	0.1 mg/kg wet weight	Some fish muscle, cereals excluding bran, potatoes, stem vegetables	Legumes, potatoes, pulses, root vegetables, stalk and stem vegetables, cereals
	0.2 mg/kg wet weight	Horsemeat, bran, soybeans, leafy vegetables, herbs, fungi;	Leafy vegetables, wheat
	0.3 mg/kg wet weight	Swordfish	
	0.4 mg/kg wet weight		Rice

(Continued)

TABLE 4. *Continued*

Element	Concentration Limit	Commission Regulation	CODEX Standard
	0.5 mg/kg wet weight	Liver, crustaceans	Salt
	1.0 mg/kg wet weight	Kidneys, bivalve molluscs, cephalopods	
	2.0 mg/kg wet weight		Molluscs and cephalopods
Mercury	0.001 mg/liter		Mineral waters (inorganic)
	0.1 mg/kg wet weight		Salt (inorganic)
	0.5 mg/kg wet weight	Nonpredatory fish	Fish (methylmercury)
	1.0 mg/kg wet weight	Predatory fish	Predatory fish (methylmercury)
Tin	50 mg/kg wet weight	Canned baby foods, infant formulae, dietary foods	Cooked cured meat in specific containers
	100 mg/kg wet weight	Canned beverages including juices	
	200 mg/kg wet weight	Canned foods other than beverages	Canned strawberries
	250 mg/kg wet weight		Canned fruits and vegetables, tomato concentrates

as "any substance not intentionally added to food which is present in such food as a result of the production (including operations carried out in crop husbandry, animal husbandry, and veterinary medicine), manufacture, processing, preparation, treatment, packing, packaging, transport, or holding of such food, or as a result of environmental contamination." Levels of contaminants should be kept as low as can reasonably be achieved, and a food exceeding the limits cannot be marketed. There are regulations setting the maximum levels for certain contaminants. Among trace elements, those mentioned by the legislation are lead, cadmium, mercury, and tin. Regulation (EC) 333/2007 establishes methods for sampling and analysis [EC, 2007a].

Commission Regulation (EC) 1881/2006 of 19 December 2006 [EC, 2006a] establishes limits for some heavy metals in food amending the previous Commission Regulation (EC) 466/2001 of 8 March 2001 [EC, 2001]. The limits for lead, cadmium, mercury, and inorganic tin are reported in Table 4.

4.3 Supplementation of Minerals to Foods

Regulation (EC) 1925/2006 concerns the addition of minerals to foods, considering that this can contribute to improve the nutritional status across the EU [EC, 2006c].

In fact, it is recognized that some population subgroups do not have an optimal intake of vitamins and minerals, especially considering iron, calcium, and zinc [Flynn et al., 2003]. The legislation envisages the preparation of positive lists of minerals allowed, only those essential for human nutrition, and of the chemical forms in which they can be supplied. Moreover, it declares that minimum and maximum amounts for them should be set, subject to revisions. The minerals that can be added are the following, as reported in the Community Register [EC, 2007b]: calcium, magnesium, iron, copper, iodine, zinc, manganese, sodium, potassium, selenium, chromium, molybdenum, fluoride, chloride, and phosphorus. Labeling describing the additions is compulsory.

5 FOOD CHAIN SAFETY

5.1 Soil and Plants

The geochemical distribution of elements impacts human health, especially in those locations in which population is still in contact with the natural environment. In developing countries the health effects of metals in the local environment are much higher, as compared with what happens in developed countries [Dissanayake and Chandrajith, 1999]. Micronutrient deficiency in agricultural soils is therefore of high significance, and data indicate that it is widespread [White and Zasoski, 1999]. A soil is considered deficient when addition of the nutrient improves plant growth.

Mineral elements are introduced into the diet of humans with food and water, and plants can be considered to be the primary source of their presence in food products. Plants take up minerals during water uptake, and it is usually considered that both essential and nonessential elements follow the same route. It means that plants can also take up noxious elements, even those of no physiological significance to them, because the transport mechanisms lack specificity. Indeed, some plants are even hyperaccumulators or accumulators for specific elements. Nickel is one element for which several hyperaccumulator plants have been reported, with levels above 1 mg/kg dry weight [WHO, 1991a].

Transfer from soil to plants is a very debated issue, and it depends on several factors leading to the so-called "bioavailability" of elements [White and Zasoski, 1999; Caussy et al., 2003]: element species, its concentration, soil pH, plant species, and cultivar/ecotype. Only a small portion of the total element content is available to plants. Deficiency can result from small concentrations in soils, but also from reduced availability to plants and from interactions among metals decreasing availability. For instance, the incidence of iodine deficiency disorders in Sri Lanka is not due to low iodine in soils, but rather to low iodine in plants [Dissanayake and Chandrajith, 1999], whereas cadmium accumulated in crop plants, especially rice, was the cause of the outbreak of Itai-Itai disease in Japan at the beginning of the twentieth century [Senesi et al., 1999]. Elements with similar electron arrangements are often antagonists in biological systems (e.g., iron, copper, and zinc).

Determination of metals and trace elements preferentially requires techniques based on physical principles, such as atomic emission spectrometry, atomic absorption spectrometry, X-ray fluorescence spectrometry, and others.

Soils may be extracted with solutions of increasing strength used in succession. In this way, called sequential extraction, nutrients in soils are subdivided in fractions which are considered to be differently available to extraction and to plant uptake— for example, soluble, exchangeable, weakly absorbed, organically bound, oxide and carbonate phases. Those which are not extracted except by complete digestion are the residual fraction [White and Zasoski, 1999]. A comprehensive recent review [Filgueiras et al., 2002] reports an extensive literature on the subject, comparing different schemes for sequential extraction.

The rhizosphere is the volume of soil around plant roots which is affected by root activity, and it is where nutrient uptake takes place [Hinsinger et al., 2006]. Root activity can change the pH and in this way increase or decrease the mobility of cations. Also secretion of compounds from the roots can affect the complexation and uptake of minerals.

For manganese it has been reported that its availability to plants is affected by the activity of microorganisms altering pH and oxidoreduction potentials [WHO, 1980b].

Once in plants, elements and metals move and reach different tissues [Broadley et al., 2007]. The element ions, acquired from the soil solution free or bound to ligands, enter the root cells cytoplasm in epidermis and cortex, moving via symplastic transport. They are then pumped into the xylem vessels in the central stele and transported to the upper portions of the plants. Movement of cations through the cell walls, in the apoplastic transport, is affected by the cation exchange capacity.

5.2 Animal Products

Another input to the food chain comes from supplementation of trace elements in livestock feed, usually to improve their availability for the animals [EC SCAN, 2003a,b]. Copper, for instance, is given to growing pigs to enhance their performance, acting probably also as antibacterial agent [Directive 70/524/EEC; EC, 1970]. Zinc is also given to pigs and other animals, up to 250 mg/kg in feedingstuff. Copper and zinc are supplemented to poultry as cofactors for enzymes.

Toxic elements like cadmium and lead are not supplemented but may be present as contaminants. Nickel and chromium may be considered essential for animal nutrition; some National legislation allows supplementation of chromium depending on the status of the animals. In EU, supplementation with these elements is not allowed in feed [AROMIS, 2005]. In some instances, supplementation exceeds the real needs, and this can have consequences on the presence of trace elements in animal products. Part of the elements is transferred to milk and meat, but most is excreted and can return to soil with manure and sludge. Consuming offal from animals usually brings many minerals to the diet. This is mainly because animals store excess metals in liver, kidney, and other organs, and they store much less in muscle.

5.3   Geological Correlates

Medical geochemistry is the discipline studying correlations between endemic disease and geological features: Concentration of element in the soil and water affects the local population through availability in the diet.

Selenium is one of the elements for which geological correlates are stronger. As previously stated, the effects of selenium deficiency have been identified in regions where the low content of selenium in the soil is a problem, starting with the observation and identification of disease in livestock. Daily intakes of selenium in European countries are often about half of the recommended dose, and the level in blood is not sufficient to guarantee full functionality of the glutathione peroxidase [Rayman, 2000]. Foods contributing selenium to the diet are quite rare, and they depend on geographic origin. For instance, wheat is a good source in the United States but not in Europe: selenium content in wheat ranges worldwide from 0.001 to 30 mg/kg [Broadley et al., 2006]. This difference is mostly due to geological features and not to agronomic practices. In China, the incidence of the Keshan disease and of the Kashin–Beck disease correlate closely with the distribution of selenium deficiency in soils, in grains, in human hair, and in human blood [Tan et al., 2002]. In particular, acid soils can bind selenium in forms that are not absorbed by crops [FAO/WHO, 2001]. On the other hand, in regions with high selenium content such as Norfolk in United Kingdom, the beneficial effect of selenium is a reduced incidence of cancer to the stomach, longer life duration, and fewer heart diseases.

Again in China, there are regions rich in selenium in which daily intake is higher than the recommended dose, up to 1338 micrograms per day [FAO/WHO, 2001]. Some plants accumulate selenium at levels which are toxic for livestock [WHO, 1987], and it has been speculated that the defeat of General Custer at Little Big Horn was due to poisoning of horses by forage with too high concentrations of selenium. Accumulating plants contain over 1 g/kg dry weight, but they are not consumed directly by humans. The levels of selenium in animal tissues, however, are somewhat buffered, so the fluctuations of the element in animal products is less, as compared to plant products. Cooking practices lead to a loss of selenium from foods. Therefore, the case of selenium exemplifies how concentration in soils, availability to plants, and entry into the food supply chain cause deficiencies, toxicities, or beneficial effects.

Also iodine deficiency is linked to its irregular distribution, with areas at risk in mountainous regions and flood plains. Leaching and erosion deplete iodine compounds from rocks and soils, and iodine accumulates in the seas and surface waters. It is reported that 1600 million people are at risk for iodine deficiency due to scarcity in the environment [EC SCF, 2002]. The only possible correction is a supply from external sources—for example, salt, bread, oil, fortified with iodine.

Zinc is the most common deficiency experienced by crops, especially in soils with high pH [Broadley et al., 2007]. On the other hand, crops growing on contaminated soils can accumulate zinc at high level. Zinc availability, however, has to be taken into consideration: Zinc bioavailability is much higher in meat as compared to plants.

Iron is a very critical element in plant nutrition: Plants can take up Fe(III) in complexed forms, or reduce it to Fe(II) [Graham and Stangoulis, 2003], using strategies that work well in soils containing adequate iron content. In case of iron deficiency, chlorosis ensues, and inducible biological mechanisms can be activated. Cereals are plants that excrete phytosiderophores, organic compounds chelating iron in soil, spending energy to increase iron uptake when supply is inadequate. These compounds generally bind other cations, such as zinc, copper, and manganese. The genetic control of this system is polygenic [Graham and Stangoulis, 2003]. However, the main impact for human nutrition is due to the scarce transport of iron to seeds, taking place in the phloem. In fact, all metals move in the phloem as complexes with organic molecules, like carboxylic acids, amino acids, amides, and amines, including nicotianamine; and in the end, loading of the grain is a complex process controlled by many genes.

Considering one toxic element such as cadmium, there are reports of soils with high natural concentrations which transfer to locally grown vegetables and then to the human food products. The extent of consumption of local products will determine the entity of contamination [WHO, 1992]. For arsenic, countries such as Bangladesh, Taiwan, India, China, Mexico, Argentina, and Chile have cases of poisoning from geochemical origin, with release of arsenic in groundwater: Concentration in water may be as high as 50 ppb, whereas the limit recommended by WHO is 10 ppb [Dissanayake and Chandrajith, 1999].

5.4 Intentional Contamination

Trace elements can also be used as poisons, and recently they have become a matter of concern as possible tools for intentional contamination of food products. In recent years, terrorist threats performed with chemical and biological agents have increased in importance, and many National and International initiatives have been developed to devise prevention and counteraction measures. It is worthwhile to remember that toxic trace elements were among the first examples of chemical agents in terrorism actions.

One well-known example, cited in many sources, is the contamination of oranges and lemons from Israel with inorganic mercury, recorded in 1978 [Smithson and Levy, 2000]. A Palestinian group claimed this action for the purpose of sabotaging Israeli economy. Few fruits spiked with mercury were found in the Netherlands, Germany, Belgium, and United Kingdom; apparently the spiking had been performed in Rotterdam, where cargoes were repackaged and shipped to European destinations. The main effect of this act was not to impair health of citizens, since very few cases of illness were reported, but instead to undermine trust in one important sector of the economy of the target country.

Arsenic is also used as intentional toxic agent. In particular arsine, arsenic hydride, AsH_3, is the most toxic form, binding to hemoglobin, affecting red blood cells and oxygen transport. At 250 ppm it is instantly lethal. Lewisite is an arsine compound in the list of potential chemical agents for terrorism and warfare [WHO, 2004]. It is a vesicant compound giving skin lesions, followed by systemic arsenic toxicity.

As an example of food contamination, in 1998 in Wakayama, Japan, four deaths and 63 illnesses followed ingestion of curry during a public festival. Within 30 min after eating the dish, the affected people manifested nausea and vomiting, and in the following days they developed a series of different symptoms and dysfunctions. The curry was found to be poisoned with arsenic trioxide, estimated in 135 g, seemingly for insurance fraud purposes [Suzuki et al., 2000]. Within 2 weeks, 56% of surviving patients developed skin lesions and enanthemas, a consequence of hepatic dysfunction: The main mechanism of arsenic poisoning is the combination with sulfhydryl-containing enzymes, which disrupts cellular oxidative processes [Uede and Furukawa, 2003]. But one of the first reports of terrorist action using arsenic dates back to 1946, when the group called DIN planned to poison Nazi officers held prisoner, to avenge the death of Jewish people in the concentration camps. What they did was to poison with arsenic about 3000 bread loaves, of the type eaten by German troops, in the bakeries of the Stalag 13 camp: The poisoning was performed by coating loaves with a mixture of arsenic and glue, using brushes. The number of people affected is not known, since reports are contradictory [Smithson and Levy, 2000].

Cadmium poisoning has also been reported [Suzuki et al., 2000]; selenium has also been considered as a chemical warfare agent [WHO, 2004].

Fluorine in gas form, F_2, can be used as irritant; the fluoride ion is very aggressive, penetrating tissues and reacting with calcium and magnesium inside cells. The effects involve local bone demineralization and systemic lowering of calcium, magnesium, and potassium with serious effects on neurotransmission and metabolism, even death. Depending on concentration, the effect may be acute and immediate, or the symptoms may appear several days after exposure [WHO, 2004].

These examples are enough to prompt particular attention to early warning systems and improved analytical capabilities as possible countermeasures to protect citizens from intentional contamination of the food chain with trace elements.

5.5   Availability of Minerals

An important issue concerning trace elements in the food supply chain is their real availability to the human consumer. Phytates and dietary fibers are present in grains and legumes, and they bind divalent cations. Their presence in the diet can decrease the bioavailability of minerals such as zinc, copper, manganese, and selenium [Gibson, 1994]. On the contrary, the presence of animal proteins can improve zinc absorption in these conditions: Zinc absorption from a nonvegetarian diet including both meats and vegetables is therefore higher. Attempts are currently being made to improve this situation by adding phytase from external sources, even through genetic engineering (see below).

Additional processing of foods is another factor modifying availability: Milling grains and removing bran decreases the amount of minerals, canning leads to leaching in water, additives may alter the chemical forms, and heat treatment can change the chemical binding in organometallic compounds. On the other hand, cooking partially breaks down proteins, making the metals linked to them more available.

Another factor affecting element availability to humans is the presence of other molecules in the intestine during absorption. It is well known that ascorbic acid decreases copper absorption in the intestine, whereas citric acid and other organic acids increase it [Wapnir, 1998]. Copper from vegetables, probably linked to glyco-proteins and lectins, is less available than copper from animal sources, because these require a less intense enzymatic digestion; soybean proteins reduce copper avail-ability by 90%. Acid conditions in the stomach can free metals from ligands, but the pH in the intestine is higher. Enterohepatic circulation also affects the balance: for instance half of the copper reaching the intestine is circulated to the bile, in strongly bound forms, and is excreted with faeces [Wapnir, 1998].

6 BIOFORTIFICATION

In many cases, as evidenced before, crop plants grow on soils deficient in some trace element, and food products are consequently deficient. Agricultural practices can increase the mineral content in crops through manipulation of soil pH and addition of amendments. For example, fertilizers or foliar sprays containing zinc can alleviate zinc deficiency symptoms in plants [Broadley et al., 2007]. Also, selenium can be administered via fertilizers to crops and to pastures to avoid diseases in livestock [Broadley et al., 2006]. The introduction of amelioration practices in Finland in the space of 14 years has increased selenium content in plants and animals, increased sel-enium intake of humans from 25 microgram per day to 124 microgram per day, and increased selenium blood levels in the population [Broadley et al., 2006]. In general, agronomic practices are considered to be effective in amelioration for the content of zinc, molybdenum, nickel, selenium, and iodine [Welch and Graham, 2005].

Plant breeding and biotechnology have also been exploited and considered for this purpose [Lönnerdal, 2003; Tucker, 2003]. Cultivars with high content of trace elements, high concentration of enhancers of bioavailability, or low content of inhibi-tors can be bred into high-yield lines [Welch and Graham, 2004]. Fortified crops are suitable for growth on micronutrient-poor soil, because their bioconcentration capacity will lead to higher content of micronutrients in edible tissues. Knowledge of mechanisms controlling metal accumulation is a prerequisite for elucidating the biochemical basis of these phenomena. In this sense, biofortification and phytoreme-diation are considered aspects of the same issue: interaction between plants and elements (metals) in the environment. Whereas phytoremediation looks for plants able to accumulate toxic metals [Pilon-Smits, 2005], biofortification looks for plants able to accumulate essential elements. Biofortification with these means can only be applied when there is a clear target nutrient. In the case of trace elements, zinc has been mostly targeted. In general, it may be argued that when in crops there is genetic variation in mineral content, then this could be exploited in breeding programs. Beans containing 50–70% more iron have been obtained through conven-tional breeding [Nestel et al., 2006]. Often, however, the genetic basis is quantitative, and Quantitative Trait Loci (QTLs) should be identified and mapped. Broadley et al. [2007] wrote a recent review paper in which existing variability for zinc uptake in plants has been collected and discussed. Identification of candidate genes involved in transport, uptake, or sequestration of minerals could provide the targets for genetic engineering approaches trying to improve the plant content for a specific element. In order to increase the content of a nutrient in plants, different processes have to be improved: uptake in the root, transport to the edible portion, and accumulation in the edible portion [Guerinot and Salt, 2001].

In rice, increase in iron content has been attempted by introducing the gene for the iron-storage protein ferritin from leguminous plants under control of an endogenous endosperm-specific promoter [Goto et al., 1999]. Alternative strategies concern the increase of siderophore production, compounds which the root exudates to chelate iron in soil; nicotianamine synthase is the enzyme required.

In the case of zinc and iron, an alternative approach to increase their availability concerns the decrease of the seed content of phytic acid, or phytate. This is a storage molecule for phosphorus, present in cereals and oil seeds. Monogastric animals cannot degrade this molecule, and one effect of the presence of phytic acid is the sequestration of cations, mainly iron, zinc, and calcium. Therefore, plants have been engineered with the phytase gene: Overexpression of the phytase can degrade phytate and therefore increase availability of zinc and iron. Phytase from *Aspergillus* has been expressed in rice and wheat [Lucca et al., 2001; Brinch-Pedersen et al., 2006]. In this case, it is important that the phytase enzyme is thermo-tolerant, to be able to function even after cooking of food.

The same target can be reached through traditional breeding by selecting low phytic acid mutants, lpa, which store phosphorus as inorganic [White and Broadley, 2005]. However, these mutants have several disadvantages for farmers, linked to their germination and growth properties. Recently, the gene impaired in the mutants has been identified in maize [Shi et al., 2007]; it encodes for a multidrug resistance-associated protein (MRP) ATP-binding cassette (ABC) transporter, expressed in seeds and at low levels in other tissues. Antisense transgenic plants have been produced which express the antisense fragments only in seeds, under control of seed-specific promoters. These plants have low phytic acid, high content of inorganic phosphorus, and normal growth behavior, as opposed to natural mutants.

Biofortification of crops per se is not sufficient to improve the nutritional status of food products; it is also necessary that the nutrient is retained in the product following processing and cooking and that bioavailability to humans is high.

7   CONCLUDING REMARKS

The impact of trace elements on human health is exerted at several different levels. Entry in the food chain can occur through plants and animal products, enhanced in some cases by supplementation and fortification, but also through instances of pollution and environmental contamination, and even through intentional contamination and fraud. Legislation and regulation enforce precise limits for most relevant trace elements, trying to safeguard populations from both deficiency and toxicity consequences.

In the future, particular attention should be paid to issues that have not have been considered adequately so far and that impact the norms and regulations dealing with trace elements: Hypersensitive populations and their protection, improvement of analytical techniques for metal speciation, and problems linked to the dual approach biofortification/phytoremediation are the ones identified in the present review.

ACKNOWLEDGMENTS

The research in this chapter has been financed with the contribution of Fondazione Cassa di Risparmio di Parma and NATO, project Science for Peace 982498 "Development of a prototype for the International Situational Centre on Interaction in Case of Ecoterrorism."

REFERENCES

Andreini C, Banci L, Bertini I, Rosato A. 2006. Counting the zinc-proteins encoded in the human genome. *J Proteome Res* **5**:196–201.

AROMIS, Concerted Action. 2005. Assessment and reduction of heavy metal input into agro-ecosystems. Darmstadt: Kuratorium fuer Technik und Bauwesen in der Landwirtschaft e.V. (KBTL).

Ballatori N. 2002. Transport of toxic metals by molecular mimicry. *Environ Health Persp* **110**(suppl 5): 689–694.

Berg T, Licht D. 2002. International legislation on trace elements as contaminants in food: A review. *Food Addit Contam* **19**:916–927.

Blunden S, Wallace T. 2003. Tin in canned food: a review and understanding of occurrence and effect. *Food Chem Toxicol* **41**:1651–1662.

Brinch-Pedersen H, Hatzack F, Stöger E, Arcalis E, Pontopidan K, Holm PB. 2006. Heat-stable phytases in transgenic wheat (*Triticum aestivum* L.): Deposition pattern, thermostability, and phytate hydrolysis. *J Agric Food Chem* **54**:4624–4632.

Broadley MR, White PJ, Bryson RJ, Meacham MC, Bowen HC, Johnson SE, Hawkesford MJ, McGrath SP, Zhao F-J, Breward N, Harriman M, Tucker M. 2006. Biofortification of UK food crops with selenium. *P Nutr Soc* **65**:169–181.

Broadley MR, White PJ, Hammond JP, Zelko I, Lux A. 2007. Zinc in plants. *New Phytol* **173**:677–702.

Caussy A, Gochfeld M, Gurzau E, Neagu C, Ruedel H. 2003. Lessons from case studies of metals: Investigating exposure, bioavailability, and risk. *Ecotox Environ Safe* **56**:45–51.

Codex Alimentarius Commission. 1985. Guidelines on nutrition labelling, CAC/GL2–1985. Rome: Codex Alimentarius Commission.

Codex Alimentarius Commission. 1995. General Standard for contaminants and toxins in foods, CODEX STAN 193–1995, Rev. 2–2006. Rome: Codex Alimentarius Commission.

Cornelis R, Crews H, Donard O, Ebdon L, Pitts L, Quevauviller. 2001. Summary paper of the EC network on trace element speciation for analysts, industry and regulators—What we have and what we need. *J Environ Monitor* **3**:97–101.

Cox DW. 1999. Disorders of copper transport. *Br Med Bull* **55**:544–555.

Dissanayake CB, Chandrajith R. 1999. Medical geochemistry of tropical environments. *Earth Sci Rev* **47**:219–258.

Eide DJ. 2006. Zinc transporters and the cellular trafficking of zinc. *Biochim Biophys Acta* **1763**:711–722.

European Commission. 1970. Directive 70/524/EEC of 23 November 1970 concerning additives in feedingstuffs. OJ L 270 14.12.1970, pp. 1–17.

European Commission. 1986. Council Directive 86/278/EEC of 12 June 1986 on the protection of the environment, and in particular of the soil, when sewage sludge is used in agriculture. OJ L181 5.7.1986, pp. 6–12.

European Commission. 1993. Council Regulation (EEC) No 315/93 of 8 February 1993 laying down Community procedure for contaminants in food. OJ L 37, 13.2.1993, pp. 1–3.

European Commission. 1996. Council Directive 96/61/EC of 24 September 1996 concerning integrated pollution prevention and control. OJ L 257, 10.10.1996, pp.26–40.

European Commission. 2001. Commission Regulation (EC) 466/2001 of 8 March 2001 setting maximum levels for certain contaminants in foodstuffs. OJ L77 16.3.2001, p. 1.

European Commission. 2002. Directive 2002/32/EC of the European Parliament and of the Council of 7 May 2002 on undesirable substances in animal feed. OJ L140 30.5.2002, pp. 10–21.

European Commission. 2003a. Commission Directive 2003/100/EC of 31 October 2003 amending Annex I to Directive 2002/32/EC of the European Parliament and of the Council on undesirable substances in animal feed. OJ L285 1.11.2003, pp. 33–37.

European Commission. 2003b. Regulation (EC) No 2003/2003 of the European Parliament and of the Council of 13 October 2003 relating to fertilizers. OJ L304 21.11.2003, pp. 1–194.

European Commission. 2003c. Report on tasks for scientific cooperation (SCOOP). Assessment of the dietary exposure to organotin compounds of the population of the EU member states. Directorate-General Health and Consumer Protection.

European Commission. 2006a. Commission Regulation (EC) 1881/2006 of 19 December 2006 setting maximum levels for certain contaminants in foodstuffs. OJ L 364, 20.12.2006, pp. 5–24.

European Commission. 2006b. Directive 2006/11/EC of the European Parliament and of the Council of 15 February 2006 on pollution caused by certain dangerous substances discharged into the aquatic environment of the Community. OJ L64 4.3.2006, pp. 52–59.

European Commission. 2006c. Regulation (EC) No 1925/2006 of the European Parliament and of the Council of 20 December 2006 on the addition of vitamins and minerals and of certain other substances to foods. OJ L404 30.12.2006, pp. 26–38.

European Commission. 2007a. Commission Regulation (EC) No 333/2007 of 28 March 2007 laying down the methods of sampling and analysis for the official control of the levels of lead, cadmium, mercury, inorganic tin, 3-MCPD and benzo(*a*)pyrene in foodstuffs. OJ L 088 29.3.2007, pp. 29–38.

European Commission. 2007b. Community Register on the addition of vitamins and minerals and of certain other substances to foods, pursuant to Regulation (EC) No 1925/2006. Directorate E—Safety of the food chain. Unit E4: Food law, nutrition and labelling.

European Commission, Scientific Committee for Animal Nutrition (SCAN). 2003a. Opinion of the Scientific Committee for Animal Nutrition on the use of copper in feedingstuffs.

European Commission, Scientific Committee for Animal Nutrition (SCAN). 2003b. Opinion of the Scientific Committee for Animal Nutrition on the use of zinc in feedingstuffs.

European Commission, Scientific Committee on Food (SCF). 2002. Opinion of the Scientific Committee on Food on the tolerable upper intake level of Iodine. Health & Consumer Protection Directorate-General, Scientific Committee on Food, Document SCF/CS/ NUT/UPPLEV/26 Final.

European Commission, Scientific Committee on Food (SCF). 2003. Opinion of the Scientific Committee on Food on the tolerable upper intake level of Zinc. Health & Consumer Protection Directorate-General, Scientific Committee on Food, Document SCF/CS/ NUT/UPPLEV/62 Final.

European Food Safety Authority. 2004. Opinion of the Scientific Panel on Contaminants in the Food Chain on a request from the Commission related to mercury and methylmercury in food (Request N° EFSA-Q-2003–030). *EFSA J* **34**:1–14.

Expert Group on Vitamins and Minerals. 2003. *Safe Upper Levels for Vitamins and Minerals.* UK: Food Standards Agency Publications.

FAO/WHO. 2001. *Human Vitamin and Mineral Requirements: Report of a Joint FAO/WHO Expert Consultation*, second edition. Bangkok: FAO/WHO [available at http://www.fao.org].

Filgueiras AV, Lavilla I, Bendicho C. 2002. Chemical sequential extraction for metal partitioning in environmental solid samples. *J Environ Monitor* **4**:823–857.

Florea A.-M, Büsselberg D. 2006. Occurrence, use and potential toxic effects of metals and metal compounds. *BioMetals* **19**:419–427.

Flynn A, Moreiras O, Stehle P, Fletcher RJ, Müller DJG, Rolland V. 2003. Vitamins and minerals: A model for safe addition to foods. *Eur J Nutr* **42**:118–130.

Gibson RS. 1994. Content and bioavailability of trace elements in vegetarian diets. *Am J Clin Nutr* **59**(suppl):1223S–1232S.

Goto F, Yoshihara T, Shigemoto N, Toki S, Takaiwa F. 1999. Iron fortification of rice seed by the soybean ferritin gene. *Nat Biotechnol* **17**:282–286.

Goyer RA, Mahaffey KR. 1972. Susceptibility to lead toxicity. *Environ Health Persp* **2**:73–80.

Graham RD, Stangoulis JCR. 2003. Trace element uptake and distribution in plants. *J Nutr* **133**:1502S–1505S.

Griffiths WJH. 2007. Review article: the genetic basis of haemochromatosis. *Aliment Pharmacol Ther* **26**:331–342.

Guerinot ML, Salt DE. 2001. Fortified foods and phytoremediation. Two sides of the same coin. *Plant Physiol* **125**:164–167.

Hillman JP, Hill J, Morgan JE, Wilkinson JM. 2003. Recycling of sewage sludge to grassland: A review of the legislation to control of the localization and accumulation of potential toxic metals in grazing systems. *Grass Forage Sci* **58**:101–111.

Hinsinger P, Plassard C, Jaillard B. 2006. Rhizosphere: A new frontier for soil biogeochemistry. *J Geochem Explor* **88**:210–213.

Kelada SN, Shelton E, Kaufmann RB, Khoury MJ. 2001. δ-aminolevulinic acid dehydratase genotype and lead toxicity: A HuGE review. *Am J Epidemiol* **154**:1–13.

Lönnerdal B. 2003. Genetically modified plants for improved trace element nutrition. *J Nutr* **133**:1490S–1493S.

Lucca P, Hurrell R, Potrykus I. 2001. Genetic engineering approaches to improve bioavailability and the level of iron in rice grains. *Theor Appl Genet* **102**:392–397.

Montenegro MF, Barbosa F, Sandrim VC, Gerlach RF, Tanus-Santos JE. 2006. A polymorphism in the delta-aminolevulinic acid dehydratase gene modifies plasma/whole blood lead ratio. *Arch Toxicol* **80**:394–398.

Mukherjee B, Patra B, Mahapatra S, Banerjee P, Tiwari A, Chatterjee M. 2004. Vanadium—An element of atypical biological significance. *Toxicol Lett* **150**:135–143.

Nestel P, Bouis HE, Meenakshi JV, Pfeiffer W. 2006. Biofortification of staple food crops. *J Nutr* **136**:1064–1067.

Oliver MA. 1997. Soil and human health: A review. *Eur J Soil Sci* **48**:573–592.

Pilon-Smits E. 2005. Phytoremediation. *Annu Rev Plant Biol* **56**:15–39.

Rayman MP. 2000. The importance of selenium to human health. *Lancet* **356**:233–241.

Reinhold JG. 1975. Trace elements—A selective survey. *Clin Chem* **21**:476–500.

Sadhra SS, Wheatley AD, Cross HJ. 2007. Dietary exposure to copper in the European Union and its assessment for EU regulatory risk assessment. *Sci Total Environ* **374**:223–234.

Senesi GS, Baldassarre G, Senesi N, Radina B. 1999. Trace element inputs into soils by anthropogenic activities and implications for human health. *Chemosphere* **39**:343–377.

Shi J, Wang H, Schellin K, Li B, Faller M, Stoop JM, Meeley RB, Ertl D, Ranch JP, Glassman K. 2007. Embryo-specific silencing of a transporter reduces phytic acid content of maize and soybean seeds. *Nat Biotechnol* **25**:930–936.

Smith CM, Wang X, Hu H, Kelsey KT. 1995. A polymorphism in the δ-aminolevulinic acid dehydratase gene may modify the pharmacokinetics and toxicity of lead. *Environ Health Persp* **103**:248–253.

Smithson AE, Levy L-A. 2000. Ataxia: the chemical and biological terrorism threat and the US response. Report No.35. Washington:The Henry L. Stimson Center.

Suzuki O, Seno H, Watanabe-Suzuki K, Ishii A. 2000. Situations of poisoning and analytical toxicology in Japan. *Forensic Sci Int* **113**:331–338.

Tan J, Zhu W, Wang W, Li R, Hou S, Wang D, Yang L. 2002. Selenium in soil and endemic diseases in China. *Sci Total Environ* **284**:227–235.

Tiffany-Castiglioni E, Venkatraj V, Qian Y. 2005. Genetic polymorphisms and mechanisms of neurotoxicity: Overview. *NeuroToxicology* **26**:641–649.

Tucker G. 2003. Nutritional enhancement of plants. *Curr Opin Biotech* **14**:221–225.

Uede K, Furukawa F. 2003. Skin manifestations in acute arsenic poisoning from the Wakayama curry-poisoning incident. *Br J Dermatol* **149**:757–762.

Uthus EO, Seaborn CD. 1996. Deliberations and evaluations of the approaches, endpoints and paradigms for dietary recommendations of the other trace elements. *J Nutr* **126**:2452S–2459S.

Wapnir RA. 1998. Copper absorption and bioavailability. *Am J Clin Nutr* **67**:1054S–1060S.

Welch RM, Graham RD. 2004. Breeding for micronutrients in staple food crops from a human nutrition perspective. *J Exp Bot* **55**:353–364.

Welch RM, Graham RD. 2005. Agriculture: The real nexus for enhancing bioavailable micronutrients in food crops. *J Trace Elem Med Bio* **18**:299–307.

White JG, Zasoski RJ. 1999. Mapping soil micronutrients. *Field Crop Res* **60**:11–26.

White PJ, Broadley MR. 2005. Biofortifying crops with essential mineral elements. *Trends Plant Sci* **10**:586–592.

World Health Organization. 1980a. *Environmental Health Criteria 15—Tin and Organotin Compounds.* Geneva: WHO.

World Health Organization. 1980b. *Environmental Health Criteria 17—Manganese.* Geneva: WHO.

World Health Organization. 1987. *Environmental Health Criteria 58—Selenium.* Geneva: WHO.

World Health Organization. 1988a. *Environmental Health Criteria 58—Chromium.* Geneva: WHO.

World Health Organization. 1988b. *Environmental Health Criteria 81—Vanadium.* Geneva: WHO.

World Health Organization. 1991a. *Environmental Health Criteria 108—Nickel.* Geneva: WHO.

World Health Organization. 1991b. *Environmental Health Criteria 118—Inorganic mercury.* Geneva: WHO.

World Health Organization. 1992. *Environmental Health Criteria 134—Cadmium.* Geneva: WHO.

World Health Organization. 1995. *Environmental Health Criteria 165—Inorganic Lead.* Geneva: WHO.

World Health Organization. 2001a. *Environmental Health Criteria 221—Zinc.* Geneva: WHO.

World Health Organization. 2001b. *Environmental Health Criteria 224—Arsenic and Arsenic Compounds.* Geneva: WHO.

World Health Organization. 2002a. *Environmental Health Criteria 227—Fluorides.* Geneva: WHO.

World Health Organization. 2002b. *Environmental Health Criteria 228—Principles and Methods for the Assessment of Risk from Essential Trace Elements.* Geneva: WHO.

World Health Organization. 2004. *Public Health Response to Biological and Chemical Weapons.* WHO *Guidance*, second edition. Geneva: WHO.

3 Trace Elements in Agro-ecosystems

SHUHE WEI

Key Laboratory of Terrestrial Ecological Process, Institute of Applied Ecology, Chinese Academy of Sciences, Shenyang 110016, People's Republic of China

QIXING ZHOU

Key Laboratory of Terrestrial Ecological Process, Institute of Applied Ecology, Chinese Academy of Sciences, Shenyang 110016, People's Republic of China; and College of Environmental Science and Engineering, Nankai University, Tianjin 300071, People's Republic of China

1 INTRODUCTION

Usually, trace elements in agro-ecosystems mainly concern zinc (Zn), iron (Fe), selenium (Se), boron (B), molybdenum (Mo), cobalt (Co), nickel (Ni), lead (Pb), cadmium (Cd), chromium (Cr), arsenic (As), copper (Cu), manganese (Mg), mercury (Hg), and so on. Except for B, these elements are also heavy metals. Though many trace elements or so-called micronutrients (such as Cu, Zn, Mn, Fe, Mo, and B) are beneficial or essential to crops some of them are harmful under much higher concentration, especially for some heavy metals such as Hg, Pb, Cd, As, Cr, and Ni [He et al., 2005]. Thus, these trace elements are often called heavy-metal contaminants or toxic trace elements in environmental science and are the main points of agro-environment protection science. Some protection measurements and remediation technologies have been studied on a large scale.

The biogeochemical cycle of trace elements is very important, which may let us know what will happen after they enter agro-systems such as input, translation, transportation, accumulation, and output. Some main objects like soil, soil micro-organism, crop, grass, animal, or human being is concerned. Usually, input or contamination paths of some trace elements like heavy metals are especially concerned. Because most heavy metals would be accumulated in soil after long-time addition, they may be harmful to human healthy through contaminated agro-production,

Trace Elements as Contaminants and Nutrients: Consequences in Ecosystems and Human Health, Edited by M. N. V. Prasad

even though some of them can be output from soil by some paths like removing agro-production from agro-systems [Sun et al., 2001]. Therefore, some background concentrations of trace elements in soil, crop, or grass are cited so as to realize the normal status of trace elements in agro-systems. Some soil environmental quality standards are also cited to determine if a land is polluted.

The most important aim of studying trace elements in agro-ecosystems is to know their benefit and harmfulness to human health, and how to control them, because most agro-productions (especially from crops) are the main food resources for human beings whether in direct or indirect [Senesi et al., 1999; Lam et al., 2004]. Basically, the effects of trace elements on crops consist of two aspects. One is element deficiency like Fe, Mn, Mo, Zn, Cu, and so on, while the other is toxicity like heavy-metal contaminants Hg, Pb, Cd, As, Cr, Ni, and so on. However, more attention is paid to agro-production qualities caused by these heavy metals. The reason partly lies in the fact that element deficiency in food is not more dangerous than those of element toxicity. Thus, the harmfulness of some elements to crops, along with implication to human health, should be considered.

Many research results showed that with the development of economics, some heavy metal concentrations in fields have been continuously increasing in the past several decades. Thus, some agro-soils have been contaminated. Phytoremediation is a promising technology for cleaning contaminated soil, especially for phytoextraction to remove toxic trace elements (i.e., some heavy metals) from huge area field by mainly using hyperaccumulators [Wei and Zhou, 2004]. Some types of research progress in phytoextraction are introduced, and some agricultural improving measurements are also discussed at the end of this chapter.

2   BIOGEOCHEMISTRY OF TRACE ELEMENTS IN AGRO-ECOSYSTEMS

2.1   Input and Contamination

Trace elements enter into agro-ecosystems through two paths: One is natural, the other is anthropogenic. Natural path means that trace elements are originally inherited from their parent materials with background concentrations. Sometimes, trace elements are being input to agro-ecosystems through air sedimentation or rainout, especially from near mine or volcano. Anthropogenic inputting processes mainly include the addition of chemicals such as fertilizers and pesticides, the use of sludge or some compost, car exhausts, and so on [Mortvedt and Beaton 1995; Yang and Yang, 2000; Chaney et al., 2001; Walker, 2001; McBride, 2004].

Soil is an important part of agro-ecosystems and also the living media of plants, microorganisms, animals, or humans. On one hand, agricultural land is the resource of materials for producing or living. On the other hand, it bears the weight of many pollutants including some trace elements, especially for toxic heavy metals. There are three phase configurations in soil (solid, liquid, and vapor), in which all kinds of physical and chemical reactions are eternally processing. Therefore, soil can self-clean pollutants to some extent—that is, output them through volatilization or eluviation to air or water, or decrease contaminant availabilities to health levels due to dilution, diffusion, or transformation by physical and chemical reactions. However, the properties of soil self-cleaning are not limitless. Once the concentrations of pollutants exceed the typical values of self-cleaning, soil will be contaminated [Zhou and Song, 2004].

2.1.1 Natural Pollution Resources Usually, some mines rich in trace elements are the main reasons for agricultural soil pollution near them. These trace elements may be concentrated by some actions like magma acting in deep geosphere, metamorphose acting, or many geochemical processions. Some trace element concentrations in soils near mines may be higher than several times or even several hundred thousand times of those far from mines. After chronically weathering conditions due to a series of complicated reactions such as dissolution–precipitation, oxidation–reduction, and adsorption–desorption, some soils originally inherited from their parent materials rich in trace elements or derived from alluvium by groundwater rich in heavy metals are contaminated. Usually, the concentrations of heavy metals in soils around mines are often higher. Sometimes, some ores are also important sources of heavy metals in soils though they are buried 200–300 m under subsurface soil [Xu and Yang, 1995; Dou and Li, 1998]. According to our studies, some trace element concentrations in soils near Qingchengzi Pb–Zn mining area (40°41′ N, 123°37′ E), Fengcheng County, Liaoning Province, China are 1.1–32.1 mg/kg for Cd, 249.2–8958.9 mg/kg for Pb, 25.2–119.4 mg/kg for Cu, and 79.2–2740.7 mg/kg for Zn, respectively, except for Cu, most of these ranges are higher than those of the National Soil Environmental Quality Standard for agricultural land [Wei et al., 2005]. Another good example is the Se contamination where soils were developed from high Se parent materials in the West-central San Joaquin Valley [Presser, 1994]. Some compounds of trace elements in ores are shown in Table 1 [He et al., 1998; Shen, 2002; Wei and Zhou, 2004].

2.1.2 Industry Pollution Resources Industry pollution sources resulted from anthropogenic activities mainly include mining, smelting, ore-processing, production of pesticides and chemical fertilizers, release of automobile exhausts, and a pile-up of municipal wastes. Among the above-referenced pollution sources, mining wastes and pile-up of wastes containing high levels of heavy metals are very prominent.

Huge excavation wastes can be brought to the soil surface when mining a metalliferous mine. Heavy metals in ore spoils enter the soil during their leaching by rainfall or dry deposits. Therefore, soils are contaminated and the concentrations of heavy metals in soils may be 100–1000 times higher than their backgrounds [Jiang et al., 2004]. More seriously, tailings of ores may also be contaminated soils due to the tailing dam broken. Tailings are often with higher heavy metal concentration and low pH, which can quickly destroy soil biology and damage crops. In addition, some large industrial emissions like metal smelters have also great impact on trace element accumulation in soil. Usually, more close to these points, the pollution of

TABLE 1. Main Pollution Sources and Compounds of Heavy Metals into Soils

Heavy Metal	Mineral	Human Activity	Main Compounds Entered into Soil
Cd	CdS, CdO	Mining, ore dressing, and smelting of nonferrous metals; Cd compound production; battery manufacturing industry; electroplate	$CdCO_3$, CdS, $Cd_3(PO_4)_2$, $Cd(OH_2)_2$, $CdSiO_3$, $CdSO_4$, $CdBr_2$, CdI_2, CdF_2, $Cd(NO_3)_2$, $CdCl_2$, $Cd(CH_3COO)_2$, $(CH_3)_2Cd$
Hg	AuHg, PdHg, HgS	Production and apply of Hg catalyst in chemical industry; Hg battery manufacturing; smelting and restoring of Hg; Hg compound production; pesticide and medicine making; production and apply of fluorescent light and Hg lamp; Hg slime resulted from caustic soda production	$HgBr_2$, HgBr, HgI, HgI_2, $Hg(NO_3)_2$, $HgNO_3$, HgO, Hg_2O, $HgSO_4$, Hg_2SO_4HgS, Hg_2Cl_2, $HgClHgCO_3$, $HgHPO_4$, $(CH_3Hg)_2S$, $(CH_3)_2Hg$, C_6H_5HgCl, $C_6H_5HgNO_3$, $C_6H_5HgOCH_3COO$, CH_3HgCl, C_2H_5HgCl, $(CH_3CH_3)_2Hg$, C_6H_5HgCl
As	AsS, As_2S_3, $FeAs_2$, FeAsS, $FeAsO_4 \cdot 2H_2O$	Mining, ore dressing, and smelting of nonferrous metals; production of As and As compounds; petroleum and chemical industry; pesticide production; dyestuff and tanning industry	As_2O_3, H_3AsO_4, H_3AsO_3, As_2S_3, AsH_3, As_4S_4, $Zn_3(AsO_4)_2$, $FeAsO_4$, $(NH_4)_3AsO_4$, Na_3AsO_4, Hg_3AsO_4, Pb_3AsO_4, $Mg_3(AsO_4)_3$, K_3AsO_4, $AsCl_3$, Zn_3As_2, Cu_3As_2, Ca_3As_2, AsS_2, $CuAs(CH_3CH_2)_5$
Cu	$CuFeS_2$, CuS, Cu_2O, $Cu_3(OH)_2(CO_3)_2$, $Cu_2(OH)_2CO_3$	Mining, ore dressing and smelting of nonferrous metals; metal and plastic electroplate; Cu compounds production	$CuBr_2$, CuBr, $Cu(OH)_2$, $CuSO_4$, Cu_2SO_4, CuI, $CuCO_3Cu(NO_3)_2$, CuS, CuF_2, Cu_2S, $CuCl_2$, CuCl, $Cu(CH_3COO)_2$, CuO, $Cu_3(PO_4)_2$
Pb	$PbCO_3$, PbS, $Pb_2CO_3Cl_2$, $PbSO_4$, $PbCrO_4$	Residue resulted from the production of lead melting and electrolying; residue resulted from the production of Pb electric accumulator; abandoned Pb electric accumulator; residue and sludge resulted from lead caster and product industry; waste resulted from the production and apply of lead compound	$Pb(CH_3COO)_2$, $PbBr_2$, $Pb(OH)_2$, PbI_2, $Pb_3(PO_4)_2$, $Pb(NO_3)_2$, PbO, $PbSO_4$, $PbCl_2$, PbF_2, PbS, Pb_2ClO_4, $Pb(CH_3)_4$

(Continued)

TABLE 1. *Continued*

Heavy Metal	Mineral	Human Activity	Main Compounds Entered into Soil
Cr	$FeOCr_2O_3$, $Mg \cdot FeCr_2O_4$, $MgFe(Cr \cdot Al)_2O_4$	Cr compound production; leather working industry; metal and plastic electroplate; dyestuff and dying by acidic medium; production and apply of dyestuff; metal Cr smelting	Cr_2O_3, CrO_3, $CrCl$, Na_2CrO_4, K_2CrO_4, $ZnCrO_4$, $CaCrO_4$, Ag_2CrO_4, $PbCrO_4$, $BaCrO_4$, $H_2Cr_2O_7$, K_2CrO_7, $NaCr_2O_7$
Zn	ZnO, $ZnCO_3$, Fe_2O_4, Zn_2SiO_4, $Zn_3(PO_4O_2)4H_2O$	Mining, ore dressing, and smelting of nonferrous metals; metal and plastic electroplate; pigment, beaded paint and rubber working; Zinoky compound production; Zinoky battery product industry	$ZnBr_2$, ZnI_2, $Zn(NO_3)_2$, $ZnSO_4$, ZnF_2, ZnS, ZnO, $Zn(CH_3COO)_2$, $Zn(CH_2COO)_2$, $ZnCrO_4$, Zn_3BrO_4, $Zn_3(PO_4)_2$, Zn_3P_2, $ZnMnO_4$
Ni	$(Ni \cdot Fe)S$, NiS, $3NiS \cdot FeS_2$, NiO, $NiAs_2$	Residue resulted from the production of nickeliferous compound; Abandoned nickeliferous catalyst; nickeliferous residue and waste from electroplate technology; nickeliferous waste resulted from analysis, assay, and testing activity	$NiBr_2$, $Ni(NO_3)_2$, $NiSO_4$, $NiCl_2$, NiS, NiO, $Ni(OH)_2$

lands are seriously. Waste gas from cars using Pb-enriched gasoline can cause Pb pollution [Krauskopf and Bird, 1995].

Table 1 lists eight heavy metals (these are preferentially controlled unorganic pollutants in China), their pollution sources, and various chemical forms [He et al., 1998; Shen, 2002; Wei and Zhou, 2004], from which we may know that heavy metal contaminations are very complicated.

2.1.3 Aricultural Pollution Resources Addition of chemical fertilizer is a very important measurement to gain high yield from crops. However, this is often a pollution resource of heavy-metal contamination. Many fertilizers contain trace amounts of heavy metals due to their origin from rocks that include high levels of trace elements. Heavy-metal concentrations in phosphorus fertilizer depend on the source of phosphate rock like triple superphosphates and Ca/Mg phosphate. Sometimes the Cd concentration in P fertilizer may be higher than 50 ppm [Mortvedt and Beaton, 1995]. Even in some common fertilizers such as urea, ammonium chloridate, ammonium sulfate, ammonium nitrate, superphosphate, and potassium sulfate, their trace element concentrations are also higher than soil background contents or soil environmental quality standards (see Table 2 [Senesi et al., 1999; Frost and Ketchum, 2000]), and have become the resource of heavy metal pollution after long-time addition of them to soil.

Manures are one type of organic material, along with biosolids or composts, which are also used to increase crop yield. Like fertilizer, they may contain higher concentrations of heavy metals than agricultural land background contents. Table 3 [Jackson and Bertsch, 2001; Bolan et al., 2003, 2004] shows some heavy-metal concentrations in some most common manures. Compared with soil background concentrations and soil environmental quality standards (see Tables 4 [Xia, 1996] and 5), obviously, manures are one of the heavy metal input resources or pollution resources too.

In order to control weeds, insects, or diseases in various crops, some pesticides such as herbicides, fungicides, and insecticides are widely used in modern agriculture. In these pesticides, very high amounts of heavy metals such as As, Cu, Zn, Mn, Hg, and Pb may be used. These pesticides may contain As $0.8-60$ ppm, Cu $4-56$ ppm, Hg $0.6-42$ ppm, Mn $1-17$ ppm, Pb $11-60$ ppm, and Zn $1-30$ ppm [Senesi et al., 1999], which are one of a potential resources of trace elements.

Sewage sludge is solid surplus of wastewater treatment or sedimentation of river. Because of containing high nutrients and organic materials, sludge is applied in agricultural soil to increase crop yield. Though some effective pretreatments have been done before sludge being added to soil, some heavy-metal concentrations such as Cd, Pb, Hg, and As in sludge are still very high, which is also an important input/pollution path of trace elements in agro-ecosystems [McBride, 1995].

In addition, irrigation is indispensable for agricultural production. Usually, heavy-metal concentration in nonpolluted natural water is extremely low. Industry, even domestic wastewaters, often contain a higher level of heavy metals than does natural water. Long-time irrigation of wastewater may significantly enhance heavy-metal contents in soil. For example, Zhangshi region farmland has been contaminated by Cd due to irrigation by using industry wastewater from Shenyang city of China during 1962 to 1975. According to survey, average Cd concentration in topsoil $(0-35$ cm$)$ is $3-5$ ppm, and some area Cd contents were $5-7$ in heavy polluted soil, which are seriously contaminated by Cd compared with China Soil Environmental Quality Standards [Zhou and Song, 2004].

2.2   Translation, Translocation, Fate, and Their Implication to Phytoremediation

Translation, translocation, and fate of trace elements in agro-ecosystems consist in three aspects. First is the geochemical process, and its main media is soil. Second

TABLE 2. Trace Element Concentrations in Some Common Chemical Fertilizers (mg/kg)

Trace Elements	Urea	Ammonium Chloridate	Ammonium Sulfate	Ammonium Nitrate	Superphosphate	Potassium Sulfate
Cd	0.22–3.20	0.00	0.08–4.80	0.08–8.50	0.10–2.20	0.06–3.80
Hg	0.72–0.75		0.30–0.75	0.75–2.88	0.40–1.20	3.20
As	5.6–33.4	1.33	4.2–29.0	2.7–119.7	2.4–28.5	2.4–8.0
Cu	1	0.06	1	3	20–135	138
Pb	5.5–48.7	0.00	2.4–13.6	1.9–27.8	3.1–17.4	13.9
Cr	1.6–8.0	20.85	1.0–4.4	1.0–4.2	39.6–134.0	2.0–2.8
Zn	1	0.93	1–6	3–7	55–235	138
Ni	7.2–10.2	0.22	25.0–32.0	7.0–34.2	19.2–38.2	44.2
Co	1.0–1.4	0.00	7.7–12.0	5.4–11.5	8.8–21.0	5.8–7.0
Mo	2	0.00	1–4	4–7	8–15	21
B	0		0	<1	140–185	20–30
Mn	2–8		3–4	3–65	10–119	197
Se	0		0	0–10	0	13.25

TABLE 3. Trace Element Concentrations in Some Common Manures (mg/kg)

Trace Elements	Swine		Dairy Cattle		Poultry	
	Total	Water-Soluble	Total	Water-Soluble	Total	Water-Soluble
Cd	0.25	0.07	0.87	0.12	0.10	0.02
As					16.79	15.45
Cu	419	130	356	112.5	656.1	314.1
Pb	13.4	1.26	7.53	0.34	0.74	0.02
Cr	13.2	1.23	11.15	0.98	2.59	0.42
Zn	1210	23.56	765	123	246.9	18.24
Ni	12.3	3.23	8.67	1.23	7.97	5.53
Co					0.68	0.31
Se					0.95	0.38
Mn	865	14.56	274.5	6.46	345	17.65

is the reaction between plant/crop or microorganism and soil each other. Third is animal or human influence on soil-plant/crop system. Figure 1 shows the complex correlations of trace elements among soil, plant/crop, and human activities [Wei and Zhou 2004; Wei et al., 2007]. Basically, after a series of complicated physical and chemical reactions, trace elements in soil will be partly absorbed by plant, and almost all of these plants are crops, whereas few are grasses. Due to food and production material requirements of human from agricultural production such as crop seed, root, stem, leaf, or inflorescence, trace elements can be removed from farmland along with the removal of these objects. As mentioned before, many human activities like fertilizer, manure addition, or wastewater irrigation may re-input trace elements into farmland again. Thus, it is a recycle process of trace elements in agro-ecosystem.

Soil is a very complicated medium that concerns (a) solid, liquid, and air phase, (b) organic and inorganic materials, and (c) animal, plant root systems, and microorganisms. Thus, trace elements will continuously react with various components in soil after they entered into soil regardless what forms they are. The main reactions include dissolution–precipitation, adsorption–desorption, complexation–dissociation, and oxidation–reduction. These reactions and soil properties such as pH, Eh, colloid content and composition, climate conditions, hydrology, and biology are the main factors affecting trace element formation and bioavailability in soil. Trace element species distribution in space and time is very complex, which can be usually classified into water-soluble, exchangeable, bound to organic matter, bound to carbonates, bound to Fe–Mn oxides, and residual forms bound in a mine crystal lattice. According to the extraction of plant, trace elements in soil can be roughly sorted into three forms including available, exchangeable, and unavailable fractions. Available trace elements mean free ions and chelating ions which can be easily extracted by plants. Residual forms are very difficult to be absorbed by a plant, called unavailable. Between available and unavailable is exchangeable fraction.

TABLE 4. Soil Environmental Quality Standards or the Maximum Permission Concentrations of Heavy Metals in Soils of Some Countries and Areas (mg/kg)

Country	pH	Cd	Hg	As	Cu	Pb	Cr	Zn	Ni
China[a]	<6.5	0.30	0.30	Rice paddy 30 / Dry field 40	Farmland 50 / Orchard 155	250	Rice paddy 250 / Dry field 150	200	40
	6.5–7.5	0.30	0.50	Rice paddy 25 / Dry field 30	Farmland 100 / Orchard 200	300	Rice paddy 300 / Dry field 200	250	50
	>7.5	0.60	1.0	Rice paddy 20 / Dry field 25	Farmland 100 / Orchard 200	350	Rice paddy 350 / Dry field 250	300	60
Germany	≥6.0	3	2	20	100	100	100	300	50
France	≥6.0	2	1	—	100	100	150	300	50
Italy	≥6.0	3	2	—	100	100	50	300	50
Canada (Ontario)	≥6.0	1.6	0.5	—	100	60	120	220	32
England[b]	Field land ≥6.0 / Pasture land ≥6.5	3.5	1	10	140/280[c]	550	600	280/560	35/70
Scotland	≥5.5	1.6	0.4	—	80	90	120	150	48

[a] If positive ion exchange capacity ≤5 cmol(+)/kg, the value of Cr is a half of the table.

[b] The values of Cu, Zn, Ni are measured by EDTA extract.

[c] Soil media is noncalcareous soil/calcareous soil.

Source: After Xia [1996], revised.

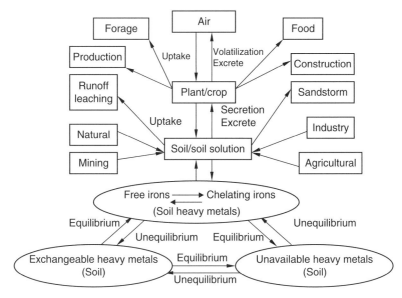

Figure 1. Interactions among of plants/crops, trace elements, soils, and human activities.

An equilibrium status is often to be arrived between available and unavailable speciation of trace elements. Once bioavailable trace elements are reduced due to plant uptake, they would be supplied from unavailable heavy metals. When bioavailable trace elements increase owing to input from external surroundings, some bioavailable and exchangeable fractions can change into unavailable trace elements as a result of an external disturbance or a change in environmental conditions such as plant uptake, organic chelation, or the fluctuation of temperature and moisture [He et al., 1998; Zhou and Huang, 2001; Wei and Zhou, 2004; Wei et al., 2007].

There is a dynamic process among of absorbing, excreting, and accumulating trace elements by a plant (Fig. 1). These processes may often reach an equilibrium point during plant growth at some moment, but which could be broken too often with large changes in environmental conditions [Zhou and Song, 2004]. When the concentration of bioavailable heavy metals in soils decrease owing to plant uptake, the process of transformation from unavailable forms into bioavailable forms will be accelerated, and the accumulation of heavy metals in a plant will increase accordingly. Once the quantity of bioavailable heavy metals released from unavailable fractionations cannot meet the accumulation in a plant, the contents accumulated by a plant will not increase. If bioavailable heavy metals from the external environment are greater than the critical concentration that a plant can accumulate maximally, the accumulation of heavy metals by the plant does not increase either. On the contrary, bioavailable heavy metals may be translated into unavailable speciation. Thus it can be seen that the accumulation of heavy metals by a plant is not unlimited. This process is also one of the main restrictive factors for phytoremediation, so some strengthening measures should be taken to improve the efficiency of phytoremediation [Wei and Zhou, 2004; Wei et al., 2007].

3 BENEFIT, HARMFULNESS, AND HEALTHY IMPLICATION OF TRACE ELEMENTS

Harmfulness of heavy metal to humans have been known from some famous pollution accidents such as Hg and Cd pollution which were called as minamata and itai-itai (ouch-ouch) disease emerged in Japan in the 1950s. However, it is relative to element harmfulness to biology. Actually, many heavy metals in some concentrations are a benefit to biology, and some of them are essential, even some of the heavy metals that are usually regarded as toxic metals. For example, Ni is indispensable to maintain the structure and function of urease. Moreover, it can also increase the activities of catalase, phlyphenol oxidase, and ascorbic acid [Zhang and Chen, 1996]. Fe, Cu, Zn, Mn, and Mo are components of enzymes which help the cell to complete oxidation–reduction reactions through their electron transportations. Therefore, to plant/crop, only when heavy-metal concentrations in soil are higher than a typical limitation which can be harmful to plant/crop growth, or their contents in productions from plant/crop are harmful to cattle/human health, are these elements called pollution elements, and the contaminated soil should be remedied [Dou and Li, 1998]. Owing to the fact that human health is indirectly impacted by trace elements through the food chain in agro-ecosystems, especially through crop food (i.e., crop is primitive biology influenced by heavy metals), the research of heavy metal impacting crop is the main point about benefit or harmfulness.

3.1 Benefit to Plant/Crop

The benefit of trace elements to crop is usually displayed by occurrences of their deficiency. This is because if trace element concentration in soil is a little higher, the crop cannot show any improvement. But once the deficiency becomes much higher, crop growth will be prohibited. Usually, Fe deficiency is a very normal problem for most crops in the world [Marschner, 1995; He et al., 2005], which easily occurs in young leaves when chloroplast chlorophyll synthesis is inhibited under low Fe nutrition status of the plant. However, Fe deficiency is not caused by short input of it in soil, but by some inducements like bicarbonate or nitrate, which may lead to the soluble Fe concentration in rhizosphere far below the level required for plant normal growth. In this case, it is easy to induce Fe deficiency chlorosis. Likewise, Zn, Cu, or Co deficiencies seriously effect crop growth too.

3.2 Harmfulness to Plant/Crop Physiology

After excessive heavy metals entered the plant, these elements may participate in some physiological and biochemical reactions, which will destroy normal growth of the plant by disturb absorb, translocation, or synthesize processes. More seriously, they may combine with some huge molecule like nucleic acid, protein, and enzyme, or may substitute special functional elements in protein or enzyme, so as to induce a series of turbulence of metabolism. Therefore, the growth and procreation of plant will be prohibited, and the plant will die [Jiang and Zhao, 2001].

Many researches have proved that As can prohibit water transportation in crop. Cr can permanently induce separation of cellular wall from membrane. Some heavy metals may impact plant transpiration by disturbing air stoma resistance. Photosynthesis can be prohibited by heavy metals through changing photosynthesis system like chloroplast and chlorophyll. For example, Cd can decrease photosynthesis intensity of rice. Hg can reduce photoelectron translocation activity or promote degradation of photosynthesis membrane components of water shield. Cu can prohibit photoelectron translocation of chloroplast and CO_2 fixation. Heavy metals can also result in turbulence of plant respiration and some decompensations. During carbohydrate metabolization, Cd pollution can decrease several concentrations of dissolvable sugars in plant. By reducing N uptake and activity of nitric acid deoxidization enzyme due to heavy-metal contamination, N element metabolization is also disturbed like changed components of amino acid, and it prohibited synthesis of protein and accelerated protein degradation. Thereby, nucleolus of plant may be broken, which results in abnormality of chromosome replication and DNA composition. Hormones are impacted after plant polluted by heavy metals. Zn can reduce the synthesis of heteroauxin and then decrease the concentration of auxin in plant, which may prohibit plant growth. In addition, heavy metals may harm plants through an impact on the micro-ecological environment of root. Much higher heavy metal concentration is very toxic to soil microorganism. Plant growth can be indirectly prohibited due to the harmfulness of heavy metal to some beneficial rhizosphere microorganism like root nodule bacteria or mycorrhiza.

3.3   Soil Environmental Quality Standards and Background of Trace Elements

Though heavy metal pollutants in soil may be removed after a series of translation, translocation, and output, their concentrations are still increased by continuously input. However, we cannot say the soil is polluted just because some heavy-metal concentrations are obviously increased. Only when heavy metal concentration in soil has prohibited crop growth or crop production quality is harmful to human or cattle health we think that the soil has been contaminated. Thus, people established soil environmental quality standards (also known as highest permitted concentration in soil) to protect farmland; that is, when heavy metal or other pollutant concentrations in soil are higher than some standard values, the soil has been contaminated [Xia, 1996; Dou and Li, 1998]. Basically, there are two methods to constitute the standard. One is the geochemical method, and the other is the ecological environment effect method. Based on application effects, the combined utilization of these two methods is better. Until now, there has not been an accordant quantity standard of soil environment in the world. The standard is influenced by many factors such as soil texture, soil usefulness method, economical status, and so on. Table 4 lists some heavy metal soil environmental quality standards or the maximum permission concentrations of heavy metals in soils of some countries and areas, which showed that there are huge differences among of them even for the same heavy metal. The standards in some Northern European countries are very strict. But for England, it is relatively loose. In China, the soil environmental quality standard is very particular, which include three grades. The second grade is applicable to food crop or vegetable, tea, orchard, and pasture soil.

To better understand clean/healthy status of agricultural soil and protect it, some heavy metal background concentrations should be known. Table 5 shows some soil background contents of trace elements in the world and in some country soils. In China, Zn background concentration range is from 3 to 790 ppm, which means that some soils are polluted compared the standard value shown in Table 4. This indicated that some agricultural soils are not safe even only with background heavy-metal concentration.

Food from crops is the main resource of human survival. High heavy-metal concentration in crop food seriously affects human health. Understanding background concentration of heavy metals in crop food is necessary to protect human health, which can let us know the natural status of heavy metal. Then, we do not need to feel panic when we discover some crop foods containing some heavy metals which may be harmful for people. Usually, such food with background heavy metal concentration can be safely applied. Table 6 shows some trace elements background concentrations in some crop seeds.

Weed species is one of the main parts of agro-ecosystems, which can also reflect the clean status of soil by their background of trace elements in soil. We determined some heavy-metal background concentrations in 55 weed species which often survive in the Shenyang Station of Experimental Ecology soil, Chinese Academy of Sciences (41°31′ N and 123°41′ E). This site is a temperate zone with a semi-moist continental climate. The soil is meadow burozem soil and is relatively unpolluted compared with

TABLE 5. Trace Elements Background Concentrations in the World and Some Country Soils (mg/kg)

Trace Elements	World	China	Japan	Brazilian (Cerrado)
Zn	10–300	3–790	23	38
Fe			80	75
Se	0.2	0.29		
Mo	1–5	0.2–6		
Co	10–40	5–40	0.78	5
Ni	40	35	5.4	14
Pb	10–150	13–42	24	26
Cd	0.06	0.097		
Cr	20–200	<100	2.4	112
As	9.36	10.38		
Cu	20	22	350	33
Hg	0.03	0.04		
Reference	Xie and Lu [2000]	Yang and Yang [2000]	Takeda et al. [2004]	Marques et al. [2004]

TABLE 6. Trace Elements Background Concentrations in Some Crop Seeds (mg/kg)

Trace Elements	Corn	Soybean	Wheat	Durox	Wild Rice	Buckwheat
Co	0.1	0.1		0.08		0.167
Cu	9	20	6.33	6.43	0.87	
Mn	5	40				
Zn	425	45	234.24	19.10	2.32	26.0
Cd			0.00	0.41	5.32	
Cr			3.56	0.51		0.138
Pb			20.36	0.07		
As				0.19	6.63	
Mo				1.33		
Ni				0.46		1.10
Fe					3.39	60.5
Se						0.019
Hg						<0.01
Reference	Bolan et al. [2004]	Bolan et al. [2004]	Frost and Ketchum [2000]	Frost and Ketchum [2000]	Nriagu and Lin [1995]	Bonafaccia et al. [2003]

the National Soil-Environmental Quality Standard of China (Table 4). The background contents of heavy metals were Cd 0.15 mg/kg, Pb 14.2 mg/kg, Cu 12.4 mg/kg, and Zn 39.9 mg/kg, respectively, and with organic matter of 1.52%. The pH of soil was 6.6. The results showed that heavy-metal concentration ranges are Cd nd (not detectable) ~1.9 mg/kg, Pb nd~13.9 mg/kg, Cu 0.4~45.4 mg/kg, and Zn nd~338.4 mg/kg, respectively (Table 7). Zn concentrations in various weed species are very different.

4   PHYTOREMEDIATION OF TRACE ELEMENT CONTAMINATION

4.1   Basic Mechanisms of Phytoremediation

Generally, phytoremediation is using a plant or a microorganism in a plant rhizosphere to remove or degrade pollutants from contaminated environment, or stabilize contaminants in soil through changing their availabilities into unavailable speciations by plant/microorganism secretions [Wei and Zhou, 2004]. These polluted environments include air, soil, and water, and the pollutants include organic materials like PAHs, pesticide, and petroleum, as well as inorganic materials like heavy metals and radio nuclides. Figure 2 lists six main ways of phytoremediation such as phytoclean, phytodegradation, rhizosphere degradation, phytoextraction, phytovolatilization, and phytostobilization. Usually, phytoextraction is widely regarded as the most promising remediation technology with many advantages that physical and chemical remediation do not hold, such as low cost, no soil-structure destruction,

TABLE 7. Trace Element Background Concentrations in 55 Farmland Weed Species (mg/kg)

Plant Species	Family	Part	Cd	Pb	Cu	Zn
Bidens tripartita L	Asteraceae	Root	0.2	0.4	11.5	12.9
		Shoot	0.3	1.4	8.7	35.3
Lactuca indica L	Asteraceae	Root	1.1	0.9	26.2	21.1
		Shoot	0.9	1.4	5.4	44.4
Artemisia lavandulaefolia DC	Asteraceae	Root	0.8	10.3	22.1	39.6
		Shoot	0.4	3.0	5.4	45.6
Silphium perfoliatum L	Asteraceae	Root	0.1	1.4	4.3	19.2
		Shoot	0.2	3.7	7.3	40.6
Cirsium pendulum Fisch	Asteraceae	Root	0.6	2.7	10.4	18.1
		Shoot	1.7	1.9	2.3	30.9
Helianthus tuberosus L	Asteraceae	Root	0.1	0.1	5.0	22.7
		Shoot	nd	0.4	3.4	15.4
Artemisia argyi Levl et Vant	Asteraceae	Root	0.1	0.2	5.0	15.6
		Shoot	0.2	3.4	5.8	30.5
Inula britannica Thunb	Asteraceae	Root	0.2	0.5	8.8	14.3
		Shoot	0.6	1.7	6.5	34.8
Sonchus brachyotus DC	Asteraceae	Root	0.3	2.7	12.1	17.6
		Shoot	0.5	1.3	5.0	22.3
Cirsium pendulum Fisch	Asteraceae	Root	0.6	2.7	10.4	18.1
		Shoot	1.7	1.9	2.3	30.9
Helianthus tuberosus L	Asteraceae	Root	0.1	0.1	5.0	22.7
		Shoot	nd	0.4	3.4	15.4
Kalimeris integrifolia Turcz. ex DC	Asteraceae	Root	0.3	0.5	9.8	14.1
		Shoot	0.2	nd	3.7	29.2
Bidens pilosa L	Asteraceae	Root	0.2	9.2	9.5	83.8
		Shoot	0.8	2.6	11.5	53.0
Ambrosia trifida L	Asteraceae	Root	0.2	13.9	13.8	17.1
		Shoot	0.2	6.4	4.2	11.4
Artemisia scoparia Waldst et Kitag	Asteraceae	Root	0.3	4.3	8.1	20.2
		Shoot	0.4	1.8	2.9	14.8
Cephalanoplos setosum (Willd) Kitam	Astcraceae	Root	0.1	0.9	12.1	20.5
		Shoot	0.2	8.0	12.8	36.2
Polygonum roseoviride L	Polygonaceae	Root	0.2	2.0	5.5	28.3
		Shoot	nd	0.4	2.1	12.6
Polygonum aviculare L	Polygonaceae	Root	0.5	1.8	1.3	9.2
		Shoot	0.4	0.1	0.4	19.8
Polygonum hydropiper L	Polygonaceae	Root	0.5	4.3	15.8	6.4
		Shoot	0.1	1.2	5.0	4.7
Polygonum lapathifolium L	Polygonaceae	Root	0.1	5.7	4.4	12.6
		Shoot	0.1	0.9	3.9	19.7
Polygonum bungeanum	Polygonaceae	Root	0.1	nd	8.7	27.3
		Shoot	nd	nd	2.9	13.4
Perilla frutescens (L) Britt	Labiatae	Root	0.2	0.6	12.6	28.0
		Shoot	0.1	0.4	4.9	34.6

(Continued)

TABLE 7. *Continued*

Plant Species	Family	Part	Cd	Pb	Cu	Zn
Salvia plebeia R.Br.	Labiatae	Root	nd	7.1	12.6	69.0
		Shoot	nd	1.7	5.6	23.9
Mentha haplocalyx Briq.	Labiatae	Root	0.1	0.6	9.5	23.7
		Shoot	0.1	2.4	2.6	33.3
Leonurus heterophyllus Sweet	Labiatae	Root	nd	1.5	4.5	11.3
		Shoot	nd	2.6	2.7	47.5
Plantago lanceolata L	Plantaginaceae	Root	0.7	9.5	31.3	338.4
		Shoot	0.3	4.7	6.6	33.6
Plantago asiatica L	Plantaginaceae	Root	0.2	2.8	19.6	31.5
		Shoot	0.1	5.4	6.8	29.7
Plantago depressa Willd	Plantaginaceae	Root	0.4	3.5	20.5	34.5
		Shoot	0.1	1.6	3.0	8.5
Polygonum viscosum Ham	Plantaginaceae	Root	nd	3.7	6.1	33.5
		Shoot	nd	3.9	2.9	30.2
Chenopodium hybridum L	Chenopodiaceae	Root	0.1	0.9	2.1	14.6
		Shoot	0.1	0.6	0.8	2.8
Chenopodium album L	Chenopodiaceae	Root	0.3	3.8	7.8	75.9
		Shoot	0.2	2.0	1.3	18.1
Carex rigescens (Franch) V Krecz	Cyperaceae	Root	0.2	3.4	4.1	40.0
		Shoot	0.1	0.8	3.4	nd
Cannabis sativa L	Moraceae	Root	nd	1.8	2.2	18.8
		Shoot	nd	1.2	2.1	19.6
Humulus scandens (Lour) Merr	Moraceae	Root	0.2	4.3	3.3	74.1
		Shoot	0.2	1.3	1.8	29.8
Echinochloa crusgalli (L.) Beauv.var mitis (Pursh) Peterm	Gramineae	Root	1.2	2.3	45.4	32.9
		Shoot	0.1	2.4	4.4	37.8
Arthraxon hispidus (Thunb) Makino	Gramineae	Root	0.2	1.3	4.7	30.7
		Shoot	0.1	5.1	3.3	43.1
Oenothera biennis L	Onagraceae	Root	nd	0.1	2.6	15.8
		Shoot	nd	2.4	3.9	16.3
Amaranthus retroflexus L	Amaranthacae	Root	0.2	0.8	3.3	21.9
		Shoot	0.1	1.6	3.1	24.1
Amaranthus lividus L.	Amaranthacae	Root	0.1	0.3	4.5	27.6
		Shoot	0.1	1.7	2.6	46.8
Glycine soja Sieb et Zucc	Leguminosae	Root	0.1	5.0	15.6	37.8
		Shoot	nd	4.2	11.3	32.7
Glycyrrhiza uralensis Fisch	Leguminosae	Root	0.1	nd	7.4	40.3
		Shoot	0.2	nd	6.5	35.5
Commelina communis L	Commelinaceae	Root	0.1	0.7	4.6	41.4
		Shoot	0.1	2.7	4.2	47.8
Peucedamum terebinthaceum L	Umbelliferae	Root	0.2	4.9	7.3	33.4
		Shoot	0.1	4.1	4.4	26.1
Potentilla paradoxa Nutt.	Rosaceae	Root	0.3	2.8	6.3	17.9
		Shoot	0.1	7.6	5.6	37.7

(Continued)

TABLE 7. *Continued*

Plant Species	Family	Part	Cd	Pb	Cu	Zn
Portulaca oleracea L	Portulacaceae	Root	0.2	nd	9.9	54.5
		Shoot	nd	nd	6.1	24.2
Polygonum orientale L	Polygonaceae	Root	0.2	4.8	2.5	4.9
		Shoot	0.1	5.6	1.4	4.4
Calystegia hederacea Wall	Convolvulaceae	Root	0.2	0.4	4.1	19.0
		Shoot	0.4	4.3	7.9	51.4
Metaplexis japonica (Thunb) Makino	Asclepiadaceae	Root	nd	0.6	4.3	26
		Shoot	nd	0.7	3.7	35.4
Abutilon theophrasti Medic	Malvaceae	Root	0.3	5.1	13.1	105.8
		Shoot	0.1	0.9	4.3	59.8
Lepidium apetalum willd	Cruciferae	Root	0.2	2.2	2.0	26.6
		Shoot	0.1	1.4	1.1	12.0
Ranunculus chinensis Bge.	Ranunculaceae	Root	0.2	3.2	18.0	27.9
		Shoot	nd	2.2	5.3	20.9
Acalypha australis L	Euphorbiaceae	Root	0.2	7.2	16.7	110.7
		Shoot	0.1	3.8	3.1	36.1

nd, not detectable.

no secondary pollution, environmental beautification, and easy acceptance by the public [Raskin et al., 1994; Salt et al., 1995; Chaney et al., 1997]. The main procedures to clean contaminated soil consist in the use of hyperaccumulators to transnormally accumulate metals from soils; and the biomass of plants, especially their above-ground parts, are seasonally harvested until metal concentrations in the soil decrease to acceptable levels [Baker et al., 1994; Chaney et al., 1997; Krämer, 2005].

Hyperaccumulators are such plants that can accumulate exceptionally high quantities of contaminants, mostly heavy metals. Although there have been different

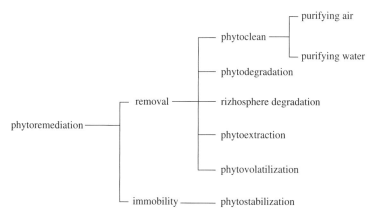

Figure 2. The main ways of phytoremediation.

definitions after Brooks et al. [1977] first coined the term, the main characteristics of hyperaccumulating plants can be summarized as follows [Wei et al., 2006]: (1) *accumulation capability*; that is, the minimum concentration in the shoots of a hyperaccumulator for As, Pb, Cu, Ni, and Co should be greater than $1000\,mg/kg$ dry mass, and Zn and Mn $10,000\,mg/kg^{-1}$, Au is $1\,mg/kg$, and Cd is $100\,mg/kg^{-1}$, respectively; (2) *translocation capability*; that is, elemental concentrations in the shoots of a plant should be higher than those in roots, that is, TF > 1 (translocation factor, concentration ratio of shoots to roots); (3) *enrichment capability* [enrichment factor (EF), concentration ratio of plant to soil]: that is, EF value in shoots of plants should be higher than 1; and (4) *tolerance capability*; that is, a hyperaccumulator should have high tolerance to toxic contaminants. In addition, for the plants tested under experimental conditions, their aboveground biomass should not decrease significantly when growing in contaminated soils.

4.2 Research Progress of Phytoextraction

According to some documents, more than 400 hyperaccumulators have been reported, but only very few of them were used to remedy contaminated soil or water [Raskin and Ensley, 2000]. Many researches are still in lab study or scale field experiment phases. The main reasons why phytoextraction is not widely used commercially consist in some shortcoming of these discovered hyperaccumulators such as small biomass, long growth period, and so on. Also, the progress of constructing ideal hyperaccumulators by using biological technologies is slow. Thus, in the meantime for screening ideal hyperaccumulators, some studies on hyperaccumulating mechanisms and improving measurements have been developed on a large scale in the world.

Some research has shown that some organic materials secreted by root systems especially for organic acid can promote plant accumulating heavy metals by chelation or rhizosphere environmental acidification [Krishnamurti et al., 1997; Cieslinski et al., 1998], but it is not clear if the mechanisms whereby hyperaccumulators accumulate heavy metals from soil are significantly different from those of non-hyperaccumulators. Compared with non-hyperaccumulators, there is obviously selection for hyperaccumulator extracting heavy metals. The possible mechanism lies in some transportation protein with specified affinity toward some heavy metals in plants [Salt and Kramaer, 2000]. After they entered root systems, heavy metals were transported into vacuoles through symplastic transport. Then these special transporters in vacuole membrane which distinguish them from those of non-hyperaccumulators load and transport heavy metals in vacuole into xylem vessels so as to benefit their transportation from the root to the above-ground part of a plant. Thus, much-higher-level heavy metals are accumulated in plant shoot, which may be the reason why heavy metal concentration in hyperaccumulator shoot is higher than those in root [Lasat et al., 1996; Chardonnens et al., 1999]. Some studies indicated that there are two reasons for hyperaccumulators with strong tolerance to heavy metals. One is compartmentation in leaves where heavy metals are localized in apoplastic and vacuole, which can prevent the harmfulness of heavy metal to cell [Vazquez et al., 1994; Kramer et al., 1997]. Another is an anti-oxidative or chelation system in plants such as superoxidase dimutase (SOD), phyto-chelatins (PCs), or histidine, which can remove free radicals caused by heavy metals. Meanwhile, the chelation of PCs or histidine to heavy metals can also decrease their toxicities and promote their uptake by plants [Krämer et al., 1996; Maitani, 1996; Boominathan and Doran, 2003]. There has been some progress regarding hyperaccu-mulative mechanisms in molecular biology in the past decade, such as the clone and identification of Zn transport protein ZNT1 [Lasat et al., 2000], the clone and screen-ing of enzyme protein genes THG1, THB1, and THD1 used to synthesize histidine, and the discovering, separating, and identifying of some function genes connected with hyperaccumulative tolerance from bacteria, fungi, plant, or animal [Chiang et al., 2006]. Furthermore, some transgene plants have been used as remediative materials in some experiments [Rugh et al., 1998; Bennett et al., 2002]. Bioactivities of heavy metals in soil and biomass regulation of hyperaccumulators are also key points of restricting phytoremediation efficiency. Remediation effect and impact to the environment of following the addition of soil additives such as EDTA and EDDS, some organic acids, and fertilizer have been tested [Meers et al., 2002; Sudova et al., 2007]. In addition, the remediation potentials of some special plants with higher biomass but without higher heavy-metal concentration (i.e., non-hyperaccumulator) are examined [Citterio et al., 2003]. The combined remediation of plant with microorganism (especially mycorrhiza) is becoming a hot topic in environmental science [White, 2001; Vogel-Mikus et al., 2006; Leung et al., 2006; Marques et al., 2006].

4.3 Discussion on Agro-Strengthen Measurements

4.3.1 Common Ground and Difference in Agriculture and Phytoextraction

Usually, the main research object in agriculture is the crop, which includes food crops like corn, rice, and wheat, vegetable crops like cabbage and eggplant, forage/fertilizer crops like clover, and so on. Nearly every part of a crop can be regarded as an agricultural production. For example, root of sweet potato, stem of potato, leaf of celery cabbage, inflorescence of daylily, and seed of corn. Basically, the aim of agriculture is to obtain the highest seed or biomass yield such as root, stem, leaf, or inflorescence from crops. In order to gain the largest quantity of agricultural product from crops, many measurements are applied such as watering, fertilizer, breeding, and so on.

Phytoextraction mainly uses plants to remove pollutants from heavy metal con-taminated soils. Thus, some special plants like hyperaccumulator are the hard core of phytoextraction, which inevitably involves plant growth and propagation, especially for plant biomass. Agricultural aims are to how to improve crop growth and propagation so as to obtain the highest economical biomass (stem, leaf, or inflor-escence biomass) or seed yield. The hard cores of agriculture and phytoextraction are accordant (Table 8); that is, plants are used by both of them [Wei and Zhou, 2004;

TABLE 8. Common Ground and Difference in Agriculture and Phytoextraction

Items	Common Grounds	Difference Research Object	Research Aim	Using Parts of Plant
Agriculture	(1) Plant growth including germination, growth, and harvest. (2) Obtain the largest biomass in limited time.	Food, forage or green fertilizer crops like corn, rice, wheat, cabbage, eggplant, clover, etc.	Obtain the highest seed yield or economical biomass (stem, leaf, etc.)	Different crops with different aim plant parts including seed, root, stem, leaf, inflorescence
Phytoextraction	(3) Need suitable environmental conditions like nutrient, water, light, temperature to grow and enhance plant biomass.	Any plant which exceptively accumulating heavy metals such as weed species, trees, sometimes, some crops, and so on	Using plant to remove heavy metals from contaminated soils	Usually, stem and leaf is main organs to remove heavy metals

Wei et al., 2007]. Thus, agricultural technologies that are used to strengthen phyto-extraction are very important and feasible.

4.3.2   Increase Unavailable Heavy-Metal Bioactivity   Many heavy metals under unavailable status are difficult to extract from contaminated soil. Some agricultural activating measures can be taken to enhance bioavailability of heavy metals.

To most of heavy metals, decreased soil pH may increase their bioavailabilities in soil due to increased H^+ exchangeable capacity [He et al., 1998]. To increase soil pH, direct acidification methods or indirect acidification methods can be considered, namely, concentrated sulfuric acid is diluted and then sprayed onto the surface of soils; or one mixes a nutrient reagent that is made of organic fertilizers, chemical fertilizers, or diluted concentrated sulfuric acid with contaminated soil. These methods are too often used in rice production when requiring a seedling culture, which usually decreases pH from 7 to 5.5 [Yang, 1998]. However, some elements are excepted such as, whose bioavailability increases with pH owing mostly to speciation of As in soil as AsO_4^{2-} or AsO_3^{3-}, which is very different from other cation elements.

4.3.3   Crop Breeding Technology   Crop breeding technology is one of the most important technologies of agriculture to gain maximal resources from crops such as food, vegetables, oil, fodder, fiber, beverage, and rubber. As we all know, it is very difficult for seed breeding to get a higher seed yield through optimizing desired crop traits such as plant height, stem thick, leaf area, growth period, and so on. But for phytoextraction that mainly uses plant stems or leaves, this point is relatively easy to achieve [Yang, 1998]. By using system breeding technology (continuously choosing), some hyperaccumulator species with huge stems and leaf biomass can be selected [Yang, 1998], which is beneficial to the accumulation of heavy metals by plants [Wei et al., 2006]. Likewise, some cultural traits like consistently ripe stage and short growing period can also be improved by crop breeding technology, which will strengthen efficiency of phytoextraction too.

4.3.4   Seed-Coating Technology   Most hyperaccumulators are wild plants whose seeds are usually small, which are very inconvenient for sowing. Seed coating is the use of a layer of materials commonly containing some fertilizers and pesticides to enclose them, which can improve germination, prevent and cure seedling diseases and pests, and dispel mice. Moreover, increased seed volume is convenient for mechanized sowing. Seed coating may be an indispensable technology for phytoextraction applied at a large, commercial scale.

4.3.5   Water and Fertilizer Utilization   Shoot biomass of hyperaccumulator is one of the important factors impacting the efficiency of phytoextraction. Some experimental results showed higher shoot biomass with higher remedying efficiency. Irrigation and fertilization are two important factors that increase shoot biomass. But overwatering and overfertilization may also prohibit plant growth. Appropriate application of water and fertilizer should be considered. Usually, the seedling and flowering stage of plants are the most sensitive periods, and plenty of water and nutrients is indispensable.

4.3.6   Shorten the Cycle of Phytoextraction   Temperature, sunlight, soil, water, air, and heat can greatly affect plant growth. Therefore, some agricultural measurements can be taken according to the reaction of a plant to these environmental conditions. A greenhouse can accelerate the growth of a plant, such as some outdoor plants that cannot survive well at low temperatures. Shade devices may promote the growth of a plant, mirroring faded light conditions. An increase in carbon dioxide concentration can enhance the plant's photosynthesis [Silva et al., 2005]. Transplanting can also shorten the cycle of phytoextraction, and hyperaccumulator seedlings that are kept under cover can be transplanted into contaminated soils once field conditions are fit for growth, which shortens the remediation time from sowing to seedling. Furthermore, taking advantage of the fact that for some hyperaccumulators the growth duration from seedling-transplantation to the flowering phase is short, and that the concentration of heavy metals accumulated in their shoots at the flowering phase is high, the efficiency of phytoextraction can be greatly improved using the method of two-phase planting—that is, harvesting the hyperaccumulator at its flowering phase, then transplanting its seedlings again [Wei et al., 2006].

ACKNOWLEDGMENTS

The research was supported by hi-tech research and development program of China (No. 2006AA06Z386), by the National Natural Science Foundation of China as the overseas young scientist grant (No. 20428707), and by the Sino-Russian Joint Research Center on Natural Resources and Eco-Environmental Sciences.

REFERENCES

Baker AJM, McGrath SP, Sidoli CMD, Reeves RD. 1994. The possibility of *in situ* heavy metal decontamination of polluted soils using crops of metal-accumulating plants. *Resour Conserv Recycl* **11**:41–49.

Bennett LE, Burkhead JL, Hale KL, et al. 2002. Analysis of transgenic Indian mustard plants for phytoremediation of metal-contaminated mine tailing. *J Environ Quality* **83**:442–451.

Bolan NS, Khan MA, Donaldson DC, Adriano DC, Matthew C. 2003. Distribution and bioavailability of copper in farm effluent. *Sci Total Environ* **309**:225–236.

Bolan NS, Adriano DC, Mahimairaja S. 2004. Distribution and bioavailability of trace elements in livestock and poultry manure by-products. *Crit Rev Environ Sci Technol* **34**:291–338.

Boominathan R, Doran PM. 2003. Organic acid complexation, heavy meal distribution and effect of ATPase inhibition in hairy roots of hyperaccumulator plant species. *J Biotechnol* **101**:131–146.

Bonafaccia G, Gambelli L, Fabjan N, Kreft I. 2003. Trace elements in flour and bran from common and tartaary buckwheat. *Food Chem* **83**:1–5.

Brooks RR, Lee J, Reeves RD. 1977. Detection of nickliferous rocks by analysis of herbarium species of indicator plants. *J Geochem Expl* **7**:49–77.

Chaney RL, Malik M, Li YM, Brown SL, Angle JS, Baker AJM. 1997. Phytoremediation of soil metals. *Curr Opin Biotechnol* **8**:279–284.

Chaney RL, Ryan JA, Kukier U, Brown SL, Siebielec G, Malik M, Angle JS. 2001. Heavy metal aspects of compost use. In: Stoffella PJ, Khan BA, editors. *Compost Utilization in Horticultural Cropping Systems*. Boca Raton, FL: CRC Press, pp. 324–359.

Chardonnens AN, Koevoets PLM, Zanten AV, et al. 1999. Properties of enhanced tonoplast zinc transport in naturally selected zinc-tolerant *Silene vulgaris*. *Plant Physiol* **120**:779–785.

Chiang HC, Lo JC, Yeh KC. 2006. Genes associated with heavy metal tolerance and accumulation in Zn/Cd hyperaccumulator Arabidopsis halleri: A genomic survey with cDNA microarray. *Environ Sci Technol* **40**:6792–6798.

Cieslinski G, Van Ress KCJ, Szmigielska AM, et al. 1998. Low-molecular-weight organic acids in rhizosphere soils of durum wheat and their effect on cadmium bioaccumulation. *Plant Soil* **203**:109–117.

Citterio S, Reprint A, Santagostino A, et al. 2003. Heavy metal tolerance and accumulation of Cd, Cr and Ni by *Cannabis sativa* L. *Plant Soil* **256**:243–252.

Dou YJ, Li CH. 1998. Principle of environment science. Nanjing: Nanjing University Press.

Frost HL, Ketchum LHJ. 2000. Trace metal concentration in durum wheat from application of sewage sludge and commercial fertilizer. *Adv Environ Res* **4**:347–355.

He ZL, Zhou QX, Xie ZM. 1998. *Chemical Equilibrium of Beneficial and Pollution Elements in Soil*. Bejing: Chinese Environ Sci Press.

He ZL, Yang XE, Stoffella PJ. 2005. Trace elements in agroecosystems and impacts on the environment. *J Trace Elem Med Biol* **19**:125–140.

Jackson BP, Bertsch PB. 2001. Determination of arsenic speciation in poultry wastes by IC-ICP-MS. *Environ Sci Technol* **35**:4868–4873.

Jiang LY, Yang XE, He ZL. 2004. Growth response and phytoextraction of copper at different levels in soils by Elsholtzia splendens. *Chemosphere* **55**:1179–1187.

Jiang XY, Zhao KF. 2001. Heavy metal damage to plant and its resisting mechanisms. *J Appl Environ Biol* **7**:92–99.

Krämer U. 2005. Phytoremediation: novel approaches to cleaning up polluted soils. *Curr Opini Biotechnol* **16**:133–141.

Krämer U, Cotter-Howells JD, Charnock JM, et al. 1996. Free histidine as a metal chelator in plants that accumulate nickel. *Nature* **379**:635–638.

Krämer U, Smith RD, Wenzel WW, et al. 1997. The role of metal transport and tolerance in nickel in hyperaccumulation by *Thlaspi goesingense* Halacsy. *Physiol Plant* **115**:1641–1650.

Krauskopf KB, Bird DK. 1995. *Introduction to Geochemistry*, third edition. New York: McGraw-Hill.

Krishnamurti GSR, Cieslinski G, Huang PM, et al. 1997. Kinetics of cadmium release from soils as influenced by organic acids: Implication in cadmium availability. *J Environ Qual* **26**:271–277.

Lam JCW, Tanabe S, Wong BSF, Lam PKS. 2004. Trace element residues in eggs of little egret (*Egretta garzetta*) and black-crowned night heron (*Nycticorax nycticorax*) from Hong Kong, China. *Marine Pollut Bull* **48**:378–402.

Lasat MM, Baker AJM, Kochian LV. 1996. Physiological characterization of root Zn^{2+} absorption and translocation to shoots in Zn hyperaccumulator and non-accumulator species of *Thlaspi*. *Plant Physiol* **112**:1715–1722.

Lasat MM, Pence NS, Garvin DE, et al. 2000. Molecular physiology of zinc transport in the Zn hyperaccumulator *Thlaspi caerulescens*. *J Exp Bot* **51**:71–79.

Leung HM, Ye ZH, Wong MH. 2006. Interactions of mycorrhizal fungi with *Pteris vittata* (As hyperaccumulator) in As-contaminated soils. *Environ Pollut* **139**:1–8.

Maitani T. 1996. The composition of metals bound to class. III metallothionein (phytochelation and its desglycyl peptide) induced by various metals in root culture of *Rubia tinctorum*. *Plant Physiol* **110**:1145–1150.

Marques APGC, Oliveira RS, Rangel AOSS, et al. 2006. Zinc accumulation in *Solanum nigrum* is enhanced by different arbuscular mycorrhizal fungi. *Chemosphere* **65**:1256–1263.

Marschner H. 1995. *Mineral Nutrition of Higher Plants*, second edition. London: Academic Press.

Marques JJ, Schulze DG, Curi N, Mertzman SA. 2004. Trace element geochemistry in Brazilian Cerrado soils. *Geoderma* **121**:31–43.

McBride MB. 1995. Toxic metal accumulation from agricultural use of sludge: Are USEPA regulations protective? *J Environ Qual* **24**:5–18.

McBride MB, 2004. Molybdenum, sulfur, and other trace elements in farm soils and forages after sewage sludge application. *Commun Soil Sci Plant Anal* **35**:517–535.

Meers E, Ruttens A, Hopgood MJ, et al. 2005. Comparison of EDTA and EDDS as potential soil amendments for enhanced phytoextraction of heavy metals. *Chemosphere* **58**:1011–1022.

Mortvedt JJ, Beaton JD. 1995. Heavy metal and radionuclide contaminants in phosphate fertilizers. In: Tiessen H, editor. *Phosphorus in the Global Environment: Transfer, Cycle and Management*. New York: John Wiley & Sons, pp. 93–106.

Nriagu JO, Lin TS. 1995. Trace metals in wild rice sold in the United States. *Sci Total Environ* **172**:223–228.

Presser TS. 1994. Geologic origin and pathways of selenium from the California coast ranges to the West-central San Joaquin Valley. In: Frankenberger WT Jr, Benson S, editors. *Selenium in the Environment*. New York: Marcel Dekker, pp. 139–156.

Raskin I, Ensley BD. 2000. *Phytoremediation of Toxic Metals*. New York: John Wiley & Sons.

Raskin I, Nanda-Kumar PBA, Dushenkov S. 1994. Removal of radionuclides and heavy metals from water and soil by plants. *Bioremediation* **3**:345–354.

Rugh CL, Senecoff JF, Meager RB, et al. 1998. Development of transgenic yellow poplar for mercury phytoremediation. *Nature Biotechnol* **16**:925–928.

Salt DE, Kramaer U. 2000. Mechanisms of metal hyperaccumulation in plants. In: Raskin H, Ensley BD, editors. *Phytoremediation of Toxic Metals: Using Plants to Clear Up the Environment*. New York: John Wiley & Sons, pp. 231–246.

Salt DE, Blaylock M, Kumar NPBA, Dushenkov V, Ensley BD, Chet I, Raskin I, 1995. Remediation: A novel strategy for the removal of toxic metals from the environment using plants. *Biotechnology* **13**:468–474.

Senesi GS, Baldassarre G, Senesi N, Radina B. 1999. Trace element inputs into soils by anthropogenic activities and implications for human health. *Chemosphere* **39**:343–377.

Shen DZ. 2002. *Bioremediation of Contaminated Soil*. Beijing: Chemical Industry Press and Environmental Sci and Engineering Press.

Silva JAT, Giang DTT, Tanaka M. 2005. Micropropagation of sweet potato (*Ipomoea batatas*) in a novel CO_2-enriched vessel. *J Plant Biotechnol* **1**:1–8.

Sudova R, Pavlikova D, Macek T, et al. 2007. The effect of EDDS chelate and inoculation with the arbuscular mycorrhizal fungus *Glomus intraradices* on the efficacy of lead phytoextraction by two tobacco clones. *Appl Soil Ecology* **35**:163–173.

Sun TH, Zhou QX, Li PJ. 2001. *Pollution Ecology*. Beijing: Science Press.

Takeda A, Kimura K, Yamasaki SI. 2004. Analysis of 57 elements in Japanese soils, with special reference to soil group and agricultural use. *Geoderma* **119**:291–307.

Vazquez MD, Poschenrieder C, Barcelo J, et al. 1994. Compartment of zinc in roots and leaves of the zinc hyperaccumulator *Thlaspi caerulescens* J & C Presl. *Bot Acta* **107**:243–250.

Vogel-Mikus K, Pongrac P, Kump P, et al. 2006. Colonisation of a Zn, Cd and Pb hyperaccumulator *Thlaspi praecox* Wulfen with indigenous arbuscular mycorrhizal fungal mixture induces changes in heavy metal and nutrient uptake. *Environ Pollut* **139**:362–371.

Walker JM. 2001. US Environmental protection agency regulations governing compost production and use. In: Stoffella PJ, Khan BA, editors. *Compost Utilization in Horticultural Cropping Systems*. Boca Raton, FL: CRC Press, pp. 381–399.

Wei SH, Zhou QX. 2004. Disscussion on basic principles and strengthening measures for phytoremediation of soils contaminated by heavy metals. *Chin J Ecol* **23**:65–72.

Wei SH, Zhou QX, Wang X. 2005. Identification of weed plants excluding the uptake of heavy metals. *Environ Int* **31**:829–834.

Wei SH, Zhou QX, Koval PV. 2006. Flowering stage characteristics of cadmium hyperaccumulator *Solanum nigrum* L. and their significance to phytoremediation. *Sci Total Environ* **369**:441–446.

Wei SH, Silva JAT, Zhou QX. 2007. Agro-improving method of phytoextracting heavy metal contaminated soil. *J Hazard Mater* doi:10.1016/j.jhazmat.2007.05.014.

White PL. 2001. Phytoremediation assisted by microorganisms. *Trends Plant Sci* **6**:502–507.

Xia JQ. 1996. *Detailed Explanation on the State Soil-Environment Quality Standard of China*. Beijing: Chinese Enviromental Science Press, China.

Xie ZM, Lu SM. 2000. Trace elements and environmental quality. In: Wu QL, editor. *Micronutrients and Biohealth*. Guiyan, China: Guizhou Sci Technol Press, pp. 208–216.

Xu JL, Yang JR. 1995. *Heavy Metal in Terrestrial Ecology Systems*. Chinese Environmental Science Press.

Yang SR. 1998. *Rice Collectanea of Yang Shouren*. Shenyang: Liaoning Sci Tech Press.

Yang WY. 2002. *An Introduction to Agronomy*. Beijing: Chinese Agriculture Press.

Yang YA, Yang XE. 2000. Micronutrients in sustainable agriculture. In: Wu QL, editor. *Mircronutrients and Biohealth*. Guiyan, China: Guizhou Science and Technology Press, pp. 120–134.

Zhang XZ, Chen FY. 1996. *Plant Physiology*. Chengchun: Jilin Science and Technology Press.

Zhou QX, Huang GH. 2001. *Environment Biogeochemistry and Global Environment Change*. Beijing: Science Press.

Zhou QX, Song YF. 2004. *Remediation of Contaminated Soils: Principles and Methods*. Beijing: Science Press.

4 Metal Accumulation in Crops—Human Health Issues

ABDUL R. MEMON and YASEMIN YILDIZHAN

Institute of Genetic Engineering and Biotechnology, 41470 Gebze,
Kocaeli, Turkey

EDA KAPLAN

Department of Biology, Istanbul University, 34134 Eminonu, Istanbul, Turkey

1 INTRODUCTION

All plants must obtain a number of inorganic mineral elements from their environment to ensure successful growth and development of both vegetative and reproductive tissues. A total of 14 mineral nutrients are required. Plants gain most of their mineral nutrients by extracting them from solution in the soil or the aquatic environment and critically depends on the activity of membrane transporters that translocate minerals from the soil into plant. Mineral nutrients are derived from complex interactions involving weathering of rock minerals, decaying organic matter, animals, and microbes. Seventeen nutrient elements are known to be essential for higher plants, among which 14 are mineral elements that plants mainly acquire from the soil [Epstein, 1972] (see Fig. 1). These are: nitrogen, phosphorus, potassium, calcium, magnesium, sulfur, iron, manganese, zinc, copper, nickel, boron, molybdenum, and chlorine. Among these nutrients, N, P, K, Ca, Mg, and S are considered macro or major nutrients, because they are required in large quantities that range between 1 and 150 g per kg of plant dry matter. Fe, Zn, Mn, Cu, Ni, B, Mo, and Cl are micronutrients or minor nutrients that are required at rates of 0.1 to 100 mg per kg of plant dry matter [Marschner, 1997]. The last two elements, molybdenum and chlorine, are present in soils as anions and undoubtedly require active transport across the plasma membrane of plant root cells for uptake. Boron is either an anion or neutral molecule in most soils, and the neutral molecule is fairly permeable across biological membranes [Welch and Graham, 2004]. Whether boron is actively transported into plants is a subject of considerable interest

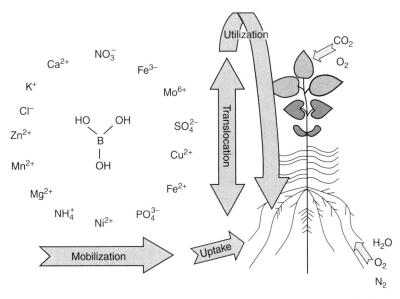

Figure 1. The 17 essential nutrients acquired by plant either from soil or air are indispensable for plant growth and development. The nutrient cycle from plant uptake to utilization is shown [Yan et al., 2006].

in current literature, but new evidence suggests that although it may enter as a neutral molecule, boron transport is facilitated when external concentrations are low [Welch and Graham, 2004] as they commonly are in acid soils everywhere. The remaining six minor elements micronutrients for higher plants are generally absorbed as divalent ions via divalent ion channels. These channels have considerable specificity for each element, or homeostasis is achieved by specific active-excretion mechanisms that are controlled by cytoplasmic concentrations [Welch, 1995]. In addition, sodium and silicon have been shown to be essential for some plants, beneficial to some, and possibly of no benefit to others. Cobalt has also been shown to be essential for the growth of legumes when relying upon atmospheric nitrogen. Claims that two other chemical elements (vanadium and selenium) may be essential micronutrients have still to be firmly established. The element aluminum is of general occurrence in plants, but seems to be without direct nutritional value, although aluminum sulfate is used, because of its acidifying properties, to change the color of hydrangeas growing on alkaline soils from pink to blue, and aluminum may also exert indirect influences on nutritional processes. Other elements often occur in plants but they are not known to serve any useful function and frequently they act as plant poisons or toxins.

Figure 2. (*Continued*) will enable us to elucidate the functions and interactions of plant nutrients at the molecular, cellular, organ, and whole-plant levels. The concept of plant nutriomics, therefore, is to integrate nutritional functions at various levels (molecular, cellular, organ, and whole-plant) with different tools (genomics, transcriptomics, proteomics and metabolomics), as schematically presented this figure [Lahner et al., 2003; Salt, 2004; Yan et al., 2006].

2 THE CONCEPT OF IONOMICS AND NUTRIOMICS IN THE PLANT CELL

Although essential minerals in plants are generally present in soils in sufficient amounts to support plant growth, most of them exist in forms that are not easily available to plants. To overcome this availability problem, plants have developed unique abilities to acquire elements from the soil and numerous studies have shown substantial genetic variation in nutrient efficiency in plants (Fig. 2). To fulfill the mineral

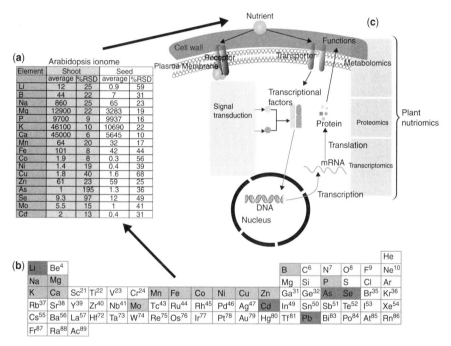

(a) Arabidopsis ionome

Element	Shoot average	%RSD	Seed average	%RSD
Li	12	25	0.9	59
B	44	22	7	31
Na	860	25	65	23
Mg	12900	22	3283	19
P	9700	9	9937	16
K	46100	10	10690	22
Ca	45000	6	5645	10
Mn	64	20	32	17
Fe	101	8	42	44
Co	1.9	8	0.3	56
Ni	1.4	19	0.4	39
Cu	1.8	40	1.6	68
Zn	61	23	59	25
As	1	195	1.3	36
Se	9.3	97	12	49
Mo	5.5	15	1	41
Cd	2	13	0.4	31

(b)

																		He
Li	Be^4											B	C^6	N^7	O^8	F^9	Ne^{10}	
Na	Mg											Mg	Si	P	S	Cl	Ar	
K	Ca	Sc^{21}	Ti^{22}	V^{23}	Cr^{24}	Mn	Fe	Co	Ni	Cu	Zn	Ga^{31}	Ge^{32}	As	Se	Br^{35}	Kr^{36}	
Rb^{37}	Sr^{38}	Y^{39}	Zr^{40}	Nb^{41}	Mo	Tc^{43}	Ru^{44}	Rh^{45}	Pd^{46}	Ag^{47}	Cd	In^{49}	Sn^{50}	Sb^{51}	Te^{52}	I^{53}	Xe^{54}	
Cs^{55}	Ba^{56}	La^{57}	Hf^{72}	Ta^{73}	W^{74}	Re^{75}	Os^{76}	Ir^{77}	Pt^{78}	Au^{79}	Hg^{80}	Tf^{81}	Pb	Bi^{83}	Po^{84}	At^{85}	Rn^{86}	
Fr^{87}	Ra^{88}	Ac^{89}																

Figure 2. A schematic diagram showing the basic concepts of ionomics and nutriomics in the plant cell. (**a**) For comparative ionomics on a genome-wide scale, *Arabidopsis* wild–type and mutant plants were grown in a uniform growth media and elemental analysis was carried out by ICP-MS. (**b**) The table represents *Arabidopsis thaliana* (Col 0) shoot and seed ionomes. All elements are presented as μg g^{-1} dry weight (taken from Salt [2004]). Elements in the Periodic Table highlighted in dark shade are essential for plant growth, and those in light shade represent nonessential trace elements. (**c**) Plant nutriomics is a new frontier of study which shows that how nutritional functions can be integrated at various levels with genomics, transcriptomics, proteomics, and metabolomics. The nutrient as either a signal or a substrate may stimulate a series of biochemical processes in the cell that can be regulated at transcriptional, translational, or metabolic levels [Yan et al., 2006]. Traditional plant nutrition studies look at nutrient efficiency mainly at the whole-plant level. Although useful, studies with whole plants cannot provide sufficient insight into the genetic nature and its specific modification of the nutritional processes. Recent progress in plant molecular biology has provided the means to tackle complex plant nutritional problems through genomic, transcriptomic, proteomic, and metabolomic approaches. All these approaches, together with phenotypic analyses,

requirement and at the same time to avoid their possible toxic effects, plants possess a complex regulatory network controlling mineral homeostasis in the cell by regulating uptake, translocation, redistribution, and sequestering nutrients in appropriate cell compartments [Welch, 2002]. Attempts have been made by breeders to develop crops adapted to either low-fertility or metal contaminated soils. However, these breeding efforts were mainly implemented through simple selection of biomass or yield in the field. Biomass or yield selections in the field are not only costly but also subject to confounding environmental interactions and spatial heterogeneity. Therefore, it would be preferable to identify and select specific traits that are directly related to a specific nutrient efficiency. Once clearly identified, these traits could be used for more efficient screening in controlled environments, or tagged with molecular markers and then improved through marker assisted selection or gene transformation. Useful traits for nutrient efficiency may be associated with altered physiological and biochemical pathways in adaptation to nutrient stress [Welch and Graham, 2004]. At present, the knowledge of the genes controlling specific steps of the mineral homeostasis network is rapidly expanding and numerous research works are being carried out [Rea, 2003]. The fact that many of the molecular and bio-chemical changes in response to nutrient deficiency occur in synchrony suggests that the genes involved are coordinately expressed and share a common regulatory system. Therefore, systematic studies are needed to understand the genomic, transcriptomic, proteomic, and metabolomic aspects of nutrient efficiency [Yan et al., 2006]. These studies will increase our understanding of the molecular basis of nutrient homeostasis and the gene/genes involved in each step of mineral regulation [Rea, 2003]. This knowledge is expected to efficiently improve crop yield, crop nutritional value, and food safety, which are nowadays a major global concern.

3 THE TRACE ELEMENT DEFICIENCIES IN THE DEVELOPING WORLD

Like plants, humans also require 17 essential minerals. Although a varied plant-based diet should supply these nutrients, mineral deficiencies are widespread throughout the world. In particular, people who consume simple diets consisting of staple foods are not supplied with adequate amounts of the essential minerals and vitamins [Grusak, 1999]. The micronutrient deficiencies are becoming a severe health problem in the developing world, and over three billion people in the world are seriously affected with micronutrient malnutrition and the numbers could increase if no preventive measures can be taken [WHO, 1999; World Bank, 1994; Welch and Graham, 2004]. Nearly two-thirds of all deaths of children are associated with nutritional deficiencies, many from micronutrients deficiencies [Caballero, 2002].

Marginal intakes of micronutrients have been shown to contribute to increased morbidity and mortality rates, diminished livelihoods, and adverse effects on learning ability, development, and growth in infants and children [Caballero, 2002; WHO, 1999]. Currently, micronutrient malnutrition in many developing nations is of alarm-ing proportions and needs urgent measures [Mason and Garcia, 1993; WHO, 2002].

Regrettably, most malnourished people living in developing countries either have no access or are not able to get access to the different strategies aimed at reducing micro-element deficiencies and have so far no possibility to improve their nutrition by diversifying the diet. The resulting poor-quality diet, characterized by high intake of staple food and low intake of other micronutrient rich products—for example, animal and fish product, fruits, and vegetables—is responsible for tremendous consequences on human health [WHO, 2002]. Several strategies are pursued to overcome micronutrient malnutrition worldwide, notably either through traditional plant breeding methods or via molecular biological techniques. These strategies should be fully exploited by the genetists, nutritiontionists, and public health officiers to combat micronutrient malnutrition [Graham et al., 2001].

Biofortifying these crops—or, in other words, increasing the micronutrient content of food crops through genetic selection and plant breeding—can significantly improve the amount of these nutrients consumed by these target populations [Welch and Graham, 2004]. In addition, increasing the micronutrient metals stored in the seeds and grains of staple food crops increases crop productivity when these seeds are sown in micronutrient-poor soils [Welch, 1999]. Several barriers to metal accumulation in food crops have to be addressed before genetically modifying plants in ways that will increase the density of micronutrient metals in staple seeds and grains [Welch, 1995]. Additionally, plant foods that contain substances (i.e., antinutrients, inhibitors, etc.) that influence the bioavailability of these nutrients to humans could be addressed in molecular genetics and traditional breeding programs in order to improve the bioavailability of microelements to humans [Graham et al., 2001].

Micronutrient-enriched staple plant foods, either through traditional plant breeding methods or through molecular biological techniques, are powerful intervention tools that target the most vulnerable people (resource-poor women, infants, and children [Bouis, 2000; Combs et al., 1996]). These tools should be fully exploited by the nutrition and public health communities to combat micronutrient malnutrition [Graham et al., 2001]. Genetically modifying plants in ways that will increase the density of micronutrients (e.g., Fe, Zn, etc.) in edible portions of seeds and grains requires that several barriers to metal accumulation within the plant be overcome [Welch, 1995].

4 IMPROVEMENT OF TRACE METAL CONTENT IN PLANTS THROUGH GENETIC ENGINEERING

The ability to manipulate the mineral content of plants through transgenic techniques and/or classical plant breeding offers an opportunity to address human mineral deficiencies by fortifying foods before harvest, thereby eliminating the need for dietary supplements or post-harvest fortification. It also offers an opportunity to improve crop productivity, which is often limited by plant mineral deficiencies. It is clear that trace elements are important for both plant and human nutrition and plant breeding; molecular biology and biotechnology hold great promise for making a significant, sustainable, low-cost contribution to the reduction of microelement deficiencies in farm animals and humans. It would be also having very significant impact on increasing farm productivity both in developing and developed countries in an environmentally beneficial way. The primary objective of plant genetists in both rich and poor countries is to improve crop productivity and quality, usually by developing crops with higher yields having good marketable characteristics such as high protein content, good taste, and good cooking qualities. Recently, genetists and breeders have put some efforts to improve the mineral nutritional content of the plants because of its importance in human nutrition [Welch and Graham, 2004]. However, this research is still lacking behind compared to other quality motivated research in agricultural products, and it becomes clear that conventional wisdom for improving micronutrients deficiencies in people in developing countries needs rethinking.

Remarkable progress has been made in understanding the nutrient ion homeostasis at molecular level in plants [Lahner et al., 2003]. The development and application of modern molecular genetic techniques and completion of the *Arabidopsis* and rice genomes has enhanced the progress in the research area of molecular plant mineral nutrition.

The discovery of DNA microarray technology has revolutionized the plant ionomics research, and the regulation of the genes by ionic changes can be identified. Interestingly, many genes are shown to be transcriptionally responsive to changes in nutrient availability, including transporters, transcription factors, and signaling factors [Thimm et al., 2001; Negishi et al., 2002; Maathuis et al., 2003; Wang et al., 2003; Wintz et al., 2003]. The efforts have been made to establish the biological function of all genes in the *Arabidopsis thaliana* and rice genome [Chory, 2000; Dangl et al., 2002]. In this connection both transgenic and mutant plants have been generated and large-scale analysis of gene products such as mRNA, proteins, and cellular metabolites are being carried out [Roessner, 2001; Ahroni and Vorst, 2002; Koller et al., 2002]. The successful profiling of 18 elements including essential macro- and microelements and various nonessential elements, in shoots of 6000 mutagenized M2 *Arabidopsis thaliana* plants, was carried out by using inductively coupled plasma spectroscopy [Lahner et al., 2003]. These 51 mutants were isolated with altered elemental profiles, and one of the mutants contained a deletion in FRD3, a gene known to control iron-deficiency responses in *A. thaliana*. It was estimated that around 2–4% of the *A. thaliana* genome is involved in regulating the plant's nutrient and trace element content, and these results demonstrate that the use of elemental profiling is one of the practical functional genomics tools for the identification of genes involved in the accumulation of mineral nutrients and trace elements in plants [Lahner et al., 2003]. The availability of *Arabidopsis* high-density gene arrays now makes it possible to simultaneously genotype plants for several hundred thousand loci, and it is possible to map a locus to approximately 12 cM [Borevitz et al., 2003]; in some cases it could be lower, down to 0.5 cM [Hazen and Kay, 2003]. The emerging high-density *Arabidopsis* genome tiling arrays would greatly facilitate the identification of genomic polymorphisms, allowing the rapid identification of deletions produced by FN mutagenesis in a single F2 bulk segregant analysis experiment. This will make it possible to identify a mutation within a short time. Positional cloning of the genes responsible for ion-profile changes in several of the ionomic mutants have been reported [Lahner et al., 2003]. Natural variation in *Arabidopsis* seed and shoot phosphate accumulation is known to exist between the Ler and Cvi accessions [Bentsink et al., 2003] and for potassium, sodium, calcium, magnesium, iron, manganese, zinc, and phosphorus in seeds of numerous ecotypes [Vreugdenhil et al., 2004]. Analyses of Ler/Cvi recombinant inbred lines revealed quantitative trait loci (QTL) that explain between 10% and 79% of this variation for the different elements [Vreugdenhil et al., 2004]. The finding of ionomic variation in various ecotypes through QTL mapping holds the promise in identifying the genes involved in regulation of the ionome not only in *Arabidopsis* but also in crop plants.

The collection of *Arabidopsis* T-DNA insertional mutants with sequenced boarders is curated at the Salk Institute, Genomic Analysis Laboratory (SIGnAL), making the forward genetic approach attractive in *Arabidopsis*, and could also be appllied to other crop plants in genus *Brassicacae*. High-throughput screening for induced point mutations (TILLING) also makes it possible to identify alternative alleles in genes of interest in many plant species, including maize (*Zea mays*), black cottonwood (*Populus balsamifera* subsp. trichocarpa), *Brassica oleracea*, and *Lotus japonicus*. The use of genomic DNA pooling and PCR also makes the identification of mutants in specific genes possible in many other species using fast neutron deletion mutagenesis [Li et al., 2001]. The ionomic project is being carried out in David Salt's laboratory in Purdue University, USA, and they have quantified the shoot ionome in approximately 1000 homozygous T-DNA insertional lines of *Arabidopsis*. Mutants with insertions in exons of genes are thought to be involved in regulating the ionome, including ion transporters and various metal-related regulatory proteins [Baxter et al., 2006]. This ionomics database can be found at http://hort.agriculture.purdue.edu/Ionomics/database.asp. Ionomic mutants reported by Lahner et al. [2003] are detailed in the ionomics database and available through the ABRC (http://www.biosci.ohiostate.edu/approximately plantbio/Facilities/abrc/abrchome.htm). Genes that appear to alter the ionome, identified in the T-DNA insertional ionomics database, can be further characterized using T-DNA insertional lines curated at the SIGnAL database (http://signal.salk.edu/cgibin/tdnaexpress).

The Purdue Ionomics Information Management System (PiiMS) provides integrated workflow control, data storage, and analysis to facilitate high-throughput data acquisition, along with integrated tools for data search, retrieval, and visualization for hypothesis development. PiiMS is deployed as a World Wide Web-enabled system, allowing for integration of distributed workflow processes and open access to raw data for analysis by numerous laboratories. PiiMS currently contains data on shoot concentrations of P, Ca, K, Mg, Cu, Fe, Zn, Mn, Co, Ni, B, Se, Mo, Na, As, and Cd in over 60,000 shoot tissue samples of *Arabidopsis thaliana*, including ethyl methanesulfonate, fast-neutron and defined T-DNA mutants, and natural accession and populations of recombinant inbred lines from over 800 separate experiments, representing over 1,000,000 fully quantitative elemental concentrations. PiiMS is accessible at www.purdue.edu/dp/ionomics [Baxter et al., 2006]. These data provide important clues for functional annotation of genes related to mineral nutrion in plants. Techniques such as functional analysis in yeast, expression studies in plants under different metal availabilities, and analysis of mutant phenotypes will provide breeders an important information regarding the high- and low-affinity transporter systems needed to cope with varying metal availability in the soil and the specific proteins required for metal transport at the different cellular and organelle membranes in plant cells [Hall and Williams, 2003]. This knowledge will help the breeders to improve the metal content in the edible part of crop plants for human nutrition.

5   GENETIC ENGINEERING APPROACHES TO IMPROVE THE BIOAVAILABILITY OF IRON AND ZINC IN CEREALS

The research efforts are underway at national and international level to pursue breeding programs for microelement-dense staple food crops. In most of the programs, Fe and Zn nutrition is being emphasized in food crops—for example, in wheat, corn, rice, and beans [Graham and Welch, 1996]. An elaborated research program at the national level in Turkey on the nutrition of Zn in wheat is being carried out for the past decade [Cakmak et al., 1996; Ozturk et al., 2006]. Deficiencies of iron and zinc are common worldwide. Iron deficiency anemia is by far the most common micronutrient deficiency exist worldwide and is estimated to affect about 30% of the world population [Lucca et al., 2001]. Besides being a major problem in developing countries, it is also highly prevalent in developed countries both in women of childbearing ages and children [Lucca et al., 2006]. The highest rise of iron deficiency occurs during rapid growth and nutritional demand, such as in children, adolescence, and pregnancy. Severe anemia in pregnant women is estimated to be responsible for up to 40% of the half million deaths associated with childbirth each year [Brotanek et al., 2007]. During infancy, lack of sufficient iron in the brain causes irreversible changes in mental and psychomotor development, seriously limiting the intellectual potential of the adult [Lozoff et al., 1991, 2006]. Abnormalities in mental performance, including apathy, irritability, impaired attentiveness, and reduced learning capacity, have been observed as the consequences of childhood anemia [Grantham-McGregor and Ani, 2001].

Zn is another micronutrient essential for both plant growth and human health and is involved in the activity of more than 300 enzymes. Studies of zinc uptake in biology are critical because zinc is essential for all organisms and humans; zinc deficiency ranks third in importance after iron and vitamin A deficiency [Hambidge, 2000]. Food zinc content is very important because the supplementation of minerals is often difficult to achieve, particularly in developing countries. Therefore, it has been suggested that increased levels of zinc in staple foods may play a role in reducing zinc deficiency [Ruel and Bouis, 1998; Graham et al., 1999; Welch and Graham, 1999]. Because zinc plays multiple roles in plant biochemical and physiological processes, even slight deficiencies will cause a decrease in growth, yield, and zinc content of edible parts. Therefore, it is essential to understand the molecular details of how plants take up, translocate, and store zinc. Zinc plays many essential unique biological roles in part because it possesses unique chemical characteristics. It is well known that a vast array of proteins use zinc for stabilizing their structures in a functional form [Christianson, 1991]. In many cases, zinc interacts with cysteines and histidines in proteins. The most conspicuous examples of such interactions are the zinc finger transcription factors, which require the binding of zinc for activation of transcription [Alberts et al., 1998].

Zinc deficiency is probably the most widespread micronutrient deficiency in cereals. Sillanpää [1990] found that, of a global sample of 190 soils in 25 countries, 49% were low in zinc. In Turkey, for example, <50% of the arable soils were found to contain <0.5 mg pentetic acid-extractable Zn/kg, which is a widely accepted critical concentration indicating zinc deficiency [Cakmak et al., 1996]. In central Anatolia, where nearly 45% of Turkey's wheat production is located, >90% of the soils sampled were below the critical concentration of zinc. The actual content of zinc in soils is fairly high, ranging from 40 to 80 mg/kg, but the availability to plants is extremely low.

Various strategies have been used to combat these deficiencies including supplementation, food fortification, and modification of food preparation and processing methods. As mentioned in Section 4, a new possible strategy is to use biotechnology to improve trace element nutrition. Genetic engineering can be used in several ways; the most obvious is to increase the trace element content of staple foods such as cereals and legumes. This may be achieved by introduction of genes that code for trace element–binding proteins, overexpression of storage proteins already present, and/or increased expression of proteins that are responsible for trace element uptake into plants [Lonnerdal, 2003]. Increasing the expression of compounds that enhance trace element absorption such as ascorbic acid is also a possibility, although this has received limited attention so far. Iron absorption may be increased by higher ascorbic or citric acid content but require overexpression of enzymes that are involved in the synthetic pathways. In addition, a combination of all of these approaches with conventional breeding techniques may prove successful.

Plant breeding research on screening for genetic variability in concentration of trace minerals is in progress [Welch, 2002]. All crops show significant genotypic variation for mineral uptake and accumulation. For example, of nearly 1000 traditional cultivars, improved commercial varieties, and elite breeding lines of brown rice that were evaluated for iron and zinc, the average iron concentration in the grain was 12 parts per million (ppm) with a range of 8–24 ppm, and zinc concentrations averaged 25 ppm with a range of 14–42 ppm. Similar ranges of iron and zinc concentrations were found in wheat; as for rice, there appeared to be more genetic diversity for zinc than for iron. In maize, the range of genotypic differences in iron and zinc concentrations in 150 improved genotypes was ∼50% of the mean value. In beans, screening of 1500 traditional varieties and wild relatives showed iron and zinc concentrations ranging from 34 to 89 ppm and from 21 to 54 ppm, respectively [Ruel and Bouis, 1998].

Genetically modifying plants for improved trace element nutrition is a complementary approach. By increasing the trace element content of traditionally grown crops and/or enhancing the bioavailability of iron and zinc in such crops, trace element nutrition of the population that consumes these crops may be improved. Genetic modification can be achieved by two fundamentally different approaches. First, by using conventional breeding and selection techniques, cultivars with the highest content of trace elements, the highest concentration of enhancers of trace element bioavailability, and/or the lowest content of inhibitors of trace element absorption can be bred into stable and high-producing lines. Second, genetic engineering techniques can be used to create novel cultivars with the desired properties (see details in Sections 4 and 6). To date, two novel iron-binding proteins have been incorporated into rice. Rice has several advantages as compared to other plants: Rice proteins have very low allergenicity (i.e., any contaminants from rice in products such as infant formula or baby foods are unlikely to cause adverse reactions), and rice contains no toxic compounds and very high expression levels can be achieved [Suzuki et al., 2003]. Wheat is another staple crop used worldwide for bread and pasta making. The two major types of wheat crops—tetrapoloid wheat (used for pasta) and hexapoloid wheat—is used for bread making and account for ~20% of all calories consumed worldwide. Grain-yield reduction of up to 80% along with reduced grain level has been observed under Zn deficiency [Cakmak et al., 1998]. This has a serious implication for human health in countries where predominant diet for daily consumption is derived from cereals. Annual worldwide production of wheat is estimated to be 620 million tons of grain, translating into approximately 62 million tons of protein [United Nations Food and Agriculture Organization, Food Outlook 4, 2005]. Uauy et al. [2006] have characterized and cloned *Gpc-B1*, a quantitative trait locus from wild emmer wheat that is associated with increased levels of grain protein, zinc, and iron as a consequence of accelerated senescence and increased nutrient mobilization from leaves to the developing grains. The recombinant lines carrying this locus senesced on average 4–5 days earlier and exhibited a 10–15% increase in GPC, Zn, and Fe content in the grain. Complete linkage of the 7.4-kb region with the different phenotypes suggests that *Gpc-B1* is a single gene encoding a NAC domain protein (NAM-B1), characteristic of the plant specific family of NAC transcription factors, and plays important roles in developmental processes, auxin signaling, defense and abiotic responses, and leaf senescence [Otegui et al., 2002; Stuppy et al., 2003; Uauy et al., 2006]. RNA interference in transgenic plants caused delayed maturation and reduced grain protein, iron, and zinc content by more than 30%. The cloning of *Gpc-B1* provides a direct link between the regulation of senescence and nutrient remobilization and an entry point to characterize the genes regulating these two processes. The ancesteral wild wheat allele encodes this transcription factor that accelerates senescence and increases nutrient remobilization from leaves to developing grains, whereas modern wheat varieties carry a nonfunctional NAM-B1 allele. These results have a significant impact on improvinig the Fe and Zn content in the grains and provides a direct link between *Gpc-B1* gene expression and nutrient remobilization from leaves to seeds. These results indicate that efficient manipulation of these transcription factors in crop plants could improve the grain quality with enhanced nutritional value. Totally another strategy could also be used to increase the metal content of the seed especially for iron and

Zn. This can be achieved by either decreasing the inhibitors or increasing the synthesis of enhancers of trace elemnt absorption in humans.

6 DECREASING THE CONTENT OF INHIBITORS OF TRACE ELEMENT ABSORPTION

Phytin is the primary storage form of phosphorus in most mature seeds and grains. It is required for early seed maturation, seedling growth, vigor, and viability [Welch and Graham, 2004]. The negative effect of the phytic acid on Fe and Zn absorption in human is well-documented and is believed to be a major contributing factor to the worldwide problem of iron and zinc deficiency (see Table 1; Welch and Graham [2004], Ghandilyan et al. [2006]). Decreasing the phytate content of the diet is shown to be strongly correlated to increased iron [Hallberg et al., 1989] and zinc absorption [Navert et al., 1985]. Thus, any reduction in the content of phytate in staple foods is likely to result in improved iron and zinc status. Current plant molecular biological and genetic modification approaches now make it possible to reduce or eliminate antinutrients from staple food plant foods or to significantly increase the levels of promoter substances in these foods [Forssard et al., 2000].

Data compiled from various studies of zinc absorption from cereal-based meals in humans show a gradual decrease in zinc absorption as phytic acid concentrations increase [Sandström and Lönnerdal, 1989]. At concentrations of 400–500 mmol (264–330 mg) phytic acid, zinc absorption is <10% and is reduced to <5% at concentrations of 1000 mmol (660 mg). This indicates that substantial reductions in phytic acid would be necessary to significantly improve zinc absorption. A similar and probably even more dramatic phenomenon was described for nonheme iron absorption. Research in humans showed that minimal amounts of phytic acid added to meals can produce a severe inhibition of nonheme iron absorption [Ruel and Bouis, 1998]. For example, the daily intake of the population whose staple diets are based on cereals, legumes, and starchy roots and tubers are estimated to range from 600 to 1900 mg—that is, 200–600 mg/meal. Cereals such as whole wheat, corn, and millet contain <800 mg phytic acid/100 g cereal [Ruel and

TABLE 1. Antinutrients in Plant Foods that Reduce Fe and Zn Bioavailability [Welch, 2002]

Antinutrients	Major Dietary Sources
Phytic acid or phytin	Whole legume seeds and cereal grains
Fiber (e.g., cellulose, hemicellulose, lignin, cutin, suberin, etc.)	Whole cereal grain products (e.g., wheat, rice, maize, oat, barley, and rye)
Certain tannins and other polyphenolics	Tea, coffee, beans, sorghum
Oxalic acid	Spinach leaves, rhubarb
Hemagglutinins (e.g., lectins)	Most legumes and wheat
Goitrogens	*Brassicas* and *Alliums*
Heavy metals (e.g., Cd, Hg, Pb, etc.)	Contaminated leafy vegetables and roots

Bouis, 1998]. It is clear from these data that considerable amount of reduction in phytic acid in staple food is necessary in order to obtain significant improvements in absorption of both zinc and nonheme iron.

Plant breeders could breed for genotypes that contain a lower amount of phytic acid, or molecular geneticists could alter genes involved in ways that reduce the amount of dietry phytic acid from plant staple foods. This strategy would affect the bioavailability of zinc and iron simultaneously and will help the population suffering from Fe and Zn deficiencies. Selective breeding for low-phytate varieties of several staple crops was shown to be successful [Raboy, 2002]. Spontaneous low–phytic acid (lpa) mutations have been found in maize, barley, and rice and result in seed phytic acid–phosphorus levels that range from 50% to 95% of controls [Raboy et al., 2001; Raboy, 2002].These plants are shown to be normal in total phosphorus content but having significantly reduced levels of phytic acid, which in turn increases the level of inorganic phosphorus. These high levels of inorganic phosphorus provide a quick, sensitive, and inexpensive test for this trait and thereby make plant breeding practical [Lonnerdal, 2003]. The first lpa mutation (lpa 1-1) in maize was introduced into several inbred lines using traditional backcrossing breeding techniques [Raboy, 2002]. Iron absorption in humans was investigated from test meals based on tortillas baked with regular maize or lpa maize, and significantly increased iron absorption was found in the low-phytate variety [Lonnerdal, 2003]. Experiments on suckling rat pups showed that lpa varieties also enhance zinc absorption as compared to wild-type varieties. Microbial phytase can also be added to feeds and allowed to act during the digestive process to improve mineral utilization in domestic animals [Yi et al., 1996].

7   INCREASING THE SYNTHESIS OF PROMOTER COMPOUNDS

Another approach that could be complementry in increasing the bioavailability of Fe and Zn is to increase the concentration of sulfur-containing amino acids in staple crops. The candidate amino acids that may increase the absorption of metals are methionine, lysine, and cysteine. Cysteine or cysteine-rich peptides are shown to have a positive effect on iron absorption [Layrisse et al., 1984; Taylor et al., 1986] and histidine is shown to facilitate zinc absorption [Lonnerdal, 2003]. Other organic acids such as fumarate, citrate, and succinate can also enhance trace element absorption. One of the best-known enhancers of iron absorption is ascorbic acid [Hallberg et al., 1989]. Many studies show that increasing dietary ascorbic acid leads to a substantial increase in iron absorption in humans. Thus, overexpression of ascorbic acid in plants is likely to have a positive effect on iron nutrition of human populations. High expression levels of α-tocopherol and β-carotene were achieved in plants [Welch, 2002; Lonnerdal, 2003], but there are limited reports on water-soluble vitamins. Most synthetic pathways for vitamins are quite complex, and it is possible that both insertion of novel genes as well as promoters for enzymes involved in these pathways are required.

8   CONCLUSIONS

Deficiencies of essential micronutrients especially Fe and Zn are common in both plants and humans, and millions of people worldwide suffer from nutrional imbalances of these essential metals. Here we describe various strategies that could be used to combat these deficiencies including the improvement of trace metal content in the plants through genetic means, supplementation, food fortification, and modification of food preparation by reducing the amount of antinutrients in grains. A new possible strategy is to use ionomics, genomics, and proteomics strategies to improve trace elemental nutrition. Genetic engineering can be used in several ways; the most obvious is to increase the trace element content of staple foods such as cereals and legumes. This may be achieved by introduction of genes that code for trace element-binding proteins, overexpression of storage proteins already present, and/or increased expression of proteins that are responsible for trace element uptake into plants. Finally, a combination of all of these approaches perhaps complemented with conventional breeding techniques may prove successful. Much work is needed to prove that these approaches will lead to grow healthy plants with increased crop yield and high trace elemental content. The consumption of these varieties as a staple food will improve the health of the population and will eliminate the trace metal disorders worldwide.

ACKNOWLEDGMENTS

We would like to thank all the researchers working in our laboratory for their support and encouragement during the writing of this chapter. The financial support by TUBITAK and ESF (Project No TOVAG- COST 859- 104 O 211) to ARM is appreciated.

REFERENCES

Aharoni A, Vorst O. 2002. DNA microarrays for functional plant genomics. *Plant Mol Biol* **48**:99–118.

Alberts IL, Nadassy K, Wodak SJ. 1998. Analysis of zinc binding sites in protein crystal structures. *Protein Sci* **7**:1700–1716.

Baxter I, Ouzzani M, Orcun S, Kennedy B, Jandhyala S, Salt DE. 2006. Purdue ionomics information management system. An integrated functional genomics platform. *Plant Physiol* **143**:600–611.

Bentsink L, Yuan K, Koorneef M, Vreugdenhil D. 2003. The genetics of phytate and phosphate accumulation in seeds and leaves of *Arabidopsis thaliana*, using natural variation. *Theor Appl Genet* **106**:1234–1243.

Borevitz JO, Liang D, Plouffe D, Chang HS, Zhu T, Weigel D, Berry CC, Winzeler E, Chory J. 2003. Large scale identification of single feature polymorphisms in complex genomes. *Genome Res* **13**:513–523.

Bouis HE. 2000. Enrichment of food staples through plant breeding: A new strategy for fighting micronutrient malnutrition. *Nutrition* **16**:701–704.

Brotanek JM, Gosz J, Weitzman M, Flores G. 2007. Iron deficiency in early childhood in the United States: Risk factors and radical/ethnic disparities. *Pediatrics* **120**:568–575.

Caballero B. 2002. Global patterns of child health: The role of nutrition. *Ann Nutr Metab* **46**(suppl 1):3–7.

Cakmak I, Yilmaz A, Ekiz H, Torun B, Erenoglu B, Braun HJ. 1996. Zinc deficiency as a critical nutritional problem in wheat production in Central Anatolia. *Plant and Soil* **180**:165–172.

Cakmak I, Erenoglu B, Gulut KY, Derici R, Romheld V. 1998. Light mediated release of phytosiderophores in wheat and barley under iron or zinc deficiency. *Plant Soil* **202**: 309–315.

Chory J, Ecker JR, Briggs S, Caboche M, Coruzzi GM, et al. 2000. National Science Foundation—sponsored workshop report: "The 2010 project" functional genomics and the virtual plant. A blueprint for understanding how plants are buit and how tho improve them. *Plant Physiol* **123**:423–425.

Christianson DW. 1991. Structural biology of zinc. *Adv Protein Chem* **42**:281–355.

Combs GF Jr, Welch RM, Duxbury JM, Uphoff NT, Nesheim MC. 1996. *Food-Based Approaches to Preventing Micronutrient Malnutrition. An International Research Agenda*, Ithaca, NY: Cornell International Institute for Food, Agriculture, and Development, Cornell University, pp. 1–68.

Dangl J, Haselkorn R, Martiensen R, McCouch S, Retzel EF, Somerville CR, Wessler S, Yales J. 2003. The national plant genomics initiative objects for 2003–2008, *Plant Physiol* **130**:1741–1744.

Epstein E. 1972. *Mineral Nutrition of Plants: Principles and Perspectives*. New York: John Wiley & Sons.

Forssard E, Bucher M, Machler F, Mozafar A, Hurrell R. 2000. Review: Potential for increasing the content and bioavailability of Fe, Zn and Ca in plants for human nutrition. *J Sci Food Agric* **80**:861–879.

Graham RD, Welch RM. 1996. Breeding for staple food crops with high micronutrient density. Working papers on agricultural strategies for micronutrients. No. 3 Washington, DC: International Food Policy Research Institute.

Graham R, Senadhira D, Beebe S, Iglesias C, Monasterio I. 1999. Breeding for micronutrient density in edible portions of staple food crops: Conventional approaches. *Field Crops Res* **60**:57–80.

Graham RD, Welch RM, Bouis HE. 2001. Addressing micronutrient malnutrition through enhancing the nutritional quality of staple foods: Principles, perspectives and knowledge gaps. *Adv Agron* **70**:77–142.

Ghandilyan A, Vreugdenhil D, Aarts MGM. 2006. Progress in the genetic understanding of plant iron and zinc nutrition. *Physiolo Plantarum* **126**(3):407–417.

Grantham-McGregor S, Ani C. 2001. A review of studies on the effect of iron deficiency on cognitive development in children, *J Nutr* **131**:6495–6685.

Grusak MA. 1999. Genomics-assisted plant improvement to benefit human nutrition and health. *Trends Plant Sci* **4**:164–166.

Hallberg L, Brune M, Rossander L. 1989. Iron absorption in man: Ascorbic acid and dose-dependent inhibition by phytate. *Am J Clin Nutr* **49**:140–144.

Hambidge M. 2000. Human zinc deficiency. *J Nutr* **130**:1344S–1349S.

Hall JL, Williams LE. 2003. Transition metal transporters in plants. *J Exp Bot* **54**:2601–2613.

Hazen SP, Kay SA. 2003. Gene arrays are not just for measuring gene expression. *Trends Plant Sci* **8**:413–416.

Koller A, Washburu MP, Lauge BM, Andon NL, Deciu C, et al. 2002. Proteomic survey of metabolic pathway in rice. *Proc Natl Acad Sci USA* **99**:11969–11974.

Lahner B, Gong J, Mahmoudian M, Smith EL, Abid KB, Rogers EE, Guerinot EML, Harper JF, Ward JM, McIntyre L, Schroeder JI, Salt DE. 2003. Genomic scale profiling of nutrient and trace elements in *Arabidopsis thaliana. Nature Biotechnol* **21**:1215–1221.

Layrisse M, Martinez TC, Leets I, Taylor P, Ramirez J. 1984. Effect of histidine, cysteine, glutathione or beef on iron absorption in humans. *J Nutr* **114**:217–223.

Li L, Luo YX, Wang X, Endonuclease G. 2001. (EndoG) is an apoptotic DNase when released from mitochondria. *Nature* **412**:95–99.

Lonnerdal B. 2003. Genetically modified plants for improved trace element nutrition. *J Nutr* **133**:1490S–1499S.

Lozoff B, Jimenez E, Wolf AW. 1991. Long-term developmental outcome of infants with iron deficiency. *N Engl J Med* **325**:687–694.

Lozoff B, Kaciroti N, Walter T. 2006. Iron deficiency in infancy applying a physiologic framework for prediction. *Am J Clin Nutr* **84**:1412–1421.

Lucca P, Hurrell R, Potrykus I. 2001. Genetic engineering approaches to improve the bio-availability and the level of iron in rice grains. *Theor Appl Genet* **102**:392–397.

Lucca P, et al. 2006. Cholinesterase inhibitor use and age in the general population. *Arch Neurol* **63**:154–155.

Maathuis FJ, Filatov V, Herzyk P, Krijger GC, Axelsen KB, Chen S, Green BJ, Li Y, Madagan KL, Sanchez-Fernandez R, et al. 2003. Transcriptome analysis of root transporters reveals participation of multiple gene families in response to cation stress. *Plant J* **35**:675–692.

Marschner H. 1997. Introduction, definition, and classification of mineral nutrients. In: *Mineral Nutrition of Higher Plants*, second edition. London: Academic Press, pp. 3–5.

Mason JB, Garcia M. 1993. Micronutrient deficiency—the global situation. *SCN News* **9**:11–16.

Navert B, Sandstrom B, Cederblad A. 1985. Reduction of the phytate content of bran by leavening in bread and its effect on absorption of zinc in man. *Br J Nutr.* **53**:47–53.

Negishi T, Nakanishi H, Yazaki J, Kishimoto N, Fujii F, Shimbo K, Yamamoto K, Sakata K, Sasaki T, Kikuchi S, et al. 2002. cDNA microarray analysis of gene expression during Fe deficiency stress in barley suggests that polar transport of vesicles is implicated in phytosiderophore secretion in Fe deficient barley roots. *Plant J* **30**:83–94.

Otegui MS, Capp R, Staehelin LA. 2002. Developing seeds of *Arabidopsis* store different minerals in two types of vacuoles and in the endoplasmic reticulum. *Plant Cell* **14**:1311–1327.

Ozturk L, Yazici MA, Yucel C, Torun A, Cekic C, Bagci A, Ozkan H, Braun HJ, Sayers Z, Cakmak I. 2006. Concentration and localization of zinc during seed development and germination in wheat. *Physiol Plantarum* **128**(1):144–152 SEP.

Raboy V. 2002. Progress in breeding low phytate crops. *J Nutr* **132**:503–505.

Raboy V, Young KA, Dorsch JA, Cook A. 2001. Genetics and breeding of seed phosphorus and phytic acid. *J Plant Physiol* **158**:489–497.

Rea PA. 2003. Ion genomics. *Nature Biotech* **21**:1149–1151.

Roessner U. 2001. Metabolic profiling allows comprehensive phenotyping of genetically or environmentally modified plant systems. *Plant Cell* **13**:11–29.

Ruel MIT, Bouis HE. 1998. Plant breeding: A long-term strategy for the control of zinc deficiency in vulnerable populations. *Am J Clin Nutr* **68**(2-S):488S–494S.

Salt D. 2004. Update on plant ionomics. *Plant Physiol* **136**:2451–2456.

Sandström B, Lönnerdal B. 1989. Promoters and antagonists of zinc absorption. In: Mills CF, editor. *Zinc in Human Biology. Human Nutrition Reviews*. Berlin: Springer-Verlag, International Life Sciences Institute, pp. 57–78.

Sillanpää M. 1990. Micronutrient assessment at the country level: An internationalstudy. FAO Soil Bulletin 63. Food and agriculture organization of the United Nations in cooperation with the Finnish International Development Agency (FINNIDA). Rome.

Stuppy WH, et al. 2003. Three-dimensional analysis of plant structure using high-resolution X-ray computed tomography. *Trends Plant Sci* **8**:2.

Suzuki YA, Kelleher SL, Yalda D, Wu L, Huang J, Huang N, Lonnerdal B. 2003. Expression, characterization and biological activity of recombinant human lactoferrin in rice. *J Pediatr Gastroenterol Nutr* **36**:190–199.

Taylor PG, Torres MC, Romano EL, Layrisse M. 1986. The effect of cysteine-containing peptides released during meat digestion on iron absorption in humans. *Am J Clin Nutr* **43**:68–71.

Thimm O, Essigmann B, Kloska S, Altmann T, Buckhout TJ. 2001. Response of *Arabidopsis* to iron deficiency stress as revealed by microarray. *Plant Physiol* **127**:1030–1043.

Uauy C, Distelfeld A, Fahima T, Blechl A, Dubcovsky J. 2006. A NAC gene regulating senescence improves grain protein, Zn, and Fe content in wheat. *Science* **314**:1298–1301.

United Nations Food and Agriculture Organization. 2005. *Food Outlook* **4**, www.fao.org/documents.

Vreugdenhil HJI, Mulder PGH, Emmen HH, Weisglas-Kuperus N. 2004. Effects of perinatal exposure to PCBs on neuropsychological functions in the Rotterdam cohort at 9 years of age. *Neuropsychology* **18**(1):185–193.

Wang L, et al. 2003. Spectropolarimetry of SN 2001el in NGC 1448: Asphericity of a normal Type Ia Supernova 1. *Astrophys J* **591**:1110–1128.

Welch RM. 1995. Micronutrient nutrition of plants. *Crit Rev Plant Sci* **14**:49.

Welch RM. 1999. Importance of seed mineral nutrient reserves in crop growth and development. In: Rengel Z, editor. *Mineral Nutrition of Crops. Fundamental Mechanisms and Implications*. New York: Food Products Press, pp. 205–226.

Welch RM. 2002. Breeding strategies for biofortified staple plant foods to reduce micronutrient malnutrition globally. *J Nutr* **132**:495S–499S.

Welch RM, Graham RD. 1999. A new paradigm for world agriculture: Meeting human needs–productive, sustainable, nutritious. *Field Crops Res* **60**:1–10.

Welch RM, Graham RD. 2004. Breeding for micronutrients in staple food crops from a human nutrition perspective. *J Exp Bot* **55**:353–364.

WHO 2002. *The World Health Report 2002. Reducing Risks, Promoting Healthy Life*. Geneva, Switzerland: World Health Organization, pp. 1–168.

WHO 1999. *Malnutrition Worldwide*. Geneva, Switzerland: World Health Organization, http://www.who.int/nut/malnutrition_worldwide.htm, pp. 1–13.

Wintz H, Fox T, Wu YY, Feng V, Chen W, Chang HS, Zhu T, Vulpe C. 2003. Expression profiles of *Arabidopsis thaliana* in mineral deficiencies reveal novel transporters involved in metal homeostasis. *J Biol Chem* **278**:47644–47653.

World Bank. 1994. The challenge of dietary deficiencies of vitamins and minerals. In: *Enriching Lives: Overcoming Vitamin and Mineral Malnutrition in Developing Countries*. Washington, DC: World Bank, pp. 6–13.

Yan X, Wu P, Ling H, Xu G, Xu F, Zhang Q. 2006. Plant nutriomics in China: An overview. *Ann Botany* **98**:473–482.

Yi Z, Kornegay ET, Denbow DM. 1996. Supplemental microbial phytase improves zinc utilization in broilers. *Poult Sci* **75**:540–546.

5 Trace Elements and Plant Secondary Metabolism: Quality and Efficacy of Herbal Products

CHARLOTTE POSCHENRIEDER, JOSEP ALLUÉ, ROSER TOLRÀ, MERCÈ LLUGANY, and JUAN BARCELÓ

Department of Plant Physiology, Bioscience Faculty, Autonomous University of Barcelona, E-08193 Bellaterra, Spain

Plant secondary metabolism is responsible for the synthesis of most of the active principles of herbal drugs. Up to date about 200,000 of these active principles or phytochemicals have been identified. Many of these substances are specifically produced only in certain plant species and are implied in the plant's defense and attraction mechanisms. The pathways for the synthesis of these products derive from the primary metabolism (Fig. 1) and they also produce many substances that are required for growth and development of all plants, like aromatic amino acids, plant hormones, or carotenes.

1 COEVOLUTIONARY ASPECTS

The medicinal use of plants by humans is as old as mankind itself. The empirical use of many ancient herbal remedies has now been supported by scientific knowledge about the chemical and pharmacological characteristics of their active principles. Moreover, new herbal drugs of immense pharmaceutical value are still being discovered, especially in the main target areas of pharmacological research such as cancer, diabetes, Alzheimer disease, and AIDS [Jachak and Saklani, 2007].

We have come a long way from the hypothesis of signatures, which in the sixteenth century postulated that plants were put on Earth with certain morphological characteristics, by the divine intent to signal their therapeutic utility. However, still

Trace Elements as Contaminants and Nutrients: Consequences in Ecosystems and Human Health, Edited by M. N. V. Prasad
Copyright © 2008 John Wiley & Sons, Inc.

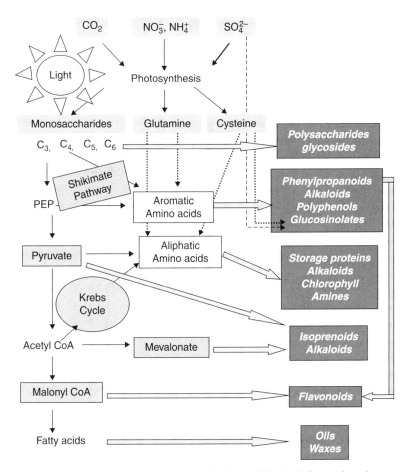

Figure 1. Main groups of plant secondary metabolites and their relation to the primary metabolism (modified from Schopfer and Brennicke [1999]).

today it is sometimes forgotten that the richness of the plant secondary metabolism is not a purpose for human convenience, but a consequence of millions of years of coevolution of plants with other organisms, competing for essential resources, establishing communications, and arming defenses [Strauss and Irwin, 2004].

The autotrophic terrestrial plant lives anchored in the substrate where the root prospects for water, oxygen, and essential mineral nutrients. Chemical energy in the form of ATP and reducing power required for growth, development, and reproduction is obtained by conversion from solar energy in leaves. These large photosynthetic surfaces are exposed to the atmosphere for light interception and CO_2 uptake. The optimization of the autotrophic function has led to a sessile lifestyle of plants, excepting in the seed phase, where dispersion is an essential factor for evolution [Levin, 2004]. The handicap of limited mobility during the main part of the life cycle is efficiently overcome by highly developed systems for perception and signaling of environmental stimuli of physical, chemical, and biological origin [Dodd, 2005; Brenner et al., 2006; Devoto and Turner, 2005; Felle and Zimmermann, 2007].

Perception of light intensity and light quality by specific photoreceptors such as phytochromes, chryptochromes, and phototropins, along with their interactions with the circadian rhythms, are among the most investigated sensorial properties of plants [Nozue and Maloof, 2006]. Temperature perception and a memory for winter [Ensminger et al., 2006; Sung and Amasino, 2006], or sensing of soil water status followed by adjustment of transpiratory water loss, growth, and development, are further essential features for withstanding seasonal changes of the environment [Davies et al., 2005].

Even more complex challenges for the sessile plants can be the competition with microorganisms and other plants for essential mineral nutrients, or the defense strategies against mobile pathogens and herbivores that attack plants for acquisition of nutrients required for their heterotrophic lifestyle. Here coevolution among microbes, plants, and animals has given rise to multiple forms of highly specific biological interactions of antibiosis and symbiosis. Amensalism, like alellopathy, commensalism, and mutualism, is based on the emission and perception of chemical signals leading to deterrence or attraction. Lack of perception of a chemical defense can also be effective against generalist herbivores due to intoxication. Specialist herbivores, in contrast, not only have evolved tolerance to the chemical defenses of certain plants, but even more they can accumulate in their bodies high concentrations of potentially toxic plant components to defend themselves against predators.

In the view of the time scale of hundred of millions of years of plant evolution, the short period of coexistence of plants and humans seems relatively short for specific human–plant co-evolution. This is certainly erroneous. On one side, humans have inherited from their animal ancestors many of the mechanisms for perception and/or tolerance of plant-derived chemical defense substances. Examples are multidrug resistance proteins, membrane transporters that actively pump out drugs from cells, thus increasing their resistance against these active principles. Human genes encoding for some of these transporters have high sequence and structural similarity to those from ruminants or mice [Longley et al., 1999].

On the other side, humans have a tremendous influence on plant evolution by domestication and breeding. Increased nutritional quality of crops, in turn, makes a decisive contribution to human development. Humans are omnivores; but a large diversity of plants makes a significant contribution to their diets. In this sense the dietary intake of plants seems to follow generalist behavior, initially guided, as in the case of animals, by a learning process that associates smell and taste with nutritional quality. Agricultural qualities of crops have had a strong additional influence of plant diversity in human diets since the origin of Agriculture. Inputs of information from scientific research on nutritional quality have been a distinctive feature in dietary choices of humans only since the nineteenth century. The consumption of specific medicinal plants for fighting infectional diseases and other pathologies falls within a specialist herbivore behavior. A therapeutic effect is achieved because the plant's active principles either are more toxic to the infective agent than to the human, or they activate defense responses in the human body that efficiently combat the disease. The cultivation of medicinal plants, the identification and isolation of their active principles, and the scientific use of herbal drugs by humans can be considered as the most developed evolutionary step of specialist herbivorism.

2   ENVIRONMENTAL FACTORS AND ACTIVE PRINCIPLES

It is well established that the concentration of active principles or phytochemicals in medicinal plants depends not only on the plant's genetics and developmental stage, but also, to a great extent, on environmental factors [Mathur et al., 2000; Rohloff et al., 2000; Gobbo-Neto and Lopes, 2007; Martins et al., 2007]. It is an often observed fact that stress favors the accumulation of secondary plant products. Heat, drought [Christiansen et al., 1997], and herbivore or pathogen attacks usually are stimulating factors, while air pollution frequently decreases the number and quantity of active principles, especially that of terpenoids [Judzentiene et al., 2007].

Soil fertility and UV radiation have contrasting influence. For example, the production of salidroside, the active principle of the Chinese medicinal herb *Rhodiola sachalinensis*, is favored by acid soil pH and low P and K availability, but also by high organic matter content and high N availability [Yan et al., 2004]. Experimental exposure to UV radiation can strongly inhibit nicotine production in *Nicotiana rustica*, while concentrations of phenolics are usually enhanced by experimental or natural UV radiation [Baztan and Barceló, 1980; Barceló et al., 1981; Spitaler et al., 2006].

The high degree of variability of the active principles in nature-collected herbal drugs limits their utility for the pharmaceutical or cosmetic industry. Moreover, the high demand of herbal drugs cannot be satisfied by collecting herbs from natural sites. Therefore, cultivation of selected genotypes under more or less environmentally controlled conditions is now a common practice.

3   INFLUENCE OF MACRONUTRIENTS

Optimization of mineral nutrition in medicinal plant crops under field conditions, in hydroponics, in aeroponics, or in cell culture is a challenge because an optimal nutrient supply for plant growth can be supraoptimal for production of secondary metabolites [Liu et al., 1999]. This holds especially true for cell cultures [Serrano and Piñol, 1991].

Medicinal plants require the same essential elements as other plants: C, H, O, N, S, P, Ca, K, Mg as macronutrients and Fe, Mn, Zn, Cu, B, Mo, Cl, Ni as micronutrients. Many studies have made approaches for optimizing nutrient supply. Table 1 shows some representative examples. Among the macronutrients, the carbon/nitrogen balance (CNB) is of special relevance for both biomass production [Lambers et al., 2008] and N-containing secondary metabolites like alkaloids [Palumbo et al.,

TABLE 1. Some Examples of the Influence of Essential Macronutrients on the Concentrations of Active Principles in Medicinal Plants

Species	Macronutrient	Effect on Phytochemicals	Cultivation	Reference
Ilex vomitoria	N surplus	↑Caffeine (A), ↓Phenolics	Potted plants	Palumbo et al. [2007]
Origanum syriacum	N surplus	↑Thymol (EO), ↑Carvacol (EO)	Field grown	Omer [1999]
Catharanthus roseus	N surplus	↑Alkaloids	Field grown	Sreevalli et al. [2004]
Hyptis suaveolens	N deficiency	↑Essential oils	Potted plants	Martins et al. [2007]
Erysimum cheiranthoides	N deficiency	↑Cardenolides (GL)	Potted plants	Hugentobler and Renwick [1995]
Hyosciamus albus	N deficiency	↓Atropine (A)	Hydroponics	Diez et al. [1983b]
Datura innoxia	NPK surplus	↑Hyosc. + scopol. (A)	Field grown	Al-Humaid [2004]
Catharanthus roseus	P deficiency	↑Triptamine, ↑indole alkaloids, ↑phenolics	Cell culture	Knobloch and Berlin [1983]
Nicotiana rustica	P deficiency	↓Nicotine	Hydroponics	Diez et al. [1983a]
Lupinus angustifolius	P deficiency	↓Lupinine (A)	Seeds potted plants	Gremigni et al. [2003]
	K deficiency	↑↑Lupinine (A)	Seeds potted plants	
Atropa acuminata	K deficiency	↑↑Total alkaloid	Hydroponics	Khan and Harborne [1991]
Brassica napus	S deficiency	↓Aliphatic glucosinolates	Field grown	Zhao et al. [1994]
Catharanthus roseus	S deficiency	↓Alkaloid	Cell culture	Arvy et al. [1995]
Brassica rapa	S surplus	↔Glucosinolate, ↑Phenolics acids	Field grown	De Pascale et al. [2007]
Datura stramonium	Ca deficiency	↓Hyoscyamine (A)	Hairy root culture	Piñol et al. [1999]
Hyoscyamus niger	Ca surplus	↓Hyoscyamine (A) ↑Scopolamine (A)	Root culture	Pudersell et al. [2003]
Brugmansia candida	Ca surplus	↑Hyoscyamine (A) ↑↑Scopolamine (A)	Hairy root culture	Pita-Alvarez et al. [2000]
Colchicum autumnale	↑Ca	↑Seed alkaloid	Field collected	Poutaraud and Girardin [2005]
Chamomilla recutita	Mg surplus	↑Farnesene (EO)	Sterile culture	Szöke et al. [2004]
Digitalis obscura	↑Mg	↑Cardenolides (GL)	Field collected	Roca-Pérez et al. [2005]

A, alkaloid; EO, essential oil; GL, glycoside; ↑, increased concentration; ↓, decreased concentration; ↔, no influence.

2007]. In contrast, N-deficiency can favor the production of essential oils [Martins et al., 2007], while the synthesis of alkaloids is strongly inhibited [Diez et al., 1983a,b]. In sulfur-deficient plants, S-fertilization enhances the production of S-containing phytochemicals, but surplus S is not stimulating in S-sufficient plants [Matula and Zukalova, 2001]. Sulfur fertilization in *Brassica rapa* enhanced the production of phenolics acids without an increase in total glucosinolates [de Pascale et al., 2007]. The supply of NPK fertilizer favored both biomass and alkaloid production in *Datura stramonium* plants [Al-Humaid, 2004], while P and especially K deficiency stress can stimulate the production of alkaloids and essential oils [Knobloch and Berlin, 1983; Khan and Harborne, 1991; Gremigni et al., 2003; Martins et al., 2007].

4 INFLUENCE OF MICRONUTRIENTS

Micronutrient deficiency has variable influences on the concentrations of active principles (Table 2). The specific roles for essential micronutrients in the production of active principles is due to their function as components or activators of enzymes of the secondary metabolism. Moreover, metals can quelate certain phytochemicals in plant tissues. Micronutrients with redox functions like Fe, Cu, Mn, and Mo are key factors for many enzymes involved in biosynthetic steps of secondary compounds. Copper, for example, is essential for the function of many oxidases and oxygenases with a key role in secondary metabolism. Copper deficiency strongly inhibits the activities of diamine oxidase, which is essential for the metabolism of the diamines putrescine and cadaverine [Cona et al., 2006], of polyphenoloxidase, and of superoxide dismutase (Cu/ZnSOD) [Marschner, 1995; Mayer, 2006].

Iron is a component of heme and non-heme oxygenases with key function in isoprenoid synthesis [Bouvier et al., 2005]. As a component of iron sulfur proteins (IspG, IspH), Fe is essential in the mevalonate-independent pathway (Fig. 1) of isopentenyl diphosphate (IPP) and dimethylallyl diphosphate (DMAPP) [Rohdich et al., 2004]. Highly substrate-specific, cytochrome P450-dependent monooxygenases are involved in alkaloid biosynthesis. Cytochrome P 450 enzymes are also acting in several steps of biochemical pathways to different types of phenolics and flavonoids. Synthesis of dihydroflavonols, a major branch point in flavonoid biosynthesis, requires Fe for the activity of the dioxygenase acting on naringenin [Buchanan et al., 2000].

Manganese is essential for Mn-SOD activity. In the shikimate pathway, Mn stimulates the pre-chorismate step catalyzed by the metalloenzyme 3-deoxy-D-arabino-heptulosonate 7-phosphate synthase [Entus et al., 2002]. Molybdenum is essential for nitrogenase and nitrate reductase and therefore plays a key role for the production of N-containing active principles like alkaloids. However, Mo deficiency seems to limit biomass production rather than alkaloid synthesis [Khan et al., 1994].

Micronutrients without redox functions are also directly or indirectly involved in plant secondary metabolism. Boron has a central function in the phenol metabolism in plants, and its deficiency strongly stimulates polyphenoloxidase [Cakmak and

TABLE 2. Variation of Concentrations of Active Principles as Influenced by Deficiency of Essential Micronutrients or Elicited by High Trace Element Concentrations

Species	Micronutrient	Effect on Photochemical	Cultivation	Reference
Digitalis obscura	↑Cu	↓Cardenolides (GL)	Field collected	Roca-Pérez et al. [2002]
Panax quinquefolium	Zn deficiency	↓Ginsenosides (S)	Hydroponics	Ren et al. [1993]
Picea abies	B deficiency	↑Protocatechuic acid (P)	Potted plants	Rummukainen et al. [2007]
Nicotiana tabacum	B deficiency	↑Chlorogenic acid (P)	Potted plants	Camacho-Cristobal et al. [2004]
Mentha arvensis	↑Fe	↑Menthol (EO)	Potted plants	Pande et al. [2007]
Cymbopogon flexuosus	Fe deficiency	↓Citral (EO), ↑Geraniol (EO)	Field grown	Misra and Khan [1992]
Beta vulgaris	↑Co	↑Betalain (AC)	Cell suspension	Trejo-Tapia et al. [2001]
Papaver bracteatum	Cu excess	↑Dihydrosanguinarine (A)	Cell suspension	Lecky et al. [1992]
Dioscorea bulbifera	Cu excess	↑Diosgenin (GL)	Tissue culture	Narula et al. [2005]
Alpinia zerumbet	Cu foliar spray	↑Camphor, borneol, 1,8-cineol, terpinene-4-ol (EO)	Potted plants	Elzaawely et al. [2007]
Rumex acetosa	Al	↓Flavonoids (root), ↑Flavonoids (shoot) ↑Anthraquinones (root)	Hydroponics	Tolrà et al. [2005]
Gloriosa superba	Al, Cd	↑Colchicine (A)	Root culture	Gosh et al. [2006]
Catharanthus roseus	Cd	↑Ajmalicine (A)	Cell suspension	Zheng and Wu [2004]
Phyllanthus amarus	Cd	↑Phyllanthine (A) ↑Hypophyllanthine (A)	Field plots	Rai et al. [2005]
Matricaria chamomilla	Cd	←→Herniarin (CU), ←→Umbelliferone (CU) ↑Glucopyranosyloxy-4-methoxycinnamic acids	Hydroponics	Kovacik et al. [2006]
Salvia miltiorrhiza	Ag	↑Tanshinone (T)	Hairy root culture	Ge and Wu [2005]
Taxus chinensis	Ag	↑Paclitaxel (A)	Cell suspension	Zhang et al. [2007]
Ocimum tenuiflorum	Cr	↑Eugenol (EO)	Potted plants	Rai et al. [2004]
Brassica oleuracea	Se	↓Glucosinolate	Hydroponics	Toler et al. [2007]
Catharanthus roseus	Se	↑Alkaloid	Cell culture	Arvy et al. [1995]
Coffea arabica	Se	↑Coffeine	Soil grown	Mazzafera [1998]
Panax ginseng	V	↑Ginsenosides (S)	Hairy root	Palazón et al. [2003]

A, alkaloid; AC, anthocyan; CU, coumarin; EO, essential oil; GL, glycoside; P, phenolics; S, saponine; T, terpenoids.

Römheld, 1997]. Zinc is required for the activity of thousands of plant proteins [Broadley et al., 2007]; a few examples are Zn finger proteins with key functions in gene regulation, glutathione transferases essential for the cell redox status, acid phosphate, MAPK, and Cu-Zn SOD.

5   TRACE ELEMENTS AS ELICITORS OF ACTIVE PRINCIPLES

Many of the active principles that justify the medical use of herbal drugs are involved in the plant's chemical defense against pathogens and herbivores. Constitutionally or preformed defenses and inducible defenses can be distinguished. The inducible defenses are synthesized in response of attack by elicitor-mediated mechanisms. Phytopathological studies on inducible defenses mainly focus on the hypersensitive response (HR), systemic acquired resistance (SAR), PR synthesis, and phytoalexin production . By definition, phytoalexins are chemical defenses that are not present in healthy plants, but are elicitor-induced after attack, especially in incompatible interactions [Agrios, 2004]. These phytochemicals fall into the classes of terpenoids (mostly sesquiterpens), glycosteroids, and alkaloids. In contrast, active principles of herbal drugs are constitutively present in the medicinal plants; that is, plants must not suffer from pathogen attack for accumulation of these chemical defenses.

However, the production of preformed phytochemicals can also be significantly enhanced by different elicitors. Among those, trace elements are being widely used to stimulate the production of active principles, especially in cell suspension and hairy root cultures (Table 2).

The mechanisms of this enhancement of phytochemical production by the supply of excess trace elements are still not fully established. Currently, however, there are considerable advances by the incorporation of this phenomenon into the integrated view of stress perception and signaling in plants [Poschenrieder et al., 2006a]. Briefly, according to this integrated view, external elicitors like oligosaccharides or membrane proteins from pathogens, chemical components of the saliva of herbivores, or even abiotic stress factors such as UV radiation or high, toxic concentrations of trace elements induce internal elicitors like jasmonate, salicylic acid, systemin, and other endogenous signaling compounds like reactive oxygen species (ROS) [Mittler et al., 2004; Devoto and Turner, 2005]. This signal transduction leads to the production of defense substances either by the directly stimulation of metabolic pathways or by activation of defense gene expression (Fig. 2). A key question, still not fully understood, is that of how specificity is achieved within this quite general system of defense activation [Poschenrieder et al., 2006b; Desender et al., 2007].

Although both excess of trace elements and biotic stress factors are able to elicit stress signaling cascades that share common signaling molecules, the modes of signal transduction and the resistance mechanisms for both types of stress are highly specific [Glombitza et al., 2004]. Receptor specificity, subcellular sites of production of ROS [Mittler et al., 2004], specificity and regulation of MAPK activities [Pedley and Martin, 2005], and differences in the activated genes and their products

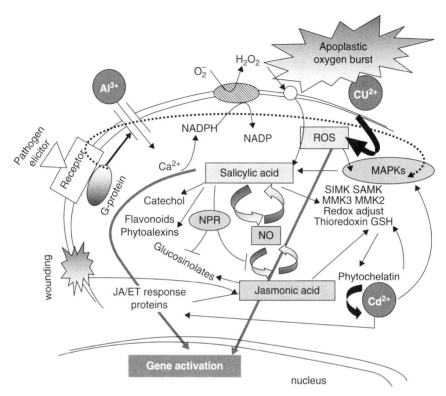

Figure 2. Specificity and cross talk in signal transduction pathways induced by pathogen elicitors, wounding and trace metal ions (modified from Poschenrieder et al. [2006b]).

are responsible for the specificity of responses. Different elicitors can induce different defense patterns, and the reaction patterns seem specific to each plant genotype/ elicitor pair [Desender et al., 2007]. Moreover, regulatory genes may play an important role in the coordination of stress-responding genes and synthesis of phytochemicals. An example for this are the *A* and *B* loci responsible for the regulation of nicotine synthesis in tobacco. Recent investigations suggest that the A-B regulon is a network of differentially expressed stress response genes, only a part of which is involved in nicotine biosynthesis [Kidd et al., 2006].

6   TRACE ELEMENTS AS ACTIVE COMPONENTS OF HERBAL DRUGS

Quality of herbal drugs depends on the presence of active principles. Among those, organic components have deserved most attention. Trace elements in medicinal plants, however, also can have a substantial influence on the therapeutic value of herbal remedies: a positive contribution as a source of essential nutrients or even as active principles, or a negative effect because of the accumulation of high concentrations of potentially toxic elements.

Elements required for human and animal health can be divided into those with a well-established biochemical function and those that are considered to be essential because deficiency symptoms appear when not supplied in sufficient amounts. The first group contains macronutrients (Ca, P, Na, Cl, K, and Mg) and micronutrients (Fe, I, Cu, Mn, Co, Mo, Se, and Zn), while in the second group only trace elements are found (F, Cr, V, Si, Ni, As, Li, Pb, B) [Epstein and Bloom, 2005].

Medicinal plants, spices, and herbal products can be rich in trace elements [Abou-Arab and Abou Donata, 2000; Basgel and Erdemoglu, 2006; Nookabkaew et al., 2006]. Despite the relatively small weight proportion of these plant products in the diet, their regular consumption can make a significant contribution to the daily recommended intake [Szentmihalyi et al., 2006]. It has to be taken into account, however, that only part of the trace elements present in the herbal product is leached into the infusions (Table 3). Moreover, organic substance present in the infusions can make part of the trace elements unavailable for absorption in the digestive tract. This is especially important for Al, a trace element that is extremely toxic when available in ionic form [Rengel, 2004]. So, the high concentrations of Al in both leaves of tea (*Comelina sinensis*), an Al hyperaccumulator, and tea infusions (Table 3) do not represent a danger for human health [Drewitt et al., 1993] because the Al is tightly bound to catechins [Barceló and Poschenrieder, 2002]. The high concentrations of Al found in some herbal teas (Table 3) are frequently due to contamination by soil and dust particles. This Al is mostly insoluble and is usually found in only very small proportions in the infusions.

Also, phytopharmaceutical formulations can have a quantitatively quite different mineral composition than the medicinal plants used for the preparations. A well-studied example is that of *Echinacea purpurea* [Razic et al., 2003]. An ethanolic extract from this plant is used as a therapeutic immunostimulant. This activity seems to be due to cynarin, the 1,4-dicaffeyl ester of quinic acid [Dong et al., 2006]. The concentrations of most mineral elements in the extract are lower than in the dry plant. A noteworthy exception is Zn, which is considerably more concentrated in the ethanolic extract [Razic et al., 2003]. Further investigations are required for showing whether the high Zn content contributes to the pharmacological activity of *Echinacea purpurea* preparations. This seems probable in the view of the well-established requirement for Zn in immune functions [Wellinghausen et al., 1997].

Enhanced intake of minerals like Ca, K, Na, and P can be useful in patients with chronic renal failure. Mineral concentrations in herbal drugs used for renal and urinary tract disorders in India have been reported to be higher than in nonmedicinal plants [Rajurkar and Damame, 1998]. However, the considerably high concentrations of Hg (e.g., 650 µg/g Hg in glandular hairs of *Mallotus philipensis*) and Cd (e.g., 16 µg/g Cd in flowers of *Michelia champak*) in some of these drugs would make a frequent ingestion extremely dangerous if these metals were available. Also, some herbal drugs used in India for cardiovascular disorders have been reported to contain extremely high Hg and Cd concentrations [Rajurkar and Damame, 1997]. Herbal preparations from Europe contain much lower metal concentrations, and even daily consumption would not exceed the

TABLE 3. Concentrations of Trace Elements in Herbal Preparations (mg/kg dry weight) and in Their Infusions[a]

Species	Preparation	Fe	Cu	Mn	Zn	Cr	Ni	Pb	Cd	Al	Reference
Matricaria chamomilla	Herb	503	8.3	60	31	1.2	1.8	0.72	0.44	356	Basgel and Erdemoglu [2006]
	Infusion	8.1	6.7	17	13	nd[b]	1.0	nd	nd	1.6	
Foeniculum vulgare	Herb	225	16	28	37	1.0	5.4	0.48	0.004	158	Basgel and Erdemoglu [2006]
	Infusion	4.9	8.1	4.3	18	nd	2.9	nd	nd	6.2	
Tilia vulgaris	Herb	228	9.6	71	36	0.34	2.5	0.26	nd	87	Basgel and Erdemoglu [2006]
	Infusion	2.4	3.8	15	9.9	nd	0.9	nd	nd	7.4	
Urtica dioica	Herb	810	5.6	80	47	1.2	3.6	4.8	0.06	596	Basgel and Erdemoglu [2006]
	Infusion	8.2	3.3	5.8	3.9	0.37	0.04	nd	nd	15	
Rosa canina	Fruit	267	4.9	244	22	0.92	2.9	0.34	0.07	157	Basgel and Erdemoglu [2006]
	Infusion	38	2.9	49	6.1	0.25	1.9	nd	nd	5.6	
Salvia officinalis	Herb	297	36	33	48	2.1	2.9	1.1	nd	178	Basgel and Erdemoglu [2006]
	Infusion	107	2.7	12	14	nd	0.85	nd	nd	7.9	
Cassia angustifolia	Herb	323	3.9	23	23	0.34	0.90	nd	nd	123	Basgel and Erdemoglu [2006]
	Infusion	6.7	2.4	9.5	15	nd	0.60	nd	nd	20	
Mentha piperita	Herb	239	12	188	51	0.94	3.0	2.4	0.09	na[b]	Lozak et al. [2002]
	Infusion	20	3.0	27	6.3	0.39	2.6	1.1	0.008		
Taraxacum officinale	Herb	853	27	101	68	na	na	7	na	2096	Queralt et al. [2005]
	Infusion	0.13	0.2	0.6	0.3			0.01		<0.1	
Eucalyptus globulus	Herb	89	10	2134	56	na	na	6	na	225	Queralt et al. [2005]
	Infusion	0.06	0.07	19	0.12			0.05		<0.1	
Plantago lanceolata	Herb	373	13	46	56	na	na	6	na	1409	Queralt et al. [2005]
	Infusion	0.11	0.11	0.50	0.50			0.007		<0.1	
Morus alba	Herb	200	8.1	156	29	0.79	0.96	0.40	0.008	180	Nookabkaew et al. [2006]
	Infusion	0.14	0.048	0.36	0.21	0.0006	0.011	<0.001	<0.001	0.07	
Camellia sinensis	Herb	167	8.1	813	32	1.5	5.6	3.9	0.035	1179	Nookabkaew et al. [2006]
	Infusion	0.08	0.04	4.10	0.22	0.0021	0.08	0.004	0.0001	5.51	
Gynostemma pentaphyllum	Herb	791	9.1	137	46	2.9	2.3	9.3	1.3	2014	Nookabkaew et al. [2006]
	Infusion	0.29	0.04	1.56	0.26	0.003	0.01	0.02	0.007	0.47	

[a]Concentration units for infusions are mg/kg in Basgel and Erdemoglu [2006] and Lozak et al. [2002] and are mg/liter in Queralt et al. [2005] and Nookabkaew et al. [2006].
[b]nd, not determined; na, not available.

FAO/WHO provisional weekly tolerable (PWTI) intake for Cd, Hg, or Pb [Falco et al., 2003, 2005].

Trace elements are also being studied as possible active principles in hypoglycemic plants. The mineral fraction of certain medicinal herbs from indigenous folk medicines (Ayurvedic, Unani, and Sidda systems) exhibits higher glucose tolerance factor action than the organic fraction of the plants [Kar et al., 1999]. The high concentrations of K, Ca, Cr, Mn, Cu, and Zn in several anti-diabetic medicinal plants have been made responsible for stimulation of insulin action [Raju et al., 2006]. Among the trace elements, Cr has a recognized function as glucose tolerance factor [Mertz, 1993]. Evaluation of herbal remedies traditionally used in mid-European folk medicine and in common adjuvant therapy for the prevention of complications in type 2 diabetes revealed high Fe, Zn, and Cr concentrations in some of them. Although infusions showed much lower concentration than the herbal remedies, a high proportion of the Cr present in the infusion made out from *Myrtilli folium, Phaseoli fructus sine seminibus*, and *Salvia folium* was able to cross membranes and to reach the plasma [Szentmihalyi et al., 2006]. This contrasts with the low membrane permeability of inorganic Cr III [Barceló and Poschenrieder, 1997].

Considerable research interest is being paid to the trace element selenium as an active principle against cancer. Early studies on the chemopreventive role of Se in carcinogenesis were confirmed by the finding that supplementation with selenized brewer's yeast was able to reduce the cancer morbidity and mortality in patients with carcinoma of the skin by nearly 50% [Clark et al., 1996]. Selenium has also been shown useful in the protection against mammary and prostate cancer [Ip et al., 1992; Chun et al., 2006]. The essential role of Se relies on its incorporation into selenoproteins, low-molecular-weight proteins that contain Se-cysteine. Examples of selenoproteins with essential functions are thioredoxine reductase and glutathione peroxidase involved in cell redox homestasis and thyroid hormone deiodinase required for thyroid hormone metabolism [Papp et al., 2007].

To enhance dietary intake of organic Se, garlic was successfully enriched in Se-methylselenocysteine by fertilization with selenium salts. In natural, non-Se fertilized garlic, Se concentrations are about 0.05 µg/g, while fertilization achieves values up to 1300 µg/g dry weight [Ip and Lisk, 1993, 1994; Cai et al., 1995]. Recent advances in analytical methods have identified γ-glutamyl-Se-methyl-selenocysteine as a further active principle against cancer in garlic and onion [Arnault and Auger, 2006]. Certain species of the genus *Astragalus* are natural Se hyperaccumulators. These plants can volatilize Se, but they also accumulate high amounts in leaves and in trichomes in the forms of Se-methylselenocysteine, γ-glutamyl-Se-methylselenocysteine, selenate, and selenite [Freemann et al., 2006]. This high Se accumulation can be involved in elemental defense against herbivores and pathogens [Poschenrieder et al., 2006a]. *Astragalus membranaceus* is used in traditional Chinese medicine. The herbal drug has anticancer activity and is employed as an adjuvant in colon cancer. Hypoglucemic activity, antiviral properties, and cardiovascular protection have also been described. The pharmacological activity is based on polysaccharides and saponines [Rios and Waterman, 1997].

Higher blood Se concentrations have also been found after administration of *A. membranaceus* [Dong et al., 2004].

Dill weed oil from *Anethum graveolens* contains potentially useful chemopreventive agents. Besides active organic compounds like monoterpenes [Zheng et al., 1992], dill accumulates Se-methylselenocysteine. In roots, moreover, a major Se compound is Se-methyl-selenomethionine [Cankur et al., 2006], a precursor for volatile Se compounds.

7 TRACE ELEMENTS IN HERBAL DRUGS: REGULATORY ASPECTS

This short overview illustrates the complexity of interactions between trace elements and organic active principles in plants with potential use in medicine. The great variety of organic and inorganic substances present in herbal drugs and the high variability due to differences in genetics, environment, and preparation procedures makes a control of safety and quality rather difficult.

At present, different national and international organisms provide recommendations or guidelines that apply to herbal remedies, although criterions and classifications are far from clear. In the United States the responsible organism is the FDA, and herbal remedies fall under Title 21 Food and Drugs of the Code of Federal Regulations, which includes food and food additives. The European Community gives metals maximum levels for vegetable foodstuffs by Commission

TABLE 4. List of Useful Authoritative Sources of Information on Herbal Drugs

- *American Herbal Pharmacopoeia* (AHP) (1997–2003)
- *European Pharmacopoeia* (Ph.Eur.), fourth edition (2003)
- *Herbal Drugs and Phytopharmaceuticals*, English fourth edition (2003)
- *Indian Herbal Pharmacopoeia* (IHP), revised new edition (2002)
- *International Organization for Standardization* (ISO): specifications, standards, and tests for various herbs and spices
- *Japanese Herbal Medicine Codex* (JHMC) (English translations available in The Japanese Standards for Herbal Medicines, 1993).
- *Japanese Pharmacopoeia* (JP XIV), 14th edition, Crude Drugs Monographs (English Edition 2001)
- *Pharmacopoeia of the People's Republic of China* (PPRC English Edition volumes I, II, and III, 2005)
- *Standardization of Single Drugs of Unani Medicine, Central Council for Research in Unani Medicine* (CCRUM), Part I (1987), Part II (1992), Part III (1997)
- *Unani Pharmacopoeia of India* (UPI), Part I
- *United States National Formulary* (USNF), 21st edition (2003)
- *United States Pharmacopeia* (USP), 26th revision (2003)
- *Australian Regulatory Guidelines for Complementary Medicines* (ARGCM)
- *World Health Organization (WHO) Monographs on Selected Medicinal Plants*, Volume 1 (1999), Volume 2 (2002), Volume 3 (2003)

Regulation (EC) No 1881/2006. The Joint FAO/WHO Expert Committee on Food Additives (JECFA) has elaborated many reports on contaminants including metals. Some recent updates are setting PTWI for Cd to 7 μg/kg body weight, PTWI for Pb to 25 μg/kg body weight, and newly established PTWI of Al to 1 mg/kg body weight [JECFA, 1993, 2006]. Other authoritative sources of information on quality and uses of herbal drugs, with rare mentions of trace elements, are listed in Table 4.

Of course, evaluation and control of safety, efficiency, and quality is much easier in synthetic medications. However, the benefits from herbal drugs by far justify the efforts to achieve sound regulations that are able to both protect consumers against toxicity and adverse effects and to preserve the pharmacological properties of the herbal remedies.

ACKNOWLEDGMENTS

Financial assistance by the Spanish Government for the Project BFU2007-60332 "Integration of responses to ionic and biotic stress in model plants: from cross-talk to trade-off " is gratefully acknowledged. Supported by the Catalonian Government SGR 2005-00785.

REFERENCES

Abou-Arab AAK, Abou Donata MA. 2000. Heavy metals in Egyptian spices and medicinal plants and the effect of processing on their levels. *J Agr Food Chem* **48**:2300–2304.

Agrios GN. 2004. *Phytopathology*, fifth edition. San Diego: Academic Press.

Al-Humaid AI. 2004. Effects of compound fertilization on growth and alkaloids of Datura plants. *J Plant Nutr* **27**:2203–2219.

Arnault I, Auger J. 2006. Seleno-compounds in garlic and onion. *J Chromatogr* **1112**:23–30.

Arvy MP, Thiersault M, Doireau P. 1995. Relation between selenium, micronutrients, carbohydrates, and alkaloid accumulation in *Catharanthus roseus* cells. *J Plant Nutr* **18**:1535–1546.

Barceló J, Poschenrieder C. 1997. Chromium in plants. In: Canali S, Tittarelli F, Sequi P, editors. *Chromium Environmental Issues*. Publ Milano: Franco Agnelli, pp. 101–129.

Barceló J, Poschenrieder C. 2002. Fast root growth responses, root exudates, and internal detoxification as clues to the mechanisms of aluminium toxicity and resistance: A review. *Environ Exp Bot* **48**:75–92.

Barceló J, Torres M, Baztan J. 1981. On the effects of the near UV radiations on the growth and content in phenolics compounds of *Petroselinum crispum* (Miller). *Anal Edafol Agrobiol* **40**:337–345.

Basgel S, Erdemoglu SB. 2006. Determination of mineral and trace elements in some medicinal herbs and their infusions consumed in Turkey. *Sci Total Environ* **359**:82–89.

Baztan J, Barceló J. 1980. Variaciones en el contenido de nicotina en plantas de *Nicotiana rustica* L. sometidas a irradiaciones de ultravioleta lejano y cercano. *Anal Edafol Agrobiol* **39**:1009–1018.

Bouvier F, Rahier A, Camara B. 2005. Biogenesis, molecular regulation and function of plant isoprenoids. *Prog Lipid Res* **44**:357–429.

Brenner ED, Stahlberg R, Mancuso S, Vivanco J, Baluska F, Van Volkenburgh E. 2006. Plant neurobiology: An integrated view of plant signalling. *Trends Plant Sci* **11**:413–419.

Broadley MR, White PJ, Hammond JP, Zelko I, Lux A. 2007. Zinc in plants. *New Phytol* **173**:677–702.

Buchanan BB, Gruissem W, Jones RL. 2000. *Biochemistry and Molecular Biology of Plants.* Rockville, MD: American Society of Plant Physiologists.

Cai XJ, Block E, Uden PC, Zhang Z, Quimby BD, Sullivan JJ. 1995. Allium chemistry: identification of selenoamino acids in ordinary and selenium-enriched garlic, onion, and broccoli using gas chromatography with atomic emission detection. *J Agric Food Chem* **43**:1754–1757.

Cakmak I, Römheld V. 1997. Boron deficiency-induced impairments of cellular functions in plants. *Plant Soil* **193**:71–83.

Camacho-Cristobal JJ, Llunart L, Lafont F, Baumert A, Gonzalez-Fontes A. 2004. Boron deficiency causes accumulation of chlorogenic acid and caffeoyl polyamine conjugates in tobacco leaves. *J Plant Physiol* **161**:879–881.

Cankur O, Yathavakilla SKV, Caruso JA. 2006. Selenium speciation in dill (*Anethum graveolens* L.) by ion pairing reversed phase and cation exchange HPLC with ICP-MS detection. *Talanta* **70**:784–790.

Christiansen J, Jornsgard B, Buskov S, Olsen CE. 1997. Effect of drought stress on content and composition of seed alkaloids in narrow-leafed lupin, *Lupinus angustifolius* L. *Eur J Agron* **7**:307–314.

Chun JY, Nadiminty N, Lee SO, Onate SA, Lou W, Gao AC. 2006. Mechanisms of selenium down-regulation of androgen receptor signalling in prostate cancer. *Mol Cancer Ther* **5**:913–918.

Clark LC, Combs GF, Turnbull BW, Slate EH, Chalker DK, Chow J, Davis LS, Glover RA, Graham GF, Gross EG, Konrad A, Lesher JL, Park K, Sanders BB, Smith CL, Taylor R. 1996. Effects of selenium supplementation for cancer prevention in patients with carcinoma of the skin. *J Am Med Assoc* **276**:1957–1985.

Cona A, Rea G, Angelini R, Federico R, Tavladoraki P. 2006. Functions of amine oxidases in plant development and defence. *Trends Plant Sci* **11**:80–88.

Davies WJ, Kudoyarova G, Hartung W. 2005. Long distance ABA signalling and its relation to other signalling pathways in the detection of soil drying and the mediation of the plant's responses to drought. *J Plant Growth Regul* **24**:285–295.

De Pascale S, Maggio A, Pernice R, Fogliano V, Barbieri G. 2007. Sulphur fertilization may improve the nutritional value of *Brassica rapa* L. subsp *sylvestris*. *Eur J Agron* **26**:418–424.

Desender S, Andrivon D, Val F. 2007. Activation of defense reactions in Solanaceae: Where is the specificity. *Cell Microbiol* **9**:21–30.

Devoto A, Turner JG. 2005. Jasmonate-regulated *Arabidopsis* stress signalling network. *Physiol Plantarum* **123**:161–172.

Diez MA, Barceló J, Lopez-Belmonte F. 1983a. Effects of N, P, K, and Mg deficiencies on growth and nicotine content of Nicotiana rustica. *Anal Edafol Agrobiol* **42**:1663–1677.

Diez MA, Barceló J, Lopez-Belmonte F. 1983b. Effects of N, P, K, and Mg deficiencies on growth, development and atropine content of *Hysocyamus albus*. *Anal Edafol Agrobiol* **42**:1693–1703.

Dodd IC. 2005. Root-to-shoot signalling: Assessing the roles of 'up' in the up and down world of long-distance signalling *in planta*. *Plant Soil* **274**:251–270.

Dong GC, Chuang PH, Forrest MD, Lin YC, Chen HM. 2006. The immuno-suppressive effect of blocking the CD28 signaling pathway in T-cells by an active component of *Echinacea* found by a novel pharmaceutical screening method. *J Med Chem* **49**:1845–1854.

Dong X, Ni Q, Shen Y. 2004. Effect of *Astragalus* injection on levels of blood selenium and immunity function in children with viral myocarditis. *Chinese J Integr Med* **10**:29–32.

Drewitt PN, Butterworth KR, Springall CD, Moorhouse SR. 1993. Plasma levels of aluminium after tea ingestion in healthy-volunteers. *Food Chem Toxicol* **31**:19–23.

Elzaawely AA, Xuan TD, Tawata S. 2007. Changes in essential oil, kava pyrones and total phenolics of *Alpinia zerumbet* (Pers.) B.L. Burtt. & R.M.Sm leaves exposed to copper sulphate. *Environ Exp Bot* **59**:347–353.

Ensminger I, Bush F, Huner NPA. 2006. Photostasis and cold acclimation: Sensing low temperature through photosynthesis. *Physiol Plantarum* **126**:28–44.

Entus R, Poling M, Herrmann KM. 2002. Redox regulation of *Arabidopsis* 3-deoxy-D-arabino-heptulosonate 7-phosphate synthase. *Plant Physiol* **129**:1866–1871.

Epstein E, Bloom AJ. 2005. *Mineral Nutrition of Plants: Principles and Perspectives*, 2nd edition. Sunderland, MA: Sinauer Associates.

Falco G, Gómez-Catalán J, Llobet JM, Domingo JL. 2003. Contribution of medicinal plants to the daily intake of various toxic elements in Catalonia, Spain. *Trace Elem Electroly* **20**:120–124.

Falco G, Llobet JM, Zareba S, Krzysiak K, Domingo JL. 2005. Risk assessment of trace elements intake through natural remedies in Poland. *Trace Elem Electroly* **22**:222–226.

Felle HH, Zimmermann MR. 2007. Systemic signalling in barley through action potentials. *Planta* **226**:203–214.

Freemann JL, Zhang LH, Marcus MA, Fakra S, McGrath SP, Pilon-Smits EAH. 2006. Spatial imaging, speciation, and quantification of selenium in the hyperaccumulator plants *Astragalus bisculatus* and *Stanleya pinnata*. *Plant Physiol* **142**:124–134.

Ge XC, Wu JY. 2005. Tanshinone production and isoprenoid pathways in *Salvia miltiorrhiza* hairy roots induced by Ag^+ and yeast elicitor. *Plant Sci* **168**:487–491.

Glombitza S, Dubois PH, Thulke O, Welzl G, Bovet L, Gotz M, Affenzeller M, Geist B, Hehn A, Asnaghi C, Ernst D, Seidlitz HK, Gundlach H, Mayer KF, Martinoia E, Werck-Reichhart D, Mauch F, Schaffner AR. 2004. Crosstalk and differential response to abiotic and biotic stressors reflected at the transcriptional level of effector genes from secondary metabolism. *Plant Mol Biol* **54**:817–835.

Gobbo-Neto L, Lopes NP. 2007. Medicinal plants: Factors of influence on the content of secondary metabolites. *Quim Nova* **30**:374–381.

Gosh S, Gosh B, Jha S. 2006. Aluminium chloride enhances colchicine production in root cultures of *Gloriosa superba*. *Biotechnol Lett* **28**:497–503.

Gremigni P, Hamblin J, Harris D, Cowling VA. 2003. The interaction of phosphorous and potassium with seed alkaloid concentrations, yield and mineral content in narrow-leafed lupin (*Lupinus angustifolius* L). *Plant Soil* **253**:1251–1257.

Hugentobler U, Renwick JAA. 1995. Effects of plant nutrition on the balance of insect relevant cardenolides and glucosinolates in *Erysimum cheiranthoides*. *Oecologia* **102**:95–101.

Ip C, Lisk DJ, Stoewsand GS. 1992. Mammary cancer prevention by regular garlic and selenium-enriched garlic. *Nutr Cancer* **17**:279–286.

Ip C, Lisk DJ. 1993. Bioavailability of selenium from selenium-enriched garlic. *Nutr Cancer* **20**:129–137.

Ip C, Lisk DJ. 1994. Enrichment of selenium in allium vegetables for cancer prevention. *Carcinogenesis* **15**:1881–1885.

Jachak SM, Saklani A. 2007. Challenges and opportunities in drug discovery from plants. *Curr Sci India* **92**:1251–1257.

JEFCA (Joint FAO/WHO Expert Committee on Food Additives) 1993. Evaluation of certain food additives and contaminants: 41st report of the Joint FAO/WHO Expert Committee on Food Additives. Geneva: World Health Organization. WHO Technical Reports Series No. 837.

JEFCA (Joint FAO/WHO Expert Committee on Food Additives) 2006. Meeting (67th: 2006: Rome, Italy). Evaluation of certain food additives and contaminants: Sixty-seventh report of the Joint FAO/WHO Expert Committee on Food Additives. WHO Technical Report Series No. 940.

Judzentiene A, Stikliene A, Kupcinskiene E. 2007. Changes in the essential oil composition in the needles of Scots pine (*Pinus sylvestris* L.) under anthropogenic stress. *The Scientific World J* **7**:141–150.

Kar A, Choudhary BK, Bandyopadhyay NG. 1999. Preliminary studies on the inorganic constituents of some indigenous hypoglycaemic herbs on oral glucose tolerance test. *J Ethnopharmacol* **64**:179–184.

Khan MB, Harborne JB. 1991. Potassium deficiency increases tropane alkaloid synthesis in *Atropa acuminata* via arginine and ornithine decarboxylase levels. *Phytochemistry* **30**:3559–3563.

Khan NA, Mulchi CL, McKee CG. 1994. Influence of soil pH and molybdenum fertilization on the productivity of Maryland tobacco. 1. Field investigations. *Commun Soil Sci Plant* **25**:2103–2116.

Kidd SK, Melillo AA, Lu R-H, Reed DG, Kuno N, Uchida K, Furuya M, Jelesko JG. 2006. The *A* and *B* loci in tobacco regulate a network of stress responses, few of which are associated with nicotine synthesis. *Plant Mol Biol* **60**:699–716.

Knobloch KH, Berlin J. 1983. Influence of phosphate on the formation of indole alkaloids and phenolics compounds in cell suspension cultures of *Catharanthus roseus*. I. Comparison of enzyme activities and product accumulation. *Plant Cell Tiss Org* **2**:333–340.

Kovacik J, Tomko J, Backor M, Repcak M. 2006. *Matricaria chamomilla* is not a hyperaccumulator, but tolerant to cadmium stress. *Plant Growth Regul* **50**:239–247.

Lambers H, Chapin FS III, Pons TL. 2008. *Plant Physiological Ecology*, second edition. New York: Springer.

Lecky R, Hook I, Sheridan H. 1992. Enhancement of dihydrosanguinarine production in suspension cultures of *Papaver bracteatum*. I. Medium modifications. *J Nat Prod* **55**:1513–1517.

Levin DA. 2004. Ecological speciation: Crossing the divide. *Syst Bot* **29**:807–816.

Liu ZJ, Adams JC, Viator HP, Constantin RJ, Carpenter SB. 1999. Influence of soil fertilization, plant spacing, and coppicing on growth, stomatal conductance, abscisic acid, and camptothecin levels in *Camptotheca acuminata* seedlings. *Physiol Plantarum* **105**:402–408.

Longley M, Phua SH, van Stijn TC, Crawford AM. 1999. Isolation and mapping of the first ruminant multidrug resistance genes. *Anim Genet* **30**:207–210.

Lozak A, Soltyk K, Ostapczuk P, Fijalek Z. 2002. Determination of selected trace elements in herbs and their infusions. *Sci Total Environ* **289**:33–40.

Marschner H. 1995. *Mineral Nutrition of Higher Plants*, second edition. London: Academic Press.

Martins FT, Santos MH, Polo M, Barbosa LCA. 2007. Effects of the interactions among macronutrients, plant age and photoperiod in the composition of *Hyptis suaveolens* (L.) Poit essential oil from Alfenas (MG), Brazil. *Flavour Frag J* **22**:123–129.

Mathur S, Verma RK, Gupta MM, Ram M, Sharma S, Kumar S. 2000. Screening of genetic resources of the medicinal vegetable plant *Centella asiatica* for herb and asiaticoside yields under shaded and full sunlight conditions. *J Hortic Sci Biotech* **75**:551–554.

Matula J, Zukalova H. 2001. Sulphur concentrations and distribution in three varieties of oilseed rape in relation to sulphur fertilization at vegetative stages. *Rostlinna Vyroba* **47**:1–6.

Mayer AM. 2006. Polyphenol oxidases in plants and fungi: Going places? A review. *Phytochemistry* **67**:2318–2331.

Mazzafera P. 1998. Growth and biochemical alterations in coffee due to selenite toxicity. *Plant Soil* **201**:189–196.

Mertz W. 1993. Chromium in human nutrition—A review. *J Nutr* **123**:626–633.

Misra A, Khan A. 1992. Correction of iron deficiency chlorosis in lemongrass (*Cymbopogon flexuosus* Steud) Watts. *Agrochimica* **36**:349–360.

Mittler R, Vanderauwera S, Gollery M, van Breusegem F. 2004. Reactive oxygen network of plants. *Trends Plant Sci* **9**:490–498.

Narula A, Kumar S, Srivastava PS. 2005. Abiotic metal stress enhances diosgenin yield in *Dioscorea bulbifera* L. cultures. *Plant Cell Rep* **24**:250–254.

Nookabkaew S, Rangkadilok N, Satayavivad J. 2006. Determination of trace elements in herbal tea products and their infusions consumed in Thailand. *J Agr Food Chem* **54**:6939–6499.

Nozue K, Maloof JN. 2006. Diurnal regulation of plant growth. *Plant Cell Environ* **29**:396–408.

Omer EA. 1999. Response of wild Egyptian oregano to nitrogen fertilization in a sandy soil. *J Plant Nutr* **22**:103–114.

Palazón J, Cusidó RM, Bonfill M, Mallol A, Moyano E, Morales C, Piñol MT. 2003. Elicitation of different *Panax ginseng* transformed root phenotypes for an improved ginsenoside production. *Plant Physiol Biochem* **41**:1019–1025.

Palumbo MJ, Putz FE, Talcott ST. 2007. Nitrogen fertilizer and gender effects on the secondary metabolism of youpon, a caffeine-containing North American holly. *Oecologia* **151**:1–9.

Pande P, Anwar M, Chand S, Yadav VK, Patra DD. 2007. Optimal level of iron and zinc in relation to its influence on herb yield and production of essential oil in menthol mint. *Commun Soil Sci Plant* **38**:561–578.

Papp LV, Lu J, Holmgren A, Khanna KK. 2007. From selenium to selenoproteins: Synthesis, identity, and their role in human health. *Antiox Redox Sign* **9**:775–806.

Pedley KF, Martin GB. 2005. Role of mitogen-activated protein kinases in plant immunity. *Curr Opin Plant Biol* **8**:541–547.

Piñol MT, Palazón J, Cusidó RM, Ribó M. 1999. Influence of calcium ion concentration in the medium on tropane alkaloid accumulation in *Datura stramonium* hairy roots. *Plant Sci* **141**:41–49.

Pita-Alvarez SI, Spollansky TC, Giulietti AM. 2000. Scopolamine and hyoscyamine production by hairy root cultures of *Brugmansia candida*: influence of calcium chloride, hemicellulase and teophylline. *Biotechnol Lett* **22**:1653–1656.

Poschenrieder C, Tolrà R, Barceló J. 2006a. Can metals defend plants against biotic stress? *Trends Plant Sci* **11**:288–295.

Poschenrieder C, Tolrà R, Barceló J. 2006b. Interactions between metal ion toxicity and defenses against biotic stress: Glucosinolates and benzoxazinoids as case studies. *Snow Landsc Res* **80**:149–160.

Poutaraud A, Girardin P. 2005. Influence of chemical characteristics of soil on mineral and alkaloid seed contents of *Colchicum autumnale*. *Environ Exp Bot* **54**:101–108.

Pudersell K, Vardja R, Vardja T, Raal A, Arak E. 2003. Plant nutritional elements and tropane alkaloid production in the roots of henbane (*Hyoscyamus niger*). *Pharm Biol* **41**:226–230.

Queralt I, Ovejero M, Carvalho ML, Marques AF, Llabrés JM. 2005. Quantitative determination of essential and trace element content of medicinal plants and their infusions by XRF and ICP techniques. *X-Ray Spectrom* **34**:213–217.

Rai V, Vajpayee P, Singh SN, Mehrotra S. 2004. Effect of chromium accumulation on photosynthetic pigments, oxidative stress defense system, nitrate reduction, praline level and eugenol content of *Ocimum tenuiflorum* L. *Plant Sci* **167**:1159–1169.

Rai V, Khatoon S, Bisht SS, Mehrota S. 2005. Effect of cadmium, ultramorphology of leaf and secondary metabolites of *Phyllanthus amarus* Schum. and Thonn. *Chemosphere* **61**:1644–1650.

Raju GJN, Sarita P, Murty GAVR, Kumar MR, Reddy SB, Vijayan V. 2006. Estimation of trace elements in some anti-diabetic medicinal plants using PIXE technique. *App Rad Isotop* **64**:893–900.

Rajurkar NS, Damame MM. 1997. Elemental analysis of some herbal plants used in the treatment of cardiovascular diseases by NAA and AAS. *J Radioanal Nucl Ch* **219**:77–80.

Rajurkar NS, Damame MM. 1998. Mineral content of medicinal plants used in the treatment of diseases resulting from urinary tract disorders. *Appl Radiat Isotopes* **49**:773–776.

Razic S, Onjia A, Potkonjak B. 2003. Trace elements analysis of *Echinacea purpurea*—Herbal medicine. *J Pharmaceut Biomed* **33**:845–850.

Ren FC, Liu TC, Liu HQ, Hu BY. 1993. Influence of zinc on the growth, distribution of elements, and metabolism of one-year old American ginseng plants. *J Plant Nutr* **16**:393–405.

Rengel Z. 2004. Aluminium cycling in the soil–plant–animal–human continuum. *BioMetals* **17**:669–689.

Rios JL, Waterman PG. 1997. A review of the pharmacology and toxicology of *Astragalus*. *Phytotherapy Res* **11**:411–418.

Roca-Pérez L, Pérez-Bermudez P, Boluda R. 2002. Soil characteristics, mineral nutrients, biomass, and cardenolide production in *Digitalis obscura* wild populations. *J Plant Nutr* **25**:2015–2026.

Roca-Pérez L, Pérez-Bermudez P, Gavidia I, Boluda R. 2005. Relationships among soil characteristics, plant macronutrients, and cardenolide accumulation in natural populations of *Digitalis obscura*. *J Plant Nutr Soil Sci* **168**:774–780.

Rohdich F, Bacher A, Eisenreich W. 2004. Perspectives in anti-infective drug design. The late steps in the biosynthesis of the universal terpenoids precursors, isopentenyl diphosphate and dimethylallyl diphosphate. *Bioorg Chem* **32**:292–308.

Rohloff J, Skagen EB, Ivensen TH. 2000. Production of yarrow (*Achillea millefolium* L.) in Norway: Essential oil content and quality. *J Agr Food Chem* **48**:6205–6209.

Rummukainen A, Julkunen-Tiitto R, Raisanen M, Lehto T. 2007. Phenolic compounds in Norway spruce as affected by boron nutrition at the end of the growing season. *Plant Soil* **292**:13–23.

Schopfer P, Brennicke A. 1999. *Pflanzenphysiologie*, fifth edition. Berlin: Springer.

Serrano M, Piñol MT. 1991. *Biotecnología Vegetal.* Madrid: Editorial Síntesis.

Spitaler R, Schlorhaufer PD, Ellmerer EP, Merfort I, Bortenschlager S, Stuppner H, Zidorn C. 2006. Altitudinal variation of secondary metabolite profiles in flowering heads of *Arnica montana* cv ARBO. *Phytochemistry* **67**:409–417.

Sreevalli Y, Kulkami RN, Baskaran K, Chandrashekara RS. 2004. Increasing the content of leaf and root alkaloids of high alkaloid content mutants of periwinkle through nitrogen fertlization. *Ind Crop Prod* **19**:191–195.

Strauss YS, Irwin RE. 2004. Ecological and evolutionary consequences of multispecies plant–animal interactions. *Annu Rev Ecol Evol* **35**:435–466.

Sung S, Amasino RM. 2006. Molecular genetic studies of the memory of winter. *J Exp Bot* **57**:3369–3377.

Szentmihalyi K, Hajdu M, Fodor J, Otai L, Blazovics A, Somogyi A, Then M. 2006. *In vitro* study of elements in herbal remedies. *Biol Trace Elem Res* **114**:143–150.

Szöke E, Maday E, Kiss SA, Sonnewend L, Lemberkovics E. 2004. Effect of magnesium on essential oil formation of genetically transformed and nontransformed chamomile cultures. *J Am Coll Nutr* **23**:763S–767S.

Toler HD, Charron CS, Sams CE, Randle WR. 2007. Selenium increases sulphur uptake and regulates glucosinolate metabolism in rapid cycling *Brassica oleracea*. *J Am Soc Hortic Sci* **132**:14–19.

Tolrà RP, Poschenrieder C, Lupi B, Barceló J. 2005. Aluminium-induced changes in the profiles of both organic acids and phenolics substances underlie Al tolerance in *Rumex acetosa* L. *Environ Exp Bot* **54**:231–238.

Trejo-Tapia G, Jimenez-Aparicio A, Rodriguez-Monroy M, De Jesus-Sanchez A, Gutierrez-Lopez G. 2001. Influence of cobalt and other microelements on the production of betalains and the growth of suspension cultures of *Beta vulgaris*. *Plant Cell Tiss Org* **67**:19–23.

Wellinghausen N, Kirchner H, Rink L. 1997. The immunobiology of zinc. *Immunol Today* **18**:519–521.

Yan X, Wu S, Wang Y, Shang X, Dai S. 2004. Soil nutrient factors related to salidrosides production of *Rhodiola sachalinensis* distributes in Chang Bai Mountain. *Environ Exp Bot* **52**:267–276.

Zhang CH, Fevereiro PS, He GY, Chen ZJ. 2007. Enhanced paclitaxel productivity and release capacity of *Taxus chinensis* cell suspension cultures adapted to chitosan. *Plant Sci* **172**:158–163.

Zhao FJ, Evans EJ, Bilsborrow PE, Syers JK. 1994. Influence of nitrogen and sulphur on the glucosinolate profile of rapeseed (*Brassica napus* L). *J Sci Food Agr* **64**:295–304.

Zheng GQ, Kenney PM, Lam LKT. 1992. Anethofuran, carvone, and limonene—Potential cancer chemopreventive agents from dill weed oil and caraway oil. *Planta Medica* **58**:338–341.

Zheng ZU, Wu M. 2004. Cadmium treatment enhances the production of alkaloid secondary metabolites in *Catharanthus roseus*. *Plant Sci* **166**:507–514.

6 Trace Elements and Radionuclides in Edible Plants

MARIA GREGER

Department of Botany, Stockholm University, 106 91 Stockholm, Sweden

1 INTRODUCTION

Plants take up, from the rhizosphere, all kinds of elements from the entire periodic system; they can also absorb these elements via their shoots. Elements such as N, P, K, S, Ca, and Mg are used in large amounts as plant nutrients, whereas the trace elements Fe, Mn, Mo, Ni, Cu, Zn, B, and Cl are used in trace amounts. In addition, plants also use H, O, and C, which are taken up as water and carbon dioxide, but may also be absorbed as organic substances. Other trace elements— for example, La and Ce [Asher, 1991]—have been discussed as possibly essential for plants. However, due to their presence at very low concentrations and to difficulties isolating the elements from the surroundings of plants, it is difficult to determine whether (1) the plant can complete its life cycle without the element, (2) the element can replace a nutrient element, and (3) the element is directly involved in plant metabolism [Marschner, 1995], and hence to determine their essentiality. There are also beneficial elements—for example, V, Na, and I—which can be essential for some plant species under specific conditions, or, though not essential, may be able to increase the growth rate of plants [Marschner, 1995].

The elements in the periodic system may occur as various isotopes, and both stable and radioactive isotopes of elements are found in plants. The uptake rates of the various isotopes can vary, and it is possible that plants may discriminate between isotopes of individual elements [Shaw and Bell, 1994; Baeza et al., 1999]. Isotopes of Cs and Sr are the most studied in relation to plant uptake, but isotopes of I, U, Ra, Tc, Th, and Rb have also been examined. Cesium and Rb are very similar to K and thus follow the uptake route of K and can interact with the K uptake; the relationship between Sr and Ca is similar.

Trace Elements as Contaminants and Nutrients: Consequences in Ecosystems and Human Health, Edited by M. N. V. Prasad
Copyright © 2008 John Wiley & Sons, Inc.

2 PLANT UPTAKE AND TRANSLOCATION OF TRACE ELEMENTS

Plants take up trace elements via their roots and shoots from soil, sediment, water, and air. Plant roots are able to release trace elements from soil particles to the soil water by means of various mechanisms, such as root excretion of organic acids, phytosidero-phores, and hydrogen ions [Marschner, 1995; Greger, 2005] (Fig. 1). This raises the concentrations of the elements in the pore water, thereby increasing the amounts of the elements available for uptake. Trace elements in the pore water enter the root surface by means of diffusion, root interception, or mass flow [Marschner, 1995] (Fig. 1). Thereafter, the elements enter the cell walls and apoplast, in which cations can be trapped and anions repelled by the negative charges of the cell wall pectin. Some of the trace elements are then translocated further in the apoplast or enter the cell symplast with the help of various carriers, pumps, and channels. Most of the trace elements and heavy elements are translocated to the greatest extent via the apoplast, while many of the plant macronutrients (except Ca) are largely taken into the symplast. In the cells, shuttles handle some of the elements, transporting them into the vacuole using various mechanisms, for example, by forming phytochelatins or organic acids to which some elements can be bound for further vacuolar transport [Steffens, 1990; Jackson et al., 1990].

The long-distance translocation of elements from roots to shoots via the xylem vessels depends on the possibility of the elements first entering the cell symplast and from there moving into the vessels. Elements with restricted entrance into the symplast often move in the apoplast; to be able to reach the xylem vessels of the

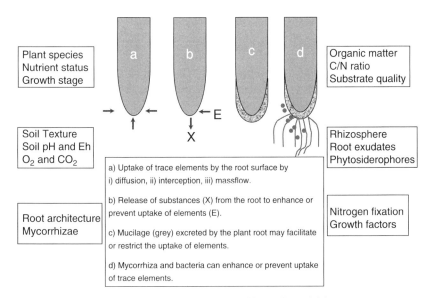

Figure 1. Trace element availability and acquisition.

roots, these elements must cross the suberinized walls of the endodermis, which is difficult. Consequently, most of the elements that reach the xylem vessels enter them via the immature parts of the roots, where the walls of the endodermis layer not yet have been suberinized. The distance from the root tip to the mature part of the root where the suberinized cell walls of the endodermis have been formed is thus important in determining the amounts of elements that can be translocated to the shoots [Lux et al., 2004]. Element translocation occurs via the xylem sap flow through the vessels; the different elements can move at different speeds due to cation exchange at the negative charges of the vessel walls, leading to the separation of cation transport from water flow in the xylem vessel (an ion-exchange column) [van de Geijn and Petit, 1979; Wolterbeek, 2005; Momoshima and Bondietti, 1990].

Plant leaves can also take up trace elements from the air, while plants that grow submersed in water can accumulate trace elements directly from the water via the whole plant body. As ions, these elements enter through the leaf cuticle, while as gases they enter the leaves through the stomata [Martin and Juniper, 1970; Lindberg et al., 1992; Marschner, 1995]. The cuticular layer functions as a weak cation exchanger due to the negative charges of the pectic material and nonesterified cutin polymers in the leaf cuticle. From the external surface toward the cell walls there is a distinct gradient from low to high charge density, which favors ion penetration across the cuticle [Yamada et al., 1964]. The uptake of ions via the cuticle can be promoted via uptake through the ectodesmata, which are less dense parts of the cuticula layer, mainly situated between the guard cells and subsidiary cells of the epidermal cell wall/cuticula.

The translocation from leaves to roots and growing shoots seems to differ between plant species and element. For example, Cd has been shown to be translocated in pea plants from the treated leaves to the stipules of the leaf stalk but no farther [Greger et al., 1993], while Cd, Cu, and Zn do not reach the roots after being applied to the leaves of pea plants [Greger, 2004]. However, there was limited translocation of Pb and Cd to the roots from the leaves in broad bean plants [Salim et al.,

TABLE 1. Distribution (%) of Cd (^{109}Cd) in Wheat Plants 14 Days after Anthesis (i.e., Onset of Grain Filling Period), 1, 6, and 96 Hours after ^{109}Cd (20–30 μL, 102 nM) Had Been Applied to the Flag Leaf Blade[a]

Plant Part	Time after Application to Flag Leaf (hours)		
	1	6	96
Ear	nd	nd	19 ± 1.0
Leaves 2 – 3	nd	5 ± 0.3	25 ± 1.6
Stem above	65 ± 3	63 ± 4.1	6 ± 0.3
Stem below	35 ± 2	30 ± 2.4	7 ± 0.2
Root	nd	1 ± 0.1	43 ± 3.0

[a]The Cd content was analyzed in roots, leaves 2 and 3, stem below and above flag leaf node, and whole ear [Greger and Landberg, unpublished data]. $n = 4 ±$ SE, nd, not detectable.

1992], while Zn and Cd applied to leaves of wheat plants are translocated both to the roots and to the grains ([Herren and Feller, 1996]; Table 1). Shinonaga and Ambe [1998] showed that various radionuclides in the atmosphere were absorbed through soybean leaves and then transported to the seeds.

3 DISTRIBUTION AND ACCUMULATION OF TRACE ELEMENTS IN PLANTS

Trace elements are usually unevenly distributed in plants, with the highest concentrations being found in the roots and the lowest in the fruit and seeds; for example, when comparing 23 crop species, it was found that Cd concentration generally decreased in the following order: fibrous roots > storage roots > stems > leaves [Jarvis et al., 1976].

Plants vary in how they accumulate specific elements, and those that can accumulate very high levels of an element in their shoots are called hyperaccumulators [Brooks, 1998]. True hyperaccumulators are unlikely to be found among edible plants; however, species accumulating high levels of specific elements can be found—for example, *Brassica juncea* can accumulate high levels of Cd and Pb, and lettuce can accumulate Cd (Table 2).

Plant cultivation over the centuries has resulted in the selection of a great many varieties of individual species. Notably, different varieties of a single species can

TABLE 2. Relative Metal Accumulations in Edible Parts of Various Plant Species [Pais and Jones, 1997]

Species	High Accumulations of	Low Accumulations of
Lettuce	Cd	—
Spinach	Cd, Zn	—
Celery	Cd, Pb,	—
Cabbage	Cd	Cu
Potato	—	Cd, Pb, Zn
Maize	—	Cd, Pb, Ni
Peas	—	Cd
Sugar beet	Cu, Ni, Zn	—
Certain barley cultivars	Cu	Pb, Ni
Certain wheat cultivars	Cd	—
Leek	—	Cu, Ni, Zn
Onion	Ni	Cu, Zn
Turnip	—	Ni
Marigold	Ni, Zn	—
Tomato	—	Zn
Beetroot	Zn	—
Kale	Pb	—
Ryegrass	Pb	—

TABLE 3. Concentration Ranges of Various Elements in Four Species from Studies in which Different Genotypes Were Screened

| Species | Element Concentration Range (mg/g DW) | | Reference |
	Cd	Pb	
Durum wheat grains	0.175–0.500	—	Greger and Löfstedt [2004]
Bread wheat grains	0.040–0.090	—	Greger and Löfstedt [2004]
Lettuce	0.084–0.405	0.06–2.90	Alexander et al. [2006]
Rice grains	0.17–1.84	—	Liu et al. [2007]

have completely different accumulation properties for particular elements, and genotypical differences in this respect have been found in many species, such as maize, non-oilseed sunflower, soybean, flax, rice, wheat, and lettuce [Grant et al., 1999; Hocking and McLaughlin, 2000; Greger and Löfstedt, 2004; Alexander et al., 2006; Table 3].

4 VEGETABLES, FRUIT, AND BERRIES

Vegetables contain various levels of metals depending on the particular metal, plant species, and plant organ. In general, legumes tend to be low accumulators, root vegetables moderate accumulators, and leafy vegetables high accumulators of metals such as Cd, Cu, Zn, and Pb [Alexander et al., 2006]. Furthermore, fruits contain lower concentrations of minerals than do vegetables, while leafy vegetables contain more minerals than other vegetables do [Varo et al., 1980]. High accumulations of some elements and low accumulations of others can be found in a single species; for example, spinach accumulates high amounts of Mn and Zn, but relatively lower concentrations of Cu and Pb, in its tissues [Intawongse and Dean, 2006].

Compared to other common vegetable species, lettuce seems to accumulate high amounts and bean low amounts of several elements, such as Pb, Zn, Cd, and Cu [Fleming and Parle, 1977; Pettersson, 1977]. Since Cd is considered one of the most dangerous elements and is the only one having specified limit values for various crops in the European Union, it has also been most frequently investigated. Not only lettuce but also beetroot, radish, and carrot are high accumulators of Cd [Lehoczky et al., 1998]. An attempt was made to classify plant families according to their Cd-accumulating properties, and it was found that Leguminosae are low Cd accumulators, Gramineae, Liliaceae, Cucurbitaceae, and Umbelliferae are moderate accumulators, and Chenopodiaceae, Cruciferae, Solanaceae, and Compositae are high accumulators [Kuboi et al., 1986].

Though a plant part may have accumulated a high amount of an element, humans consuming the plant part will not necessarily absorb all its content of the element;

bioaccessibility could depend on both the particular element and the nature of the plant tissue. A study of the gastrointestinal bioavailable concentrations of metals found that the availability varied greatly from element to element and according to plant type. The greatest metal availability was found in lettuce for Mn (63.7%), followed by radish for Cu (62.5%), Cd (54.9%), and Mn (45.8%), and lettuce for Zn (45.2%) [Intawongse and Dean, 2006].

As discussed earlier in this chapter, different varieties of a crop species may accumulate elements differently [Crews and Davies, 1985; Alexander et al., 2006]. However, these differences do not depend solely on the genotype, but also on the soil concentrations and the availability of the elements. When investigating different genotypes from various fields, the differences in accumulation found can thus be much greater than those found in an ordinary screening test using a similar substrate. McLaughlin et al. [1994] compared the uptake of Cd by 14 commonly grown potato cultivars in South Australia. They found significant differences between the cultivars at most sites; at some sites, individual cultivars had Cd concentrations exceeding the Australian maximum permissible concentration of 50 mg Cd/kg. According to the authors, it would be possible to reduce Cd concentrations in potato tubers by more than 50% by selecting low-accumulating cultivars.

Different berries also contain different levels of elements, depending on the species. An investigation of wild berries in northern Sweden found high levels of Tl, Sr, and Ba in lingonberries (*Vaccinium vitis-idaea*), while the highest levels of Cl and Re were found in blueberries (*Vaccinium myrtillus*) [Rodushkin et al., 1999]. The collection site can also influence element concentrations in berries, which may reflect the environmental pollution in a given area. The highest concentrations of Ag, As, Be, Bi, Br, Cd, Hg, I, Ni, Pb, Sb, and Tl are found in berries from mine sites, while high levels of Li, V, Hf, W, Ta, lanthanides, and actinides are found in the vicinity of high-traffic roads.

Not only polluted soils but also soils naturally high in trace elements can transfer high amounts of undesired elements to cultivated plants. For example, a study on element content in plants grown in alum shale soil displayed higher levels of Cd, U, V, As, Mo and Ba, in lettuce and U, V, and Mo in potato (Table 4) than are reported in the literature [Kabata-Pendias and Pendias, 1992]; moreover, the Cd concentration exceeded the limit values for leaf vegetables (0.2 mg Cd/kg) and for potato (0.1 mg Cd/kg) according to European Commission Regulation no. 466/2001.

In areas highly contaminated with radioactive isotopes, one can assume that elements such as Rb and Cs, which resemble the nutrient element K and also Sr (which in turn is similar to Ca), should be easily taken up and translocated and that high levels of them should be found in plants. Man-made isotopes are also easy for plants to take up and translocate; for example, technetium (^{99}Tc) is an element with a very high transfer factor (Table 5). It is probably taken up via a route similar to that of NO_3^- or other anions [van Loon, 1986] and has been detected in the form of $^{99}TcO_4^-$ in spinach cells [Lembrechts et al., 1985].

TABLE 4. Concentrations of Several Elements in a Studied Alum Shale Soil and in Soil Described in the Literature, as Well as in Lettuce and Potato Cultivated in the Alum Shale Soil [Greger and Landberg, unpublished data]

| Element | Concentration in Soil (mg/kg) | | Concentration in Potato (mg/kg DW) | Acc Factor × 10^{-3} | Concentration in Lettuce (mg/kg DW) | Acc Factor × 10^{-3} |
	Alum Shale	Literature Data[a]				
U	54	0.1–11	0.02–0.03	0.4	0.14–0.25	3
V	176	23–230	nd[b]–0.06	0.1	0.38–0.87	4
Ba	125	100–3000	2–3	20.7	16–20	143
Cd	6	0.01–3	0.22–0.26	42.6	2.5–3.3	519
Mo	67	0.5–40	1.6–1.9	25.5	0.80–0.94	13
As	117	0.1–48	nd	—	nd–0.7	6

[a]Pais and Jones [1997].
[b]nd, not detectable.

TABLE 5. Soil–Plant Transfer Factors for Selected Radionuclides [IUR, 1989]

Radionuclide	Transfer Factor (min–max)
^{99}Tc	$3 \times 10^{-1} - 3 \times 10^{3}$
^{90}Sr	$8 \times 10^{-3} - 4 \times 10^{1}$
$^{134/137}Cs$	$2 \times 10^{-4} - 3 \times 10^{1}$

5   CEREALS AND GRAINS

Problems arise when a plant species accumulates high levels in the plant part used by humans and when this specific part is much consumed. For example, high concentrations of Cd have been found in wheat grains, soybean, sunflower grains, linseed, and rice grains [e.g., Hocking and McLaughlin, 2000; Greger and Löfstedt, 2004; Arao and Ishikawa, 2006]; in comparison, Cd concentrations in the shoot tissues of wheat and sunflowers have been found to be low [Pettersson, 1977].

5.1   Cadmium in Wheat

Compared with most cereals, wheat has been demonstrated to have among the highest Cd contents in its grains. Cadmium uptake varies among cultivars of bread and durum wheat [Oliver et al., 1995; Wenzel et al., 1996; Li et al., 1997; Greger and Löfstedt, 2004], and the general pattern of Cd accumulation in different wheat types is as follows: winter bread wheat < spring bread wheat < durum wheat [Hellstrand and Landner, 1998]. However, there are greater differences between *cultivars* than between different wheat types [Greger and Löfstedt, 2004; Greger and Landberg, unpublished]. Unfortunately, the high-protein content cultivars are usually also those with high Cd concentration in the grains. This is not, however, the case in *Triticum spelta* (spelt wheat), in which the grains have a high protein content but a low Cd content [Greger and Landberg, unpublished]. Recently, we found that winter spelt wheat had a very low Cd content, 18 μg/kg DW, in its grains, compared to 40–90 μg/kg DW in bread wheat grains and approximately 175–500 μg/kg DW in durum wheat grains ([Greger and Löfstedt, 2004; Greger and Landberg, unpublished]; Swedish limit values 100 μg Cd/kg DW).

The mechanisms underlying the different Cd accumulations in grains of different cultivars are found in (a) uptake from soil, (b) translocation within the plant, and (c) binding within the grains [Greger and Landberg, unpublished results]. Cultivars with high Cd accumulation in their grains release Cd from soil colloids by means of a high root cation exchange capacity, which increases the Cd concentration in pore water and thus the uptake and the translocation of Cd into the grains; however, there is no difference between cultivars in terms Cd uptake from *soil solution*. Within the plants, Cd is translocated to the shoots to a greater degree in cultivars that accumulate high amounts of Cd in the grains, and the Cd reaching the grains is more likely to become bound there in high-accumulating varieties. Such varieties accumulate Cd in the peripheral part of the grain, possibly in the aleurone layer, and bound to an albumin or globulin protein fraction to a greater extent than do low accumulators.

5.2 Arsenic in Rice

Rice, *Oryza sativa* L., is a very important crop, particularly in Southeast and East Asia. Rice fields are often irrigated with contaminated water, increasing the As content of the soil and of all parts of the rice plant, including the grains [Abedin et al., 2002a; Williams et al., 2005]. Rice takes up As easily, and transports it to the straw and grains [Xie and Huang, 1998]. In the submerged soil environments in which rice is cultivated, *arsenite*, As(III), is the predominant species [Marin et al., 1992; Abedin et al., 2002b]. In such environments, iron plaque is commonly formed on the roots; the plaque strongly binds to As(V), enhancing As uptake by rice roots, but reducing As translocation from roots to shoots [Liu et al., 2004]. Formation of iron plaque can vary between rice genotypes [Liu et al., 2006], and a high plaque level may thus be advantageous. Inorganic As and dimethylarsinic acid are the main As species found in rice grains [Liu et al., 2006], and As concentrations and species can differ between rice cultivars [Liu et al., 2004]. In addition, As(V) uptake is strongly suppressed by phosphate, while As(III) transport is unaffected by it [Abedin et al., 2002b].

6 AQUATIC PLANTS

Plants cultivated in aquatic environments for food can accumulate high levels of trace elements from the water. The aquatic macrophyte water spinach (*Ipomea aquatica*) is a popular vegetable rich in vitamin A, vitamin C, and iron [Chughtai, 1995]. It grows rapidly by approximately 10 cm/day and requires little labor input for its production [Ruskin and Shipley, 1976]. In Southeast Asia, the plant is cultivated in freshwater watercourses, which often receive wastewaters, and this plant consequently accumulates Hg, Cd, and Pb [Göthberg et al., 2002]. The ambient nutrient concentration affects the uptake of these elements, and their accumulation declines with high ambient nutrient status [Göthberg et al., 2004]. This is likely caused by biological dilution resulting from increased growth and by interaction between nutrient elements and heavy metals at the uptake site. Part, or sometimes all, of the analyzed Hg was in the form of methyl-Hg, and this plant species is able to form methyl mercury in its new shoots [Göthberg and Greger, 2006]. Concentrations of methyl-Hg, total Hg, Cd, and Pb up to 221, 2590, 530, and 123 µg/kg DW were found in water spinach from the Bangkok area [Göthberg et al., 2002].

Macroalgae are traditionally consumed in Asia as sea vegetables, for example, the brown algae *Hizikia fusiforme*, *Kjellmaniella crassifolia*, *Laminaria japonica*, *Laminaria setchellii*, and *Undaria pinnatifida*, the green algae *Ulva lactuca* and *Caulerpa racemosa*, and the red algae *Gracilaria lemaneiformis* and *Porphyra* sp. In terms of nutrition, they are low-calorie foods with high concentrations of minerals (i.e., Mg, Ca, P, K, and I), vitamins, proteins, indigestible carbohydrates, a low lipid content, and relatively high contents of essential amino acids and unsaturated fatty acids [Jimenez-Escriq and Cambrodon, 1999]. However, macroalgae can also accumulate other elements. *Caulerpa racemosa*, for example, has high As, B, and Ti and low Fe contents [Misheer et al., 2006]. Arsenic seems to occur in high concentrations in many edible algae and is present as As(III), As(V), methylarsonic acid, dimethylarsinic acid, and arsenosugars [Misheer et al., 2006]. Arsenic concentration is lowered by soaking in water and boiling [Ichikawa et al., 2006], but cooking does not affect the bioaccessibility of arsenosugar [Almela et al., 2005]. Other elements found in algae are I, which is high in *Laminaria japonica* and *L. setchellii*, and ^{226}Ra, which is found in algae from Japan [van Netten et al., 2000]. Heavy metals are also found in macroalgae, and brown algae can bind metals such as Ca, Cd, and Cu to alginates and polysaccharides in the algal tissue. For example, the brown alga *Kjellmaniella crassifolia* contains 27% alginates [Nishide et al., 1996]. The macromolecular conformation of the alginate polymers determines the binding behavior of metals, and binding capacities of 1.8 mmol Cd/g alginates have been measured [Davis et al., 2004].

7   FUNGI

Although fungi are no longer considered part of the plant kingdom, it is still worthwhile considering them as food crops. Mushrooms contain higher amounts of heavy metals and Se than do edible plants [Varo et al., 1980]. It is known that some fungal species accumulate high amounts of elements such as Cd, Hg, Pb, V, Cu, and As [Vetter, 2003]. Decomposing fungi have higher concentrations of Hg, Pb, and Cd in their fruiting bodies than do mycorrhizal fungi [Lodenius et al., 1981], probably because decomposing fungi use plant cell walls, where most metals are bound in the plant tissue, as their nutrient source. The highest concentrations of Hg and Cd have been found in *Agaricus* sp.; up to 130 mg Cd/kg DW has been found [Lodenius et al., 1981] and *A. arvensis* is a known high accumulator of Hg [Vetter and Berta, 2005]. Other Hg accumulators are *Macrolepiota* sp., *Lycoperdon perlatum*, and *Lepista* sp., all of which contain approximately 3 mg Hg/kg DW [Vetter and Berta, 1997]. Bioaccumulation of vanadium was found to be negligible in most cases, but concentrations up to 1 mg/kg DW could be found in some species [Vetter, 1999]. The highest concentration of Cr was found in *Lactarius deliciosus* (4.51 mg/kg DW) and of Ni (9.9 mg/kg DW) in *Tricholoma terreum*; however, the Ni and Cr concentrations found in mushrooms were under toxicological limits [Vetter, 1997]. The majority of mushroom species have low As levels, though high levels are found in, for example, *Agaricus* sp. and *Macrolepiota rhacodes* [Vetter, 1994].

Metal contents may differ due to growth conditions, and the Hg content in cultivated *Agaricus* sp. was low while that of wild *Agaricus* was high [Vetter and Berta, 2005], probably due to the growth substrate. Furthermore, metal content also differs depending on how the fungi were treated. Vetter [2003] studied the metal content in preserved and fresh samples of *Agaricus bisporus* and found no difference in their Cd, Mn, and Zn contents, indicating that these elements were

TABLE 6. Different Rates of Accumulation of Radiocesium in Several Mushroom Species [Kalac, 2001]

Low	Medium	High
Armillariella mellea	*Agaricus silvaticus*	*Cantharellus lutescens*
Boletus edulis	*Leccinum aurantiacum*	*Cantharellus tubaeformis*
Calocybe gambosa	*Leccinum scabrum*	*Rozites caperata*
Cantharellus cibarius		*Russula cyanoxantha*
Lycoperdon perlatum		*Suillus variegatus*
Macrolepiota procera		
Pleurotus ostreatus		

hard bound to the fungal tissue. However, the Cu and Se contents decreased with preservation, indicating that soluble forms of these elements were present in the fungus.

In response to the Chernobyl accident, many studies have examined radioactivity in mushrooms, which should not exceed 10 kBq/kg DW according to IAEA [1994]. Mushrooms do not accumulate ^{90}Sr or radioisotopes of plutonium at toxicological levels [Mascanzoni, 1992; Mietelski et al., 1993]. *Macrolepiota procera* displayed the highest ^{103}Ru activity (1.48 kBq/kg DW) in an Austrian study [Teherani, 1987], while negligible levels of ^{110m}Ag were found in Austrian fungal species [Heinrich et al., 1989]. According to Eckl et al. [1986], levels of ^{226}Ra and ^{210}Pb were below the detection limits in edible fungal species; however, calculations regarding ^{210}Pb in a French study of mushrooms indicated that ^{210}Pb contributed to the total effective radioactive dose for humans [Kalac, 2001]. The natural radionuclide ^{40}K is highly accumulated in mushrooms, and accumulation factors ranging from 20 to 40 have been found [Seeger, 1978]. It is likely due to its similarity to K that radiocaesium (^{134}Cs and ^{137}Cs) also accumulates well in mushrooms. Radiocaesium is found to a greater extent in mushrooms from coniferous than from deciduous forests [Heinrich, 1992], and a high soil mobility of radiocaesium due to higher soil moisture increases the level of the isotope in mushrooms. Some accumulators of radiocaesium are shown in Table 6.

8   HOW TO COPE WITH LOW OR HIGH LEVELS OF TRACE ELEMENTS

To avoid high trace element concentrations in cultivated plants, one can change cultivation sites or change the species or genotype cultivated, selecting less-accumulating plants. If possible, one can even decontaminate the soil before cultivation. One way to do so is by phytoextraction, in which a high-accumulating species, preferably an energy crop, is first grown for some years; thereafter, food crops can be grown in the less-contaminated and remediated soil [Greger and Landberg, 1999]. However, this method does not work in soils with high natural background levels of the elements of concern, in which case the only option is to change to a less-accumulating crop or variety.

On the other hand, plants can sometime be deficient in elements; for example, low Zn levels in wheat grains is a problem, in which case growing genotypes efficient at Zn uptake can then be a possibility [Imtiaz et al., 2006]. Another problem is too low levels of Fe in soil causing Fe-deficient plants and lower biomass production; it is then wise to change to a grass crop, as grasses can release hard-bound Fe using by excreting phytosiderophores [Römheld, 1991]. Low selenium levels in soil are a problem in some areas around the world, though there are other areas with very high Se levels. Although Se is not an essential element for plants, it is essential for animals and humans, so Se concentrations in food and fodder plants are important. In the United States, plants cultivated on selenium-rich soil were fed to cattle living in selenium-deficient sites [Banuelos and Mayland, 1999].

REFERENCES

Abedin MJ, Cotter-Howells J, Mehrag AA. 2002a. Arsenic uptake and accumulation in rice (*Oryza sativa* L.) irrigated with contaminated water. *Plant Soil* **240**:311–319.

Abedin MJ, Feldmann J, Meharg AA. 2002b. Uptake kinetics of arsenic species in rice plants. *Plant Physiol* **128**:1120–1128.

Alexander PD, Alloway BJ, Dourado AM. 2006. Genotypic variations in the accumulation of Cd, Cu, Pb and Zn exhibited by six commonly grown vegetables. *Environ Pollut* **144**:726–745.

Almela C, Laparra JM, Velez D, Barbera R, Farre R, Monotoro R. 2005. Arsenosugars in raw and cooked edible seaweed: Characterization and bioaccessibility. *J Agric Food Chem* **53**:7344–7351.

Arao T, Ishikawa S. 2006. Genotypic differences in cadmium concentration and distribution of soybean and rice. *Jarq-Japan Agric Res Quarterly* **40**:21–30.

Asher CJ. 1991. Beneficial elements, functional nutrients, and possible new essential elements. In: Mortvedt JJ, Cox FR, Shuman LM, Welch RM, editors. *Micronutrients in Agriculture*, second edition. Madison, WI: Soil Science Society of America Book Series No. 4, pp. 703–723.

Baeza A, Barandica J, Paniagua JM, Rufo M, Sterling A. 1999. Using ^{226}Ra/^{228}Ra disequilibrium to determine the residence half-lives of radium in vegetation compartments. *J Environ Radioact* **43**:291–304.

Banuclos GS, Mayland HF. 1999. Disposal option for plants used in the phytoremediation of Se-laden soils. Vienna, Proceedings, 5th International Conference on the Biogeochemistry of Trace Elements.

Brooks RR. 1998. *Plants that Hyperaccumulate Heavy Metals.* Wallingford: CAB International.

Chughtai MA. 1995. Effects of water spinach (*Ipomea aquatica*) on nutrient regime and fish growth. AIT Thesis AE-95-38. Bangkok, Asian Institute of Technology.

Crews HM, Davies BE. 1985. Heavy metal uptake from contaminated soils by six varieties of lettuce (*Lactuca sativa* L.). *J Agric Sci* **105**:591–595.

Davis TA, Ramirez M, Mucci A, Larsen B. 2004. Extraction, isolation and cadmium binding of alginate from *Sargassum* spp. *J Appl Phycol* **16**:275–284.

Eckl P, Hofmann W, Türk R. 1986. Uptake of natural and man-made radionuclides by lichens and mushrooms. *Radiat Environ Biophys* **25**:43–54.

Fleming GA, Parle PJ. 1977. Heavy metals in soils, herbage and vegetables from an industrialised area west of Dublin city. *Irish J Agric Res* **16**:35–48.

Göthberg A, Greger M. 2006. Formation of methyl mercury in an aquatic macrophyte. *Chemosphere* **65**:2096–2105.

Göthberg A, Greger M, Bengtsson B-E. 2002. Accumulation of heavy metals in water spinach (*Ipomoea aquatica*) cultivated in the Bangkok region, Thailand. *Environ Toxicol Chem* **21**:1934–1939.

Göthberg A, Greger M, Holm K, Bengtsson B-E. 2004. Influence of nutrient levels on uptake and effects of mercury, cadmium, and lead in water spinach. *J Environ Qual* **33**:1247–1255.

Grant CA, Bailey LD, McLaughlin MJ, Singh BR. 1999. Management factors which influence cadmium concentrations in crops. In: McLaughlin MJ, Singh BR, editors. *Cadmium in Soils and Plants*. Dordrecht: Kluwer Academic Publishers, pp. 151–198.

Greger M, Johansson M, Hamza K, Stihl A. 1993. Foliar uptake of cadmium in sugar beet (*Beta vulgaris*) and Pea (*Pisum sativum*). *Physiol Plant* **88**:563–570.

Greger M. 2004. Metal availability, uptake, transport and accumulation in plants. In: Prasad MNV, editor. *Heavy Metal Stress in Plants—From Molecules to Ecosystems*, second edition. Heidelberg: Springer Verlag, pp. 1–27.

Greger M. 2005. Influence of willow (*Salix viminalis* L.) roots on soil metal chemistry: Effects of clones with varying metal uptake potential. In: Huang PM, Gobran GR, editors. *Biogeochemistry of Trace Elements in the Rhizosphere*. Amsterdam: Elsevier, pp. 301–312.

Greger M, Landberg T. 1999. Use of willow in phytoextraction. *Int J Phytorem* **1**:115–123.

Greger M, Löfstedt M. 2004. Comparison of uptake and distribution of cadmium in different cultivars of bread and durum wheat. *Crop Sci* **44**:501–507.

Heinrich G. 1992. Uptake and transfer factors of ^{137}Cs by mushrooms. *Radiat Environ Biophys* **31**:39–49.

Heinrich G, Müller HJ, Oswald K, Gries A. 1989. Natural and artificial radionuclides in selected Styrian soils and plants before and after reactor accident in Chernobyl. *Biochem Physiol Pflanz* **185**:55–67.

Hellstrand S, Landner L. 1998. Cadmium in fertilisers, soil, crops and food—The Swedish situation. In: *Cadmium Exposure in the Swedish Environment*. Stockholm, KEMI Report.

Herren T, Feller U. 1996. Effect of locally increased zinc contents on zinc transport from the flag leaf lamina to the maturing grains of wheat. *J Plant Nutr* **19**:379–387.

Hocking PJ, McLaughlin MJ. 2000. Genotypic variation in cadmium accumulation by seed of linseed, and comparison with seeds of some other crop species. *Austr J Agric Res* **51**:427–433.

IAEA. 1994. *Intervention Criteria in a Nuclear or Radiation Emergency*. Vienna: International Atomic Energy Agency Safety Series No 109.

Ichikawa S, Kamoshida M, Hanaoka K, Hamano M, Maitani T, Kaise T. 2006. Decrease of arsenic in edible brown algae *Hijikia fusiforme* by the cooking process. *Appl Organometallic Chem* **20**:585–590.

Imtiaz M, Alloway BJ, Khan P, Memon MY, Siddiqui SUH, Aslam M, Shah SKH. 2006. Zinc deficiency in selected cultivars of wheat and barley as tested in solution culture. *Com Soil Sci Plant Anal* **37**:1703–1721.

Intawongse M, Dean JR. 2006. Uptake of heavy metals by vegetable plants grown on contaminated soil and their bioavailability in the human gastrointestinal tract. *Food Add Cont* **23**:36–48.

IUR. 1989. VIth report of the working group on soil-to-plant transfer factors, 24–25 May 1989, Grimselpass Switzerland. p. 65. Internal Union of Radio Ecologists.

Jackson PJ, Unker PJ, Delhaize E, Robinson NJ. 1990. Mechanisms of trace metal tolerance in plants. In: Katterman F, editor. *Environmental Injury to Plants*. San Diego: Academic Press, pp. 231–258.

Jarvis SC, Jones LHP, Hopper MJ. 1976. Cadmium uptake from solution by plants and its transport from roots to shoots. *Plant Soil* **44**:179–191.

Jimenez-Escriq A, Cambrodon IG. 1999. Nutritional evaluation and physiological effects of edible seaweeds. *Arch Latinoamer Nutr* **49**:114–120.

Kabata-Pendias A, Pendias H. 1992. *Trace Elements in Soils and Plants*. Boca Raton, FL: CRC Press.

Kalac P. 2001. A review of edible mushroom radioactivity. *Food Chem* **75**:29–35.

Kuboi T, Noguchi A, Yazaki J. 1986. Family-dependent cadmium accumulation in higher plants. *Plant Soil* **92**:405–415.

Lehoczky É, Szabó L, Horváth Sz. 1998. Cadmium uptake by lettuce in different soils. *Com Soil Sci Plant Anal* **29**:1903–1912.

Lembrechts JF, Desmet GM, Overbeck H. 1985. Molecular mass distribution of technetium complexes in spinach leaves. *Environ Exp Bot* **25**:355–360.

Li YM, Chaney RL, Schneiter AA, Miller JF, Elias EM, Hammond JJ. 1997. Screening for low grain cadmium phenotypes in sunflower, durum wheat and flax. *Euphytica* **94**:23–30.

Lindberg SE, Meyers TP, Taylor GE Jr, Turner RR, Schroeder WH. 1992. Atmosphere-surface exchange of mercury in a forest: Results of modelling and gradient approaches. *J Geophys Res* **97**:2519–2528.

Liu JG, Qian M, Cai GL, Yang JC, Zhu QS. 2007. Uptake and translocation of Cd in different rice cultivars and the relation with Cd accumulation in rice grain. *J Hazard Mate* **143**:443–447.

Liu W-J, Zhu Y-G, Smith FA, Smith SE. 2004. Do iron plaque and genotypes affect arsenate uptake and translocation by rice seedlings (*Oryza sativa* L.) grown in solution culture? *J Exp Bot* **55**:1707–1713.

Liu WJ, Zhu YG, Hu Y, Williams PN, Gault AG, Meharg AA, Charnock JM, Smith FA. 2006. Arsenic sequestration in iron plaque, its accumulation and speciation in mature rice plants (*Oryza sativa* L.). *Environ Sci Technol* **40**:5730–5736.

Lodenius M, Kuusi T, Laaksovirta K, Liukkonen-Llja H, Piepponen S. 1981. Lead, cadmium and mercury contents of fungi in Mikkeli, SE Finland. *Ann Bot Fennici* **18**:183–186.

Lux A, Sottniková A, Opatrná J, Greger M. 2004. Differences in structure of adventitious roots in Salix clones with contrasting characteristics of Cd accumulation and sensitivity. *Physiol Plant* **120**:537–545.

Marin AR, Masscheleyn PH, Patrick WH. 1992. The influence of chemical form and concentration of arsenic on rice growth and tissue arsenic concentration. *Plant Soil* **139**:175–183.

Marschner H. 1995. *Mineral Nutrition of Higher Plants*, second edition. London: Academic Press.

Martin TJ, Juniper EB. 1970. *The Cuticles of Plants*. Edinburgh: Arnold.

Mascanzoni D. 1992. Determination of ^{90}Sr and ^{137}Cs in mushrooms following the Chernobyl fallout. *J Radioanal Nuclear Chem* **161**:483–488.

McLaughlin MJ, Williams CMJ, McKay A, Kirkham R, Gunton J, Jackson KJ, Thompson R, Dowling B, Partington D. 1994. Effect of cultivar on uptake of cadmium by potato tubers. *Austr J Agric Res* **45**:1483–1495.

Mietelski JW, LaRosa J, Ghods A. 1993. [90]Sr and [239+240]Pu, [238]Pu, [241]Am in some samples of mushrooms and forest soils from Poland. *J Radioanal Nuclear Chem* **170**:243–258.

Misheer N, Kindness A, Jonnalagadda SB. 2006. Seaweeds along KwaZulu-Natal coast of South Africa—4: Elemental uptake by edible seaweed Caulerpa racemosa (Sea grapes) and the arsenic speciation. *J Environ Sci Health Part A—Toxic/Hazard Subst Environ Eng* **41**:1219–1235.

Momoshima N, Bondietti EA. 1990. Cation binding in wood: Applications to understanding historical changes in divalent cation availability in red spruce. *Can J Forest Res* **20**:1840–1849.

Nishide E, Anzai H, Uchida N, Nisizawa K. 1996. Changes in M:G ratios of extracted and residual alginate fractions on boiling with water the dried brown alga *Kjellmaniella crassifolia* (Laminariales, Phaeophyta). *Hydrobiology* **327**:515–518.

Oliver DP, Gartrell JW, Tiller KG, Correll R, Cozens GD, Youngberg BL. 1995. Differential responses of Australian wheat cultivars to cadmium concentration in wheat grain. *Aust J Agric Res* **46**:873–886.

Pais I, Jones JB. 1997. *The Handbook of Trace Elements*. Boca Raton, FL: St. Lucie Press.

Pettersson O. 1977. Differences in cadmium uptake between plant species and cultivars. *Swed J Agric Res* **7**:21–24.

Rodushkin I, Ödman F, Holmström H. 1999. Multi-element analysis of wild berries from northern Sweden by ICP techniques. *Sci Total Environ* **231**:53–65.

Römheld V. 1991. The role of phytosiderophores in acquisition of iron and other micronutrients in graminaceous specie: An ecological approach. *Plant Soil* **130**:127–134.

Ruskin FR, Shipley DW. 1976. *Making Aquatic Weeds Useful: Some Perspectives for Developing Countries*. Washington, DC: National Academy of Sciences.

Salim R, Al-Subu MM, Douleh A, Khalaf S. 1992. Effects on growth and uptake of broad beans (*Vicia fabae* L.) by root and foliar treatments of plants with lead and cadmium. *J Environ Sci Health A* **27**:1619–1642.

Seeger R. 1978. Kaliumgehalt höherer Pilze. *Z Lebensm Untersuch Forsch* **167**:23–31.

Shaw G, Bell JNB. 1994. Plants and radionuclides. In: Farago ME, editor. *Plants and the Chemical Elements. Biochemistry, Uptake, Tolerance and Toxicity*. Weinheim: VCH, pp. 179–220.

Shinonaga T, Ambe S. 1998. Multitracer study on absorption of radionuclides in atmosphere—plant model system. *Water Air Soil Pollut* **101**:93–103.

Steffens JC. 1990. The heavy metal-binding peptides of plants. *Annu Rev Plant Physiol Plant Mol Biol* **41**:553–575.

Teherani DK. 1987. Accumulation of [103]Ru, ^{137}Cs and [134]Cs in fruitbodies of various mushrooms from Austria after the Chernobyl incident. *J Radioanal Nuclear Chem* **117**:69–74.

van de Geijn SC, Petit CM. 1979. Transport of divalent cations. Cation exchange capacity of intact xylem vessels. *Plant Physiol* **64**:954–958.

van Loon L. 1986. Kinetic aspects of the soil-to-plant transfer of technetium. Doctoratproefschrift NO 150 aan de Fakultieit der Landbouwwetenschappen van de Katholieke Universiteit te Leuven Belgium.

van Netten C, Cann SAH, Morley DR, van Netten JP. 2000. Elemental and radioactive analysis of commercially available seaweed. *Sci Total Environ* **255**:169–175.

Varo P, Lähelmä O, Nuurtamo M, Saari E. 1980. Mineral element composition of Finnish foods. VII Potato, vegetables, fruits, berries, nuts and mushrooms. *Acta Agric Scand Suppl* **22**:89–113.

Vetter J. 1994. Data on arsenic and cadmium contents of some common mushrooms. *Toxicon* **32**:11–15.

Vetter J. 1997. Chromium and nickel contents of some common edible mushroom species. *Acta Alimentaria* **26**:163–170.

Vetter J, Berta E. 1997. Mercury content of some wild edible mushrooms. *Food Res Techn* **205**:316–320.

Vetter J. 1999. Vanadium content of some common edible, wild mushroom species. *Acta Alimentaria* **28**:39–48.

Vetter J. 2003. Chemical composition of fresh and conserved *Agaricus bisporus* mushroom. *Eur Food Res Techn* **217**:10–12.

Vetter J, Berta E. 2005. Mercury content of the cultivated mushroom *Agaricus bisporus*. *Food Control* **16**:113–116.

Wenzel WW, Blum WEH, Brandstetter A, Jockwer F, Kochl A, Oberforster M, Oberlander HE, Riedler C, Roth K, Vladeva I. 1996. Effects of soil properties and cultivar on cadmium accumulation in wheat grain. *Z Pflantzenernahr Bodenkd* **159**:609–614.

Williams PN, Price AH, Raab A, Hossain SA, Feldmann J, Meharg AA. 2005. Variation in arsenic speciation and concentration in paddy rice related to dietary exposure. *Environ Sci Technol* **39**:5531–5540.

Wolterbeek HT. 1987. Cation exchange in isolated xylem cell walls of tomato. I. Cd^{2+} and Rb^{2+} exchange in adsorption experiments. *Plant Cell Environ* **10**:30–44.

Yamada Y, Bukovac MJ, Wittwer SH. 1964. Ion binding by surfaces of isolated cuticular membranes. *Plant Physiol* **39**:978–982.

Xie ZM, Huang CY. 1998. Control of arsenic toxicity in rice plants grown on an arsenic-polluted paddy soil. *Com Soil Sci Plant Anal* **29**:2471–2477.

7 Trace Elements in Traditional Healing Plants—Remedies or Risks

M. N. V. PRASAD

Department of Plant Sciences, University of Hyderabad, Hyderabad 500 046, India

1 INTRODUCTION

The medicinal and aromatic plants (MAP) sector has traditionally occupied an important position in the sociocultural, spiritual, and ethnopharmacology of rural and tribal lives in different parts of the world. MAPs constitute an effective source of traditional, alternative, and complementary medicine. The World Health Organization (WHO) estimated that 80% of the population of developing countries still relies on traditional medicines, mostly plant drugs for primary health care. Modern pharmacopeia contains at least 25% drugs derived from plants. Many synthetic analogues are based on prototype compounds isolated from plants. Today approximately 70% of "synthetic" medicines are derived from plants. Millions of rural households use medicinal plants in a self-help mode. In recent years, there has been tremendous hype for commercialization of medicinal herbs. More and more people are seeking remedies and health approaches free from the side effects caused by synthesized chemicals. The growing demand for herbal products has led to a quantum jump in the volume of plant materials traded within and across the countries. There were several valid reasons far the modern world of medicine to turn away from traditional herbal products to embrace synthetic drugs. The supply of herbs was subjected to the vagaries of nature, and the potency of different patches of herbs could vary depending on when, where, and how they are procured. The actual dose taken by the patient was also subjected to variation depending on the method of preparation. But the new scientific research in the world of phytomedicine is fueling respect for herbs and leading toward a "Greener economics" and "back to nature in health care" mindset. The Western world has shown an

Trace Elements as Contaminants and Nutrients: Consequences in Ecosystems and Human Health, Edited by M. N. V. Prasad
Copyright © 2008 John Wiley & Sons, Inc.

unprecedented resurgence in herbalism in recent times. Thus, there is a great future for the new plant-based pharmaceuticals.

2 THE INDIGENOUS SYSTEM OF MEDICINE

The traditional system of medicine has survived all vicissitudes and is being practiced in rural and urban areas in biodiversity-rich nations. Historically, the traditional system and the Western system of medicine share the same root; but during the course of civilization and development, these systems diverged. The term "traditional, complementary, and alternative systems of medicine" covers more than 600 unconventional therapies; some of the commonly used Indian medicinal systems are listed below.

Ayurveda: This traditional system of medicine is based on ancient Vedic scriptures, practiced all over India. This system combines holistic therapies, naturopathy, and herbalism for the betterment of the body, the mind, and the spirit.

Siddha: This traditional system of medicine originated from Tamil Nadu with the help of the holistic healers called "Siddars," and this system is extensively practiced in the southern states of India.

Unani: This system of medicine was brought to India by Muslim conquerors and was practiced throughout India for several hundred years. This system includes all natural medicines from plants, animals, and minerals.

Homeopathy: This is also a foreign system of medicine introduced in India many decades ago, and it is characterized by administration of small quantities of medicines with varying concentrations depending upon the symptoms of the disease being treated.

Aromatherapy: This involves the use of essential oils, normally administered by inhalation or massage.

Phytotherapy: This involves the use of medicinal plants and phytoconstituents in treating diseases.

According to an all-India ethnobiological survey carried out by the Ministry of Environment and Forests, Government of India, there are over 8000 species of plants being used by the people of India. Approximately 1800 plant species are used in Ayurveda, 600 in Siddha, 400 in Unani, and 400 in the homoeopathic systems of medicine.

The herbal drugs are derived from the plants of either terrestrial or aquatic origin. Herbal medicines, labeled medicated products that contain active ingredients come from a plant or a combination of plants. The herbal medicines may contain excipients in addition to the active ingredient. Demand for herbal medicine is increasing in both developing and developed countries due to growing recognition of natural products, which are nontoxic, have minimum side effects, and are easily available at affordable prices.

3 HERBAL DRUG INDUSTRY

The abundant sources of many varieties of medicinal plants for herbal medicine formulation have opened many opportunities for the development of unique pharmaceutical products to supply the domestic and export market. The health care systems are turning out to be more and more expensive, and development of technologies to introduce and to integrate herbal medicine systems into our health care is urgently needed. An adequate quality control of herbal drugs is a significant concern for consumers and manufacturers. The herbal drug market is growing at a rapid pace, with an annual rise of 20–30%. The medicinal and aromatic plants play a greater role as alternate sources of income and agriculture in certain nations. The medicinal plants are finding use as pharmaceuticals, neutraceuticals, cosmetics, and food supplements [Chizzola et al., 2003]:

- Plant drugs are used in Indian systems of medicine, including Ayurveda, Unani, and Siddha.
- Over-the-counter, nonprescription products contain plant extracts, such as those used in the essential oil industry, Phytopharmaceuticals, and the cosmaceutical industry.
- Plant extracts are used for natural health products such as (a) health food, (b) neutraceuticals, and (c) recombinant proteins.

The herbal drug industry is expanding globally (Fig. 1).

Asian medical systems (Ayurveda, Siddha, Folk, Unani, Tibetan, Chinese, and some modern systems of medicinal practices) use a wide range of plant products such as flowers, leaves, roots, rhizomes, whole plant, wood, bark, stem, seeds, fruits, and so on, in the preparation of herbal formulations. These are often contaminated with toxic levels of trace elements of natural or anthropogenic origin [Ajasa et al., 2004; Ang and Lee, 2006; Arzani et al., 2007; Caldas and Machado, 2004; Chan, 2003; Dampare et al., 2006; Ernst, 2002; Ernst and Coon, 2001; Haider

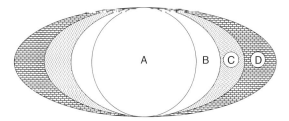

Figure 1. Expanding horizons of herbal medicines and global importance. **A**: Indigenous market for indian systems of medicine (ISM). **B**: Market for ISM where traditional medicine is predominant form of health care (east and central european countries). **C**: Developed country market where ISM is on rising trend (Western Europe, Canada, Australia, New Zealand, Japan. **D**: QC and QA needed to comply the stringent regulations (United States, Germany, Switzerland).

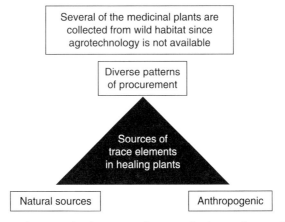

Figure 2. Technogenic contamination, geogenic contamination, and trace element contaminated groundwater irrigation are the major sources of toxic trace elements (arsenic, lead, cadmium) in healing plants collected and cultivated in periurban areas.

et al., 2004; Kabelitz, 1998; Melo et al., 2004; Mohanta et al., 2007; Obi et al., 2006; Obiajunwa et al., 2002; Sathiyamoorthy et al., 1997] (Fig. 2):

(a) Healing plants are often collected from wild habitats that are exposed to trace element pollution.

(b) Trace elements contaminated groundwater irrigation for cultivation of medicinal plants.

Use of plants for primary health care is known since time immemorial. Interactions of plants with trace elements (metals, metalloids, and radionuclide) have several economic and environmental implications and had gained global significance. In wealthy countries, contamination is often highly localized. Heavy metal pollution and contamination is widespread in eastern European and some Asian countries and is dramatically increasing in the developing world—for example, India and China. Hence, in order to develop a herbal drug industry successfully, it must cope with the increasing demand for quality herbal medicine in the domestic and export market and must utilize standard operating procedures (Fig. 3):

1. Several of the trace elements serve as constituents of biomolecules as cofactors for various enzymes and in a variety of metabolic functions.

2. Trace elements also take part in neurochemical transmission (Zn on synaptic transmission).

3. Trace elements accumulated in medicinal plants have healing power in numerous ailments and disorders.

4. Plants with Cr and Mn can be correlated with therapeutic properties against diabetic and cardiovascular diseases.

Implementing standard operating procedures (SOP)
These include:
Good Agricultural Practice (GAP)
Good Laboratory Practice (GLP)
Good Sourcing Practice (GSP)
Good Manufacturing Practice (GMP)
Good Harvesting Practices (GHP)
Good Clinical Trial Practice (GCTP)

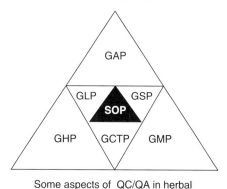

Some aspects of QC/QA in herbal
medicines taking over-the-counter (OTC)

Figure 3. Good management practices for quality control (QC) and quality assurance (QA) of herbal products.

5. Ca^{2+} ions accumulation plays an important role in urinary tract. Hypercalcemia is reported to cause renal failure and forms calcium deposits as stones in the urinary tract.
6. Na- and K-enriched plant parts are partially responsible for the diuretical action (osmotic diuretics).
7. Iron deficiency is common in uremic patients.
8. Zinc supplementation improves the health of uremic patients.
9. Certain transition group elements regulate hepatic synthesis of cholesterol.

4   NOTABLE MEDICINAL AND AROMATIC PLANTS THAT HAVE THE INHERENT ABILITY OF ACCUMULATING TOXIC TRACE ELEMENTS

Plants exhibit four different strategies when exposed to trace elements (Table 1).

Indicators: Plants in which uptake and translocation reflect metal concentration in interstetial water and show toxic symptoms.

Excluders: Plants that restrict the uptake of toxic metals over a wide range of background concentrations.

TABLE 1. Selected Examples of Healing Plants Reported to Accumulate Trace Elements

Botanical Name Family	Common Name	Useful Parts	Chemical Constituents	Medicinal Value	Reference for Metal Accumulation
Allium sepa Liliaceae	Onion	Bulbs	Glycosides	Aromatic, stimulant, expectorant	Wierzbica [1998]
Alternanthera pungense	Ponnganti	Leaves	Vitamin A	Nutraceutical galactagogue	Prasad [2001, 2006]
Alternanthera philoxeroides	Alligator weed	Leaves	Minerals	Nutraceutical	Naqvi and Rizvi [2000]
Amaranthus spinosus Amaranthaceae	Pigweed	Leaves	Sterols	Emetic, laxative, antiseptic	Prasad [2001]
Artemisia vulgaris Asteraceae	Artemisia	Leaves, flowers	Artemesin and cineole	Anti-inflammatory Antimalarial Antispasmodic	Bashmakov et al. [2006]
Bacopa monnieri	Brahmi	Leaves	Bacosides	Brain tonic	Sinha [1999], Singh et al. [2006]
Brassica juncea Brassicaceae	Indian mustard	Seeds	Glycoside, sinigrin, gluconapin	Stimulant, anthelmintic	Palmer et al. [2001]
Calotropis gigantean Asclepiadaceae	Madar	Root, flower, bark, latex	Calotropic	Abortfacient, purgative, diarrhea	Eapen et al. [2006]
Camellia sinensis Theaceae	Tea	Leaves	Theophylline, caffeine	Anti-inflammatory	Flaten [2002]
Cassia siamea	Senna	Leaves		Purgative, liver tonic, jaundice	Kumar et al. [2002], Tripathi et al. [2004]
Centella asiatica Apiaceae	Brahmi	Leaves	Saponin, glycosides	Memory enhancer	Glass [1999]

Plant / Family	Common name	Part used	Constituent	Use	Reference
Chenopodium album Chenopodiaceae	Wild spinach	Leaves	Volatile oils	Hemorrhoids	Bashmakov et al. [2006]
Datura innoxia Solanaceae	Thorn apple	Whole plants	Alkaloids, malic acid	Asthma, antispasmodic	DeWitt et al. [1993]
Eucalyptus citridora Myrtaceae	Eucalyptus	Leaves	Volatile oil	Antiseptic, expectrorant	Bhati and Singh [2003]
Fagopyrum esculentum Polygonaceae	Buck wheat	Grains	Coloring principles	Circulatory problems, diarrhea	Tani and Barrington [2005]
Healianthus annuus Asteraceae	Sunflower	seeds	Oleic and Stearic acids	Diuretic, cough	Prasad [2006]
Hemidesmus indicus	Indain Sarparilla	Roots	—	Demulcent, diaphoretic, and diuretic	Chandrasekhar et al. [2005]
Hypericum perforatum	St. John's wort	—	—	Antidepressant	Gomeza et al. [2004]
Ipomea aquatica Convolvulaceae	Water spinach, Morning glory	Seeds, leaves	Resins	Scabies, purgative	Costa-Pierce [1998]
Lawsonia inermis Lythraceae	Henna	Leaves	Tannins, flavonoids, xanthones	Natural dye, henna Skin disorders and beautification	Nnorom et al. [2005]
Mentha piperata Lamiaceae	Mint	Leaves	Menthol	Antiseptic, carminative	Arzani et al. [2007]
Nerium odoratum Apocynanceae	Nerium	Leaves	Glycosides	Cardiac arrythmiasis	de Jesus et al. [2000], Aksoy and Öztürk [1997]

(Continued)

143

TABLE 1. *Continued*

Botanical Name Family	Common Name	Useful Parts	Chemical Constituents	Medicinal Value	Reference for Metal Accumulation
Ocimum sanctum Lamiaceae	Holy basil	Leaves	Flavonoids	Genitourinary disorder, immunomodulatory agent, anthelmintic, antipyretic	Veeranjaneyulu and Das [1982]
Pelargonium graveolens Geraniaceae	Geranium	Leaves	Geranial	Astringent	Saxena et al. [1999], Dan et al. [2000]
Phyllanthus niruri Euphorbiaceae	Phyllanthus	Fruits	Glycosides	Hepatoprotective	
Salix tetraseperma Salicaceae	Indian willow	Bark	Phenolics, flavoinoids, glycosides	Anti-inflammatory	Unterbrunner et al. [2007]
Solanum nigrum Solanaceae	Black night-shade	Whole plant	Alkaloids	Emollient, laxative	Rezek et al. [2007], Kotrba et al. [1996], Marques et al. [2007]
Taraxacum officinale Asteraceae	Dandelion	Roots and leaves	Phenolic acids, terpenes	Diuretic, liver disorders	Keane et al. [2001], Bhamakov et al. [2006]
Typha latifolia Vetiveria zizanioides Poaceae	(Cat-tail) Vetiver	Rhizome, pollen roots	— Volatile oils	Medicinal value for pollen Refrigerant, stimulants, anti-AIDS	Mirka et al. [1996] Chiu et al. [2006]
Viola Violaceae	Viola	Aerial parts	Flavonoids	Eczema, expectorant	
Zea mays Poaceaea	Corn	Kernals	Starch	Ulcer, diuretic	Prasad [2006]

144

Accumulators: Plants in which uptake and translocation reflect metal concentration in soil or interstetial water without showing toxic symptoms.

Hyperaccumulators: Plants that accumulate elevated levels of trace elements. This would vary for different trace elements (Figs. 4 and 5). Plants will accumulate trace elements, essential nutrients, or contaminants if they are compartmentalized in its parts, depending upon the chemical ligands and transportation mechanisms [Prasad, 1996; Siedlecka, 1995]. The bioaccumulated trace element bioavailability is regulated by various processes (Fig. 6) [Chen et al., 2006; Mehra and Baker, 2007; Zheljazkov et al., 2004].

***Pelargonium* (Scented Geranium):** Saxena et al. [1999] reported that lemon-scented geraniums (*Pelargonium* sp. "Frensham") accumulated large amounts of Cd, Pb, Ni, and Cu from soil in greenhouse experiments.

***Vetiveria zizanioides* (Vetiver Grass):** It is known to have multiple uses. This plant had several popular names such as "the miracle grass," "a wonder grass," "a magic grass," "a unique plant," "an essential grass," "an amazing plant," "an amazing grass," "a versatile plant," "a living barrier," "a living dam," "a living nail," "a living wall," and "an eco-friendly grass." This extraordinary grass is adaptable to multiple environmental conditions, and it is globally recognized as an easy and economical alternative to control soil erosion and to solve a variety of environmental problems. It has been used for restoration, conservation, and protection of land disrupted by human activities like agriculture, mining,

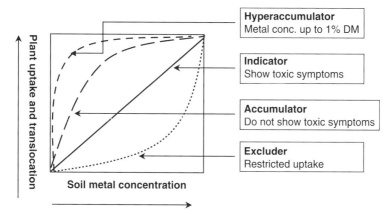

Figure 4. Plants exhibiting four different strategies when exposed to elevated concentration metals. Indicators: Plants in which uptake and translocation reflect metal concentration in interstitial water and show toxic symptoms. Excluders: Plants restrict the uptake of toxic metals into shoot over a wide range of background concentrations. Accumulators: Plants in which uptake and translocation reflect metal concentration in soil or interstitial water without showing toxic symptoms. Hyperaccumulators: Plants that show elevated levels of trace element. This would vary for different trace elements.

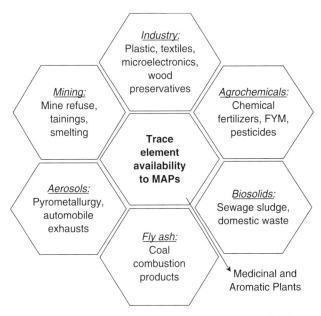

Figure 5. Trace element sources as contaminants to MAPs.

construction sites, oil exploration and extraction, and infrastructure corridors, and it has been used for water conservation in watershed management, disaster mitigation, and treatment of contaminated water and soil. Research at the global level has proved the relevance of vetiver in multiple applications [Chiu et al., 2005, 2006].

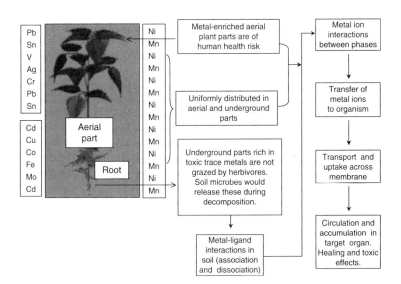

Figure 6. Bioavailability of trace elements.

The vetiver grass has rather unique morphological and physiological characteristics: It has the ability to control erosion and sedimentation, to withstand extremely soil and climatic variations, and to tolerate elevated heavy metals concentrations in water and soil. Vetiver is reportedly nonevasive and fosters the voluntary return of native plants. It is widely used in environmental remediation projects all over the world. Vetiver grass is known for its effectiveness in erosion and sediment control and is known to accumulate toxic trace elements [Pichai et al., 2001].

In Australia, *V. zizanioides* has been successfully used to stabilize mining overburden and highly saline, sodic, magnesic, and alkaline (pH 9.5) tailings of coalmines, as well as highly acidic (pH 2.7) arsenic tailings of gold mines. In China, it has been demonstrated that *V. zizanioides* is one of the best choices for revegetation of Pb/Zn mine tailings due to its high metal tolerance [Truong, 2000].

Hemidesmus indicus: Roots of this plant have several medicinal and economic uses. Recently, Chandrasekhar et al. [2005] reported its phytoextraction function for recovery of lead from industrially contaminated soils of Hyderabad, India.

Hydrocotyle umbellata (**Pennywort**): It is reported to remove heavy metals from soil and aqueous effluents [Glass, 1999].

Alternanthera philoxeroides (**Alligator weed**): It is one of the most common aquatic weeds in contaminated/polluted ecosystems. This is native to South America and is naturalized in India. Iron content is very high (2%) and is used as a leaf vegetable. It is reported to accumulate elevated doses of lead and mercury from polluted waters [Prasad, 2001]. *A. pungense* grows on sewage sludge-laden soils and is accredited with galactagogue properties, which increase the flow of milk in the cattle. It is used for treating night blindness and contains carotene ($\sim$200 μg/100 g). The leaves are reported to contain protein 5% and iron $\sim$16 mg/100 g.

Bacopa monnieri: (Brahmi) is a rooted emergent aquatic macrophyte, known for its application as brain tonic. In natural ecosystems it is reported to bioconcentrate cadmium [Singh et al., 2006].

Some plants accumulate trace elements through leaves, not necessarily through only roots (Fig. 7).

Agrotechnologies Available for the Following Medicinal and Aromatic Plants: *Artemisis annua* (Artemisin), *Cymbopogon winterianus* (Citronella), *C. martinii* (Palmarosa), *C. flexuosus* (Lemongrass), *Cassia angustifolia* (Senna), *Chrysanthemum cineraefolium* (Pyrethrum), *Mentha arvensis* (Menthol mint), *M. citrata* (Bergamot mint), *M. piperate* (Pepper mint), *M. spicata* (Spear mint), *M. cardiaca* (Scotch spear mint), *Ocimum sanctum* (Tulsi), *Pelargonium graveolens* (Geranium), *Phyllanthus amarus* (Bhumi amlai), *Vetiveria zizanioides* (Vetiver), and *Withania somnifera* (Ashwagandha). However, cultivation in peri-urban areas usage of

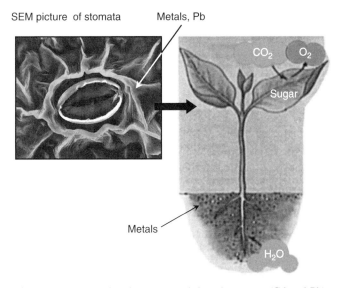

Figure 7. Uptake of toxic trace metal though stomata (Cd and Pb).

Figure 8. (**a**, **b**) *Hemidesmus indicus*, (**c**) *Centella asiatica*, (**d**) *Vetiveria zizanioides*. Vetiver grass has multiple uses: the miracle grass," "a wonder grass," "a magic grass," "a unique plant," "an essential grass," "an amazing plant," "an amazing grass," "a versatile plant," "a living barrier," "a living dam," "a living nail," "a living wall," "an eco-friendly grass." (**e**) *Ipomea sp.* and *Typha latifolia* (Cattail).

Figure 9. (**a, b**) *Alternanthera philoxeroides* (alligator weed). (**c, d**) Nutraceuticals sold in leafy vegetables in market—for example, *Alternanthera sessiles* and *A. pungens*.

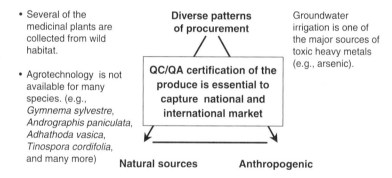

- Several of the medicinal plants are collected from wild habitat.

- Agrotechnology is not available for many species. (e.g., *Gymnema sylvestre*, *Andrographis paniculata*, *Adhathoda vasica*, *Tinospora cordifolia*, and many more)

Diverse patterns of procurement

QC/QA certification of the produce is essential to capture national and international market

Natural sources   **Anthropogenic**

Groundwater irrigation is one of the major sources of toxic heavy metals (e.g., arsenic).

Figure 10. Trace element contamination is a major impediment to herbal drug industry.

contaminated groundwater for irrigation involve risk of trace element contamination (Figs. 8–10) [Prasad, 20007].

5 CLEANUP OF TOXIC METALS FROM HERBAL EXTRACTS

Toxic trace metals from aqueous extracts of two most commonly used plants, namely, *Vitis vinifera* (raisin) and *Nordostachys jatamansi* (Jatamansi), have been cleaned using granulated *Cladosporium cladosporioides* # 2 as biosorbent [Pethkar et al., 2001]. Analysis of herbal extracts before and after contact with biosorbent significantly removed the toxic metal such as lead and cadmium (Fig. 11). Likewise, *Azadirachta indica* has been used as an effective biosorbent for removal of lead(II) from aqueous solutions [Athar et al., 2007].

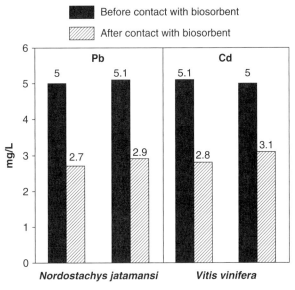

Figure 11. Toxic trace metals from aqueous extracts of two most commonly used herbal plants; that is, *Nordostachys jatamansi* (Jatamansi) and *Vitis vinifera* (raisin) have been removed using granulated *Cladosporium cladosporioides* #2 as biosorbent. Analysis of herbal extracts before and after contact with biosorbent significantly removed the toxic metal such as lead and cadmium. (Data source: Pethkar et al. [2001].)

The traditional medicinal system practiced in India for several centuries is well known as Ayurveda. Ayurveda is the first of its kind and hopes to promote global participation and generate ideas for further research. Ayurveda is now being honored as the wisdom of the great sages and has evolved greatly since ancient times. It has to grow into a system appropriate to the present age and pertinent to global vision. Ayurvedic cures and therapies are used for various diseases and disorders.

Standardization, quality control, and quality assurance of herbal formulations are difficult to achieve. Traditional healers who had the knowledge of medicinal herbs that are effective in certain diseases are not able to obtain a manufacturing license for it. There is a Wildlife Protection Act that prevents the use of certain herbal materials. There is a scarcity of widely used raw materials, numbering around 500–600. No efforts have been made to promote medicinal plant cultivation.

6   POLYHERBAL PREPARATION AND TRADITIONAL MEDICINE PHARMACOLOGY

Ayurvedic pharmacology (Dravyaguna) cannot be explained in terms of modern pharmacological evaluation methods. In Ayurveda, each drug has distinct features like rasa, guna, veerya, vipaka, and prabhava, and these qualities vary from plant to plant and from part to part of the plant. The therapeutic potency of individual drugs varies according to the place, season, and time of collection of the raw drug and according to the form in which the medicines are dispensed (i.e., kashaya, gutika arishta, etc.). According to their therapeutic efficacy and features, the drugs are grouped in different categories or vargas.

A particular varga is selected for the treatment of diseases of a particular system. "There should be compatibility between the dravyagunas and doshas predominant in a patient to obtain maximum therapeutic efficacy of the drug. Hence ayurvedic pharmacotherapy is *more* individualized according to the doshic predominance of the patient and not generalized as in the case of modern medicine. Modern medical practitioners criticize the polyherbal formula for their irrational combination of several in gredients. But these polyherbal formulations have synergistic, potentiative, agnostic, and antagonistic pharmacological action within itself due to the incorporation of plant medicines with diverse pharmacological action [Siedlecka, 1995]. These pharmacological principles work together in a dynamic way to produce maximum therapeutic efficacy with minimum side effects. The food–drug and drug–drug interactions have also been criticized by modern medical practitioners. This food–drug theory is as old as the ayurvedic system itself in ayurvedic pharmacotherapy. This is known as "pathyam," which means that a patient is put on a particular drug regime. The patient is advised to totally avoid consumption of certain foods. It is well established in science that phytate decreases the absorption of dietary zinc and iron; milk and calcium given with tetracyclines reduce the absorption of antibiotics; and so on. In certain cases of allergic diseases when the identity of the allergen is not known, medical practitioners advise patients to avoid certain foods in their diet. Ayurveda restricts intake of tamarind with mineral preparations since tamarind contains tannins and tartaric acid, which will reduce the absorption of minerals. According to this medicinal system, metal-based drugs known as "bhasma" involve the conversion of a metal into its mixed oxides. During these transformations, the zero valent metal state gets converted into a form with higher oxidation state, and the most important aspect of this synthesis (known traditionally as "bhasmikarana") is that the toxic nature (i.e., systemic toxicity causing nausea, vomiting, stomach pain, etc.) of the resulting metal oxide is completely destroyed while inducing the medicinal properties into it. The drugs known as "bhasmas" (ash) are well known in the traditional Indian Ayurveda, and these are chemically mixed oxides of one or more metals. Their traditional preparation involves conversion of a pure metal into its oxide form following a typical procedure, available in the ancient literature of Ayurveda. Such a procedure is believed to eliminate the harmful properties of metal oxides while inducing the medicinal properties into it. The critical steps in the preparation of "bhasma" are (i) trituration of a pure metal with various plant juices for several hours and (ii) repeated cycles of high-temperature calcination in a sealed earthen pot. In this process the phytochemicals lose their

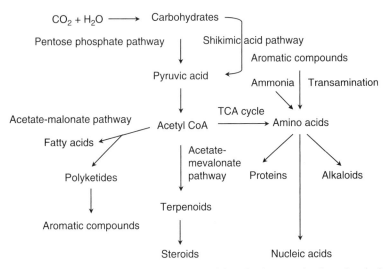

Figure 12. Major metabolic pathways synthesizing the therapeutic phytochemicals.

organic nature and are convered into inorganic form or a concentrate of trace elements (Fig. 12).

7   CONCLUSIONS

Global market for herbal medicines, functional foods (nutraceuticals), and a variety of biofortified agricultural products is a subject of contemporary interest. Plants that accumulate metals and metalloids have gained considerable significance and are implicated with healing function. Herbal drugs and functional foods that are known to have healing function require quality control and quality assurance that makes legislation set limits regarding heavy metal concentration.

Collection and cultivation of healing plants in periurban area often elevated levels of have trace elements that would pose health risks [Chiu et al., 2006; Ernst, 2002; Kabelitz, 1998] or may serve as remedy for which detailed investigations and standardization of pharmacopeias is needed. Trace element contamination in healing plants is risk as well as remedy. Therefore, standardization of herbal drugs, quality control, and quality assurance and certification is of paramount significance for the growing global herbal drug industry.

Plants that accumulate essential trace elements are implicated in propelling metabolic processes (metallomics) (Fig. 13) [Preuss et al., 1998]. Since the bioavailability depends critically on the actual species of an element present, it is becoming more and more evident that information on the concentration of an element in an herbal drug formulation, functional foods, or nutraceutical would explain the importance of a given heavy metal and its role in metabolic processes. Therefore, the precise

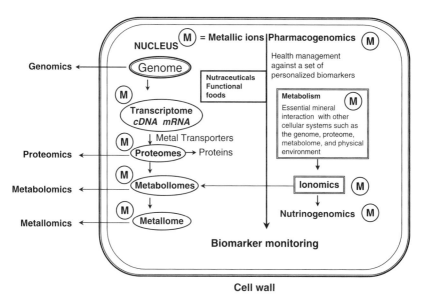

Figure 13. "-Omics" is a multidisciplinary metal-assisted functional biogeochemistry for understanding the combo therapeutic functions of herbo-minerals.

information regarding the nature and concentration of individual heavy metal present in a sample is required. Thus species-related information is of crucial value. Lack of sophisticated analytical instruments to analyze the concentration of several species in a given sample of herbal formulation is an impediment. Therefore, species-selective and sensitive analysis of heavy metals in herbal preparations—for instance, "bhasmas" (ash of the medicinal herbs, sometimes mixture of several herbs)—is of paramount importance. These "bhasmas" (traditional preparation process is known as "bhasmikarana") are chemically mixed oxides of one or more metals; in other words, these are metal-based drugs. During "bhasmikarana," chemical transformations take place so that the toxic nature of the resulting metal oxide is suppressed (i.e., toxicity causing nausea, vomiting, stomach pain, etc.) while express the healing functions [Razic et al., 2005; Wadekar et al., 2005]. Herbal preparations need to be scientifically validated and standardized by emerging tools such as metallomics, reverse pharmacology, and systems biology.

Traditional cosmetics used in Saudi Arabia, India, Morocco, Mauritania, Pakistan, India, and many other countries are reported to contain trace metals [Nnorom et al., 2005]. There is an increased consciousness regionally and globally in the production and use of plants with healing properties. Hence, analytical instrumentation is required for species-selective analysis for trace elements in these commodities (Table 2). Prominent lacunae in quality control and quality assurance related to trace elements in medicinal plant products and functional foods, along with standardization of relevant pharmacopeia, are urgently needed (Fig. 14).

TABLE 2. Sophisticated Analytical Techniques Required for Quality Control and Quality Assurance of Herbal Drugs and for Metallomics Research [Chow et al., 1995; Chuang et al., 2000; Dim et al., 2004; Drăsara and Moravcovaa, 2004; Ozcan et al., 2007; Obaiajunwa et al., 2002; Pichai et al., 2001; Prasad, 2006; Raju et al., 2006; Razic et al., 2003, 2005; Kelly et al., 2002; Salvador et al., 2003; Xia and Rayson, 2000; Yamashitaa et al., 2005]

Analytical technique	QC/QA of herbal drugs and for metallomics research
Ultratrace analysis, all-elements analysis, one atom detection, one-molecule detection	Distributions of the elements in the biological fluids, cell, organs, etc.
Hyphenated methods (LC-ICP-MS, GC-ICP-MS, MALDI-MS, ES-MS)	Chemical speciation of the elements in the biological samples and systems
X-ray diffraction analysis, EXAFS	Structural analysis of metallomes (metal-binding molecules)
NMR, XPS, laser-Raman spectroscopy, DNA sequencer, amino acids sequencer, time-resolution and spatial-resolution fluorescence detection	Elucidation of reaction mechanisms of metallomes using model complexes (bioinorganic chemistry)
LC-ES-MS, LC-MALDI-MS, LC-ICP-MS	Identification of metalloproteins and metalloenzymes
LC, GC, LC-MS, GC-MS, ES-MS, API-MS (atmospheric pressure chemical ionization mass spectrometry)	Metabolisms of biological molecules and metals (metabollomes, metabolites)
ICP-AES, ICP-MS, graphite-furnace AAS, autoanalyzer, spectrophotometry	Medical diagnosis of health and disease related to trace metals on a multielement basis
LC-MS, LC-ICP-MS, stable isotope tracers	Elucidation of phytotherapeutics
Isotope ratio measurement (chronological techniques, DNA sequencer)	Chemical evolution of the living systems and organisms on the earth
In situ analysis, immunoassay, bioassay, food analysis, clinical analysis	Metal-assisted biological functions biosciences in medicine, toxicology, and clinical and nutritional biochemistry
Mössbauer spectroscopy, X-ray photoelectron spectroscopy (XPS), electron spin resonance spectroscopy (ESR), mass or tandem mass spectrometry (MS or MS/MS) (chromatography or electrophoresis). nonspecific detectors [UV, flame ionization detector (FID)], atomic absorption spectrometry (AAS), microwave-induced plasma (MIP), inductively coupled plasma (ICP), capillary zone electrophoresis (CZE), matrix-assisted laser desorption ionization (MALDI)	Species-selective analysis of trace metals in herbal products for QC/QC

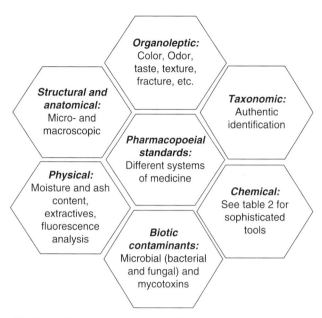

Figure 14. Selected parameters to standardize herbal products for QC and QA.

REFERENCES

Ajasa AMO, Bello MO, Ibrahim AO, Ogunwande IA, Olawore NO. 2004. Heavy trace metals and macronutrients status in herbal plants of Nigeria. *Food Chemistry* **85**:67–71.

Aksoy A, Öztürk MA. 1997. *Nerium oleander* L. as a biomonitor of lead and other heavy metal pollution in Mediterranean environments. *Sci Total Environ* **205**:145–150.

Ang HH, Lee KL. 2006. Contamination of mercury in tongkat Ali hitam herbal preparations. *Food Chem Toxicol* **44**:1245–1250.

Arzani A, Zeinali H, Razmjo K. 2007. Iron and magnesium concentrations of mint accessions (*Mentha* spp.) *Plant Physiol Biochem* **45**:323–329.

Athar M, Farooq U, Hussain B. 2007. *Azadirachta indica* (Neem): An effective biosorbent for the removal of lead (II) from aqueous solutions. *Bull Environ Contam Toxicol* **79**:288–292.

Bashmakov DI, Lukatkin AS, Prasad MNV. 2006. Temperate weeds in Russia serve as sentinels for monitoring trace element pollution. In: Prasad MNV, Sajwan KS, Naidu R, editors. *Trace Elements in the Environment: Biogeochemistry, Biotechnology and Bioremediation.* Boca Raton, FL: CRC Press (Taylor and Francis), Chapter 23, pp. 439–450.

Bhati M, Singh G. 2003. Growth and mineral accumulation in *Eucalyptus camaldulensis* seedlings irrigated with mixed industrial effluents. *Bioresource Technol* **88**:221–228.

Cala V, Cases MA, Walter I. 2005. Biomass production and heavy metal of *Rosmarinus officinalis* grown on organic waste-amended soil. *J Arid Environ* **62**:401–412.

Caldas ED, Machado LL. 2004. Cadmium, mercury and lead in medicinal herbs in Brazil. *Food Chem Toxicol* **42**:599–603.

Chakraborty D, Maji S, Bandyopadhyay A, Basu S. 2007. Biosorption of cesium-137 and strontium-90 by mucilaginous seeds of *Ocimum basilicum*. *Bioresource Technol* **98**:2949–2952.

Chan K. 2003. Some aspects of toxic contaminants in herbal medicines. *Chemosphere* **52**:1361–1371.

Chandrasekhar K, Kamala CT, Chary NS, Balaram V, Garcia G. 2005. Potential of *Hemidesmus indicus* for phytoextraction of lead from industrially contaminated soils. *Chemosphere* **58**:507–514.

Chen YM, Kuang M, Zhuang SY, Chiang PN. 2006. Chemical and physical properties of rhizosphere and bulk soils of three tea plants cultivated in Ultisols. *Geoderma* **136**:378–387.

Chiu KK, Ye ZH, Wong MH. 2005. Enhanced uptake of As, Zn, and Cu by *Vetiveria zizanioides* and *Zea mays* using chelating agents. *Chemosphere* **60**:1365–1375.

Chiu KK, Ye ZH, Wong MH. 2006. Growth of *Vetiveria zizanioides* and *Phragmities australis* on Pb/Zn and Cu mine tailings amended with manure compost and sewage sludge: A greenhouse study. *Bioresource Technol* **97**:158–170.

Chizzola R, Michitsch H, Franz C. 2003. Monitoring of metallic micronutrients and heavy metals in herbs, spices and medicinal plants from Australia. *Eur Food Res Technol* **216**:407–411.

Chow PYT, Chua TH, Tang KF. 1995. Dilute acid digestion procedure for the determination of lead, copper, and mercury in traditional Chinese medicines by atomic absorption spectrometry. *Analyst* **120**:1221–1223.

Chuang IC, Chen KS, Huang YL, Lee PN, Lin TH. 2000. Determination of trace elements in some natural drugs by atomic absorption spectrometry. Biol Trace Elem Res **76**:235–244.

Costa-Pierce BA. 1998. Preliminary investigation of an integrated aquaculture–wetland ecosystem using tertiary-treated municipal wastewater in Los Angeles County, California. *Ecol Eng* **10**:341–354.

Dampare SB, Ameyaw Y, Adotey DK, Osae S, Serfor-Armah Y, Nyarko BJB, Adomako D. 2006. Seasonal trend of potentially toxic trace elements in soils supporting medicinal plants in the eastern Region of Ghana. *Water, Air, and Soil Pollution* **169**:185–206.

Dan TV, Raj KS, Saxena PK. 2000. Metal tolerance of scented geranium (*Pelargonium* sp. "Frensham"): Effects of cadmium and nickel on chlorophyll fluorescence kinetics. *Int J Phytoremediation* **2**:91–104.

de Jesus EFO, Simabuco SM, dos Anjos MJ, Lopes RT. 2000. Synchrotron radiation X-ray fluorescence analysis of trace elements in *Nerium oleander* for pollution monitoring. *Spectrochimica Acta Part B: Atomic Spectroscopy* **55**:1181–1187.

DeWitt JG, Kuske CR, Jackson PJ. 1993. Characterization of metal binding sites of cell wall components from *Datura innoxia*. *J Inorg Biochem* **51**:568.

Dim IA, Funtua II, Oyewale AO, Grass F, Umar IM, Gwozdz R, Gwarzo US. 2004. Determination of some elements in *Ageratum conyziodes*, a tropical medicinal plant, using instrumental neutron activation analysis. *J Radioanal Nucl Chem* **261**:225–228.

Drăsara P, Moravcovaa J. 2004. Recent advances in analysis of Chinese medical plants and traditional medicines. *J Chromatogr B* **812**:3–21.

Eapen S, Singh S, Thorat V, Kaushik CP, Raj K, D'Souza SF. 2006. Phytoremediation of radiostrontium (^{90}Sr) and radiocesium (^{137}Cs) using giant milky weed (*Calotropis gigantea* R.Br.) plants, *Chemosphere* **65**:2071–2073.

El-Rjoob AWO, Massadeh AM, Omari MN. 2008. Evaluation of Pb, Cu, Zn, Cd, Ni and Fe levels in *Rosmarinus officinalis* labaiatae (Rosemary) medicinal plant and soils in selected zones in Jordan. *Environ Monit Assess* **140**:61–68.

Ernst E. 2002. Toxic heavy metals and undeclared drugs in Asian herbal medicines. *Trends in Pharmacol Sci* **23**:36–139.

Ernst E, Coon TJ. 2001. Heavy metals in traditional Chinese medicines: A systematic review. *Clin Pharmacol Ther* **70**:497–504.

Feng-Ying Z, Li Shun-Xing L, Lu-Xiu L. 2007. Assessment of bioavailability and risk of iron in phytomedicines *Aconitum carmichaeli* and *Paeonia lactiflora*. *J Trace Elements Med Biol* **21**:77–83.

Flaten TP. 2002. Aluminium in tea—concentrations, speciation and bioavailability. *Coordination Chem Rev.* **228**:385–395.

Glass DJ. 1999. *US and International Markets for Phytoremediation*, 1999–2000. Needham, MA: DJ Glass Associates, Inc., p. 266.

Gomez MR, Cerutti S, Sombra L, Silva MF, Martinez LD. 2007. Determination of heavy metals for the quality control in argentinian herbal medicines by ETAAS and ICP-OES. *Food Chem Toxicol* **45**:1060–1064.

Gomeza MR, Cerutti S, Olsina RA, Silva MF, Martınez LD. 2004. Metal content monitoring in *Hypericum perforatum* St. John's Wort antidepressant pharmaceutical derivatives by atomic absorption and emission spectrometry. *J Pharmaceut Biomed Anal* **34**:569–576.

Haider S, Naithani V, Barthwal J, Kakkar P. 2004. Heavy metal content in some therapeutically important medicinal plants. *Bull Environ Contam Toxicol* **72**:119–127.

Jonak C, Nakagami H, Hirt H. 2004. Heavy metal stress—activation of mitogen-activated protein kinase by copper and cadmium. *Plant Physiol* **136**:3276–3283.

Jyothi NVV, Mouli PC, Reddy SRJ. 2003. Determination of zinc, copper, lead and cadmium in some medicinally important leaves by differential pulse anodic stripping analysis. *J. Trace Elem Med Biol* **17**:79–83.

Kabelitz L. 1998. Heavy metals in herbal drugs. *Eur J Herb Med* **4**:25–29.

Keane B, Collier MH, Shann JR, Rogstad SH. 2001. Metal content of dandelion (*Taraxacum officinale*) leaves in relation to soil contamination and airborne particulate matter. *Sci Total Environ* **281**:63–78.

Kelly RA, Andrews JC, DeWitt JG. 2002. An X-ray absorption spectroscopic investigation of the nature of the zinc complex accumulated in *Datura innoxia* plant tissue culture. *Microchem J* **71**:231–245.

Kotrba P, Macek T, Skacel F, Ruml T. 1996. Accumulation of cadmium by hairy-root culture of *Solanum nigrum* from nutrient medium. *International Biodeterioration & Biodegradation* **37**:240.

Kumar P, Vajpayee MB, Ali RD, Tripathi N, Singh UN, Rai SN, Singh. 2002. Biochemical responses of *Cassia siamea* Lamk grown on coal combustion residue (fly-ash). *Bull Environ Contam Toxicol* **68**:675–683.

Lin S, Drake LR, Rayson GD. 2002. Affinity distributions of lead ion binding to an immobilized biomaterial derived from cultured cells of *Datura innoxia*. *Adv Environ Res* **6**:523–532.

Mamani MVC, Aleixom LM, de Abreu MF, Susanne Rath S. 2005. Simultaneous determination of cadmium and lead in medicinal plants by anodic stripping voltammetry. *J Pharmaceut Biomed Anal* **37**:709–713.

Marques APGC, Rui S, Oliveira RS, Samardjieva KA, Pissarra J, Rangel AOSS, Castro PML. 2007. *Solanum nigrum* grown in contaminated soil: Effect of arbuscular mycorrhizal fungi on zinc accumulation and histolocalisation. *Environ Pollut* **145**:691–699.

Mehra A, Baker CL. 2007. Leaching and bioavailability of aluminium, copper and manganese from tea (*Camellia sinensis*) *Food Chem* **100**:1456–1463.

Melo JS, D'Souza SF. 2004. Removal of chromium by mucilaginous seeds of *Ocimum basilicum*. *Bioresource Technol* **92**:151–155.

Melo JS, D'Souza SF, Ajasa AMO, Bello MO, Ibrahim AO, Ogunwande IA, Olawore NO. 2004. Heavy trace metals and macronutrients status in herbal plants of Nigeria. *Food Chem* **85**:67–71.

Mirka MA, Clulow FV, Davé NK, Lim TP. 1996. Radium-226 in cattails, *Typha latifolia*, and bone of muskrat, *Ondatra zibethica* (L.), from a watershed with uranium tailings near the city of Elliot Lake, Canada. *Environ Pollut* **91**:41–51

Mohanta B, Chakraborty A, Sudarshan M, Dutta RK, Baruah M. 2003. Elemental profile in some common medicinal plants of India. Its correlation with traditional therapeutic usage. *J Radioanal Nucl Chem* **258**:175–179.

Morita A, Horie H, Fujii Y, Takatsu S, Watanabe N, Yagi A, Yokota H. 2004. Chemical forms of aluminum in xylem sap of tea plants (*Camellia sinensis* L.) *Phytochemistry* **65**:2775–2780.

Murch SJ, Haq K, Rupasinghe HPV, Saxena PK. 2003. Nickel contamination affects growth and secondary metabolite composition of St. John's wort (*Hypericum perforatum* L.) *Environ Exp Bot* **49**:251–257.

Naidu GRK, Denschlag HO, Mauerhofer E, Porte N, Balaji T. 1999. Determination of macro, micro nutrient and trace element concentrations in Indian medicinal and vegetables leaves using instrumental neutron activation analysis. *Appl Radiat Isot* **50**:947–953.

Naqvi SM, Rizvi SA. 2000. Accumulation of chromium and copper in three different soils and bioaccumulation in an aquatic plant, *Alternanthera philoxeroides*. *Bull Environ Contam Toxicol* **65**:55–61.

Nnorom IC, Igwe JC, Oji-Nnorom CG. 2005. Trace metal contents of facial (make-up) cosmetics. *Afr J Biotech* **4**:1133–1138.

Obi E, Akunyili DN, Ekpo B, Orisakwe OE. 2006. Heavy metal hazards of Nigerian herbal remedies. *Sci Total Environ* **369**:35–41.

Obaiajunwa EI, Adebajo AC, Omobuwajo OR. 2002. Essential and trace element contents of some Nigerian medicinal plants. *J Radioanal Nuclear Chem* **252**:473–476.

Olabanji SO, Omobuwajo OR, Ceccato D, Buoso MC, De Poli M, Moschini G. 2006. Analysis of some medicinal plants in South-western Nigeria using PIXE. *J Radioanal Nuclear Chem* **270**:515–521.

Özcan MM and Akbulut M. 2008. Estimation of minerals, nitrate and nitrite contents of medicinal and aromatic plants used as spices, condiments and herbal tea. *Food Chem* **106**:852–858.

Özcan MM, Ünver A, Uçar T, Arslan D. 2007. Mineral content of some herbs and herbal teas by infusion and decoction. *Food Chem* **120**:1221–1223.

Palmer CE, Warwick S, Keller W. 2001. Brassicaceae (Cruciferae) family—plant biotechnology and phytoremediation. *Int J Phytoremed* **3**:245–287.

Pichai NMR, Samjiamjiaras R, Thammanoon H. 2001. The wonders of a grass, Vetiver and its multifold applications. *Asian Infrastruct Res Rev* **3**:1–4.

Pethkar AV, Gaikaiwari RP, Paknikar KM. 2001. Biosorptive removal of contaminating heavy metals from plant extracts of medicinal value. *Curr Sci* **80**:1216–1219.

Prasad MNV. 1996. Variable allocation of heavy metals in phytomass of crop plants—Human health implications. In: Sobotka P, Eybl V, editors. *Plzensky Lékarsky Sbornik*, Suppl. 71 Prague: Czech Republic Medical Association. Charles University Press, pp. 19–22.

Prasad MNV. 2001. Bioremediation potential of amaranthaceae. In: Leeson A, Foote EA, Banks MK, Magar VS, editors. *Phytoremediation, Wetlands, and Sediments,* Vol. 6(5), *Proceedings 6th International In Situ and On-Site Bioremediation Symposium.* Columbus, OH: Battelle Press, pp. 165–172.

Prasad MNV. 2006. Trace elements in medicinal plants: Environmental quality and human health concerns. *Ecol Eng* **16**:56–58.

Prasad MNV. 2007. Trace element accumulation in medicinal and aromatic plants collected and cultivated in periurban area risk or remedy. In: Zhu Y, Lepp N, Naidu R, editors. *Biogeochemistry of Race Elements: Environmental Protection, Remediation And Human Health.* China: Tsinghua University Press, pp. 105–106.

Preuss HG, Jarrell ST, Scheckenbach R, Lieberman S, Anderson RA. 1998. Comparative effects of chromium, vanadium and *Gymnema sylvestre* on sugar-induced blood pressure elevations in SHR. *J Am College Nutr* **17**:116–123.

Raju GJN, Sarita P, Murty GAVR, Kumar MR, Reddy BS, Charles MJ, Lakshminarayana S, Reddy TS, Reddy SB, Vijayan V. 2006. Estimation of trace elements in some anti-diabetic medicinal plants using PIXE technique. *Appl Radiation Isotopes* **64**:893–900.

Razic S, Onjia A, Potkonjak B. 2003. Trace elements analysis of *Echinacea purpurea*—Herbal medicinal. *J Pharmaceut Biomed Anal* **33**:845–850.

Razic S, Onjia A, Dogo S, Slavkovic L, Popovic A. 2005. Determination of metal contant in some herbal drugs—Empirical and chemomimetric approach. *Talanta* **67**:233–239.

Rezek J, Macek T, Mackova M, Triska J. 2007. Plant metabolites of polychlorinated biphenyls in hairy root culture of black nightshade *Solanum nigrum* SNC-9O. *Chemosphere* **69**: 1221–1227.

Salvador MJ, Dias DA, Moreira S, Zucchi O. 2003. Analysis of medicinal plants and crude extracts by synchroton radiation total reflection X-ray fluorescence. *J Trace Microprobe Tech* **21**:377–388.

Sathiyamoorthy P, Van Damme P, Oven M, Golan-Goldhirsh A. 1997. Heavy metals in medicinal and fodder plants of the Negev desert. *J Environ Sci Health* **32**:2111–2123.

Saxena PK, Raj KS, Dan T, Perras MR, Vettakkorumakankav NN. 1999. Phytoremediation of metal contaminated and polluted soils. In: Prasad MNV, Hagemeyer J, editors. *Heavy Metal Stress in Plants. From Molecules to Ecosystems.* New York: Springer, pp. 305–329.

Serfor-Armah Y, Nyarko BJB, Akaho EHK, Kyere AWK, Osae S, Oppong-Boachie K, Osae EKJ. 2001. Activation analysis of some essential elements in five medicinal plants used in Ghana. *J Radioanal Nuclear Chem* **250**:173–176.

Sertie JAA, Carvalho JCT, Panizza S. 2000. Antiulcer activity of the crude extract from the leaves of *Casearia sylvestris*. *Pharamceut Biol* **38**:112–119.

Siedlecka A. 1995. Some aspects of interactions between heavy metals and plant mineral nutrients. *Acta Soc Bot Pol* **2**:265–272.

Singh S, Eapen S, D'Souza SF. 2006. Cadmium accumulation and its influence on lipid peroxidation and antioxidative system in an aquatic plant, *Bacopa monnieri* L. *Chemosphere* **62**:233–246.

Sinha S, Gupta AK, Bhatt K. 2007. Uptake and translocation of metals in fenugreek grown on soil amended with tannery sludge: Involvement of antioxidants. *Ecotoxicol Environ Safety* **67**:267–277.

Sinha S. 1999. Accumulation of Cu, Cd, Cr, Mn, and Pb from artificially contaminated soil by *Bacopa monnieri*. *Environ Monit Assess* **57**:253.

Srivastava SK, Rai V, Srivastava M, Rawat AKS, Mehrotra S. 2006. Estimation of heavy metals in different *Berberis* species and its market samples. *Environ Monit Assess* **116**:315–320.

Tani FH, Barrington S. 2005. Zinc and copper uptake by plants under two transpiration rates. Part II. Buckwheat (*Fagopyrum esculentum* L.) *Environ Pollut* **138**(3):548–558.

Teske M, Trentini AMM. 1997. The flip side of Ayurveda. *J Postgrad Med* **39**:179–182.

Tolonen M. 1990. *Vitamins and Minerals in Health and Nutrition*. Chichester, England: Ellis Horwood Ltd.

Tripathi RD, Vajpayee P, Singh N, Rai UN, Kumar A, Ali MB, Kumara B, Yunus M. 2004. Efficacy of various amendments for amelioration of fly-ash toxicity: growth performance and metalcomposition of *Cassia siamea* Lamk. *Chemosphere* **54**:1581–1588.

Truong P. 2000. Vetiver grass technology for environmental protection. In: *The 2nd International Vetiver Conference: Vetiver and the Environment.* Cha Am, Thailand.

Unterbrunner R, Puschenreiter M, Sommer P, Wieshammer G, Tlustoš P, Zupan M, Wenzel WW. 2007. Heavy metal Accumulation in trees growing on contaminated sites in Central Europe. *Environ Pollut* **148**:107–114.

Vaillant N, Monnet F, Hitmi A, Sallanon H, Coudret A. 2005. Comparative study of responses in four *Datura* species to a zincstress. *Chemosphere* **59**:1005–1013.

Veeranjaneyulu K, Das VSR. 1982. *In vitro* chloroplast localization of [65]Zn and [63]Ni in a Zn-tolerant plant *Ocimum basilicum* Benth. *J Exp Bot* **33**:1161–1165.

Wadekar MP, Rode CV, Bendale YN, Patil KR, Prabhune AA. 2005. Preparation and characterization of a copper based Indian traditional drug: Tamra bhasma. *J Pharmaceut Biomed Anal* **39**:951–955.

Wierzbica M. 1998. Lead in the apoplast of *Allium cepa* L. root tips—Ultrastructural studies. *Plant Sci* **133**.

Williams PA, Rayson GD. 2003. Simultaneous multi-element detection of metal ions bound to a *Datura innoxia* material. *J Hazardous Mater* **99**:277–285.

Xia H, Rayson GD. 2000. Solid-state [113]Cd NMR studies of metal-binding to a *Datura innoxia* biomaterial. *Adv Environ Res* **4**:67–74.

Yamashitaa CI, Saikia M, Vasconcellosa MBA, Sertié JAA. 2005. Characterization of trace elements in Casearia medicinal plant by neutron activation analysis. *Appl Rad Isotopes* **63**:841–846.

Zheljazkov VD, Craker LE, Xing B. 2006. Effects of Cd, Pb, and Cu on growth and essential oil contents in dill, peppermint, and basil. *Environ Exp Bot* **58**:9–16.

Zheljazkov VD, Phil R, Warman PR. 2004. Phytoavailability and fractionation of copper, manganese, and zinc in soil following application of two composts to four crops. *Environ Pollut* **131**:187–195.

8 Biofortification: Nutritional Security and Relevance to Human Health

M. N. V. PRASAD

Department of Plant Sciences, University of Hyderabad, Hyderabad 500 046, India

1 INTRODUCTION

Trace elements, such as selenium, iron, zinc, and calcium, are essential for human as well as animal health. Nutritionally important trace elements are deficient in soils in many regions of the world. Health problems associated with excess/deficiency or uneven distribution of these essential trace elements in soils have become major public health issues in many developing countries. Therefore, the development of "foods and animal feeds" with fortified essential nutrients has been one of the most attractive research fields globally. In order to achieve the traditional forms of agriculture, conservation, and greater use of native bio-geo-diversity, genetic diversity analysis of the cultivable crops is a must.

Several of the trace elements serve as constituents of biomolecules as a cofactors for various enzymes and in a variety of metabolic functions. Trace elements accumulated in medicinal plants have the healing power in numerous ailments and disorders. Trace elements are implicated in healing function and neurochemical transmission (Zn on synaptic transmission); Cr and Mn can be correlated with therapeutic properties against diabetic and cardiovascular diseases. Certain transition group elements regulate hepatic synthesis of cholesterol (see Chapter 7).

Low mineral bioavailability (mineral deficiency) impairs a metabolic function in which a biomolecule catalytic function is dependent on that element (Tables 1–4). The fruits and vegetables that our grandparents ate when they were growing up were more nutritious than the ones that we and our children are consuming. In brief, the produce grown now has lower levels of several minerals than it did

Trace Elements as Contaminants and Nutrients: Consequences in Ecosystems and Human Health, Edited by M. N. V. Prasad
Copyright © 2008 John Wiley & Sons, Inc.

TABLE 1. Trace Element Deficiency-Induced Disease or Symptoms

Element	Deficiency Disease
Se	Keshena (China), bone, arthritis, cardiovascular and cancer
Zn	Dwarfness (Iran and Egypt) infertility, impaired taste and smell
Cu	Anemia, skeletal defects
Cr	Glucose metabolism, kidney function, cholesterol disorders
I	Goiter
F	Dental caries
Mg	Depression, nervous system
Mo	Mouth/esophageal cancer
Co	Anemia
Fe	Anemia
Na	Coma

before the "green revolution." Mineral concentrations in wheat are reportedly decreased over the last 160 years [McGrath et al., 2007].

It is estimated that about three billion people suffer the sinister effects of micronutrient deficiencies because they lack money to buy enough meat, poultry, fish, fruits, legumes, and vegetables. Women and children in Sub-Saharan Africa, South and Southeast Asia, and Latin America and the Caribbean are especially at risk for disease, premature death, and impaired cognitive abilities because of diets lacking essential micronutrients—particularly iron, iodine, zinc, and vitamin A. Therefore, it is necessary to alleviate malnutrition.

According to Davis [2005], nutrient content of 43 fruit and vegetable crops between 1950 and 1999 revealed that six out of 13 nutrients had declined in these crops over the 50-year period (the seven other nutrients showed no significant, reliable changes). Three minerals—phosphorus, iron, and calcium—declined between 9% and 16%. A study of the mineral content of fruits and vegetables grown in Great Britain between 1930 and 1980 shows similar decreases in nutrient density. The British study found significantly lower levels of calcium, magnesium,

TABLE 2. Diseases/Symptoms Associated with Excess of Certain Elements

Element	Disease Due to Excess Availability (Natural/Anthropogenic Reasons)
Hg	Mina-Mata, neurological disorders
Cd	Itai-Itai, bone crippling, cancer → prostate, lung, renal, and heart problems
F	Yellowing of teeth, skeletal deformities
Pb	Lung cancer, neurological disorders
Mn	Paraplegia = skeletal deformations → paralysis
As	Skin, lung cancer

TABLE 3. Essential Trace Elements and Their Beneficial Functions to Human Health

Trace Element	Function
Calcium	Essential for bones, teeth, healthy gums, and bone growth and mineral density in children. Calcium helps regulate the heart rate and nerve impulses, lower cholesterol, prevent atherosclerosis, develop muscles, and prevent muscle cramping. It keeps the heart beating regularly, metabolizes body iron, and is a natural sedative. It is essential for a healthy nervous system, blood clotting, and helps in the prevention of colon cancer.
Magnesium	Essential for utilization of calcium and potassium. It functions in enzyme reactions to produce energy. Magnesium protects the lining of arteries and helps form bones. It is important to enhance enzyme activity and for effective nerve and muscle function. It helps convert blood sugar to energy, and it prevents depression and stress. A deficiency can cause irritability, high blood pressure, muscle weakness, dizziness and heart disease.
Sodium	Essential for controlling control fluid balance through a mechanism called "the sodium/potassium pump."
Potassium	Essential for healthy nervous system and a steady heart rate, helps to prevent stroke and, with sodium, it is critical in maintaining fluid balance.
Boron	Essential for healthy bones, brain function, alertness, and the metabolism of bulk minerals such as calcium, phosphorus, and magnesium. This mineral is needed for calcium uptake. It also helps women on estrogen replacement therapy to keep estrogen in the blood longer. Boron retards bone loss in women after menopause.
Chromium	Essential for maintaining energy levels. Chromium helps metabolize glucose and stabilize glucose levels. It helps the body manufacture and use cholesterol and protein. Helps the body lose fat and keep muscle. It also reduces appetite and helps provide energy. Needed to normalize blood sugar levels.
Copper	Essential for healthy bones, joints, and nerves as well as hemoglobin and red blood cells. Copper contributes to healing, energy production, taste, and hair and skin color. It is essential in forming collagen for healthy bones and connective tissue, and it helps prevent osteoporosis. Essential to the formation of hemoglobin in red blood cells and bone. It is necessary for healthy nerves and energy production. Needed for healthy protein and enzyme formation (see Chapter 25).

(Continued)

TABLE 3. *Continued*

Trace Element	Function
Germanium	Essential for delivery of oxygen to tissues and remove toxins and poisons from the body.
Iodine	Essential for promoting healthy physical and mental development in children. Iodine is required for thyroid gland function and metabolizing fats. Iodine deficiency is a public health problem in parts of the world that have iodine-deficient soils. Iodine is needed to make thyroid hormone, which has a variety of roles in human embryo development.
Iron	Essential for production of hemoglobin, the oxygen-carrying protein in red blood cells, and myoglobin found in muscle tissue. Iron is essential for important enzyme reactions, growth, and maintaining a healthy immune system. In the blood, iron is found in larger amounts than any other mineral. Essential in the formation of red blood cells, promotes disease resistance and improves skin tone. Iron deficiency causes hair loss, dizziness, and anemia. Women lose twice as much iron as men (see Chapters 11 and 19).
Iodine	Iodine helps to metabolize fat, is needed for thyroid health, and prevents goiters. Iodine deficiency has been linked to breast cancer and mental retardation in children. It also promotes healthy hair, skin, and teeth (see Chapter 12).
Manganese	Essential for metabolizing fat and protein, regulating blood glucose, and supporting immune system and nervous system function. Manganese is necessary for normal bone growth and cartilage development. It is involved in reproductive functions and helps produce mother's milk. Along with B vitamins, manganese produces feelings of well-being. For healthy nerve and immune system, improves blood sugar regulation. Necessary for healthy bones, fights fatigue, improves memory, and reduces irritability. Helps lactation and produces enzymes needed to oxidize fats.
Molybdenum	Only extremely small amounts are needed to metabolize nitrogen and promote proper cell function. Molybdenum is present in beans, peas, legumes, whole grains, and green leafy vegetables. A diet low in these foods can lead to mouth and gum problems and cancer. This mineral is needed in nitrogen metabolism. It is found in the liver, bones, and kidneys. A low intake is associated with mouth and gum disorders. A low intake is also associated with cancer, and a lack of this mineral may cause sexual impotence in males.

(Continued)

TABLE 3. *Continued*

Trace Element	Function
Selenium	Important antioxidant that works with vitamin E to protect the immune system, heart, liver; may help prevent tumor formation. Selenium deficiency occurs in regions of the world where soils are selenium-poor and low-selenium foods are produced. Premature infants are naturally low in selenium with no known serious effects. This works with vitamin E. They are both antioxidants. It keeps tissue elastic, helps with dandruff, improves circulation, protects against some cancers, and alleviates menopausal distress (see Chapter 23).
Silicon	Essential for preventing cardiovascular disease. Helps to form bones and connective tissue, nails, skin, and hair.
Vanadium	For reducing cholesterol and form healthy bones and teeth. Vanadium functions in reproduction. Needed for cellular metabolism and for the formation of bones and teeth. It plays a role in growth and reproduction and inhibits cholesterol synthesis. A vanadium deficiency may be linked to cardiovascular and kidney disease, impaired reproductive ability, and increased infant mortality. Vanadium has been known to help with reversal and control of diabetes.
Zinc	For the growth of reproductive organs and regulation of oil glands. Zinc is required for protein synthesis, immune system function, protection of the liver, collagen formation, and wound healing. A component of insulin and major body enzymes, zinc helps vitamin absorption, particularly vitamins A and E. This mineral builds and supports the immune system. It is involved in protein synthesis and the formation of insulin, helps normalize the prostate, and helps restore smell and taste loss. Recommended for healing skin and acne. It accelerates healing and promotion of growth and mental alertness (see Chapters 9 and 10).

copper, and sodium in vegetables, as well as significantly lower levels of magnesium, iron, copper, and potassium in fruit. Also, another British study found that mineral concentrations decreased in wheat over the past 160 years [McGrath et al., 2007]. The decline in agriculture produce's nutritional value corresponds to the period of increasing industrialization of our farming systems, focusing on cash crops, substituting farm inputs with chemical fertilizers, pesticides, and monoculture farming for the natural cycling of nutrients and on-farm biodiversity. Organic farmers fertilize their crops by adding organic matter to the soil in the form of composts, cover crops, and manures. The organic matter feeds microorganisms in the soil that, in the process of

TABLE 4. Significant Depletion or Deficiency of Essential Trace Elements Symptoms and/or Disease

Trace Element	Deficiency Symptom/Disease
Calcium	Osteoporosis.
Magnesium	Dietary magnesium deficiency may occur in chronic alcoholics, in persons taking diuretic drugs, and as a result of severe, prolonged diarrhea.
Sodium	Prolonged imbalances/deficiency can contribute to heart disease.
Potassium	Prolonged imbalances/deficiency can lead to impairment of nervous and heart functions.
Boron	Deficiency results in vitamin D deficiency. Boron supplements can improve calcium levels as well as vitamin D levels and can help prevent osteoporosis in postmenopausal women by promoting calcium absorption.
Chromium	Altered glucose metabolism, cholesterol, and protein sythesis. Symptoms include fatigue, anxiety, poor protein metabolism, and glucose intolerance (as in diabetes). In adults, chromium deficiency can be a sign of coronary artery disease.
Copper	Copper deficiency is rare, significant changes in copper metabolism occur in two serious genetic diseases: Wilson disease and Menkes' disease.
Iodine	Iodine deficiency is a public health problem in all parts of the world that have iodine-deficient soils. A deficiency during pregnancy can cause serious birth defects. Deficiency in adults can result in an enlarged thyroid gland (goiter) in the neck (see Chapter 12).
Manganese	Deficiency can lead to convulsions, vision and hearing problems, muscle contractions, tooth-grinding, and other problems in children; and atherosclerosis, heart disease, and hypertension in older adults. Symptoms of manganese deficiency include faulty transmission of nerve and muscle impulses, irritability, nervousness, and tantrums. Confusion, poor digestion, rapid or irregular heartbeat (arrhythmia), and seizures can also result. Magnesium deficiency is associated with cardiac arrest, asthma, chronic fatigue syndrome, chronic pain, depression, insomnia, irritable bowel syndrome, and lung conditions.
Selenium	Selenium deficiency occurs in regions of the world where soils are selenium-poor and low-selenium foods are produced. Selenium deficiency occurs in regions of the world containing low-selenium soils, including parts of China, New Zealand, and Finland. In Keshan Province, China, a condition (Keshan disease) occurs that results in deterioration of regions of the heart and the development of fibers in these areas. Keshan disease, which may be fatal, is thought to result from a combination of selenium deficiency and a virus.
Iron	Iron deficiency causes anemia (low hemoglobin and reduced numbers of red blood cells), which results in tiredness and shortness of breath because of poor oxygen delivery (see Chapters 11 and 19).
Molybdenum	Lead to mouth and gum problems and cancer.
Vanadium	Vanadium deficiency may be associated with infant mortality.

eating and living and dying, recycle the nutrients embedded in the organic matter. The microbes slowly release not only nitrogen, phosphorus, and potassium, in available form to the plants, but also release a host of other nutrients in ratios difficult to replicate with synthetic fertilizers. The large populations of microorganisms that typically inhabit organically managed fields also produce substances that combine with minerals in the soil and make them more available to plants, a function that can be especially important for iron absorption. Iron is usually present in soil, but it is often in an unavailable form.

Biorganic farming takes total micronutrient care and is an important soil management practice for fortification of crop produce with essential nutrients for the health of humans as well as livestock. In this case, degradable waste generated from agroforestry activity is subjected to biocomposting and applied to crop husbandry. Application of farm yard manure (FYM) to agricultural fields is a well-established traditional practice in India. FYM generated from an animal source such as cattle, poultry, and swine is often fortified with various levels of trace metals that are of nutritional value to livestock.

Geomedicine (medicial geology or geomedical science) is the reflection of natural conditions in the health and well-being of living creatures. Geomedicine considers the importance of naturally occurring elements in the soil and parent rock for the health and well-being of humans and animals. Traditionally, geomedicine deals with the patterns in the prevalence of diseases, where the geographic variation in disease is reflective of the variation in geological characteristics. For example, molybdenosis, fluorosis, and arsenic-enriched groundwater constitute classic examples of this fascinating and emerging field. This branch of science is rapidly progressing with connections to organic farming.

Organic farmers do not use synthetic formulations of fertilizers, and this restriction is part of the reason why organic produce has relatively higher nutrient values. These farmers feed their crops only indirectly. The relatively larger root-balls of organic plants is another reason why organically grown plants can absorb a wider variety of nutrients than chemically fertilized plants. Because organic plants don't have macronutrients spoon-fed to them, they grow larger root systems out of necessity. Roots on organic plants have to range farther to access sufficient nitrogen, phosphorus, and potassium. In the process, they come into contact with more trace minerals and micronutrients than the smaller root-balls of conventional plants. "When plants are growing, they sense how big a root system they have to produce to draw from the soil the nutrients and moisture they need to grow and reach maturity and reproduce," "On a conventional farm where there are high levels of fertilizer nutrients in the soil, along with lots of water, there is little incentive for roots to penetrate far."

The role that antioxidants play in plant health probably also contributes to the higher antioxidant levels found in organic produce. Many antioxidants help a plant resist diseases, deter pests, and recover from insect damage. Because organically grown plants do not "benefit" from the protection of pesticides, they must be able to muster their own defenses and therefore produce high levels of antioxidants (Fig. 1).

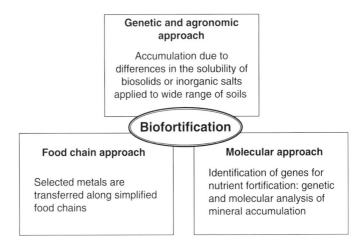

Figure 1. Biofortification: Possible approaches and achievements [Banziger and Long, 2000; Cakmak, 2007; Storckdieck et al., 2007; Graham et al., 1999; Gregorio and Htut, 2007; Long et al., 2003; Lyons et al., 2007; Mesjasz-Przybylowicz et al., 2007; Slingerland, 2007; Bañuelos, 2007; House et al., 2002; Monasterio and Graham, 2000; Nesamvuni et al., 2005; Welch et al., 2000].

2 BIOAVAILABLITY OF MICRONUTRIENTS

Current efforts to combat micronutrient malnutrition in the developing world focus on fortifying foods with these nutrients through postharvest processing. These approaches have accomplished much. In regions with adequate infrastructure and well-established markets for delivering processed foods such as salt, sugar, and cereal flours, food fortification can greatly improve the micronutrient intake of vulnerable populations. Therefore, appropriate measures are needed to complement existing interventions (Fig. 2).

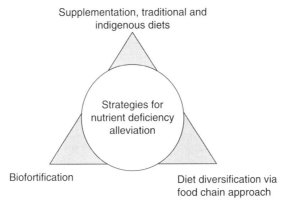

Figure 2. Different strategies for nutrient deficiency. Efficacy and effectiveness of these measures is being researched for cost-effectiveness and acceptability.

3 SOCIAL ACCEPTABILITY OF BIOFORTIFIED CROPS

The crops fortified by the HarvestPlus program are widely produced and consumed by poor households in the developing world, but farmers need to be convinced to grow biofortified varieties and consumers must be persuaded to add them to their diets. The need for maintaining superior agronomic traits is paramount with biofortified varieties. Where scientists can combine high micronutrient content with high yield, farmer adoption and market success of nutritionally improved varieties is virtually guaranteed. Research has shown that high levels of minerals in seeds can also aid plant nutrition and thus improve yield. HarvestPlus, a pioneering agency in the field of "Biofortification," employs participatory plant breeding techniques, in which scientists take farmers' perspectives and preferences are taken into account during the breeding process. Improved mineral content, which generally does not alter appearance, taste, texture, or cooking quality, is far less complicated than improving provitamin A content that may alter the color of food.

4 DEVELOPMENT AND DISTRIBUTION OF THE NEW VARIETIES

A common problem in many developing countries is the lack of delivery systems to get products—be they from health or agronomic inputs—to the poorest people. HarvestPlus is overcoming this problem by implementing an end-user strategy that identifies and addresses constraints to seed and extension systems, overcomes market impediments, and recognizes that substantive behavior change and health communication activities that must accompany the development of the biofortified varieties.

Through their ongoing work with seed systems and their contributions to disaster response, CGIAR centers have gained valuable experience in building and promoting local seed-distribution systems. These established systems offer a natural route for disseminating biofortified seed. HarvestPlus in association with leading universities is evolving strategies to create demand for nutrient-dense crops. Local and regional agricultural committees and small-farmer seed enterprises, in particular, will play a crucial role in getting micronutrient-rich varieties into the hands of growers. Local marketing organizations and health communication specialists and local civic bodies need to be integrated effectively to reach vulnerable undernourished children and their mothers.

5 SELECTED EXAMPLES OF BIOFORTIFIED CROPS TARGETED BY HARVESTPLUS IN COLLABORATION WITH A CONSORTIUM OF INTERNATIONAL PARTNERS*

The HarvestPlus has contributed an extraordinary range of knowledge in biofortification, including expertise in plant breeding, genomics, human nutrition in coalition

*Reproduced with permission from HarvestPlus.

TABLE 5. International Consortium Contributing to the Field of Biofortification[a]

CIAT	International Centre for Tropical Agriculture
CIMMYT	International Maize and Wheat Improvement Centre
CIP	International Potato Centre
ICARDA	International Centre for Agricultural Research in the Dry Areas
ICRISAT	International Crops Research Institute for the Semi-Arid Tropics
IFPRI	International Food Policy Research Institute
IITA	International Institute of Tropical Agriculture
IRRI	International Rice Research Institute
WARDA	Africa Rice Centre
IPGRI	International Plant Genetic Resources Institute
INIBAP	International Network for the Improvement of Banana and Plantain
NARES	National Agricultural Research and Extension Systems
NHRIS	National Health Research and Implementation Systems

[a]This is not an exhaustive list.

with a number of international organizations (Table 5). Selected examples of crops for better nutrition are mentioned below.

5.1   Rice

Rice is the most dominant cereal crop in many developing countries and is the staple food for more than half of the world's population. In several Asian countries, rice provides 50–80% of the energy intake of the poor. Because of the high per-capita consumption of rice in these countries, increasing its nutritive value could have significant positive health outcomes for millions of people.

Strategy. Micronutrient-dense rice varieties are being developed using the best traditional breeding and modern biotechnology methods to achieve increases in nutrient concentrations that will have measurable benefits for people consuming these new varieties. Initially, the introduction of improved varieties will be targeted to Bangladesh, Indonesia, Vietnam, India, and the Philippines.

Breeding programs aimed at producing varieties with high iron and zinc concentrations also seek to combine the high mineral content with other seed and food characteristics attractive to farmers or consumers. Studies by HarvestPlus and others have shown considerable losses of iron and zinc during the polishing of rice. For this reason, HarvestPlus breeding work is focused on increasing mineral levels in white rice. Initial germ plasm screening and field evaluations for iron and zinc have included breeding lines from Korea, Bangladesh, Indonesia, India, and the Philippines. Commercial varieties of rice normally contain 2 mg/kg iron. Thirty lines of rice with more than 5 mg/kg grain iron were initially selected from germ plasm banks and evaluated in multi-location trials in the wet and dry seasons in the Philippines to determine agronomic and nutritional performance, assess genotype by environment interactions for iron and zinc, and identify parent materials and candidates for fast-track breeding. Results show that by selective breeding, iron content in polished rice can be increased by a factor of 2–4. Zinc concentrations of 20–25 mg/kg have been identified in several varieties, suggesting the high feasibility of developing high-zinc rice. A human efficacy study using high-iron rice, recently published in the *Journal of Nutrition*, demonstrated a 20% increase in iron blood stores attributable to the consumption of biofortified high-iron rice.

5.2 Wheat

In developing countries, particularly those in South and West Asia, about half a billion people are iron-deficient. In many of these same regions, wheat is a major staple food. The main objective of biofortifying wheat is to develop nutritionally enhanced wheat to increase people's intake of iron and zinc. The potential impact of iron-enhanced wheat could be dramatic, given that spring wheat varieties developed by CIMMYT and its partners are used in 80% of the global spring wheat area.

Strategy. The initial target countries for improving the micronutrient content in wheat will be Pakistan and India, specifically the area around the Indo-Gangetic plains. This region has high population densities and high micronutrient malnutrition. As more resources become available, other countries will be included in the project. The strategy is to improve iron and zinc levels in high-I-yielding, disease-resistant varieties being developed by the national research institutes of India and Pakistan and by international agricultural research centers. The highest recorded levels of iron and zinc in wheat grain have been found in land races and wild relatives of wheat such as *Triticum dicoccon* and *Aegilops taushii*. Hence, the search for better sources of high iron and zinc within CIMMYT genebanks will be centered on screening wild relatives. Because the best sources identified to date derive from wild relatives of wheat that cannot be crossed directly with modern wheat, researchers developed a variety of hexaploid wheat that could be crossed directly with current modern varieties of wheat. These biofortified varieties contain 4–5% higher levels of iron and zinc in the grain compared to modern wheat.

Researchers are using hexaploid wheat to support breeding efforts in India and Pakistan for the development of high-yielding, disease-resistant 6 biofortified wheat. More than 2400 sources of genetic variation have been screened for iron and zinc at CIMMYT and in India, Pakistan, and Turkey. In addition, testing of local germplasm for minerals was initiated in China and Kazakhstan in 2005. The first zinc-biofortified lines were delivered to India and Pakistan in 2005. These initial advanced materials, which were planted in on-farm and on-station trials, represented broadly adapted, high-yielding, disease-resistant wheat lines that demonstrate potential under target region conditions. These trials generated genotypic and genotype x environment data on agronomic-, micronutrient-, and end-use performance. Results from Pakistan revealed that for zinc, once planted in target regions, the differences among various genotypes were minimal and concentrations of zinc were low compared with results from those planted in Mexico. Researchers are now investigating ways to minimize the impact of microenvironmental factors on grain nutrient density. For iron, by contrast, advanced lines surpassed the local varieties, on average, across eight locations in the same region in Pakistan. HarvestPlus molecular biologists, genomicists, and human nutritionists are exploring the feasibility of increasing the concentration of iron in wheat through the introduction of the ferritin gene. Molecular markers for the genes that control iron and zinc concentrations in the grain are being identified to facilitate their transfer. (See Chapters 9 and 10 for detailed information on zinc.)

5.3 Maize

Maize is the preferred staple food of more than 1.2 billion people in Sub-Saharan Africa and Latin America. Over 5 million people in these regions are vitamin A-deficient. Maize-based diets, particularly those of extremely poor individuals, often lack essential vitamins such as vitamin A. Dietary sources occur either as preformed vitamin A, as in dairy and other foods from animal sources, or as provitamins A, as found in plant foods, including maize. Identifying and increasing the supply of maize cultivars rich in provitamins A may greatly improve the health and longevity of people around the world. Researchers are currently screening germ plasm samples of existing maize varieties to identify micronutrient-dense varieties for development through conventional breeding. International research institutes, primarily the International Maize and Wheat Improvement Center (CIMMYT) and the International Institute for Tropical Agriculture (IITA), are conducting most of the preliminary breeding work. Adaptive breeding for local conditions will be carried out in partnership with national agricultural research and extension systems (NARES) in Latin America and Africa to ensure that the new varieties are competitive in terms of grain yield and other important traits compared with currently grown varieties.

Strategy. HarvestPlus maize research originally concentrated on increasing provitamins A carotenoids in maize. Efforts to increase the concentration of iron in maize through conventional breeding programs were hampered by limited natural variation. In 2005, however, potential new sources for high iron and zinc were identified. Initial target countries for dissemination of biofortified maize are Brazil, Ethiopia, Ghana, Guatemala, and Zambia. Research in the United States and at CIMMYT has focused on (a) finding genetic markers to facilitate selection in breeding programs aimed at enhancing provitamin A levels and (b) dissemination of the new varieties so as to encourage widespread acceptance and create demand for biofortified varieties. A key feature of this strategy will be the formation of alliances in each target country or in subregions within countries to accomplish specific tasks.

Evaluation of biofortified varieties is being carried out with extension services, non-governmental organizations, participation of farmers, universities, industries, or government nutritionists, food technologists, to assess the effect of local food processing on bioavailability of vitamins and other minerals. HarvestPlus project teams, nutritionists, and ministries of health and education, formulate and implement a nutrition advocacy program, including promoting adoption of biofortified maize varieties. Seed producers, retailers, and food and feed industries are involved in developing strategies to ensure access to the selected biofortified varieties by the farmers and consumers who will benefit most from their use. Micronutrient concentrations in yellow and white maize germ plasm is being assessed. HarvestPlus scientists have discovered germ plasm with maximum iron and zinc concentrations of 40–45 mg/kg and 50–62 mg/kg, respectively. Based on nutritionists' estimates, iron and zinc levels of 60 and 55 mg/kg, respectively, are required for biofortified maize have an Impact on nutrition. For provitamins A, screening revealed maximum levels between 5.0 and 8.6. In 2006, national program teams will continue to screen local germ plasm and conduct adaptive breeding research, including farmer-participatory variety evaluations.

5.4 Beans

For more than 300 million people, an inexpensive bowl of beans is the centerpiece of their daily diet. The common bean, *Phaseolus vulgaris*, is the world's most important food legume, far more so than chickpeas, faba beans, lentils, and cowpeas. Given the widespread consumption of beans throughout the world, efforts to improve their micronutrient content could potentially benefit a great many people. Biofortifying the common bean will produce the greatest returns in areas where these beans supply a significant proportion of the nutrients in the diet. These areas include parts of East, Central, and Southern Africa and all of Central America and Brazil. Improved common bean varieties have been widely accepted in Central America, where their resistance to viral diseases made them appealing. In East and Central Africa, yields of mid-altitude climbing beans are nearly triple those of standard bush varieties.

Strategy. HarvestPlus research on the development of biofortified beans focuses on increasing the concentrations of iron and zinc in agronomically superior varieties. Key productivity traits are being developed which, when combined with improved nutritional value, will create new, more nutritious common bean varieties that are even more attractive to farmers—specifically, tolerance to both low-fertility soils and drought. Early maturity is also being incorporated. Conventional breeding techniques and marker-assisted selection, combined with functional genomics, will reveal the underlying mechanisms of the genes that control these traits.

The average iron concentration in common bean is about 55 mg/kg; through germ plasm screening, however, researchers have found varieties with iron concentrations in excess of 100 mg/kg *Phaseolus polyanthus*, a sister species that crosses readily with *P. vulgaris*, which contain iron concentrations of up to g 127 mg/kg. The long-term goals are to obtain a favorable combinations of productivity and nutritional traits in bean varieties with a range of colors and plant types that farmers and consumers want. In particular, research will be aimed at doubling the iron concentration and increasing the zinc concentration by at least 4%.

Breeding is well underway, and crosses have been produced that were designed to combine varieties high in iron with those that have a high tolerance to environmental stresses. The first progeny in the red mottled bean class, which is very popular in Latin America and Africa, presented iron and zinc levels in excess of

80 and 3 mg/kg, respectively, when grown on farms in Colombia. In addition, climbing beans bred for better heat tolerance contained 4% more iron than standard varieties. As of 2006, the HarvestPlus project has entered wide-scale deployment with the production of several tons of seed in DR Congo, Rwanda, Kenya, and Ethiopia. Promising micronutrient-rich bean lines have entered on-farm trials. Breeding for high mineral content has already been incorporated into the national agricultural research agendas of many African and Latin American countries.

5.5   *Brassica juncea* (Indian Mustard)

Brassica juncea is used as vegetable, salad, condiment, and fodder oilseed, is widely consumed as a leafy vegetable in northern parts of India, and is amenable to genetic transformation.

Strategy. The present study was carried out using 35S constitutive overexpression of pea seed ferritin cDNA (gifted by Professor J. F. Briat, France, which the author thankfully acknowledges). The methodology includes construction of vector, transfer of binary vector to *Agrobacterium tumefaciens*, plant transformation, and confirmation of integration and expression of the transgene [Nirupa et al., 2007] (Fig. 3).

Putative transgenics obtained in T_0 and T_1 have been confirmed by Polymerase chain reaction (PCR) with ferritin and hptII primers. A progeny test on the seeds of the T_0 plants indicated the Mendelian pattern of inheritance of the marker gene. The

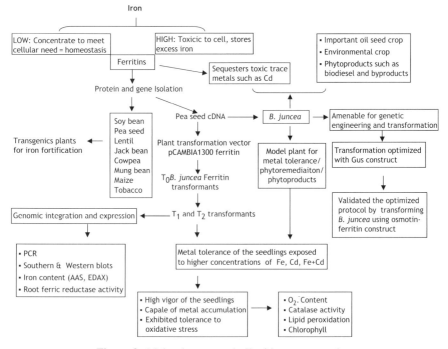

Figure 3. Molecular approach: Ferritin overexpression.

integration was confirmed by southern hybridization of digested genomic DNA of T_0 and T_1 generations with ferritin and hygromycin probes. The expression of transgene in the T_1 generation was confirmed by RT PCR. Ferritin subunits expression in T_1 generation transformants was further confirmed by immunoblot using antiferritin antibody and other established procedures [Nirupa et al., 2007].

6 SELENIUM-FORTIFIED PHYTOPRODUCTS

Selenium is essential trace element involved in etiology of several diseases. At normal concentration, Se will have protective effect against several diseases, including cancer. This protection is lost at low concentrations, and Se is highly toxic at high concentrations. Therefore, it is important that Se levels be monitored and maintianed at a desirable range [Banuelos, 2007]. GPx is Se-dependent enzymes. At low concentrations, there is a correlation between GPx and Se status. Therefore, GPx levels can be used to assess Se status than measuring Se itself because concentration of the enzyme is less easily affected by diet.

7 SOURCES OF SELENIUM IN HUMAN DIET

Selenium is a nonmetallic element that naturally occurs in different components of the environment. Selenate and selenite are common ions in natural waters and soils. Inorganic-reduced Se forms include mineral selenides and hydrogen selenide. Reduced Se compounds are the methylated species such as dimethyleselenide (DMSe) and dimethylediselenide (DMDSe), and sulfur substitution in amino acids include selenocysteine, selenocystine, and selenomethionine. Selenium generally enters the food chain through both surface and underground waters which are used for irrigation and drinking purposes. According to water quality guidelines for Se of the USEPA, the maximum contamination level (MCL) of Se in water for drinking purposes is 10 mg/liter and the maximum permissible level (MPL) for water used for irrigation is 20 mg/liter. Wheat, mustard, sunflower, oats, and rice are some of the sources of selenium in food (in descending order of the concentration).

Free radicals (FR) and reactive oxygen species (ROS) are the signaling molecules. Abiotic and biotic stresses are reported to generate FR and ROS and have been implicated in a number physiological disorders in plants and other organisms. In humans, this involves several diseases ranging from rheumatoid arthritis to hemorrhagic shock to AIDS. Overproduction of these species cause membrane damage and accelerate the tissue disease process in organisms (Tables 1–3).

8 SELENIUM (Se) AND SILICA (Si) MANAGEMENT IN SOILS BY FLY ASH AMENDMENT

Many areas of the world are deficient in Se. Therefore, forage grown on such soils does not supply adequate amounts (0.05–2 mg/kg) of Se for satisfactory animal performance. Small applications of fly ash can significantly increase the concentration of Se in the forage and thus overcome the deficiency [Gissel-Nielsen et al., 1984]. On the other hand, some areas have excessive levels of Se [Dhillon and Dhillon, 2003, El-Bayoumy et al., 2006]. The Se level can be reduced substantially by application of gypsum where the added S competes with the Se for uptake by the plants [Arthur et al., 1993]. Many acid soils subtend low levels of Si in solution, while certain crops such as rice and sugarcane have a requirement for this element. Fly ash application significantly increased yield. Application of Si also increased levels of available P, probably due to the exchange of silicate for phosphate on soil surfaces.

9   CHROMIUM FOR FORTIFICATION DIABETES MANAGEMENT

For over 150 years, evidence of the benefits of chromium—a nutrient found in brewer's yeast—in treating diabetes has been steadily accumulating. In people with most types of diabetes, chromium has been shown to improve the way the body handles glucose. Supplementing with chromium may also reduce cholesterol and triglyceride levels in people with diabetes, reducing their risk of blocked arteries

Figure 4. (**a**) *Prosopis juliflora* colonized on metal-loaded soils. Leaves and twigs accumulated large concentrations of minerals. *Prosopis juliflora* pods accumulate high Cr concentration (up to 150 ppb). This supplements Cr in fodder for animals (Please note that goats depend upon tender foliage and pods of *P. juliflora*). Cr requirement for animals was >0.1 ppm, while toxic level was 1000 ppm [McDowell et al., 1978]. (**b**) Goats grazing tender pods. (**c**) Tender shoots. (**d**) Pods.

(atherosclerosis). Despite an abundance of evidence, however, many diabetes experts have been reluctant to recommend chromium supplementation. According to the most recent 2006 Clinical Practice Recommendations from the American Diabetes Association, "The existence of a relationship between chromium picolinate and either insulin resistance or type 2 diabetes is highly uncertain." It is reported that chromium picolinate is effective in improving glycemic control and normalizing lipid levels in people with type 2 diabetes [Hsing-Hsien et al., 2004] (Fig. 4).

10 SILICA MANAGEMENT IN RICE—BENEFICIAL FUNCTIONS

Silica is the second most abundant element in the earth's crust, so it would be useful for amelioration of Al toxicity in acidic soils. Subsoil acidification in agro-ecosystems is a serious global environmental concern. Acid soils occupy nearly 30% (3.95 ha) of the arable land area in both tropical and temperate belts. In addition to the natural processes, farming and management practices such as high use of nitrogen fertilizers, removal of cations by harvested crops, leaching, and runoff of cations resulted acidification of soils. In many industrialized areas, the atmospheric deposition of sulfur and nitrogen compounds is a major source of proton influx to soils. More than the low pH of the soils, the major problem associated with acid soils is the toxicity of aluminum and manganese and the deficiency of phosphorus, calcium, magnesium, and potassium. In addition to these nutritional factors, the acid soils are also characterized by low water-holding capacity due to compaction of soils. Apart from mineral toxicities, soil acidification is also known to change the microbial activity of the soils. Acidification is also known to reduce the degradation of soil organic matter and alters

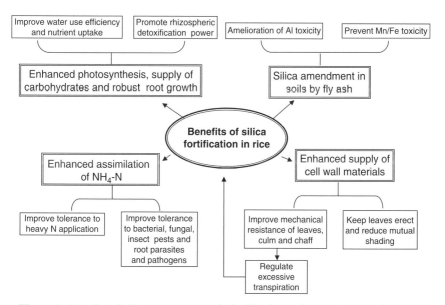

Figure 5. Benefits of silica management via fortification and amendment to rice crop.

cation and nutrient flow in the ecosystem. In general, most of the acid soils have low exchangeable bases (e.g., Ca, Mg, and K) mainly because of the low cation exchange capacity (CEC) of the soils. The oxides of Al and Fe in wet acidic soils fix large fraction of the phosphate, making it unavailable to plants leading to lower crop yields. Therefore, Al toxicity is the major agronomic problem in acid soils, and it is reported that silica had ameliorative function [Cocker et al., 1998] (Fig. 5).

11   CONCLUSIONS

Genetic engineering of crops for biofortification may involve the risks of transferring fitness-enhancing traits into the noxious weeds (sorghum versus johnson grass). Biofortification may also contribute to soil degradation and genetic erosion or alter a crop's susceptibility to pathogens. The relevance of these cutting edge fields *pros* and *cons* *vis a vis* Biosafety are presented in Figs. 6 and 7. Trace elements, such

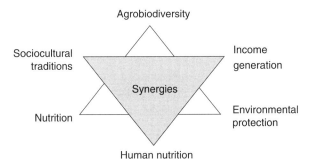

Figure 6. Linking biofortification with socioeconomic, environmental and possible impact on agrobiodiversity.

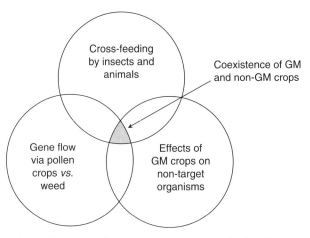

Figure 7. Transgenic crops and environmental safety issues.

as selenium, iron, zinc, and calcium, are essential for human as well as animal health. Nutritionally important trace elements are deficient in soils in many regions of the world. Bioorganic farming takes total micronutrient care and is an important soil management practice for fortification of crop produce with essential nutrients for the health of humans as well as livestock. In this case, degradable waste generated from agro forestry activity is subjected to biocomposting and applied to crop husbandry. Application of farm yard manure (FYM) to agricultural fields a well-established traditional practice in India. FYM generated from animal source such as cattle, poultry, and swine is often fortified with various levels of trace metals that are required for plant nutrition. The World Summit on Sustainable Development (WSSD), Johannesburg, South Africa, 2002, the Conference of the Parties (COP7) of the Convention on Biodiversity in Kuala Lampur, 2004, and the World Health Organization (WHO) Global Strategy on Diet, Physical Activity and Health (WHO, 2004) explicitly acknowledged the importance of traditional and indigenous diets. CBD Decision VII/32 recommended a strengthened focus on biodiversity for food and nutrition.

ACKNOWLEDGMENTS AND DISCLAIMER

This review has been prepared from information obtained from authentic and highly regarded sources such as internet, national, and international conference deliberations. Thanks are due to HarvestPlus for allowing permission to cite examples of rice, wheat, maize, and beans (see Section 5) investigated for biofortification of mineral nutrients in association with leading universities in different parts of the world. The author thankfully acknowledges these resources. The author and the publisher cannot assume responsibility for any adverse consequences of the use of material cited in this article in field or laboratory.

REFERENCES

Aarts MGM, Ghandilyan A, van de Mortel JE, Talikdar S, Wu J, Du J, Schat H, Ling HQ, Koorneef M. 2007. Identification of genes for biofortification: Genetic and molecular analysis of mineral accumulation in *Arabidopsis thaliana*. In: Zhu Y, Lepp N, Naidu R, editors. *Biogeochemistry of Trace Elements: Environmental Protection, Remediation and Human Health*. China: Tsinghua University Press, pp. 254–255.

Arthur MA, Rubin G, Woodbury PB, Weinstein LH. 1993. Gypsum amendment to soil can reduce selenium uptake by alfalfa in the presence of coal fly ash. *Plant Soil* **148**:83–90.

Bañuelos GS. 2007. Prodution of unique bio-based products from vegetation management of selenium. In: Zhu Y, Lepp N, Naidu R, editors. *Biogeochemistry of Trace Elements: Environmental Protection, Remediation and Human Health*. China: Tsinghua University Press, pp. 256–257.

Banziger M, Long J. 2000. The potential for increasing the iron and zinc density of maize through plant breeding. *Food Nutr Bull* **21**:397–400.

Cakmak I. 2007. Genetical and agronomical biofortification of cereals with Zn for better nutrition In: Zhu Y, Lepp N, Naidu R, editors. *Biogeochemistry of Trace*

Elements: Environmental Protection, Remediation and Human Health. China: Tsinghua University Press, pp. 260–261.

Cocker KM, Evans DE, Hodson MJ. 1998. The amelioration of aluminium toxicity by silicon in higher plants: Solution chemistry or an in planta mechanisms? *Physiol Plant* **104**:608.

Davis DR. 2005. Trade-offs in agriculture and nutrition. *Food Technol* **59**:3.

Dhillon KS, Dhillon SK. 2003. Quality of underground water and its contribution towards selenium enrichment of the soil–plant system for a seleniferous region of northwest India. *J Hydrol* **272**:120–130.

El-Bayoumy K, Sinha R, Pinto JT, Rivlin RS. 2006. Cancer chemoprevention by garlic and garlic-containing sulfur and selenium compounds. *J Nutr* **136**(3):864S–869S.

Gissel-Nielsen G, Gupta UC, Lamand M, Westyermarck T. 1984. Selenium in soils and plants and its importance in livestock and human nutrition. *Adv Agron* **37**:397–460.

Graham R, Senadhira D, Beebe S, Iglesias C, Monasterio I. 1999. Breeding for micronutrient density in edible portions of staple food crops: Conventional approaches. *Field Crops Res* **60**:57–80.

Gregorio GB, Htut T. 2007. Improving micronutrient value of rice through breeding. In: Zhu Y, Lepp N, Naidu R, editors. *Biogeochemistry of Trace Elements: Environmental Protection, Remediation and Human Health*. China: Tsinghua University Press, pp. 268–269.

House WA, Ross M, Welch RM, Beebe S, Cheng Z. 2002. Potential for increasing the amounts of bioavailable zinc in dry beans (*Phaseolus vulgaris* L.) through plant breeding. *J Sci Food Agric* **82**:1452–1457.

Hsing-Hsien C, Ming-Hoang L, Wen-Chi L, Chen-Ling H. 2004. Antioxidant effects of chromium supplementation with type 2 diabetes mellitus and euglycemic subjects, *J Agric Food Chem* **52**:1385–1389.

Long J, Ortiz X, Monasterio I, Banziger M. 2003. Improving the nutritional quality of maize and wheat for human consumption. In: Cakmak I, Welch RM, editors. Impacts of agriculture on human health and nutrition. *Encyclopedia of Life Support Systems*. Oxford, UK: UNESCO-EOLSS.

Lyons G, Stangoulis J, Genc Y, Liu F, Graham R. 2007. Biofortification in the food chain, and use of selenium and phyto-compounds in prevention and treatment of chronic diseases. In: Zhu Y, Lepp N, Naidu R, editors. *Biogeochemistry of Trace Elements: Environmental Protection, Remediation and Human Health*. China: Tsinghua University Press, pp. 279–280.

McDowell LR, Fick KR, Ammerman CB, Miller SM, Houser RH. 1978. Minerals for grazing ruminants in tropical regions. In: Conrad JH, McDowell LR, editors. *Proceedings of Latin American Symposium on Mineral Nutrition Research with Grazing Ruminants*. University of Florida.

McGrath SP, Zhao FJ, Fan MS. 2007. Decreasing mineral concentrations in wheat grain over the last 160 years. In: Zhu Y, Lepp N, Naidu R, editors. *Biogeochemistry of Trace Elements: Environmental Protection, Remediation and Human Health*. China: Tsinghua University Press, pp. 281–282.

Mesjasz-Przybylowicz J, Migula P, Przybylowicz WJ, Augustyniak M, Nakonieczny X, Orlowska E. 2007. Transfer of selected metals along simplified food chains of ultramafic ecosystem in Mpumalanga Province, South Africa. In: Zhu Y, Lepp N, Naidu R, editors. *Biogeochemistry of Trace Elements: Environmental Protection, Remediation and Human Health*. China: Tsinghua University Press, pp. 1020–2022.

Monasterio I, Graham RD. 2000. Breeding for trace minerals in wheat. *Food Nutr Bull* **21**:392–396.

Nesamvuni AE, Vorster HH, Margetts BM, Kruger A. 2005. Fortification of maize meal improved the nutritional status of 1- to 3-year-old African children. *Public Health Nutr* **8**:461–467.

Nirupa N, Prasad MNV, Kirti PB. 2007. Expression of pea seed ferritin cDNA in Indian mustard: Nutritional value and oxidative stress tolerance of the transformants. In: Zhu Y, Lepp N, Naidu R, editors. *Biogeochemistry of Trace Elements: Environmental Protection, Remediation and Human Health*. China: Tsinghua University Press, pp. 285–286.

Slingerland MA. 2007. Biofortification in a food chain approach for rice in China. In: Zhu Y, Lepp N, Naidu R, editors. *Biogeochemistry of Trace Elements: Environmental Protection, Remediation and Human Health*. China: Tsinghua University Press, pp. 293–295.

Storckdieck S, Bonsmann G, Hurrell R. 2007. The impact of trace elements from plants on human nutrition: a case for biofortification. In: Zhu Y, Lepp N, Naidu R, editors. *Biogeochemistry of Trace Elements: Environmental Protection, Remediation and Human Health*. China: Tsinghua University Press, pp. 258–259.

Welch RM, House WA, Beebe S, Cheng Z. 2000. Genetic selection for enhanced bioavailable levels of iron in bean (*Phaseolus vulgaris* L.) seeds. *J Agric Food Chem* **48**:3576–3580.

9 Essentiality of Zinc for Human Health and Sustainable Development

M. N. V. PRASAD

Department of Plant Sciences, University of Hyderabad, Hyderabad 500 046, India

Zinc is indispensable for life. It is a constituent of several enzymes, namely, carbonic anhydrase dehydrogenases, aldolases, Cu/zinc superoxide dismutase, isomerases, transphosphorylases, and RNA and DNA polymerases. The zinc deficiency decreases the activity of these enzymes. Zinc-metalloproteins (zinc-finger motif) are regulators of gene expression (DNA-binding transcription factors). Without them, RNA polymerase cannot complete its function of transcribing genetic information from DNA into RNA. Zinc is a co-factor of more than 200 enzymes, such as oxidoreductases, hydrolases, transferases, lyases, isomerases, and ligases. Many of the metalloenzymes are involved in the synthesis of DNA and RNA and protein synthesis and metabolism. Zinc is important for nitrogen metabolism in plants. Zinc regulates the activity of the carbonic anhydrase (CA, carbonatelyase, carbonate dehydratase), which is a ubiquitous enzyme (found in animals, terrestrial plants, eukaryotic algae, cyanobacteria) and is localized both in the cytosol and in the chloroplasts. Zinc-metalloproteins (zinc-finger motif) are regulators of gene expression (DNA-binding transcription factors). In the absence of these, RNA polymerase cannot complete its function of transcribing genetic information from DNA into RNA [Prasad and Hagemeyer, 1999; Prasad, 2001; Prasad and Strzałka, 2002].

Several important food crops can be seriously affected by zinc deficiency. Maize and rice are the most sensitive with wheat being moderately sensitive. If the zinc supply status is inadequate, the crops will be affected by deficiency (Table 1).

Trace Elements as Contaminants and Nutrients: Consequences in Ecosystems and Human Health, Edited by M. N. V. Prasad
Copyright © 2008 John Wiley & Sons, Inc.

TABLE 1. Zinc Sources for Alleviating Zinc Deficiency

Zinc sulfate	$(ZnSO_4 \cdot H_2O)$	35% Zn
Zinc sulfate	$(ZnSO_4 \cdot 7H_2O)$	22% Zn
Basic zinc sulfate	$(ZnSO_4 \cdot Zn(OH)_2)$	55% Zn
Zinc oxide	(ZnO)	67–80% Zn
Zinc carbonate	$(ZnCO_3)$	52–65% Zn
Zinc chloride	$(ZnCl_2)$	45% Zn
Zinc frits	(Zn silicate)	Variable Zn contents
Zinc EDTA (chelate)		12–14% Zn
Zinc polyflavonoid (chelate)		10% Zn
Zinc lignosulfonate (chelate)		5% Zn
Zincated fertilizers		Variable Zn contents
Multi-micronutrient mixtures		Variable Zn contents

Source: Based on Alloway B. Zinc—the vital micronutrient for healthy, high-value crops, with permission from the International Zinc Association (IZA), Belgium.

According to the WHO Report 2002, zinc deficiency ranks in the 11th position among the 20 leading risk factors, a global situation listed in the following order of priority:

1. Underweight
2. Unsafe sex
3. Blood pressure
4. Tobacco
5. Alcohol
6. Unsafe water, sanitation, hygiene
7. Cholesterol
8. Indoor smoke from solid fuels
9. Iron deficiency
10. Overweight
11. Zinc deficiency
12. Low fruit and vegetable intake
13. Vitamin A deficiency
14. Physical inactivity
15. Risk factors for injury
16. Lead exposure
17. Illicit drugs
18. Unsafe health care injections
19. Lack of contraception
20. Childhood sexual abuse

Source: *The WHO Report 2002: Reducing Risks, Promoting Healthy Life*. WHO, ISBN 92 4 256207

Selected traditional uses of zinc (*Source*: Based on IZA 2002. Zinc and sustainable development—a consultative document, Belgium. Prasad [2002])

(a) Food supplement for human health and cattle/livestock health; fertilizer to overcome zinc deficiency in soil and to improve crop productivity.

(b) Glavanized zinc is important material for industrial, commericial, and residential sectors besides highway barriers, guard rails, transmission towers, and bridges.

(c) Protects vehicle tires from degradation by UV rays.

(d) Batteries for hearing aids, portable computers, mobile phones, zero-emission vehicles, etc.

(e) Zinc alloys are in high demand for water purification systems.

(f) Zinc compounds are used as a protective shield on the surface of satellites and space vehicles from variable temperatures.

(g) Zinc compounds are used for electrical leakage circuit breakage systems for circuit protection against power surges and lightning strikes.

(h) Zinc cladding protects undersea and drilling wells communication cables.

1   BIOGEOCHEMICAL CYCLING OF ZINC[a]

Zinc, like all metals, is a natural component of the earth's crust and is an inherent part of our environment. Zinc is present not only in rock and soil, but also in air, water, the biosphere, plants, animals, and humans.

Zinc is constantly being transported by nature, a process called *natural cycling*. Rain, snow, ice, sun, and wind erode zinc-containing rocks and soil. Wind and water carry minute amounts of zinc to lakes, rivers, and the sea, where it collects as sediment or is transported further. Natural phenomena such as volcanic eruptions, forest fires, dust storms, and sea spray all contribute to the continuous cycling of zinc through nature.

Sea salt and the movement of soil dust particles in the air are the principal sources of natural zinc emissions in the atmosphere. Forest fires and volcanoes also contribute in a minor way to zinc's natural cycling. These natural emissions amount to 5.9 million metric tons each year.

Anthropogenic emissions (including industrial emissions, urban waste streams, agriculture, corrosion, tire wear, etc.) are estimated at 57,000 metric tons worldwide each year.

During the course of evolution, all living organisms have adapted to the zinc in their environment and have used it for specific metabolic processes.

---

[a]Source: IZA 2002. Zinc and sustainable development—a consultative document, Belgium.

The amount of zinc present in the natural environment varies from place to place and from season to season. For example, the amount of zinc in the earth's crust ranges between 10 and 300 milligrams per kilogram, and zinc in rivers varies from less than 10 µg per liter to over 200 µ. Similarly, falling leaves in autumn lead to a seasonal increase in zinc levels in soil and water.

Zinc is absolutely essential for the normal healthy growth and reproduction of all higher plants, animals, and humans and is therefore called an "essential trace element" or a "micronutrient." A very wide range of crops are affected by zinc deficiency, including cereals (rice, wheat, and maize), fodder crops (sorghum and sudangrass), pulses (beans, chickpeas and soybeans), bush and tree fruits (apples, citrus, peaches), nuts (pecans), vegetables (potato, onion and sugarcane), and non-food crops such as cotton and tobacco. When crops have a deficient supply of zinc, yield will be reduced and quality of the crop product may also suffer.

2   DISTRIBUTION OF ZINC DEFICIENCY IN SOILS ON A GLOBAL LEVEL[a]

Australia: Most of the coastal land (cultivated belt) in the country: Victoria, South Australia, Queensland, Northern New South Wales, South West of Western Australia, coastal areas and islands of Tasmania.

New Zealand: Shallow Niue soils.

United States of America: Most of the cultivated areas in California, Arizona, Midwest states, Florida (but most of the country affected to a certain extent in 40 states).

Central and South America: Central Plateau of Brazil, Llanos Orientales in Colombia and Venezuela, Costa Rica, Guatemala, Mexico, Peru, NW Puerto Rico, French Guinea.

East Asia: Widespread deficiencies throughout India and Pakistan. In Bangladesh, 2 million ha of paddy rice soils are zinc-deficient. China, Japan and Philipiines.

China: Calcareous soils, including Yellow River, Yangtze, and rice soils throughout Northern China, which is more deficient than Southern China.

Japan: Reported in rice.

Philippines: 500,000 ha of rice soils deficient.

Africa: Common in West Africa, NE, SW, and W Nigeria, Zimbabwe, Malagasy, Congo, Sierra Leone, Gezira region of Sudan, SW Cape Province in South Africa.

Middle east: Iran, Iraq, Syria, and Turkey, with calcareous course textured soils or arid/semi-arid areas most at risk. Current guestimate is that 50% of the area is zinc-deficient.

---

[a]Based on Alloway B. Zinc—the vital micronutrient for healthy, high-value crops, with permission from the International Zinc Association (IZA), Belgium.

TABLE 2. Content of Zinc and Phytate Content, Phytate-to-Zinc Molar Ratios of Commonly Consumed Foods[a]

Food Groups	Zinc (mg/100 g)	Phytate (mg/100 g)	Phytate:Zinc (Molar Ratio)
Seeds, nuts (sesame, pumpkin, almond, etc.)	2.9–7.8	1,760–4,710	22–88
Beans, lentils (soy, kidney bean, chickpea, etc.)	1.0–2.0	110–617	19–56
Whole-grain cereal (wheat, maize, brown rice, etc.)	0.5–3.2	211–618	22–53
Refined cereal grains (white flour, white rice, etc.)	0.4–0.8	30–439	16–54
Bread (white flour, yeast)	0.9	30	3
Fermented cassava root	0.7	70	10
Tubers	0.3–0.5	93–131	26–31
Vegetables	0.1–0.8	0–116	0–42
Fruits	0–0.2	0–63	0–31

[a]Dairy (milk, cheese) 0.4–3.1; liver, kidney (beef, poultry) 4.2–6.1; meat (beef, pork) 2.9–4.7; poultry (chicken, duck, etc.) 1.8–3.0; seafood (fish, etc.) 0.5–5.2; eggs (chicken, duck) 1.1–1.4. Eggs contain zinc but lack phytate. Data derived from the International Mini List.

Phytate:zinc molar ratio is an important factor for zinc bioavailability. Mere intake of zinc-containing foods may not be sufficient to ameliorate the zinc deficiency. Phytate in food would interfere with zinc availability and hence phytate:zinc molar ratio determines the zinc absorption in the diet.

$$\frac{\text{mg phytate per day}/660}{\text{mg zinc per day}/65.4}$$

For example, if phytate intake is 883 mg/day and zinc intake is 7 m/day, then the phytate:zinc molar ratio is 12.5.

Source: Based on data from International Zinc Nutrition Consultative Group Technical Document #1—Assessment of the risk of zinc deficiency in populations and options for its control. Hotz C, Brown KH, editors. 2004. With permission from the United Nations University, Japan.

> Europe: Unevenly distributed but mainly in southern Europe (with calcareous soils). Problems have been reported in sandy soils in South Western France and Northern Germany on sandy soils and various soil types in Romania.

Zinc deficiency in crops of different types is a major problem in many parts of the world and leads to serious inefficiencies in crop production, especially with new, high-yielding varieties of crops that require relatively expensive inputs of fertilizers, pesticides, and irrigation in order to attain their optimum yield [Cakmak et al., 1996d]. Where zinc has been shown to be deficient by soil tests, plant analysis, or diagnostic symptoms, the problem can easily be rectified (Table 2). Relatively small amounts of zinc compounds such as zinc sulfate can cure the deficiency and last for several years before they need to be repeated. This treatment is highly cost-effective when the costs of the zinc application and the value of the extra yield are considered.

The human body needs zinc (Fig. 1). The beneficial effects of zinc have been known for a long time, especially in the healing of wounds. Our body rapidly

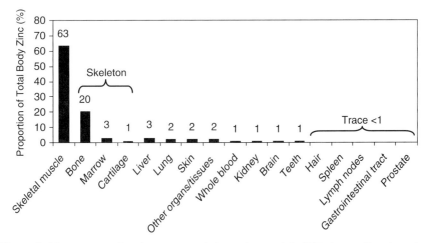

Figure 1. Zinc content of major organs and tissues in an adult (70 kg) man. Based on data from International Zinc Nutrition Consultative Group Technical Document #1—Assessment of the risk of zinc deficiency in populations and options for its control. Hotz C, Brown KH, editors. 2004. With permission from the United Nations University, Japan. Adapted from Iyengar [1998].

concentrates zinc around a fresh wound and inside the newly formed scab. Zinc is active in the healing process. The highest concentrations of zinc in the human body are found in our reproductive organs. Zinc is present in more than 300 hormones and enzymes that contribute to the human body life processes. None of the remaining metals is so important for these vital functions. These hormones and enzymes regulate wound healing, digestion, reproduction, sight, respiration, kidney function, sugar balance, taste, and many other functions.

Additional information on zinc function is needed in the following areas:

- Zinc and infection.
- Effects of zinc on specific etiologies of infection, including malaria and other parasitic diseases, tuberculosis, and HIV.
- The role of zinc in reproductive health and the consequences of zinc deficiency during pregnancy, including fetal development, delivery complications, postpartum maternal health, and infant health.
- Risk assessment of zinc toxicity.

3 ZINC INTERVENTION PROGRAMS

See Chapter 8.

Myo-inositol hexaphosphate (phytic acid) consists of a ring of six phosphate ester groups. Phytate is the magnesium, calcium, or potassium salt of phytic acid. The term

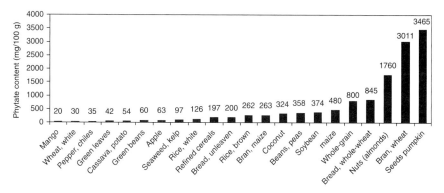

Figure 2. Phytate content of selected foods. Broccoli, cabbage, carrots, eggplant, lettuce, mushrooms, onions, squash, sweet corn, tomatoes, turnip, berries, citrus, melons, stonefruit, beef, pork, other game, poultry, organ meat, fish, shellfish, grubs, locusts, milk, cheese, yogurt, and eggs are known to have no phytate content. (Data Source: World Food Dietary Assessment Program, 2.0; University of California, Berkeley, USA.) Based on data from International Zinc Nutrition Consultative Group Technical Document #1—Assessment of the risk of zinc deficiency in populations and options for its control. Hotz C, Brown KH, editors. 2004. With permission from the United Nations University, Japan.

phytate represents phytic acid molecule, as well as its salt forms. Phytate is a phosphorus storage molecule with a high natural content in seeds, including cereal grains, nuts, and legumes, and a lower content in other plant foods, such as fruits, leaves, and other vegetables. In legumes, phytate is uniformly distributed and associated with protein, whereas in cereal grains it is generally concentrated in the bran; in maize, the majority of phytate exists in the germ. Phytate is a strong chelator of minerals, including zinc. Because phytate cannot be digested or absorbed in the human intestinal tract, minerals bound to phytate also pass through the intestine unabsorbed. The inhibitory effect of phytate on zinc absorption appears to follow a dose-dependent response, and the phytate:zinc molar ratio of the diet has been used to estimate the proportion of absorbable zinc.

The phytate content and the phytate:zinc molar ratio of some commonly consumed foods and the zinc deficiency high-risk countries are shown in Fig. 2 and Table 3, respectively. In general, seeds, nuts, legumes, and unrefined cereal grains have the highest phytate:zinc molar ratios, which range from 22 to 88, while other plant foods have phytate:zinc molar ratios in the range of 0–42. Animal source foods do not contain phytate and therefore have a phytate:zinc molar ratio equivalent to zero. Calcium also has an inhibitory effect on zinc absorption, although this may only occur when phytate is also present in the diet. The inhibitory effect of calcium may result from the formation of insoluble calcium–zinc–phytate complexes in the intestinal tract. Both the total amount and type of protein in the diet influence zinc absorption. Increasing protein content results in a greater percent absorption of dietary zinc. Animal protein, such as from meat and eggs, including

TABLE 3. Zinc Deficiency—The High-Risk Countries

Country	Region	Zinc	Phytate	P:Zn Molar Ratio
Tajikistan	Eastern Europe	5.5	811	14.3
Zerbaijan	Eastern Europe	6.9	1020	14.7
Djibouti	Sub-Saharan Africa	6.2	955	15.4
Ecuador	Latin America	8.0	1323	16.4
Kiribati	Western Pacific	9.1	1519	16.5
Philippines	Southeast Asia	7.8	1344	17.1
Guyana	Latin America	8.3	1445	17.2
Congo, Republic of	Sub-Saharan Africa	6.2	1163	18.7
Peru	Latin America	7.6	1547	20.2
Madagascar	Sub-Saharan Africa	7.4	1554	20.8
Sao Tome and principe	Sub-Saharan Africa	7.0	1489	21.0
Liberia	Sub-Saharan Africa	5.4	1162	21.2
Viet Nam	Southeast Asia	9.2	2008	21.6
Guinea	Sub-Saharan Africa	7.3	1592	21.7
Comoros	Sub-Saharan Africa	6.0	1348	22.4
Senegal	Sub-Saharan Africa	9.1	2184	23.9
Cambodia	Southeast Asia	7.1	1722	24.0
Guinea-Bissau	Sub-Saharan Africa	8.9	2163	24.2
Angola	Sub-Saharan Africa	6.5	1616	24.6
Congo, Dem Republic of	Sub-Saharan Africa	5.7	1421	24.7
Sri Lanka	Southeast Asia	8.6	2221	25.4
Cameroon	Southeast Asia	9.0	2339	25.6
Laos	Southeast Asia	7.9	2031	25.6
India	South Asia	10.9	2906	26.3
Keny	Sub-Saharan Africa	8.1	2195	26.7
Eritrea	Sub-Saharan Africa	8.2	2223	26.8
Gambia	Sub-Saharan Africa	8.1	2200	26.9
Haiti	Latin America	6.8	1870	27.3
Bangladesh	South Asia	7.4	2064	27.7
Myanmar	Southeast Asia	9.3	2612	27.9
Indonesia	Southeast Asia	10.0	2859	28.4
Korea Dem People's Rep	Western Pacific	9.8	2834	28.8
Mozambique	Sub-Saharan Africa	6.2	1861	29.6
Tanzania	Sub-Saharan Africa	7.9	2392	30.1
Honduras	Latin America	7.9	2466	30.7
Lesotho	Sub-Saharan Africa	10.2	3500	33.8
Zambia	Sub-Saharan Africa	8.3	2874	34.3
Burundi	Sub-Saharan Africa	7.6	2668	34.6
El Salvador	Latin America	8.9	3132	34.7
Zimbabwe	Sub-Saharan Africa	8.3	2953	35.2
Nicaragua	Latin America	7.5	2696	35.4
Guatemala	Latin America	8	2950	36.3
Malawi	Sub-Saharan Africa	8.9	3370	37.3

Source: Based on data from International Zinc Nutrition Consultative Group Technical Document #1—Assessment of the risk of zinc deficiency in populations and options for its control. Hotz C, Brown KH, editors. 2004. With permission from the United Nations University, Japan.

whey protein, appear to have further enhancing effects on zinc absorption, although casein may be inhibitory.

4   ZINC-TRANSPORTING GENES IN PLANTS

Use of plants well beyond food and fiber is the beginning of environmental biogeo-technology that would have drastic influence on our resource management. Global industrialization has resulted in large areas of soils worldwide that remain polluted with toxic levels of trace elements.

Plants that hyperaccumulate metals occur on metal-rich soils and accumulate metals in their tissues. Hyperaccumulation has been confirmed for several metals including zinc (up to 4% zinc in shoot dry biomass [Baker and Brooks, 1989; Lombi et al., 2000] (Fig. 3). Zinc is a constituent of several enzymes, namely, carbonic anhydrase dehydrogenases, aldolases, $Cu/zinc$ superoxide dismutase, isomerases, transphos-phorylases, and RNA and DNA polymerases. Therefore, zinc deficiency results in mal-function or no function of these enzymes. Zinc-metalloproteins (zinc-finger motif) are regulators of gene expression (DNA-binding transcription factors). In the absence of these, RNA polymerase cannot complete its function of transcribing genetic information from DNA into RNA. Zinc is a cofactor of more than 200 enzymes, such as oxidoreductases, hydrolases, transferases, lyases, isomerases, and ligases. Many of the metalloenzymes are involved in the synthesis of DNA and RNA and protein synthesis and metabolism.

In Brassicaceae, about 21 species belonging to belong to three genera, namely, *Cochlearia, Arabidopsis* and *Thlaspi*, are reported to be zinc hyperaccumulators (Fig. 3, Table 4). Zinc is reported to inhibit photosynthetic CO_2 fixation, Hill reaction, and photosynthetic electron transport in different plants. Zinc is also known to inter-act with the donor side of PSII. The identified sites of zinc action in light reactions of photosynthesis are PSI, PSII, and the plastoquinone pool. In *Phaseolus vulgaris*, zinc affected Rubisco. Inhibition of net photosynthetic CO_2 fixation rate and increase of the CO_2 compensation point were observed in intact *Phaseolus vulgaris* after zinc treatment. Zinc also inhibited Rubisco activity without affecting its oxygenase

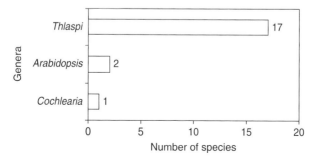

Figure 3. Zinc hyperaccumuting taxa of Brassicaceae.

TABLE 4. Brassicaceae that Accumulate Zn > 1 mg g⁻¹ day wt in Ascending Order[a]

Metal	Taxa	Distribution
1150	*Thlaspi idahoense*	United States
1400	*Cochlearia pyrenaica*	Belgium
1680	*Thlaspi violascens*	Turkey
1800	*T. goesingense*	Austria
3000	*T. montanum*	United States
3000	*T. ochroleucum*	Greece
3090	*T. parvifolium*	United States
3520	*T. liaceum*	Turkey
3890	*T. magellanicum*	Argentina, Chile
10500	*T. bulbosum.*	Greece
21000	*T. praecox*	Bulgaria
26700	*Arabidopsis thaliana*	Pennsylvania, United States
13600	*Arabidopsis halleri*	Germany
16000	*Thlaspi stenocarpum*	Spain
21000	*T. rotundifolium* subsp. *cepaeifolium*	Italy
11000	*T. rotundifolium*	France
39600	*T. caerulescens*	Germany

[a]For details, see Palmer et al. [2001].

activity. Zinc ions changed the hydrolytic activity of chlorophyllase in rice leaves. The activity of carbonic anhydrase (CA) decreased in a number of plant species as a consequence of zinc deficiency. The concomitant effect of it is reduced photosynthesis [Rengel, 1999]. The higher photosynthetic efficiency of crops growing in zinc deficiency is attributed to the genotype and the efficient function of CA activity [Rengel, 1999]. Metal transporting genes have been identified in *Arabidopsis* (Brasicaceae). Overexpression of an *Arabidopsis* zinc transporter, cation diffusion facilitator (CDF) gene-enhanced resistance to zinc accumulation. Transgenic plants showed increased zinc uptake, and tolerance and antisense of this gene led to wild-type zinc tolerance in transgenic plants. Zinc transporters can be manipulated to increase selectivity and accumulation of metal ions.

Zhao and Eide [1996a,b] isolated and identified the gene *ZINCT1* as one of the micronutrient transport genes, with high sequence homology with other zinc transport genes isolated from yeast. A family of zinc transporter genes have also been identified in *Arabidopsis* that respond to zinc deficiency. Zinc hyperaccumulation in *Thlaspi caerulescens* occurs because of the *ZINCT1* gene, which encodes a high-affinity zinc transporter. This gene is constitutively expressed at a much higher level in *T. caerulescens* than in *T. arvense*, where its expression is stimulated by zinc deficiency. In fact, plant zinc status is shown to alter the normal regulation of zinc transporter genes in *T. caerulescens*. An important aspect of zinc hyperaccumulation and tolerance in *T. caerulescens* is also the production of low-molecular-weight compounds involved in zinc detoxification in

TABLE 5. Zinc for Man and Biosphere—Human Health and Sustainable Development[a]

Function	Authors	Year
Accumulation		
Accumulation of free tryptophan and tryptamine in zinc-deficient maize seedlings	Takaki and Kushizaki	1970
Contamination of *Quercus robur* L. and *Pinus silvestris* L. foliage by smelter	Godzik et al.	1970
Bioaccumulation in plants growing on mine deposits	Rascio	1977
Zinc and Cd accumulation by corn inbreds grown on sludge amended soil	Hinesly et al.	1978
Accumulation by the aquatic liverwort *Scapania*	Whitton et al.	1982
Concentrations in soil extracts and crops grown on four soils treated with metal-loaded sewage sludges	Sanders et al.	1987
Bioconcentrations in salt marsh plants along the Dutch coast	Otte et al.	1991
Bioconcentrations in aquatic macrophytes from the Sudbury and Muskoka regions of Ontario	Reimer and Duthie	1993
Accumulation in soil and cultivated vegetables and weeds grown in industrially polluted areas	Barman and Lal	1994
Accumulation in the rhizosphere of wetland plants	Otte et al.	1995
Accumulation in plant seeds growing in metalliferous habitats in Bulgaria	Stefanov et al.	1995
Accumulation in wetland soils	Doyle and Marinus	1997
Internal zinc accumulation is correlated with increased growth in rice suspension culture	Hossain et al.	1997
Zinc distribution and excretion in the leaves of the gray mangrove, *Avicennia marina* (Forsk.) Vierh	MacFarlane and Burchett	1999
Somatic hybridization between the zinc accumulators *Thlaspi caerulescens* and *Brassica napus*	Brewer et al.	1999
Accumulation in Gametophytes and Sporophytes of the moss *Funaria hygrometrica* (Funariales)	Basile et al.	2001
Bioaccumulation in foliage mine flora	Pugh et al.	2002
Zinc and copper accumulation and tolerance in populations of *Paspalum distichum* and *Cynodon dactylon*	Shu et al.	2002
Accumulation by 12 wetland plant species thriving in contaminated sites	Deng et al.	2004
Accumulation in plants in mine area	Zu et al.	2004
Biochemical tolerance		
The role of malate, oxalate, and mustard oil glucosides in the evolution of zinc resistance in herbage plants	Mathys	1977
Tryptophan and tryptamine in zinc-deficient plant	Takaki and Kushizaki	1972
Effect of fallout of windborne zinc from a fuming kiln on soil geochemical prospecting at Berg Aukas, South West Africa	Marchant	1974
Biogeochemical prospecting of zinc using *Juniperus virginiana*	Adams and Hood	1976

(Continued)

TABLE 5. *Continued*

Function	Authors	Year
Supplementation through sewage	Stoveland et al.	1979
Biogeochemicals cycling and bioavailability, uptake, and translocation in *Phragmites australis*	Schierup and Larsen	1981
Induction of enzyme capacity in plants as a result of heavy metal toxicity: Dose–response relations in *Phaseolus vulgaris* L., treated with zinc	Van Assche et al.	1988
Zinc fertilization on different varieties of soybean (*Glycine max*)	Saxena and Chandel	1992
Zn-enriched organic manures on Zn nutrition of wheat and residual effect on soybean	Gupta et al.	1992
Free tryptophan and indoleacetic acid in zinc-deficient radish shoots	Domingo et al.	1992
Influence of zinc on the growth and distribution of elements on ginseng plants	Fu-Cheng et al.	1993
The role of low-molecular-weight organic acids in the mechanism of increased zinc tolerance in *Silene vulgaris*	Harmens et al.	1994
Genotypic differences in the production and partitioning of carbohydrates between roots and shoots of wheat grown under zinc deficiency	Perason and Rengel	1997
Expression of the Zn^{2+}-containing hydroxynitrile lyase from flax (*Linum usitatissimum*) in *Pichia pastoris*—utilization of the recombinant enzyme for enzymatic analysis and site-directed mutagenesis	Trummler et al.	1998
The zinc violet and its colonization by arbuscular mycorrhizal fungi	Hildebrandt et al.	1999
The tri-trophic transfer of Zn from the agricultural use of sewage sludge	Winder et al.	1999
In vitro selection and biochemical characterisation of Zn-adapted callus lines in *Brassica* spp	Rout et al.	1999
Toxicity level for phytoavailable zinc in compost-peat	Bucher and Schenk	2000
Indole-3-acetic acid and zinc on the growth, osmotic potential, and soluble carbon and nitrogen components of soybean plants growing under water deficiency	Gadallah	2000
Arbuscular mycorrhiza and plant succession on zinc smelter spoil heap in Katowice-Welnowiec	Gucwa-Przepiora and Turnau	2001
Changes in soil solution zinc and pH and uptake of zinc by arbuscular mycorrhizal red clover in zinc-contaminated soil	Li and Christie	2001
The redox status of plant cells (AsA and GSH) is sensitive to zinc-imposed oxidative stress in roots and primary leaves of *Phaseolus vulgaris*	Cuypers et al.	2001
The use of zinc finger peptides to study the role of specific factor binding sites in the chromatin environment	Segal et al.	2003

(Continued)

TABLE 5. *Continued*

Function	Authors	Year
Bioleaching of zinc and aluminum from industrial waste sludges by means of *Thiobacillus ferrooxidans* waste management	Solisio et al.	2002
The role of arbuscular mycorrhiza on uptake by red clover growing in a calcareous soil spiked with zinc	Chen et al.	2003
Arbuscular mycorrhiza can depress translocation of zinc to shoots of host plants in soils moderately polluted with zinc	Christie et al.	2004
Cd–Zn interaction		
Zinc–copper interaction affecting plant growth on a metal contaminated soil	Luo and Rimmer	1995
Cadmium–zinc interactions on hydroponically grown bean (*Phaseolus vulgaris* L.)	Chaoui et al.	1997
Cadmium and zinc interactions and their transfer in soil–crop system under actual field conditiona	Nan et al.	2002
Zinc alleviates cadmium–induced oxidative stress in *Ceratophyllum demersum* L.: a free-floating freshwater macrophyte plant	Aravind and Prasad	2003
Interactions with Cd and tolerance in two populations of *Holcus lanatus* L. grown in solution culture	Coughtrey and Martin	1979
Interaction with Cd on metal uptake and regeneration of tolerant plants in linseed	Chakravarty and Srivastava	1997
Deficiency		
Effects of zinc deficiency on the nitrogen metabolism of meristematic tissues of rice plants with reference to protein synthesis	Kitagishi and Obata	1986
Increase in membrane permeability and exudation in roots of zinc-deficient plants	Cakmak and Marschner	1988a,b
Zinc deficiency in mango	Agarwala et al.	1988
Identification and inheritance of inefficient zinc absorption in soybean	Hartwig et al.	1991
Zinc deficiency in wheat genotypes grown in conventional and chelator-buffered nutrient solutions	Rengel	1999
Detoxification		
Deposition of zinc phytate in globular bodies in roots of *Deschampsia caespitosa* ecotypes: A detoxification mechanism	Van Steveninck et al.	1987
Diagnosis of deficiency		
Diagnosing zinc deficiency in rapeseed and mustard by seed analysis	Rashid et al.	1994
Elemental defense		
Zinc nutrition of *Nicotiana tabacum* L. in relation to multiplication of tobacco mosaic virus	Helms and Glenn	1955

(*Continued*)

TABLE 5. *Continued*

Function	Authors	Year
Zinc nutrition in higher plants	Takaki and Kushizaki	1976
Influence of zinc smelter emissions on leaves of *Pinus sylvestris* and *Vaccinium* species as revealed by some morphological and ecophysiological indices	Czuchajowska	1987
Zinc nutritional status on growth, protein metabolism, and levels of indole-3-acetic acid and other phytohormones in bean	Cakmak et al.	1989
Selecting zinc-efficient genotypes for soils of low zinc status	Graham et al.	1992
Effects of plants and soil microflora on leaching of zinc from mine tailings	Banks et al.	1994
Importance of seed zinc content for wheat growth on zinc-deficient soil	Rengel and Graham	1995a
Wheat genotypes differ in zinc efficiency when grown in chelate-buffered nutrient solution	Rengel and Graham	1995b,c
Zinc efficiency of oilseed rape (*Brassica napus* and *B. juncea*) genotypes	Grewal et al.	1997
Impact of arsenic and zinc from mine tailing to grazing cattle	Bruce et al.	2003
Integrated assessment of Zn highway pollution: bioaccumulation in soil, grasses, and land snails	Viard et al.	2004
Exudates		
Nitrogenous constituents of the zinc-deficient barley leaves and the material exudated from them	Fujiwara and Tsutsumi	1959
Release of zinc- and iron-mobilizing root exudates	Römheld and Marschner	1990
Release of zinc mobilizing root exudates	Zhang et al.	1991
Hyperaccumulation		
Hyperaccumulation in metallophytes from mining areas of Central Europe	Reeves and Brooks	1983
Compartmentalization of zinc in roots and leaves of the zinc hyperaccumulator *Thlaspi caerulescens* J. & C. Presl	Vazquez et al.	1994
Zinc hyperaccumulation in *Thlaspi caerulescens*	Tolrà et al.	1996
Altered zinc compartmentation in the root symplasm and stimulated zinc absorption into the leaf as mechanisms involved in zinc hyperaccumulation	Lasat et al.	1998a,b
Interaction with organics		
2-(4-20, 40-dichlorophenoxy-phenoxy)-methyl propanoate enhanced uptake and utilization by wheat	Robson and Snowball	1989
Influence of chlorosulfuron on the uptake and utilization zinc in wheat	Robson and Snowball	1990

(*Continued*)

TABLE 5. *Continued*

Function	Authors	Year
The role of organic matter in association with zinc in selected arable soils from Kujawy Region	Dabkowska-Naskret	2003
Localization		
Localization in plant cells	De Filippis and Pallaghy	1976
The distribution of cadmium, copper, lead, and zinc in eelgrass (*Zostera marina* L.)	Brix and Lyngby	1982
Localization of zinc in *Thlaspi caerulescens* (Brassicaceae), a metallophyte that can hyperaccumulate cadmium	Vazquez et al.	1992
Metallothioneins		
Isolation and characterisation of two divergent type 3 metallothioneins from oil palm, *Elaeis guineensis*	Abdullah et al.	2002
Mycorrhizal uptake		
Influence of zinc on development of the endomycorrhizal fungus *Glomus mosseae* and its mediation of phosporus uptake by glycine max "Amsoy 71"	McIlveen and Cole	1979
The role of mycorrhizas in the interactions of phosphorus with zinc, copper, and other elements	Lambert et al.	1979
Effects of mycorrhizae and fertilizer amendments on zinc tolerance of plants	Shetty et al.	1995
Localization in mycorrhizal *Euphorbia cyparissias* from zinc wastes in southern Poland	Turnau et al.	1998
Acquisition of Cu, zinc, Mn, and Fe by mycorrhizal maize (*Zea mays* L.) grown in soil at different P and micronutrient levels	Liu et al.	2000
Solubilization of insoluble inorganic zinc compounds by ericoid mycorrhizal fungi derived from heavy-metal-polluted sites	Martino et al.	2003
Phytoavailability		
Effects of farmyard manure and phosphorus on zinc transformations and phytoavailability in two alfisols of India	Rupa et al.	2003
Phytoextraction		
Phytoextraction from a contaminated soil	Ebbs et al.	1997
Phytoextraction of zinc by oat (*Avena sativa*), barley (*Hordeum vulgare*) and Indian mustard (*Brassica juncea*)	Ebbs and Kochian	1998
Measurement of in situ phytoextraction of zinc by spontaneous metallophytes growing on a former smelter site	Schwartz et al.	2001

(*Continued*)

TABLE 5. *Continued*

Function	Authors	Year
Phytomanagement		
Lime and pig manure as ameliorants for revegetating lead/zinc mine tailings: a greenhouse study	Ye et al.	1999
Long-term sewage sludge applications on Zn status in a clay loam soil under pasture grass in Zimbabwe	Nyamangara and Mzezewa	1999
Phytoremediation		
Phytoremediation potential of *Thlaspi caerulescens*	Brown et al.	1994
Phytotoxicity		
Zinc phytotoxicity	Chaney	1994
Tolerance		
Differential tolerance to Cd and zinc of arbuscular mycorrhizal (AM) fungal spores isolated from heavy metal-polluted and unpolluted	Weissenhorn et al.	1994
Essentiality throughout the root zone for optimal plant growth and development	Nable and Webb	1993
Response of tomato (*Lycopersicon esculentum* L.) cultivars to foliar application of zinc when grown in sand culture at low zinc	Kaya and Higgs	2002
Nuclear microprobe studies of elemental distribution in seeds of *Biscutella laevigata* L. from zinc wastes in Olkusz	Mesjasz-Przybyłowicz et al.	2001
Relationship between zinc and auxin in the growth of higher plants	Skoog	1940
Zinc in culture solution on the growth of rice plants	Ishizuka and Tanaka	1962
Plant availability of Zn	Schnappingeran and Martens	1974
Growth response of the green algae *Chlorella vulgaris* to selective concentrations of zinc	Rachlin and Farran	1974
Soluble zinc containing extracts from two australian species	Forago and Pitt	1977
Zinc sensitivity in *Phaseolus*: Expression in cell culture	Christianson	1979
Zinc influenced water relations in *Phaseolus vulgaris*	Rauser and Dumbroff	1981
Zinc tolerance of mycorrhizal *Betula*	Brown and Wilkins	1985
Vesicular arbuscular mycorrhizae decrease zinc-toxicity to grasses growing in zinc-polluted soil	Dueck et al.	1986
Phosphorus and zinc fertilization on the solubility of zinc in two alkaline soils	Norvell et al.	1987
Effect of zinc deficiency in wheat on the release of zinc and iron mobilizing root exudates	Zhang et al.	1989
Potato response to zinc as influenced by genetic variability	Sharma and Grewal	1990
The influence of zinc on the distribution and evolution of metallophyte	Baker and Proctor	1990

(Continued)

TABLE 5. *Continued*

Function	Authors	Year
Zinc tolerance and the binding of zinc as zinc phytate in *Lemna minor*. X-ray microanalytical evidence	Van Steveninck et al.	1990
Low-molecular-weight metal complexes in the freshwater moss *Rhynchostegium riparioides* exposed to elevated concentrations of Zn, Cu, Cd, and Pb in the laboratory and field	Jackson et al.	1991
Survival of the indigenous population of *rhizobium leguminosarum biovar trifolii* in soil spiked with Zn	Chaudri et al.	1992
Is the release of phytosiderophores in zinc-deficient wheat plants a response to impaired iron utilization?	Walter et al.	1994
Effect of zinc and iron deficiency on phytosiderophore release in wheat genotypes differing in zinc efficiency	Cakmak et al.	1994
Characters of root geometry of wheat genotypes differing in zinc efficiency	Dong et al.	1995
Phytosiderophore release in bread and durum wheat genotypes differing in zinc efficiency	Cakmak et al.	1996a
Phytosiderophore release in bread and durum wheat genotypes differing in zinc efficiency	Cakmak et al.	1996b
Zinc-efficient wild grasses enhance release of phytosiderophores under zinc deficiency	Cakmak et al.	1996a
Dry matter production and distribution of zinc in bread and durum wheat genotypes differing in zinc efficiency	Cakmak et al.	1996b
Physiological characterization of root zinc^{2+} absorption and translocation to shoots in zinc hyperaccumulator and nonaccumulator	Lasat et al.	1996
Toxicity of zinc and copper to *Brassica* species: Implications for phytoremediation	Ebbs and Kochian	1997
Rhizobium in soils contaminated with copper and zinc following the long-term application of sewage sludge and other organic wastes	Smith and Read	1997
Phytosiderophore release by sorghum, wheat, and corn under zinc deficiency	Hopkins et al.	1998
Rhizobia-early leaf acacia (*Acacia auriculaeformis*) symbiotic association	Zhang et al.	1998
Concerted action of antioxidant enzymes and curtailed growth under zinc toxicity in *Brassica juncea*	Prasad et al.	1999
Micro-PIXE studies of elemental distribution in seeds of *Silene vulgaris* from a zinc	Mesjasz-Przybyłowicz et al.	1999
In vitro selection and regeneration of zinc tolerant calli from *Setaria italica*	Samantaray et al.	1999
Differential tolerance to Fe and zinc deficiencies in wheat germplasm	Rengel and Römheld	2000

(*Continued*)

TABLE 5. *Continued*

Function	Authors	Year
Ecotypes of *Holcus lanatus* tolerant to zinc toxicity also tolerate zinc deficiency	Rengel	1999
Life-cycle phases of a zinc- and cadmium-resistant ecotype of *Silene vulgaris* in risk assessment of polymetallic mine soils	Ernst and Nelissen	2000
Growth response of *Sesbania rostrata* and S. cannabina to sludge-amended lead/zinc mine tailings: A greenhouse study	Ye et al.	2001
Strong induction of phytochelatin synthesis by zinc in marine green alga, *Dunaliella tertiolecta*	Hirata et al.	2001
Evaluation of major constraints to revegetation of zinc mine tailings using bioassay	Ye et al.	2002
Tolerance of rice germplasm to zinc deficiency	Quijano-Guerta et al.	2002
Zinc complex accumulated in *Datura innoxia* plant tissue culture	Rebekah et al.	2002
Comparison of two hyperspectral imaging and two laser-induced fluorescence instruments for the detection of zinc stress and chlorophyll concentration in bahia grass (*Paspalum notatum* Flugge)	Schuerger et al.	2003
Evaluation of cytological effects of Zn^{2+} in relation to germination and root growth of *Nigella sativa* L. and *Triticum aestivum* L.	El-Ghamery et al.	2003
Growth and metal accumulation in vetiver and two *Sesbania* species on lead/zinc mine tailings	Yang et al.	2003
Regulation of phytochelatin synthesis by zinc and cadmium in marine green alga, *Dunaliella tertiolecta*	Tsuji et al.	2003
Shoot biomass and zinc/cadmium uptake for hyperaccumulator and nonaccumulator *Thlaspi* species in response to growth on a zinc-deficient calcareous soil	Ozturk et al.	2003
Zinc and lead sequestration in an impacted wetland system	Peltier et al.	2003
Translocation		
Absorption and translocation of zinc in eelgrass (*Zostera marina* L.)	Lyngby et al.	1982
Transfer of cadmium, lead, and zinc from industrially contaminated soil to crop plants	Dudka et al.	1996
Absorption and translocation of sludge-borne zinc in field-grown maize (*Zea mays* L.)	Jarausch-Wehrheim et al.	1999
Mobility of heavy metals in self-burning waste heaps of the zinc smelting plant in Belovo (Kemerovo Region, Russia)	Sidenko et al.	2001

(*Continued*)

TABLE 5. *Continued*

Function	Authors	Year
Soil–water distribution coefficients and plant transfer factors for ^{65}Zn under field conditions in tropical Australia	Twining et al.	2004
Transport		
Yeast *ZRT1* gene encodes the zinc transporter protein of a high-affinity uptake system induced by zinc limitation	Zhao and Eide	1996a
ZRT2 gene encodes the low-affinity zinc transporter	Zhao and Eide	1996b
Transport of cadmium via xylem and phloem in maturing wheat shoots: Comparison with the translocation of zinc, strontium, and rubidium	Herren and Feller	1997
Identification of a family of zinc transporter genes from *Arabidopsis thaliana* that respond to zinc deficiency	Grotz et al.	1998
Zinc transport in naturally selected zinc-tolerant *Silene vulgaris*	Chardonnens et al.	1999
Molecular physiology of zinc transport in the zinc hyperaccumulator *Thlaspi caerulescens*	Lasat et al.	2000
The molecular physiology of heavy metal transport in the zinc hyperaccumulator (i.e., *Thlaspi caerulescens*)	Pence et al.	2000
Transport rate of arsenic, cadmium, copper, and zinc in *Potamogeton pectinatus*	Wolterbeek and van der Meer	2002
Uptake		
VAM-enhanced uptake		
Uptake patterns by genotypes of sorghum and maize differ amongst hybrids and their parents	Ramani and Kannan	1985
Zinc uptake by corn as affected by vesicular–arbuscular mycorrhizae	Faber et al.	1990
Release of zinc mobilizing root exudates in different plant species as affected by zinc nutritional status	Zhang et al.	1991
Contribution of the VA mycorrhizal hyphae in acquisition of phosphorus and zinc by maize grown in a calcareous soil	Kothari et al.	1991
Role of zinc in the uptake and root leakage of mineral nutrients	Welch and Norvell	1993
^{65}Zn uptake in subterranean clover (*Trifolium subterraneum* L.) by three vesicular–arbuscular mycorrhizal fungi in a root-free sandy soil	Bürkert and Robson	1994
Uptake by hyperaccumulator *Thlaspi caerulescens* and metal tolerant *Silene vulgaris* grown on sludge amended soils	Brown et al.	1995a
Uptake by hyperaccumulator *Thlaspi caerulescens* grown in nutrient solution	Brown et al.	1995b

(Continued)

TABLE 5. *Continued*

Function	Authors	Year
Uptake by water hyacinth (*Eichhornia crassipes*) in Mukuvisi and Manyame Rivers	Zaranyika and Ndapwadza	1995
Influence of arbuscular mycorrhizae on heavy metal (zinc and Pb) uptake and growth of *Lygeum spartum* and *Anthyllis cytisoides*	Diaz et al.	1996
Regulation of zinc uptake in wheat plants	Santa-María and Cogliatti	1998
Uptake of zinc and iron by wheat genotypes differing in zinc efficiency	Rengel et al.	1998
Uptake of zinc by arbuscular mycorrhizal white clover from zinc-contaminated soil	Zhu et al.	2001
Zinc mobility in wheat: Uptake and distribution of zinc applied to leaves or roots	Haslett et al.	2001
Effect of mixed cadmium, copper, nickel and zinc at different pHs upon alfalfa growth and heavy metal uptake	Peralta-Videa et al.	2002
Foliar and root uptake of ^{65}Zn in tomato plants (*Lycopersicon esculentum* Mill.)	Brambilla et al.	2002
Uptake and release of zinc by aquatic bryophytes (*Fontinalis antipyretica* L. ex. Hedw.)	Martins and Boaventura	2002
Influence of early stages of arbuscular mycorrhiza on uptake of zinc and phosphorus by red clover from a low-phosphorus soil amended with zinc and phosphorus	Bi et al.	2003

[a]This listed information on zinc in plant cell and environment pertains to its presence in environment as a contaminant, uptake, accumulation, phytotoxicity, physiology, and biochemistry of toxicity and tolerance in plants, biochemical chraracterization, and sustainable management and development.

the cell (cytoplasm and vacuole) and in the long-distance transport of zinc in the xylem vessels. Citrate was not shown to play an important role in zinc chelation, and malate had constitutively high concentrations in the shoots of both the accumulator *T. caerulescens* and the nonaccumulator *T. ochroleucum*. More recently, direct measurements of the *in vivo* speciation of zinc in *T. caerulescens* using the noninvasive technique of X-ray absorption have revealed that histidine is responsible for the transport of zinc within the cell, whereas organic acids (citrate and oxalate) chelate zinc during long-distance transport and storage. Another constitutive aspect of *T. caerulescens* is the high zinc requirement for maximum growth, compared to other species. This probably depends on the strong expression of the metal sequestration mechanism, which would subtract a large amount of intracellular zinc to normal physiological processes even when the zinc supply is low.

Arabidopsis thaliana has multiple zinc transporters designated ZIP1, ZIP2, ZIP3, and ZIP4. Grotz et al. [1998] demonstrated the specificities of each of the ZIP genes with their experiments. They tested other metal ions for their ability to inhibit zinc uptake mediated by these proteins. Zinc uptake by ZIP1 was not inhibited by a 10-fold excess of Mn, Ni, Fe, and Co. Zinc was the most potent competitor, demonstrating that ZIP1 prefers zinc as its substrate over these metal ions. Cd and Cu also inhibited zinc uptake, but to a lesser extent. This suggests that Cd and Cu may also be substrates for ZIP1.

Pence et al. [2000] reported on the cloning and characterization of a high-affinity zinc^{2+} transporter cDNA, *ZINCT1*, from the zinc/Cd-hyperaccumulating plant, *Thlaspi caerulescens*. Through comparisons to a closely related, nonaccumulator species, *Thlaspi arvense*, the researchers determined that the elevated ability of *T. caerulescens* to take up zinc and Cd was due, in part, to an enhanced level of expression of zinc transporters. Previous physiological studies by the group indicated that the hyperaccumulating ability of *T. caerulescens* was linked to zinc transport at a number of sites in the plant. The researchers transformed a zinc-transport-deficient strain of yeast, ZHY3, with a cDNA library from *T. caerulescens*. By screening for growth on low-zinc medium, they were able to isolate seven clones, five of which represented a 1.2-kb cDNA designated *ZINCT1* (for Zinc transporter) that restored the yeast's ability to grow under low-zinc conditions. The *ZINCT1* gene displayed considerable identity to two *Arabidopsis thaliana* metal transporter genes, ZIP4 (for transporting zinc) and *IRT1* (for transporting Fe). For purposes of comparison, they then cloned the homolog of *ZINCT1* (designated *ZINCT1*-arvense) from the non-hyperaccumulator species *T. arvense*. Expression studies using Northern blots of RNA isolated from the roots and shoots of both *Thlaspi* species revealed that *ZINCT1* is expressed in *T. caerulescens* at extremely high levels. In contrast, expression of *ZINCT1* in *T. arvense* could only be detected at a very low level in both shoots and roots, and then only when the plants had been exposed to conditions of zinc deficiency. To further explore the role of zinc status on *ZINCT1* expression, Pence et al. (2000) exposed both species to a range of zinc concentrations. They found that when *T. caerulescens* was grown in a nutrient solution containing an excess of zinc (50 μM), the transcript level of *ZINCT1* decreased, indicating that *ZINCT1* is not expressed constitutively in *T. caerulescens*. *ZINCT1* transcript levels in *T. arvense* appeared to be unaffected by exposure to excess zinc.

Transport studies in *T. caerulescens* show that *ZINCT1* mediates high-affinity zinc transport. In many plant species, the induction of a high-affinity transporter is characteristic of a nutrient-deficiency response and would correlate with the expression pattern observed for *ZINCT1* in *T. arvense*. The authors speculate that the hyperaccumulation phenotype in *T. caerulescens* may then be due to a mutation in the plant's ability to sense or respond to zinc levels; that is, these plants may be functioning as if they are constantly experiencing zinc deficiency. They propose this is likely the result of a change in global regulation linked to the plant's overall zinc status, supporting the concept that the zinc hyperaccumulation phenotype, at least in this species, is due to a change in the regulation, and not the constitutive expression, of a single gene.

5   ADDRESSING ZINC DEFICIENCY WITHOUT ZINC FORTIFICATION

Zinc is recycled in the body through pancreas, and metallothionein I is implicated in this function.

Dietary phytate is complexed in the intestine by the ionic zinc endogenously secreted in the small intestine and duodenum. A significant portion of the zinc is attached to the carboxylpeptidases A and B. However, large pool of duodenal zinc pool must be reused to sustain homeostasis. It is rather difficult to achieve zinc homeostatis in diets very rich in phytate. However, phytases are available from sprouts and from avariety of sources such as teats, bacteria, and fungi. Supplemenation of phytatse would hydrolyze phytate to provide adequate reabsorbale zinc. Phytase products are commonly administered for monogastric animals to optimize phosphate utilization, but there is a scanty amount of information and research to explore for human health. Phytase application for zinc homeostasis without zinc fortification may be another feasible option for human health.

6   ZINC DEFICIENCY IS A LIMITATION TO PLANT PRODUCTIVITY

Zinc deficiency is one of the most critical factors in plants growing on calcareous soils with high pH values. It is well known that zinc is an important component of many vital enzymes and is a structural stabilizer for proteins, membranes, and DNA-binding proteins [Vallee and Auld, 1990]. Under conditions of Zn deficiency, several metabolic processes are impaired, such as plant development and growth [Cakmak et al., 1989, 1996a–d], RNA metabolism, and protein synthesis [Falchuk et al., 1978] (Fig. 4). Zn is known to have a stabilizing and protective effect on the biomembranes against oxidative and peroxidative damage, loss of plasma membrane integrity [Aravind and Prasad, 2003; Cakmak, 2000; Powell, 2000], and alteration of the permeability of membrane [Bettger and O'Dell, 1981]. It was suggested that Zn may have a role in modulating free radicals and its related processes through anti-oxidant properties [Zago and Oteiza, 2001]. Lack of Zn causes overproduction of oxy-radicals and disorders of the structural and functional integrity of cell membranes [Cakmak and Marschner, 1988a,b].

In many species (corn, sorghum, beans) there may be interveinal chlorosis of the leaves, followed by the development of white necrosis spots when they were grown without zinc. The reduction of chlorophyll level and the destruction of chloroplasts ultrastructure led to decrease in photosynthesis in Zn-deficient plants. The activity of some photosynthetic enzymes is also affected by Zn deficiency. This was shown for carbonic anhydrase (CA), and for ribulose, 1,5-bisphosphate carboxylase (RuBPC, Rubisco) [Jyung et al., 1972]. CA is a ubiquitous enzyme among living organisms which catalyzes the reversible interconversion of CO_2 and HCO_3^-. Zn has a catalytic role in plant and animal CA, being coordinated to the imidazole rings of three histidines close to the active site. Since CA activity is affected by Zn

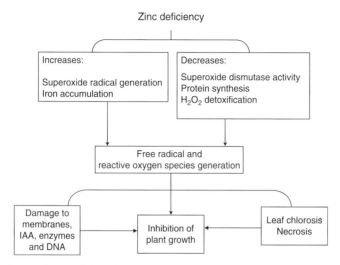

Figure 4. Zinc-deficiency-induced symptoms in plants. Redrawn based on Cakmak [2000], Cakmak and Marschner [1988a,b], Cakmak et al. [1989, 1994, 1996a–d].

deficiency and also it plays a very vital role in maintaining the inorganic carbon levels, this enzyme is used as a biochemical marker to diagnose Zn deficiency [El-Fouly et al., 1997; Salama et al., 1997]. Zinc for man and biosphere and sustainable development is detailed in Table 5.

ACKNOWLEDGMENTS AND DISCLAIMER

The author gratefully acknowledges the United Nations University, Japan, the International Zinc Nutrition Consultative Group (IZiNCG) and the International Zinc Association (IZA) for allowing permission to use copyright information in this review. The author and the publisher cannot assume responsibility for any adverse consequences of the use of material cited in this chapter in field or laboratory.

REFERENCES

Abdullah SNA, Cheah SC, Murphy DJ. 2002. Isolation and characterisation of two divergent type 3 metallothioneins from oil palm, *Elaeis guineensis. Plant Physiology and Biochemistry* **40**(3):255–263.

Adams SC, Hood WC. 1976. Biogeochemical prospecting for copper and zinc using *Juniperus virginiana* L. *Journal of Geochemical Exploration* **6**(1–2):163–175.

Agarwala SC, Nautiyal BD, Chatterjee C, Sharma CP. 1988. Manganese, zinc and boron deficiency in mango. *Scientia Horticulturae* **35**(1–2):99–107.

Antonovics J, Bradshaw AD, Turner RG. 1971. Heavy metal tolerance in plants. *Adv Ecol Res* **7**:1–85.

Aravind P, Prasad MNV. 2003. Zinc alleviates cadmium-induced oxidative stress in *Ceratophyllum demersum* L.: A free floating freshwater macrophyte. *Plant Physiol Biochem* **41**:391–397.

Baker AJM, Brooks RR. 1989. Terrestrial higher plants which hyperaccumulate metallic elements. A review of their distribution, ecology, and phytochemistry. *Biorecovery* **1**:81–126.

Baker AJM, Proctor J. 1990. The influence of cadmium, copper, lead, and zinc on the distribution and evolution of metallophytes in the British Isles. *Plant Syst Evol* **173**:91–108.

Baker AJM, McGrath SP, Sidoli CMD, Reeves RD. 1994a. The possibility of *in situ* heavy metal decontamination of polluted soils using crops of metal-accumulating plants. *Resour Conserv Recycl* **11**:41–49.

Baker AJM, Reeves RD, Hajar ASM. 1994b. Heavy metal accumulation and tolerance in British populations of the metallophyte *Thlaspi caerulescens* J. & C. Presl (Brassicaceae). *New Phytol* **127**:61–68.

Banks MK, Schwab AP, Fleming GR, Hetrick BA. 1994. Effects of plants and soil microflora on leaching of zinc from mine tailings. *Chemosphere* **29**:1691–1699.

Barman SC, Lal MM. 1994. Accumulation of heavy metals (Zn, Cu, Cd and Pb) in soil and cultivated vegetables and weeds grown in industrially polluted fields. *J Environ Biol* **15**:107–115.

Basile A, Cogoni AE, Bassi P, Fabrizi E, Sorbo S, Giordano S, Castaldo Cobianchi R. 2001. Accumulation of Pb and Zn in gametophytes and sporophytes of the moss *Funaria hygrometrica* (Funariales). *Ann Bot* **87**:537–543.

Bettger WJ, O'Dell BL. 1981. A critical physiological role of zinc in the structure and function of biomembranes. *Life Sci* **28**:1425–1438.

Bi YL, Li XL, Christie P. 2003. Influence of early stages of arbuscular mycorrhiza on uptake of zinc and phosphorus by red clover from a low-phosphorus soil amended with zinc and phosphorus. *Chemosphere* **50**(6):831–837.

Boyd RS. 1998. Hyperaccumulation as a plant defensive strategy. In: Brooks RR, editor. *Plants that Hyperaccumulate Heavy Metals*. Oxford, UK: CAB International, pp. 181–201.

Boyd RS, Martens SN. 1998a. The significance of metal hyperaccumulation for biotic interactions. *Chemoecology* **8**:1–7.

Boyd RS, Martens SN. 1998b. Nickel hyperaccumulation by *Thlaspi montanum* var. *montanum* (Brassicaceae): a constitutive trait. *Am J Bot* **85**:259–265.

Bradley R, Burt AJ, Read DJ. 1981. Mycorrhizal infection and resistance to heavy metal toxicity in *Calluna vulgaris*. *Nature (Lond.)* **292**:335–337.

Brambilla M, Fortunati P, Carini F. 2002. Foliar and root uptake of [134]Cs, [85]Sr and [65]Zn in processing tomato plants (*Lycopersicon esculentum* Mill.). *Journal of Environmental Radioactivity* **60**(3):351–363.

Brewer EP, Saunders JA, Angle JS, Chaney RL, McIntosh MS. 1999. Somatic hybridization between the zinc accumulator *Thlaspi caerulescens* and *Brassica napus*. *Theor Appl Genet* **99**(5):761–771.

Brix H, Lyngby JE. 1982. The distribution of cadmium, copper, lead, and zinc in eelgrass (*Zostera marina* L.). *Sci Total Environ* **24**(1):51–63.

Brown MT, Wilkins DA. 1985. Zinc tolerance of mycorrhizal Betula. *New Phytol.* **99**:101–106.

Brown SL, Chaney RL, Angle JS, Baker AJM. 1994. Phytoremediation potential of *Thlaspi caerulescens* and Bladder Campion for zinc and cadmium contaminated soil. *J Environ Qual* **23**:1151–1157.

Brown SL, Chaney RL, Angle JS, Baker AM. 1995a. Zinc and cadmium uptake by hyperaccumulator *Thlaspi caerulescens* and metal tolerant *Silene vulgaris* grown on sludge amended soils. *Environ Sci Technol* **29**:1581–1585.

Brown SL, Chaney RL, Angle JS, Baker AM. 1995b. Zinc and cadmium uptake by hyperaccumulator *Thlaspi caerulescens* grown in nutrient solution. *Soil Sci Am J* **59**:125–133.

Bruce SL, Noller BN, Grigg AH, Mullen BF, Mulligan DR, Ritchie PJ, Currey N, Ng JC. 2003. A field study conducted at Kidston Gold Mine, to evaluate the impact of arsenic and zinc from mine tailing to grazing cattle. *Toxicology Letters* **137**(1–2):23–34.

Bucher AS, Schenk MK. 2000. Toxicity level for phytoavailable zinc in compost-peat substrates. *Scientia Horticulturae* **83**(3–4):339–352.

Bürkert B, Robson A. 1994. ^{65}Zn uptake in subterranean clover (*Trifolium subterraneum* L.) by three vesicular–arbuscular mycorrhizal fungi in a root-free sandy soil. *Soil Biol Biochem* **26**(9):1117–1124.

Cakmak I. 2000. Possible roles of zinc in protecting plant cells from damage by reactive oxygen species. *New Phytol* **146**:185–205.

Cakmak I, Marschner H. 1988a. Increase in membrane permeability and exudation in roots of zinc deficient plants. *J Plant Physiol* **132**:356–361.

Cakmak I, Marschner H. 1988b. Enhanced superoxide radical production in roots of zinc-deficient plants. *J Exp Bot* **39**:1449–1460.

Cakmak I, Marschner H, Bangerth F. 1989. Effect of zinc nutritional status on growth, protein metabolism and levels of indole-3-acetic acid and other phytohormones in bean (*Phaseolus vulgaris* L.). *J Exp Bot* **40**:405–412.

Cakmak I, Gülüt KY, Marschner H, Graham RD. 1994. Effect of zinc and iron deficiency on phytosiderophore release in wheat genotypes differing in zinc efficiency. *J Plant Nutr* **17**:1–17.

Cakmak I, Ozturk L, Karanlik S, Marschner H, Ekiz H. 1996a. Zinc-efficient wild grasses-enhance release of phytosiderophores under zinc deficiency. *J Plant Nutr* **19**:551–563.

Cakmak I, Sari N, Marschner H, Kalayci M, Yilmaz A, Ekei S, Gülüt KY. 1996b. Dry matter production and distribution of zinc in bread and durum wheat genotypes differing in zinc efficiency. *Plant and Soil* **180**:173–181.

Cakmak I, Sari N, Marschner H, Ekiz H, Kalayci M, Yilmaz A, Braun HJ. 1996c. Phytosiderophore release in bread and durum wheat genotypes differing in zinc efficiency. *Plant and Soil* **180**:183–189.

Cakmak I, Yalmaz A, Kalayci M, Eekiz H, Torun B, Erenoglu B, Braun HJ. 1996d. Zn deficiency as a critical problem in wheat production in central Antalia. *Plant Soil* **180**:165–172.

Chakravarty B, Srivastava S. 1997. Effect of cadmium and zinc interaction on metal uptake and regeneration of tolerant plants in linseed. *Agriculture, Ecosystems and Environment* **61**(1):45–50.

Chaney RL. 1994. Zinc phytotoxicity. In: Robson AD, editor. *Zinc in Soils and Plants*. Dordrecht: Kluwer Academic Publishers, pp. 135–150.

Chaoui A, Ghorbal MH, El Ferjani E. 1997. Effects of cadmium–zinc interactions on hydroponically grown bean (*Phaseolus vulgaris* L.). *Plant Sci* **126**(1):21–28.

Chardonnens AN, Koevoets PLM, van Zanten A, Schat H, Verkleij J. 1999. Properties of enhanced tonoplast zinc transport in naturally selected zinc-tolerant *Silene vulgaris*. *Plant Physiol* **20**:779–785.

Chaudri AM, McGrath SP, Giller KE. 1992. Survival of the indigenous population of *rhizobium leguminosarum* biovar *trifolii* in soil spiked with Cd, Zn, Cu and Ni salts. *Soil Biol Biochem* **24**:625–632.

Chen BD, Li X, Tao HQ, Christie P, Wong MH. 2003. The role of arbuscular mycorrhiza in zinc uptake by red clover growing in a calcareous soil spiked with various quantities of zinc. *Chemosphere* **50**:839–846.

Christianson ML. 1979. Zinc sensitivity in *Phaseolus*: Expression in cell culture. *Environmental and Experimental Botany* **19**(3):217–221.

Christie P, Li X, Chen B. 2004. Arbuscular mycorrhiza can depress translocation of zinc to shoots of host plants in soils moderately polluted with zinc. *Plant Soil* **261**:209–217.

Coughtrey PJ, Martin MH. 1979. Cadmiun, lead and zinc interactions and tolerance in two populations of *Holcus lanatus* L. grown in solution culture. *Environmental and Experimental Botany* **19**(4):285–290.

Cuypers A, Vangronsveld J, Clijsters H. 2001. The redox status of plant cells (AsA and GSH) is sensitive to zinc imposed oxidative stress in roots and primary leaves of *Phaseolus vulgaris*. *Plant Physiol Biochem* **39**:657–664.

Dabkowska-Naskret H. 2003. The role of organic matter in association with zinc in selected arable soils from Kujawy Region, Poland. *Organic Geochemistry* **34**(5):645–649.

De Filippis LF, Pallaghy CK. 1976. The effect of sub-lethal concentrations of mercury and zinc on *Chlorella* III. *Z. Pflanzenphysiol* **79S**:332–335.

Deng H, Ye ZH, Wong MH. 2004. Accumulation of lead, zinc, copper and cadmium by 12 wetland plant species thriving in metal-contaminated sites in China. *Environmental Pollution* **132**:29–40.

Diaz G, Azcon-Aguilar C, Honrubia M. 1996. Influence of arbuscular mycorrhizae on heavy metal (zinc and Pb) uptake and growth of *Lygeum spartum* and *Anthyllis cytisoides*. *Plant Soil* **180**:241–249.

Domingo AL, Nagatomo Y, Tamai M, Takaki H. 1992. Free tryptophan and indoleacetic acid in zinc-deficient radish shoots. *Soil Sci Plant Nutr* **38**:262–267.

Dong B, Rengel Z, Graham RD. 1995. Characters of root geometry of wheat genotypes differing in zinc efficiency. *J Plant Nutr* **18**:2761–2773.

Dudka S, Piotrowska M, Terelak H. 1996. Transfer of cadmium, lead, and zinc from industrially contaminated soil to crop plants: A field study. *Environ Pollut* **94**:181–188.

Dueck TA, Visser P, Ernst WHO, Schat H. 1986. Vesicular arbuscular mycorrhizae decrease zinc-toxicity to grasses growing in zinc-polluted soil. *Soil Biol Biochem* **18**:331–333.

Ebbs SD, Kochian LV. 1997. Toxicity of zinc and copper to *Brassica* species: Implications for phytoremediation. *J Environ Qual* **26**:776–781.

Ebbs S, Kochian LV. 1998. Phytoextraction of zinc by oat (*Avena sativa*), barley (*Hordeum vulgare*) and Indian mustard (*Brassica juncea*). *Environ Sci Technol* **32**:802–806.

Ebbs SD, Lasat MM, Brady DJ, Cornish J, Gordon R, Kochian LV. 1997. Phytoextraction of cadmium and zinc from a contaminated soil. *J Environ Qual* **26**:1424–1430.

El-Fouly MM, Amberger A, Shabana MR, Abou El Nour EAA. 1997. Tolerance of faba bean genotypes to low zinc using carbonic anhydrase (CA) activity as a screening technique for sustainable food production and environment. *Plant Nutr* **273**:274.

El-Ghamery AA, El-Kholy MA, Abou El-Yousser MA. 2003. Evaluation of cytological effects of Zn^{2+} in relation to germination and root growth of *Nigella sativa* L. and *Triticum aestivum* L. *Mutation Research/Genetic Toxicology and Environmental Mutagenesis* **537**(1):29–41.

Ernst WHO, Nelissen HJM. 2000. Life-cycle phases of a zinc- and cadmium-resistant ecotype of *Silene vulgaris* in risk assessment of polymetallic mine soils. *Environ Pollut* **107**:329–338.

Faber BA, Zasoski RJ, Burau RG, Uriu K. 1990. Zinc uptake by corn as affected by vesicular–arbuscular mycorrhizae. *Plant Soil* **129**:121–130.

Falchuk KH, Hardy C, Ulpino L, Vallee BL. 1978. RNA metabolism, manganese, and RNA polymerases of zinc sufficient and zinc deficient *Euglena gracilis*. *Proc Nat Acad Sci USA* **75**:4175–4179.

Farago ME, Pitt MJ. 1977. Plants which accumulate metals. Part III. A further investigation of two australian species which take up zinc. *Inorg Chim Acta* **24**:211–214.

Fu-Cheng R, Tie-Ching L, Liu HQ, Hu BY. 1993. Influence of zinc on the growth and distribution of elements on ginseng plants. *J Plant Nutr* **16**:393–405.

Fujiwara A, Tsutsumi M. 1959. Biochemical studies of microelements in green plants. 3. On the composition of nitrogenous constituents of the zinc-deficient barley leaves and the material exudated from them. *Tohoku J Agric Res* **10**:327–332.

Gadallah MAA. 2000. Effects of indole-3-acetic acid and zinc on the growth, osmotic potential and soluble carbon and nitrogen components of soybean plants growing under water deficit. *Arid Environ* **44**:451–467.

Godzik ST, Florkowski S, Piorek M, Sassen MA. 1970. An attempt to determine the tissue contamination of *Quercus robur* L. and *Pinus silvestris* L. Foliage by particulates from zinc and lead smelters. *Environmental Pollution* **18**:97–106.

Graham RD, Ascher JS, Hynes SC. 1992. Selecting zinc-efficient genotypes for soils of low zinc status. *Plant Soil* **146**:241–250.

Grewal HS, Stangoulis JCR, Potter TD, Graham RD. 1997. Zinc efficiency of oilseed rape (*Brassica napus* and *B. juncea*) genotypes. *Plant Soil* **191**:123–132.

Grotz N, Fox T, Connolly E, Park W, Guerinot ML, Eide D. 1998. Identification of a family of zinc transporter genes from *Arabidopsis thaliana* that respond to zinc deficiency. *Proc Natl Acad Sci USA* **95**:7220–7225.

Gucwa-Przepiora E, Turnau K. 2001. Arbuscular mycorrhiza and plant succession on zinc smelter spoil heap in Katowice-Welnowiec. *Acta Soc Bot Poloniae* **70**:153–158.

Gupta VK, Singh CP, Relan PS. 1992. Effect of Zn-enriched organic manures on Zn nutrition of wheat and residual effect on soyabean. *Biores Technol* **42**:155–157.

Harmens H, Koevoets PLM, Verkleij JACS, Ernst WHO. 1994. The role of low molecular weight organic acids in the mechanism of increased zinc tolerance in *Silene vulgaris* (Moench) Garcke. *New Phytol* **126**:615–621.

Hartwig EE, Jones WF, Kilen TC. 1991. Identification and inheritance of inefficient zinc absorption in soybean. *Crop Sci* **31**:61–63.

Haslett BS, Reid RJ, Rengel Z. 2001. Zinc mobility in wheat: Uptake and distribution of zinc applied to leaves or roots. *Annals of Botany* **87**(3):379–386.

Helms K, Glenn S. 1955. Pound zinc nutrition of *Nicotiana tabacum* L. in relation to multiplication of tobacco mosaic virus. *Virology* **1**(4):408–423.

Herren T, Feller URS. 1997. Transport of cadmium via xylem and phloem in maturing wheat shoots: Comparison with the translocation of zinc, strontium and rubidium. *Annals of Botany* **80**:623–628.

Hildebrandt U, Kaldorf M, Bothe H. 1999. The zinc violet and its colonization by arbuscular mycorrhizal fungi. *J Plant Physiol* **154**:709–717.

Hinesly TD, Alexander DE, Ziegler EL, Barrett GL. 1978. Zinc and Cd accumulation by corn inbreds grown on sludge amended soil. *Agron J* **70**:425–428.

Hirata K, Tsujimoto Y, Namba T, Ohta T, Hirayanagi N, Miyasaka H, Zenk MH, Miyamoto K. 2001. Strong induction of phytochelatin synthesis by zinc in marine green alga, *Dunaliella tertiolecta*. *Journal of Bioscience and Bioengineering* **92**(1):24–29.

Hopkins BG, Whitney DA, Lamond RE, Jolley VD. 1998. Phytosiderophore release by sorghum, wheat, and corn under zinc deficiency. *J Plant Nutr* **21**:2623–2637.

Hossain B, Hirata N, Nagatomo Y, Akashi R, Takaki H. 1997. Internal zinc accumulation is correlated with increased growth in rice suspension culture. *J Plant Growth Regul* **16**:239–243.

Ishizuka Y, Tanaka A. 1962. Effect of boron, zinc, and molybdenum in culture solution on the growth of rice plants. *Soil Sci Plant Nutr* **33**:93–99.

Iyengar GV. 1998. Revaluation of the trace element content in reference man. *Radiat Phys Chem* **51**:545–560.

Jackson PP, Robinson NJ, Whitton BA. 1991. Low molecular weight metal complexes in the freshwater moss *Rhynchostegium riparioides* exposed to elevated concentrations of Zn, Cu, Cd, and Pb in the laboratory and field. *Environmental and Experimental Botany* **31**(3):359–366.

Jarausch-Wehrheim B, Mocquot B, Mench M. 1999. Absorption and translocation of sludge-borne zinc in field-grown maize (*Zea mays* L.). *European Journal of Agronomy* **11**(1):23–33.

Jyung WH, Camp ME, Polson DE, Adams MW, Wittwer SH. 1972. Differential response of two bean varieties to zinc as revealed by electrophoretic protein pattern. *Crop Sci* **12**:26–29.

Kaya C, Higgs D. 2002. Response of tomato (*Lycopersicon esculentum* L.) cultivars to foliar application of zinc when grown in sand culture at low zinc. *Scientia Horticulturae* **93**:53–64.

Kitagishi K, Obata H. 1986. Effects of zinc deficiency on the nitrogen metabolism of meristematic tissues of rice plants with reference to protein synthesis. *Soil Sci Plant Nutr* **32**:397–405.

Kothari SK, Marschner H, Römheld V. 1991. Contribution of the VA mycorrhizal hyphae in acquisition of phosphorus and zinc by maize grown in a calcareous soil. *Plant Soil* **131**:177–185.

Lambert DH, Baker DE, Cole H. 1979. The role of mycorrhizas in the interactions of phosphorus with zinc, copper and other elements. *Soil Sci Soc Am J* **43**:976–980.

Lasat MM, Baker AJM, Kochian LV. 1996. Physiological characterization of root Zn^{2+} absorption and translocation to shoots in Zn hyperaccumulator and nonaccumulator species of *Thlaspi*. *Plant Physiol* **112**(4):1715–1722.

Lasat MM, Baker AJM, Kochian LV. 1998a. Altered zinc compartmentation in the rootsymplasm and stimulated zinc absorption into the leaf as mechanisms involved in zinc hyper accumulation in *Thlaspi caerulescens*. *Plant Physiol* **118**:875–883.

Lasat MM, Fuhrmann M, Ebbs SD, Cornish JE, Kochian LV. 1998b. Phytoremediation of a radiocesium-contaminated soil: evaluation of cesium-137 bioaccumulation in the shoots of three plant species. *J Environ Qual* **27**:165–169.

Lasat MM, Pence NS, Garvin DF, Ebbs SD, Kochian LV. 2000. Molecular physiology of zinc transport in the zinc hyperaccumulator *Thlaspi caerulescens*. *J Exp Bot* **51**:71–79.

Li XL, Christie P. 2001. Changes in soil solution zinc and pH and uptake of zinc by arbuscular mycorrhizal red clover in zinc contaminated soil. *Chemosphere* **42**:201–207.

Liu A, Hamel C, Hamilton RI, Ma BL Smith DL. 2000. Acquisition of Cu, Zn, Mn and Fe by mycorrhizal maize (*Zea mays* L.) grown in soil at different P and micronutrient levels. *Mycorrhiza* **9**:331–336.

Lombi E, Zhao FJ, Dunham SJ, McGrath SP. 2000. Cadmium accumulation in populations of *Thlaspi caerulescens* and *Thlaspi goesingense*. *New Phytologist* **145**:11–20.

Luo Y, Rimmer DL. 1995. Zinc–copper interaction affecting plant growth on a metal contaminated soil. *Environ Pollut* **88**:79–93.

Lyngby JE, Brix H, Schierup H-H. 1982. Absorption and translocation of zinc in eelgrass (*Zostera marina* L.). *Journal of Experimental Marine Biology and Ecology* **58**(2–3): 259–270.

Marchant JW. 1974. Effect of fallout of windborne zinc, lead, copper and cadmium from a fuming kilnon soil geochemical prospecting at Berg Aukas, South West Africa. *Geochem Explor* **3**(2):191–198.

Martins RJE, Boaventura RAR. 2002. Uptake and release of zinc by aquatic bryophytes (*Fontinalis antipyretica* L. ex. Hedw.). *Water Research* **36**:5005–5012.

Martino E, Perotto S, Parsons R, Gadd GM. 2003. Solubilization of insoluble inorganic zinc compounds by ericoid mycorrhizal fungi derived from heavy metal polluted sites. *Soil Biol Biochem* **35**:133–141.

Mathys W. 1977. The role of malate, oxalate, and mustard oil glucosides in the evolution of zinc-resistance in herbage plants. *Physiol Plant* **40**:130–136.

MacFarlane GR, Burchett MD. 1999. Zinc distribution and excretion in the leaves of the grey mangrove, *Avicennia marina* (Forsk.) Vierh. *Environ Exp Bot* **41**:167–175.

McIlveen WD, Cole H. 1979. Influence of zinc on development of the endomycorrhizal fungus *Glomus mosseae* and its mediation of phosphorus uptake by glycine max 'Amsoy 71'. *Agriculture and Environment* **4**:245–256.

Mesjasz-Przybyłowicz J, Grodzińska K, Przybyłowicz WJ, Godzik B, Szarek-Łukaszewska G. 1999. Micro-PIXE studies of elemental distribution in seeds of *Silene vulgaris* from a zinc dump in Olkusz, southern Poland. *Nuclear Instruments and Methods in Physics Research Section B: Beam Interactions with Materials and Atoms* **158**(1–4):306–311.

Mesjasz-Przybyłowicz J, Grodzińska K, Przybyłowicz WJ, Godzik B, Szarek-Łukaszewska G. 2001. Nuclear microprobe studies of elemental distribution in seeds of *Biscutella laevigata* L. from zinc wastes in Olkusz, Poland. *Nuclear Instruments and Methods in Physics Research Section B: Beam Interactions with Materials and Atoms* **181**:634–639.

Nable RO, Webb MJ. 1993. Further evidence that zinc is required throughout the root zone for optimal plant growth and development. *Plant Soil* **150**:247–253.

Nan Z, Li J, Zhang J, Cheng G. 2002. Cadmium and zinc interactions and their transfer in soil-crop system under actual field conditions. *Sci Total Environ* **285**(1–3):187–195.

Norvell WA, Dabkovska-Naskret H, Cary EE. 1987. Effect of phosphorus and zinc fertilization on the solubility of Zn^{2+} in two alkaline soils. *Soil Sci Soc Am J* **51**:584–588.

Nyamangara J, Mzezewa J. 1999. The effect of long-term sewage sludge application on Zn, Cu, Ni and Pb levels in a clay loam soil under pasture grass in Zimbabwe. *Agricul Ecosyst Environ* **73**:199–204.

Otte ML, Bestebroer SJ, van der Linden JM, Rozema J, Broekman RA. 1991. A survey of zinc, copper and cadmium concentrations in salt marsh plants along the Dutch coast.

Otte ML, Kearns CC, Doyle MO. 1995. Accumulation of arsenic and zinc in the rhizosphere of wetland plants. *Bull Environ Contam Toxicol* **55**:154–161.

Ozturk L, Karanlik S, Ozkutlu F, Cakmak I, Kochian LV. 2003. Shoot biomass and zinc/cadmium uptake for hyperaccumulator and non-accumulator *Thlaspi* species in response to growth on a zinc-deficient calcareous soil. *Plant Sci* **164**(6): 1095–1101.

Peltier EF, Webb SM, Gaillard J-F. 2003. Zinc and lead sequestration in an impacted wetland system. *Advances in Environmental Research* **8**(1):103–112.

Pence NS, Larsen PB, Ebbs SD, Letham DL, Lasat MM, Garvin DF, Eide D, Kochian LV. 2000. The molecular physiology of heavy metal transport in the Zn/Cd hyperaccumulator *Thlaspi caerulescens. Proc Natl Acad Sci USA* **97**:4956–4960.

Peralta-Videa JR, Gardea-Torresdey JL, Gomez E, Tieman KJ, Parsons JG, Carrillo G. 2002. Effect of mixed cadmium, copper, nickel and zinc at different pHs upon alfalfa growth and heavy metal uptake. *Environ Pollut* **119**:291–301.

Powell SR. 2000. The antioxidant properties of zinc. *J Nutr* **130**:1447–1454.

Prasad KVSK, Saradhi PP, Sharmila P. 1999. Concerted action of antioxidant enzymes and curtailed growth under zinc toxicity in *Brassica juncea. Environ Exper Bot* **42**(1):1–10.

Prasad MNV, editor. 2001. *Metals in the Environment—Analysis by Biodiversity.* New York: Marcel Dekker.

Prasad MNV. 2002. Zinc is the friend and foe of life. Zeszyty Naukowe PAN 33, Com. "*Man and Biosphere.*" Kabata-Pendias A, Szteke Warsaw B, editors. pp. 49–54.

Prasad MNV, Hagemeyer J, editors. 1999. *Heavy Metal Stress in Plants: From Molecules to Ecosystems.* Heidelberg: Springer-Verlag, pp. xiii +401.

Prasad MNV, Strzałka K. 2002. *Physiology and Biochemistry of Metal Toxicity and Tolerance in Plants.* Dordrecht: Kluwer Academic Publishers.

Pugh RE, Dick DG, Fredeen AL. 2002. Heavy metal (Pb, Zn, Cd, Fe, and Cu) contents of plant foliage near the Anvil Range lead/zinc mine, Faro, Yukon Territory. *Ecotoxicol Environ Saf* **52**:273–279.

Ramani S, Kannan S. 1985. An examination of zinc uptake patterns by genotypes of sorghum and maize; differences amongst hybrids and their parents. *J Plant Nutr* **8**:1199–1210.

Rascio W. 1977. Metal accumulation by some plants growing on zinc mine deposits. *Oikos* **29**:250–253.

Rashid A, Buthio N, Rafique E. 1994. Diagnosing zinc deficiency in rapeseed and mustard by seed analysis. *Commun Soil Sci Plant Anal* **25**:3405–3412.

Rauser WE, Dumbroff EB. 1981. Effects of excess cobalt, nickel and zinc on the water relations of *Phaseolus vulgaris*. *Environ Exper Bot* **21**:249–255.

Reddy KB, Ashalatha M, Venkaiah K. 1993. Differential response of groundnut genotypes to iron-deficiency stress. *J Plant Nutr* **16**:523–531.

Reeves RD, Brooks RR. 1983. European species of *Thlaspi* L. (*Cruciferae*) as indicators of nickel and zinc. *Geochem Explor* **18**:275–283.

Rengel Z. 1999. Physiological mechanisms underlying differential nutrient efficiency of crop genotypes. In: Rengel Z, editor. *Mineral Nutrition of Crops: Fundamental Mechanisms and Implications*. New York: Haworth Press, pp. 227–265.

Rengel Z, Graham RD. 1995a. Importance of seed zinc content for wheat growth on zinc-deficient soil. I. Vegetative growth. *Plant Soil* **173**:259–266.

Rengel Z, Graham RD. 1995b. Wheat genotypes differ in zinc efficiency when grown in chelate-buffered nutrient solution. I. Growth. *Plant Soil* **176**:307–316.

Rengel Z, Graham RD. 1995c. Wheat genotypes differ in zinc efficiency when grown in chelate-buffered nutrient solution. II. Nutrient uptake. *Plant Soil* **176**:317–324.

Rengel Z, Römheld V, Marschner H. 1998. Uptake of zinc and iron by wheat genotypes differing in zinc efficiency. *J Plant Physiol* **152**:433–438.

Rengel Z, Römheld V. 2000. Differential tolerance to Fe and zinc deficiencies in wheat germplasm. *Euphytica* **113**:219–225.

Robson AD, Snowball K. 1989. The effect of 2-(4-20, 40-dichlorophenoxy-phenoxy)-methyl propanoate on the uptake and utilization of zinc by wheat. *Aust J Agric Res* **40**:981–990.

Robson AD, Snowball K. 1990. The effect of chlorosulfuron on the uptake and utilization of copper and zinc in wheat. *Aust J Agric Res* **41**:19–28.

Römheld V, Marschner H. 1990. Genotypical differences among gramineaceous species in release of phytosiderophores and uptake of iron phytosiderophores. *Plant Soil* **123**:147–153.

Rout GR, Samantaray S, Das P. 1999. In vitro selection and biochemical characterisation of zinc and manganese adapted callus lines in *Brassica* spp. *Plant Sci* **146**:89–100.

Salama ZA, Amberger AA, El-Fouly MM. 1997. Aldolase, carbonic anhydrase and catalase activity in faba bean leaves as affected by iron and zinc. *Egypt J Physiol Sci* **21**:31–39.

Samantaray S, Rout GR, Das P. 1999. In vitro selection and regeneration of zinc tolerant calli from *Setaria italica* L. *Plant Sci* **143**:201–209.

Sanders JR, McGrath SP, McM. Adams T. 1987. Zinc, copper and nickel concentrations in soil extracts and crops grown on four soils treated with metal loaded sewage sludges. *Environ Pollut* **44**(3):193–210.

Saxena SC, Chandel AS. 1992. Effect of zinc fertilization on different varieties of soybean (*Glycine max*). *Indian J Agric Sci* **62**:695–697.

Santa-María GE, Cogliatti DH. 1998. The regulation of zinc uptake in wheat plants. *Plant Sci* **137**:1–12.

Schierup HH, Larsen VJ. 1981. Macrophyte cycling of zinc, copper, lead and cadmium in the littoral zone of a polluted and a non-polluted lake. I. Availability, uptake and translocation of heavy metals in *Phragmites australis* (Cav.) Trin. *Aquat Bot* **11**:197–210.

Schuerger AC, Capelle GA, Die Benedetto JA, Mao C, Thai CN, Evans MD, Richards JT, Blank TA, Stryjewski EC. 2003. Comparison of two hyperspectral imaging and two laser-induced fluorescence instruments for the detection of zinc stress and chlorophyll concentration in bahia grass (*Paspalum notatum* Flugge). *Remote Sensing Environ* **84**:572–588.

Schwartz C, Gérard E, Perronnet K, Morel JL. 2001. Measurement of in situ phytoextraction of zinc by spontaneous metallophytes growing on a former smelter site. *Sci Total Environ* **279**:215–221.

Segal DJ, Stege JT, Barbas CF III. 2003. Zinc fingers and a green thumb: manipulating gene expression in plants. *Current Opinion in Plant Biology* **6**:163–168.

Sharma UC, Grewal JS. 1990. Potato response to zinc as influenced by genetic variability. *J Indian Potato Assoc* **17**:1–5.

Shetty KG, Hetrick BAD, Schwab AP. 1995. Effects of mycorrhizae and fertilizer amendments on zinc tolerance of plants. *Environ Pollut* **88**:307–314.

Shu WS, Ye ZH, Lan CY, Zhang ZQ, Wong MH. 2002. Lead, zinc and copper accumulation and tolerance in populations of *Paspalum distichum* and *Cynodon dactylon*. *Environmental Pollution* **120**:445–453.

Sidenko NV, Gier R, Bortnikova SB, Cottard F, Pal'chik NA. 2001. Mobility of heavy metals in self-burning waste heaps of the zinc smelting plant in Belovo (Kemerovo Region, Russia). *J Geochem Explor* **74**(1–3):109–125.

Skoog F. 1940. Relationship between zinc and auxin in the growth of higher plants. *Am J Bot* **27**:939–951.

Smith SE, Read DJ. 1997. *Mycorrhizal Symbiosis*, second edition. London: Academic Press.

Solisio C, Lodi A, Veglio F. 2002. Bioleaching of zinc and aluminium from industrial waste sludges by means of *Thiobacillus ferrooxidans*. *Waste Manage* **22**(6):667–675.

Stoveland S, Astruc M, Lester JN, Perry R. 1979. The balance of heavy metals through a sewage treatment works II. Chromium, nickel and zinc. *Sci Total Environ* **12**(1):25–34.

Takaki H, Kushizaki M. 1970. Accumulation of free tryptophan and tryptamine in zinc-deficient maize seedlings. *Plant Cell Physiol* **11**:793–804.

Takaki H, Kushizaki M. 1972. Tryptophan and tryptamine in zinc deficient plant. *J Sci Soil Manure Jpn* **43**:81–85.

Takaki H, Kushizaki M. 1976. Zinc nutrition in higher plants. *Bull Natl Inst Agric Sci* **28**:75–118.

Trummler K, Roos J, Schwaneberg U, Effenberger F, Frster S, Pfizenmaier K, Wajant H. 1998. Expression of the Zn^{2+}-containing hydroxynitrile lyase from flax (*Linum usitatissimum*) in *Pichia pastoris*—utilization of the recombinant enzyme for enzymatic analysis and site-directed mutagenesis. *Plant Sci* **139**(1):19–27.

Tsuji N, Hirayanagi N, Iwabe O, Namba T, Tagawa M, Miyamoto S, Miyasaka H, Takagi M, Hirata K, Miyamoto K. 2003. Regulation of phytochelatin synthesis by zinc and cadmium in marine green alga, *Dunaliella tertiolecta*. *Phytochemistry* **62**(3):453–459.

Turnau K. 1998. Heavy metal content and localization in mycorrhizal *Euphorbia cyparissias* from zinc wastes in southern Poland. *Acta Soc Bot Poloniae* **67**:105–113.

Twining JR, Payne TE, Itakura T. 2004. Soil-water distribution coefficients and plant transfer factors for 134Cs, 85Sr and 65Zn under field conditions in tropical Australia. *J Environ Radioactiv* **71**:71–87.

Vallee BL, Auld DS. 1990. Zn coordination, function and structure of zinc enzymes and other proteins. *Biochemistry* **29**:5647–5659.

Van Stevenink RFM, Van Stevenink ME, Fernando DR, Horst WJ, Marschner H. 1987. Deposition of zinc phytate in globular bodies in roots of *Deschampsia caespitosa* ecotypes: A detoxification mechanism? *J Plant Physiol* **131**:247–257.

Van Stevenink RFM, Van Stevenink ME, Wells AJ, Fernando DR. 1990. Zinc tolerance and the binding of zinc as zinc phytate in *Lemna minor*. X-ray microanalytical evidence. *J Plant Physiol* **137**:140–146.

Vazquez MD, Barcelo J, Poschenrieder C, Madico J, Hatton P, Baker AJM, Cope GH. 1992. Localization of zinc and cadmium in *Thlaspi caerulescens* (Brassicaceae), a metallophyte that can hyperaccumulate both metals. *J Plant Physiol* **140**(3):350–355.

Vazquez MD, Poschenreider C, Barcelo J, Baker AJM, Hatton P, Cope GH. 1994. Compartmentalization of zinc in roots and leaves of the zinc hyperaccumulator *Thlaspi caerulescens* J. & C. Presl. *Bot Acta* **107**:243–250.

Viard BN, Pihan F, Promeyrat S, Pihan J-C. 2004. Integrated assessment of heavy metal (Pb, Zn, Cd) highway pollution: bioaccumulation in soil. Graminaceae and land snails *Chemosphere* **55**(10):1349–1359.

Walter A, Römheld V, Marschner H, Mori S. 1994. Is the release of phytosiderophores in zinc-deficient wheat plants a response to impaired iron utilization? *Physiol Plant* **92**:493–500.

Weissenhorn I, Glashoff A, Leyval C, Berthelin J. 1994. Differential tolerance to Cd and zinc of arbuscular mycorrhizal (AM) fungal spores isolated from heavy metal-polluted and unpolluted soils. *Plant Soil* **167**:189–196.

Welch RM, Norvell WA. 1993. Growth and nutrient uptake of barley (*Hordeum vulgare* L. cv Herta). Studies using an *n*-(2-hydroxyethyl)ethylenedinitrilotriacetic acid-buffered nutrient solution technique. 1. Role of zinc in the uptake and root leakage of mineral nutrients. *Plant Physiol* **101**:627–631.

Whitton BA, Say PJ, Jupp BP. 1982. Accumulation of zinc cadmium and lead by the aquatic liverwort *Scapania*. *Environ Pollut B* **3**:299–316.

Winder L, Merrington G, Green I. 1999. The tri-trophic transfer of Zn from the agricultural use of sewage sludge. *Sci Total Environ* **229**(1–2):73–81.

Wolterbeek HT, van der Meer AJGM. 2002. Transport rate of arsenic, cadmium, copper and zinc in *Potamogeton pectinatus* L.: radiotracer experiments with 76As, 109,115Cd, 64Cu and 65, 69Zn. *Sci Total Environ* **287**(1–2):13–30.

Ye ZH, Shu WS, Zhang ZQ, Lan CY, Wong MH. 2002. Evaluation of major constraints to revegetation of lead/zinc mine tailings using bioassay techniques. *Chemosphere* **47**(10):1103–1111.

Ye ZH, Wong JWC, Wong MH, Lan CY, Baker AJM. 1999. Lime and pig manure as ameliorants for revegetating lead/zinc mine tailings: a greenhouse study. *Bioresource Technol* **69**(1):35–43.

Ye ZH, Yang ZY, Chan GYS, Wong MH. 2001. Growth response of *Sesbania rostrata* and *S. cannabina* to sludge-amended lead/zinc mine tailings: a greenhouse study. *Environ Int* **26**(5–6):449–455.

Zago MP, Oteiza PI. 2001. The antioxidant properties of zinc: Interactions with iron and antioxidants. *Free Rad Bio Med* **31**:266–274.

Zaranyika MF, Ndapwadza T. 1995. Uptake on Ni, Zn, Fe, Co, Cr, Pb, Cu, and Cd by water hyacinth (Eichhornia crassipes) in Mukuvisi and Manyame Rivers, Zimbabwe. *J Environ Sci Health* **A30**(1):157–170.

Zhang F, Romheld V, Marschner H. 1989. Effect of zinc deficiency in wheat on the release of zinc and iron mobilizing root exudates. *Z Pflanzenernaehr Bodenkd* **152**:205–210.

Zhang F, Romheld V, Marschner H. 1991. Release of zinc mobilizing root exudates in different plant species as affected by zinc nutritional status. *J Plant Nutr* **14**:675–686.

Zhao H, Eide D. 1996a. The yeast *ZRT1* gene encodes the zinc transporter protein of a high affinity uptake system induced by zinc limitation. *Proc Natl Acad Sci USA* **93**:2454–2458.

Zhao H, Eide D. 1996b. The *ZRT2* gene encodes the low affinity zinc transporter in *Saccharomyces cerevisiae*. *J Biol Chem* **271**:23203–23210.

Zhu YG, Christie P, Laidlaw AS. 2001. Uptake of zinc by arbuscular mycorrhizal white clover from zinc-contaminated soil. *Chemosphere* **42**:193–199.

Zhang ZQ, Wong MH, Nie XP, Lan CY. 1998. Effects of zinc (Zinc sulfate) on Rhizobia-earleaf acacia (*Acacia auriculaeformis*) symbiotic association. *Bioresource Technol* **64**(2):97–104.

10 Zinc Effect on the Phytoestrogen Content of Pomegranate Fruit Tree

FATEMEH ALAEI YAZDI

Department of Agronomy, Yazd Agricultural and Natural Resources Research Center, Yazd, Yazd Province, I.R. of Iran

FARHAD KHORSANDI

Department of Agronomy, Islamic Azad University—Darab Branch, Darab, Fars Province, I.R. of Iran

1 INTRODUCTION

Pomegranate (*Punica granatum* L.) is a fruit crop well adapted to arid regions, where the summer is hot and dry and the winter is cold. It is grown extensively in Iran, India, and the United States, but it is also grown in most Mediterranean and Far East countries. This fruit tree is an important cash crop in different arid regions. It has an extensive market around the world as fresh fruit, concentrated fruit juice, and different pharmaceutical and medicinal products. The main use of pomegranate is as table fruit, but large amounts are used in the beverage industry [Nagy et al., 1990], as well as paste, fruit roll, and natural dye (for carpet industry). The peel, containing up to 30% tannins, is used for tanning leather [Duke and Ayensu, 1985]. In a study with 69 strains of 19 different species of gram-positive and gram-negative bacteria, pomegranate was reported to possess good antibacterial effect [Saeed and Tariq, 2006]. Ghasemian et al. [2006], working with two pomegranate cultivars of Iran (Malas Saveh and Aghamohammadali cv.), reported that the peel extracts had both antioxidant and antimutagenic properties, which may be used as biopreservative in food and neutracutical industries. Pomegranate is also a rich source of different kinds of polyphenols as well as microelements, particularly in the edible parts

Trace Elements as Contaminants and Nutrients: Consequences in Ecosystems and Human Health, Edited by M. N. V. Prasad
Copyright © 2008 John Wiley & Sons, Inc.

[Poyrazoglu et al., 2002; Mirdehghan and Rahemi, 2007]. The biological uniqueness of pomegranate is that it has no close botanical relatives, and consequently is a potential source for several physiological factors, which could have significant effect on human health and diseases [Lansky et al., 2000].

Zinc deficiency is common in calcareous soils of arid regions. Countries with widespread Zn-deficiency problems include China, India, Pakistan, Iran, Syria, Australia, Sudan, many states in the United States, and parts of Europe [Alloway, 2004]. The main causes of Zn deficiency in Iran are calcareous soils with high pH, excessive application of phosphorus fertilizers, high bicarbonate concentration in irrigation water, and prolonged water logging and flooding [Alloway, 2004]. Zinc deficiency is commonly observed in pomegranate orchards of Iran [Taghavi, 2000], India [Balakrishnan et al., 1996] and USA [LaRue, 1980]. The main symptoms of Zn deficiency are short internodes (rosetting), narrowing and decrease in leaf size, irregular green bands along the midrib and main veins on a light yellow background, reduced fruit size and delayed maturity [Alloway, 2004; Srivastava and Singh, 2005].

Zinc fertilization of the plant is the most practical solution to correction of Zn deficiency. In experiments at 25 sites with zinc, iron, manganese, and copper in Iran, Ziaeian and Malakouti [2001] reported that zinc treatments significantly increased wheat grain yield and the yield components. Bybordi and Malakouti [2004], in experiments with the same nutrients under saline condition, reported that zinc treatment significantly increased wheat grain yield and yield components. Asadi Kangarshahi and Malakouti [2004] reported significant increase in soybean seed yield and quality in experiments at 20 sites in Iran with soil incorporation of 40 kg/ha zinc sulfate. Increasing supply of zinc significantly increased dry matter production of potatoes in India [Tiwari et al., 1982]. Foliar application of Zn overcame negative effects of Zn deficiency on the growth of two varieties of tomatoes [Kaya and Higgs, 2002]. Seilsepour and Ghaemi [2006] reported significant increase in tomato yield and water use efficiency by foliar application of Zn.

Although zinc fertilization of cereal crops in saline and non-saline, calcareous soils of Iran is well reported, the literature on zinc fertilization of fruit trees, particularly pomegranate, is scarce. Zn fertilization increased hazelnut yield and quality in Turkey [Serdar et al., 2005]. Fall application of urea and zinc sulfate to sweet cherry (Bing cv.) decreased floral bud death and increased fruit set compared to untreated trees [Glozer and Grant, 2006]. Apple leaf area, leaf chlorophyll index, current year's branch growth, and quality indices were significantly improved by zinc sulfate fertilization in West Azerbaijan, Iran [Rasouli Sadaghiani et al., 2002]. Foliar application of Zn on date palms (Shahani cv.) significantly increased fruit yield, fruit length, and pulp weight, without significantly affecting seed characteristics [Khayyat et al., 2007]. Foliar application of 0.25% each of zinc sulfate, iron sulfate, and manganese sulfate along with 0.15% boric acid significantly increased the yield of pomegranate from 18.5 kg/plant in control to 26.37 kg/plant and also increased the juice content from 65.6% to 74.8% [Balakrishnan et al., 1996]. In a field study with pomegranate (Ghajagh cv.) in Iran, soil application of N, P, and K, based on soil analysis, along with foliar application of 0.5% zinc sulfate, increased pomegranate yield over 1.3 t/ha compared to control, which was fertilized according to farmer's routine program [Taghavi, 2000]. The farmers do not include micronutrients such as Zn in their routine fertilization program. Raghupathi and Bhargava [1998b], by using DRIS (Diagnosis and Recommendation Integrated System) and CND (Composition Nutrient Diagnosis) methods, showed that zinc and nitrogen were the most limiting plant nutrients in pomegranate orchards of Bijapour, India. Recently, Daryashenas and Dehghani [2006] determined DRIS reference norms of pomegranate in Yazd Province (Iran) for several nutrients, including Zn, which is useful in determining nutritional problems of pomegranate trees. They concluded that Zn deficiency is more severe in pomegranate orchards of Yazd than Bijapour.

The interest in phytoestrogens has increased dramatically due to the ineffectiveness and unsafety of hormone replacement therapy [Cornwell et al., 2004], along with the emergence of several lines of evidence about their possible role in preventing a range of diseases, including hormonally dependent cancers [van Elswijk et al., 2004; Albrecht et al., 2004], and cardiovascular diseases [Larkin et al., 2000]. Phytoestrogens are a group of naturally occurring phenolic compounds present in legumes, whole grains, fruits, and vegetables [Rohrdanz et al., 2002]. Currently, four different families of plant-derived phenolic compounds are considered phytoestrogens: isoflavonoids, stilbenes, lignans, and coumestans [Cornwell et al., 2004]. Isoflavonoids and coumestans have been identified as the most common estrogenic compounds in plants [Price and Fenwick, 1985]. Leguminous plants, such as soybean, chickpeas, and beans, are the most abundant in isoflavonoid phytoestrogen [Mazur, 2000]. Grains, fiber-rich fruits and vegetables, and tea are rich in lignans [Mazur, 2000; Cornwell et al., 2004]. The commercial possibility of recovering high amounts of phenolics with antioxidant properties from the residues of eleven fruits and vegetables for food and cosmetic applications has been demonstrated [Peschel et al., 2006]. Pomegranate (*Punica granatum*) is one of the rich sources of phytoestrogen compounds among horticultural and fruit crops. Pomegranate peel had the highest antioxidant activity among the peel, pulp, and seed fractions of 28 kinds of commonly consumed fruits in China [Li et al., 2006].

Pomegranate trees have been used extensively for indigenous medicine in many cultures, at least as far back as 1550 B.C. [Wren, 1988]. The oil from the seeds contains about 80% of a rare 18-carbon fatty acid, or punicic acid [Longtin, 2003]. Also present in the oil are isoflavonic phytoestrogens: genistein and daidzein, and the phytoestrogenic coumestan: coumestrol [Moneam et al., 1988]. Pomegranate is one of the only plants in nature known to contain the sex steroid estrone, and it has the highest botanical concentration of the steroid estrone at 14 mg/kg dried seed [Heftmann et al., 1966]. Pomegranate juice and seed oil contain phytoestrogenic compounds that have been shown to exhibit antioxidant activity [Schubert et al., 1999], exert antiproliferative effects on human breast cancer cells *in vitro* [Kim et al., 2002], exhibit significant antitumor activity against human prostate cancer [Albrecht et al., 2004], and prevent heart disease [Aviram et al., 2004]. Also, pomegranate has been recommended as medicinal food for treatment of acquired immune deficiency syndrome (AIDS) patients [Lee and Watson, 1998; Neurath et al., 2005].

As was mentioned before, zinc deficiency is common in calcareous soils of arid regions [Alloway, 2004], where pomegranate production is extensive. Zinc is essential for the healthy growth of plants, animals, and humans. It plays an essential role in several key plant metabolic and physiological pathways that are concerned with photosynthesis and sugar formation, protein and hormone synthesis, growth regulation, seed production, and disease resistance [Alloway, 2004]. It may also have an important role in the synthesis of phytoestrogenic compounds in plants.

Several *in vitro* and *in vivo* studies have shown positive antagonistic relationship between zinc and phytoestrogens. Isoflavonoid effects on bone loss prevention were enhanced by zinc in both *in vitro* and *in vivo* studies with elderly and young rats [Gao and Yamaguchi, 1998; Yamaguchi et al., 2000; Ma et al., 2000]. In a glasshouse experiment, foliar application of zinc increased isoflavones levels in 47-day-old clover (*Trifolium subterraneum*) seedlings [Rossiter, 1967]. Venkatesan et al. [2005] reported positive and highly significant correlation between Zn and polyphenols contents of mature leaves of tea.

It is speculated that zinc fertilization of agronomic and horticultural crops may increase their phytoestrogen content. Both zinc and phytoestrogen play an important role in human health. Pomegranate, an important source of phytoestrogen, is grown extensively as a cash crop in many zinc-deficient, calcareous soils of arid regions. Therefore, the main objective of this study was to evaluate the effects of zinc fertilization on the phytoestrogen contents of four commercial cultivars of pomegranate.

2 MATERIALS AND METHODS

Field experiments were carried out in 2002 and 2003 on a sandy loam soil at the Yazd Agricultural and Natural Resources Research Center (YANRC), Yazd Province, Iran. The soil moisture and temperature regimes at the site are Aridic and Thermic, respectively. The experiment was factorial with randomized complete blocks design and four replications. The pomegranate cultivars were Togh Gardan (TG), Shahvar Dane Ghermez (SDG), Malas Yazdi (MY), and Zagh Yazdi (ZY), which are the most popular cultivars among the local growers. These cultivars are sweet–sour, sweet, sweet–sour and sour, respectively; and based on their taste and color, they have their particular market.

The area of the experimental field was 5000 m^2. The fruit trees were all planted at the same time; and at the start of the experiment, they were 18 years old. The distance between trees was 4 m in a row and 6 m between rows (420 trees/ha). The irrigation method was border irrigation, and the irrigation depth was 180 cm (18,000 m^3/ha) in both years.

Soil analysis of the experimental field indicated zinc deficiency (Table 1). The type and amount of fertilizers per tree were potassium sulfate (450 g), triple super phosphate (375 g), urea (twice, 300 g each time), iron sulfate (200 g), manganese sulfate (80 g), and cupper sulfate (50 g). The required fertilizers were deep, placed about 1 m from the tree trunks at a depth of 50 cm—except for urea, which was surface-applied.

TABLE 1. Selected Properties of the Soil Before the Start of the Experiment

Depth (cm)	Texture	EC$_e$ (dS/m)	pH	CaCO$_3$ (%)	OC (%)	P	K	Cu	Mn	Fe	Zn
						(mg/kg soil)					
0–30	SL	3.85	7.9	23.2	0.19	9.8	110	0.34	1.8	4.2	0.64
30–60	SL	4.90	7.8	22.6	0.09	11.3	150	0.86	3.4	5.8	0.70
60–90	SL	6.18	7.8	21.7	0.17	11.7	165	0.87	3.8	5.8	0.76

OC, organic carbon; SL, sandy loam; ECe, electrical Conductivity of the saturated soil paste.

Zinc sulfate was foliar, applied twice (April 15 and May 15) at a rate of 0.4% in both years of the experiment on designated plots. Since Zn moves very slowly through the soil, and a large proportion of the fruit tree root system is located at deep soil layers, foliar application of Zn are generally more effective and economical than soil application [Swietlik, 2002; Zekri and Obreza, 2003] or fertilation [Kallsen et al., 2000]. Therefore, foliar method of Zn application was used in this experiment. Furthermore, zinc does not readily translocate within plants; thus, Zn deficiency symptoms occur mainly in new growth. Zhang and Brown [1999] reported more Zn transport out of immature leaflets than mature leaves in Pistachio, sprayed with Zn, as was detected by stable ^{68}Zn isotope. Post-bloom Zn sprays are also replacing dormant and post-harvest sprays as the primary means for applying Zn in commercial apple orchids of Washington State, USA [Peryea, 2006]. Therefore, maximum benefits are obtained if spray is applied to young leaves before hardening off, preferably in the spring when most of the new growth flushes occur [Zekri and Obreza, 2003]. Thus, the pomegranate trees in the experimental plots were foliar, treated twice in the spring when the branches had developed young leaves. The poor mobility of Zn in plants also suggests the need for a constant supply of available Zn for optimum growth. In a four-year field study with apple (Golden Delicious cv.), Peryea [2006] observed no detectable lasting effect of three seasons of Zn sprays on leaf Zn in the fourth year. Similarly, there was no detectable effect in any year of the Zn spray treatments on bud Zn concentration in the following winter [Peryea, 2006]. That is why the pomegranate trees were Zn-sprayed in both years of this experiment.

The fruits were harvested on September 7–10 each year. Before harvest, eight fruits from the four sides of each tree were randomly picked and were used for determination of seed zinc and phytoestrogen contents in the laboratory. To determine seed dry weights, 100 g of fresh seed (after removal of juice and fruit meat) was oven-dried at 80°C for 72 h and then weighed again. Afterwards, the percent dry weight was calculated. Zinc content of the leaves, fruit juice, and seeds were determined by dry ash procedure as described by Saini et al. [2001], using an atomic absorption spectrometer.

The total content of the major phytoestrogens (daidzin, daidzein, genistin, genistein, and coumestrol) in pomegranate seeds were determined according to the method described by Moneam et al. [1988]. The dried seeds from each of the four cultivars and both Zn treatments were dried and grinded separately. Then, 10 g of grinded seeds were placed in HPLC-grade methanol and filtered. The total phytoestrogens were determined by reversed-phase HPLC apparatus (Jasco FP-920), using a non-linear methanol-deionized water gradient on a C-18 column (250 mm × 4.6 mm), at a flow rate of 0.3 ml/min for a total of 17 min per sample. The detection was analyzed by spectrophotometer at 254 nm.

The data were evaluated by analysis of variance procedures, using the SAS software [SAS Institute, 1989]. A Duncan Multiple Range Test at 0.05 level of probability was used for separation of mean differences.

3   RESULTS AND DISCUSSIONS

3.1   Pomegranate Yield

The effect of year on total yield was not significant, but it was highly significant on unmarketable yield (data not shown). This indicates the importance of environmental conditions on yield characteristics. The maximum, minimum, and average temperatures were warmer than the 50-year average in both years of the experiment (data not shown). The average yearly rainfalls were also less than the 50-year average by 24% and 12% in the years 2002 and 2003, respectively. The 2003 season was in general cooler than 2002 season and had more rainfall, particularly during the fall and winter months. Cooler temperatures indicated less evapotranspiration in 2003 than 2002; and since the rainfall was also more, the fruit trees were under less water and heat stresses.

The total fruit yield did not significantly increase in any of the cultivars due to zinc fertilization, and they were not different among cultivars (Fig. 1a). ZY cultivar had the highest total fruit yield in both Zn-fertilized and unfertilized treatments (Fig. 1a).

Zn fertilization significantly decreased unmarketable fruit yields (Table 2, Fig. 1b). There were also significant differences among cultivars in either control or Zn-fertilized treatment (Table 2, Fig. 1b). Among cultivars, TG had the lowest

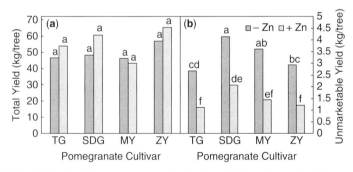

Figure 1. Total fruit yield (**a**) and unmarketable fruit yield (**b**) as affected by Zn treatment (0% and 0.4%) for the pooled data of the two study years. Bars with the same letters at each graph are not significantly different at $p < 0.05$ according to Duncan's multiple range test.

TABLE 2. Mean Unmarketable Fruit Yields of the Pomegranate Cultivars

Year	Cultivar	Fertilizer Treatment				Reduction in Unmarketable Yield (%)	
		Control		0.4% Zn			
2002	TG	3.98	ab	1.85	d	47.4	c
	SDG	4.75	a	3.50	b	26.7	d
	MY	4.15	ab	2.20	cd	49.3	c
	ZY	4.18	ab	1.93	d	54.5	bc
2003	TG	1.38	def	0.38	f	73.5	ab
	SDG	3.55	b	0.63	ef	83.1	a
	MY	3.05	bc	0.68	ef	77.1	a
	ZY	1.70	de	0.48	f	69.8	ab

Means with the same letters are not significantly different at $p < 0.05$ according to Duncan's multiple range test.

unmarketable yield, but it was not significantly different from ZY cultivar in control and from ZY and MY in Zn treatment (Fig. 1b). SDG had the highest amount of unmarketable yield in both treatments, but it was not significantly different from MY cultivar (Fig. 1b).

Unmarketable yields were lower in 2003 than 2002 in both control and Zn-fertilized treatments (Table 2). This was due to better weather conditions and less heat and water stresses in 2003. Also, the percent reductions in unmarketable yields due to Zn fertilization were greater in 2003 (Table 2). Therefore, in addition to a more favorable weather condition, the beneficial effects of first-year Zn application on general tree health was another reason for lower production of unmarketable yield in the second year. Therefore, it seems that Zn may have increased the ability of the pomegranate trees to resist diseases and environmental stresses. Graham [1983] has reviewed evidences for linkages between Zn nutrient stress and plant diseases. Zn-efficient wheat genotypes were less susceptible to crown rot disease [Crewal et al., 1996] and *Rhizoctonia* root rot [Thongbai et al., 2001].

3.2 Pomegranate Zinc Content

The effect of year on Zn concentrations of leaf, fruit juice, and seed was not significant (data not shown). In addition, they were not significantly different among cultivars in Zn-fertilized or unfertilized treatments (Fig. 2). Zinc application significantly increased leaf Zn concentration in all four pomegranate cultivars (Fig. 2a). Normal range of Zn in most plants (mg Zn/kg leaf dry weight) is 20–100, and below 15 is considered Zn deficiency [Plank, 1989]. Leaf Zn-deficiency threshold values (mg Zn/kg leaf dry weight) has been reported to be 14 for apple [Shear and Faust, 1980], 30 for pecan [Plank, 1989], and 20 for avocado [Goodall et al., 1979]. Also, Zn sufficiency range (mg Zn/kg leaf dry weight) has been reported to be 30–75 for woody ornamentals [Plank, 1989], 30–150 for avocado [Goodall et al., 1979],

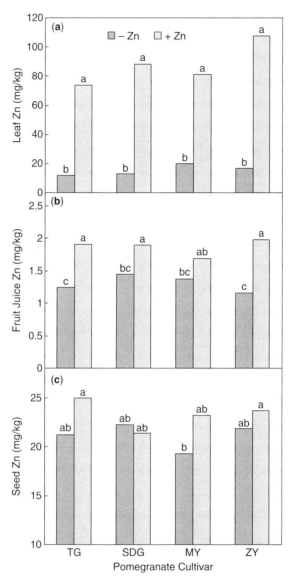

Figure 2. Zinc concentration (mg/kg dry weight) of leaf (**a**), fruit juice (**b**), and seed (**c**) as affected by Zn treatment (0% and 0.4%) for the pooled data of the two study years. Bars with the same letters at each graph are not significantly different at $p < 0.05$ according to Duncan's multiple range test.

and 20–50 for apple and pear and 15–50 for peach [Plank, 1989]. Sufficiency range of Zn in pomegranate leaves has been reported to be 38–45 mg/kg [Raghupathi and Bhargava, 1998a] and 14–72 mg/kg [El Kassas et al., 1993]. Daryashenas and Dehghani [2006] reported that leaf Zn concentration in high (>14 ton/ha) and low

(<14 ton/ha) yielding pomegranate orchards of Yazd were between 11–24 and 10–25 mg/kg, respectively. In this experiment, the average Zn concentration of leaf in four pomegranate cultivars was between 12.0 and 19.8 mg/kg in the control (Fig. 2a). Thus, pomegranate trees had Zn deficiency. Zinc foliar application significantly increased average leaf Zn concentrations to 73.5–107.4 mg/kg and corrected Zn deficiency of the trees, which was evident in highly significant reduction in unmarketable (rotted) fruit yields.

Foliar application of zinc significantly increased Zn concentration of pomegranate juice in three out of four cultivars (Fig. 2b). The average Zn concentration of fruit juice in control was between 1.16 and 1.45 mg/kg juice dry weight (Fig. 2b). Pomegranate pulp (arils without the seeds) has been reported to contain 1.2 mg Zn per 100 g fresh weight [USDA-NDB, 2006]. Mirdehghan and Rahemi [2007] studied the seasonal changes of mineral nutrients in pomegranate (Malas Yazdi cv., one of the four cultivars studied in this experiment) and reported the aril (pulp and seed) Zn concentration at harvest to be about 11.75 mg Zn/kg aril dry weight. The reason for differences in our results and the results of Mirdehghan and Rahemi [2007] was that they measured Zn concentration in the whole dried aril (pulp and seed), while we measured Zn concentration in only the juice (pulp) fraction of the aril. The average Zn concentration of fruit juice in Zn foliar application treatment was between 1.69 and 1.98 mg/kg juice dry weight (Fig. 2b). It was increased significantly in TG, SDG, and ZY cultivars; and although it was increased in MY cultivar by more than 23%, the increase was not significant (Fig. 2b). The results showed that fruit juice nutritional value in terms of Zn has been improved by foliar application of Zn to the pomegranate trees.

Foliar application of zinc did not have a significant effect on seed Zn concentration in any of the four pomegranate cultivars (Fig. 2c). The average Zn concentration of seeds in control was between 19.3 and 22.3, and in Zn-fertilized treatment it was between 21.4 and 25.0 mg/kg seed dry weight (Fig. 2c). These values were higher than the ones reported by Mirdehghan and Rahemi [2007] for the whole aril. Considering our Zn concentrations in dry weights of fruit juice (Fig. 2b) and seed (Fig. 2c), the results of Mirdehghan and Rahemi [2007] seems reasonable, since they analyzed the whole aril (pulp and seed fractions) for Zn concentration, while we analyzed each fraction separately. In control treatment, Zn concentration of seeds was higher than Zn concentrations of leaf and juice (Fig. 2). However, in Zn-fertilized trees the order of Zn concentration from highest to the lowest was leaf > seed > juice (Fig. 2). In both Zn treatments, fruit juice had the lowest concentration of Zn in the three fractions studied. The main reason is that metabolic activities in fruit juice are less than those in leaves and seeds. Nutrients within the plants tend to move more to the sites where metabolic activities are for any reason high [Faust, 1989].

3.3 Phytoestrogen Content

Phytoestrogen content of pomegranate seeds was significantly different among cultivars (Table 3). In control treatment, TG cultivar significantly had the lowest seed content of phytoestrogen, but it was not significantly different among other three

TABLE 3. The ANOVA *F* Values for Some of the Measured Characteristics of Pomegranate Seed for the Pooled Two-Year Data

Treatment	Phytoestrogen (mg/kg)	Zinc (mg/kg)
Year (Y)	NS	22.05**
Cultivar (C)	15.35**	NS
Y × C	NS	3.29*
Fertilizer (Zn)	19.38**	NS
Zn × C	NS	NS
Zn × Y	NS	NS
Zn × C × Y	NS	NS

* and **: significant at 0.05 and 0.01 probability levels, respectively.
NS, not significant at 0.05 probability level.

cultivars (Fig. 3). Zinc fertilization effect on seed phytoestrogen was highly significant (Table 3). Seed phytoestrogen increased in all cultivars between 3.6% and 14.8%, but it was significant only in TG and MY cultivars (Fig. 3). TG cultivar also had the highest increase in seed phytoestrogen content due to Zn fertilization (Fig. 3). Although it was increased by about 3.6 mg/kg (14.8%) in TG cultivar, still it was the lowest among cultivars in Zn fertilization treatment.

Phytoestrogen content of seeds had a significant and positive correlation with leaf Zn content at 0.05 probability level, in the pooled two-year data (Table 4). Within cultivars, only in TG cultivar, seed phytoestrogen content had positive and highly significant correlation with Zn contents of seed and leaf (Table 4). TG and MY cultivars had the highest increase in seed Zn content due to zinc fertilization (18% and 20.3%,

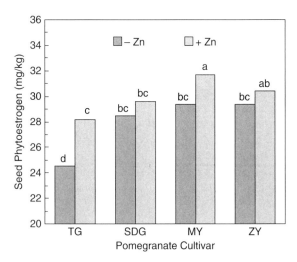

Figure 3. Seed phytoestrogen content of pomegranate cultivars as affected by Zn treatment (0% and 0.4%) for the pooled data of the two study years. Bars with the same letters are not significantly different at $p < 0.05$ according to Duncan's multiple range test.

TABLE 4. Correlation Coefficient of Seed Phytoestrogen Content with Zn Contents of Leaf, Juice, and Seed

Data Set	Leaf Zn	Juice Zinc	Seed Zn
Pooled	0.248*	0.173	0.011
TG cv.	0.790**	0.367	0.624**

* and **: significant at 0.05 and 0.01 probability levels, respectively.

respectively) (Fig. 2c). The highest increase in seed phytoestrogen content was also observed in these two cultivars (14.8% and 8.1%, respectively) (Fig. 2). Zinc is not a part of phytoestrogen structures. However, it is reported that the DNA-binding domain of estrogen receptors and other steroid hormone receptors contain two zinc fingers [WGPH, 2003]. Zinc was used as an effective catalytic agent for synthesis of genistein [Whalley et al., 2000]. Therefore, it seems that Zn acts as a catalytic substance for biosynthesis of phytoestrogen in pomegranate seed.

4   SUMMARY AND CONCLUSIONS

In this two-year field study, foliar zinc fertilization of four pomegranate cultivars significantly increased seed phytoestrogen content of two cultivars, although the concentration increased in all four cultivars. MY cultivar had the highest phytoestrogen content among the four cultivars, either with or without Zn fertilization. TG cultivar had the highest response to Zn fertilization, and its phytoestrogen content increased by 3.6 mg/kg. Therefore, different pomegranate cultivars respond differently to Zn fertilization in terms of seed phytoestrogen content. A further research with more number of Zn treatments and cultivars is recommended.

Improvement of phytoestrogen concentration in pomegranate seed has a positive effect on the harvestable portion of phytoestrogen from the seed, which has a good market in European and Far East countries. The significant increase in these two components has a double effect on the harvestable portion of phytoestrogen from the seed and increases the value of pomegranate fruit tree as a rich source of phytoestrogen. The results suggest that Zn may act as a catalytic agent in physiological pathways for biosynthesis of the phytoestrogen in pomegranate seeds. Zinc fertilization of Zn-deficient pomegranates cultivated in calcareous soils of arid regions is recommended for correction of Zn deficiency, reduction of unmarketable yield, and enhancement of fruit quality in terms of phytoestrogen content.

ACKNOWLEDGMENTS

The authors wish to express their gratitude to Mr. H. Kheradmand from the Laboratory of Rafsanjan Pistachio Company, for conducting the HPLC analysis of the phytoestrogen content of the seeds. Our gratitude is extended to Mr. M. R. Vazifehshenas, the horticulture specialist at YANRC, for his helpful suggestions and assistance in performing the statistical analysis of the data sets.

REFERENCES

Albrecht M, Jiang W, Kumi-Diaka J, Lansky EP, Gommersall LL, Patel A, Mansel RE, Geldof AA, Campbell MJ. 2004. Pomegranate extracts potently suppress proliferation, xenograft growth, and invasion of human prostate cancer cells. *J Med Food* **7**(3):274–283.

Alloway BJ. 2004. *Zinc in Soils and Crop Nutrition*. Brussels, Belgium: International Zinc Association (ZA).

Asadi Kangarshahi A, Malakouti MJ. 2004. Zinc calibration in the field and its effects on soybean yield. *Iranian J Soil Water Sci* **17**(2):115–122.

Aviram M, Rosenblat M, Gaitini D, Nitecki S, Hoffman A, Dorufeld L, Volkova N, Presser D, Attias J, Liker H. 2004. Pomegranate juice consumption for 3 years by patients with carotid artery stenosis reduces common carotid intima-media thickness, blood pressure and LDL oxidation. *Clin Nutr* **23**(3):423–433.

Balakrishnan K, Vekatesan K, Sambandamurthis S. 1996. Effect of foliar application of Zn, Fe, Mn and B on yield quantity of pomegranate, cv. Ganesh. *Orissa J Hortic* **24**(1–2):33–35.

Bybordi A, Malakouti MJ. 2004. Effects of iron, manganese, zinc and copper on wheat yield and quality under saline condition. *Iranian J Soil Water Sci* **17**(2):140–150.

Cornwell T, Cohick W, Raskin I. 2004. Dietary phytoestrogens and health. *Phytochemistry* **65**:995–1016.

Crewal HS, Graham RD, Rengel Z. 1996. Genotypic variation in zinc efficiency and resistance to crown rot disease (*Fusarium graminaearum* Schw. Group 1) in wheat. *Plant Soil* **186**:219–226.

Daryashenas AM, Dehghani F. 2006. Determination of DRIS reference norms for pomegranate in Yazd Province. *Iranian J Soil Water Sci* **20**(1):1–10.

Duke AJ, Ayensu SE. 1985. *Medicinal Plants of China*. Algonac, MI: Reference Publications.

El Kassas SE, Amen KIA, Hussein AA, Osman SM. 1993. Effect of certain methods of weed control on some micronutrient concentrations in Manfaloury pomegranate leaves. *Assiut J Agric Sci* **24**:85–96.

Faust M. 1989. *Physiology of Temperate Zone Fruit Trees*. New York: John Wiley & Sons.

Gao YH, Yamaguchi M. 1998. Zinc enhancement of genistein's anabolic effect on bone components in elderly female rats. *Gen Pharmacol* **31**(2):199–202.

Ghasemian A, Mehrabanian S, Majd A. 2006. Peel extracts of two Iranian cultivars of pomegranate (*Punica granatum*) have antioxidant and antimutagenic activities. *Pakistan J Biol Sci* **9**(7):1402–1405.

Glozer K, Grant JA. 2006. Effects of fall applications of urea and zinc sulfate to Bing sweet cherry spring bud break. *HortScience* **41**:1030–1031.

Goodall GE, Embleton TW, Platt RG. 1979. Avocado Fertilization. University of California Cooperative Extension Bulletin 2024, Davis, CA.

Graham RD. 1983. Effects of nutrient stress on susceptibility of plants to disease with particular reference to the trace elements. In: Woolhouse H, editor. *Advances in Biological Research*, Vol. 10. London: Academic Press, pp. 221–276.

Heftmann E, Ko ST, Bennet RD. 1966. Identification of estrone in pomegranate seeds. *Phytochemistry* **5**:1337–1340.

Kallsen CE, Holtz B, Villaruz L, Wylie C. 2000. Leaf zinc and copper concentrations of mature pistachio trees in response to fertigation. *HortTechnology* **10**:172–176.

Kaya C, Higgs D. 2002. Response of tomato (*Lycopersicon esculentum* L.) cultivars to foliar application of zinc when grown in sand culture at low zinc. *Sci Hortic* **93**(1):53–64.

Khayyat M, Tafazoli E, Eshghi S, Rajaee S. 2007. Effect of nitrogen, boron, potassium and zinc sprays on yield and fruit quality of date palm. *American-Eurasian J Agric Environ Sci* **2**(3):289–296.

Kim ND, Mehta R, Yu WP, Neeman I, Livney T, Amichay A, Poirier D, Nicholls P, Kirby A, Jiang W, Mansel R, Ramachandran C, Rabi T, Kaplan B, Lansky EP. 2002. Chemopreventive and adjuvant therapeutic potential of pomegranate (*Punica granatum*) for human breast cancer. *Breast Cancer Res Treat* **71**:203–217.

Lansky EP, Shubert S, Neeman I. 2000. Pharmacological and therapeutic properties of pomegranate. In: *Proceedings of the Symposium on Production, Processing and Marketing of Pomegranate in the Mediterranean Region: Advances in Research and Technology.* Orihuela, Spain, October 15–17, 1998, p. 231.

Larkin TA, Astheimer L, Price WE. 2000. Analysis of phytoestrogens in foods and biological samples. Agro-Food-Industry-Hi-Tech, November/December, pp. 24–27.

LaRue JH. 1980. Growing pomegranates in California. Available online: http://rics.ucdavis.edu/fnric2/crops/pomegranate_factsheet.shtml

Lee J, Watson RR. 1998. Pomegranate: A role in health promotion and AIDS? In: Watson RR, editor. *Nutrition, Food and AIDS*. Boca Raton, FL: CRC Press, p. 179.

Li Y, Guo C, Yang J, Wei J, Xu J, Cheng S. 2006. Evaluation of antioxidant properties of pomegranate peel extract in comparison with pomegranate pulp extract. *Food Chem* **96**(2):254–260.

Longtin R. 2003. The pomegranate: Nature's power fruit? *J Natl Cancer Inst* **95**(5):346–348.

Ma ZJ, Igarashi A, Inagaki M, Mitsugi F, Yamaguchi M. 2000. Supplemental intake of isoflavones and zinc-containing mineral mixture enhances bone components in the femoral tissue of rats with increasing age. *J Health Sci* **46**(5):363–369.

Mazur W. 2000. Phytoestrogens: Occurrence in foods, and metabolism of lignans in man and pigs. Dissertation, Medical Faculty of the University of Helsinki.

Mirdehghan SH, Rahemi M. 2007. Seasonal changes of mineral nutrients and phenolics in pomegranate (*Punica granatum* l.) fruit. *Sci Hortic* **8**(2):120–127.

Moneam NMA, El-Sharasky AS, Badereldin MM. 1988. Oestrogen content of pomegranate seeds. *J Chromatogr* **438**(2):438–442.

Nagy P, Shaw PE, Wardowski WF. 1990. *Fruits of Tropical and Subtropical Origin*. Lake Alfred, FL: Florida Science Source.

Neurath AR, Strick N, Li Y, Debnath AK. 2005. *Punica granatum* (pomegranate) juice provides and HIV-1 entry inhibitor and candidate topical microbicide. *Ann NY Acad Sci* **1056**:311–327.

Peryea FJ. 2006. Phytoavailability of zinc in postbloom zinc sprays applied to "Golden Delicious" apple trees. *HortTechnology* **16**:60–65.

Peschel W, Sanchez-Rabaneda F, Diekmann W, Plescher A, Gartzia I, Jimenez D, Lamuela-Raventos R, Buxaderas S, Codina C. 2006. An industrial approach in the search of natural antioxidants from vegetable and fruit wastes. *Food Chem* **97**(1):137–150.

Plank CO. 1989. *Plant Analysis Handbook for Georgia*. University of Georgia Cooperative Extension Services, Athens, GA, Available online: http://aesl.ces.uga.edu/docbase/publications/plant/plant.htm

Poyrazoglu E, Gokmen V, Artik N. 2002. Organic acids and phenolic compounds in pomegranates (*Punica granatum* l.) grown in Turkey. *J Food Comp Anal* **15**:567–575.

Price KR, Fenwick GR. 1985. Naturally occurring oestrogens in foods: A review. *Food Addit Contam* **2**:73–106.

Raghupathi HB, Bhargava BS. 1998a. Leaf and soil nutrient diagnostic norms for pomegranate (*Punica granatum* l.). *J Indian Soc Soil Sci* **46**:412–416.

Raghupathi HB, Bhargava BS. 1998b. Diagnosis of nutrient imbalance in pomegranate by Diagnosis and Recommendation Integrated System and Compositional Nutrient Diagnosis. *Comm Soil Sci Plant Anal* **29**(19&20):2881–2892.

Rasouli Sadaghiani MH, Malakouti MJ, Samar SM, Sepehr E. 2002. The effectiveness of different application methods of zinc sulfate on nutritional conditions of apple in calcareous soils of Iran. Paper S13-P-154-A, 16th International Horticultural Congress, August 11–17, Toronto, Canada, p. 23.

Rohrdanz E, Ohler S, Tran-Thi Q, Kahl R. 2002. The phytoestrogen daidzein affects the antioxidant enzyme system of rat hepatoma H4IIE cells. *J Nutr* **132**:370–375.

Rossiter RC. 1967. Physiological and ecological studies on the oestrogenic isoflavones in subterranean clover (*T. subterraneum* L.) IV. Effects of zinc deficiency in clover seedlings. *Aust J Agric Res* **18**(1):39–46.

Saeed S, Tariq P. 2006. Effects of some seasonal vegetables and fruits on the growth of bacteria. *Pakistan J Biol Sci* **9**(8):1547–1551.

Saini RS, Sharma KD, Dhankhar OP, Kaushik RA. 2001. *Laboratory Manual of Analytical Techniques in Horticulture.* Jodhpur, India: Agrobios.

SAS Institute. 1989. *User's Guide: Statistics.* Version 6. Cary, NC: SAS Institute.

Schubert SY, Lansky EP, Neeman I. 1999. Antioxidant and eicosanoid enzyme inhibition properties of pomegranate seed oil and fermented juice flavonoids. *J Ethnopharmacol* **66**:11–17.

Seilsepour M, Ghaemi MR. 2006. Effects of different irrigation water quantities and use of Fe and Zn on yield and water use efficiency of tomato. *Iranian J Soil Water Sci* **20**(2):309–318.

Serdar U, Horuz A, Demir T. 2005. The effects of B–Zn fertilization on yield, cluster Drop and nut traits in hazelnut. *J Biol Sci* **5**(6):786–789.

Shear CB, Faust M. 1980. Nutritional ranges in deciduous tree fruits and nuts. *Hortic Rev* **2**:142–163.

Srivastava AK, Singh S. 2005. Zinc nutrition, a global concern for sustainable citrus production. *J Sust Ag* **25**(3):5–42.

Swietlik D. 2002. Zinc nutrition of fruit crops. *HortTechnology* **12**:45–50.

Taghavi GR. 2000. The effects of macronutrients and foliar application of zinc sulfate on the yield and quality of pomegranate. In: *Proceedings of the Second National Conference on the Optimum Utilization of Chemical Fertilizers and Pesticides in Agriculture*, January 24–26, 2000, Karaj I.R. of Iran, pp. 230–231.

Thongbai P, Hannam RJ, Graham RD, Webb MJ. 2001. Interaction between zinc nutritional status of cereals and *Rhizoctonia* root rot severity. *Plant Soil* **153**:207–214.

Tiwari KN, Nigam V, Pathak AN. 1982. Effect of potassium and zinc applications on dry matter production and nutrient uptake by potato variety "Kufri chandramukhi" (*Solanum tuberosum* L.) in an alluvial soil of Uttar Pradash. *Plant Soil* **65**(1):141–147.

USDA Nutrient Data Base (NDB). 2006. USDA National Nutrient Database for Standard References, Release 19. Available online: http://riley.nal.usda.gov/NDL/cgi-bin/list_nut_edit.pl

van Elswijk DA, Schobel UP, Lansky EP, Irth H, van der Greef J. 2004. Rapid dereplication of estrogenic compounds in pomegranate (*Punica granatum*) using on-line biochemical detection coupled to mass spectrometry. *Phytochemistry* **65**(2):233–241.

Venkatesan S, Murugesan S, Senthur Pandian VK, Ganapathy MNK. 2005. Impact of sources and doses of potassium on biochemical and Greenleaf parameters of tea. *Food Chem* **90**(4):535–539.

Whalley JL, Oldfield MF, Botting NP. 2000. Synthesis of [4–13C] isoflavonoid phytoestrogens. *Tetrahedron* **56**:455–460.

Working Group on Phytoestrogen and Health (WGPH). 2003. *Phytoestrogens and Health*. London: Food Standard Agency.

Wren RC. 1988. *Potter's New Cyclopedia of Botanical Drugs and Preparations*. Essex: C. W. Daniel Company.

Yamaguchi M, Gao YH, Ma ZJ. 2000. Synergistic effect of genistein and zinc on bone components in femoral-metaphyseal tissues of female rats. *J Bone Miner Metab* **18**(2):77–83.

Zekri M, Obreza TA. 2003. Micronutrient deficiencies in citrus: Iron, zinc and manganese. Fact sheet SL 204, University of Florida Cooperative Extension Series, Gainesville, FL.

Zhang Q, Brown PH. 1999. Distribution and transport of foliar applied zinc in pistachio. *J Am Soc Hortic Sci* **124**:433–436.

Ziaeian AH, Malakouti MJ. 2001. Effects of Fe, Mn, Zn, and Cu fertilization of wheat in the calcareous soils of Iran. In: Horst WJ, editor. *Plant Nutrition-Food Security and Sustainability of Agro-ecosystems through Basic and Applied Research*. London: Kluwer Academic, p. 840.

11 Iron Bioavailability, Homeostasis through Phytoferritins and Fortification Strategies: Implications for Human Health and Nutrition

N. NIRUPA and M. N. V. PRASAD

Department of Plant Sciences, University of Hyderabad, 500 046 Hyderabad, India

1 INTRODUCTION

Iron is one of the essential micronutrients for plant and human nutrition. About 30% of the world's population suffers from iron deficiency anemia. Uptake and transport of iron by the plant is an integrated process of membrane transport, reduction, and trafficking between chelator species, whole-plant allocation, and genetic regulation [Wei and Theil, 2000; Schmidt, 2003]. Therefore, iron homeostasis is regulated through complex mechanisms involving various pathways to reduce the toxic effects of the iron [Hell and Stephan, 2003].

Ferritin, a ubiquitous class of iron storage nuclear-encoded protein, plays a major role in eukaryotic iron homeostasis [Harrison and Arosio, 1996; Theil and Briat, 2004]. It is composed of 24 subunits, which can store up to 4000 iron atoms in the central cavity as solid oxo mineral in soluble bioavailable form. Ferritin is the only protein capable of solving iron/oxygen chemistry with cellular concentration requirements of $\sim 10^{-4}$ M compared to the 10^{-18} M solubility of the iron, a gradient of 100 trillion-fold [Theil, 2000; Liu and Theil, 2005]. Intracellularly most of the metabolic iron is sequestered in ferritin [Marentes and Grusak, 1998]. Besides enzymatic scavenging, ferritin controls the concentration of transition metals, which have a prime role in oxygen activation [Rama Kumar and Prasad, 1999b]. It has been demonstrated that

Trace Elements as Contaminants and Nutrients: Consequences in Ecosystems and Human Health, Edited by M. N. V. Prasad
Copyright © 2008 John Wiley & Sons, Inc.

ferritin plays a key role in alleviating oxidative damage and pathogens [Deak et al., 1999; Rama Kumar and Prasad, 1999a; Mata et al., 2001; Dellagi et al., 2005].

Ferritin in legumes is one of the dietary nonheme iron sources in human nutrition [Goto et al., 1998; Goto et al., 1999; Murray-Kolb et al., 2003]. Because ferritin iron is separated from the chelating components such as phytates by its protein coat, it is more stable rendering iron bioavailability [Theil, 2000]. The role of ferritin as a transient iron buffer has been documented in the developmental processes of plants [Strozycki et al., 2003].

Plant products that deliver increased levels of essential minerals or vitamins are termed "fortified" foods. The introduction of genes that code for trace elements binding proteins or storage proteins produce fortified foods. Biofortification is a sustainable approach to alleviate malnutrition [Foyer et al., 2006]. A notable example of biofortification was the creation of iron-fortified rice and "Golden Rice" (vitamin A fortified) [Goto et al., 1999; Ye et al., 2000]. Edible parts also include vegetative tissues, and regulation of plant ferritins occurs predominantly at the level of transcription overexpression of ferritin and might serve as a strategy for fortification [Prasad and Nirupa, 2007; Ragland and Theil, 1993]. Ferritin can also sequester nonferrous metals along with iron, reducing the availability of metals that catalyze cell damage [Sczekan and Joshi, 1989]. It is also considered an important cytoprotectant. Since ferritin plays a major role in diminishing the toxic effects generated due to free radicals [Rama Kumar and Prasad, 1999a; Fourcroy et al., 2004] and also heavy metals by sequestering excess metal, the behavior of the transgenic plants overexpressing ferritin with regard to cadmium toxicity is analyzed in some experiments [Sappin-Didier et al., 2005].

2 IRON IMPORTANCE

Iron is an essential element for all forms of life. Iron takes part in photosynthesis, respiration, DNA synthesis, and hormone structure and action. Many regulatory mechanisms are involved in iron uptake, transport, and storage, which are still to be elucidated [Briat and Lobreaux, 1997; Curie and Briat, 2003; Hell and Stephan, 2003; Schmidt, 2003; Graziano and Lamattina, 2005]. The vital properties of iron are counteracted by its problematic chemistry as a transitional metal affecting the productivity of photosynthetic organisms. Although iron is the fourth most abundant element in the earth's crust (after oxygen, silicon, and aluminum), it is largely unavailable since iron has a tendency to combine with other constituents in the soil to form products, which are not available for plant uptake [Gueirnot and Salt, 2001]. The problem is not abundance but mainly solubility. Iron is present in two oxidation states: Fe^{3+} ($Ar3d^5$) or ferric and Fe^{2+} ($Ar3d^6$) or ferrous. Fe(II) is relatively soluble but is readily oxidized by atmospheric oxygen. The solubility of Fe(III) decreases dramatically with increasing pH values due to hydroxylation as $Fe(OH)_3$, polymerization, and finally precipitation with inorganic ions. While the free Fe(III) is soluble up to 10^{-6} M at pH 3.3, this concentration is only 10^{-17} M Fe(III), and thus the solubility of iron ranges orders of magnitude lower than the required optimal concentration for plant growth in well-aerated soils with pH values more than 7. Iron is easily dissolved into the acidic soil solution, which the plant can then use, while in basic soil solutions it is relatively insoluble. The availability of iron in the soil can be decreased up to 1000-fold for each unit increase in pH. High concentrations of carbonate, bicarbonate, and phosphates in the soil and irrigation water may also lower the availability of iron because low-solubility salts are formed [Yadav and Singh, 1990]:

$$CaCO_3 + CO_2 + H_2O \rightarrow Ca_2^+ + 2HCO_3^+$$

Apart from reducing the iron solubility, bicarbonate ions also reduce the mobility in the plants vascular tissue. The lack of mobility of iron in the presence of bicarbonate causes deficiency in the leaves without affecting the iron in the stems and petioles. High concentration of soil CO_2, a byproduct of respiration by soil microbes and plant roots in less aerated soils, increases the bicarbonate concentration. High-pH soils (8 or greater) are prone to be iron-deficient for many plant species.

3 IRON TOXICITY

Excess Fe generates peroxidative damage of lipids, loss of K^+ to the external medium (suggesting damage of the membranes), diminution of the reduced form of [glutathione] and increase of the oxidized form, increase in the activity of superoxide dismutase, and diminution of the chlorophyll content [Sinha et al., 1997]. Similar observation was done by Caro and Puntarulo [1996] that the addition of Fe-EDTA *in vivo* up to an exogenous concentration of 5×10^{-4} M (500,000 times the minimum of 10^{-9} M) gives rise to an increase in the Fe content of the tissues accompanied by oxidative stress in the roots of soy (*Glycine max*). At the subcellular level, the Fe content and the rate of reduction of Fe-EDTA increased in the roots exposed to Fe when compared to a control without Fe. There was a 50% increase in production of the superoxide radical, along with a fourfold increase in the production of hydroxide in the root endoplasmic reticulum in the medium with iron compared to control without Fe. But there was no affect on the thiol content or antioxidants except for alpha tocopherol when supplemented with Fe. A report by Caris et al. [1995] is interesting in this aspect. They indicate that Fe chelated with a synthetic siderophore (O-Trensox) did not generate oxidative damage, as occurred with Fe-EDTA and Fe-citrate. The ability of a plant or other organism to assimilate iron depends on the conjunctive action of a series of mechanisms involving several rate-limiting steps. This assimilation varies between species, cultivars, or varieties within the same species [Bienfait, 1988].

4 INTERACTIONS WITH OTHER METALS

Understanding the interactions of metals will serve to design more effective strategies for dissecting the mechanisms of metal homeostasis. Many reports hypothesize that

TABLE 1. Iron Interactions with Other Metals

Metal	Type of Interaction	Experimental System (Other Nutrients/Metals Involved are Shown in Parentheses)	Toxicity	Reference
Boron	B levels influence Fe absorption and translocation paralleling the dry matter production.	*Lycopersicon esculentum* Hydroponic culture (manganese)	—	Alvarez-Tinaut et al. [1980]
Copper	When excess Cu was applied, Cu competed with Fe uptake.	*Glycine max* Hydroponic culture (zinc)	Lower chlorophyll content, decreased oxygen evolution at high concentrations	Bernal et al. [2007]
Chromium	Action of Cr is dependent on a low Fe/Cr ratio.	*Plantago lanceolata* L. roots	Decreased chlorophyll content of iron-deficient plants	Schmidt [1996]
	Chromium has been shown to increase leaf iron.	Spinach		Misra and Jaiswal [1982]
Molybdenum	Interaction between [4Fe-4S] cluster and the molybdenum of the Mo-bis MGD cofactor.	*Escherichia coli*	—	Rothery et al. [1999]
Manganese	No interaction was observed for the rate of root absorption of iron.	Manganese-sensitive *Glycine max* cultivar Hydroponic culture *Lycopersicon esculentum*	Fe deficiency correlated poorly, high levels of Mn in leaves displaying Mn toxicity	Heenan and Campbell [1983]
	Deficient or normal Mn levels antagonize Fe absorption.	Hydroponic culture (boron)		Alvarez-Tinaut et al. [1980]
Zinc	Translocation of Zn decreased with increasing levels of Fe.	*Oryza sativa* L. Hydroponic culture	—	Brar and Sekhon [1975]

Element				
	When more Zn was applied, less Zn was translocated to grains; when more Fe was applied, more Zn was translocated to grains. The effects of Fe and Zn on Fe distribution at maturity were opposite to that of Zn distribution.	*Oryza sativa* L. Green house	—	Verma and Tripathi [1983]
Arsenic	Presence of iron-plaque-enhanced arsenite and decreased arsenate uptake.	*Oryza sativa* hydroponic culture (Phosphate)	Chlorosis resulted from Fe deficiency induced by As.	Hu et al. [2007] Chen et al. [2005]
	As was sequestered in roots when arsenate was supplied and most As concentrated in Fe plaque when arsenate was supplied.	*Oryza sativa* Hydroponic culture (manganese)		Liu et al. [2005]
Aluminum	There are strong interactions between Al toxicity and concentrations of Fe and reduced toxicity in presence of Fe.	*Aspergillus nidulans, Neurospora crassa* and *Hymenoscyphus ericae*	Reduced biomass production were compensated by high Fe availability.	Illmer and Buttinger [2006]
Cadmium	Root Cd concentrations was higher and directly related to solution Cd concentrations added.	*Oryza sativa* Hydroponic culture	Cd toxicity symptoms resembled Fe chlorosis and affected plant growth, yield, and Cd accumulation.	Adhikari et al. [2006]
Iodine	Correlation between the oxidizing power of the rice roots and the amount of iodine absorbed.	*Oryza sativa*	Reduced growth.	Yamada et al. [2006]
Lead	Iron deficiency has been shown to increase lead absorption from the intestinal tract.	Rat and humans	Inhibition of heme synthesis and in anemia.	Six and Goyer [1970] Ahamed et al. [2007]

iron has an impact on health outcomes from essential and nonessential metal exposures. Some toxic metals like cadmium and nutritionally essential metals like zinc share common chemical characteristics competing for the sites of enzyme binding, thereby disturbing the mineral metabolism and distribution in the tissues and cells. Experimental evidence suggests that essential elements may protect from the effects of heavy metal exposure, while their deficiency may increase toxicity [Chowdary and Chandra, 1987]. Iron deficiency increases absorption of cadmium, lead, and aluminum. Some of the interactions are summarized in the Table 1.

5 IRON ACQUISITION BY PLANTS

Precise mechanisms are adapted by plants to obtain adequate amounts of this essential nutrient. In response to iron deficiency, plants obtain iron from the soil using one of two diverse processes [Connolly and Guerinot, 2002]. All plants except grasses induce a set of responses that lower the pH in their rhizosphere for the solublising and reduction of Fe(III) to Fe(II) termed "strategy I" [Guerinot and Yi, 1994]. Cloning of the genes with putative roles in iron signaling has begun to untie the nature of the sensor(s) and its downstream targets in plants at a molecular level [Rogers and Guerinot, 2002; Ling, 2002], but most pieces of the puzzle are still missing. Genes encoding the enzymes that function in response to iron deficiency in roots such as Fe(III) chelate reductase (*fro2*) and the Fe(II) transporter (*irt1*) are reported to be up-regulated [Robinson et al., 1999; Waters, 2002]. The IRT1 protein is the major iron transporter responsible for iron uptake from the soil [Connolly et al., 2002; Vert et al., 2002].

Grasses employ the strategy II response for iron acquisition from the soil. In response to iron deficiency, the grasses secrete Fe(III) chelators called phytosidero-phores which are specific plasma-membrane transporters. In barley, phytosidero-phores belong to the mugineic acid (MA) family [Mori, 1999; Curie et al., 2001; Neigishi et al., 2002]. Their functions collectively include iron uptake into the plant. The most obvious symptom of an iron deficiency is foliar iron chlorosis. Development of plants that are capable of growth on the one-third of the world's soils that are iron-deficient or that accumulate elevated levels of bioavailable iron in aid of human nutrition is essential for better utilization of iron.

6 TRANSLOCATION OF IRON IN PLANTS

Once taken up by the root, iron is loaded in the xylem sap and translocated into the plants aerial parts through the transpiration stream. Organic acids, especially citrate, are the main metal chelators in the xylem. The mechanism implies that active root transporters must load iron from the cortical cells into the xylem. Efflux ion transporters have not yet been characterized at the molecular level in plants. Once in the leaves, Fe^{3+} citrate is likely to be the substrate of leaf ferric chelate reductase since such an enzymatic reaction has been described in leaf mesophyll cells. Transporters among the ZIP, NRAMP, and YSL families will probably prove to be involved in iron transport through the plasma membrane of leaf cells. Mobility of iron from sink tissues via the phloem sap is poorly documented. However, it is well established that phloem sap contains iron arising from its mobilization in source organs [Grusak, 1995]. Unloading of the iron in sink organs is supposed to take place by a symplastic route [Hell and Stephan, 2003]. One of the molecules identified as a phloem metal transporter is nicotinamine. Nicotinamine and NRAMP handle iron in its ferrous form. Since most of iron in the phloem is chelated in the Fe^{3+} form by ITP [Curie and Briat, 2003], it is assumed that nicotinamine plays a role of shuttle by chelating Fe^{2+} from ITP-bound Fe^{3+} during uptake [Shingles et al., 2002]. In plants iron homeostasis is maintained primarily through transcriptional control of gene expression. Expression of the iron deficiency response genes *fro2*, *irt1*, and *irt2* is controlled at the level of transcript accumulation by iron status. Plants adapt to the iron conditions by activating a shoot-to-root signal. Loading and unloading of phloem, the putative YSL transporters represent good candidates for performing this task [Curie et al., 2001]. Very little information is available in the intracellular iron movement in plant cells. Plant vacuoles are likely to play an important role in handling excess iron. The best evidence for such vacuole function comes from the finding that upon iron overload in pea mutants overaccumulating iron, nicotinamine concentration increases and the bulk of iron chelator is relocated in the vacuoles, where nicotinamine is observed in the cytoplasm under normal or deficient iron conditions [Pich et al., 2001]. The bulk of the iron in leaves is found within the chloroplasts where it is engaged in the photosynthetic process. Plastids contain ferritin, and ferritin iron represents more than 90% of the iron found in a pea embryo axis. Phytoferritin serves to store excess iron to avoid oxidative stress [Savino et al., 1997; Deak et al., 1999; Rama Kumar and Prasad, 1999a]. Iron transport to the plastids is therefore of primary importance in plant physiology, and this subcellular iron transport activity is poorly documented. Iron transport into the chloroplasts appears to be mediated via a potential energized uniport mechanism in a nonspecific manner, since other micronutrients also compete with Fe^{2+} [Rogers et al., 2002].

7 IRON DEFICIENCY IN HUMANS

About two-thirds of the world's population is at risk of iron-deficiency induced anemia (http://www.who.int/nut/ida.htm). Iron deficiency is the one of the most widespread micronutrient deficiency in humans. It has been estimated by WHO that nearly 3.7 billion people are iron-deficient, and the problem is severe enough to cause anemia in 2 billion people. Among them, 40% are nonpregnant women and 50% were pregnant women. It has also been estimated that 31% of children fewer than 5 years are anemic, with mostly iron-deficiency anemia. According to WHO [2001] and FAO, micronutrient deficiencies appear to increase in prevalence, and the diet diversity declines as population pressure influences patterns of land use. One of the major contributing factors for maternal deaths is iron deficiency anemia in Asia and Africa [Conway and Toenniessen, 1999]. About 400 million women of childbearing age suffer as a result, and they are more prone to stillborn or underweight children and to mortality at childbirth.

Iron deficiency anemia can increase fatigue, reduce work competence in adults, reduce attention span, decrease resistance to infection, reduce intellectual performance, and impair cognitive development in children. Frequently aggravated by malaria and worm infections, iron deficiency anemia can endanger the health of pregnant women, and in their most severe forms can be a direct cause of death. Even mild iron deficiency anemia might decrease work capacity across all sectors of the population, which is generally reversible on the elimination of iron deficiency. For infants, permanent cognitive deficits often result from iron deficiency anemia [Hulthén, 2003]. An accurate and recent method for diagnosing iron deficiency is the measurement of the serum ferritin concentration [Brugnara, 2002; Crichton et al., 1978].

The main suggested causes for iron deficiency anemia include

- Poor eating habits
- Large intake of inhibitors
- Decline in purchasing power
- Poor weaning practices
- Chronic illnesses such intestinal parasites
- Thalassemia
- Insufficient consumption of vegetables and food
- Low dietary intake of iron
- Poor bioavailability of iron

The primary cause of iron deficiency apart from other causes is that most individuals, especially in developing countries, subsist on diets comprised mainly of plant foods that do not supply adequate daily intakes of Fe. The bioavailability of iron is fairly low in the vegetable foods almost about 10%. Although plants can provide almost all the essential nutrients required by humans [Grusak, 1999], plant Fe concentration is low relative to that of most animal-derived foods (Table 2), and

TABLE 2. Iron Content in Some Common Edible Food

Food Group	mg Fe/100 g Edible Portion
Cereal grains	0.4–2.1
Legume seeds	1.5–2.9
Pea seeds	1.5
Leafy vegetables	0.7–3.6
Root crops	0.2–0.8
Fruit	0.1–0.3
Red meat	2.7–10.7

Source: Adapted from Grusak [2000].

the form of Fe in plants is poorly absorbed in the gut [Theil, 2004]. Less consumption of meats that are rich in highly-bioavailable haeme iron is an important contributing factor as the body also absorbs less of the nonheme iron found in grains and vegetables due to the low haeme iron which otherwise enhances absorption of nonheme iron form.

8 AMELIORATION OF IRON DEFICIENCIES

Various approaches have been used to eliminate iron deficiencies which include modification of food processing methods which supplement and fortify foods (Fig. 1). In spite of the value of the advances, a number of factors like infrastructure, acceptance by people, and economics play a key role in limiting their utility. Programs to control iron deficiency and anemia should follow an integrated, long-term approach. Consequently, it is rational to attempt to employ all approaches accessible to improve the iron nutrition. Biofortification is the process of improving the nutritional value of staple food crops by the development of genotypes with high bioavailable nutrients. Ferritin, a protein that can bind up to 4500 atoms of iron ions per molecule present mostly in legumes, potentially can provide a substantial amount of iron content to meet the demanding iron requirements. Experimental evidence indicates that the bioavailability of ferritin iron may be like iron from ferrous sulfate [Murray-Kolb et al., 2003; Davila-Hicks et al., 2004].

One potential way is through the introduction of transgenes that code for iron-binding proteins, storage proteins already present and/or increased expression of proteins that are responsible for trace element uptake into plants [Lucca et al., 2002,

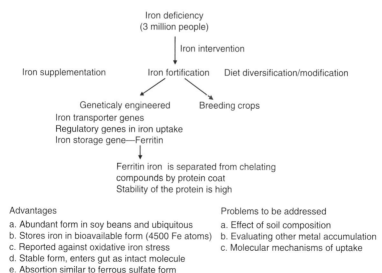

Figure 1. Various approaches for eliminating iron deficiencies.

2006]. However, one possible limitation to increase the bioavailable iron content even with very high levels of expression may not considerably increase the iron content in the presence of naturally occurring inhibitors; if iron content is not increased, many atoms of trace elements will be bound per protein molecule. Introducing ferritin in plants which can bind as many as 4500 atoms of iron may prove beneficial [Murray-Kolb et al., 2002; Lönnerdal, 2003].

9   FERRITIN

Ferritin, a ubiquitous class of iron storage nuclear-encoded protein, plays a major role in eukaryotic iron homeostasis [Harrison and Arosio, 1996] (Fig. 2). It is composed of 24 subunits, which can store up to 4000 iron atoms in the central cavity as a solid oxo mineral in a soluble bioavailable form. Ferritin is the only protein capable of solving iron/oxygen chemistry with cellular concentration requirements of $\sim 10^{-4}$ M compared to the 10^{-18} M solubility of the iron, a gradient of 100 trillion-fold [Theil, 2000; Liu and Theil, 2005]. Intracellularly, most of the metabolic iron is sequestered in ferritin [Marentes and Grusak, 1998]. Besides enzymatic scavenging, ferritin controls the concentration of transition metals, which have a prime role in oxygen activation [Rama Kumar and Prasad, 1999a]. It has been demonstrated that ferritin plays a key role in alleviating oxidative damage and pathogens [Deak et al., 1999; Rama Kumar and Prasad, 1999a; Mata et al., 2001]. Mata et al. [2001] reported a reduction of infection and ROS in the *Phytophthora infestans*-infected leaves of *Solanum tuberosum*, upon addition of iron chelator deferoxamine. Reactive oxygen species in leaves inoculated with *P. infestans* were also reduced after adding deferoxamine. Ferritin mRNA accumulated in response to pathogen attack in the leaves and upon treatment with the elicitor eicosapentaenoic acid in tubers, suggesting a role of

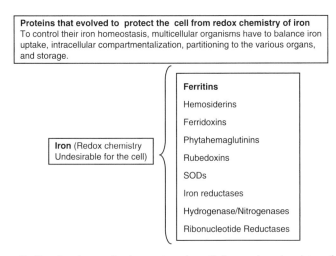

Figure 2. Proteins that evolved to protect the cell from redox chemistry of iron.

ferritin iron chelation in pathogen attack. Pro-oxidant (H_2O_2) and ABA treatment resulted in induction of ferritin in *Vigna mungo*. Pretreatment of iron-deficient de-rooted seedlings with free radical scavengers and antioxidants followed by co-treatment with ferric citrate inhibited ferritin induction indicating the antioxidant role of ferritin [Rama Kumar and Prasad, 1999a]. Iron sequestration in ferritins was found to be a part of an iron-withholding defense system induced in response to bacterial invasion, when *Arabidopsis thaliana* was used as a susceptible host for the pathogenic bacterium *Erwinia chrysanthemi* [Dellagi et al., 2005]. Ferritin in legumes is one of the dietary nonheme iron sources in human nutrition [Goto et al., 1998, 1999; Murray-Kolb et al., 2003]. Because ferritin iron is separated from chelating components such as phytates by its protein coat, it is more stable rendering iron bioavailability [Theil, 2000]. The role of ferritin as a transient iron buffer has been documented in the developmental processes of plants [Strozycki et al., 2003]. Developmental degradation of ferritin has been previously described for different parts of germinating pea seeds; degraded fragments were observed in the radicle, but not in other parts of the seed [Lobreaux and Briat, 1991]. Ferritin is accumulated in seed maturation and degraded during germination, indicating its role as a transient buffer iron supply. Ferritin levels increase in developing leaves [Theil and Hase, 1993], indicating that ferritin synthesis in leaves is developmentally controlled. Plant ferritin mRNA has been shown to accumulate during the early stages of nodule development [Kimata and Theil, 1994]. In senescing nodules of *Lupinus luteus*, ferritin is resynthesized through the expression of two out of the three lupine ferritin genes [Strozycki et al., 2003]. Deleting ferritin genes is detrimental to life in animals [Ferreira, 2000], and its importance is indicated by the presence even in strictly anaerobic bacteria [Da-Costa et al., 2001]. In humans, diseases related to ferritin mutations were discovered and are relatively benign or appear late in life [Cazzola et al., 1997].

10   FERRITIN STRUCTURE

Ferritin is a large protein (12-nm diameter, 480,000 Da) with a large cavity (256 nm^3) for the mineral that is created by the spontaneous assembly of 24 ferritin polypeptides folded into four-helix bundles bound to each other by hydrogen and salt (ionic) bonds. Ferritins managing iron–oxygen biochemistry in animals, plants, and micro-organisms belong to the di-iron carboxylate protein family and concentrate iron as ferric oxide $\approx 10^{14}$ times above the ferric K_s [Liu and Theil, 2004]. Iron is a cofactor in the dioxygenases, but in ferritin it is a substrate. Protein mineralization surfaces, Fe entry and exit sites in ferritin share properties with ion channel/pore proteins that form biominerals and may be progenitors of ferritin pores. The number of functional sites in ferritin is as follows: one mineral cavity in the protein center, eight entry and exit pores on the outer surface, 12 mineral attachment sites on the protein cavity surface, and a variable number of catalytic ferroxidase sites in the center of each subunit [Lobreaux et al., 1992b; Theil, 2000] depending on the number of H and L subunits per ferritin molecule. H-ferritin subunits have an active ferroxidase site

TABLE 3. Ferritin Genes Isolated from Various Sources[a]

Gene	Source of Gene	Ferritin Induction	Gene Size (Base Pairs)	Distinct Features	Gen Bank Accession	Reference
Sof 35	*Glycine max* cell cultures	Iron-regulated ferritin mRNA and protein synthesis	986	Transit peptide and the extension peptide are conserved in the iron-induced mRNA.	M72894	Lescure et al. [1991]
pfe	*Phaseolus vulgaris* young leaves and shoot meristem tissue	Ferritin was purified from seedlings that had been treated with 0.8 mM ferric sodium EDTA	1246	Substantial similarity with other ferritin sequences; 5' untranslated region contains two out-of-frame AUG codons, a region of extreme pyrimidine composition bias, and potentially stable secondary structure.	X58274	Spence et al. [1991]
PeSd 1	*Pisum sativum* seed	Iron induction, Fe EDTA 100 μM	1023	Lacks 5' UTR.	X64417	Lobreaux and Briat [1991]
FM1 and *FM2*	*Zea mays* root and seed	Iron treatment (500 μM) Fe-EDTA/75 μM) Fe-citrate-induced ferritin protein accumulation in roots and leaves	1292	Both were identical, except in their 3' UTRs.	X61391	Lobreaux et al. [1992]
Pe Sd2	*Pisum sativum* seed	Recombinant protein was expressed in *E.coli*	1023	ΔTP/ΔEP; ΔTP (transit peptide, extension peptide), 6 base differences. compared to *PeSd 1* (consensus ferrooxidase site).	X73369	Van Wuytswinkel and Briat [1995]
Zm fer 1	*Zea mays* seedling	Accumulation of *Zm fer 1* transcripts in response to iron	3294	Eight exons and seven introns.	X83076	Fobis-Loisy et al. [1996]
Zm fer 2	*Zea mays* seedling	Accumulation of *Zm fer 2* and transcripts in response to ABA	2902	Eight exons and seven introns.	X83077	Fobis-Loisy et al. [1996]

Gene	Source	Description	bp	Notes	Accession	Reference
At fer 1	*Arabidopsis thaliana* cell suspension	*AtFer1* transcript abundance in response to iron and not to ABA.	1413	Localized on chromosome 5	X94248	Gaymard et al. [1996]
LSC30	*Brassica napus* leaves	Enhanced expression during leaf senescence.	977	Identified from cDNA subtractive hybridization study in *Brassica*.	U68217	Buchanan-Wollaston and Ainsworth [1997]
Cp2 Cp3	*Vigna unguiculata* leaves	mRNA was detected from developing leaves.	958	Significantly divergent from other ferritins (only 77% identical to soybean ferritin. No similarity of transit peptide in *Cp2*. *Cp 1 Transit peptide shares similarity*	AF052058 AF052057	Wardrop et al. [1999]
MsFer	*Medicago sativa* Somatic embryo library	Transgenic tobacco plants accumulating ferritin in their leaves exhibited tolerance to necrotic damage.	1036	89% identity with pea ferritin.	X97059	Deak et al. [1999]
At fer 2 At fer 3 At fer 4	*Arabidopsis thaliana* Analysis of the *A. thaliana* EST database with BLASTN	*AtFer1* and *AtFer3* transcript abundance in response to iron and not to ABA. *At fer 2* transcript abundance in response to ABA and not to iron, found mainly in seeds.	1006 1042 985	*AtFer2* and *AtFer3* are on chromosome 3 and *AtFer4* is on chromosome 2. All 4 genes have 7 introns located at same place. *cis*-IDRS shares similarity in the four genes.	AC009991 AL163763 AF085279	Petit et al. [2001]
SFerH-2	*Glycine max* seedlings		1135	Corresponding region in the 28-kDa soybean ferritin subunit identified in this study was not susceptible to cleavage.	AB062754	Masuda et al. [2001]

(Continued)

TABLE 3. *Continued*

Gene	Source of Gene	Ferritin Induction	Gene Size (Base Pairs)	Distinct Features	Gen Bank Accession	Reference
StF1	*Solanum tuberosum* leaves	Ferritin mRNA accumulated in response to pathogen attack	826	No presence 5'UTR.25 amino acids of the plastid transit peptide are missing.	AF133814	Mata et al. [2001]
Apf1	*Malus xiaojinensis*	—	771	—	AF315505	Zhou et al. [2001]
LlFer1, *LlFer2,* *LlFer3*	*Lupinus luteus*	*LlFer2* class) was transcribed in response to ABA *LlFer3* gene was repressed by ABA, but up-regulated by light. *LlFer2* and *LlFer3* induced on symbiotic interaction.	1032 1118 1039	Amino acid sequence identity of mature polypeptides (86–90%)	—	Strozycki et al. [2003]
Ferritin 2	*Conyza canadensis* seedlings	Up-regulated by paraquat.	765	Exhibit similarity and possess all the structural characteristics of known plant ferritin genes.	AJ786262	Soós et al. [2006]
NtFer1 *NtFer2*	*Nicotiana tobacum* seedlings	Iron loading of tobacco plantlets increased the ferritin mRNA abundance in both. *NtFer1*, was expressed in both leaves and roots.	1214 1125	Share the same characteristics as the other plant ferritins.	AY083924 AY141105	Jiang [2005]

Source: Prasad and Nirupa [2007].

[a]Reproduced with permission from Global Science Books Journal, The Asian and Australasian Journal of Plant Science and Biotechnology.

and occur in multiple forms in humans, animals, plants, and bacteria. L-Ferritin subunits have a degenerate ferroxidase site, and the gene duplication to encode L-ferritin subunits is found only in vertebrate animals. Plant and bacterial ferritins have only a single type of subunit, which possibly performs both functions. Masuda et al. [2001] reported two types of ferritin subunits in soybean having different functions from each other.

Subunits of pea and many other leguminous ferritins are synthesized as precursor proteins with a unique N terminal sequence that is composed of two domains [Ragland et al., 1990; Van Wuytswinkel and Briat, 1995]. These N-terminal domains are not present in the mammalian or other ferritins. The N-terminal transit peptide is the first domain that contains 40–50 residues and is presumed to facilitate transport of the ferritin precursor to plastids [Briat et al., 1995]. The second domain, the extension peptide, is a part of the mature protein [Laulhere et al., 1988] whose function is still unclear.

Plant ferritin genes have been obtained from many different plants (Table 3)—for example, ferritin gene from *Lens esculenta* [Crichton, 1990], *Glycine max* cell suspensions, cotyledon [Sczekan and Joshi, 1987; Lescure et al., 1991; Ragland et al., 1990], *Pisum sativum* seed [Lobreaux et al., 1992b; Van Wuytswinkel and Briat, 1995], *Vigna ungiculata* [Wicks and Entsch, 1993], *Phaseolus vulgaris* [Rama Kumar and Prasad, 2000; Spence et al., 1991], maize [Lobreaux et al., 1992], *Medicago truncatula* [Gyorgyey et al., 2000], *Medicago sativa* [Deak et al., 1999], and *Chlorella protothecoides* [Horstensteiner et al., 2000].

11 MINERAL CORE FORMATION

There are eight iron entry sites, three 24 iron oxidation sites, 24 iron translocation sites, 12 iron mineral attachment/mineralization sites, and eight iron exit sites. When Fe^{2+} enters the ferritin protein pores, it rapidly (in milliseconds) readies the ferroxidase site buried within each H-ferritin subunit [Chasteen and Harrison, 1999]. The ferroxidase site activity that oxidizes Fe^{2+} and forms diferroxo-mineral precursors ($Fe^{3+}-O-Fe^{3+}$) is the function best characterized in ferritin [Liu and Theil, 2004]. Mossbauer, resonance, Raman, and EXAFS spectroscopies (Hwang et al., 2000; Theil, 2000] have been used to analyze the fast ferroxidase reaction intermediates (in milliseconds) by rapid mixing and freezing to trap enough of a blue (A650 nm) diferric peroxo transition state for the spectroscopic studies. Trapping the diferric complex in protein crystals has not been possible. The peroxo intermediate also forms in di iron peroxo oxygenases. Iron is a substrate converted to the diferric peroxo intermediate in ferritin and is released into the interior of the protein as a diferric oxo mineral precursor. Ferritin releases H_2O_2 [Zhao et al., 2001] and diferric oxo mineral precursors, leaving behind an active site that is altered for a fairly long time, presumably to allow peroxide to diffuse away before binding the next Fe(II) substrate atoms [Liu and Theil, 2003] (Fig. 3).

Ferritin also concentrates phosphate in the central core [Theil, 2000]. Various amounts of phosphate are trapped inside the mineral core and seem to reflect the

$$2Fe^{2+} + (H_2O)_6 + O_2 \longrightarrow Fe^{3+}-O-O-Fe^{3+}(H_2O)_2$$

$$(F^{3+}-O-Fe^{3+})\cdot X \quad \text{Minerale} + H_2O_2$$

Figure 3. Ferroxidase site reaction in H-ferritin subunits. (Adapted from Theil [2003].)

cellular concentrations of phosphate. The mineral formed in ferritin of animals has much less phosphate than those formed in phosphate-rich cytoplasm of organisms without nuclei (prokaryotes) or in plants where the ferritin mineral is formed inside the prokaryotic like plasmid [Waldo et al., 1995; Briat, 1996; Polanams et al., 2005].

12 FERRITIN GENE FAMILY AND REGULATION

Plant ferritins are usually the products of a small gene family, and all plant ferritin genes so far reported are single-copy genes (*Zea mays*, Fobis-Loisy et al. [1996]; *Vigna unguiculata*, Wardrop et al. [1999]; *Arabidopsis*, Petit et al. [2001]; *Glycine max*, and Yoshihara Goto et al. [2001]). In *Arabidopsis*, *V. ungiculata* there are four genes belonging to the ferritin family, while three were detected in *Lupinus luteus* and two in maize. Expression of individual family members of the known plant ferritin genes is always under the differential regulation [Strozycki et al., 2003; Jiang, 2005]. The differential expression of ferritin genes was detected by the induction of iron and abscisic acid (ABA) in *Lupinus* [Strozycki et al., 2003], and by that of iron, ABA and H_2O_2 in *A. thaliana* [Fobis-Loisy et al., 1996; Harrison and Arosio, 1996; Petit et al., 2001]. Paraquat-induced expression of the *ferritin 2* gene was also reported in *A. thaliana* [Camp et al., 2003; Soós et al., 2005]. Recently, nitric oxide-mediated ferritin regulation has been shown in *Arabidopsis* [Murgia et al., 2002]. Nitric oxide was shown to act downstream of iron through the iron-dependent regulatory sequence [Petit et al., 2001] of the *AtFer1* promoter, suggesting that Nitric Oxide plays an important role in the regulation of iron homeostasis in plants [Murgia et al., 2002]. Transcriptional control was achieved through transcriptional repression for the *ZmFer1* and *AtFer1* ferritin genes from maize and *A. thaliana*, respectively [Petit et al., 2001].

Studies of plant ferritins have revealed several important differences in the structure, localization, and regulation of plant ferritins as compared to animal ferritins [Connolly and Guerinot, 2002]. For example, while animal ferritins are found in the cytosol, plant ferritins contain transit peptides for delivery to organelles called plastids [Proudhon, 1996]. Moreover, while iron-regulated expression of animal ferritin is controlled mainly at the level of translation by a system of iron-responsive elements (IREs) and iron regulatory RNA-binding proteins (IRPs) [Eisenstein, 2000], experiments in soybean and maize have shown that iron regulates expression of plant ferritins both transcriptionally through iron regulatory element and iron-dependent regulatory sequence [Lescure et al., 1991; Lobreaux et al., 1992b; Wei and Theil, 2000; Petit et al., 2001] and posttranscriptionally [Fobis-Loisy, 1996]. Transcriptional control was achieved through transcriptional repression for the *ZmFer1* and *AtFer1* ferritin genes from maize and *A. thaliana*, respectively [Petit et al., 2001].

13 DEVELOPMENTAL REGULATION

Ferritin is a developmentally regulated protein that is detected only during specific stages of the life cycle of the plant [Lobreaux and Briat, 1991]. Ferritin subunit abundance increases in seeds during their maturation and is detected in dry seeds. Ferritins are degraded during germination, and the pattern observed by immunodetection is suggestive of free radical degradation of ferritin during iron exchanges *in vitro*. This suggests that the pool of iron stored in seed ferritins is mobilized during germination and that the protein shell is degraded. The determinate nature of maize leaf development allowed it to be shown that ferritins are present in the young leaf section and in the tip section containing senescent cells, but not in the central part of the leaf, where mature chloroplasts are abundant [Theil and Hase, 1993; Buchanan-Wollaston and Ainsworth, 1997]. Leaf ferritin content correlates with the differentiation stage of chloroplasts. This would serve as an iron buffer for iron-containing proteins in photosynthesis and would allow the sequestration of iron when chloroplasts degenerate.

Plant ferritins are more likely than animal ferritins to be the source of ferritin in natural foods, and their mineral has a higher ratio of phosphate to iron (usually 4:1) than does that of animal ferritins (usually 1:8; Davila-Hicks et al. [2004]). Studies of plant ferritins have revealed several important differences in the introns/exons organization, structure, localization, and regulation of plant ferritins as compared to animal ferritins. Two different ferritin subunits, H and L, encoded by different genes have been described in animals. The H subunits contain conserved amino acids defining a ferroxidase site responsible for rapid Fe (II) oxidation, leading to a rapid uptake of iron inside the protein cavity; L subunits lack this site but are enriched in E residues facing the central cavity of the protein, thus enabling better nucleation of Fe (III) for its long-term storage [Harrison and Arosio, 1996; Connolly and Guerinot, 2002]. One type of plant ferritin subunit has been described, sharing the characteristics of both the H and L subunits, namely a ferroxidase center and additional E residues facing the protein cavity [Lobreaux et al., 1992b]. Animal ferritins are found in the cytosol; plant ferritins contain transit peptides for delivery to specific organelles, the plastids [Proudhon, et al., 1996]. An N-terminal extension signal was found in all the genes cloned from *Arabidopsis thaliana*, which shares characteristics with plant-specific transit peptides responsible for the targeting of precursor proteins to plastids [Petit et al., 2001]. More recently, ferritins were reported to occur in mitochondria of both animals and plants with a possible role of protection against oxidative stress [Levi and Arosio, 2004; Zancani et al., 2004].

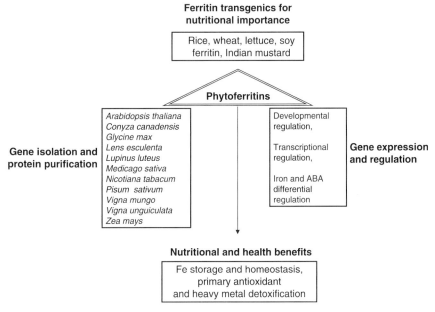

Figure 4. Biotechnological advancements in the field of phytoferritins.

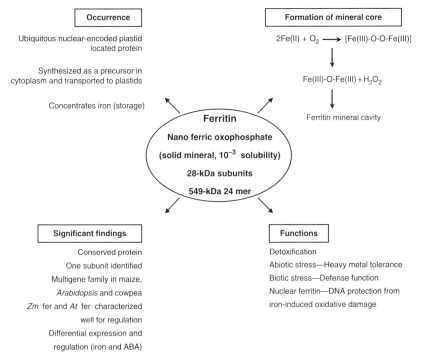

Figure 5. Salient features of ferritins.

Moreover, while iron-regulated expression of animal ferritin is controlled mainly at the level of translation by a system of iron-responsive elements (IREs) and iron regulatory RNA-binding proteins (IRPs) [Eisenstein, 2000], experiments in soybean and maize have shown that iron regulates expression of plant ferritins by both transcriptional and posttranscriptional regulation, through an iron regulatory element (FRE in soybean), which is a *cis*-acting element (identified in soybean ferritin gene), and by an iron-dependent regulatory sequence (IDRS in maize and *Arabidopsis*) [Lescure et al., 1991; Wei and Theil, 2000; Petit et al., 2001] (Fig. 3). Post-transcriptional regulation was reported in maize mutant *ys1* where ferritin protein and mRNA abundance does not correlate in *ys1* leaves upon iron induction, indicating that iron also controls plant ferritin accumulation posttranscriptionally [Fobis-Loisy et al., 1996]. Transcriptional control was achieved through transcriptional repression for the *ZmFer1* and *AtFer1* ferritin genes from maize and *A. thaliana*, respectively [Petit et al., 2001]. The biotechnological advancements and nutritional importance of phytoferritins is depicted in Figs. 4 and 5 respectively.

14   ROLE OF FERRITIN

The role of ferritin is to concentrate iron in the cells to an effective level that matches cellular need [Goto and Yoshihara, 2001]. Concentration of iron leads to iron storage function. When iron concentrations are very high, ferritin also has a protective function by sequestering the iron inside the protein [Rama Kumar and Prasad, 1999a; Connolly et al., 2002; Fourcroy et al., 2004], thus performing detoxification function. The possible role of ferritin acting against biotic stresses is also reported [Deak et al., 1999; Hegedûs et al., 2002; Dellagi et al., 2005). Recent reports indicate the potential role of ferritin as a protector of the genome (Surguladze et al., 2005).

Ferritin, a rich form of nonheme iron in many plant foods particularly in legumes, has been little considered as a dietary iron source until recently. Beard et al. [1996] have demonstrated that the iron in pure ferritin, the major form of iron in soybeans [Ambe et al., 1987], can be absorbed by iron-deficient rats to correct anemia. Hallberg [1981] reported that iron from ferritin was well-absorbed and did not vary notably from ferrous sulfate iron (high bioavailabile form) in humans with varied iron status when given in meals with a low content of inhibitors. However, when added to foods for fortification, ferrous sulfate causes rancidity and discoloration [Hurrell, 2002], which make the product inedible. Ferritin iron, highly bioavailable to humans, may not likely affect the food in which it is consumed and may serve as a form that can be used as dietary source. Effects of inhibitors and enhancers of nonheme iron absorption on the absorption of iron from ferritin should be evaluated for further insights. Ferritin is reported to resist low pH, heat denaturation (temperatures up to $85°C$), urea, many proteolytic enzymes, and *in vitro* digestion [Theil, 2000; Liu et al., 2003; Lönnerdal, 2003; Davila-Hicks et al., 2004]. Experiments scrutinizing ferritin absorption from food have been initiated, but the potential impact of ferritin, "nature's iron," on global iron deficiency could be immense [Theil, 2003].

TABLE 4. Examples of Transgenic Plants Overexpressing Ferritin[a]

Origin of the Gene (Gene Name)	Promoter Expression	Target Plant	Expressed Function	Reference
Soybean ferritin cDNA *SoF 35*	CaMV 35S	*Nicotiana tabacum*	Iron accumulation	Goto et al. [1998]
	CaMV 35S P6 (chloroplastic) C5 (cytoplasmic)	*Nicotiana tabacum*	Iron accumulation	Van Wuytswinkel et al. [1998]
Soybean ferritin cDNA	Seed-specific *Glu B*	*Oryza sativa*	Iron accumulation	Goto et al. [1999]
Alfalfa ferritin cDNA	Constitutive CaMV 35S	*Nicotiana tabacum*	Iron accumulation; tolerance to oxidative damage and biotic stress	Deak et al. [1999]
Soybean ferritin cDNA	Constitutive CaMV 35S	*Triticum estivum* and *Oryza sativa*	Increased iron levels in vegetative tissues but not in seeds	Drakakaki et al. [2000]
Soybean ferritin cDNA	Constitutive CaMV 35S	*Lactuca sativa*	Iron accumulation and improved growth rate	Goto et al. [2000]
Soybean ferritin cDNA	Plastid and cytoplasm expressors using CaMV 35S	*Nicotiana tabacum*	Soil-dependent variability in iron accumulation	Vansuyt et al. [2000]
Phaseolus vulgaris ferritin, phytase gene (*Phy A*) from *Aspergillus* and metallothionein-like protein (*rgMT*)	Seed-specific	*Oryza sativa*	Iron accumulation	Lucca et al. [2001]
Soybean ferritin cDNA	CaMV 35S P6 (chloroplastic) C5 (cytoplasmic)	*Nicotiana tabacum*	No protection against photoinhibition and ozone stress	Murgia et al. [2002]
Alfalfa ferritin cDNA	Constitutive CaMV 35S	*Nicotiana tabacum*	Abiotic stress tolerance	Hegedüs et al. [2002]

Gene	Promoter/Expression	Plant species	Effect	Reference
Soybean ferritin cDNA	Plastid and cytoplasm expressors using CaMV 35S	*Nicotiana tabacum*	Increased root ferric reductase and H^+-ATPase activities and iron content, phosphate-regulated iron accumulation	Vansuyt et al. [2003]
Soybean ferritin cDNA	Endosperm-specific	*Oryza sativa*	Iron and zinc accumulation	Vasconcelos et al. [2003]
Soybean ferritin cDNA	Constitutive CaMV 35S	*Nicotiana tabacum*	Iron and other metals accumulation	Yoshihara et al. [2003]
Soybean ferritin cDNA	Overexpression in plastids or in cytoplasm CaMV 35S	*Nicotiana tabacum*	Heavy metal (Cd) accumulation	Sappin-Didier et al. [2005]
Soybean ferritin cDNA *Aspergillus* phytase (*Phy A*)	Endosperm-specific	*Zea mays*	Increased bioavailable iron	Drakakaki et al. [2005]
Soybean ferritin cDNA	Seed-specific Globulin and glutelin promoter	*Oryza sativa*	Imbalance of ferritin expression and iron accumulation	Qu et al. [2005]
Soybean ferritin cDNA	Plastid expressor CaMV 35S	*Nicotiana tabacum*	Rhizosphere bacteria of transgenics less susceptible to iron stress than wild type in spite of increased iron content in overexpressors	Robin et al. [2005]
Soybean ferritin cDNA	Plastid expressor CaMV 35S	*Nicotiana tabacum*	Study of structure of bacterial and pseudomonads in soil and roots in ferritin overexpressors	Robin et al. [2006]

Source: Prasad and Nirupa [2007].

[a]Reproduced with permission from Global Science Books Journal, The Asian and Australasian Journal of Plant Science and Biotechnology.

15 METAL SEQUESTRATION BY FERRITIN: HEALTH IMPLICATIONS

Knowledge of plant–metal interactions is important for socioeconomic reasons and also for reducing the risks associated with the introduction of trace metals into the food chain. Transitional elements, such as iron and copper, react with reduced forms of oxygen and through Haber–Weiss and Fenton's reaction to generate free radicals and lead to oxidative stress. The transfer of one electron from the electron transport chain to oxygen (univalent reaction) generates superoxide anion ($O_2^-\cdot$), which then dismutates, spontaneously or enzymatically, to hydrogen peroxide (H_2O_2). The latter can react with iron(II) ion (Fenton reaction), generating the highly reactive hydroxyl radical (OH^-). This metal-dependent conversion to the highly toxic —OH via the Haber–Weiss reaction is thought to be responsible for the majority of the biological damage associated with these molecules. Heavy metals such as mercury, lead, and cadmium have no known beneficial effect on organisms, and their accumulation over time can cause serious problems. These elements do not break down or change into other forms and therefore persist in the environment and can accumulate to toxic levels in people or plants. In order to cope with these toxic effects and to maintain the essential metals within the physiological range, plants have evolved complex mechanisms that serve to control the uptake, accumulation, and detoxification of metals [Prasad, 2004a]. Besides enzymatic scavenging, control of the concentrations of metals (known for their prime role in oxygen activation and enzyme inactivation) by sequestering them could form an important complementary way in the prevention of toxic effects. Ferritin is also capable of binding cations such as aluminum, beryllium, cadmium, and zinc apart from iron in the mineral core [Sczekan and Joshi, 1989; Rama Kumar and Prasad, 1999b; Polanams et al., 2005]. It is suggested that the phosphate anion in the iron core of ferritin is necessary to bind with such nonferrous metals. Wade et al. [1993] showed that pea ferritin contains about one-third phosphate atoms.

Genetic engineering has already been used successfully to enhance plant metal tolerance and accumulation [Lupino et al., 2005]. This was achieved either by overproducing metal-chelating molecules such as ferritin [Goto et al., 1999] or by overexpression of metal transporter proteins [Hirschi et al., 2000]. Sappin-Didier et al. [2005] have reported increased accumulation of cadmium in the ferritin overexpressors grown in the soil containing iron and other metals along with cadmium (Table 4). Hence the plants overexpressing ferritin should be assessed for the toxic metal accumulation under different soil conditions.

16 OVEREXPRESSION OF FERRITIN

Knowledge of molecular genetics obtained from one organism can be readily utilized for the improvement of another. Moreover, a large variety of techniques are available which enhance the power and speed of genetic manipulation. The mechanisms underlying iron transport and deposition in the different tissues are of particular importance

TABLE 5. Ferritin Genes Expressed in Transgenic Crops[a]

Gene	Gene Source	Plasmid Used	Method of Gene Introduction	Promoter Used	Level of Expression	Result of Transformation	Reference
Lettuce							
Soybean ferritin cDNA	*Glycine max*	pBG1	*Agrobacterium*-mediated transformation	*CaMV 35 S* promoter	1.2 to 1.7 times in leaves	Enhanced growth, high photosynthesis rates	Goto et al. [2000]
Rice							
Soybean ferritin cDNA	*Glycine max*	pGPTV	*Agrobacterium*-mediated transformation	*GluB-1* Glutelin promoter	Threefold increase in seeds	Normal growth and development	Goto et al. [1999]
SoyferH-1	*Glycine max*	pGPTV	Biolistic method	*GluB-1* Glutelin promoter	Threefold increase in seeds	Iron concentrations increased even after polishing; Zinc also detected	Vasconcelos et al. [2003]
Soy ferH-1	*Glycine max*	pGPTV	*Agrobacterium*-mediated transformation	*GluB-1 Glb-1* Glutelin, Globulin promoters	Threefold increase in seeds	No significant morphological changes; increase of iron accumulation did not parallel ferritin mRNA	Qu et al. [2005]
pfe + Phytase (phyA) and the cysteine-rich protein metallothionein. (rgMT)	*Phaseolus vulgaris*	pCAMBIA 1390	*Agrobacterium*-mediated transformation	*G1* glutelin promoter	Twofold increase in seeds	Normal growth and development	Lucca et al. [2001]

(Continued)

TABLE 5. *Continued*

Gene	Gene Source	Plasmid Used	Method of Gene Introduction	Promoter Used	Level of Expression	Result of Transformation	Reference
Soybean ferritin cDNA	*Glycine max*	pSF1	Particle bombardment	Constitutive maize *ubiquitin-1* promoter 2-10	Twofold increase in leaves but not in seeds	Chlorotic, reduced fertility	Drakakaki et al. [2000]
Soybean ferritin cDNA	*Phaseolus limensis*	pCAMIBA1301	*Agrobacterium*-mediated transformation	Glutelin *GluB-1* promoter	64% increase in seeds	Normal growth and development; no zinc content traced	Liu and Theil [2004]
				Wheat			
Soybean ferritin cDNA	*Glycine max*	pSF1 and pACH20	particle bombardment	Constitutive maize *ubiquitin-1* promoter 2-10	50% in leaves but not in seeds	Ferritin mRNA and protein levels decreased during seed maturation; normal growth and development	Drakakaki et al. [2000]
Soybean ferritin cDNA + Aspergillus phytasePhy A-encoding phytase	*Glycine max*	pSF2	Particle bombardment	Rice seed-specific Gt1 promoter	20–70%	Increased level of iron, and bioavailability of iron in transgenic maize seeds	Drakakaki et al. [2005]

Source: Prasad and Nirupa [2007].
[a]Reproduced with permission from Global Science Books Journal, The Asian and Australasian Journal of Plant Science and Biotechnology.

since the regulatory mechanisms of iron homeostasis can be manipulated to increase the iron content of plants [Ghandilyan et al., 2006]. Classical breeding and biotechnology could contribute together to the required improvement [Foyer et al., 2006].

Constitutive expression of ferritin has been done in various crops like wheat, rice, lettuce, and maize [Goto et al., 1998; Van Wuytswinkel et al., 1998; Deak et al., 1999; Drakkaki et al., 2000, 2005; Goto et al., 2000, Nirupa et al., 2007] where there was increase of iron content in the vegetative parts but not in the seed when expresses under constitutive promoter. The endosperm-specific expression of a *Glycine max* [Goto et al., 1999] or *Phaseolus vulgaris* [Lucca et al., 2001] ferritin gene in rice resulted in a threefold increase or doubling, respectively, of the iron content in the seed (Table 5).

A complete perception of the chemistry and biology of iron metabolism in plants would serve in developing plants with improved performance on soils with poor iron availability and thus contribute to increased yields. Biotechnological appliance of the basic research in the iron biology contributes developing elite varieties of crops aimed at the alleviation of iron deficiency with a positive impact on agricultural and human health problems. Food security and nutritional value of the foods might be addressed explicitly with the concerted efforts of nutritionists and of plant and soil scientists using the ferritin as a model system to develop strategies for iron fortification. The increasing demand for nutraceuticals and fortified foods makes ferritin an ideal model to a beneficial effect on human health.

ACKNOWLEDGMENTS

This review has been prepared from information obtained from authentic and highly regarded sources such as internet, national and international conference deliberations. The authors thankfully acknowledge all these resources, particularly the authors listed in the bibliography. The authors thank Prof. Jaime A. Teixeira da Silva, Editor-in-Chief, Global Science Books, Editorial Office, Miki cho Post Office, Kagawa ken, Kita gun, Miki cho, Ikenobe 3011-2, P.O. Box 7, 761-0799, Japan, for permission to use some of our previously published data, particularly Tables 3–5, in this chapter.

REFERENCES

Adhikari T, Tel-Or E, Libal Y, Shenker M. 2006. Effect of cadmium and iron on rice (*Oryza Sativa* L.) plant in chelator-buffered nutrient solution. *J Plant Nutr* **29**:1919–1940.

Ahamed M, Singh S, Behari JR, Kumar A, Siddiqui MKJ. 2007. Interaction of lead with some essential trace metals in the blood of anemic children from Lucknow, India. *Clin Chim Acta* **377**:92–97.

Alvarez-Tinaut MC, Leal A, Martínez RL. 1980. Iron-manganese interaction and its relation to boron levels in tomato plants. *Plant Soil* **55**:377–388.

Ambe S, Ambe F, Nozuki T. 1987. Mössbauer study of iron in soybean seeds. *J Agric Food Chem* **35**:292–296.

Beard JL, Burton JW, Theil EC. 1996. Purified ferritin and soybean meal can be sources of iron for treating iron deficiency in rats. *J Nutr* **126**:154–160.

Bernal M, Cases R, Picorel R, Yruela I. 2007. Foliar and root Cu supply affect differently Fe- and Zn-uptake and photosynthetic activity in soybean plants. *Environ Exp Bot* **60**:145–150.

Bienfait HF. 1988. Mechanisms in Fe-efficiency reactions of higher plants. *J Plant Nutr* **11**:605–629.

Brar MS, Sekhon GS. 1975. Interaction of zinc with other micronutrient cations. *Plant Soil* **45**:137–143.

Briat JF. 1996. Roles of ferritin in plants. *Plant Nutr* **19**:1331–1342.

Briat JF, Lobreaux S. 1997. Iron transport and storage in plants. *Trends Plant Sci* **2**:187–192.

Briat JF, Laboure AM, Laulhere JP, Lescure AM, Lobreaux S, Pessey H, Proudhan D, Van Wuytswinkel OP. 1995. Molecular and cellular biology of plant ferritins. In: Abadia J, editor. *Iron Nutrition in Soil and Plants*. Netherlands: Kluwer Academic Publishers, pp. 265–276.

Brugnara C. 2002. A hematologic "Gold Standard" for iron-deficient states? *Clin Chem* **48**:981–982.

Buchanan-Wollaston V, Ainsworth C. 1997. Leaf senescence in *Brassica napus:* cloning of senescence related genes by subtractive hybridization. *Plant Mol Biol* **33**:821–834.

Camp RGL, Przybila D, Oschenbein C, Laloi C, Kim C, Danon A. 2003. Rapid induction of stress responses after release of singlet oygen in *Arabidopsis. Plant Cell* **15**:2320–2332.

Caris C, Baret P, Beguin C, Serratrice G, Pierre JL, Laulhère JP. 1995. Metabolization of iron by plant cells using O-Trensox, a high-affinity abiotic iron-chelating agent. *Biochem J* **312**:879–885.

Caro A, Puntarulo S. 1996. Effect of in vivo iron supplementation on oxygen radical production by soybean roots. *Biochim Biophys Acta* **1291**:245–251.

Cazzola M, Bergamaschi G, Tonon L, Arbustini E, Grasso M, Vercesi E, Barosi G, Bianchi PE, Cairo G, Arosio P. 1997. Hereditary hyperferritinemia-cataract syndrome: relationship between phenotypes and specific mutations in the iron-responsive element of ferritin light-chain mRNA. *Blood* **9**:814–821.

Chasteen ND, Harrison PM. 1999. Mineralization in ferritin: an efficient means of iron storage. *J Struct Biol* **126**:182–194.

Chen Z, Zhu YG, Liu WJ, Meharg AA. 2005. Direct evidence showing the effect of root surface iron plaque on arsenite and arsenate uptake into rice (*Oryza sativa*) roots. *New Phytol* **165**:91–97.

Chowdhury BA, Chandra RK. 1987. Biological and health implications of toxic heavy metal and essential trace element interactions. *Prog Food Nutr Sci* **11**:55–113.

Connolly EL, Guerinot ML. 2002. Iron stress in plants. *Genome Biol* **3**:1024.1–1024.4.

Connolly EL, Fett JP, Guerinot ML. 2002. Expression of the IRT1 metal transporter is controlled by metals at the levels of transcript and protein accumulation. *Plant Cell* **14**:1347–1357.

Conway G, Toenniessen G. 1999. Feeding the world in the twenty-first century. *Nature* **402**:55–58.

Crichton RR. 1990. Proteins of iron storage and transport. *Adv Protein Chem* **40**:281–363.

Crichton RR, Ponze-Ortiz Y, Koch MHJ, Parfait R, Stuhrman HB. 1978. Isolation and characterization phytoferritin from pea (*Pisum sativum*) and lentil (*Lens esculenta*). *Biochem J* **171**:349–356.

Curie C, Briat JF. 2003. Iron transport and signaling in plants. *Annu Rev Plant Biol* **54**:183–206.

Curie C, Panaviene Z, Loulergue C, Dellaporta SL, Briat JF, Walker EL. 2001. Maize yellow stipe1 encodes a membrane protein directly involved in Fe(III) uptake. *Nature* **409**:346–349.

da Costa PN, Romao C, Le-Gall J, Xavier AV, Melo E, Teixeira M, Saraiva LM. 2001. The genetic organization of Desulfovibrio desulphuricans ATCC 27774 bacterioferritin and rubredoxin-2 genes: Involvement of rubredoxin in iron metabolism. *Mol Microbial* **4**:217–227.

Davila-Hicks P, Theil EC, Lönnerdal BL. 2004. Iron in ferritin or in salts (ferrous sulfate) is equally available in non-anemic women. *Am J Clin Nutr* **81**:1179.

Deak M, Horvarth GV, Davletova S, Török K, Vass I, Barna B, Kiraly, Dudits D. 1999. Plants ectopically expressing the iron binding protein, ferritin are tolerant to oxidative damage and pathogens. *Nat Biotechnol* **17**:192–196.

Dellagi A, Rigault M, Segond D, Roux C, Kraepiel Y, Cellier F, Briat JF, Gaymard F, Expert D. 2005. Siderophore-mediated upregulation of *Arabidopsis* ferritin expression in response to *Erwinia chrysanthemi* infection. *Plant J* **43**:262–272.

Drakakaki D, Christou P, Stöger E. 2000. Constitutive expression of soybean ferritin cDNA in transgenic wheat and rice results in increased iron levels in vegetative tissues but not in the seeds. *Transgenic Res* **9**:445–452.

Drakakaki G, Sylvain M, Raymond G, Elizabeth L, Sandra P, Rainer F, Paul C, Eva S. 2005. Endosperm-specific co-expression of recombinant soybean ferritin and *Aspergillus* phytase in maize results in significant increases in the levels of bioavailable iron. *Plant Mol Biol* **59**:869–880.

Eisenstein RS. 2000. Iron regulatory proteins and the molecular control of mammalian iron metabolism. *Annu Rev Nutr* **20**:627–662.

Ferreira F, Bucchini D, Martin ME, Levi S, Arosio P, Grandchamp B, Beaumont C. 2000. Early embryonic lethality of H ferritin gene deletion in mice. *J Biol Chem* **275**:3021–3024.

Fobis-Loisy I, Aussel L, Briat JF. 1996. Post-transcriptional regulation of plant ferritin accumulation in response to iron as observed in the maize mutant ys1. *FEBS Lett* **397**:149–154.

Fourcroy P, Vansuyt G, Kushnir S, Inze D, Briat JF. 2004. Iron-regulated expression of a cytosolic ascorbate peroxidase encoded by the APX1 gene in *Arabidopsis* seedlings. *Plant Physiol* **134**:605–613.

Foyer CH, Dean DellaPenna C, Straeten DVD. 2006. A new era in plant metabolism research reveals a bright future for bio-fortification and human nutrition. *Physiol Plant* **126**:289–290.

Ghandilyan A, Vreugdenhil D, Aarts MGM. 2006. Progress in the genetic understanding of plant iron and zinc nutrition. *Physiol Plant* **126**:407–417.

Goto F, Yoshihara T. 2001. Improvement of micronutrient contents by genetic engineering development of high iron content crops. *Plant Biotechnol* **18**:7–15.

Goto F, Yoshihara T, Saiki H. 1998. Iron accumulation in tobacco plants expressing soybean ferritin gene. *Transgenic Res* **7**:173–180.

Goto F, Yoshihara T, Shigemoto N, Toki S, Takaiwa F. 1999. Iron fortification of rice seed by the soybean ferritin gene. *Nat Biotechnol* **17**:282–286.

Goto F, Yoshihara T, Saiki H. 2000. Iron accumulation and enhanced growth in transgenic lettuce plants expressing the iron binding protein ferritin. *Theor Appl Gen* **100**:658–664.

Gratão PL, Prasad MNV, Lea PJ, Azevedo RA. 2005. Phytoremediation: Green technology for the clean up of toxic metals in the environment. *Braz J Plant Physiol* **17**:53–64.

Graziano M, Lamattina L. 2005. Nitric oxide and iron in plants: an emerging and converging story. *Trends Plant Sci* **10**:4–8.

Grusak MA. 1995. Whole-root iron (III)-reductase activity throughout the life cycle of iron-grown *Pisum sativum* L. (Fabaceae): Relevance to the iron nutrition of developing seeds. *Planta* **197**:111–117.

Grusak MA. 1999. Improving the nutrient composition of plants to enhance human nutrition and health. *Annu Rev Plant Physiol Plant Mol Biol* **50**:133–161.

Grusak MA. 2000. Strategies for improving the iron nutritional quality of seed crops: lessons learned from the study of unique iron-hyperaccumulating pea mutants. *Pisum Genet* **32**:1–6.

Guerinot ML, Salt DE. 2001. Fortified foods and phytoremediation. Two sides of the same coin. *Plant Physiol* **125**:164–167.

Guerinot ML, Yi Y. 1994. Iron: Nutritious, noxious, and not readily available. *Plant Physiol* **104**:815–820.

Gyorgyey J, Vaubert D, Jimenez-Zurdo JI, Charon C, Troussard L, Kondorosi A, Kondorosi E. 2000. Analysis of *Medicago truncatula* nodule expressed sequence tags. *Mol Plant Microbe Interact* **13**:62–71.

Hallberg L. 1981. Bioavailability of dietary iron in man. *Annu Rev Nutr* **1**:123–147.

Harrison PM, Arosio P. 1996. The ferritins: Molecular properties, iron storage function and cellular regulation. *Biochim Biophys Acta* **127**:161–203.

Hegedûs A, Erdei S, Janda T, Szalai J, Dudits D, Horváth G. 2002. Effects of low temperature stress on ferritin or aldose reductase overexpressing transgenic tobacco plants. *Acta Biol Szegediensis* **46**:97–98.

Heenan DP, Campbell LC. 1983. Manganese and iron interactions on their uptake and distribution in soybean (*Glycine max* (L.) Merr.) *Plant Soil.* **30**:317–326.

Hell R, Stephan UW. 2003. Iron uptake, trafficking and homeostasis in plants. *Planta* **216**:541–551.

Hirschi KD, Korenkov V, Wilganowski N, Wagner G. 2000. Expression of *Arabidopsis* CAX2 in tobacco: Altered metal accumulation and increased manganese tolerance. *Plant Physiol* **124**:125–134.

Hortensteiner S, Chinner J, Matile P, Thomas H, Donnison IS. 2000. Chlorophyll breakdown in chlorella protothecoides: Characterization of degreening and cloning of degreening related genes. *Plant Mol Biol* **42**:439–450.

Hu Z-Y, Zhu Y-G, Li M, Zhang L-G, Cao Z-H, Smith FA. 2007. Sulfur (S)-induced enhancement of iron plaque formation in the rhizosphere reduces arsenic accumulation in rice (*Oryza sativa* L.) seedlings. *Environ Pollut* **147**:387–393.

Hulthén L. 2003. Iron deficiency and cognition. *Scand J Food Nutr* **47**:152–156.

Hurrell RF. 2002. Fortification: Overcoming technical and practical barriers. *J Nutr* **132**:806–812.

Hwang J, Krebs C, Huynh BH, Edmondson DE, Theil EC, Penner-Hahn JE. 2000. A short Fe–Fe distance in peroxo diferric ferritin: Control of Fe substrate versus cofactor decay? *Science* **287**:122–125.

Illmer P, Buttinger R. 2006. Interactions between iron availability, aluminium toxicity and fungal siderophores. *BioMetals* **19**:367–377.

Jiang TB. 2005. Isolation and expression pattern analysis of two ferritin genes in tobacco. *J Integr Plant Biol* **47**:477–478.

Kimata Y, Theil EC. 1994 Posttranscriptional regulation of ferritin during nodule development in soybean. *Plant Physiol* **104**:263–270.

Laulhere JP, Lescure AM, Briat JF. 1988. Purification and characterization of ferritin from maize, pea and soybean seeds. *J Biol Chem* **26**:10289–10294.

Lescure AM, Proudhon D, Pesey H, Ragland M, Theil EC, Briat JF. 1991. Ferritin gene transcription is regulated by iron in soybean cell cultures. *PNAS* **88**:8222–8226.

Levi S, Arosio P. 2004. Mitochondrial ferritin. *Int J Biochem Cell Biol* **36**:1887–1889.

Ling HQ. 2002. The tomato fer gene encoding a bHLH protein controls iron-uptake responses in roots. *PNAS* **99**:13938–13943.

Liu WJ, Zhu YG, Smith FA. 2005. Effects of iron and manganese plaques on arsenic uptake by rice seedlings (*Oryza sativa* L.) grown in solution culture supplied with arsenate and arsenite. *Plant Soil* **277**:127–138.

Liu X, Jin W, Theil EC. 2003. Opening protein pores with chaotropes enhances Fe reduction and chelation of Fe from the ferritin biomineral. *PNAS* **100**:3653–3658.

Liu X, Theil EC. 2004. Ferritin reactions: direct identification of the site for the diferric peroxide reaction intermediate. *PNAS* **101**:8557–8562.

Liu X, Theil EC. 2005. Ferritins: Dynamic management of biological iron and oxygen chemistry. *Acc Chem Res* **38**:167–75.

Lobreaux S, Briat JF. 1991. Ferritin accumulation and degradation in different organs of pea (*Pisum sativum*) during development. *Biochem J* **274**:601–606.

Lobreaux S, Massenet O, Briat JF. 1992a. Iron induces ferritin synthesis in maize plantlets. *Plant Mol Biol* **19**:563–575.

Lobreaux S, Yewdall SJ, Briat JF, Harrison PM. 1992b. Amino-acid sequence and predicted three-dimensional structure of pea seed (*Pisum sativum*) ferritin. *Biochem J* **288**:931–939.

Lönnerdal B. 2003. Genetically modified plants for improved trace element nutrition. *J Nutr* **133**:1490–1493.

Lucca P, Hurrell RF, Potrykus I. 2001. Genetic engineering approaches to improve the bioavailability and the level of iron in rice grains. *Theor Appl Genet* **10**:392–397.

Lucca P, Hurrell RF, Potrykus I. 2002. Fighting iron deficiency anemia with iron rich rice. *J Am Coll Nutr* **21**:84–190.

Lucca P, Poletti S, Sautter C. 2006. Genetic engineering approaches to enrich rice with iron and vitamin A. *Physiol Plant* **126**:291–303.

Marentes E, Grusak MA. 1998. Iron transport and storage within the seed coat and embryo of developing seeds of pea (*Pisum sativum* L.). *Seed Sci Res* **8**:367–375.

Masuda T, Goto F, Yoshihara T. 2001. A novel plant ferritin subunit from soybean that is related to a mechanism in iron release. *J Biol Chem* **276**:19575–19579.

Mata CG, Lamattina L, Cassia RO. 2001. Involvement of iron and ferritin in the potato—*Phytophera infestans* interaction. *Eur J Plant Pathol* **107**:557–562.

Mori S. 1999. Iron acquisition by plants. *Curr Opin Plant Biol* **2**:250–253.

Misra SG, Jaiswal PC. 1982. Absorption of Fe by spinach on chromium(VI) treated soil. *J Plant Nutr* **5**:755–760.

Murgia I, Delledonne M, Soave C. 2002. Nitric oxide mediates iron-induced ferritin accumulation in *Arabidopsis*. *Plant J* **30**:521–528.

Murray-Kolb LE, Theil EC, Takaiwa F, Goto F, Yoshihara T, Beard JL. 2002. Transgenic rice as a source of iron for iron-depleted rats. *J Nutr* **132**:957–960.

Murray-Kolb LE, Welch R, Theil EC, Beard JL. 2003. Women with low iron stores absorb iron from soybeans. *Am J Clin Nutr* **77**:180–184.

Negishi T, Nakanishi H, Yazaki J, Kishimoto N, Fujii F, Shimbo K, Yamamoto K, Sakata K, Sasaki T, Kikuchi S. 2002. cDNA microarray analysis of gene expression during Fe-deficiency stress in barley suggests that polar transport of vesicles is implicated in phytosiderophore secretion in Fe-deficient barley roots. *Plant J* **30**:83–94.

Nirupa N, Prasad MNV, Kirti PB. 2007. Expression of pea seed ferritin cDNA in indian mustard: Nutritional value and oxidative stress tolerance of the transformants. In: Zhu Y, Lepp N, Naidu R, editors. *Biogeochemistry of race elements: Environmental protection, remediation and human health.* China: Tsinghua University Press, pp. 285–286.

Petit JM, Van Wuytswinkel O, Briat JF, Lobreaux S. 2001. Characterization of an iron-dependent regulatory sequence involved in the transcriptional control of AtFer1 and ZmFer1 plant ferritin genes by iron. *J Biol Chem* **276**:5584–5590.

Pich A, Manteuffel R, Hillmer S, Scholz G, Schmidt W. 2001. Fe homeostasis in plant cells: Does nicotianamine play multiple roles in the regulation of cytoplasmic Fe concentration? *Planta* **213**:967–976.

Polanams J, Ray AD, Watt RK. 2005. Nano iron phosphate, iron arsenate, iron vandate and, iron molybdate minerals synthesized within the protein cage of ferritin. *Inorg Chem* **44**:3203–3209.

Prasad MNV. 2004a. *Heavy Metal Stress in Plants: From Biomolecules to Ecosystems*, second edition. Heidelberg: Springer-Verlag, p. 462.

Prasad MNV, Nirupa N. 2007. Phytoferritins—Implications for human health and nutrition. *Asian and Australasian J Plant Sci Biotechnol* **1**:1–9.

Proudhon D, Wei J, Briat JF, Theil EC. 1996. Ferritin gene organization: Differences between plants and animals suggest possible kingdom-specific selective constraints. *J Mol Evol* **42**:325–336.

Qu L, Yoshihara T, Ooyama A, Goto F, Takaiwa F. 2005. Iron accumulation does not parallel the high expression level of ferritin in transgenic rice seeds. *Planta* **222**:225–233.

Ragland M, Theil EC. 1993. Ferritin mRNA, protein and iron concentration during soybean nodule development. *Plant Mol Biol* **21**:555–560.

Ragland M, Briat JF, Gagnon J, Laulhere JP, Massenet O, Theil EC. 1990. Evidence for conservation of ferritin sequences among plants and animals and for transit peptide in soybean. *J Biol Chem* **265**:18339–18344.

Rama Kumar T, Prasad MNV. 1999a. Ferritin induction by iron mediated oxidative and ABA in *Vigna mungo* (L.) Hepper seedlings: Role of antioxidants and free radical scavengers. *J Plant Physiol* **155**:652–655.

Rama Kumar T, Prasad MNV. 1999b. Metal binding properties of ferritin in vitro in *Vigna mungo* (L). Hepper (Black gram). Possible role in heavy metal detoxification. *Bull Environ Contam Toxicol* **62**:502–507.

Rama Kumar T, Prasad MNV. 2000. Partial purification and characterization of ferritin from *Vigna mungo* (Blackgram) seeds. *J Plant Biol* **27**:241–246.

Robin A, Vansuyt G, Corberand T, Briat JF, Lemanceau P. 2006a. The soil affects both the differential accumulation of iron between wild type and ferritin over-expressor tobacco plants and the sensitivity of their rhizosphere bacterioflora to iron stress. *Plant and Soil* **283**:73–81.

Robin A, Mougel C, Siblot S, Vansuyt G, Mazurier S, Lemanceau P. 2006b. Effect of ferritin overexpression in tobacco on the structure of bacterial and pseudomonad communities associated with the roots. *FEMS Microbiol Ecol* **58**:492–502.

Robinson NJ, Proctor CM, Connolly EL, Guerinot ML. 1999. A ferric chelate reductase for iron uptake from soils. *Nature* **397**:694–697.

Rogers EE, Guerinot ML. 2002. FRD3, a member of the multidrug and toxin efflux family, controls iron deficiency responses in *Arabidopsis*. *Plant Cell* **1**:1787–1799.

Rothery RA, Trieber CA, Weiner JH. 1999. Interactions between the molybdenum cofactor and iron–sulfur clusters of *Escherichia coli* dimethylsulfoxide reductase. *J Biol Chem* **274**:13002–13009.

Sappin-Didier V, Vansuyt G, Mench M, Briat JF. 2005. Cadmium availability at different soil pH to transgenic tobacco overexpressing ferritin. *Plant Soil* **270**:189–197.

Savino G, Briat JF, Lobréaux S. 1997. Inhibition of the iron-induced ZmFer1 maize ferritin gene expression by antioxidants and serine/threonine phosphatase inhibitors. *J Biol Chem* **272**:33319–33326.

Schmidt W. 1996. Influence of chromium (lll) on root-associated Fe(lll) reductase in *Plantago lanceolata* L. *J Exp Bot* **47**:805–810.

Schmidt W. 2003. Iron homeostasis in plants: Sensing and signaling pathways. *J Plant Nutr* **26**:2211–2230.

Sczekan SR, Joshi JG. 1987. Isolation and characterization of ferritin from soybean (glycine max). *J Biol Chem* **262**:13780–13788.

Sczekan SR, Joshi JG. 1989. Metal binding properties of phytoferritin and synthetic iron cores. *Biochim Biophyis Acta* **990**:8–14.

Shingles R, North M, McCarty RE. 2002. Ferrous ion transport across chloroplast inner envelope membranes. *Plant Physiol* **128**:1022–1030.

Sinha S, Gupta M, Chandra P. 1997. Oxidative stress induced by iron in *Hydrilla verticillata* (l.f.) Royle: Response of antioxidants. *Ecotoxicol Environ Safety* **38**:286–291.

Six KM, Goyer RA. 1972. The influence of iron deficiency on tissue content and toxicity of ingested lead in the rat. *J Lab Clin Med* **79**:128–136.

Soós V, Jóri B, Páldi E, Szegő D, Szigeti Z, Rácz IL, Lásztity DS. 2005. Ferritin2 gene in paraquat-susceptible and resistant biotypes of horseweed *Conyza canadensis* (L.) Cronq. *J Plant Physiol* **163**:979–982.

Spence MJ, Henzl MT, Lammers PJ. 1991. The structure of a *Phaseolus vulgaris* cDNA encoding the iron storage protein ferritin. *Plant Mol Biol* **17**:499–504.

Strozycki PM, Skapska A, Szczesniak K, Sobieszczuk E, Briat JF, Legocki AB. 2003. Differential expression and evolutionary analysis of the three ferritin genes in the legume plant *Lupinus luteus*. *Physiol Plant* **118**:380–389.

Surguladze N, Patton S, Cozzi A, Fried MG, Connor JR. 2005. Characterization of nuclear ferritin and mechanism of translocation. *Biochem J* **388**:731–740.

Theil EC, Briat JF. 2004. *Plant Ferritin and Non-heme Iron Nutrition in Humans.* Harvest Plus Technical Monograph 1. Washington, DC and Cali, International Food Policy Research Institute and International Center for Tropical Agriculture.

Theil EC. 1990. Regulation of ferritin and transferring receptor mRNA. *J Biol Chem* **265**:4771–4774.

Theil EC. 2000. Ferritin. In: Messerschmidt A, Huber R, Poulos T, Wieghardt K, editors. *Handbook of Metalloproteins.* Chichester: John Wiley & Sons, pp. 771–781.

Theil EC. 2003. Ferritin: At the cross roads of Iron and oxygen metabolism. *J Nutr* **133**:1549–1553.

Theil EC, Hase T. 1993. Plant and microbial ferritins. In: Barton LL, Hemming BC, editors. *Iron Chelation in Plants and Soil Microorganisms.* San Diego: Academic Press, pp. 133–156.

Theil EC. 2004. Iron, ferritin, and nutrition. *Ann Rev Nutr* **24**:327–343.

Vansuyt G, Souche G, Straczek A, Briat JF, Jaillard B. 2003. Flux of protons released by wild type and ferritin over-expressor tobacco plants: effect of phosphorus and iron nutrition. *Plant Physiol Biochem* **41**:27–33.

Van Wuytswinkel O, Briat JF. 1995. Conformational changes and in vitro core formation modification induced by sitedirected mutagenesis of the specific N-terminus of pea seed ferritin. *Biochem J* **305**:959–965.

Van Wuytswinkel O, Vansuyt G, Grignon N, Fourcroy P, Briat JF. 1998. Iron homeostasis alteration in transgenic tobacco overexpressing ferritin. *Plant J* **17**:93–97.

Vasconcelos M, Datta K, Oliva N, Mohammad M, Torrizo L, Krishnan S, Oliveira M, Goto F, Datta SK. 2003. Enhanced iron and zinc accumulation in transgenic rice with the ferritin gene. *Plant Sci* **164**:371–378.

Verma TS, Tripathi BR. 1983. Zinc and iron interaction in submerged paddy. *Plant Soil* **72**:101–116.

Vert G, Grotz N, Dedaldechamp F, Gaymard F, Guerinot ML, Briat JF, Curie C. 2002. IRT1, an *Arabidopsis* transporter essential for iron uptake from the soil and for plant growth. *Plant Cell* **14**:1223–12334.

Wade VJ, Treffry A, Laulhere JP, Baminger ER, Cleton MI, Mann S, Briat JF, Harrison PM. 1993. Structure and composition of ferritin cores from pea seed (*Pisum sativum*). *Biochim Biophys Acta* **1161**:91–96.

Waldo GS, Wright E, Whang ZH, Briat JF, Theil EC, Sayers DE. 1995. Formation of the ferritin iron mineral occurs in plastids (an X-ray absorption spectroscopy study). *Plant Physiol* **109**:797–802.

Wardrop AJ, Wicks RE, Entsch B. 1999. Occurrence and expression of members of the ferritin gene family in cowpeas. *Biochem J* **337**:523–530.

Waters BW. 2002. Characterization of FRO1,a pea ferric chelate reductase involved in iron acquisition. *Plant Physiol* **129**:85–94.

Wei J, Theil EC. 2000. Identification and characterization of the iron regulatory element in the ferritin gene of a plant (soybean). *J Biol Chem* **275**:17488–17493.

WHO/UNICEF/UNU. 2001. *Iron Deficiency Anaemia: Assessment Prevention and Control.* Geneva: World Health Organization, (WHO/NHD/01.3). http://www.who.int/nut/documents/ida_assessment_prevention_control.pdf

Wicks R E, Entsch B. 1993 Functional genes found for three different plant ferritin subunits in the legume, *Vigna unguiculata. Biochem Biol Chem* **275**:17488–17493.

Yadav DV, Singh K. 1990. Lime-induced iron chlorosis in sugarcane. *Nutr Cyc Agro ecosyst* **16**:119–136.

Yamada H, Takeda C, Mizushima A, Yoshino K, Yonebayashi K. 2005. Effect of oxidizing power of roots on iodine uptake by rice plants. *Soil Sci Plant Nutr* **51**:141–145.

Ye X, Al-Babili S, Kloti A, Zhang J, Lucca P, Beyer P, Potrykus I. 2000. Engineering the provitamin A (beta-carotene) biosynthetic pathway into (carotenoid-free) rice endosperm. *Science* **87**:303–305.

Yoshihara T, Masuda T, Jiang T, Goto F, Mori S, Nishizawa N. 2003. Analysis of some divalent metal contents in tobacco expressing the exogenous soybean ferritin gene. *J Plant Nutr* **26**:2253–2265.

Zancani M, Peresson C, Biroccio A, Federici G, Urbani A, Murgia I, Soave C, Micali F, Vianello A, Macrì F. 2004. Evidence for the presence of ferritin in plant mitochondria. *Eur J Biochem* **271**:3657–3664.

Zhao G, Bou-Abdallah F, Yang X, Arosio P, Chasteen ND. 2001. Is hydrogen peroxide produced during iron (II) oxidation in mammalian apoferritins? *Biochemistry* **40**:10832–10838.

Zhou ZQ, Cheng MH, Zhou ZY, Pei P, Yang WG. 2001. The ferritin gene of apple trees (*Malus xiaojinensis*). *China J Biotechnol* **17**:342–344.

12 Iodine and Human Health: Bhutan's Iodine Fortification Program

KARMA LHENDUP

Faculty of Agriculture, College of Natural Resources, Lobesa,
PO Box Wangduephodrang, Bhutan

Iodine (I) is one of the essential trace elements required in small amounts for human and animal nutrition. Other essential trace elements include zinc (Zn), selenium (Se), iron (Fe), fluorine (F), chromium (Cr), boron (B), copper (Cu), manganese (Mn), and molybdenum (Mo) [Deckers et al., 2000]. These elements are required to be present at certain concentrations in the diets for normal life processes.

1 ROLE OF IODINE

Iodine is required in the body in minute amounts, mainly in the thyroid gland for the production of thyroid hormones, thyroxin (T4), and triiodo-thyronine (T3). These hormones are essential for the normal development and functions of nervous system and body metabolism [Dunn and Van der Haar, 1990]. The normal thyroid contains between 2 and 20 mg of iodine. Under normal circumstances, the thyroid takes about 20% of the available iodine, which is adequate for normal physiological life processes. When the intake is less than the physiological requirements of the body, thyroid function is impaired and is unable to synthesize sufficient amounts of thyroid hormone, as a result a series of functional and developmental abnormalities occurs, which includes endemic goiter and cretinism, mental retardation, decreased fertility rate, and increased prenatal death and infant mortality, which are referred to as iodine deficiency disorders or IDD [Hetzel, 1983].

In addition, when too much is ingested, the major epidemiological consequence is iodine-induced hyperthyroidism (IIH), whereby the thyroid may shut down or under certain circumstances it might become overactive [Todd et al., 1995; Stanbury, 1998; Delange and Lecomte, 2000].

2   IODINE DEFICIENCY DISORDERS (IDD)

As mentioned earlier, IDD refers to all the ill effects (Table 1) of iodine deficiency in a population that can be prevented by ensuring an adequate intake of iodine for all people [Hetzel, 1983; WHO, UNICEF, ICCIDD, 2001]. These spectrums of ill effects are caused mainly by the deficiency of iodine in soils, which leads to deficiency in all forms of plant life, including all cereals grown in the deficient soil (Fig. 1). Therefore, populations living in systems of subsistence agriculture on soils deprived of iodine are "locked into" iodine deficiency [Koutras et al., 1985; Hetzel and Mano, 1989].

Iodine deficiency is a global public health problem with large population exposed to iodine deficiency [WHO, UNICEF, ICCIDD, 1994; WHO, 1999]. Studies have shown that about 130 countries (out of 191 member states) are affected by iodine deficiency disorders, with a total of 807 million population equivalent to 13% of the world's population [WHO, 1999]. Brain damage—that is, endemic cretinism (neurological and myxedematous) and irreversible mental retardation—is the most common disorder of iodine deficiency in the world as established by clinical and epidemiological studies [WHO, 1994; Delange, 1994].

TABLE 1. Iodine Deficiency Disorders at Different Stages

Stages in Life	Ill Effects
Fetus	Abortions
	Still births
	Congenital anomalies
	Increased perinatal mortality
	Endemic cretinism; deaf mutism
Neonate	Neonatal goiter
	Neonatal hypothyroidism
	Endemic mental retardation
	Increased susceptibility of the thyroid gland to nuclear radiation
Child and adolescent	Goiter
	Subclinical hypothyroidism
	Subclinical hyperthyroidism
	Impaired mental function
	Retarded physical development
	Increased susceptibility of the thyroid gland to nuclear radiation
Adult	Goiter, with its complications
	Hypothyroidism
	Impaired mental function
	Spontaneous hyperthyroidism in the elderly
	Iodine-induced hyperthyroidism
	Increased susceptibility of the thyroid gland to nuclear radiation

Source: WHO [2004].

Iodine Deficiency – A Disease of the Soil

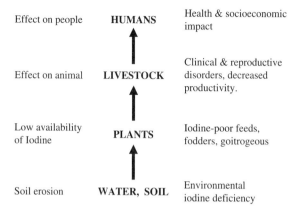

Figure 1. Iodine deficiency in soils and related effets to the food chain. (*Source*: ICCIDD Report [2004].)

Studies conducted in severely iodine deficient areas have shown that iodine deficiency is responsible for a mean intelligence quotient (IQ) loss of 13.5 points in children as compared to those from comparable communities where there is no iodine deficiency [Bleichrodt and Born, 1994]. This mental deficiency affects the learning ability of children, women's health, and the quality of life of communities, which altogether puts a great strain on social, cultural, and economic productivity [WHO, 2001].

3 SOURCES OF IODINE

The bulk of iodine required for human consumption is derived from food (80%) and water (20%) [BGS, 2000]. Seafood has the highest iodine content, in which it can be

TABLE 2. Average Iodine of Various Foods (μg/kg)

Food	Fresh Basis		Dry Basis	
	Mean	Range	Mean	Range
Fish (fresh water)	30	17–40	116	68–194
Fish (marine)	832	163–3180	3715	471–4591
Shellfish	798	308–1300	3866	1292–4987
Meat	50	27–97	—	—
Milk	47	35–56	—	—
Eggs	93	—	—	—
Cereal grains	47	22–72	65	34–92
Fruits	18	10–29	154	62–277
Legumes	30	23–36	234	223–245
Vegetables	29	12–201	385	204–1636

Source: Koutras et al. [1985].

as high as 3 mg/kg. Ocean water contains about 50 μg of iodine per liter and is the primary store and source of iodine in the world [Koutras et al., 1985]. The other main dietary sources of iodine are milk, eggs, cereals, and meat. The average iodine content of some of the foods (fresh and dry basis) as reported by Koutras et al. [1985] is shown in Table 2. According to Koutras et al. [1985], these average iodine values in foods cannot be used universally for estimating iodine intake because the content varies with geographical location and environmental media.

4 RECOMMENDED INTAKE OF IODINE

The total requirement for iodine is quite small (teaspoonful over a life time), which is also indicated by the recommended daily allowance (RDA) endorsed by WHO [DeLange, 1994]. The recommended daily intake of iodine are 90 μg for infants and children up to 6 years, 120 μg for children aged 7–10 years, 150 μg for adults above 12 years, and 200–250 μg for pregnant and lactating women [WHO, UNICEF, ICCIDD, 2001]. These amounts are proposed to allow normal T4 production without stressing the thyroid iodide trapping mechanism or raising TSH levels.

5 INDICATORS FOR ASSESSMENT OF IODINE STATUS AND EXPOSURE

Several indicators are used to assess the iodine status of a population: thyroid size by palpation or ultrasound, urinary iodine concentration or excretion (UIC/E) and the blood constituents, TSH (thyroid stimulating hormone or thyrotropin), and thyroglobulin (Tg). Until the 1990s, total goiter prevalence (TGP) or total goiter rate (TGR) was recommended as the main indicator to assess IDD prevalence. However, TGP has been found to have limited role because it reflects a population's history of chronic iodine deficiency but not its present iodine status or the impact of salt iodization. Nevertheless, it remains useful in the baseline assessment of IDD severity [WHO, 2004].

Urinary iodine concentration (UIC) or urinary iodine excretion (UIE) is a more sensitive indicator to the daily dietary intake of iodine, and it is now recommended over TGP. It is also a widely used biochemical marker of iodine status since most dietary iodine is excreted in urine (usually over 90% of the daily intake) and only a minor fraction in feces [Lamberg, 1993; Dunn et al., 1993]. UIC/UIE greater than 100 μg/liter is considered normal. A range of excretion of 50–99 μg/liter indicates mild deficiency, while 20–49 μg/liter indicates moderate deficiency. Severe iodine urinary excretion is defined for concentration lower than 20 μg/liter [WHO, 1996].

Thyroid stimulating hormone (TSH) levels in neonates are particularly sensitive to iodine deficiency where TSH screening program exists; however, it has limited applicability in most developing countries due to cost and difficulties in interpretation. The value of thyroglobulin as an indicator of IDD status has yet to be fully explored and to gain wide acceptance [WHO, 2004].

6 CONTROL OF IDD

Since 1990, tremendous progress has been made in the control and elimination of iodine deficiency disorders with the implementation of various IDD Control Programmes (IDDCP): iodine supplementation and fortification; monitoring and evaluation; inter-sectorial collaboration and advocacy; and communication to educate the public.

Iodine supplementation and fortification through food items such as bread, milk, water, and salt were generally used as a means to control IDD in various countries. Among these, iodized salt or salt fortified with iodine was found the most promising and appropriate iodine supplementation [WHO, UNICEF, ICCIDD, 1994]. According to WHO [2004] (i) salt is one of the few commodities consumed by everyone; (ii) salt consumption is fairly stable throughout the year; (iii) salt production is usually in the hands of few producers; (iv) salt iodization technology is easy to implement and available at a reasonable cost; (v) the addition of iodine to salt does not affect its color, taste, or odor; (vi) the quality of iodized salt can be monitored at the production, retail, and household levels; and (vii) salt iodization programs are easy to implement.

The recommended level of iodination of salt was 20–40 parts per million (ppm) of iodine to salt, which satisfies the iodine requirements of a population (assuming an average salt intake of 10 g per capita/day) [WHO, UNICEF, ICCIDD, 1996]. This quantity of iodination was also enough to cover the losses from the point of production to the point of consumption. Since then, the adoption of universal salt iodization (USI) (the iodization of salt for both human and livestock consumption) by the World Health Assembly and the World Summit for children as the method of choice to virtual elimination of IDD has been successful.

Other components of the IDD control program such as (a) monitoring and evaluation, (b) multi-sectorial collaboration and advocacy, and (c) educating the public have been vital to a greater extent in the reduction, control, and elimination of IDD. These have been a suitable mechanism responsible for coordinating the sectors involved in the control of iodine deficiency disorders and for overseeing the program. For instance, monitoring of the iodine concentration in salt at production level, retail shops, and households periodically by using titration methods—or, in the case of imported salt, by using reliable test kits at the point of entry—was taken care of. In addition, conducting goiter prevalence surveys and measuring urinary iodine occasionally helped to assess and understand the adequate and continuing coverage of USI. Furthermore, information dissemination through education and communication has contributed to creating awareness of taking iodized salt daily to the general public and schoolchildren and in turn has been crucial in controlling IDD.

The review carried out by WHO in collaboration with UNICEF and ICCIDD on the global situation of the IDD in 1999 had a remarkable progress because of the IDD control program in place. As per WHO, UNICEF, and ICCIDD [1999], of the 131 countries with IDD, 98 (75%) had legislation on salt iodization in place, and an additional 12 countries had it in draft form. The figures would have gone up since then. Bhutan, a small Himalayan country, is one of the countries that have achieved virtual elimination of IDD, which was set for the year 2005 [ICCIDD Report, 2004].

7 IDD SCENARIO IN BHUTAN: PAST AND PRESENT

Iodine deficiency disorders have long been a major public health problem in Bhutan [RGOB, ICCIDD, AIIMS, UNICEF, WHO, MI, 1996]. The prevalence of goiter was observed in majority of people by health personnel since early 1960s and 1970s. Among the several studies conducted on the prevalence of IDD in Bhutan, the first known published report is by two English doctors in 1964, which stated that goiter was almost universally prevalent.

Bhutan is located in the high Himalayan mountain range and is a land-locked country surrounded by mountains. Bhutan falls in the iodine-deficient belt of the Himalayas as reflected in the world distribution of recognized iodine deficiency map [Dunn and van der Haar, 1990]. The long history of glacial activity and the heavy monsoon rains might have resulted in the removal the iodine from the soils. Additionally, being far away from the nearest sea, the rains are normally deprived of iodine, which could otherwise replenish the superficial layer of the soil with iodine. As a result, none of the foods produced in Bhutan have significant iodine levels. As a matter of fact, iodine deficiency is purely a disease of the soils and Bhutan had some of the world's most severe iodine deficiency in the past few decades.

In view of the seriousness of the situation, the 63rd National Assembly (Parliament) of the Royal Government of Bhutan (RGOB) convened in 1964 passed a resolution to use only imported iodized salt. The 1975 study report by Dr. Mahendra in nine districts across Bhutan indicated a goiter prevalence of 47–60% in schoolchildren, and 50–53% among the adult population [RGOB, ICCIDD, AIIMS, UNICEF, WHO, MI, 1996]. Since then, the Royal Government of Bhutan has prioritized its effort towards preventing and eliminating IDD in the country. Some of the key events of the IDD control programs that helped in virtual elimination of IDD in Bhutan are provided in Table 3.

TABLE 3. Key Events in Bhutan's IDD Control Program

1. Iodine deficiency disorders in Bhutan: Extent and severity—first nationwide IDD study	1983
2. Production and distribution of iodized salt to control iodine deficiency disorders in Bhutan	1983
3. National policy, strategy and Plan of Action to Control Iodine Deficiency Disorders in Bhutan: The National IDD Control Program in Bhutan	1985
4. Situational analysis of the salt iodization program in Bhutan	1986
5. A nationwide internal program evaluation of IDDCP	1992
6. Iodine deficiency disorders in Bhutan: Extent and severity—External evaluation	1996
7. Statement signed by His Holiness the Je Khenpo	1997
8. Introduction of annual cyclic monitoring	1998
9. Sustainable elimination of IDD	2002
10. Sustainable elimination of IDD: External evaluation	2003

Source: Adapted from Sithey [2006].

8 TOWARD IDD ELIMINATION IN BHUTAN: HIGHLIGHTS OF THE IDD CONTROL PROGRAM

8.1 IDD Survey

The first nationwide IDD survey was undertaken in 1983 by a team from All India Institute of Medical Sciences (AIIMS), New Delhi and with financial support from WHO and UNICEF. The survey was carried out in 11 of the then 18 districts and included adult males and females, primary school students, and preschool children [UNICEF, WHO, 1983; RGOB, ICCIDD, AIIMS, UNICEF, WHO, MI, 1996]. The team reported a total goiter rate (TGR) of 60.0% (Table 4). Cretinism was reported in all districts, reaching 10% or higher in those districts most severely affected by iodine deficiency. The trend was that districts in the southern belt were found to be affected the most. Urinary iodine estimation showed that 62% of school-children and 77% of the general population had urinary iodine concentration lower than 50 µg iodine/g cretinine (approximately equivalent to 50 µg/liter) [RGOB, ICCIDD, AIIMS, UNICEF, WHO, MI, 1996].

As a result of the findings of the first nationwide IDD survey, the Health Division initiated the IDDCP in 1984. Salt iodization and distribution was one of the top priorities of the IDDCP, followed by monitoring of iodine in salt at production site, retail shops, and households, evaluation of the program, and community level education. In 1985, a salt iodization plant was commissioned at Phuntsholing on Bhutan's southern border, to iodize all salt entering the country. It was a joint firm of the RGOB and a private individual. With a capacity of 6 tons/hr, the plant is of standard continuous spray mixing type of UNICEF design. The imported crushed "Kurkutch" variety of common salt from Gujarat, India is powdered uniformly at the plant before iodization—that is, addition of iodine as potassium or sodium iodate. The recommended iodization level at the plant was maintained at 60 ppm of iodine, which ensured that salts reaching the household level have the acceptable iodine content of ≥ 15 ppm. The loss of iodine from production site to the households was taken care of. The iodized salt is then packed into a woven high-density polyethylene (HDPE) or laminated jute bags supplied by the Food Corporation of Bhutan (FCB), which is then distributed through the network of nine FCB depots, 64 FCB Commission agents, and 198 centers of World Food Programme. Legislation was in force to prevent the import or sale of salt that has not been iodized at this plant. Since 1994, Bhutan Salt Enterprise (BSE) was privatized and has been responsible for procurement of common salt from India, its iodization and packaging at Phuntsholing, and subsequent distribution in Bhutan. However, monitoring of salt at the consumer and retailer level, as well as at the production site, was done by health workers, district hospitals and laboratories, and the Public Health Laboratory, respectively. Together with the commissioning of the plant, other components of the IDDCP as mentioned above were undertaken.

As a result, the second nationwide evaluation program was conducted in late 1991 and early 1992, to assess the IDD situation and the iodine content in salt, the impact of the IDDCP, and also the knowledge and attitudes of the population concerning the

TABLE 4. Results of the National IDD Survey in 1983, 1991–1992, and 1996

| Indicator | Nationwide IDD Survey, 1983 | Nationwide IDD Survey, 1991–1992 | | | | Nationwide IDD Survey, 1996 |
| | | Northern | | Southern | | |
		Children (6–11)	Women (15–45)	Children (6–11)	Women (15–45)	Children aged 6–11 years
Total goiter rate (%)	60.0 (12,045)	18.4 (1443)	28.5 (1581)	32.5 (992)	45.9 (988)	14.0 (1200)
Visible goiter prevalence (%)	—	0.3 (1443)	3.8 (1581)	1.0 (992)	7.1 (988)	1.2 (1200)
Cretinism prevalence (%)	≥10	0.4 (1443)	0.9 (1581)	0.4 (992)	0.8 (988)	Not done
UIE (%) (≥100 μg/liter)	3.1* (822)	87.4 (864)	85.2 (893)	83.8 (871)	81.9 (878)	76 (333)
Median UIE (μg/liter)	—	283 (864)	239 (893)	244 (871)	233.5 (878)	230 (333)
TSH (%) (>5 mU/liter)	—	20.38 (1423)	10.8 (—)	22.34 (985)	7.4 (—)	Not done

Figures in parentheses are the number of people examined or samples collected.

*UIE ≥ 100 (μg/g of creatinine).

Source: Adapted from RGOB, ICCIDD, AIIMS, UNICEF, WHO, MI [1996].

nature and causes of IDD [DHS, 1992]. The study was undertaken by Directorate of Health Services, Ministry of Social Services, RGOB using a 30-cluster sample survey method in two districts, one in the north and another in the south. Children aged 6–11 years and women aged 15–45 years were examined for iodine deficiency indicators such as goiter and cretinism; urine and blood samples were collected for estimation of urinary iodine and thyroid stimulating hormone (TSH), respectively.

The result of the study is presented in Table 4. The main findings were that goiter prevalence in children and women was 18.4% and 28.5% in the north and 32.5% and 45.9% in the south, respectively [DHS, 1992]. The prevalence of cretinism was 0.4% for children in both north and south, while for women it was 0.9% and 0.8% in north and south, respectively.

A total of 87.4% of children in the north and 83.8% in the south had acceptable urinary iodine excretion (UIE)—that is, urinary iodine $\geq 100 \, \mu g$/liter. The UIE for women was 85.2% in the north and 81.9% in the south. The mean and median UIE for children and women were found all above $200 \, \mu g$/liter. Also, 79.6% of children in the north and 77.6% in the south had acceptable thyroid stimulating hormone (TSH) values of $\geq 5 \, \mu g$/liter. The corresponding figures for the women were 89.2% and 92.6%, respectively.

Furthermore, 96.6% of the 146 household salt samples collected from the north and 95% of the 140 household salt samples from the south had acceptable iodine content of ≥ 15 ppm. The impact of salt iodization and distribution was quite apparent in both the districts. Since then, the emphasis was on achieving sustainable elimination of IDD in the country.

Four years after the completion of internal evaluation in 1991–1992, a national assessment was undertaken in mid-1996 to track progress toward the sustainable elimination of IDD in Bhutan. It was conducted jointly by Nutrition Section, Directorate of Health Services, Royal Government of Bhutan, the International Council for Control of Iodine Deficiency Disorders (ICCIDD), New Delhi (India), UNICEF, WHO, and The Micronutrient Initiative (MI), Ottawa (Canada). The

TABLE 5. Age-wise Prevalence of Goiter in Schoolchildren

Age	Total Children Examined	Goiter Grade#			
		0	1	2	TGR (1 + 2)
6–7	278 (100%)	238 (85.6%)	37 (13.3%)	3 (1.1%)	40 (14.4%)
8–9	491 (100%)	423 (86.2%)	66 (13.4%)	2 (0.4%)	68 (13.8%)
10–11	431 (100%)	370 (85.8%)	51 (11.8%)	10 (2.3%)	61 (14.1%)
Total	1200 (100%)	1031 (86.0%)	154 (12.8%)	15 (1.2%)	169 (14.0%)

#Goiter classification by palpation as per WHO, UNICEF, ICCIDD [2001]:
Grade 0 No palpable or visible goiter.
Grade 1 A goiter that is palpable but not visible when the neck is in the normal position (i.e., the thyroid is not visibly enlarged).
Grade 2 A swelling in the neck that is clearly visible when the neck is in a normal position and is consistent with an enlarged thyroid when the neck is palpated.
Source: Directorate of Health Services [1992]; RGOB, ICCIDD, AIIMS, UNICEF, WHO, MI [1996].

study on 1200 schoolchildren revealed a further drop in total goiter rate to 14%, indicating mild iodine deficiency as per the recommended criteria of WHO/UNICEF/ICCIDD [Health Division, 1998]. The age-wise prevalence of goiter in the samples (schoolchildren) is presented in Table 5.

The median level of urinary iodine excretion of 230 μg/liter indicated no iodine deficiency. However, the frequency distribution of urinary iodine levels showed 24% of urine samples having less than 100 μg/liter. The distribution was as follows: 3% of children had less than or equal to 20 μg/liter; 9.9% had between 21 and 50 μg/liter and 11.1% had between 51 and 100 μg/liter. Furthermore, 82% and 73.7% of the salt samples collected from 273 households and 137 retail shops, respectively, had adequate levels of iodine [Health Division, 1998].

The low levels of iodine in salt at household level and retail shops in this study as compared to only 13% to 16% in 1991–92 survey, very well explains the current observation that 24% of schoolchildren had urinary iodine excretion less than 100 μg/liter. Other reasons provided are the breakdown in monitoring of the iodine content of salt at the production level at the Salt Iodization Plant, Phuntsholing, problems faced by BSE in the regular procurement of common salt from India, and failure of the salt crusher. These led the retailers and consumers to buy salt from across the Indo-Bhutan border directly, with consequent loss of control over the iodine content of salt [RGOB, ICCIDD, AIIMS, UNICEF, WHO, MI, 1996].

Comparison of the three nationwide IDD studies in 1983, 1992, and 1996 are presented in Table 4. The comparative results revealed a significant decrease in the prevalence and severity of iodine deficiency disorders, which demonstrated the considerable impact of the IDDCP. The prevalence of TGR reduced from 60% in 1983 to 14% in 1996, while the percentage of urinary iodine excretion (<100 μg/liter) decreased from 96.9% in 1983 to 24% in 1996. In addition, the iodized salt coverage at the household level increased from 0% to 95.8% in 1992 (Fig. 2). The results provided more confidence to the various sectors involved in eliminating the IDD and to bringing about further progress, which ensued yearly monitoring of the progress.

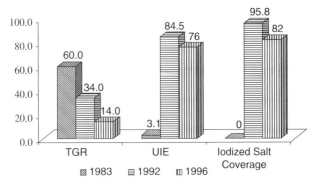

Figure 2. Comparative of percent prevalence of total goiter rate (TGR), urinary iodine excretion (UIE), and iodized salt coverage at the household level in 1983, 1992, and 1996.

9   1996 ONWARD: INTERNAL EVALUATION OF THE IDDCP THROUGH CYCLIC MONITORING

Bhutan initiated internal evaluation of the IDDCP through cyclic monitoring as per the recommendation of the 1996 external evaluation. The result of the four years of cyclic monitoring conducted annually in five different districts and 30 schools with a sample size of 1200 is presented in Table 6.

TABLE 6. Comparative Survey Results from Four Years of Cyclic Monitoring of IDDCP

	Year			
Indicator	1998	1999	2000	2001
Total goiter rate (%)	17	12	8	5
Iodized salt coverage in % (>15 ppm)	95.8	71.1	92.2	95.0
Median UIE (µg/liter)	277.5	170.0	261.0	298.0
UIE (%) (≥ 100 µg/liter)	87.7	76.8	93.07	88.0

Source: Adapted from Sithey [2006].

Figure 3. (**a**) Pema Gatshel district; (**b**) Wangduephodrang district; (**c**) A young man with goiter symptom from a remote village in Chukha district; (**d**) An old lady with grade 2 goiter from a village in Wangduephodrang district.

The implementation of these cyclic studies by the Royal Government of Bhutan further saw the reduction of the prevalence of goiter to 5% in 2000. The recent evaluation on Bhutan's IDD Control Programme against WHO/UNICEF/ICCIDD ten point indicators has declared Bhutan's achievement of total elimination of IDD by all standards in 2003 [Sithey, 2006]. In recognition of the RGOB's commitment in controlling and eliminating IDD, the AIIMS presented a citation to the RBG the same year. It took over two decades for Bhutan to achieve universal salt iodization and elimination of IDD from being a public health problem in sixties. Goiter, which is one of the most visible manifestations of IDD, can still be seen in some of the older people in remote villages of Bhutan (Fig. 3).

10   CONCLUSION

Iodine deficiency, through its effects on the developing brain, has an enormous impact on communities and nations. However, there has been substantial progress in the last decade toward the elimination of iodine deficiency disorders due to the availability and existence of cost-effective prevention measures such as iodization of salt. Increased iodine intake through food fortification with iodine and/or iodine supplementation has improved the Iodine nutrition worldwide. It reflects the efforts made by countries to implement effective IDD control programs and is proof of successful collaboration between all partners in IDD control; the health authorities and the salt industry in particular.

In Bhutan, increased availability and consumption of adequately iodized salt at the household level was accompanied by an almost complete normalization of urinary iodine levels and a dramatic reduction in the prevalence of goiter. Tight market control and legislation for iodization of salt has been a major step forward to a nationwide strategy to alleviate iodine deficiency. Sustained government commitment, regular monitoring of iodine at various levels, information dissemination through communication, and education on IDD, private–public partnership in iodization, and distribution are other key factors that have played a significant role in Bhutan's progress toward sustainable elimination of IDD.

Thus far, strategies to control IDD in Bhutan had been very effective and successful. Nevertheless, should the IDD resurface due to unforeseen reasons, monitoring mechanisms existing in the country should be strictly continued to be implemented such that only iodized salts are imported and made readily available to the Bhutanese populations. Furthermore, people residing in the bordering areas having easier access to markets across the border should be given more attention on the use of iodized salt by the concerned authority.

REFERENCES

Bleichrodt N, Born MA. 1994. Meta-analysis of research on iodine and its relationship to cognitive development. In: Stanbury JB, editor. *The Damaged Brain of Iodine Deficiency.* New York: Cognizant Communication Corporation, pp. 195–200.

British Geological Survey (BGS). 2000. Water quality fact sheet: Iodine. WaterAid, p. 4.

Deckers J, Laker M, Van Herreweghe S, Vanclooster M, Swennen R, Cappuyns V. 2000. State of art on soil-related geo-medical issues in the World. In: Låg J, editor. *Geomedical Problems in Developing Countries*. Olso: The Norwegian Academy of Science and Letters, pp. 23–42.

DeLange F. 1994. The disorders induced by iodine deficiency. *Thyroid* **4**:107–128.

Delange F, Lecomte P. 2000. Iodine supplementation: Benefits outweigh risks. *Drug Safety* **22**:89–95.

Directorate of Health Services (DHS) Ministry of Social Services, Royal Government of Bhutan. 1992. *Iodine Deficiency Disorders: The Bhutan Story*. Thimphu.

Dunn JT, van der Haar F. 1990. *A Practical Guide to the Correction of Iodine Deficiency*. Report of the International Council for the Control of Iodine Deficiency Disorders. Netherlands, p. 62.

Dunn JT, Crutchfield HE, Gutekunst R, Dunn AD. 1993. Two simple methods for measuring iodine in urine. *Thyroid* **3**:119–123.

Hetzel BS. 1983. Iodine deficiency disorders (IDD) and their eradication. *Lancet* **2**:1126–1129.

Hetzel BS, Mano M. 1989. A review of experimental studies of iodine deficiency during foetal development. *J Nut* **119**:145–151.

ICCIDD Report. 2004. Tracking Progress Towards Sustainable Elimination of Iodine Deficiency Disorders in Rajasthan. Report submitted by ICCIDD to UNICEF India & MI. p. 49.

Health Division, Ministry of Social Services, Royal Government of Bhutan. 1998. *Internal Evaluation of Iodine Deficiency Disorder Control Programme (IDDCP) by Cyclic Monitoring*. Thimphu.

Koutras DA, Matovinovic J, Vought R. 1985. The ecology of iodine. In: Stanbury JB, Hetzel BS, editors. *Endemic Goiter and Endemic Cretinism*. New Delhi: Wiley Eastern Limited, pp. 185–195.

Lamberg BA. 1993. Iodine deficiency disorders and endemic goitre. *Eur J Clin Nutr* **47**:1–8.

RGOB, ICCIDD, AIIMS, UNICEF, WHO, MI. 1996. Tracking progress towards sustainable elimination of Iodine Deficiency Disorders in Bhutan, p. 97.

Sithey G. 2006. Progress towards Sustainable Elimination of IDD in Bhutan—Lessons to be shared. *IDD News* Vol. IV, Issue 1, pp. 10–12.

Stanbury JB. 1998. Prevention of iodine deficiency. In: Howson CP, Eileen TK, Abraham H, editors. *Prevention of Micronutrient Deficiencies: Tools for Policymakers and Public Health Workers*. Washington, DC: Institute of Medicine, National Academies Press, p. 224.

Todd CH, Allain T, Gomo ZAR, Hasler JA, Ndiweni M, Oken E. 1995. Increase in thyrotoxicosis associated with iodine supplements in Zimbabwe. *Lancet* **346**:1523–1564.

UNICEF, WHO. 1983. Iodine Deficiency Disorders in Bhutan: Extent and Severity. Report of a study team from the All India Institute of Medical Sciences. New Delhi.

WHO. 1994. *Iodine and Health: A Statement by World Health Organization*. Geneva: WHO/NUT/94.4.

WHO. 1996. *Recommended Iodine Levels and Salt and Guidelines for Monitoring Their Adequacy and Effectiveness*. Geneva: WHO.

WHO. 1999. *Prevention and Control of Iodine Deficiency Disorders.* 52nd World Health Assembly. Provisional agenda item 13, A52/11. Geneva: World Health Organization.

WHO. 2001. *Assessment of Iodine Deficiency Disorders and Monitoring Their Elimination.* World Health Organization Geneva: WHO.

WHO. Bruno de Benoist et al., editors. 2004. *Iodine Status Worldwide: WHO Global Database on Iodine Deficiency.* Department of Nutrition for Health and Development. Geneva: World Health Organization, p. 58.

WHO, UNICEF, ICCIDD. 1994. *Indicators for Assessing Iodine Deficiency Disorders and Their Control Through Salt Iodization.* Geneva: World Health Organization. WHO/NUT/94.6, pp. 1–55.

WHO, UNICEF, ICCIDD. 1996. *Recommended Iodine Levels in Salt and Guidelines for Monitoring Their Adequacy and Effectiveness.* Geneva: World Health Organization (WHO/NUT/96.13).

WHO, UNICEF, ICCIDD. 1999. *Progress Towards Elimination of Iodine Deficiency Disorders.* Geneva: WHO/NHD/99.4.

WHO, UNICEF, ICCIDD. 2001. *Assessment of the Iodine Deficiency Disorders and Monitoring Their Elimination.* Geneva: World Health Organization, WHO/NHD/01.1, pp. 1–107.

13 Floristic Composition at Kazakhstan's Semipalatinsk Nuclear Test Site: Relevance to the Containment of Radionuclides to Safeguard Ecosystems and Human Health

K. S. SAGYNDYK and S. S. AIDOSSOVA

Botany Department, Biology Faculty, Kazakh National al-Farabi University, Almaty 050040, Republic of Kazakhstan

M. N. V. PRASAD

Department of Plant Sciences, University of Hyderabad, Hyderabad 500 046, India

1 INTRODUCTION

Nuclear technology was initiated during the mid-20th century for economic and military purposes [Negri and Hinchman, 2000]. As a result, nuclear bombs and nuclear reactor facilities were developed with radionuclides as the main source of energy [Zumdahl, 1989]. Radionuclides pose environmental risks when exposed and/or deposited in soil and water [Negri and Hinchman, 2000]. Sources of radionuclides are fallout from above-ground nuclear testing, accidents at nuclear reactor facilities, and fission by-products from nuclear bombs [Entry et al., 1997; Negri and Hinchman, 2000].

Once radionuclides are deposited on the soil surface, they eventually are incorporated into the soil structure. Physical and biological nutrient cycles can distribute radionuclides throughout soil and water [Entry et al., 1997]. Natural weathering and chemical leaching can also release radionuclides into the soil [Dushenkov et al., 1999]. Soils with

Trace Elements as Contaminants and Nutrients: Consequences in Ecosystems and Human Health, Edited by M. N. V. Prasad

clay particles have a strong adsorption for radionuclides due to clay's large surface area [Dushenkov et al., 1999; Negri and Hinchman, 2000]. Soil constituents, such as organic matter and oxides, can temporarily bind radionuclides, allowing their release under specific environmental conditions [Negri and Hinchman, 2000]. The amount of bound radionuclides is usually linearly correlated with the amount of time that has passed since the radionuclide deposition. If radionuclides are adsorbed or bound in the soil, then they are not bioavailable to plants, microorganisms, or soil invertebrates. If radionuclides remain in the soil solution, then they are bioavailable to soil biota and plants. Radionuclide bioavailability mostly depends on the type of radionuclide deposition, the time of deposition, and the soil characteristics [Dushenkov et al., 1999]. Radionuclides ^{137}Cs and $^{238-241}$Pu tightly bind to soil particles, decreasing their bioavailability [Negri and Hinchman, 2000].

Four common radionuclides of environmental concern are ^{137}Cs, ^{90}Sr (strontium), ^{234,235,238}U (uranium), and $^{238-241}$Pu (plutonium). Cesium-137 and ^{90}Sr come from fission by-products, while Pu is released from nuclear weapon testing and nuclear fuel facilities [Negri and Hinchman, 2000]. Uranium is released from nuclear fuel cycles, but is the only one of the four radionuclides that occurs naturally [Negri and Hinchman, 2000]. If any of these radionuclides are in the soil solution and bioavailability, they could pose a risk to environmental and human health. Radionuclides cannot be degraded, so they have the potential to accumulate in plant species and

TABLE 1. Tree, Grass, and Forb Species Able to Accumulate Radionuclides

Tree Species	Radionuclide
Red maple (*Acer rubrum*)	^{137}Cs, ^{238}Pu, ^{90}Sr
Liquidambar stryaciflua	^{137}Cs, ^{238}Pu, ^{90}Sr
Tulip tree (*Liriodendron tulipifera*)	^{137}Cs, ^{238}Pu, ^{90}Sr
Coconut palm (*Cocos nucifera*)	^{137}Cs
Montery pine (*Pinus radiata*)	^{137}Cs, ^{90}Sr
Ponderosa pine (*P. ponderosa*)	^{137}Cs, ^{90}Sr
Forest redgum (*Eucalyptus tereticornis*)	^{137}Cs, ^{90}Sr
Black spruce (*Picea mariana*)	^{137}Cs
Oak (*Quercus*)	^{137}Cs
Juniper (*Juniperus*)	^{137}Cs
Grass/Forb Species	
Tall fescue (*Festuca arundinacea*)	^{137}Cs
F. rubra	^{137}Cs
Perennial ryegrass (*Lolium perenne*)	^{137}Cs
White clover (*Trifolium repens*)	^{137}Cs
Chickweed (*Cerastium fontanum*)	^{137}Cs
Bentgrass (*Agrostis* plant communities)	^{137}Cs
Redroot (*Amaranthus retroflexus* cv. *belozernii*, *aureus*, and Pt-95)	^{137}Cs
Beet, quinoa, and Russian thistle (*Chenopodiaceae*)	^{137}Cs, ^{90}Sr
Umbelliferae and Legume family (a)	^{90}Sr
Spiderwort (*Tradescantia bracteata*)	^{137}Cs, ^{90}Sr

Source: Entry et al., 1997; Dushenkov et al., 1999; Negri and Hinchman, 2000.

increase in concentration as they make it higher in the food chain [Entry et al., 1997]. For example, a plant that has accumulated ^{137}Cs, if consumed by a deer, it gets bio-accumulated in humans when they hunt for food. Humans exposed to radionuclides or who ingest contaminated food accumulate radionuclide Table 1. Humans can also be directly exposed to the radionuclide through contact in soil or water. Some human health effects of radionuclide contamination are forms of cancer and mutations [Entry et al., 1997].

Cleanup of soil radionuclides would be possible. The first one is soil excavation, exorbitantly expensive, $1600 per cubic yard to excavate, package, transport, and disposal [Ensley, 2000; Negri and Hinchman, 2000]. In the United States alone, $200–300 billion is estimated to be spent on remediation of radionuclides in water and soil [Ensley, 2000; Entry et al., 1997]. The second is soil washing [Huang et al., 1998; Entry et al., 1997]. Both these methods have only relocated the problem.

Nitric acid, ammonium salt, ammonium nitrate, ammonium chloride, and ammonium acetate enhanced the bioavailability and accumulation of ^{137}Cs [Dushenkov et al., 1999; Hossner et al., 1998; Negri and Hinchman, 2000] in cabbage, tepary beans, Indian mustard, and reed canary grass *Amaranthus retroflexus* (redroot pigweed). Other prominent chleators are (K^+, NH_4^+, Rb^+, Na^+, and Cs^+) citric acid, natural organic matter, dissolved organic compounds, microbial exudates, and sulfur ($>10\,\mathrm{mmol/kg}$) (Dushenkov et al., 1999; Huang et al., 1998; Negri and Hinchman, 2000].

2 KAZAKHSTAN: SEMIPALATINSK NUCLEAR TEST SITE

The Republic of Kazakhstan (northeastern part) consists of 470 nuclear test sites comprising an area of $430\,\mathrm{km}^2$ conducted during 1949–1989; 26 of the sites were surface, 87 air, and 357 underground. An area of about $430\,\mathrm{km}^2$ is polluted with radionuclides. The main contribution to radioactive pollution was made by above-ground nuclear tests conducted until 1963. After 1963, radioactive pollution was caused by radionuclide splinters of nuclear fuel fission, unreacted nuclear fuel, and radionuclide products of explosive devices activated by neutrons. Nuclear decay of the underground nuclear tests and the subsequent decay products eventually accumulated in these test sites. Dangerous radionuclides such as ^{239,240}Pu, ^{90}Sr, and ^{137}Cs (^{137}Cs-has a half-life of 30 years and ^{90}Sr has a half-life of 28 years) leached to surface waters and groundwaters [Entry et al., 1997; Freiling, 1962; Izrael et al., 1970; Killham, 1995]. ^{137}Cs acts similarly to potassium and ^{90}Sr behaves similarly to calcium [Killham, 1995]. Vertical and horizontal migration of these decay products of radionuclides enter plants and reach humans through the food chain and is thus a health hazard [Aidossova, 2003; Westnoff, 1997]. Unauthorized activities of local people to recover cables and radioactive scrap [Akhmetov et al., 2000] further aggravate the situation. Presently, this contaminated territory is a grazing land and is partly mined for coal and gold.

Radioecological investigations of the territory, migration of radionuclides from underground water into soil and into the roots of plants, uptake and translocation to aerial parts, the transfer factor, and concomitant affects on the food chain have not been investigated. Strategies for rehabilitation of ecosystems destroyed by nuclear tests have not been planned, although radiophytoremediation is gaining considerable progress in other parts of the world [Vandenhove, 2006].

The vegetation of the study area in Kazakhstan represents steppes predominated by xerophytic grasses such as *Stipa capillata* L., *Festuca valesiaca* Gaud, and *Agropyron cristatum* L [Sagyndyk et al., 2007]. This region is contaminated by radionuclides, products of 357 underground nuclear explosions, and tests carried out during 1949 to 1989. This area is being used as grazing land and partly for mining (coal and gold). Focused radioecological and radiophytoremediation investigations on the migration of radionuclides from underground waters into soils, translocation to plant aerial parts, the transfer factor, and concomitant effects on the food chain (namely, underground water → soils → plants → animals) are rather scanty. This chapter discusses the Floristic composition at Kazakhstan's Semipalatinsk nuclear test site.

Semipalatinsk Test Site facilities are under the jurisdiction of the National Nuclear Center of the Republic of Kazakhstan, which is involved in civilian activities and conversion of the site to nondefense uses (Fig. 1). The National Nuclear Center includes three research institutes in the town of Kurchatov and three research reactors at the Semipalatinsk test site. The Semipalatinsk test range covered an area of 18,000 km². Between 1949 and 1989, 470 nuclear tests, including 357 underground and 113 atmospheric tests, were conducted at Semipalatinsk Test Site facilities (Balmukhanov et al., 2000). Semipalatinsk'sDegelen Mountain nuclear test facility (Figs. 2 and 3) was the largest underground nuclear test site in the world, consisting of 186 separate tunnels in natural mountain formations. Between October 11, 1961 and October 10, 1989, 224 tests were conducted there. In addition to Degelen Mountain, underground tests were also conducted at Balapan, in vertical holes drilled in the ground rather than in tunnels. These holes are about 500–600 m

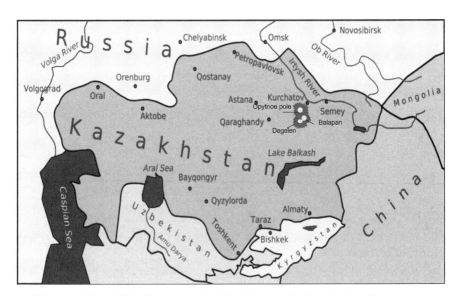

Figure 1. Map of Kazakhstan showing Kazakhstan's Semipalatinsk nuclear test site.

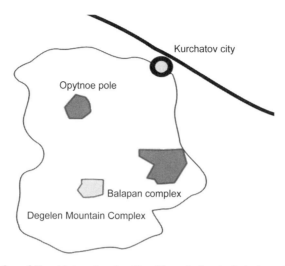

Figure 2. Map of Kazakhstan showing Kazakhstan's Semipalatinsk nuclear test site.

deep, and the bottoms of the holes are up to 900 m in diameter. The last nuclear test conducted at the Semipalatinsk test site took place at Balapan in November 1989 (Fig. 2). From 1997 to 2000, a series of calibrated explosions destroyed testing infrastructure at Degelen and Balapan as part of a joint US–Kazakhstan effort under the Weapons of Mass Destruction Elimination Initiative, Cooperative Threat Reduction Program.

Figure 3. Filed spots: Degelen, Balapan, and Opytnoe nuclear test sites for comparative floristic study. mcR/h is microRoentgen per hour. This measurement is used for environmental gamma-radiation measurements and also can be converted to Sievert/hour. 100 mcR/h = 1 Sievert/h.

There are no fences around the perimeter of Semipalatinsk; people and animals roam freely across the former test site. Only the Baykal-1 and Institute of Geophysical Research (IGR) reactor complexes are cordoned off, according to National Nuclear Center officials. According to Kazakhstani nuclear physicists, extensive mining operations are underway at the test site: Beryllium, coal, and gold are mined, and table salt is produced from a lake located near the main test field. In addition, scrap metal is gathered illegally from the site. According to some reports, bore holes at Degelen Mountain have been breached by scrap metal gatherers, although National Nuclear Center officials deny this [Carlsen, 2000].

3   FLORA OF NUCLEAR TEST SITE

On the territory of the Semipalatinsk nuclear site, 530 species of vascular plants were found Table 2. They belong to 72 families and 281 genera. Within the bounds of the Semipalatinsk nuclear test, the floristic variety of each test sites greatly distinguishes. Among the three nuclear test sites, the highest floristic diversity is on the Degelen: 490 species, which belong to 70 families and 271 genera. It is caused by a variety of natural conditions of Degelen mountain range. The flora of Balapan is presented by 213 species (42 families and 141 genera), and flora of Opytnoe pole contain 194 species (39 family and 120 genera) [Sultanova, 2000].

Natural flora of "Opytnoe pole" test sites was completely destroyed during air and above-ground nuclear explosions. Now it is presented by regenerated zonal flora. The changes in the floristic composition are depicted in Fig. 4–13. Asteraceae and Poaceae families have kept their positions. On the "Balapan" and "Opytnoe pole" test sites the phytocenotic role of Brassicaceae and Chenopodiaceae families have grown. On the Degelen, Fabaceae and Chenopodiaceae families moved to 7–8 places, but Brassicaceae took third place. On the natural and low destroyed ecotopes of Degelen, the Lamiaceae family is in eighth place, but on the technogenic nuclear ecotopes it moves to fourth place.

On the technogenic ecotopes the number of xerophytes (from 8.9% up to 11%) and mesoxerophytes increase (from 24.5% up to 33.3%), but number of mesophytes decrease (from 31.5% to 20.5%).

By frequency of occurrence, the plants can be divided into five groups:

1. Omnipresent. Mostly weeds and adventitious plants: *Artemisia scoparia*, *A. dracunculus*, *A. sieversiana*, *Acroptilon repens*, *Amaranthus retroflexus*, *Berteroa incana*, *Cannabis ruderalis*, *Ceratocarpus arenarius*, *Chenopodium botrys*, *Erigeron canadensis*, *Fumaria vaillantii*, *Kochia scoparia*, *Lepidium densiflorum*, *L. latifolium*, *Polygonum aviculare*, *Setaria viridis*, *Solanum dulcamara*, *Urtica dioica*, *Xanthium strumarium*, and so on. There are 63 species in this group (11.9%).

2. Natural Phytocenosis. *Stipa capillata*, *S. sareptana*, *Festuca valesiaca*, *Artemisia frigida*, *A. gracilescens*, *Spiraea hypericifolia*, *Caragana pumila*, *Cleistogenes squarrosa*, *Helictotrichon desertorum*, *Galium ruthenicum*, *Ephedra distachya*, *Potentilla acaulis*, *Ancathia igniaria*, *Dianthus rigidus*,

TABLE 2. On the Territory of the Semipalatinsk Nuclear Site, 530 Species of Vascular Plants Were Found[a]

Plant Species	Medicinal Value
Achillea nobilis	Hemostatic remedy
Acroptilon repens	Malaria, epilepsy
Artemisia dracunculus	Radiculitis, gastritis, epilepsy, taeniacide drug
Artemisia scoparia	Use for "Tarhkun" spice
Berberis sibirica	Rheumatism, goiter, hypertension, diarrhea, dysentery
Berteroa incana	Diuretic, sudorific, debilitant, wound-healing remedy, asphyxia, singultus
Betula pendula	Eczema
Cannabis ruderalis	Hashish
Delphinium dictyocarpum	Painkiller, wound-healing and cholagogue remedy, antiseptic
Ephedra distachya	Rheumatism, digestive organ and respiratory tract disease
Fragaria viridis	Antipyretic
Fumaria vaillantii	Cholagogue, diuretic remedy, hepatitis, stomach, bowel disease, suscitate appetite, infectious disease, loss of blood
Juniperus sabina	Abscess, dermatoses
Lepidium latifolium	Rheumatism, wound healing remedy
Matricaria recutita	Stomach, laxative
Melica transsilvanica	Sedative
Padus avium	Anticeptic, stomach and lung disease
Pinus sylvestris	Tuberculosis, bronchitis, pleuritis
Populus nigra	Antipyretic, antiinflammatory, diaphoretic drug
Populus tremula	Rheumatism, antipyretic, sedative remedy, antiseptic, diaphoretic drug
Polygonum aviculare	Hepatitis, stomach, rheumatism, malaria, dysentery, pertussis, tuberculosis, bronchitis, pleuritis, cholecystitis, hemorrhage
Ribes saxatile	Antipyretic
Sanguisorba officinalis	Spasmolytic, uterus and heart disease, burn, stomatitis
Urtica dioica	Hemorrhoids, excessive menstruation, edema, wound, hurts
Viburnum opulus	Gastric and duodenal ulcers, hemostatic drug
Xanthium strumarium	Eczema, dermatitis, dysentery, bronchitis, rheumatism, chill, goiter, antipyretic, sedative remedy
Ziziphora clinopodioides	Rheumatism, chill, spasmolytic

[a]They belong to 72 families and 281 genera; 68 species belong to medicinal plants.

Phragmites australis, Calamagrostis epigeios, Atriplex cana, and so on. This group is represented by 160 species (30.2%).

3. Sporadic Occurrence. *Achillea nobilis, Agrostis vinealis, Berberis sibirica, Betula pendula, Delphinium dictyocarpum, Fragaria viridis, Juniperus sabina, Stipa kirghisorum, Populus tremula, Ziziphora clinopodioides*, and so on. This is the largest group, with a total of 189 species (35.7%).

4. Rare Occurrence. *Cotoneaster melanocarpus, Crataegus clorocarpa, Hieracium virosum, Limonium chrysocomum, Lonicera pallasii, Matricaria*

recutita, Melica transsilvanica, Padus avium, Parnassia palustris, Potentilla nudicaulis, Reaumuria soongorica, Ribes saxatile, Typha laxmannii, Viola montana, and so on. This group consists of 91 species (17.2%).

5. Unit Occurrence Group. *Pinus sylvestris, Dianthus versicolor, Viburnum opulus, Populus nigra, Brachypodium pinnata, Sagittaria sagittifolia, Psoralea degelenica, Campanula sibirica, Fritillaria meleagroides, Valeriana altaica, Angelica sylvestris,* and so on. This group has 27 species (5%).

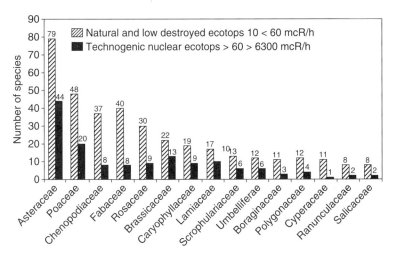

Figure 4. Comparative floristic composition of prevalent genera of Degelen nuclear test site. (*Data source*: Sultanova [2000].)

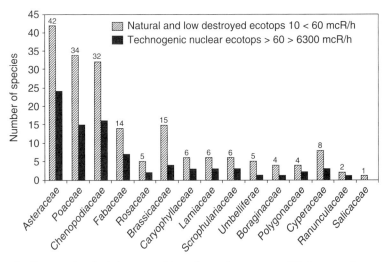

Figure 5. Comparative floristic composition of prevalent genera of Balapan nuclear test site. (*Data source*: Sultanova [2000].)

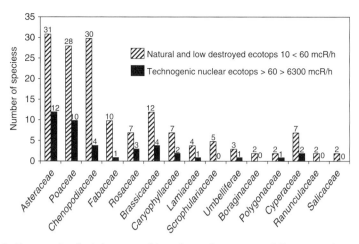

Figure 6. Comparative floristic composition of prevalent genera of Opytnoe pole nuclear test site. (*Data source*: Sultanova [2000].)

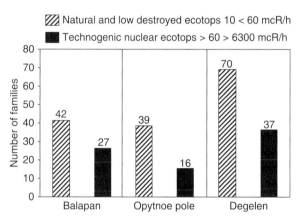

Figure 7. Comparison of nuclear test site's flora—Number of families. (*Data source*: Sultanova [2000].)

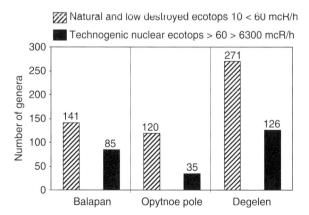

Figure 8. Comparison of nuclear test site's flora—Number of genera. (*Data source*: Sultanova [2000].)

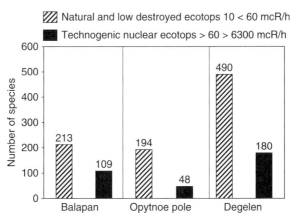

Figure 9. Comparison of nuclear test site's flora—Number of species. (*Data source*: Sultanova [2000].)

Plants that grow at equivalent dose rate 6300 microR/h are represented by following species (highly tolerant): *Artemisia frigida*, *Chamaenerion angustifolium*, *Ephedra distachya*, *Festuca valesiaca*, *Kochia scoparia*, *Lotus angustissimus*, *Phragmites australis*, *Psathyrostachys juncea*, *Silene suffrutescens*, *Stipa sareptana*, and *Typha angustifolia*. Plants that grow on technogenic ecotopes with equivalent dose rate between 2900 and 6300 microR/h are: *Artemisia scoparia*, *Dianthus rigidus*, *Iris scariosa*, *Kochia sieversiana*, *Lepidium latifolium*, *Limonium suffrutescens*, *Rumex*

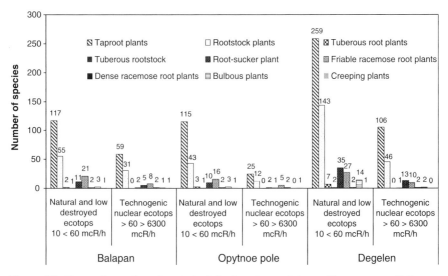

Figure 10. Comparison of nuclear test site's flora by root type. (*Data source*: Sultanova [2000].)

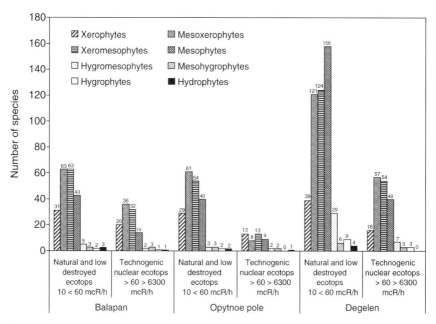

Figure 11. Comparison of nuclear test site's flora by water factor ecomorphs. (*Data source*: Sultanova [2000].)

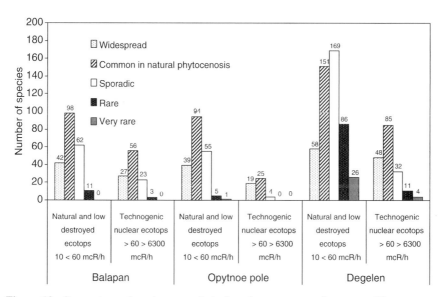

Figure 12. Comparison of nuclear test site's flora by occurrence frequency. (*Data source*: Sultanova [2000].)

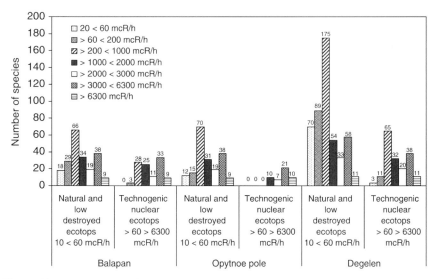

Figure 13. Floristic comparison by dose rate of nuclear test site. (*Data source*: Sultanova [2000].)

confertus, Salix cinerea, Sanguisorba officinalis, Stipa capillata, Urtica urens, and so on. Plants that grow on "potentially dangerous" ecotopes with equivalent dose rate between 60 and 3000 microR/h were divided into the following groups: 200 microR/h, 1000 microR/h, 2000 microR/h, 3000 microR/h. This is the most numerous group (389 species, 73.4%), including such species as *Atriplex cana*.

4   FODDER PLANTS

Sheep: Sheep use pascual plants better than do cows and horses. They willingly eat not only grasses and legumes, but even different types of miscellaneous herbs containing fair quantity of bitter substances (wormwood, milfoil), ashes (the goosefoot family), and barbs (in this case they eat soft parts such as leaves and draws). Mesophyte plants are more preferable plants (alfalfa, couch grass, etc.) than xerophytes (fescue, June grass, mat-grass, wormwood, etc.). Steppe vegetation is more frequently eaten by sheep than by other livestock; that is why sheep is more preferable than cattle in the steppe region.

Cow: Cows eat lush and palatable grasses (mesophytes and gygrophytes). Bitter-tasting, poignant-smelling, high-ash, barbly grasses are either badly eaten or not eaten by cows (wormwood, milfoil, the goosefoot family, camel's-thorn, *Spiraea hypericifolia, Caragana pumila*). Cows eat Compositae with semifloscules, some Cruciferae, Umbelliferae, Rosaceae, and Cyperaceae very well.

Horse: Horses are less fastidious than cows, but more fastidious than sheep and camel. They eat dry xerophytes plants (fescue, feather grass, couch grass, June grass), bitter-tasting and poignant-smelling grasses (wormwood), high-ash goosefoot grasses (kokpek, tas-buirgun, leban (Vitex), eurotia, glasswort (Salsola)), mesophyte grasses (alfalfa (Medicago), clover (Trifolium), esparcet (Onobrychis), quack grass (*Agropyron repens*), randall (*Festuca pratense*)), gygrophyte grasses very well.

5   CONCLUSIONS

Nuclear technology was initiated by the onset of the Cold War in the middle of the 20th century [Negri and Hinchman, 2000]. As a result, nuclear bombs and nuclear reactor facilities were developed with radionuclides as the main constituent. The vegetation of the Semipalatinsk nuclear test site is mostly represented by xerophytic plants. Five hundred thirty species of vascular plants that belong to 72 families and 281 genera were found on this territory. Among the three nuclear test sites, the highest floristic diversity is on the Degelen: 490 species. The flora of Balapan is represented by 213 species, and the flora of Opytnoe pole consists of 194 species [Sultanova, 2000]. Natural flora of Opytnoe pole test sites was completely destroyed during air and above-ground nuclear explosions. Now it is represented by regenerated zonal flora.

Floristic investigation shows that number of families, genera, and species, number of species of lifetime, root type, and water factor ecomorphs were decreased as result of radiation effects in the Degelen nuclear test site as well as in the Opytnoe pole and Balapan test sites.

ACKNOWLEDGMENTS AND DISCLAIMER

One of the authors, MNVP acknowledges Kazakh National al-Farabi University, Almaty, Republic of Kazakhstan for supporting short academic exchange visits in 2006 and 2007 "under the auspices of invited visiting professors scheme." This academic exchange resulted in preparation of this chapter. This review has been prepared from information obtained from authentic and highly regarded sources such as internet, national, and international conference deliberations. The authors thankfully acknowledge all these resources. The authors and the publisher cannot assume responsibility for any adverse consequences of the use of material cited in this article in field or laboratory.

REFERENCES

Aidossova SS. 2003. The morphological changes of different life form plants in condition of radioactive pollution. *Bull Kazakh National al-Farabi University, Ecological Series* **1**:67–71.

Akhmetov MA, Artem'ev OI, Ptitskaia LD. 2000. Radiological monitoring of currents and problems of rehabilitation on Degelen mountain range of Semipalatinsk nuclear test site. *Bull National Nuclear Centre of Republic of Kazakhstan. Radioecology. Environment Control* **3**:23–28.

Alexander LT. 1950. Radioactive materials as plant stimulants-field results. *Agron J* **42**:252–255.

Balmukhanov SB, Abdrakhmanov JN, Balmukhanov TS, Gusev BI, Kurakinaand NN. 2000. Kazakhstan Closes Nuke Test Facility, Associated Press, Saturday, July 26.

Carlsen TM. 2000. *Young Investigator Program in Kazakhstan: The Health and Environmental Impacts of Nuclear Testing in the Semipalatinsk Region*. Livermore, CA: Lawrence Livermore National Laboratory (UCRL-AR-141293).

Dushenkov S, Mikheev A, Prokhnevsky A, Ruchko M, Sorochinsky B. 1999. Phytoremediation of radiocesium-contaminated soil in the vicinity of Chernobyl, Ukraine. *Environ Sci Technol* **33**:469–475.

Ensley BD. 2000. Rationale for use of phytoremediation. In: Raskin I, and Ensley BD, editors. *Phytoremediation of Toxic Metals: Using Plants to Clean Up the Environment*. New York: Wiley-Interscience, pp. 1–12.

Entry JA, Watrud LS, Manasse RS, Vance NC. 1997. Phytoremediation and reclamation of soils contaminated with radionuclides. In: *Phytoremediation of Soil and Water Contaminants*. New York: American Chemical Society, №667, pp. 299–306.

Entry JA, Watrud LS, Reeves M. 1999. Accumulation of 137Cs and 90Sr from contaminated soil by three grass species inoculated with mycorrhizal fungi. *Environ Pollut* **104**:449–457.

Freiling EC. 1962. *Radioactive Fallout from Nuclear Weapon Tests*, V.I., TID-7632. Washington, DC, 47 pp.

Hossner LR, Loeppert RH, Newton RJ, Szanislzo PJ, Moses A Jr. 1998. *Literature Review: Phytoaccumulation of Chromium, Uranium, and Plutonium in Plant Systems*. Amarillo National Resource Center for Plutonium, pp. 1–8.

Huang JW, Blaylock MJ, Kapulnik Y, Ensley BD. 1998. Phytoremediation of uranium contaminated soils: Role of organic acids in triggering uranium hyperaccumulation in plants. *Environ Sci Technol* **32**:2004–2008.

Izrael Yu A, Petrov VN, Presman A Ya. 1970. *Radiactive Pollution of Environment Under Underground Nuclear Explosions and Methods of Forecast*. Leningrad: Hydrometeorological Publishing House, 213 pp.

Killham K. 1995. *Ecology of Polluted Soils. Soil Ecology*. Cambridge: Cambridge University Press, pp. 175–181.

Negri CM, Hinchman RR. 2000. The use of plants for the treatment of radionuclides. In: Raskin I, Ensley BD, editors. *Phytoremediation of Toxic Metals: Using Plants to Clean Up the Environment*. New York: Wiley-Interscience, pp. 1–7.

Sagyndyk KS, Aidossova SS, Prasad MNV. 2007. Grasses tolerant to radionuclides growing in Kazakhstan nuclear test sites exhibit structural and ultrastructural changes—implications for phytoremediation and involved risks. *Terrestrial and Aquatic Ecotoxicology* **1**:70–77.

Sultanova BM. 2000. Assessment of restoration potential of Semipalatinsk nuclear test site's flora. *Bull Radioecol Environ Protection* **3**:62–70.

Vandenhove H. 2006. Phytomanagement of radioactively contaminated sites. In: Prasad MNV, Sajwan KS, Naidu R, editors. *Trace Elements in the Environment: Biogeochemistry, Biotechnology and Bioremediation*. Boca Raton, FL: CRC Press, pp. 583–609.

Westnoff A. 1999. Mycorhizal plants for phytoremediation of soils contaminated with radionuclides. *Restoration and Reclamation* **5**:114–119.

Zumdahl SS. 1989. *Atoms, molecules, and ions*. Chapter 2 of *Chemistry*, second edition. Lexington, MA: D. C. Health and Company, pp. 292–359.

14 Uranium and Thorium Accumulation in Cultivated Plants

IRINA SHTANGEEVA

Chemical Department, St. Petersburg University, St. Petersburg 199034, Russia

1 INTRODUCTION: URANIUM AND THORIUM IN THE ENVIRONMENT

The transfer of artificial radionuclides along terrestrial food chains has been studied extensively over the last 30 years, with understandable emphasis on cesium-137 since 1986. Naturally occurring radionuclides have not been studied to the same extent. However, in recent years the interest in assessment of the impacts of these radioactive elements on arable soils, soil microbiota, edible plants, and humans has been increasing constantly. Many investigations have been carried out in different countries, especially in those where concentrations of naturally occurring radionuclides in soils are particularly high [Vera Tomé et al., 2002; Termizi Rampli et al., 2005; Chen et al., 2005].

There are two main sources of radiation arising from the decay of naturally occurring radioactive elements. These are cosmic and terrestrial radiation. Besides, 200 years of industrialization has produced and redistributed increasing amounts of radioactive matter via release of additional radiation through mining (especially uranium), coal combustion, cement production, street construction, and other human activities.

The most common terrestrial radioisotopes are uranium-238, thorium-232, and potassium-40. In this chapter, we will discuss the biogeochemical behavior of two of them: thorium and uranium. Uranium and thorium are major energy sources, which drive the evolution of the Earth and the planets. Both of these radionuclides are components of the biosphere, and thus they occur naturally in all soils and plants.

Trace Elements as Contaminants and Nutrients: Consequences in Ecosystems and Human Health, Edited by M. N. V. Prasad
Copyright © 2008 John Wiley & Sons, Inc.

Uranium (U) is a naturally occurring radioactive element with atomic number 92. It has the highest atomic weight among the naturally occurring radioactive elements. U is commonly found in very small amounts in rocks, soil, water, plants, and animals (including humans). This metal is weakly radioactive and contributes to low levels of natural background radiation in the environment. U has three isotopes, ^{238}U, ^{235}U, and ^{234}U. Although uranium is present in the environment at low concentrations, it is even more abundant than such metals as cadmium, mercury, and lead. Uranium had only limited use prior to 1939. However, after the development of the atomic bomb, it was mined extensively.

Thorium (Th) is surprisingly abundant in the Earth's crust, being almost three times more abundant than uranium. Thorium is found in small amounts in most rocks and soils. Thorium occurs in several minerals, the most common being the rare earth–thorium–phosphate mineral, monazite, which contains up to 12% of thorium oxide. Granitile contains up to 80 mg/kg of Th. In the environment, thorium exists in various combinations with other minerals, such as silica. It is commonly accepted that most Th compounds found in the environment do not dissolve easily in water and do not evaporate from soil or water into the air.

The important characteristic of each radioactive element is its half-life. The half-life of a radionuclide is time period during which one-half of the initial number of atoms undergoes decay to the daughter product. Uranium and thorium isotopes have extremely long half-lives that range from 4,468,000,000 years for ^{238}U to 13,900,000,000 years for ^{232}Th.

U and Th have big cation radii and small electronegativities and ionization potentials (U^{4+}: 9.7 nm, 1.4, 6.08 eV, respectively; Th^{4+}: 11.4 nm, 1.0, 6.95 eV, respectively). Therefore, in this respect, their ions are similar to K^+, Na^+, and the ions of rare earth elements (REE) [Yingjun, 1984]. They are usually enriched in rocks that are also rich with K, Na, and REE [Bao and Zhang, 1998]. Because of a high valence state, both U and Th can rather easily combine with other elements or ionic groups to form complexes. The complexes of U and Th are smaller in density, and thus they can easier migrate in the environment. Soil U and Th adsorption and complexation processes apparently influence the uptake of these metals by plants growing in the soils.

2   URANIUM AND THORIUM IN SOIL

In general, soils and plants contain all naturally occurring radioactive elements with half-lives comparable to the age of the Earth, although their concentrations in plants may be rather low. The biogeochemistry of U and Th in the environment is of major environmental concern [Airey and Ivanovich, 1986; Buck et al., 1996]. Fundamental knowledge of biogeochemical cycles of these metals and processes involved in their environmental migration are also of practical importance for power generation, water supply, agriculture, sewage disposal, and environmental protection and remediation. All previous experimental results demonstrated that distribution of U and Th in soil is highly variable. For example, activity concentrations of ^{238}U in soil can vary by around three orders of magnitude depending on various factors [Ewers et al.,

2003]. Therefore, an assessment of the radionuclide distributions in the environment (first of all, in the soil–plant system) may be rather complicated.

Although there are numerous reports in literature on biogeochemistry of both elements [Sheppard and Eveden, 1988; Mazor, 1992; Mortvedt, 1994; Voigt et al., 2000; Yoshida et al., 2000; Edmands et al., 2001; Vera Tomé et al., 2002; Ehlken and Kirchner, 2002; Rodrigues et al., 2002; Vera Tomé et al., 2003; Thiry et al., 2005; Tsuruta, 2006; Galindo et al., 2007], the number of publications on uranium prevails. This is not surprising because uranium is an important element for nuclear production. Over 16% of the world's electricity is generated from uranium in nuclear reactors and this consumption amounts to be 65,434 tons of uranium per year [Hore-Lacy, 2003]. The published data on the fate and transport of thorium in the environment are more limited. However, recently the biogeochemistry of thorium has generated more interest to environmental scientists [Higgy and Pimpl, 1998; Zararsiz et al., 1997; Morton et al., 2001; Sar and D'Souza, 2002; Morton et al., 2002; Larson et al., 2005; Höllriegl et al., 2007].

Thorium is widely distributed in the environment. Th is a typical lithophilic element, and its geochemical behavior is very similar to that of rare earth elements (especially cerium), zirconium, hafnium, and uranium [Huist, 1997]. The geochemistry of Th is simplified by the existence of just one valence state, $+4$. Since Th in solution is a highly charged cation, it undergoes extensive interaction with water and many anions to form complex compounds [Wright, 1999].

Uranium exists naturally in the Earth crust at mean content of 2.5 mg/kg. The behavior of U in soils is mostly influenced by physicochemical processes (speciation in soil solution and binding with mineral/organic particles) and biological processes (interactions with soil microorganisms, root exudates). Uranium is present in the soil primarily (80–90%) in the $+6$ oxidation state as the uranyl (UO_2^{2+}) cation [Shahandeh and Hossner, 2002]. U(VI) is the most mobile form of U. It exists in solution predominantly as UO_2^{2+} and as soluble carbonate complexes, $(UO_2)_2 CO_3(OH)_3^-$, $UO_2CO_3^0$, $UO_2(CO_3)_2^{2-}$, $UO_2(CO_3)_3^{4-}$, and possibly $(UO_2)_3(CO_3)_6^{6-}$ [Duff and Amrhein, 1996]. Within the pH range of 4.0 and 7.5 and in the absence of dissolved inorganic ligands (carbonate, fluoride, sulfate, and phosphate), the hydroxyl species UO_2OH^+, $UO_2(OH)_2^0$, and $(UO_2)_2(OH)^{2+}$ dominate U(VI) speciation [Meinrath et al., 1996]. Uranium in soil and in water forms complexes with sulfate and phosphate as well as with carbonate and hydroxide. These complexes may increase the total solubility of U.

The variety of the molecular forms in which uranium is present in the environment provides the ability of the uranium atom to form complex connections. Uranium in the uranyl form can be absorbed in the organism, for example, with phosphate or carbonate complexes [Schott et al., 2006]. All these different forms have different biological activities and, thus, also different toxicities.

2.1 Soil Characteristics Affecting Uranium and Thorium Plant Uptake

The study of U and Th transfer from soil to edible vegetation through root uptake is very important, especially considering accumulation of these radionuclides in the food chains. An understanding of the mobility of U and Th in soils and their transfer to different plants requires a detailed knowledge of U and Th interactions with soil composed of abiotic and biotic components. Despite numerous studies on U and Th content in vegetation, there is little information related to the rate of their uptake and storage by different plant species, especially in field conditions.

Concentrations of Th and U in soil (one of the main factors affecting plant uptake of the radionuclides) differ significantly depending on soil type, parent rocks, climate, relief, vegetation season (if we say about rhizosphere soil), and many other reasons. Typical concentration range of Th in soils is $2-12$ mg/kg with an average value of 6 mg/kg [Kabata-Pendias and Pendias, 2000]. The worldwide mean U concentration in noncontaminated soils ranges from 0.4 to 6.0 mg/kg [Shacklette and Boerngen, 1984]. In general, distribution of natural radionuclides in background soils follows Gaussian distribution (Fig. 1). However, in contaminated soils a highly heterogeneous distribution of these metals may be observed (Fig. 2).

Among other factors, radionuclide uptake by plants depends on soil characteristics [Pulhani et al., 2000]. Table 1 illustrates mean concentrations of U and Th in different soils. Concentrations of Th and U in natural soils may differ, for instance, depending on the soil particle size. As was reported [Frindik and Vollmer, 1999], in the range below 150 μm, total alpha activities of soil particles have been found to increase continuously with decreasing particle size. The activity increase was due to the growing share of uranium in this fine-particle range. Thorium isotopes demonstrated even a higher contribution to the activity compared to uranium isotopes.

Certain differences in bioavailability of radionuclides among soils may or may not be based on just quantitative properties of the soils [Sheppard and Evenden, 1992].

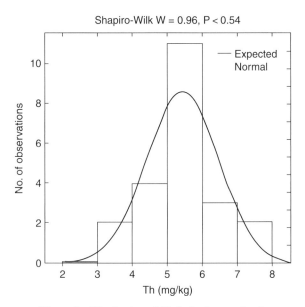

Figure 1. Distribution of Th in background soil.

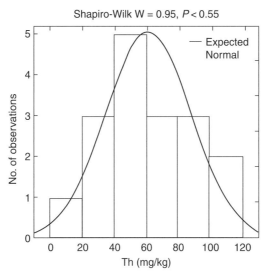

Figure 2. Distribution of Th in soil taken from a contaminated area.

**TABLE 1. Mean Concentrations of U and Th in Different Soils Taken
from Background Areas**

Type of Soil	U (mg/kg)	Th (mg/kg)	Source
Podzolic	0.12	4.8	Tynybekov and Hamby [1999]
Gray forest soil	0.12	6.0	Tynybekov and Hamby [1999]
Wetland soil	0.4	8.9	Knox et al. [2006]
Red soil	0.35	2.6	Tynybekov and Hamby [1999]
Sandy soil	0.6	5.8	Bednar et al. [2004]
Limed sand	0.6		Sheppard and Evenden [1992]
Chernozem	0.93	5.2	Tynybekov and Hamby [1999]
Ophiolitic soil	1.0	2.4	Tzortis and Tsertos [2005]
Loam	2.1		Sheppard and Evenden [1992]
Loam	2.2	5.7	Shtangeeva et al. [2005]
Sandy loam	2.2	9.6	Shtangeeva et al. [2006]
Coral sand soil	3.0		Robison et al. [2005]
Alluvial soil	3.7	5.7	Pulhani et al. [2000]
Acid sand	4.1		Sheppard and Evenden [1992]
Orthic ferralsols	4.9	14.9	Termizi Ramli et al. [2005]
Calcerous		7.6	Orvini et al. [2000]
Silt		8.5	Chaosheng [2002]
Silt mud		11.7	Chaosheng [2002]
Mud		13.4	Chaosheng [2002]
Agricultural soil	5.8		Miljevic et al. [2001]

Unfortunately, a large part of the available literature refers to the studies either in soils of high background areas or pot experiments with spiked radioactivity in nutrient solution. The experiments using nutrient solutions as a growth medium give only approximate estimates of the processes occurring in the complex soil solution under ordinary conditions and do not reflect the real situation existing in a field.

The basic premise of many radioecological assessments is the assumption that the transfer of radionuclides from soil to plants and hence animals derived food products is a positive linear relationship for a given set of ecological/agricultural conditions [Beresford and Wright, 2005]. Numerous publications have reported such a linear relationship between total radionuclide concentration in the hydroponic solution and total amount of the radionuclide in the plant roots, even no matter what is the pH of the growth medium [Ebbs et al., 1998; Shtangeeva and Ayrault, 2004; Laroche et al., 2005; Blanco Rodríguez et al., 2006].

However, plant radionuclide concentrations are not so often linearly related to soil radionuclide concentrations [Diebold and McGrath, 1985; Sheppard and Sheppard, 1985; Sheppard and Evenden, 1988; Sheppard et al., 1989]. Nonlinearity can complicate the measurement of bioavailability, because each plant and soil combination may have a unique curvilinear relationship. We may assume that it would be hardly possible to use experimental results of greenhouse tests conducted with plants grown in various nutrient solutions to predict uptake and translocation of metals by plants grown in soil. Soil and liquid media are absolutely different systems, and mechanisms of metal uptake by plants growing in nutrient solutions and in soils may be rather different. As was shown, uranium uptake by soil-grown plants is not related to simple bioavailability parameters, and only complex models considering several soil characteristics can help to predict uranium uptake [Vandenhove et al., 2007]. It is not surprising that nutrition of plants growing in hydroponic solution may be different compared with that in soil. In liquid media, trace metals, including U and Th, are already present in bioavailable forms. In soil, many factors will affect mobility of these metals and, thus, their availability to plants. In particular, soil type can greatly affect the sorption and subsequent desorption of metals. As an example, we can compare here uptake and translocation of radionuclides by plants grown in soil and in a hydroponic medium. Ramaswami et al. [2001] reported results of experiments performed simultaneously in hydroponics and in two soils (a sandy-loam soil and an organic-rich soil). They found that efficiency of uranium extraction decreased sharply from hydroponic to sandy and, especially, organic soil, indicating that soil organic matter sequestered uranium, rendering it largely unavailable for plant uptake. These results indicate that detailed description of site-specific soils must be done to screen plants for radionuclide extraction capability.

2.2 Effects of Soil Amendments

The data on plant uptake of radionuclides under natural field conditions, especially from background areas, are scarce. The soil properties such as pH, clay minerals, Ca, K, organic matter contents, and fertilizer application can strongly affect the retention, uptake, and distribution of radionuclides in plants. As a result, it may be rather difficult to predict the rates of uptake of radionuclide in a particular site.

Uranium and thorium concentrations are often higher in phosphate-rich soils. Application of certain fertilizers with higher concentrations of these radionuclides—for example, common mineral fertilizer super phosphate—can potentially result in accumulation of U and Th in food crops grown in the amended soils. Fertilizers can contain elevated levels of the uranium series, the thorium series and potassium-40 [Akhtar and Tufail, 2007]. The enormous utilization of phosphate rock and super phosphate derived from it has the potential of being an important factor in the soil contamination with these radionuclides. Both rock phosphate and super phosphate contain substantial levels of natural uranium, amounting to hundreds of ppm. It was shown [Hamamo et al., 1995] that whereas the uranium series in phosphate rock is nearly in equilibrium, in super phosphate the ^{226}Ra and its progeny are depleted by 60–70%. This is a result of the chemical processing of the rock phosphate. On the other hand, the super phosphate is much more soluble and may be expected to release radionuclides to the environment more rapidly than rock phosphate.

Cooper et al. [1995] published results of field trials on application of RMD (bauxite mining residues termed "red mud" combined with gypsum) to sandy loam soils. They found that such a soil treatment resulted in a linear increase in concentrations of radionuclides of uranium and thorium series in the soil. It is interesting that tests with application of super phosphate to these soils did not show such significant increase of radionuclide levels in the soils as was observed by the authors with RMG.

3 RADIONUCLIDES IN PLANTS

In general, accumulation of radionuclides in plants depends upon many factors, including plant species, tissue type (e.g., leaf versus fruit), soil–water–plant relationships, soil type, and the amount and chemical form of the radionuclide in the soil. Until now for many soil–crop combinations, only a few datasets could be found. The radionuclide concentrations in plants vary significantly with type of crop and the part(s) harvested. A method describing the radionuclide accumulation in a plant is called concentration ratio (CR). CR is the ratio of radionuclide concentration in a plant to concentration of the radionuclide in soil. The CR can vary throughout several orders of magnitude. Published data indicates that under ordinary conditions, concentrations of radionuclides found in plant tissues are quite low compared with those in soil. For example, CR values reported by Mortvedt [1994] ranged from 0.00148 to 0.00006.

We will not cite here concentrations of U and Th in various cultivated plant species. Unfortunately, reported values differ significantly. We may assume that natural difference in the uptake of radionuclides by plant is not the only reason for such a situation. Certain contribution may also be provided by significant differences in the methodologies of plant sampling and preparation of the plant material for elemental analysis as well as quite expectable differences in accuracy of various analytical techniques used for determination of very low Th and U concentrations in the plants. Here we will just consider some of main factors affecting U and Th uptake and transport in a plant.

3.1 Accumulation of Uranium and Thorium in Plant Roots

It has been reported in numerous publications that concentrations of U and Th in roots are much higher than in leaves and in seeds [Shtangeeva, 1993; Pulhani et al., 2000; Shtangeeva and Ayrault, 2004; Chang et al., 2005]. Typical distribution of U and Th in roots and leaves of wheat grown in Podzolic soil is demonstrated in Fig. 3.

In general, roots serve as a natural barrier preventing the transport of many trace metals, including radionuclides to upper plant parts. Moreover, the rate of radionuclide translocations from roots to shoots is probably species-dependent. It may be different for different species and even cultivars. For example, Shahandeh and Hossner [2002] reported that U concentration in roots of different plants collected from the same site was 30–50 times higher than U concentration in shoots. Among other plant species tested by the authors, sunflower and Indian mustard had the highest root U concentrations, and wheat and ryegrass had the lowest U concentrations in roots.

Uptake of U and Th is likely to be influenced by the type of the plant roots. It was reported [Sheppard et al., 1985; Apps et al., 1988] that U content in root samples was consistently higher for fine roots than for bigger roots. For each root diameter class, the root barks (including periderm and living phloem) were shown to contribute greatly to the U accumulation in roots. Since the adhering soil particles may be easily washed from the roots, incomplete removal of substrate from the root surface hardly could explain the high concentration of U in the root bark, which accounted for 19–60% of the average soil concentration. But some doubt remains

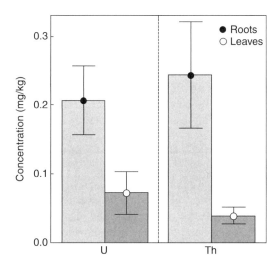

Figure 3. Mean concentrations of U and Th in roots and leaves of wheat.

on the form and exact localization of U in the root tissues. Mineral precipitates (autunite, $Ca(UO_2\text{-}PO_4)_2 \cdot 10H_2O$) have been found in the tips of plant roots [Jones et al., 1990]. It is probable that a major part of U is simply sorbed on the outer root tissues. A striking fission track image of the U distribution in root samples of trees on abandoned tailings showed that U was concentrated outside the vascular cambium forming a uranium-rich sheath [Apps et al., 1988]. This surface accumulation could result from the formation of sparingly soluble carbonates or hydroxy-carbonates as suggested by Trüby [1995] for other metals.

It is clear that uptake of U and Th by plants depends significantly on the plant species. We will demonstrate this in the example of two plants—rye and wheat—grown simultaneously under the same conditions in soil where were added small amounts of U and/or Th.

3.2 Differences in U and Th Uptake by Different Plant Species (in the example of wheat *Triticum aestivum* and Rye *Secale cereale*)

Five-day-old germinated seedlings of wheat (*Triticum aestivum* L.) and rye (*Secale cereale* L.) were transferred to large ceramic pots filled with soil. Before sowing, soil in the pots was watered with 500 ml of four different water solutions: control (ordinary water); $Th(NO_3)_4$; $(UO_2)(NO_3)_2$; mixture of nitrates of U and Th (concentrations of U and Th in the solutions were 50 mg/l). Plants and soil (from the root surface) were collected three times: within 3, 7, and 11 days after sowing.

Tables 2 and 3 present mean concentrations of elements in roots and leaves of rye and wheat. Although rye is botanically related to wheat (both belong to the tribe *Hordeae*), these two plants differ markedly in concentrations of many elements. In most cases, wheat has higher amounts of elements than rye. This is typical both for roots and for leaves of the plants. The only exception is Na; its concentration in the rye leaves is statistically significantly higher than in the leaves of wheat. In roots of rye and wheat, concentration of Na is very similar.

Concentrations of many elements (As, Co, Eu, Fe, Na, Sc, Ta, U, and Zn) in roots of both these plants are statistically significantly higher than in leaves. This indicates that roots prevent penetration of large amounts of different elements to upper plant parts. Plants have evolved highly specific mechanisms to take up, translocate, and store various elements. The uptake mechanism is selective; plants preferentially acquire some elements over others. For example, concentrations of essential plant nutrients such as K (and Rb, a chemical analogue of K) and Sr in leaves of rye and wheat are higher than in roots. This means plants use the mechanism of active transport of these elements from roots to leaves. It was a bit strange, however, that concentration of Sb in leaves of rye and wheat was higher than in roots. Sb is not an essential plant nutrient; at the moment, however, we have no explanation of this fact.

After adding U and Th to soil, concentration (total amount) of these radionuclides in the soil increased significantly. As compared to control, U concentration was 6 times higher after treatment with U and 3 times higher after treatment with U+Th. Soil Th concentration was 5 times higher after treatment with Th and 3 times higher after treatment with U+Th. We did not observe leaching of these

TABLE 2. Mean Concentrations of Elements (mg/kg) in Rye

Element	Control		+U		+Th		U+Th	
	Roots	Leaves	Roots	Leaves	Roots	Leaves	Roots	Leaves
As	0.43 ± 0.11	<0.4	0.57 ± 0.06	<0.4	0.45 ± 0.23	<0.4	0.46 ± 0.14	<0.4
Au	0.09 ± 0.02[a]	0.19 ± 0.11	0.12 ± 0.03	0.16 ± 0.10	0.16 ± 0.12	0.13 ± 0.12	0.14 ± 0.02[b]	0.19 ± 0.07
Ba	25.3 ± 3.0[c]	35.1 ± 0.5	20.5 ± 7.4	28.8 ± 12.2	33.9 ± 8.0	34.9 ± 8.9	18.8 ± 2.0[b]	38.0 ± 8.5
Br	18.9 ± 18.2[c]	4.84 ± 1.58	7.12 ± 1.84	3.15 ± 0.31	14.9 ± 10.4	3.56 ± 0.43	12.5 ± 10.7	3.33 ± 1.19
Ca (%)	0.49 ± 0.03[c]	0.78 ± 0.24	0.56 ± 0.27	0.84 ± 0.27	0.43 ± 0.05	0.78 ± 0.20	0.46 ± 0.13	0.72 ± 0.33
Ce	1.71 ± 0.49	1.72 ± 0.37	<5	1.82 ± 0.70	2.21 ± 1.69	2.10 ± 0.82	3.01 ± 1.99	1.60 ± 0.63
Co	0.31 ± 0.01[a,c]	0.08 ± 0.01[a]	0.45 ± 0.06[b]	0.09 ± 0.01	0.42 ± 0.09	0.10 ± 0.01	0.37 ± 0.08	0.08 ± 0.04
Cr	8.66 ± 2.23	6.16 ± 0.63	22.0 ± 16.7	6.88 ± 1.63	12.8 ± 8.5	6.16 ± 1.23	15.3 ± 8.4	5.76 ± 2.11
Cs	1.50 ± 1.08	1.43 ± 0.91	0.96 ± 0.10	1.26 ± 0.83	3.09 ± 1.76	0.53 ± 0.13	1.27 ± 0.65	2.07 ± 1.66
Eu	0.08 ± 0.02[a,c]	0.03 ± 0.01[a]	0.12 ± 0.01[b]	0.03 ± 0.02	0.13 ± 0.10	0.03 ± 0.03	0.35 ± 0.42	0.03 ± 0.01
Fe	610 ± 109[a,c]	317 ± 86	855 ± 160	319 ± 88	779 ± 14	304 ± 29	628 ± 179	287 ± 122
Hf	1.16 ± 0.40	0.94 ± 0.19[a]	1.70 ± 0.49	1.08 ± 0.35	1.62 ± 1.05	0.90 ± 0.29	1.38 ± 0.04	0.94 ± 0.41
K (%)	1.92 ± 0.78[c]	5.09 ± 2.16	2.45 ± 1.16	5.63 ± 2.89	2.13 ± 0.69	5.75 ± 1.72	2.00 ± 0.97	5.68 ± 1.75
La	0.59 ± 0.05	0.72 ± 0.17[a]	<1	0.77 ± 0.21	0.79 ± 0.61	0.82 ± 0.26	0.35 ± 0.23	0.67 ± 0.22

Lu	$0.02 \pm 0.01^{a,c}$	0.05 ± 0.01	0.05 ± 0.01^{b}	0.05 ± 0.02	0.04 ± 0.03	0.04 ± 0.02	0.04 ± 0.01^{b}	0.04 ± 0.01
Na (%)	0.83 ± 0.21^{c}	0.13 ± 0.02^{a}	0.82 ± 0.03	0.13 ± 0.01	1.04 ± 0.36	0.12 ± 0.01	0.94 ± 0.32	0.13 ± 0.03
Rb	22.3 ± 3.6^{c}	36.4 ± 14.0	25.6 ± 6.63	38.4 ± 17.8	23.7 ± 4.6	38.3 ± 10.4	19.6 ± 6.4	36.9 ± 8.9
Sb	0.73 ± 0.05^{c}	1.83 ± 0.97^{a}	1.20 ± 0.57	2.08 ± 1.69	1.94 ± 1.31	3.01 ± 3.27	0.84 ± 0.42	4.44 ± 3.13
Sc	$0.11 \pm 0.03^{a,c}$	0.03 ± 0.01^{a}	0.17 ± 0.06	0.04 ± 0.01	0.15 ± 0.01	0.03 ± 0.01	0.12 ± 0.03	0.03 ± 0.01
Sm	0.07 ± 0.02^{a}	0.28 ± 0.02^{a}	0.99 ± 0.38^{b}	0.11 ± 0.03	0.13 ± 0.02^{b}	0.09 ± 0.03	0.60 ± 0.35	0.07 ± 0.03
Sr	<25	16.6 ± 5.7	<36	60.5 ± 42.8	<21	80.2 ± 55.3	<21	124 ± 7
Ta	0.06 ± 0.03	<0.05	0.14 ± 0.09	<0.05	0.09 ± 0.03	<0.05	0.08 ± 0.03	<0.05
Th	0.21 ± 0.02^{a}	0.18 ± 0.01	0.32 ± 0.06^{b}	0.19 ± 0.06	9.56 ± 2.30^{b}	0.22 ± 0.06	5.68 ± 2.67^{b}	0.27 ± 0.09
U	<0.40	0.77 ± 0.37	34.2 ± 13.8^{b}	0.35 ± 0.05	1.26 ± 0.10	0.42 ± 0.08	17.5 ± 12.3	0.45 ± 0.23
Yb	0.12 ± 0.05	0.18 ± 0.03^{b}	0.20 ± 0.07	0.21 ± 0.08	0.19 ± 0.10	0.19 ± 0.06	0.14 ± 0.04	0.20 ± 0.09
Zn	171 ± 63^{c}	68.2 ± 8.6	179 ± 46	65.8 ± 3.1	181 ± 54	71.4 ± 15.4	148 ± 51	71.9 ± 12.3

[a] Differences between rye and wheat are statistically significant ($P < 0.05$).

[b] Differences between control and different treatments are statistically significant ($P < 0.05$).

[c] Differences between roots and leaves of control plants are statistically significant ($P < 0.05$).

TABLE 3. Mean Concentrations of Elements (mg/kg) in Wheat

Element	Control Roots	Control Leaves	+U Roots	+U Leaves	+Th Roots	+Th Leaves	U+Th Roots	U+Th Leaves
As	0.86 ± 0.09	<0.4	0.75 ± 0.26	<0.4	0.71 ± 0.15	<0.4	0.45 ± 0.07	<0.4
Au	0.31 ± 0.03	0.21 ± 0.16	0.19 ± 0.10	0.17 ± 0.09	0.20 ± 0.07	0.28 ± 0.11	0.29 ± 0.13	0.20 ± 0.15
Ba	27.3 ± 3.2	32.7 ± 15.1	19.7 ± 3.9	30.7 ± 9.8	29.6 ± 16.9	32.3 ± 19.6	32.2 ± 15.8	33.5 ± 13.3
Br	18.7 ± 6.5[a]	7.52 ± 1.99	10.5 ± 5.5	5.98 ± 1.46	24.6 ± 20.0	5.92 ± 1.77	18.6 ± 13.0	5.51 ± 1.77
Ca (%)	0.66 ± 0.37	0.82 ± 0.22	0.70 ± 0.04	0.88 ± 0.26	0.91 ± 0.17	0.85 ± 0.51	0.79 ± .022	0.88 ± 0.40
Ce	2.30 ± 1.17	2.42 ± 0.67	3.01 ± 2.63	2.42 ± 0.78	2.03 ± 0.66	3.23 ± 0.42	3.84 ± 1.67	2.24 ± 0.39
Co	0.47 ± 0.01[a]	0.14 ± 0.04	0.43 ± 0.04	0.13 ± 0.02	0.51 ± 0.02	0.13 ± 0.02	0.41 ± 0.04	0.14 ± 0.02
Cr	29.9 ± 23.0[a]	9.56 ± 2.68	15.8 ± 0.4	8.67 ± 2.42	22.7 ± 10.5	10.9 ± 2.9	25.7 ± 19.8	8.47 ± 1.01
Cs	1.23 ± 0.93	1.53 ± 1.09	0.82 ± 0.35	1.63 ± 1.34	1.73 ± 0.21	0.91 ± 0.07	1.98 ± 1.50	1.22 ± 0.39
Eu	0.15 ± 0.01[a]	0.05 ± 0.01	0.17 ± 0.05	0.04 ± 0.01	0.21 ± 0.08	0.07 ± 0.06	0.14 ± 0.01	0.04 ± 0.01
Fe	978 ± 9[a]	409 ± 120	625 ± 209	364 ± 87	915 ± 304	429 ± 101	614 ± 63[b]	355 ± 66
Hf	2.06 ± 0.52	1.66 ± 0.40	1.85 ± 0.40	1.59 ± 0.47	2.87 ± 1.45	1.96 ± 0.59	3.37 ± 2.68	1.37 ± 0.18
K (%)	2.19 ± 0.75[a]	6.23 ± 0.34	2.80 ± 0.98	6.47 ± 2.19	3.23 ± 1.02	6.63 ± 1.35	3.24 ± 1.22	7.39 ± 2.32
La	0.94 ± 0.20	1.22 ± 0.24	0.95 ± 0.05	1.11 ± 0.24	0.95 ± 0.30	1.39 ± 0.60	0.44 ± 0.59	1.09 ± 0.18

Lu	0.05 ± 0.01	0.08 ± 0.03	0.09 ± 0.03	0.08 ± 0.02	0.08 ± 0.04	0.10 ± 0.03	0.09 ± 0.07	0.07 ± 0.01
Na (%)	0.80 ± 0.14[a]	0.06 ± 0.01	0.54 ± 0.13	0.06 ± 0.01	0.91 ± 0.42	0.06 ± 0.01	0.83 ± 0.15	0.05 ± 0.01
Rb	28.8 ± 3.5[a]	40.1 ± 3.7	30.6 ± 11.1	40.8 ± 18.8	37.2 ± 12.4	40.3 ± 10.1	33.9 ± 14.5	45.9 ± 13.8
Sb	0.82 ± 0.45[a]	4.13 ± 0.62	0.85 ± 0.34	4.64 ± 3.50	1.99 ± 1.29	2.68 ± 2.18	1.51 ± 0.48	1.17 ± 0.60
Sc	0.20 ± 0.01[a]	0.05 ± 0.01	0.13 ± 0.05	0.04 ± 0.01	0.18 ± 0.08	0.08 ± 0.03	0.11 ± 0.05	0.04 ± 0.01
Sm	0.14 ± 0.02	0.15 ± 0.02	0.76 ± 0.10[b]	0.15 ± 0.03	0.16 ± 0.05	0.19 ± 0.06	0.56 ± 0.28	0.13 ± 0.01
Sr	<40	15.0 ± 5.0	<32	27.6 ± 15.2	<19	28.9 ± 12.1	<25	24.7 ± 10.0
Ta	0.10 ± 0.01	<0.05	0.09 ± 0.02	<0.05	0.13 ± 0.06	<0.05	0.14 ± 0.10	<0.05
Th	0.32 ± 0.04	0.28 ± 0.09	0.26 ± 0.04	0.24 ± 0.09	8.59 ± 3.90[b]	0.79 ± 0.46	9.92 ± 12.1	0.32 ± 0.04
U	0.91 ± 0.04[a]	0.25 ± 0.10	23.2 ± 3.9[b]	0.32 ± 0.13	1.06 ± 0.47	0.32 ± 0.13	11.6 ± 2.9[b]	0.21 ± 0.06
Yb	0.25 ± 0.07	0.35 ± 0.07	0.37 ± 0.30	0.30 ± 0.09	0.29 ± 0.15	0.44 ± 0.13	0.39 ± 0.38	0.29 ± 0.06
Zn	170 ± 2[a]	87.6 ± 11.9	195 ± 87	83.2 ± 8.8	163 ± 17	82.3 ± 4.3	147 ± 50	76.9 ± 0.8

[a]Differences between roots and leaves of control plants are statistically significant ($P < 0.05$).
[b]Differences between control and different treatments are statistically significant ($P < 0.05$).

metals to deeper soil layers. Concentrations of Th and U in the soil from the bottom
of all pots were approximately the same, regardless of the treatments.

Rye and wheat grown in the radionuclide-enriched soils demonstrated significant
increase in concentrations of Th and especially U in roots. After addition of mixture
of U and Th to soil, concentrations of these metals in the plant roots were also
increased. However, this increase was not so marked as it was in the case where U
or Th was added to soil as a single element, probably because of competition
between these metals during uptake process. It is significant that concentrations of
U and Th in leaves remained unchanged. This means that transfer of these radionuclides from soil to leaves via root uptake was minimal.

Figure 4 shows distribution of U in control soil, in soil where was added U and in
roots of rye and wheat grown in the soils. In roots of both plants grown in clean

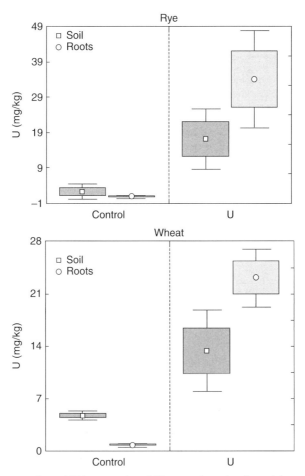

Figure 4. Concentration of U in control and U-contaminated soils and in roots of rye and
wheat grown in the soils.

(control) soil, concentration of U was lower than U content in the soil. However, in roots of the plants grown in U-enriched soil, U concentration was much higher than that in roots of the control plants, and it was higher than U content in the rhizosphere soil. Another situation was observed in the plants grown in Th-enriched soil (Fig. 5). In this case, Th concentration in roots was always lower than that in soil. This indicates that Th is less available to plant uptake than U.

Figure 6 shows ratios of U and Th concentrations in roots of wheat and rye to concentrations of these elements in the soil where the plants were grown. The ratios are higher for U than for Th for all the treatments and for both plant species. It was reported that mobility of uranium in soil is higher than thorium mobility, regardless of the soil type [Titaeva, 1992; Morton et al., 2001]. Th^{4+} is readily soluble, but at the same time it may be quickly adsorbed or precipitated as hydrolysate. It was suggested [Bednar et al., 2004] that Th could migrate in soil differently than U: either as a negatively charged particle or as anionic complex with organic matter. Since soil metal

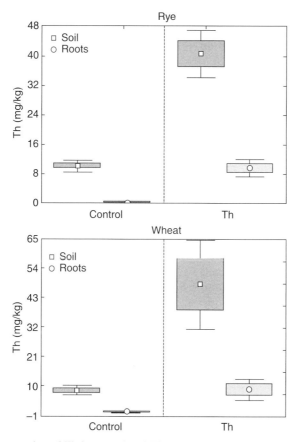

Figure 5. Concentration of Th in control and Th-contaminated soils and in roots of rye and wheat grown in the soils.

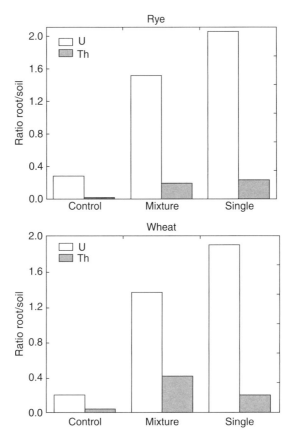

Figure 6. Ratios of U and Th concentrations in roots of rye and wheat grown in control and contaminated soils to U and Th concentrations in the control soil and in the soil where U or Th (single) and U+Th (mixture) were added.

mobility correlates with availability of the metal to plant, we can expect certain differences in the bioavailability of U and Th.

$U_{[root]}/U_{[soil]}$ ratios in the experiments with U and U+Th are higher than 1. One might expect that such a significant uptake of U by plants could result in a decrease of U concentration in the rhizosphere soil. As an example, Fig. 7 illustrates dynamics of U in the soil where rye was grown (control, after adding U, and after adding U+Th). With time, soil U concentration slightly decreased in the experiment with mixture of U and Th and decreased rather significantly in the experiment where only U was added to soil. Such a decrease of soil U concentration resulted from uptake of U by the plant roots. The decrease was lower in the experiment with U+Th because plants could uptake less U than was observed in the experiment with U alone.

Rye could uptake more U than wheat. This was typical for all treatments (U, Th, and U+Th). As a result, there was no such a significant decrease of U content in the rhizosphere soil of wheat as we observed in the rhizospere soil of rye. The ratios of

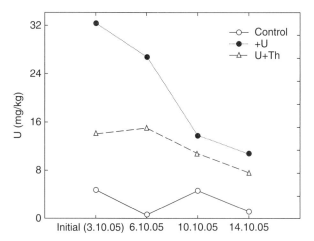

Figure 7. Dynamics of U in control soil and in soil where U and U+Th were added.

radionuclide concentrations in roots of the plants grown in contaminated soil to those in roots of the control plants were ~45 (rye) and ~26 (wheat). Despite such a significant increase of U and Th concentrations in the plants, U content in the contaminated soils decreased only two times. Soil Th concentration remained unchanged. Moreover, U was not transferred from roots to leaves; therefore, it was not actually removed from the contaminated soil.

Main conclusions from these observations are the following: (1) Biomass of the young seedlings was probably too small to remove large amounts of radionuclides from the contaminated soil; (2) due to higher mobility of U in soil and, as a result, higher level of bioavailability of this metal to plants, U may be more easily removed from soil than Th; and (3) rye can uptake more U than wheat, and thus it may be a promising plant species for U phytoextraction from contaminated soils.

3.3 Effects of U and Th Bioaccumulation on Distribution of Other Elements in Rye and Wheat

Bioaccumulation of any metal in a plant can result in certain variations in concentrations of some other elements in the plant. In our experiments, roots of wheat suffered less than roots of rye. Compared to control, in roots of wheat grown in U-contaminated soil, concentration of Sm was statistically significantly higher, and in roots of wheat grown in soil where were added U+Th, concentration of Fe was lower ($P < 0.01$). In roots of rye, more variations in element concentrations were observed after the treatment of soil with U. In this case, concentrations of Co, Eu, Lu, Sm, and Th were statistically significantly higher than those in roots of the control plants. Concentration of Sm in roots of rye grown in Th-enriched soil was higher ($P < 0.05$) than Sm content in roots of the control plants. In roots of rye grown in the soil treated with U+Th, concentrations of Au and Lu increased and Ba content decreased. Elemental composition of leaves of rye and wheat seedlings

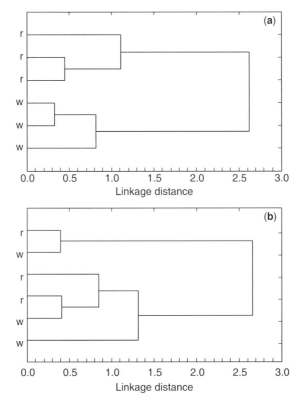

Figure 8. Cluster analysis (Ward's method) of leaves of rye (r) and wheat (w) grown in control soil (**a**) and in soil enriched with U (**b**).

remained rather stable for all the treatments (as we remember, there was no increase of U and Th concentrations in leaves of the plants grown in contaminated soils).

Figure 8 illustrates results of cluster analysis performed on the basis of element concentrations in leaves of rye and wheat (control plants and plants grown in U-contaminated soil). Leaves of the control plants are well-separated into two groups: rye and wheat. We could expect such a separation; it was due to differences in concentrations of several elements in leaves of wheat and rye (Table 2). However, there is no such a good separation of the plants grown in contaminated soil. Considering the fact that concentrations of elements in leaves of the plant species grown in control and contaminated soils are similar, these variations, probably, may be explained by certain changes in the relationships between different elements in the plants grown in contaminated soils. Absolutely the same situation is observed in the plant roots.

3.4   Relationships Between U and Th in Soils and in Different Plant Parts

Figures 9 and 10 show correlations between U (Fig. 9) and Th (Fig. 10) concentrations in roots of rye and wheat and in soil where the plants were grown.

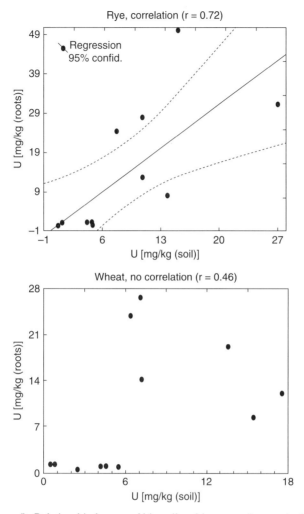

Figure 9. Relationship between U in soil and in roots of rye and wheat.

Although rye and wheat were grown and harvested simultaneously, we can see that both U and Th were more easily transferred from soil to roots of rye as compared to wheat (correlation between U and Th concentrations in soil and in roots of rye was statistically significant, while no correlation between concentrations of these metals in soil and in roots of wheat was found).

The relationships between U and Th may differ significantly, depending on soil type. If we compare relationships between U and Th in loam soil with total carbon concentration 8.4% and in plants grown in the soil, we will see that the correlation between these radionuclides in soil and in the plant roots is statistically significant and positive (Fig. 11). On the other hand, there is no correlation between U and Th in sandy loam soil with a lower amount of total carbon (3.6%) and U and Th

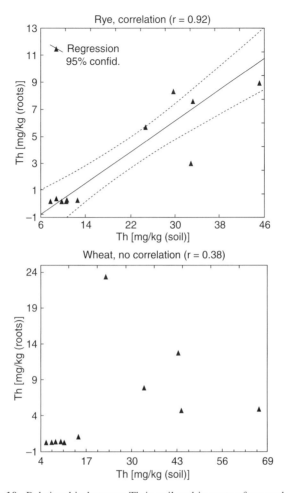

Figure 10. Relationship between Th in soil and in roots of rye and wheat.

in roots of rye and wheat grown in the soil (Fig. 12). It is interesting that there is also no correlation between U and Th in leaves of the plants grown in both soils (Fig. 13).

U and Th have similar chemical properties. But we may assume that behavior of these metals in soil and in plants (more exactly, in different plant parts) may be rather different, thus suggesting that there are additional factors influencing biochemistry of these metals in soil and in plants.

3.5   Phytotoxicity of U and Th

Little is known about phytotoxicity of thorium and uranium. The chemical toxicity of actinides may be similar to that of other metals. However, the ionizing radiation associated with radioactive decay can result in additional toxic effects on a plant.

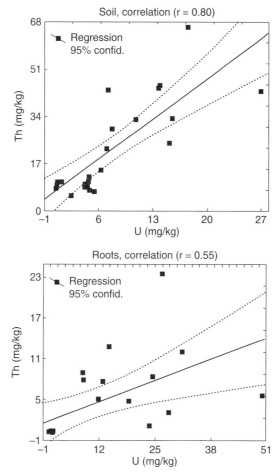

Figure 11. Relationship between U and Th in loam soil and in roots of the plants grown in the soil.

Natural concentrations of U and Th in plants hardly can pose any problem for the vegetation [Bowen, 1966]. The effects may certainly be significant when the radio-nuclide concentration exceeds micromolar levels in the biomass. In this case the higher Th and U concentrations can affect certain biochemical and physiological processes in the plants.

A review of the literature on the toxicological effects of U (and Th, Pu, etc.) generally leads back to reviews on metal toxicity and chemistry [Paquet et al., 1998]. There are many contradictory publications in the literature on phytotoxicity of natural radionuclides to plants [Sheppard and Evenden, 1992]. It was reported that the same concentration of a radionuclide may be harmful in one soil but may have no impact in another soil [Marquenie, 1985]. The toxicity effects may be rather different for different plant species. In fact, the effects are actually species-dependent.

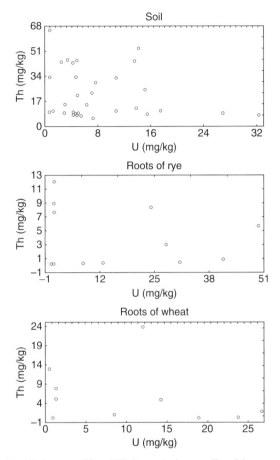

Figure 12. Relationship between U and Th in sandy loam soil and in roots of rye and wheat grown in the soil.

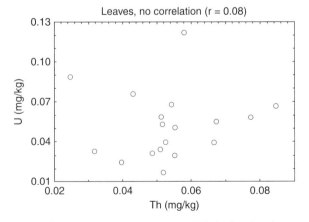

Figure 13. Relationship between U and Th in the plant leaves.

For example, Gulati et al. [1980] reported that the highest yield of wheat was obtained at 3.0 mg U per kilogram of soil and that tomato yield decreased continuously when the U level in the soil was increased from 1 to 6 mg/kg. Levels as low as 5 mg U per kilogram of soil, well within the normal background range, have been cited as toxic, whereas other studies reported no toxicity at levels 100- to 1000-fold higher. The authors also studied the effects of irrigation on uptake of U. It was found that U uptake by crops increased significantly with increase in quantity of applied water. Thus, it may be possible to regulate the plant uptake of U by controlling application of water.

We may assume (and will try to show this in more detail below) that U and Th, similarly to other trace metals, may be both toxic and—within certain concentration range—favorable for plants and humans. These effects depend on various factors.

To assess the impacts of U and Th on soil biota, plant nutrition, and some physiological characteristics of plants, the following field experiment was carried out. Four small (1 m × 1.5 m) plots were used for the trial. The soil was classified as Ferric Podzol [FAO UNESCO, 1988] with a loam texture. Plants (couch-grass and plantain) were watered with solutions of $Th(NO_3)_4$ (test 1), $(UO_2)(NO_3)_2$ (test 2), and a mixture of nitrates of U and Th (test 3). Concentrations of U and Th in the solutions were 60 mg/kg. Each site was watered with 1 liter of the solution. The control (test 4) plants were watered with the same amount of ordinary water. After two days, the plants and soil from the root surface were collected during the day (from 6:00 to 22:00, every four hours). The plant samples were washed carefully just after sampling while they keep turgor to remove dust and the smallest particles of soil from the surface of leaves and especially from roots.

The mean concentrations of elements in soils and in different parts of couch-grass and plantain are shown in Tables 4–6. Concentrations of U and Th in control plants were low. The treatments significantly affected radionuclide concentrations in soils and in plants. As we expected, concentrations of U and Th in the rhizosphere soil of the plants treated with U, Th, and U+Th were higher than in the control soil.

The treatments also affected concentrations of some other elements in the soils. The most significant variations were observed as a result of Th treatment. It is interesting that variations of element concentrations in soil were well-correlated with variations of the element concentrations in couch-grass. For example, content of Zn in the rhizosphere soil of U- and U+Th-treated plants decreased, and concentrations of Zn in roots of couch-grass grown in the soil treated with U and Th increased. Thus, we assume that U and Th treatments could transfer Zn to more mobile form and stimulate uptake of Zn by couch-grass. A decrease of soil Rb content (after treatment with Th) was followed by an increase of Rb concentration in couch-grass roots. We also observed a significant increase of K concentration both in roots and in leaves of couch-grass resulted from the treatment with Th. Interestingly, there was no such an effect of Th on K content in plantain (Fig. 14). We hardly could expect a statistically significant decrease of the total amount of K in soil since the amount of K taken by roots of couch-grass was relatively small (considering the differences between the biomass of roots and the mass of soil). Nevertheless, taking into account that Rb is a chemical analogue of K, we may suggest that Th treatment stimulated transfer of certain part of soil K to a more bioavailable form. This means that thorium

TABLE 4. Mean Concentrations of Elements (mg/kg) in Soil

Element	Control	+U	+Th	U+Th
Na (%)	1.2 ± 0.1	1.2 ± 0.1	1.2 ± 0.1	1.1 ± 0.1
K (%)	2.4 ± 0.4	2.5 ± 0.2	2.4 ± 0.4	2.2 ± 0.4
Ca (%)	1.1 ± 0.2	1.0 ± 0.2	0.9 ± 0.3	1.1 ± 0.4
Sc	8.1 ± 0.7	8.1 ± 0.8	7.0 ± 0.7^a	7.2 ± 1.2
Cr	53 ± 5	52 ± 8	50 ± 8	49 ± 7
Fe (%)	2.4 ± 0.2	2.3 ± 0.3	2.1 ± 0.2^a	2.2 ± 0.4
Co	8.8 ± 0.6	9.0 ± 0.9	8.0 ± 0.7^a	8.2 ± 1.5
Zn	135 ± 20	106 ± 21^a	136 ± 19	114 ± 18^a
As	2.8 ± 0.3	2.8 ± 0.5	2.9 ± 0.5	2.7 ± 0.7
Br	5.0 ± 0.5	5.2 ± 1.2	6.1 ± 0.6^a	5.1 ± 1.2
Rb	111 ± 9	115 ± 8	99 ± 9^a	104 ± 12
Sr	125 ± 55	143 ± 60	152 ± 63	124 ± 37
Sb	1.1 ± 0.4	1.0 ± 0.6	1.1 ± 0.3	0.8 ± 0.1^a
Cs	2.8 ± 0.2	2.7 ± 0.3	2.1 ± 0.4^a	2.4 ± 0.6
Ba	604 ± 83	617 ± 56	585 ± 42	573 ± 38
La	32 ± 3	32 ± 5	30 ± 3	40 ± 20
Sm	5.1 ± 0.5	5.2 ± 0.8	4.7 ± 0.5	5.6 ± 1.8
Eu	1.1 ± 0.2	1.0 ± 0.1	1.0 ± 0.1	1.0 ± 0.1
Yb	2.0 ± 0.5	2.0 ± 0.2	1.8 ± 0.3	1.9 ± 0.3
Lu	0.4 ± 0.1	0.39 ± 0.05	0.35 ± 0.05	0.38 ± 0.07
Hf	5.7 ± 1.7	7.1 ± 2.0	6.2 ± 2.5	6.0 ± 1.9
Ta	0.6 ± 0.2	0.7 ± 0.2	0.6 ± 0.1	0.7 ± 0.3
Th	9.6 ± 3.1	8.5 ± 1.7	28 ± 20^a	19 ± 8^a
U	2.2 ± 0.3	25 ± 21^a	2.2 ± 0.4	12 ± 7^a

[a]Differences between control and treated with U, Th, and U+Th samples are statistically significant at $P < 0.05$.

could provide favorable effects on the plant nutrition, thereby stimulating uptake of potassium, which is an essential plant nutrient.

Couch-grass and plantain belong to two different orders: *Graminaceae* and *Plantaginaceae*, respectively. Comparison of plants grown in control plots showed that concentrations of K and Ca (and its chemical analogue Sr) in roots of couch-grass were higher than those in roots of plantain, and concentrations of Na and Zn were significantly lower ($P < 0.05$). Concentrations of K, Ca, Sr, and Ba in leaves of couch-grass were also higher ($P < 0.05$) than those in leaves of plantain.

Cluster analysis of the plant samples performed on the basis of element concentrations in roots and leaves of the plants treated with the radionuclides showed that roots of couch-grass treated with U, Th, and U+Th formed separated groups (Fig. 15a). Roots of the plants treated with Th and U+Th were closer to each other and well-separated from roots of the plants treated with U. Such a tendency was also observed in the leaf samples of couch-grass (Fig. 15b). We may assume that this resulted from more significant effects of U on the plant as compared to the effects of Th. Besides, simultaneous addition of U and Th could reduce the effects,

TABLE 5. Mean Concentrations of Elements (mg/kg) in Roots and Leaves of Wheat-Grass

Element	Control Roots	Control Leaves	+U Roots	+U Leaves	+Th Roots	+Th Leaves	U+Th Roots	U+Th Leaves
Na	2460 ± 390	612 ± 194	1820 ± 200[a]	550 ± 152	2050 ± 370	514 ± 65	2750 ± 350	614 ± 442
K (%)	1.1 ± 0.4	2.8 ± 0.7	1.3 ± 0.2	3.0 ± 0.2	2.3 ± 0.3[a]	4.0 ± 0.4[a]	1.2 ± 0.4	3.6 ± 0.03
Ca (%)	0.42 ± 0.10	0.47 ± 0.06	0.40 ± 0.12	0.54 ± 0.08	0.39 ± 0.10	0.40 ± 0.06	0.38 ± 0.14	0.48 ± 0.09
Sc	0.34 ± 0.01	0.06 ± 0.01	0.52 ± 0.15[a]	0.06 ± 0.02	0.37 ± 0.16	0.04 ± 0.01[a]	0.23 ± 0.11	0.06 ± 0.01
Cr	8.0 ± 2.3	3.5 ± 0.6	7.4 ± 2.1	4.2 ± 1.0	9.9 ± 2.8	3.1 ± 0.8	9.1 ± 5.7	2.8 ± 0.3
Fe	1700 ± 237	307 ± 28	1900 ± 430	332 ± 80	1330 ± 450	226 ± 46[a]	965 ± 313[a]	313 ± 49
Co	0.96 ± 0.27	0.19 ± 0.07	1.4 ± 0.3[a]	0.19 ± 0.06	0.78 ± 0.25	0.08 ± 0.02[a]	0.76 ± 0.23	0.13 ± 0.03
Zn	80 ± 10	49 ± 3	103 ± 13[a]	49 ± 8	101 ± 17[a]	50 ± 8	78 ± 17	58 ± 12
Br	3.8 ± 0.9	1.4 ± 1.3	4.1 ± 0.9	1.3 ± 0.4	5.0 ± 1.2	1.1 ± 0.6	18 ± 30	1.2 ± 0.6
Rb	4.4 ± 0.8	4.1 ± 2.4	5.7 ± 1.4	2.9 ± 0.9	6.7 ± 1.0[a]	2.9 ± 0.4	4.2 ± 1.3	4.7 ± 0.9
Sr	16 ± 2	11 ± 4	10 ± 1	11 ± 4	13 ± 6	13 ± 3	13 ± 3	17 ± 11
Sb	0.42 ± 0.11	0.15 ± 0.02	0.26 ± 0.06[a]	0.25 ± 0.08	0.31 ± 0.08	0.16 ± 0.03	0.80 ± 0.69	0.19 ± 0.06
Cs	0.67 ± 0.30	0.09 ± 0.02	0.37 ± 0.19	0.11 ± 0.08	0.47 ± 0.15	0.08 ± 0.02	0.41 ± 0.25	0.07 ± 0.02
Ba	30 ± 11	14 ± 3	41 ± 12	14 ± 5	33 ± 13	22 ± 6[a]	21 ± 8	12 ± 1
La	1.7 ± 0.2	0.59 ± 0.14	3.1 ± 1.9	0.54 ± 0.14	1.5 ± 0.7	0.48 ± 0.09	2.4 ± 0.9	0.43 ± 0.07
Sm	0.27 ± 0.06	0.07 ± 0.01	0.35 ± 0.08	0.05 ± 0.02[a]	0.23 ± 0.10	0.07 ± 0.02	0.30 ± 0.12	0.06 ± 0.01[a]
Eu[b]	220 ± 108	20 ± 5	70 ± 23[a]	20 ± 7	160 ± 106	20 ± 3	80 ± 28[a]	20 ± 7
Yb[b]	0.18 ± 0.08	0.08 ± 0.02	0.14 ± 0.06	0.12 ± 0.03	0.12 ± 0.07	0.09 ± 0.02	0.10 ± 0.05	0.08 ± 0.02
Lu[b]	40 ± 16	20 ± 6	10 ± 11[a]	30 ± 7	30 ± 11	20 ± 4	30 ± 5	20 ± 3
Hf	0.60 ± 0.04	0.53 ± 0.10	0.57 ± 0.06	0.61 ± 0.13	0.80 ± 0.47	0.52 ± 0.14	0.67 ± 0.16	0.38 ± 0.06[a]
Ta	0.09 ± 0.03	0.03 ± 0.01	0.05 ± 0.02	0.02 ± 0.01	0.07 ± 0.02	0.03 ± 0.01	0.05 ± 0.02	0.02 ± 0.01
Th	0.43 ± 0.04	0.14 ± 0.01	0.8 ± 0.4	0.19 ± 0.05	13 ± 6[a]	8.2 ± 7.8[a]	1.8 ± 0.8[a]	8.2 ± 3.8[a]
U	0.16 ± 0.19	0.14 ± 0.08	162 ± 96[a]	29 ± 11[a]	0.31 ± 0.20	0.13 ± 0.06	3.2 ± 2.7[a]	4.9 ± 1.7[a]

[a]Differences between control and treated with U, Th, and U+Th plants are statistically significant at $P < 0.05$.
[b]Concentrations of the elements are shown in μg/kg.

319

TABLE 6. Mean Concentrations of Elements (mg/kg) in Roots and Leaves of Plantain

Element	Control		+U		+Th		U+Th	
	Roots	Leaves	Roots	Leaves	Roots	Leaves	Roots	Leaves
Na	1300 ± 220	953 ± 475	1670 ± 250[a]	535 ± 233	1220 ± 220	582 ± 191	1630 ± 570	488 ± 255
K (%)	5.4 ± 0.6	4.0 ± 0.6	6.1 ± 0.6	3.6 ± 0.6	5.9 ± 0.4	3.8 ± 0.5	6.3 ± 1.3	3.5 ± 1.3
Ca (%)	0.66 ± 0.10	1.6 ± 0.2	0.60 ± 0.09	1.7 ± 0.3	0.67 ± 0.04	2.2 ± 0.6	0.66 ± 0.03	1.9 ± 0.4
Sc	0.36 ± 0.08	0.13 ± 0.07	0.24 ± 0.09	0.09 ± 0.02	0.30 ± 0.07	0.08 ± 0.04	0.35 ± 0.12	0.11 ± 0.02
Cr	5.8 ± 1.2	5.3 ± 0.9	7.7 ± 3.4	5.0 ± 0.7	5.6 ± 1.4	4.7 ± 0.5	9.5 ± 3.6	5.0 ± 2.4
Fe	1870 ± 840	562 ± 239	1140 ± 270	436 ± 52	1030 ± 210	397 ± 135	1170 ± 340	474 ± 93
Co	0.77 ± 0.10	0.39 ± 0.13	0.65 ± 0.11	0.21 ± 0.04[a]	1.1 ± 0.4	0.38 ± 0.14	0.87 ± 0.22	0.28 ± 0.11
Zn	48 ± 5	49 ± 3	60 ± 14	55 ± 7	72 ± 26	61 ± 10[a]	54 ± 2	60 ± 6[a]
Br	4.3 ± 3.1	4.6 ± 6.1	1.9 ± 0.3	1.3 ± 0.4	2.4 ± 1.3	2.2 ± 0.7	2.4 ± 1.1	1.2 ± 0.4
Rb	6.7 ± 2.8	3.9 ± 2.3	4.9 ± 0.9	2.8 ± 0.7	6.6 ± 0.8	3.3 ± 0.6	9.1 ± 2.4	4.4 ± 1.3
Sr	42 ± 7	40 ± 8	67 ± 17[a]	42 ± 7	58 ± 15	58 ± 19	47 ± 19	59 ± 16[a]
Sb	0.23 ± 0.04	0.26 ± 0.05	0.24 ± 0.05	0.29 ± 0.23	0.22 ± 0.03	0.57 ± 0.80	0.38 ± 0.16	0.22 ± 0.04
Cs	0.26 ± 0.06	0.09 ± 0.07	0.20 ± 0.08	0.06 ± 0.02	0.27 ± 0.14	0.09 ± 0.01	0.42 ± 0.24	0.07 ± 0.01
Ba	56 ± 18	51 ± 9	51 ± 22	56 ± 6	85 ± 23	69 ± 34	51 ± 4	35 ± 25
La	1.6 ± 0.3	1.1 ± 0.5	1.2 ± 0.3	0.99 ± 0.25	1.5 ± 0.3	1.0 ± 0.3	1.6 ± 0.4	0.95 ± 0.40
Sm	0.20 ± 0.07	0.15 ± 0.07	0.15 ± 0.05	0.12 ± 0.03	0.20 ± 0.04	0.21 ± 0.15	0.22 ± 0.07	0.13 ± 0.05
Eu[b]	60 ± 14	30 ± 7	19 ± 18	30 ± 3	80 ± 28	30 ± 4	22 ± 19	30 ± 1
Yb[b]	0.06 ± 0.02	0.14 ± 0.06	0.07 ± 0.02	0.14 ± 0.05	0.10 ± 0.04	0.12 ± 0.06	0.20 ± 0.11[a]	0.09 ± 0.07
Lu[b]	20 ± 2	40 ± 18	20 ± 12	40 ± 8	20 ± 8	30 ± 8	40 ± 20	40 ± 20
Hf	0.56 ± 0.21	0.76 ± 0.15	0.63 ± 0.11	0.82 ± 0.14	0.59 ± 0.16	0.70 ± 0.11	0.81 ± 0.12	0.80 ± 0.44
Ta	0.04 ± 0.01	0.04 ± 0.02	0.06 ± 0.03	0.04 ± 0.01	0.08 ± 0.09	0.03 ± 0.01	0.07 ± 0.02[a]	0.07 ± 0.07
Th	0.38 ± 0.08	0.25 ± 0.13	0.27 ± 0.09	0.27 ± 0.15	4.4 ± 7.9	159 ± 193	3.2 ± 4.5	44 ± 33[a]
U	0.11 ± 0.08	0.09 ± 0.01	35 ± 30[a]	8.6 ± 7.7[a]	0.22 ± 0.15	0.68 ± 0.74	5.8 ± 10.8	23 ± 17[a]

[a]Differences between control and treated with U, Th, and U+Th plants are statistically significant at $P < 0.05$.
[b]Concentrations of the elements are shown in μg/kg.

320

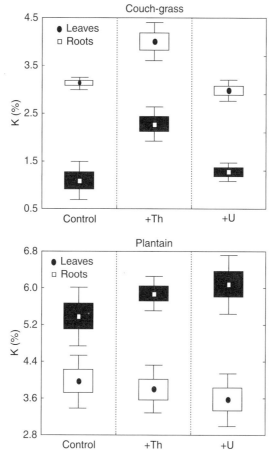

Figure 14. Potassium concentration in roots and leaves of couch-grass and plantain grown in control soils and in soils enriched with Th and U.

probably owing to competition between these two metals during uptake process. In plantain, the differences between different treatments were not so clear (Fig. 16).

Comparison of the data presented in Tables 5 and 6 shows that couch-grass probably is more affected by the treatments than plantain. At least the number of statistically significant variations in macro- and micronutrient concentrations in roots and leaves of couch-grass resulted from the treatments was higher than that in plantain. It seems likely that plantain was more tolerant to the treatments than couch-grass.

3.6 Effects of U and Th on Leaf Chlorophyll Content and the Rhizosphere Microorganisms

As was reported, some physiological characteristics of plants, including such an important parameter as leaf chlorophyll content, may be negatively affected by radio-nuclides [Aery and Jain, 1997; Jain and Aery, 1997]. In our experiment, after the

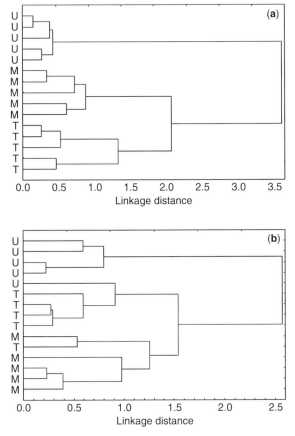

Figure 15. Cluster analysis (Ward's method) of element concentrations in roots (**a**) and leaves (**b**) of couch-grass. U, T, and M: The plants were treated with U, Th, and mixture of U and Th, respectively.

treatments with U and Th, concentration of chlorophyll in leaves decreased compared to that in leaves of the control plants (Fig. 17). The decrease was most significant in the plants treated with U (the analyses were performed only for plantain). One mechanism of U toxicity may relate to the disruption of the first step in glycolysis by uranyl displacing Mg in the enzyme hexokinase [Van Horn and Huang, 2006]. It is known that Mg is an important component of chlorophyll molecule [Nikiteshen, 2003].

Microbial biomass that represents the soil fraction responsible for the energy and nutrient cycling can influence significantly the plant nutrition and the yield of the plants. Microorganisms appear to be very sensitive and predictive tools in soil health monitoring. It was reported that microorganisms play an important role in regulating the mobility of U in soil [Ticknor, 1994; Fowle et al., 2000; Francis et al., 2004; Ohnuki et al., 2005]. It is known that microorganisms produce into soil extracellular compounds abundant with proteins [Wright and Upadhyaya, 1996]. Measurements of the amounts of the rhizosphere proteins are supposed to

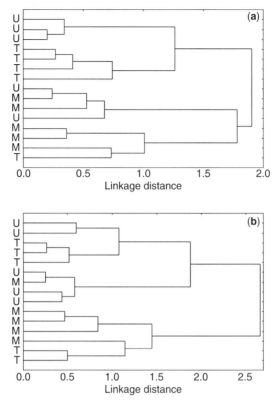

Figure 16. Cluster analysis (Ward's method) of element concentrations in roots (**a**) and leaves (**b**) of plantain. U, T, and M: The plants were treated with U, Th, and mixture of U and Th, respectively.

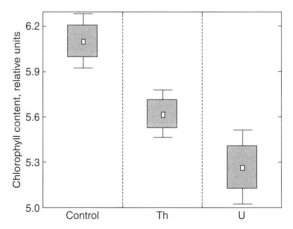

Figure 17. Chlorophyll content in leaves of plantain grown in control soil and in soil enriched with Th and U.

be accurate indirect method to assess soil microbial biomass because of a relatively high protein content in the microbial cells [Baatn, 1998]. Our results showed that total amount of proteins in the rhizosphere of the plants treated with U and Th increased (Fig. 18). This indicates that addition of U and Th to soil can enhance growth of the rhizosphere microorganisms. Furthermore, it was found that the higher the amount of proteins observed in the rhizosphere of plantain, the higher the amount of U taken by the plant roots and translocated to leaves. On the other hand, there was no such correlation registered for couch-grass.

In most previous experiments the negative effects of radionuclides on plant nutrition and growth of soil microorganisms were observed. We may assume that this inconsistence between published materials and our experimental results might be explained by the following reasons:

- The plants could release into the rhizosphere-specific organic compounds that might stimulate the microbial growth.

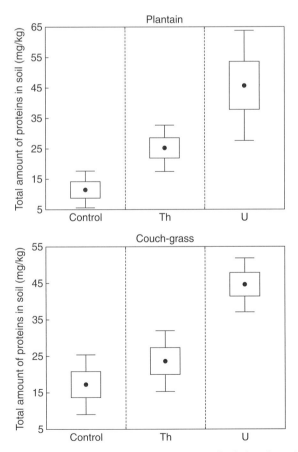

Figure 18. Total amount of proteins in the rhizosphere soil of plantain and couch-grass.

- As a result of U poisoning, part of the microorganisms usually presented in the rhizosphere died due to the induced stress to the microbes. In this situation the rate of reproduction of the microorganisms, which is normally suppressed in the rhizosphere soil, might increase.

Based on the results of this experiment, we cannot say for a sure whether U and Th actually contributed significantly to the variations or part of the variations was provided by the nitrate component of the U and Th compounds. However, we can conclude that both Th and U treatments induced significant changes in plant nutrition and physiological activity of the plants and soil biota. Numerous studies describing uptake of metals by plants often completely ignore the impact of soil microbes on the process of the metal uptake. Meanwhile, soil microorganisms may be effective mediators of mobilization and immobilization of the metals in the rhizosphere. The role of bacteria in the rhizosphere chemistry is particularly interesting. The root-colonizing bacteria excrete their own exudates into the surrounding soil. These organic compounds can catalyze specific oxidation–reduction reactions that will change mobility of certain metals [Wielinga et al., 1999]. The microorganisms are able to facilitate movement of the metals that would otherwise be unavailable to the plants [Domelly and Fletcher, 1994]. In some cases the bacteria are perhaps more proficient than root exudates at solubilization and absorbing of some metals and radionuclides [Jackson, 1993].

Although now a lot of data are available both on ion uptake by different plant species and on attraction of radionuclides to bacterial cell walls, the system "soil–bacteria–plant" as a network is not fully understood. Besides, the rhizosphere colonization is specific to certain plant species. We may suggest that more information may be obtained if we use microbial pretreatment of seeds instead of addition of microorganisms to soil. Such an inoculation of seeds with microorganisms may be more effective for enhancement of metal bioavailability than microbiological treatment of soil. Seed germination is an important stage of the plant growth, and it may have a crucial significance for further life of the plant.

Figure 19 illustrates uptake and translocation of U and Th by wheat seedlings infected with a culture of phosphate-mobilizing bacteria *Cellulomonas sp.*32 SPBTI. It is important that not only uptake of Th and U by the plant roots was increased after the seed treatment with *Cellulomonas*, but also these metals were more easily translocated from roots to leaves. This indicates that exudates of *Cellulomonas* could assist in transferring U and Th to more available to plants forms.

3.7   Temporal Variations of U and Th in Plants

It is well known that element concentrations in the plant tissues can vary with time, for example, during the vegetation season [Savari and Lockwood, 1991; Myung and Thornton, 1997; Otero and Macias, 2002]. Meanwhile, we can also expect certain variations in the plant element concentrations over shorter time (days or even hours). This assumption is based on the circadian rhythms of the plant development [Carter et al., 1991; Behrenfeld et al., 2004]. There are publications reporting that

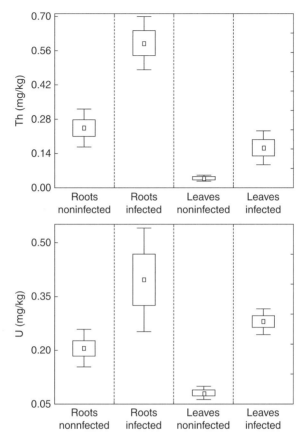

Figure 19. Th and U in roots and leaves of wheat seedlings noninfected and infected with bacteria.

uptake of K by plants is regulated by light [Becker et al., 1989; Kim et al., 1992; Suh et al., 2000] and the circadian clock [Kim et al., 1993]. We may assume that not only concentration of K, but also concentrations of other elements in plants, are controlled by light and the biological clock, and these variations are species-specific.

Figure 20 shows diurnal dynamics of U and Th concentrations in leaves of couch-grass grown in soil enriched with U and Th. During the day the leaf U and Th concentrations changed significantly. These variations were regular and very similar for both these radionuclides. There was a clear maximum at 14:00 that might be explained by the highest soil temperature at this time (Fig. 21).

As is seen from Fig. 22a, concentration of U in roots and leaves of couch-grass grown in U-rich soil was also the highest at 14:00. However, the highest U concentration in roots and leaves of plantain grown in the same plot was registered 4 hr later, at 6:00 (Fig. 22b). Thus, the short-term variations in U concentration in plantain could not be explained by the changes in soil temperature. But, as was mentioned

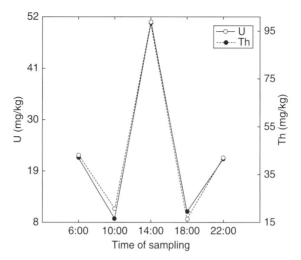

Figure 20. Dynamics of U and Th concentrations in leaves of couch-grass grown in soil enriched with U and Th.

above, diurnal dynamics of U concentration in roots and leaves of plantain was well correlated with total amount of proteins in the rhizosphere of this plant.

It has been reported [Walter and Schurr, 2005] that, in spite of certain interspecies differences, diurnal changes in the plant growth rate are generally larger than the changes in mean growth rate of the plant from day to day. This shows that processes controlling the plant growth variations within 24 hr are stronger than processes acting on a day-to-day scale. Thus, such big variations in root and leaf element concentrations may result from the significant short-term changes in the rates of the plant growth. Different plants have different sensitivity to temperature and photoperiod.

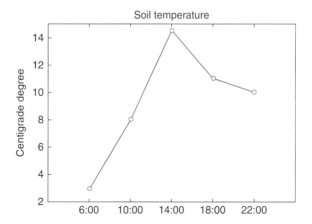

Figure 21. Diurnal dynamics of soil temperature.

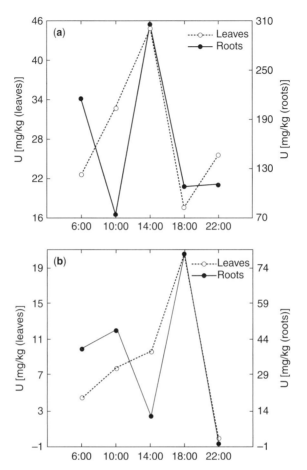

Figure 22. Dynamics of U concentrations in leaves and roots of couch-grass (**a**) and plantain (**b**) grown in U-rich soil.

Therefore, the differences in diurnal dynamics of radionuclide concentrations in plantain and couch-grass may be expected.

3.8   Effects of Thorium on a Plant During Initial Stages of the Plant Growth

The seedling stage is considered to be crucial for overall further life of the plant [Bajji et al., 2002]. To assess the effects of radionuclides on seed germination, plant nutrition, and development during initial stages of the plant growth, the following experiment has been carried out.

Seeds of wheat *Triticum vulgare (vill) Horst* were divided into two groups and germinated for 6 days on a moist filter paper at room temperature. Thorium nitrate was added to the medium where one group of seeds was germinated (concentration of Th in the solution was 20 ng/ml), and the second group served as a control.

The same volume of water (just water in the first case and water supplemented with Th in the second case) was added to both germination media. No more water was added during the 6-day germination. After germination, a portion of seedlings from both groups was taken for elemental analysis, and rest of the seedlings were transferred to pots filled with soil. The wheat seedlings were then grown in a naturally illuminated greenhouse. After 7 days the plants were harvested. The soil samples from the surface of the plant roots (rhizosphere soil) were taken simultaneously with the plants.

Transfer of the control 6-day-old seedlings after germination to soil favored the uptake of different elements by roots and their translocation from roots to other plant parts (Tables 7 and 8). Compared to the 6-day-old seedlings (before their transfer to soil), concentrations of all elements, except Na, Sc, Zn, and Br in leaves and Ca, Sc, Br, Rb, and Au in roots in the plants grown in soil for next 7 days, were significantly higher.

Germination of seeds in Th-supplemented medium resulted in a significant increase of Th content in all parts of the 6-day-old wheat seedlings (Table 7). Transfer of the seedlings germinated in Th-supplemented medium to soil and the seedling growth in the soil for 7 days led to a significant decrease of Th concentration in roots and in leaves (Table 8). In particular, in leaves Th content decreased up to the level found in the control plants. However, concentration of Th in seeds did not change and was comparable to Th content found in the 6-day-old seedlings just after the end of germination in the Th-enriched medium. This might be a result of a strong absorption/adsorption of Th in the very beginning of germination on the surface of the seeds. In the control plants grown in soil, the highest concentration of Th was observed in roots, and the lowest Th content was found in leaves. The accumulation of Th in wheat seedlings was associated with variations in uptake of other elements. Although Th concentration in leaves was more stable, the most strongly affected part of the plants was the leaf. In leaves of the plants germinated in Th-amended medium and then grown in soil, concentrations of Na and Rb were higher ($P < 0.05$), and concentrations of Ca, Fe, and La were lower ($P < 0.05$) than those in the control plants. The effects of Th treatment on concentrations of other elements in seeds and in roots were not so significant as in leaves. Seeds and roots of the plants germinated in Th medium and subsequently grown in soil had significantly lower ($P < 0.05$) Ca than did seeds and roots of the control plants.

Thus, the first and the most important reaction of the young wheat seedlings on the germination in Th enriched medium was a decrease of Ca in all parts of the seedlings. Calcium is a cell wall structure component, and it is also an effective regulator of various processes in the plant cells [Frausto da Silva and Williams, 1993]. Cell membranes are usually less permeable for calcium, and its concentration is fairly constant in plants [Evans et al., 1991]. The decrease of Ca may have harmful consequences to a plant.

The variations in uptake of main mineral nutrients such as potassium, calcium, and sodium may provide valuable information on the toxic effects of thorium. In particular, the K^+/Na^+ ratio is an important characteristic of physiological state of a plant [Santa-Maria and Epstein, 2001]. Table 9 shows the K/Na ratio in leaves, seeds, and

TABLE 7. Mean Concentrations (mg/kg) of Elements in Roots, Leaves, and Seeds of 6-Day-Old Wheat Seedlings Germinated in Ordinary Conditions (1) and After Addition of Th to the Germination Medium (2)

Element	Roots 1	Roots 2	Leaves 1	Leaves 2	Seeds 1	Seeds 2
Na (%)	0.38 ± 0.04	0.45 ± 0.05	0.025 ± 0.004	0.037 ± 0.006	0.0036 ± 0.0007	0.0061 ± 0.0010
K (%)	0.63 ± 0.04	0.78 ± 0.02	1.9 ± 0.1	1.7 ± 0.1	0.10 ± 0.01	0.11 ± 0.01
Ca (%)	0.15 ± 0.02	<0.12	0.058 ± 0.004	<0.12	0.067 ± 0.007	0.28 ± 0.09
Sc	0.18 ± 0.03	0.09 ± 0.02	0.15 ± 0.02	0.06 ± 0.4	0.005 ± 0.002	0.03 ± 0.01
Cr	0.88 ± 0.08	1.4 ± 0.5	0.52 ± 0.04	<0.4	<0.4	<0.4
Fe	251 ± 18	239 ± 23	150 ± 15	149 ± 17	44.8 ± 7.3	48.4 ± 11.0
Co	0.10 ± 0.03	0.11 ± 0.04	0.04 ± 0.02	0.03 ± 0.02	0.03 ± 0.03	0.04 ± 0.02
Zn	25.9 ± 3.7	36.2 ± 4.1	52.2 ± 8.3	55.6 ± 7.5	23.9 ± 3.9	58.9 ± 15.0
Br	12.4 ± 6.0	19.8 ± 7.1	9.3 ± 2.6	21.1 ± 9.5	0.54 ± 0.10	1.8 ± 0.25
Rb	5.2 ± 1.1	4.7 ± 1.2	6.8 ± 2.0	6.5 ± 1.6	0.44 ± 0.09	1.1 ± 0.5
Sr	11.2 ± 4.3	<8.0	<8.0	<8.0	9.0 ± 3.2	13.0 ± 4.2
Sb	0.10 ± 0.08	0.68 ± 0.09	0.04 ± 0.03	0.33 ± 0.25	0.009 ± 0.008	0.02 ± 0.02
Cs	<0.008	0.51 ± 0.40	<0.008	0.24 ± 0.18	<0.008	<0.008
Ba	<3.0	<3.0	<3.0	<3.0	3.3 ± 1.0	13.0 ± 2.5
Au	0.008 ± 0.004	0.02 ± 0.01	0.005 ± 0.005	0.008 ± 0.004	0.001 ± 0.001	0.001 ± 0.001
Th	<0.02	2.4 ± 1.3	<0.02	0.17 ± 0.04	<0.02	1.6 ± 0.6

TABLE 8. Mean Concentrations of Elements (mg/kg) in 13-Day Old Seedlings

Element	Roots		Leaves		Seeds	
	1	2	1	2	1	2
Na (%)	$0.65 \pm 0.14^{a,b}$	0.64 ± 0.11	0.03 ± 0.01	0.06 ± 0.01^d	0.06 ± 0.03	0.07 ± 0.05
K (%)	$1.26 \pm 0.41^{a,b}$	0.95 ± 0.15	4.9 ± 0.4^c	5.7 ± 2.3	0.35 ± 0.03	0.25 ± 0.34
Ca (%)	0.19 ± 0.09	0.12 ± 0.03^d	0.18 ± 0.02	0.13 ± 0.03^d	0.22 ± 0.05	0.12 ± 0.05^d
Sc	0.10 ± 0.07	0.12 ± 0.07	0.03 ± 0.03	0.01 ± 0.01	0.03 ± 0.02	0.27 ± 0.17
Cr	2.33 ± 1.00^b	1.3 ± 0.2	1.0 ± 0.3	0.85 ± 0.19	0.55 ± 0.25	0.49 ± 0.25
Fe	614 ± 262^b	308 ± 33	225 ± 28	155 ± 22^d	166 ± 70	146 ± 25
Co	$0.88 \pm 0.12^{a,b}$	0.95 ± 0.10	0.08 ± 0.02	0.11 ± 0.05	0.14 ± 0.06	0.25 ± 0.08
Zn	$146 \pm 8^{a,b}$	140 ± 15	57 ± 6^c	55.0 ± 3.7	70.1 ± 3.1	53.9 ± 30.9
Br	20.1 ± 9.4^b	20.4 ± 6.2	8.6 ± 4.0	14.1 ± 5.9	7.5 ± 5.2	6.2 ± 1.4
Rb	8.1 ± 2.4^a	6.5 ± 0.6	23.3 ± 1.7^c	26.8 ± 0.3^d	2.6 ± 0.8	2.2 ± 1.2
Sr	31.2 ± 6.0^a	28.4 ± 6.9	13.6 ± 7.8	17.4 ± 14.5	21.0 ± 16.6	<8.0
Ag	0.24 ± 0.11	0.20 ± 0.09	0.11 ± 0.05	0.070 ± 0.003	0.070 ± 0.005	0.27 ± 0.18
Sb	0.27 ± 0.12^b	0.33 ± 0.32	0.08 ± 0.06	0.22 ± 0.16	0.05 ± 0.02	0.07 ± 0.06
Cs	0.12 ± 0.09	0.20 ± 0.28	0.03 ± 0.01^c	0.12 ± 0.15	0.011 ± 0.004	<0.008
Ba	17.9 ± 4.7	9.7 ± 2.9	17.1 ± 5.1	20.7 ± 6.8	14.2 ± 3.6	18.0 ± 17.3
La	0.30 ± 0.24	0.19 ± 0.17	0.07 ± 0.05	0.01 ± 0.02^d	0.09 ± 0.07	0.05 ± 0.04
Sm	0.04 ± 0.03	33.7 ± 58.3	0.010 ± 0.002	<0.01	0.013 ± 0.005	0.02 ± 0.01
Au	0.02 ± 0.01^b	0.02 ± 0.01	0.010 ± 0.006	0.010 ± 0.004	0.004 ± 0.001	0.007 ± 0.002
Th	0.09 ± 0.04	0.87 ± 0.19^d	<0.02	<0.02	0.03 ± 0.02	2.0 ± 1.7^a

1, Control. 2, Th was added to soil; 3, the seedlings were germinated in Th-amended medium and then grown in soils.

$a-c$Differences between roots and leaves (a), roots and seeds (b), and leaves and seeds (c) were statistically significant at $P < 0.05$.

dDifferences between control and treatments with Th for particular plant parts were statistically significant at $P < 0.05$.

TABLE 9. Ratios of K/Na in Different Parts of Wheat
Seedlings Grown in Soil for Seven Days

Leaves		Seeds		Roots	
1	2	1	2	1	2
151	96	5.9	3.8	1.9	1.5

1—Control; 2—Th was added to germination medium.

roots of both control seedlings and seedlings germinated in the presence of Th. Potassium was preferentially accumulated in all parts of the plants. The K/Na ratio in leaves was much higher than in seeds and in roots because of the higher K content and lower concentration of Na in the leaves. These results are consistent with the previously published data [Mahmood, 1998; Schachtman and Liu, 1999]. As was mentioned above, leaves of the seedlings germinated in the presence of Th showed higher Na values that consequently decreased significantly the K/Na ratio in the leaves.

This raises the question of whether the sodium influx is toxic to the plants. The earlier studies [Marschner, 1995; Chartzoulakis et al., 2002] pointed out that the increasing concentration of Na in a plant could result in a decrease of the leaf area, the shoot dry weight, the leaf length, and the leaf width. In contrast, we did not observe any significant variation in the leaf biomass when the seedlings were germinated in the presence of thorium. However, it is necessary to note that in our experiment the growth time was rather short, and this does not exclude the possibility of certain physiological changes at the further stages of the plant development.

To obtain a more detailed estimate of the effects of the treatments on the wheat seedlings, we also performed principal component analysis of the plant samples (Fig. 23). For the calculations we used Varimax normalized rotation of our data. Two first PCs explained 49% of the total variance of the system. The first factor (PC1) was responsible for separation of leaves, seeds, and roots into three different groups. Leaves were far separated from roots and seeds, while groups of seeds and roots were located closer to each other. A highly correlated group of elements Cr, Fe, Na, Au, Zn, Sm, Co, and Br was the dominant factor in the first PC. Comparison of the data on control wheat seedlings (Table 8) revealed significant differences between concentrations of these elements (except Sm) and also K and Sb in roots and in seeds. Concentrations of all these elements in roots were significantly higher than in seeds. However, differences between seeds and leaves were not so great as those between seeds and roots. Zn concentration was statistically significant lower and concentrations of alkaline elements (K, Rb, and Cs) were much higher in leaves than in seeds. Lastly, concentration of Sr was lower, while Na, K, Co, and Zn contents were higher in roots as compared with those in leaves.

It is known that, in spite of even careful washing of plants after sampling, the problem of possible contamination of plants by dust and soil still exists. This is particularly important for root samples. It has been reported on the example of such elements like Sc and REE that surface contamination of the plant roots may represent a serious problem for elemental analysis of the plant samples [Schleppi et al., 2000].

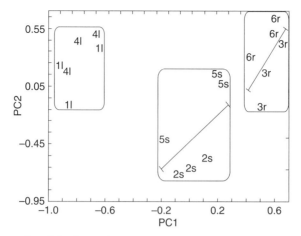

Figure 23. Score plot of the first and second principal components. l, s, and r: leaves, seeds, and roots, respectively. 1 (leaves), 2 (seeds), and 3 (roots) of the control plants; 4 (leaves), 5 (seeds), and 6 (roots) of the plants germinated in Th-enriched medium and then grown in soil.

Therefore, it would appear reasonable that in the case of contamination of roots by soil, one might also expect higher concentrations of the listed above elements in the root samples. However, we did not observe increased concentrations of Sc and REE in roots as compared to those in leaves and in seeds. Thus, we can conclude that the possible effect of contamination of plant roots was negligible.

As seen in Fig. 23, inside each group of root and seed samples, control plants and plants germinated in Th-enriched medium formed their own groups. The second PC (PC2) was responsible for the separation. Th had a significant negative loading value in the PC2; K and Na demonstrated a high positive correlation with the second PC. We may suggest that thorium, probably, has affected the behavior of these two plant nutrients at the early stage of the plant growth. Interestingly, there was not a good separation between leaves of the plants germinated in control and contaminated with Th media.

4 POTENTIAL HEALTH EFFECTS OF EXPOSURE TO U AND TH

The search for acute and chronic health effects from uranium exposure dates to 1824 when Christian Gmelin published a report on the effects of uranyl nitrate, chloride, and sulfate on dogs and rabbits [Hodge et al., 1973]. Early studies of humans were conducted from about 1860 to the early 1900s. During that time, uranium was administered as a therapeutic agent for diabetes because it had been shown to increase glucose excretion [Hodge et al., 1973].

People always experience exposure to a certain amount of U and Th from food, air, soil, and water, because they are naturally present in all these components. Food, such as root vegetables, and water will provide us with small amounts of natural U and Th and we will breathe minimal concentrations of these radionuclides with air. People that live near hazardous waste sites or near mines or have to eat crops grown on contaminated soil, and people that work in the phosphate industry may experience a higher exposure than other people.

Two potential ingestion exposure routes may be considered: inadvertent ingestion of soil and ingestion of radionuclides incorporated in the food chain. It is possible that radionuclides deposited on the ground or on the surface of plants may be incorporated into the edible portion of the plant and consumed directly by humans.

Published data on the effects of U and Th on human health is rather contradictory. There is no data available on long-term effects of ingestion of these radionuclides on humans. The information on the possible health effects of U and Th intake relies mainly on the experiments with animals that have digestive systems similar to that of humans (e.g., rats, dogs, pigs, and monkeys). However, it is clear that direct transfer of the experimental findings based on the experiments with animals to humans is hardly possible.

Uranium and thorium may present both chemical and radiological hazards. Unfortunately, it is difficult to divide the chemical and radiological components and to assess the exact contribution of each hazard to the harmful effects. Until now we have insufficient data on chemical toxicity of uranium and thorium. There is no data available on long-term effects of ingestion of these radionuclides on humans; all information available came from intermediate-term studies on animals.

Today, an extensive literature has evolved which traces the health effects of natural and enriched uranium. The main routes of natural uranium into the human body are by inhalation and ingestion. A very small percentage of the inhaled uranium is retained by the lymph nodes for a long time, and another small fraction is solubilized and goes to the blood, where most is subsequently excreted. A fraction of the blood content is deposited in the kidneys, liver, other organs, and the skeleton.

Extensive information is available about the occupational exposure of workers in the uranium industry. No increase in overall deaths has been observed in several epidemiological studies of workers exposed to uranium. Because natural uranium produces very little radioactivity, the health effects from exposure of humans to uranium are usually attributed to the chemical properties of uranium. Based on the studies with experimental animals, the most likely adverse health effect on humans from ingesting large amounts of uranium is on the kidneys. Studies of humans exposed to abnormally high levels of uranium in drinking water (averaging 100–600 μg/liter) for many years suggest that there may be minor damage to kidney tissue. The effects are a very mild decrease in the kidney's ability to hold onto proteins, sugar, and other compounds. However, this damage does not cause major effects on kidney function and is reversible after the exposure to uranium stops. There are no specific symptoms after long-term consumption of drinking water containing high levels of uranium. Studies of workers with occupational exposure to uranium have not shown any evidence of serious kidney disease or other health effects [http://www.epa.gov/safewater/index.html, 2004]. Only limited evidences suggest that even chronic exposure to natural uranium in food or water, except presumably at extraordinary concentrations, is associated with morbidity of man or animal.

The following conclusions may be made from the available scientific literature [McDiarmid and Keogh, 1998]:

- Although any increase in radiation to the human body can be calculated to be harmful from extrapolation from higher levels, there are no peer-reviewed published reports of detectable increases of cancer or other negative health effects from radiation exposure to inhaled or ingested natural uranium at levels far exceeding those typical for background areas. This is mainly because the body is very effective at eliminating ingested and inhaled natural uranium. The mass of uranium needed for significant internal exposure is virtually impossible to obtain.

- External radiation takes the primary form of alpha radiation, but there are also beta and gamma radiation. Alpha radiation is not capable of penetrating cloth and would therefore have no negative health effect. Beta and gamma radiation, which can have negative health effects, have been measured at levels below those expected to be of concern.

- Large variations in exposure to natural uranium in the normal environment have not been associated with negative health effects.

- Exposure to uranium in large doses can cause changes in renal function and, at very high levels, result in renal failure. However, in spite of these findings, no increased morbidity or frequency of end-stage renal disease has been observed in relatively large occupational populations chronically exposed to natural uranium at concentrations above normal level.

Compared to uranium, relatively less information is available on health effects of thorium. The available data are rather contradictory. Animal studies have shown that breathing thorium dust may result in lung damage. Other studies in animals suggest that drinking massive amounts of thorium can cause death from metal poisoning. The presence of large amounts of thorium in the environment could result in exposure to more hazardous radioactive decay products of thorium, such as radium and thoron. We can say that we know very little about specific exposure levels of thorium that result in harmful effects on people or animals. High levels of exposure have been shown to cause death in animals associated only with thorium.

Najem and Voyce [1990] performed a case–control study of 112 households residing in the vicinity of a thorium waste disposal site and found a higher prevalence of birth defects and liver diseases among the exposed group than among the unexposed group. However, the numbers were quite small and the confidence intervals were wide, so that no definite conclusions can be drawn from the data.

Chen et al. [2004] reported results of a 20-year study carried out at Baiyun Obo Rare-Earth Iron Mine in China. The ore contained 0.04% of ThO_2 and 10% of SiO_2. The purpose of this study was to investigate possible health effects in dust-exposed miners following long-term exposure to thorium-containing dusts and thoron progeny. An epidemiological study showed that the lung cancer mortality of the dust-exposed miners was significantly ($P < 0.005$) higher than that of the non-exposed group. It was suggested that this might result from the long-term exposure to thorium-containing dusts (carcinogens are ThO_2 and SiO_2) and thoron progeny. However, from this research we cannot conclude for sure about the exact contribution of thorium to the phenomenon.

Another study of thorium workers [Terry and Hewson, 1995] showed that breathing thorium dust might cause an increased chance of developing lung disease and cancer of the lung or pancreas many years after being exposed. Changes in the genetic material of body cells have also been shown to occur in workers who breathed thorium dust. Liver diseases and effects on the blood have been found in people injected with thorium in order to take special X-rays. Many types of cancer have also been shown to occur in these people many years after thorium was injected into their bodies. Since thorium is radioactive and may be stored in bone for a long time, bone cancer is also a potential concern for people exposed to thorium.

Finally, the risks associated with radionuclides in foods are considered to be insignificant or *de minimis*. We may assume that in certain situations, U and Th may provide even positive effects for humans as we observed in the example of certain improvements in the plant nutrition resulting from the radionuclide treatments. We should remember that without the warmth of radiation, life would not have evolved in the first place.

REFERENCES

Aery NC, Jain GS. 1997. Effect of uranyl nitrate on seed germination and early seedling growth of *Triticum aestivum. Biologia* **52**:115–119.

Airey PL, Ivanovich M. 1986. Geochemical analogues of highlevel radioactive waste repositories. *Chem Geol* **55**:203–213.

Akhtar N, Tufail M. 2007. Natural radioactivity intake into wheat grown on fertilized farms in two districts of Pakistan. *Radiat Protect Dosim* **123**:103–112.

Apps MJ, Duke MJM, Stephens-Newsham LG. 1988. A study of radionuclides in vegetation on abandoned uranium tailings. *J Radioanal Nucl Chem* **123**:133–147.

Baatn MN. 1998. Growth rates of bacterial communities in soils varying pH: A comparison of thymidine and leucine incorporation techniques relatively high protein content in microbial cells. *Microbial Ecol.* **36**:316–327.

Bajji M, Kinet J-M, Luts S. 2002. Osmotic and ionic effects of NaCl on germination, early seedling growth, and ion content of *Atriplex halimus* (Chenopodiaceae). *Can J Bot* **80**:297–304.

Bao X, Zhang A. 1998. Geochemistry of U and Th and its influence on the origin and Evolution of the Earth's crust and the biological evolution. *Acta Petrol. Mineralogica* **17**:160–172.

Becker D, Moshelion M, Czempinski K, Moran N, Lowen CZ, Satter RL. 1989. Light-promoted changes in apoplastic K^+ activity in the *Samanea saman* pulvinus, monitored with liquid membrane microelectrodes. *Planta* **179**:421–427.

Bednar AJ, Gent DB, Gilmore JR, Sturgis TC, Larson SL. 2004. Mechanisms of thorium migration in a semiarid soil. *J Environ Qual* **33**:2070–2077.

Behrenfeld MJ, Prasil O, Babin M, Bruyant F. 2004. In search of a physiological basis for covariations in light-limited and light-saturated photosynthesis. *J Phycol* **40**:4–25.

Beresford NA, Wright SM. 2005. Non-linearity in radiocaesium soil to plant transfer: Fact or fiction? *Radioprotection* **40**(suppl 1):67–72.

Blanco Rodríguez P, Vera Tomé F, Pérez Fernández M, Lozano JC. 2006. Linearity assumption in soil-to-plant transfer factors of natural uranium and radium in *Helianthus annuus* L. *Sci Tot Environ* **361**:1–7.

Bowen HJM. 1966. The biogeochemistry of the elements. In: Bowen HJM, editor. *Trace Elements in Biochemistry*. New York: Academic Press, pp. 173–210.

Buck EC, Brown NR, Dietz NL. 1996. Contaminant uranium phases and leaching at the Fernald site in Ohio. *Environ Sci Technol* **30**:81–88.

Carter PJ, Nimmo HG, Fewson CA, Wilkins MB. 1991. Circadian rhythms in the activity of a plant protein kinase. *EMBO J* **10**:2063–2068.

Chang P, Kim K-W, Yoshida S, Kim S-Y. 2005. Uranium accumulation of crop plants enhanced by citric acid. *Environ Geochem Health* **27**:529–538.

Chaosheng Z. 2002. Grain size effect on multi-element concentrations in sediments from intertidal flats of Bohai Bay, China. *Appl Geochem* **17**:59–68.

Chartzoulakis K, Loupassaki M, Bertaki M, Androulakis I. 2002. Effects of NaCl salinity on growth, ion content and CO_2 assimilation rate of six olive cultivars. *Sci Horti* **96**:235–247.

Chen SB, Zhu YG, Hu QH. 2005. Soil to plant transfer of ^{238}U, ^{226}Ra and ^{232}Th on a uranium mining-impacted soil from southeastern China. *J Environ Radioactivity* **82**:223–236.

Chen XA, Cheng YE, Xiao H, Feng G, Deng YH, Feng ZL, Chen L, Han XM, Yang YJ, Dong ZH, Zheng R. 2004. Health effects following long-term exposure to thorium dusts: A twenty-year follow-up study in China. *Radioprotection* **39**:525–533.

Cooper MB, Clarke PC, Robertson W, McPharlin IR, Jeffrey RC. 1995. An investigation of radionuclide uptake into food crops grown in soils treated with bauxite mining residues. *J Radioanal Nucl Chem* **194**:379–387.

Diebold FE, McGrath S. 1985. Investigation of *Artemisia tridentata* as a biogeochemical uranium indicator. *J Geochem Explor* **23**:1–12.

Domelly PK, Fletcher JS. 1994. Potential use of mycorrhizal fungi as bioremediation agents. In: Anderson TA, Coats JR, editors. *Bioremediation through Rhizosphere Technology*. Washington, DC: American Chemical Society, pp. 93–99.

Duff MC, Amrhein C. 1996. Uranium(VI) adsorption on goethite and soil in carbonate solutions. *J Soil Sci Soc Am* **60**:1393–1400.

Ebbs SD, Brady DJ, Kochian LV. 1998. Role of uranium speciation in the uptake and translocation of uranium by plants. *J Exp Bot* **49**:1183–1190.

Edmands JD, Brabander DJ, Coleman DS. 2001. Uptake and mobility of uranium in black oaks: Implications for biomonitoring depleted uranium-contaminated groundwater. *Chemosphere* **44**:789–795.

Ehlken S, Kirchner G. 2002. Environmental processes affecting plant root uptake of radioactive trace elements and variability of transfer factor data: A review. *J Environ Radioactivity* **58**:97–112.

Evans DE, Briars SA, Williams LE. 1991. Active calcium transport by plant cells membranes. *J Exp Bot* **42**:285–303.

Ewers LW, Ham GJ, Wilkins BT. 2003. Review of the transfer of naturally occurring radionuclides to terrestrial plants and domestic animals. National Radiological Protection Board Report, NRPB-W49, Chilton, pp. 1–64.

FAO-UNESCO. 1988. *Soil Map of the World*, Revised Legend. World Soil Resource Report 60, Rome.

Fowle DA, Fein JB, Martin AM. 2000. Experimental study of uranyl adsorption onto Bacillus subtilis. *Environ Sci Technol* **34**:3737–3741.

Francis AJ, Gillow JB, Dodge CJ, Harris R, Beveridge TJ, Papenguth HW. 2004. Uranium association with halophilic and non-halophilic bacteria and archaea. *Radiochim Acta* **92**:481–488.

Frausto da Silva JJR, Williams RJP. 1993. *The Biological Chemistry of the Elements. The Inorganic Chemistry of Life*. Oxford: Clarendon Press.

Frindik O, Vollmer S. 1999. Particle-size dependent distribution of thorium and uranium isotopes in soil. *J. Radioanal Nucl Chem* **241**:291–296.

Galindo C, Mougin L, Fakhi S, Nourreddine A, Lamghari A, Hannache H. 2007. Distribution of naturally occurring radionuclides (U, Th) in Timahdit black shale (Morocco). *J Environ Radioactivity* **92**:41–54.

Gulati KL, Oswal MC, Nagpaul KK. 1980. Assimilation of uranium by wheat and tomato plants. *Plant Soil* **55**:55–59.

Hamamo H, Landsberger S, Harbottle G, Panno S. 1995. Studies of radioactivity and heavy metals in phosphate fertilizer. *J Radioanal Nucl Chem* **194**:331–336.

Higgy RH, Pimpl M. 1998. Natural and man-made radioactivity in soils and plants around the research reactor of Inshass. *Appl Rad Isot* **49**:1709–1712.

Hodge HC, Stannard JN, Hursh JB. 1973. *Handbook of Experimental Pharmacology*. Berlin: Springer-Verlag.

Hore-Lacy I. 2003. *Nuclear Electricity*, seventh edition. Melbourne: Uranium Information Centre Ltd.

Höllriegl V, Greiter M, Giussani A, Gerstmann U, Michalke B, Roth P, Oeh U. 2007. Observation of changes in urinary excretion of thorium in humans following ingestion of a therapeutic soil. *J Environ Radioactivity* **95**:149–160.

http://www.epa.gov/safewater/index.html, 2004.

Huist W. 1997. Thorium. *Gmelin Handbook of Inorganic and Organometallic Chemistry*, Supplement v. B1, eighth edition. Berlin: Springer-Verlag.

Jackson WR. 1993. Humic, fulvic, and microbial balance: Organic soil conditioning, an agricultural text and reference book. Evergreen, CO: Jackson Research Center.

Jain GS, Aery NC. 1997. Effect of uranium additions on certain biochemical constituents and uranium accumulation in wheat. *Biologia* **52**:599–604.

Jones KC, Lepp NM, Obbard JP. 1990. Other metals and metalloids. In: Alloway BJ, editor. *Heavy Metals in Soils*. Glasgow: Blackie, pp. 280–321.

Kabata-Pendias A, Pendias H. 2000. *Trace Elements in Soil and Plants*, third edition. Boca Raton, FL: CRC Press.

Kim HY, Cote GG, Crain RC. 1992. Effects of light on the membrane potential of protoplasts from *Samanea saman* pulvini. Involvement of the H-ATPase and K channels. *Plant Physiol* **99**:1532–1539.

Kim HY, Cote GG, Crain RC. 1993. Potassium channels in *Samanea saman* protoplasts controlled by phytochrome and the biological clock. *Science* **260**:960–962.

Laroche L, Henner P, Camilleri V, Morello M, Garnier-Laplace J. 2005. Root uptake of uranium by a higher plant model (*Phaseolus vulgaris*)—Bioavailability from soil solution. *Radioprotection* **40**(suppl 1):33–39.

Larson SL, Bednar AJ, Ballard JH, Shettlemore MG, Gent DB, Christodoulatos C, Manis R, Morgan GC, Fields MP. 2005. Characterization of a military training site containing [232]thorium. *Chemosphere* **59**:1015–1022.

Mahmood K. 1998. Effects of salinity, external K^+/Na^+ ratio and soil moisture on growth and ion content of *Sesbania rostrata*. *Biol Plant* **41**:297–302.

Marquenie JM. 1985. Bioavailability of micropollutants. *Environ Technol Lett.* **6**:351–358.

Marschner H. 1995. *Mineral Nutrition of Higher Plants*. London: Academic Press.

Mazor E. 1992. Uranium in plants of Southern Sinai. *J Arid Environ* **22**:363–368.

McDiarmid MA, Keogh JP. 1998. *The Depleted Uranium Follow-Up Program*. Baltimore: VA Medical Center, AFRRI Special Publication. **98-3**:29–31.

Meinrath G, Kato Y, Kimura T, Yoshida Z. 1996. Solid–aqueous phase equilibria of uranium(VI) under ambient conditions. *Radiochim. Acta* **75**:159–167.

Miljevic NR, Markovic MM, Todorovic DJ, Cvijovic MR, Goloboeanin DD, Orlic MP, Vevelinovic DS, Bioeanin RN. 2001. Uranium content in the soil of the Federal Republic of Yugoslavia after NATO Intervention. *Arch Oncol* **9**:245–249.

Morton LS, Evans CV, Harbottle G, Estes GO. 2001. Pedogenic fractionation and bioavailability of uranium and thorium in naturally radioactive spodosols. *J Soil Sci Soc Am* **65**:1197–1203.

Morton LS, Evans CV, Estes GO. 2002. Natural uranium and thorium distributions in podzolized soils and native blueberry. *J Environ Qual* **31**:155–162.

Mortvedt JJ. 1994. Plant and soil relationships of uranium and thorium decay series radionuclides—A review. *J Environ Qual* **23**:643–650.

Myung CJ, Thornton I. 1997. Environmental contamination and seasonal variation of metals in soils, plants and waters in the paddy fields around a Pb–Zn mine in Korea. *Sci Tot Environ* **198**:105–121.

Najem GR, Voyce LK. 1990. Health effects of a thorium waste disposal site. *Am J Public Health* **80**:478–480.

Nikiteshen VI. 2003. *Ecologo-agrochemical Basis of Balanced Application of Fertilisers*. Moscow: Nauka (in Russian).

Ohnuki T, Yoshida T, Ozaki T, Samadfam M, Kozai N, Yubuta K, Mitsugashira T, Kasama T, Francis AJ. 2005. Interactions of uranium with bacteria and kaolinite clay. *Chem Geol* **220**:237–243.

Orvini E, Speziali M, Salvini A, Herborg C. 2000. Rare earth element determination in environmental matrices by INAA. *Microchem J* **67**:97–104.

Otero XL, Macias F. 2002. Variation with Depth and Season in Metal Sulfides in Salt Marsh Soils. *Biogeochemistry* **61**:247–268 .

Paquet F, Ramounet B, Métivier H, Taylor DM. 1998. The bioinorganic chemistry of Np, Pu and Am in mammalian liver. *J Alloys Compounds* **271–273**:85–88.

Pulhani V, Kayasth S, More AK, Mishra UC. 2000. Determination of traces of uranium and thorium in environmental martices by neurton activation analysis. *J Radioanal Nucl Chem* **243**:625–629.

Ramaswami A, Carr P, Burkhardt M. 2001. Plant-uptake of uranium: Hydroponic and soil system studies. *Int J Phytorem* **3**:189–201.

Robison WL, Hamilton TF, Martinelli RE, Kehl SR, Lindman TR. 2005. Concentration of beryillim (Be) and depleted uranium (DU) in marine fauna and sediment samples from Illeginni and Boggerik Islands at Kwajalein Atoll. Report of the US Department of Energy, UCRL-TR-210057, pp. 1–18.

Rodrigues PB, Vera Tomé F, Lozano JC. 2002. About the assumption of linearity in soil-to-plant transfer factors for uranium and thorium isotopes and ^{226}Ra. *Sci Tot Environ* **284**:167–175.

Santa-Maria GE, Epstein E. 2001. Potassium/sodium selectivity in wheat and the amphiploid cross wheat X *Lophopyrum elongatum*. *Plant Sci* **160**:523–534.

Sar P, D'Souza SF. 2002. Biosorption of thorium (IV) by a *Pseudomonas* biomass. *Biotechnol Lett* **24**:239–243.

Savari A, Lockwood APM, Sheder M. 1991. Effects of season and size (age) on heavy metal concentrations of the common cockle (*Cerastoderma edule* (L.)) from Southampton water. *J Moll Stud* **57**:45–57.

Schachtman DP, Liu W. 1999. Molecular pieces to the puzzle of the interaction between potassium and sodium uptake in plants. *Trends Plant Sci* **4**:281–287.

Schleppi P, Tobler L, Bucher JB, Wyttenbach A. 2000. Multivariate interpretation of the foliar chemical composition of Norway spruce (*Picea abies*). *Plant Soil* **219**:251–262.

Schott A, Brand RA, Kaiser J, Schmidt D. 2006. Depleted uranium (DU)—chemo- and radiotoxicity. In: Merkel J, Hasche-Berger A, editors. *Uranium in the Environment. Mining Impact and Consequences*. Berlin: Springer, pp. 165–174.

Shacklette HT, Boerngen JG. 1984. Element concentrations in soils and other surficial materials of the conterminous United States. *US Geol Surv Prof Paper* **1270**:1–105.

Shahandeh H, Hossner LR. 2002. Role of soil properties in phytoaccumulation of uranium. *Water Air Soil Pollut* **141**:165–180.

Sheppard MI, Sheppard SC. 1985. The plant concentration ratio concept as applied to natural U. *Health Phys* **48**:494–500.

Sheppard MI, Thibault DH, Sheppard SC. 1985. Concentrations and concentration ratios of U, As and Co in Scots pine grown in a waste-site soil and an experimentally contaminated soil. Water *Air Soil Pollut* **26**:85–94.

Sheppard SC, Evenden WG. 1988. The assumption of linearity in soil and plant concentration ratios: An experimental evaluation. *J Environ Radioactivity* **7**:221–247.

Sheppard SC, Evenden WG. 1992. Bioavailability indices for uranium: Effect of concentration in eleven soils. *Arch Environ Contain Toxicol* **23**:117–124.

Sheppard SC, Evenden WG, Pollock RW. 1989. Uptake of natural radionuclides by field and garden crops. *Can J Soil Sci* **69**:751–767.

Shtangeeva IV. 1993. Chemical element distribution in soils and some species of plants. In: Frontasieva M, editor. *Activation Analysis in Environmental Protection*. Dubna: JINR, pp. 340–351.

Shtangeeva I, Ayrault S. 2004. Phytoextraction of thorium from soil and water media. *Water Air Soil Pollut* **154**:19–35.

Shtangeeva I, Ayrault S, Jain J. 2005. Thorium uptake by wheat at different stages of plant growth. *J Environ Radioactivity* **81**:283–293.

Shtangeeva I, Lin X, Tuerler A, Rudneva E, Surin V, Henkelmann R. 2006. Thorium and uranium uptake and bioaccumulation by wheat-grass and plantain. *Forest Snow Landscape Res* **80**:181–190.

Suh S, Moran N, Lee Y. 2000. Blue light activates depolarization-dependent K channels in flexor cells from *Samanea saman* motor organs via two mechanisms. *Plant Physiol* **123**:833–843.

Termizi Ramli A, Wahab A, Hussein MA, Khalik Wood A. 2005. Environmental ^{238}U and ^{232}Th concentration measurements in an area of high level natural background radiation at Palong, Johor, Malaysia. *J Environ Radioactivity* **80**:287–304.

Terry KW, Hewson GS. 1995. Thorium lung burdens of mineral sands workers. *Health Phys* **69**:233–242.

Thiry Y, Schmidt P, Van Hees M, Wannijn J, Van Bree P, Rufyikiri G, Vandenhove H. 2005. Uranium distribution and cycling in Scots pine (*Pinus sylvestris* L.) growing on a revegetated U-mining heap. *J Environ Radioactivity* **81**:201–219.

Ticknor KV. 1994. Uranium sorption on geological materials. *Radiochim Acta* **64**:229–236.

Titaeva NA. 1992. *Nuclear Geochemistry.* Moskow: MGU (in Russian).

Trüby P. 1995. Distribution patterns of heavy metals in forest trees on contaminated sites in Germany. *Angew Bot* **69**:135–139.

Tsuruta T. 2006. Bioaccumulation of uranium and thorium from the solution containing both elements using various microorganisms. *J Alloys Compounds* **408–412**:1312–1315.

Tynybekov AK, Hamby DM. 1999. A screening assasment of external radiation levels on the shore of lake Issyk-Kul in the Kyrgyz Republic. *Health Phys* **77**:427–430.

Tzortzis M, Tsertos H. 2005. Natural radioelement concentration in the Troodos Ophiolite Complex of Cyprus. *J Geochem Expl* **85**:47–54.

Vandenhove H, Van Hees M, Wannijn J, Wouters K, Wang L. 2007. Can we predict uranium bioavailability based on soil parameters? Part 2: Soil solution uranium concentration is not a good bioavailability index. *Environ Pollut* **145**:577–586.

Van Horn JD, Huang H. 2006. Uranium(VI) bio-coordination chemistry from biochemical, solution and protein structural data. *Coord Chemi Revi* **250**:765–775.

Vera Tomé F, Rodrigues PB, Lozano JC. 2002. Distribution and mobilization of U, Th and ^{226}Ra in the plant-soil compartments of a mineralized uranium area in South-west Spain. *J Environ Radioactivity* **59**:223–243.

Vera Tomé F, Rodrigues PB, Lozano JC. 2003. Soil-to-plant transfer factors for natural radionuclides and stable elements in a Mediterranean area. *J Environ Radioactivity* **65**:161–175.

Voigt G, Scott EM, Bunzl K, Dixon P, Sheppard SC, Whicker WF. 2000. Radionuclides in the environment: Radiological quantities and sampling designs. *Radiat Protection Dosimetry* **92**:55–57.

Walter A, Schurr U. 2005. Dynamics of leaf and root growth: Endogenous control versus environmental impact. *Ann Bot* **95**:891–900.

Wielinga B, Lucy JK, Moore JN, Gannon JE. 1999. Microbial and geochemical characterization of fluvially deposited sulfidic mine tailing. *Appl Environ Microbiol* **65**:1548–1585.

Wright SF, Upadhyaya A. 1996. Extraction of an abundant and unusual protein from soil and comparison with hyphal protein of arbuscular mycorrhizal fungi. *Soil Sci* **161**:575–586.

Wright RJ. 1999. Thorium. In: Fairbridge RW, editor. *Encyclopaedia of Geochemistry and Environmental Sciences*. New York: Van Nostrand, pp. 1183–1189.

Yingjun L. 1984. *Element Geochemistry*. Beijing: Science Press.

Yoshida S, Muramatsu Y, Tagami K, Uchida S, Ban-nai T, Yonehara H, Sahoo S. 2000. Concentrations of uranium and $^{235}U/^{238}U$ ratios in soil and plant samples collected around the uranium conversion building in the JCO campus. *J Environ Radioactivity* **50**:161–172.

Zararsiz A, Kizmar R, Arikan P. 1997. Field study on thorium uptake by plants within and around of a thorium ore deposit. *J Radioanal Nucl Chem* **222**:257–262.

15 Exposure to Mercury: A Critical Assessment of Adverse Ecological and Human Health Effects

SERGI DÍEZ

Environmental Geology Department, ICTJA-CSIC, 08028 Barcelona, Spain; and Environmental Chemistry Department, IIQAB-CSIC, 08034 Barcelona, Spain

CARLOS BARATA

Environmental Chemistry Department, IIQAB-CSIC, 08034 Barcelona, Spain

DEMETRIO RALDÚA

Laboratory of Environmental Toxicology (UPC), 08220 Terrassa, Spain

1 HUMAN HEALTH EFFECTS

1.1 Introduction

Mercury (Hg) has caused a variety of documented, significant adverse impacts on human health and the environment throughout the world. Mercury and its compounds are highly toxic, especially to the developing nervous system. The toxicity to humans and other organisms depends on the chemical form, the amount, the pathway of exposure, and the vulnerability of the person exposed. Human exposure to mercury can result from a variety of pathways, including consumption of fish, occupational and household uses, dental amalgams, and mercury-containing vaccines.

Because Hg has become a proved cause of concern due to its high toxicity, special attention has paid to its determination. In particular, methylmercury (MeHg), the most toxic form of mercury, can cause severe neurological damage to humans and wildlife [Grandjean et al., 1999; Clarkson et al., 2003]. Mercury levels in both drinking water and ambient air are usually irrelevant, and then the two major ways of exposure to mercury in humans are related to dental amalgam (mercury vapor, Hg^0) and fish consumption (MeHg). Lately, an increasing concern has been expressed by the American Academy of Pediatrics and the Public Health Service of the USA regarding the safety of a Hg preservative (ethyl mercury compound, thimerosal) in many vaccine preparations routinely administered to infants. Although patterns of human usage have changed over the centuries, occupational exposure still occurs especially to Hg vapor in the mining operations, including gold mining [Malm, 1998; Grandjean et al., 1999], the chlor-alkali industry [Calasans and Malm, 1997; Montuori et al., 2006], and dentistry [Harakeh et al., 2002; Morton et al., 2004].

Nowadays, the general population is primarily exposed to three different forms of mercury: mercury vapors emitted by dental amalgam fillings [Goering et al., 1992; Razagui and Haswell, 2001; Hansen et al., 2004], MeHg naturally bioaccumulated in fish [Bjornberg et al., 2003; Hightower and Moore, 2003; Canuel et al., 2006], and an ethyl mercury compound, thimerosal, which is used as a preservative in certain commonly used childhood vaccines [Halsey, 1999; Pichichero et al. 2008].

1.2   Sources and Cycling of Mercury to the Global Environment

Mercury is a natural element of the earth, a silver-colored, shiny, liquid metal that is found a variety of chemical forms in rocks, soil, water, air, plants, and animals. Mercury is found usually combined with other elements in various compounds, which may be inorganic (e.g., the mineral cinnabar, a combination of mercury and sulfur) or organic (e.g., MeHg), although occasionally Hg occurs in its elemental, relatively pure form, as a liquid or vapor.

The global cycling of mercury begins with the evaporation of Hg vapor from land and sea surfaces. Most of the mercury in the atmosphere is elemental Hg vapor, a chemically stable monatomic gas that circulates in the atmosphere for up to a year and hence can be widely dispersed and transported thousands of miles from likely sources of emission. This vapor is oxidized in the upper atmosphere to a water-soluble ionic mercury, which is returned to the earth's surface in rainwater via wet deposition. Even after it deposits, mercury commonly is emitted back to the atmosphere either as a gas or associated with particles, to be re-deposited elsewhere. As it cycles between the atmosphere, land, and water, mercury undergoes a series of complex chemical and physical transformations, many of which are not completely understood. In fact, about 90% of the total Hg input to oceans is recycled to the atmosphere and less than 10% reaches the sediments. However, a small percentage (about a 2%) is methylated in the biota, resulting in accumulation in the food chain, while only a small fraction is lost to the atmosphere, mainly as highly volatile dimethyl mercury [Fitzgerald et al., 1998]. Mercury accumulates most efficiently in the aquatic food web, where predatory organisms at the top of the food chain generally have higher mercury concentrations. Almost all of the mercury that accumulates in fish tissue is MeHg; and inorganic mercury (IHg), which is less efficiently absorbed and more readily eliminated from the body than MeHg, does not tend to bioaccumulate.

The releases of mercury can be grouped into four categories: (i) natural sources; (ii) current anthropogenic releases from the mobilization of mercury impurities in raw materials; (iii) current anthropogenic releases resulting from mercury used intentionally in products and processes; and (iv) remobilization of historic anthropogenic mercury releases previously deposited everywhere.

Natural sources include volcanoes, evaporation from soil and water surfaces, degradation of minerals, and forest fires. Available information indicates that natural sources account for less than 50% of the total releases. On average around the globe, there are indications that anthropogenic emissions of mercury have resulted in deposition rates today that are 1.5 to 3 times higher than those during preindustrial times. In and around industrial areas, the deposition rates have increased by 2–10 times during the last 200 years [EPA US, 1997; Bergan et al., 1999]. Global natural emissions at about 1650 metric tons/year was estimated in 1994 [Mason et al., 1994] or at 1400 metric tons/year in an update performed later [Lamborg et al., 2002]. The Programme for Monitoring and Evaluation of the Long-Range Transmission of Air Pollutants in Europe (EMEP) refers for an estimated global natural emission of about 2400 metric tons, of which 1320 were emitted from land and 1100 were emitted from oceans [Bergan and Rodhe, 2001]. Emission inventories indicate that Asian Hg sources account for more than 50% of the global anthropogenic total Hg [Jaffe et al., 2005]. In the next few decades, due to rapid economic and industrial development in Asia, a significant increase in anthropogenic Hg emissions is expected to occur, unless drastic measures will be taken [Wong et al., 2006].

Some of the more important anthropogenic processes that mobilize mercury impurities include: coal-fired power and heat generation; cement production; and mining and other metallurgic activities involving the extraction and processing of mineral materials, such as production of iron and steel, zinc, and gold. Some important sources of anthropogenic releases that occur from the intentional extraction and use of mercury include: mercury mining; small-scale gold and silver mining; chloralkali production; (breakage during) use of fluorescent lamps, auto headlamps, manometers, thermostats, thermometers, and other instruments; dental amalgam fillings; manufacturing of products containing mercury; waste treatment and incineration of products containing mercury; landfills; and cremation.

The atmospheric residence time of elemental mercury is in the range of months to roughly one year. This makes transport on a hemispherical scale possible, and emissions in any continent can thus contribute to the deposition in other continents. For example, based on modeling of the intercontinental mercury transport performed by EMEP/MSC-E [Travnikov, 2005], up to 67% of total depositions to the continent are from external anthropogenic and natural sources. Even knowing that the Arctic region has no significant local sources of mercury emission on its territory at all, about half the mercury deposition is due to the atmospheric transport from foreign anthropogenic emission sources, of which the greatest contribution is made by Asian (33%) and European sources (22%).

Speciation influences the transport of mercury within and between environmental compartments. Mercury adsorbed onto particles and ionic (e.g., divalent) mercury compounds will fall on land and water mainly in the vicinity of the sources (local to regional distances), while elemental mercury vapor is transported on a hemispherical/global scale, making mercury emissions a global concern. Another example is the so-called "polar sunrise mercury depletion incidence," a special phenomenon that has been shown to influence the deposition of mercury in Polar regions. It has also been termed "the mercury sunrise," because a highly elevated deposition of mercury is taking place during the first few months of the Polar sunrise. It seems that solar activity and the presence of ice crystals influence the atmospheric transformation of elemental gaseous mercury to divalent mercury, which is more rapidly deposited. Mercury depletion has now been observed in Alert, Canada [Schroeder et al., 1998], in Barrow, Alaska, USA [Lindberg et al., 2002], in Svalbard, Norway [Berg et al., 2003], in Greenland [Ferrari et al., 2004] and in the Antarctic [Ebinghaus et al., 2002].

1.3 Methylmercury

1.3.1 Pathways of Human Exposure. As a result of the global cycling of mercury mostly described above, IHg reaches the aquatic environment and is transformed to MeHg via methylation by microbial communities present in sediments of aquatic environments. This is believed to be a protective mechanism by microorganisms, because inorganic mercuric mercury is more toxic. In aquatic systems, microorganisms responsible for methylation of mercury are mainly the sulfate-reducing bacteria. This conversion to MeHg produced is able to enter the aquatic food chain. Accordingly, human exposure to MeHg nowadays occurs almost exclusively from consumption of fish and marine mammals.

The mercury concentrations in various fish species are generally from about 0.01 to 4 mg/kg, depending on factors such as pH and redox potential of the water, and species, age and size of the fish. Since mercury biomagnifies in the aquatic food web, fish of higher trophic level tend to have higher levels of mercury. Hence, large predatory fish, such as king mackeral, pike, shark, swordfish, walleye, barracuda, large tuna, scabbard, and marlin, as well as seals and toothed whales, contain the highest concentrations. Available data indicate that mercury is present throughout the globe in concentrations that adversely affect human beings and wildlife. These levels have led to consumption advisories in a number of countries (for fish and sometimes marine mammals), warning people, especially sensitive subgroups (pregnant women and young children), to limit or avoid consumption of certain types of fish from water bodies. Moderate consumption of fish (with low mercury levels) is not likely to result in exposures of concern. However, people who consume higher amounts of fish or marine mammals may be highly exposed to mercury and are therefore at risk [Hightower and Moore, 2003].

The most dramatic case of severe MeHg poisoning is from Minamata Bay, Japan, in the 1950s, when a fishermen community consumed fish containing high levels of MeHg. The source was a chemical plant manufacturing acetaldehyde, where IHg mercury was used as a catalyst. The amount of mercury discharged from 1932 to 1968 was large and was estimated at 456 tons. As a result, hundreds of people died, and thousands were affected, many with permanent damage [Harada, 1995; Akagi et al., 1998]. Recent studies in the Brazilian Amazon demonstrated that the population had increased exposure to MeHg because of their consumption of fish contaminated by upstream gold-mining activities [Lebel et al., 1998; Grandjean et al., 1999; Dolbec et al., 2000].

1.3.2   Toxicokinetics. Several recent reviews have given extensive details on the absorption, disposition, and excretion of MeHg in the human body [ATSDR, 1999; EPA US, 2001; Clarkson and Magos, 2006]. It is well known that about 95% of MeHg ingested in fish is absorbed by humans in the gastrointestinal tract, although the exact site of absorption is unknown. It is distributed to all tissues, a process that takes some 30–40 hr, where about 5% is found in the blood and 10% in brain. In blood, most MeHg is bound to the hemoglobin in red cells, with a concentration of about 20 times the concentration in plasma. The concentration in brain is about 5 times and in scalp hair about 250 times the corresponding concentration in blood. Human hair is a widely accepted biomarker for MeHg exposure [NRC, 2000; Diez and Bayona, 2002; Montuori et al., 2004; Diez et al., 2007]. Whereas results from the other body specimens can only be used as a measure of recent exposure, hair, in contrast, can provide a historical exposure record. Mercury levels in hair have been found to be a suitable indicator of dietary, environmental and occupational human exposure to mercury [Carrington and Bolger, 2002; Montuori et al., 2006; Diez et al., 2008]. Hair levels closely follow blood levels. MeHg crosses the blood–brain and placental barriers. Moreover, levels in cord blood are proportional to, but slightly higher than, levels in maternal blood. Levels in the fetal brain are about five to seven times higher than levels in maternal blood [Cernichiari et al., 1995a].

MeHg is slowly metabolized to IHg by microflora in the intestines; nevertheless, the biochemical mechanisms are still not known. Although MeHg is the predominant form of mercury during exposure, IHg slowly accumulates and resides for long periods in the central nervous system. IHg is believed to be in an inert form probably in the form of insoluble mercury selenide [Clarkson et al., 2003].

Urinary excretion is negligible, of the order of 10% or less of total elimination from the body. Most of the MeHg is eliminated from the body by demethylation and excretion of the inorganic form in the feces. The processes of biliary excretion and demethylation by microflora do not occur in suckling animals. The failure of neo-nates to excrete MeHg may be associated with the inability of suckling infants to perform these two processes [Rowland et al., 1977; Ballatori and Clarkson, 1982].

The formation of complexes with thiol-containing small molecules such as cysteine and glutathione plays a major role in the processes of transport. MeHg enters the endothelial cells of the blood–brain barrier as a complex with L-cysteine. This complex is structurally similar to methionine and is transported into cells via a widely distributed neutral amino acid carrier protein [Kerper et al., 1992].

1.3.3   Toxicity and Effects on Humans. Once MeHg is dispersed throughout the body by blood and enters the brain, it may cause structural damage. The critical target for MeHg toxicity is the central nervous system. The physical lesions may lead to tingling and numbness in fingers and toes, loss of coordination, difficulty in walking, generalized weakness, impairment of hearing and vision, tremors, and finally loss of consciousness and death. The developing fetus may be at particular risk from MeHg exposure. Infants born to mothers exposed to MeHg during pregnancy have exhibited a variety of developmental neurological abnormalities, including the following: delayed onset of walking and talking, cerebral palsy, altered muscle tone and deep tendon reflexes, and reduced neurological test scores. Maternal toxicity may or may not have been present during pregnancy for those offspring exhibiting adverse effects. For the general population, the critical effects observed following MeHg exposure are multiple central nervous system effects including ataxia and paresthesia.

1.3.4 Risk Evaluations of Exposure. As mentioned in Section 1.3.1, intake of MeHg in fish and/or other aquatic foods is considered the most serious general impact on humans. Based on risk assessments, several countries and international organizations have used risk evaluation tools such as levels of daily or weekly MeHg or mercury intakes considered safety (RfD, reference dose; and PTWI, provisional tolerable weekly intake), limits/guidelines for maximum concentrations in fish, and fish consumption advisories. The RfD is defined by the United States Environmental Protection Agency (US EPA) as an estimate (with uncertainty spanning perhaps an order of magnitude) of a daily exposure to the human population (including sensitive subgroups) that is likely to be without an appreciable risk of deleterious effects during a lifetime. On the other hand, the tolerable intake represents the maximum acceptable level of a contaminant in the diet; the goal should be to limit exposure to the maximum feasible extent, consistent with the PTWI.

Recent prospective epidemiologic studies from the Faroe Islands, the Seychelles Islands and New Zealand cohorts assessed the developmental effects of lower MeHg exposure in fish-consuming populations resulting from maternal and fetal exposures to MeHg [Cernichiari et al., 1995b; Grandjean et al., 1997; Crump et al., 1998]. The US EPA decided to rely on the Faroe Islands study [EPA U.S., 1997] and a benchmark analysis dose was established and converted into a maternal intake of 1.1 μg mercury per kilogram body weight (bw) per day. Finally, applying a safety factor of 10, a reference dose (RfD) of 0.1 μg/kg bw/day was recommended. The reference dose will be exceeded if a substantial amount of fish, contaminated with mercury, is ingested. As an example, if the weekly intake is about 100 g (one typical fish meal per week) of fish with >0.4 mg/kg, the RfD will be exceeded. This suggests that fish mercury levels should be kept below this limit.

During the 67th meeting of the Joint FAO/World Health Organization (WHO) Expert Committee on Food Additives (JECFA), held in Rome from June 20 to June 29, 2006, to evaluate certain food additives and contaminants, resulted in the confirmation of the PTWI of 1.6 μg MeHg/kg bw/week for the general population [JECFA, 2006]. The Committee confirmed the existing PTWI, based on the most sensitive toxicological endpoint developmental neurotoxicity in the most susceptible species (humans). However, the Committee noted that life stages other than the embryo and fetus may be less sensitive to the adverse effects of MeHg. The US Food and Drug Administration (FDA) and EPA recommend to eat no more than

12 oz per week of fish and shellfish lower in mercury. This is two average 6-oz meals. Fish lower in mercury include shrimp, canned light tuna (not albacore tuna), salmon, pollock, and catfish. If no advice is available, eat no more than two meals per week. In Europe, following a request from the European Commission, the European Food Safety Authority's (EFSA) Scientific Panel on Contaminants in the Food Chain (CONTAM) has evaluated the possible risks to human health from the consumption of foods contaminated with mercury—in particular, MeHg—based on intake estimates for Europe. The Panel considered the PTWI established recently by the JECFA as well as the intake limits established by the U.S. National Research Council (US-NRC).

2 ADVERSE ECOLOGICAL EFFECTS

2.1 Laboratory Toxicity Studies

Despite the existence of substantial information on bioaccumulation patters of inorganic and organic Hg species to aquatic plant and invertebrate species, little is known about their bioavailability and toxicity in both laboratory and field studies [Boening, 2000].

In an exhaustive review of mercury, Boening [2000] reported that MeHg was about 10 times more toxic to fish than the inorganic forms. However, due to the lack of data, it was not possible to establish if organic or inorganic forms of mercury were more toxic to aquatic plants, algae, and invertebrate spacies. Furthermore, Boening [2000] also showed that due to the speciation of IHg, its toxicity varied largely across studies within species. Thus both the rather high disparity of environmental water conditions and the lack of information in organic mercury forms prevented Boening from obtaining conclusive results.

In Figs. 1 and 2 we presented updated information on Hg and MeHg toxicity on fish, aquatic plants, and invertebrate species using the EPA ECOTOX Database (http://cfpub.epa.gov/ecotox/) and the LC50s as the measured endpoint.

For fish considering only 10 species (tilapia, trout, catfishes, striped bass, carp, white perch, eels, mummichog, pumpkinseed, goldfish), we obtained 370 and 150 LC_{50} values for inorganic and methyl mercury, respectively (Fig. 1). For plant and invertebrate aquatic taxa, from almost 1000 case studies found, only 50 were conducted with methyl mercury (Fig. 2), which are still indicating an alarming lack of toxicity studies on the most toxic form of mercury. The analysis of the data indicated a log normal distribution of the species sensitivities to inorganic Hg but not to MeHg (Figs. 1 and 2), with the latter probably being related to the lack of data. The median sensitivity to Hg and MeHg were 400 and 40 µg/liter for fish and 180 and 16 µg/liter, for plant and invertebrate species, respectively, which indicates about 10-fold greater toxicity of MeHg compared to that of Hg. These results agree with the 10-fold differences in toxicity reported with regard to fish [Boening, 2000]. Within fish species, trout and salmons were the most sensitive organisms and catfishes, with Tilapia and carps being the most tolerant ones (Fig. 1). For invertebrate and plants, small plant species like water celery, crustacean and molusc larvae and

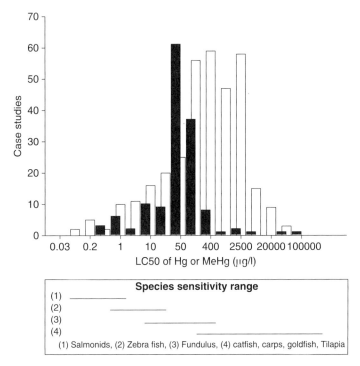

Figure 1. Toxicity, reported in log scale of LC_{50} values, IHg (mercuric chloride, white bars), and of organic mercury (MeHg, filled black bars) to fish. The range of sensitivity of the principal taxonomic groups considered is shown in the box below the graph using lines.

small cladocerans (*Ceriodaphnia*) were the most sensitive taxa, followed by the water flea (*Daphnia*, Fig. 2). The most tolerant plant and invertebrate taxa included big decapoda species such as lobsters and crayfish, as well as clamps and microalgae. It is also important to notice the quite large variability in tolerance reported within species across studies (i.e., for the water flea, trout and salmons), which agree with the fact that mercury has a complex speciation and hence its toxicity will vary dramatically across small changes in environmental water conditions. Although reported information on mercury toxicity and speciation in water is not conclusive, there is strong evidence that factors such as organic matter may affect dramatically its bioavailability and toxicity [Verslycke et al., 2003; Ravichandran, 2004].

It is important to point out that the data reported here and in most databases are lethal responses and hence of limited ecological value. Future laboratory work—in particular, within non-fish species—should be focused on studying and modeling how mercury speciation may affect sublethal responses of different sensitive aquatic organisms such as small aquatic plants, larvae, and cladocera. In relation to this, sublethal assays of sensitive species such as the sea urchin larvae [Fernandez and Beiras, 2001], and duckweed [Li and Xiong, 2004] and the use of short-term demographic or sublethal responses using cladoceran species [Barata and Baird,

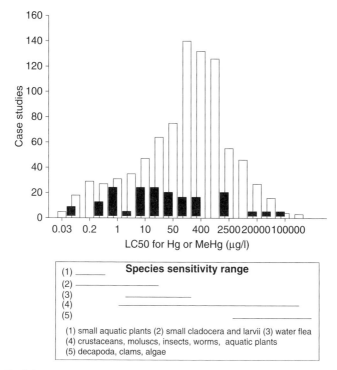

Figure 2. Toxicity, reported in log scale of LC_{50} values, IHg (mercuric chloride, white bars), and of organic mercury (MeHg, filled black bars) to aquatic plants, algae, and invertebrates. The range of sensitivity of the principle taxonomic groups considered is shown in the box below the graph using lines.

2000; Barata et al., 2002] are promising ways to relate Hg speciation with ecological-related and cost-effective toxicity responses.

2.2   Biochemical Approaches to Study Bioavailability and Effects

Another important hallmark of most studies performed in the field is the lack of existing information on both bioavailability and effects of mercury in nonvertebrate species inhabiting impacted locations [Wolfe et al., 1998; Boening, 2000; Ullrich et al., 2001; Wang et al., 2004]. In most cases the studies are limited to report mercury and methyl mercury levels in the organism. For instance, not all metal measured in an organism can react with biological macromolecules and hence be susceptible to impair organism's health [Barata et al., 2002]. This problem is especially important for most invertebrate species since due to their small size, metal body burdens are often measured in whole-animal tissues and not on target organs or tissues. A plausible alternative, however, is to measure simultaneously both metal levels in organism tissues and specific metal-toxicological responses [Canesi et al., 1999]. Metal complexation with glutathione, induction of metalothionein proteins, and metal sequestration into lysosomes can be considered good biochemical responses to Hg exposure. Reduced glutathione (GSH) and the cysteine-rich protein metallothionein (MT) have been suggested to play a cooperative protective roles against metal toxicity, with the former acting as initial defense and the latter acting at a second stage [Viarengo, 1994; Ochi et al., 1988]. GSH, the most abundant cellular thiol, is involved in metabolic and transport processes and in the protection of cells against the toxic effects of metals. The Hg cation (Hg^{2+}) is characterized by an extremely high affinity for SH residues [Viarengo, 1994], forming GS–Hg complexes through its thiolate sulfur atom [Rabestein et al., 1985]. Methyl mercury also has high affinity for SH groups, including those of GSH; MeHg–SG complexes have been identified in different mammalian tissues and in the bile of fish, and tissue GSH levels are an important determinant of both renal and hepatic MeHg uptake [Ballatori, 1991]. Although the main role of heavy metal detoxication can be ascribed to MT, the induction of MT biosynthesis by metal cations is relatively slow and considerable toxicity may occur before establishment of effective levels of MT [Roesijadi, 1992; Ochi et al., 1988; Viarengo, 1994]. Rates of MT synthesis induction and related metal binding to MT can become limiting at different times of exposure, depending on the tissue and on the kind of metal cation, thus resulting in differential binding of metals to other structures, including target sites for metal toxicity and detoxication [Roesijadi, 1992]. In marine invertebrates it has been reported that Hg^{2+} can rapidly displace other cations such as Zn^{2+} from preexisting metallothioneins and induce the synthesis of apothioneins.

The role of lysosomal vacuolar system in heavy metal compartmentalization and detoxification in marine invertebrate cells is also widely recognized. In the digestive gland of mussels, copper is detoxified into lysosomes trapped in the form of both oxidized insoluble Cu-thioneins and lipofucsin, which are subsequently eliminated by exocytosis of residual bodies [Regoli et al., 1998; Viarengo et al., 1989; Marchi et al., 2004]. Although lysosomal mercury accumulation and MT induction have been demonstrated in MeHg-treated organisms [Schionning et al., 1991; Rising et al., 1995], information on aquatic invertebrate species is scarce. Nevertheless, studies on mussels suggest that complexation with MeHg may be one of the detoxification pathways of this metal species [Canesi et al., 1999].

Inorganic and organic mercury forms are also known to exert toxicity causing oxidative stress or disrupting the metabolism and physiological role of essential metals such as Zn, Cu, and Ca. To minimize oxidative damage to cellular components, organisms have developed antioxidant defenses. Important antioxidant enzymes are the enzymes superoxide dismutase (SOD, EC 1.15.1.1 converts $O_2^{-\cdot}$ to H_2O_2), catalase (CAT; EC 1.11.1.6—reduces H_2O_2 to water), and glutathione peroxidase (GPX; EC 1.11.1.9—detoxifies H_2O_2 or organic hydroperoxides produced, for example, by lipid peroxidation) [Di Giulio et al., 1995; Halliwell and Gutteridge, 1999]. Glutathione S-transferases (GST; EC 2.5.1.18) catalyzes the conjugation of glutathione (GSH) with various electrophilic substances, and it plays a role in preventing oxidative damage by conjugating breakdown products of lipid peroxides to GSH [Ketterer et al., 1983]. Some GST isozymes display peroxidase activity [Ketterer et al., 1983; Hayes and Pulford, 1995]. Finally, glutathion reductase

(GR, EC 1.6.4.2) catalyze the reduction of oxidized glutathione GSSG to GSH and hence contribute to reestablish the glutathione-mediated antioxidant defensive system.

Organisms can adapt to increasing reactive oxygen species (ROS) production by up-regulating antioxidant defences, such as the activities of antioxidant enzymes [Livingstone, 2003]. Failure of antioxidant defenses to detoxify excess ROS production can lead to significant oxidative damage including enzyme inactivation, protein degradation, DNA damage, and lipid peroxidation [Halliwell and Gutteridge, 1999]. In particular, lipid peroxidation is considered to be a major mechanism, by which oxyradicals can cause tissue damage, leading to impaired cellular function and alterations in physicochemical properties of cell membranes, which, in turn, disrupt vital functions [Rikans and Hornbrook, 1997]. Lipid peroxides are known to decompose and produce a variety of substances, the most important of which is malondialdehyde (MDA) [Leibovitz and Siegel, 1980]. MDA is incorporated into various large fluorescent biomolecules, which accumulate in the cells. These lipid-soluble fluorescent products, also known as lipofucsin, are regarded as good biomarkers of age and of the degree of oxidative stress in animals [Sohal, 1981].

Most studies on ROS production and effects have been conducted at the suborganismal level and limited to vertebrate species. Studies that have demonstrated potential for ROS production, antioxidant defense responses, and oxidative damage in aquatic invertebrates are restricted to few species, and information relating biochemical effects with individual level responses is scarce or absent (see reviews of Di Giulio et al. [1995] and Livingstone [2001]). Therefore, validation of pro-oxidant and antioxidant biochemical biomarkers as early indicators of Hg to aquatic biota require those systems to be characterized in more invertebrate species and require their ecological relevance to be established.

Here we are presenting original data on the freshwater mussels species, *Dreissena polymorpha*, exposed to inorganic and organic forms of Hg in the laboratory and in the field.

2.3   Methods

2.3.1   Laboratory Trials.   Speciments of *D polymorpha*, 2 cm long, were obtained from Riba-Roja reservoir (Ebro River, NE Spain, Fig. 3). Before use, mussels were kept in an aquarium in static tanks containing aereated ASTM synthethic hard

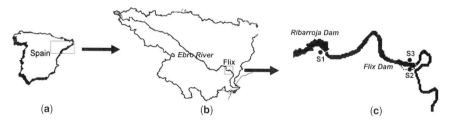

Figure 3. Field-studied sites showing the location of the Ebro River basin (**a**), that of the studied area (**b**), and that of stations 1, 2, and 3 in the lower part of the river basin.

water at 20°C for 10 days and fed with a suspension 1:1 of *Scenedesmus subspicatus* and *Chlorella vulgaris* (10^6 cells/ml, daily). During the experiments, mussels were exposed for 4 days to $HgCl_2$, CH_3HgCl at 0.2 μmol/liter Hg. Metals were added every day to the water as $HgCl_2$ aqueous standard solutions and as CH_3HgCl from an acetone stock solution. Acetone was never higher than 0.5 ml/liter. The water (50 liters per 50 mussels) was changed every day; 2 hr before changing the medium, mussels were allowed to feed on the suspension of algae (see above). After the exposure period, digestive glands were dissected and pools of five individuals were frozen in liquid N_2 and preserved at $-80°C$ until analysis.

2.3.2 Field Studies. Mussels (2 cm long) were collected from three sites that include upstream reference location (S1, Ribaroja) and two historically impacted chlor-alkali Hg sites in Flix reservoir (Ebro River, NE Spain): S2, right beside the factory, and S3 in the opposite river margin. Once collected, they were allowed to depurate for 2 hr and then their digestive glands were dissected and preserved as above until analysis (Fig. 3). Chemical analysis of total Hg (THg) in those mussels evidence levels of 0.028, 0.525, and 0.066 mg/kg, w.w. (wet weight) for sites S1, S2, and S3, respectively.

2.3.3 Biochemical Analysis. Pools of five digestive glands (wet 0.2–0.3 g) were homogenized at 4°C in 1:4 w.w./buffer volume ratio in 100 mM phosphate buffer, pH 7.4, containing 100 mM KCl and 1 mM EDTA. Homogenates were centrifuged at 10,000–30,000 g for 10–30 min, and the supernatants were immediately used as enzyme GSH sources or to determine metalothionein levels, respectively. Biochemical measurements were carried out using a microplate fluorimeter Synergy HT (BioTec) and a spectrophotometer reader (Uvikon 941) at $25 \pm 0.5°C$. Assays were run at least in quadruplicate. Enzyme activities including SOD, CAT, GPX, GR, GST, and levels of lipid peroxidation were measured following Barata et al. [2005] and Carlberg and Mannervik [1985], GSH levels were determined following the fluorimetric assay of Kamencic [2000], DNA strand breaks were measured according to Lafontaine [2000] and MT levels were determined following Viarengo et al. [1997] procedures. A detailed procedure for the previously mentioned assays can be found in Faria et al. [2007].

2.4   Results and Discussion

Laboratory exposures results are depicted in Figs. 4 and 5 and showed different patterns of response of mussels exposed to inorganic and MeHg. For instance, MeHg (but not Hg) significantly reduced ($P < 0.05$) GSH levels (Fig. 4). Alternatively, Hg (but not MeHg) induced metallothionein levels (Fig. 3). Regarding antioxidant defensive systems, both compounds affected antioxidant enzyme activities similarly, inhibiting CAT and GPx-Se and increasing the activity of SOD (Fig. 4). Tissue damage, measured as peroxidation levels of lipids and of DNA strand breaks in digestive gland, were only evident for Hg (Fig. 5). Nevertheless, it is important to consider that short-term (4 days) exposures in the laboratory may have not allowed substantial damage to be produced.

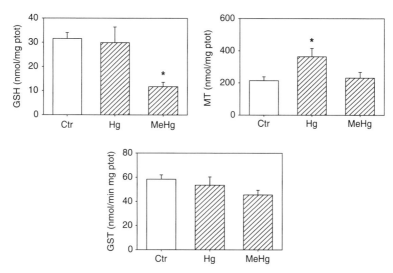

Figure 4. Biotransformation biomarkers (mean $\pm$ SEM, $n = 6$–7) in mussels exposed to 0.2 μmol/liter Hg of Hg Cl$_2$ (Hg) and CH$_3$HgCl (MeHg). Levels of reduced glutathione (GSH) and of metallothionein proteins (MT) and activities of glutathione S transferase (GST). *Significant ($P < 0.05$) different from control (Ctr) values following ANOVA and Tukey's post hoc comparisons.

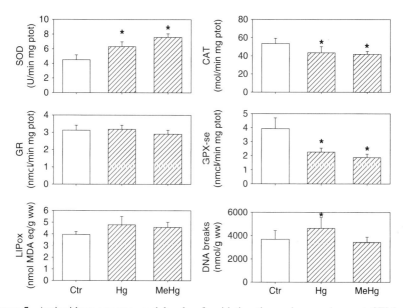

Figure 5. Antioxidant enzymes and levels of oxidative tissue damage (mean $\pm$ SEM $n = 6$–7) in mussels exposed to 0.2 μmol/liter Hg of HgCl$_2$ (Hg) and CH$_3$HgCl (MeHg). SOD, CAT, GPX-Se, LPox, and DNA brakes are superoxide dismutasa, catalase, Se-dependent glutathione peroxidase, and glutathione reductase activities, and levels of lipid perodides and of DNA strand brakes. *Significant ($P < 0.05$) different from control values following ANOVA and Tukey's post hoc comparisons.

Results for field exposed mussels are reported in Figs. 6 and 7. Mussels from the two affected locations (sites 2 and 3) showed low levels of GSH and of MT and increased activities of GST relative to the upstream location (site 1). Activities of the antioxidant enzymes SOD and CAT, however, were higher in the most exposed mussels (site 2), followed by those collected in the affected area but in the other margin (site 3). Tissue damage (lipid peroxidationa and DNA brakes) were only apparent in mussels from site 2.

Overall, the results reported here indicate that the use of a battery of biomarkers can be useful to detect exposure to and effects of Hg. In the laboratory under controlled and high exposure levels, two biotransformation biomarkers (levels of GSH and of MT) were able to distinguish inorganic and organic Hg forms. However, in the field, where MeHg is the dominant form of mercury and where other contaminant sources (other persistent contaminants for instance) may co-occur as well, the results are less apparent. For example, as expected from laboratory experiments, activities of SOD and tissue damage levels (LPox, DNA brakes) in the field were higher in exposed mussels. However, contrary to laboratory findings, MT levels decreased and so did activities of CAT in field-exposed mussels.

In summary, the data presented here indicate that the combined use of chemical and specific biochemical markers will improve our ability to detect environmental hazards of mercury to biota, provided that the latter are characterized and "calibrated" in laboratory assays.

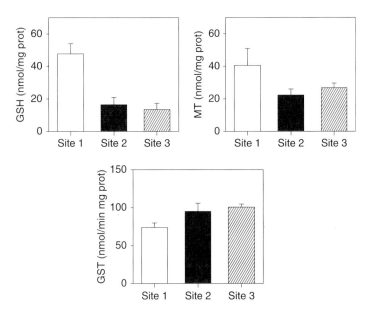

Figure 6. Biotransformation biomarkers (mean $\pm$ SEM, $n = 8-10$) in field collected mussels. Levels of reduced glutathione (GSH) and of metallothionein proteins (MT) and activities of glutathione S transferase (GST). *Significant ($P < 0.05$) different from control values following ANOVA and Tukey's post hoc comparisons.

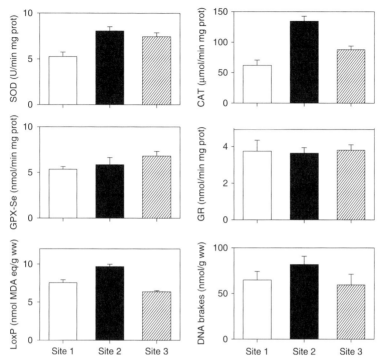

Figure 7. Antioxidant enzymes and levels of oxidative tissue damage (mean $\pm$ SEM, $n =$ 6–7) in field-collected mussels. SOD, CAT, GPX-Se, LPox, and DNA brakes are superoxide dismutase, catalase, Se-dependent glutathione peroxidase and glutathione reductase activities, and levels of lipid perodides and of DNA strand brakes, respectively. *Significant ($P < 0.05$) different from control values following ANOVA and Tukey's post hoc comparisons.

3 CASE STUDY: MERCURY-CELL CHLOR-ALKALI PLANTS AS A MAJOR POINT SOURCES OF MERCURY IN AQUATIC ENVIRONMENTS—THE CASE OF CINCA RIVER, SPAIN

3.1 Introduction

Mercury pollution in aquatic ecosystems has received great attention since the discovery of mercury as the cause of Minamata disease in Japan in the 1950s. Large quantities of mercury are released to the aquatic environment, and one of the major point sources of mercury has been its use in the electrolytic preparation of chloride and caustic soda in mercury-cell chlor-alkali plants. The effluents from these plants may contain high quantities of mercury in the inorganic form [Arribére et al., 2003]. Once in the aquatic ecosystems, part of the IHg can be microbiologically converted into MeHg and uptaken up by aquatic organisms.

Fishes accumulate mercury directly from food and the surrounding water [Rainbow, 1985], and they can bioconcentrate large amounts of this metal. In fact, the half-life of elimination of MeHg is among the longest known for metals, with values of 640–1200 days in different freshwater species [Moore and Ramamoorthy, 1983]. Temperature accelerates uptake of mercury from the water by increasing the metabolic rate and the respiratory volume. Because the gills are the primary surface for absorption of waterborne substances by fish, uptake increases as respiratory volume increases. Increased metabolic rate also increases energy demand and thus increases food consumption and exposure to mercury through the food chain. Distribution of mercury within a fish can be considered as a flowing system in which the flow pattern moves from the absorbing surfaces (gills, skin, and gastrointestinal tract), into the blood, then to the internal organs and eventually either to the kidney and bile for excretion or to the muscle for long-term storage.

One of the first adaptive and protective biological responses to mercury exposure is induction of metallothionein synthesis. Metallothioneins are low-molecular-weight proteins that serve as a storage depot for copper and zinc and that "scavenge" sulfhydryl-reactive metals that enter the cell. Metallothioneins are rich in cysteine (~30%) and have higher affinities for Hg and Cd than for Zn [Coyle et al., 2002]. Studies on the regulation of the MT gene expression provided evidence that induction by metals is a direct response to increased intracellular metal concentration, mediated through the action of metal-binding regulatory factors [Thiele, 1992]. When the amount of accumulated metal exceeds the fish ability to synthesize the metallothionein, toxicity of the Hg occurs. The pro-oxidant properties of the Hg are exacerbated by their inhibitory effect on antioxidant processes. Hg promotes lipid peroxidation, depletes glutathione (GSH), and inhibits antioxidative processes, with the consequent disruption of the membrane structure and the mitochondrial function (Stohs and Bagchi, 1995). In addition, mercury disrupts the structure and function of numerous important proteins through direct binding to free sulfhydryl groups. Histopathological findings in liver of fish exposed to mercury include reduced lysosomal membrane stability, lipid peroxidation with an enhaced content of ceroid/lipofucsin, apoptosis and/or necrosis, increased number and size of macrophage aggregates, or alterations of cytoplasmatic organization [Meinelt et al., 1997; Oliveira-Ribeiro et al., 2002].

3.2   The Case of Mercury Pollution in Cinca River, Spain

One well-documented case of feral fish chronically exposed to mercury through the effluent of a chlor-alkali plant is the Cinca River, a tributary of the Ebro river, downstream of the mercury-cell chlor-alkali plant at Monzón [Raldúa and Pedrocchi, 1996; Lavado et al., 2006; Quiros et al., 2007; Raldúa et al., 2007]. Here we review all the data available about the biological effects on the feral fish of the Cinca River downstream of this plant.

Monzón (NE Spain) is a small and highly industrialized city in the middle course of the Cinca River (Fig. 8). This industrial activity has caused the historical release of organic and inorganic compounds to the river due the pollution coming from the mercury-cell chlor-alkali industry [Raldúa and Pedrocchi, 1996], production and utilization of solvents and chlorinated pesticides [Raldúa et al., 1997], and usage of flame retardants in electrical plants [Eljarrat et al., 2004]. Although chlorine production of the Monzón plant was held from 1998 to 2002 (ca. 31,000 tonnes), mercury

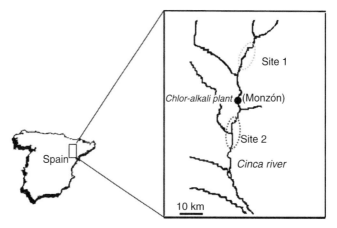

Figure 8. Map with the sampling areas in the Cinca River (NE Spain), upstream (S1) and downstream (S2) of the chlor-alkali plant of Monzón.

losses of this plant have decreased from 1999 to 2002 [OSPAR Commission, 2004]. In spite of the reported decrease in the mercury losses, the performance of this mercury-based chlor-alkali plant is still one of the worst in Europe (1.545 g Hg/t Cl_2, in 2002). Since 1993 we have been assessing impact of the mercury released by this chlor-alkali plant on the fish population of the Cinca River, using the barbel as sentinel species.

3.2.1 Mercury Levels in Sediments. THg levels in sediment downstream of Monzón (0.4 mg/kg, d.w.) are about 25 times higher than those in sediment upstream (0.015 mg/kg, d.w.). The level of mercury in sediments from S2 is smaller than that in the commonly reported sediment downstream of chlor-alkali plants (>1 mg/kg; Abreu et al., [2000], Alonso et al., [2000]). In fact, the mercury levels reported in 1993 in sediments from S2 were 2.69 mg/kg d.w. [Raldúa and Pedrocchi, 1996]. Mercury losses of this plant decreased, but chlorine production did not, from 1999 to 2002, increasing during this period the amount of mercury in safely deposited wastes [OSPAR Commission, 2004]. Figure 9 shows that the important decrease observed in the content of THg in the sediments of Cinca River downstream of the plant during the period 1993–2002 (http://oph.chebro.es/DOCUMENTACION/Calidad/CalidadDeAguas.html), about 7 times in 9 years, reflects this improvement in the production system of the plant to minimize the disposal of metallic Hg into the Cinca River. Sediments from the S1 are under the Interim Sediment Quality Guideline (ISQG) for mercury, 0.170 mg/kg-d.w., but S2 is quite close to the probability effect level (PEL), 0.460 mg/kg-d.w. [EC, 1999].

3.2.2 Mercury Levels in Fish. *Barbus graellsii* is a bottom-feeding fish species endemic to the western Mediterranean and very common in the Ebro River

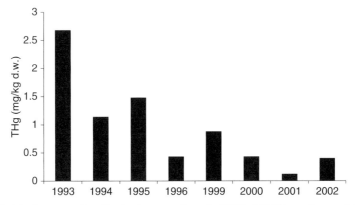

Figure 9. Total mercury (mg/kg d.w.) temporal trends (1993–2002) in sediment of the Cinca River downstream of the chlor-alkali plant of Monzón.

[Doadrio et al., 1991]. The muscle and liver of the barbel downstream of the plant show average concentrations 10- and 30-fold higher than those found upstream, respectively, with a maximal concentration of 3.64 mg/kg in the liver of an 8 years old barbel from S2 (Fig. 10).

Mercury levels in muscle and liver in barbel from the Cinca River show a very significant correlation ($r^2 = 0.83$, $P < 0.005$). Nevertheless, the liver/muscle ratio of mercury concentration is significantly different ($P < 0.0005$) in barbel sampled upstream and downstream of the factory, with values of 0.4 ± 0.07 and 1.32 ± 0.24 (average $\pm$ standard error), respectively. Muscle is the main target for organic mercury [Hakanson, 1984] and liver for inorganic and metallic mercury

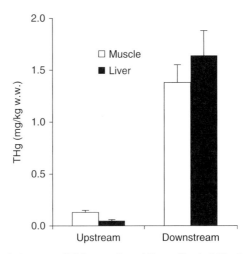

Figure 10. THg levels (mean $\pm$ SE) in muscle and liver of barbel (*Barbus graellsli*) upstream and downstream of the chlor-alkali plant in the Cinca River.

[Burrows and Krenkel, 1973]. A high liver/muscle ratio of mercury concentration have been reported in fish exposed to the effluents from chlor-alkali plant containing high quantities of Hg in inorganic form [Abreu et al., 2000; Arribére et al., 2003]. Liver of fish downstream of a chlor-alkali factory has, also, a very high inorganic to THg ratio [Arribére et al., 2003].

Despite the fact that the improvements in the production system of the chlor-alkali plant during the last 9 years have been reflected in a decrease of 70% in the mercury levels in sediments, the decrease in the content of mercury in the muscle of adult barbels has been only 30% [Raldúa and Pedrochhi, 1996; Raldúa et al., 2007]. These data are consistent with the fact that the effects of the mercury in fish populations can be observed long after the chlor-alkali plants have closed down [Arribére et al., 2003].

An evaluation of several studies about the critical body burden of mercury in fish suggests that whole-body concentrations of $10-20$ mg/kg could be lethal to fish, and whole-body concentrations of $1-5$ mg/kg could have chronic effects on fish [Niimi and Kissoon, 1994]. Downstream of Monzón, whole-body concentrations of 9 of the 11 barbel exceeded 1 mg/kg. All the barbel samples collected downstream of the chlor-alkali plant showed mercury levels higher than 0.3 mg/kg wet flesh, with the quality objective of the EU in the areas affected by discharges of mercury from the chlor-alkali electrolysis industry [Council Directive 82/176/CEE].

In order to address human health risks from fish consumption, we have targeted adult fish, which typically contain higher concentrations of Hg and are more commonly sought by anglers. All the barbel from downstream had Hg concentrations above 0.5 mg/kg wet weight. Historically, 0.5 mg/kg was a commonly reported human advisory Hg fish concentration applicable to nonpredatory fish [EC, 2005]. Recently, the U.S. EPA reduced the human health Hg fish criteria to 0.3 mg/kg w.w. [National Listing of Fish and Wildlife Advisories, 2001]. Local authorities are aware of the high levels of mercury in fish downstream of Monzón, and since June 2000 they have recommended that anglers catch and release in this area.

3.2.3 Molecular Effects. Recently, the group of Dr. Piña in the Department of Molecular Biology-CSIC developed molecular tools in barbel for environmental monitoring [Quiros et al., 2007]. Comparison between the levels of MT mRNA and the MT protein in barbels from the Cinca River shows a bimodal correlation (Fig. 11), with a linear correlation at low MT expression levels (up to about 2 μg protein per gram of cytosolic total protein; solid symbol in Fig. 11) and a plateau at high expression levels (empty symbols). A simple explanation for this is the accumulation of MT protein at high levels of MT transcripts, likely occurring after chronic exposures to high levels of the inductor(s). Both parameters correlate with mercury content in liver (Fig. 12). A similar good correlation was found between hepatic MT mRNA and muscle mercury in feral largemouth bass [Schlenk et al., 1995]. Although it is well known that most of the mercury present in fish is MeHg, upon chronic exposures as much as 50% of the hepatic MeHg is demethylated to IHg [Barghigiani et al., 1989], which is capable of inducing MT transcription. However, it is unclear if MeHg is able to induce MT transcription

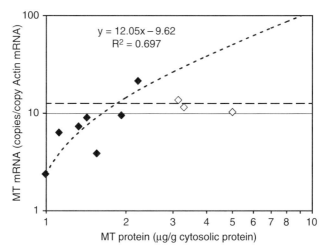

Figure 11. Correlation between metalothionein mRNA and MT protein levels in liver from adult barbels from S1–S3. Samples with low MT protein levels, which follow a linear correlation to mRNA levels, are indicated with solid diamonds; linear regression parameters for these samples are indicated at the top of the graph. The correlation was statistically significant ($P = 0.019$). The putative maximum level of mRNA transcript resulting in protein accumulation is indicated by a horizontal dashed line.

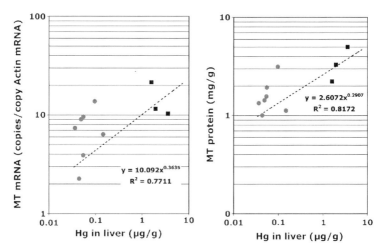

Figure 12. Correlations between Hg content and metallothionein mRNA levels (**left**) and between Hg content and metallothionein protein levels (**right**) in the liver of adult barbels from S1 (gray circles) and S2 (black squares). The graph also shows power regression fitting parameters for the latter samples. Both regressions are statistically significant ($P = 0.021$ for mRNA and $P = 0.013$ for protein).

in fish. Subdermal exposure of rats during 14 days to different doses of MeHg (25–200 µg/kg/day) resulted in 2.5- to 7-fold increases in MT I mRNA in testis, without changes in either MT II or MT III mRNA levels [Dufresne and Cyr, 1999]. Atlantic salmon fed for 4 months on fish-meal-based diet supplied with MeHg showed significant induction of hepatic MT after 2 months of exposure, with a nonobserved effect level (NOEL) of 0.5 mg MeHg/kg [Berntssen et al., 2004]. Nevertheless, catfish exposed to dietary MeHg for 30 days showed unchanged MT mRNA levels, despite the fact that hepatic and muscle mercury were significantly increased [Schlenk et al., 1997]. From the current study it is not possible to discriminate which form of the mercury, inorganic or organic, is involved in the induction of the MT transcription.

Chemical mixtures have been reported to affect biomarkers' response differently from compounds administered alone. For example, the sensitivity of biomarkers like MT and CYP1A are likely to be influenced by pollutants other than their primary inducers. For instance, pretreatment of the European flounder (*Platichthys flesus L.*) with benzo[a]pyrene (BaP) and PCB-156, two well-known inducers of the CYP1A, inhibited the MT induction by cadmium (30–50%). Furthermore, CYP1A induction caused by BaP was inhibited (40% compared with BaP treatment alone) in flounder pretreated with cadmium [Sandvik et al., 1997]. Similar results have been found using fish hepatocyte cultures [De Faverney et al., 2000; Ghosh et al., 2001]. Nevertheless, total TEQ values, including PCDDs, PCDFs, and PCBs, in sediment of the Cinca River downstream of the chlor-alkali plant are 9.92 pg/g d.w. [Eljarrat et al., 2008], a value below the safe sediment value (20 pg TEQ/g dw), and we have found a relatively uniform EROD levels in liver of barbel from S1 and S2 (between 30 and 50 pmol/min/mg protein). Thus, the relative small induction of metallothionein found in barbels downstream of Monzón in comparison with the high levels of mercury present in their livers could not be mediated by the inhibitory action of PCBs or other dioxin-like compounds.

3.2.4 Histopathology. Histological analysis has shown a significant higher prevalence of histological alterations in liver of fish downstream of the chlor-alkali plant ($P < 0.05$, Fig. 8). These anomalies included proliferation of macrophage aggregates (MA), disruption of the normal morphology, and degeneration of hepatocytes with nuclear pycnosis, and they occasionally included cytoplasmatic condensation, decreased cytoplasm/nucleus ratio, and an enhanced content in ceroid/lipofucsin in liver [Raldúa et al., 2007]. It is worth noting that barbels with the highest hepatic mercury levels showed also the more severe liver pathologies.

Mercury is a well-known pro-oxidant that exerts oxidative stress via H_2O_2 production and causes lipid peroxidation, and ceroid/lipofucsin are lipogenic pigments arising from the oxidation of unsatured lipids [Agius and Robert, 2003]. The enhanced content of lipofuscin may have been stimulated by mercury accumulation because lipofuscin, the end product of lipid peroxidation, contributes to the detoxication of heavy metals that are trapped both chemically and mechanically and then eliminated by excretion of residual bodies [Regoli, 1992]. Pathological findings in liver of *Salvelinus alpinus* associated with MeHg exposure included severe necrosis and alterations of cytoplasmatic organization [Olivira-Riveiro et al., 2002]. Nonetheless, an earlier study on prolonged dietary MeHg exposed fish did not find pathological liver damage [Wobeser, 1975]. Although the suitability of the macrophage aggregates as a possible bioindicator for mercury pollution has been discussed in the literature and positive correlations between mercury and MA response in the liver of pikes have been found [Meinelt et al., 1997], there are many other factors that can affect MA. Larger fish, fish with nutritional deficiencies, or fish in poor health tend to have more or larger MA [Wolfe, 1992; Teh et al., 1997]. In addition, the number and/or size of macrophage aggregates increase with age [Brown and George, 1985].

REFERENCES

Abreu SN, Pereira E, Vale C, Duarte AC. 2000. Accumulation of mercury is sea bass from a contaminated lagoon (Ria de Aveiro, Portugal). *Mar Pollut Bull* **40**:293–297.

Agius C, Robert RJ. 2003. Melano-macrophage centres and their role in fish pathology. *J Fish Dis* **26**:499–509.

Akagi H, Grandjean P, Takizawa Y, Weihe P. 1998. Methylmercury dose estimation from umbilical cord concentrations in patients with Minamata disease. *Environ Res* **77**:98–103.

Alonso D, Pineda P, Oliveiro J, Gonzalez H, Campos N. 2000. Mercury levels in muscle of two fish species and sediments from the Cartagena Bay and the Ciénaga Grande de Santa Marta, Colombia. *Environ Pollut* **109**:157–163.

Arribére MA, Ribeiro Guevara S, Sanchez RS, Gil MI, Román Ross G, Daurade LE, Fajon V, Horvat M, Alcalde R, Kestelman AJ. 2003. Heavy metals in the vicinity of a chlor-alkali factory in the upper Negro River ecosystem, Northern Patagonia, Argentina. *Sci Total Environ* **301**:187–203.

ATSDR. 1999. *Toxicological Profile for Mercury*. Atlanta, GA: Agency for Toxic Substances and Disease Registry.

Ballatori N, Clarkson TW. 1982. Developmental changes in the biliary excretion of methylmercury and glutathione. *Science* **216**:61–63.

Ballatori N. 1991. Mechanisms of metal transport across liver cell plasma membranes. *Drug Metab Rev* **23**:83–132.

Barata C, Baird DJ, Soares AMVM. 2002. Food supply on density-dependent effects on demographic responses of the cladoceran *Moinodaphnia macleayi* to heavy metal exposure. *Ecol Appl* **12**:552–564.

Barata C, Baird DJ. 2000. Determining the ecotoxicological mode of action of toxicants from measurements on individuals: Results from short duration chronic tests with *Daphnia magna* Straus. *Aquat Toxicol* **48**:195–209.

Barata C, Lekumberri I, Vila-Escalé M, Prat N, Porte C. 2005. Trace metal concentration, antioxidant enzyme activities and susceptibility to oxidative stress in the tricoptera larvae *Hydropsyche exocellata* from the Llobregat riven basin (NE Spain). *Aquat Toxicol* **74**:3–19.

Barata C, Markich SJ, Baird DJ, Soares AMVM. 2002. The relative importance of water and food as cadmium sources to *Daphnia magna* Straus. *Aquat Toxicol* **61**:143–154.

Barghigiani C, Pellegrini D, Carpene E. 1989. Mercury binding proteins in liver and muscle of flat fish from the northern Tyrrhenian sea. *Comp Biochem Physiol C* **94**:309–312.

Berg T, Sekkesaeter S, Steinnes E, Valdal AK, Wibetoe G. 2003. Springtime depletion of mercury in the European Arctic as observed at Svalbard. *Sci Total Environ* **304**:43–51.

Bergan T, Gallardo L, Rodhe H. 1999. Mercury in the global troposphere: A three-dimensional model study. *Atmos Environ* **33**:1575–1585.

Bergan T, Rodhe H. 2001. Oxidation of elemental mercury in the atmosphere; Constraints imposed by global scale modelling. *J Atmos Chem* **40**:191–212.

Berntssen MHG, Hylland K, Julshamn AK, Waagbø R. 2004. Maximum limits of organic and inorganic mercury in fish feed. *Aquaculture Nutr* **10**:83–97.

Bjornberg KA, Vahter M, Petersson-Grawe K, Glynn A, Cnattingius S, Darnerud PO, Atuma S, Aune M, Becker W, Berglund M. 2003. Methyl mercury and inorganic mercury in Swedish pregnant women and in cord blood: Influence of fish consumption. *Environ Health Perspect* **111**:637–641.

Boening DW. 2000. Ecological effects, transport, and fate of mercury: A general review. *Chemosphere* **40**:1335–1351.

Brown CL, George CJ. 1985. Age-dependent accumulation of macrophage aggregation in the yellow perch, *Perca flavescens* (Mitchill). *J Fish Dis* **8**:135–138.

Burrows WD, Krenkel PA. 1973. Studies on uptake and loss of methylmercury-203 by bluegills (*Lepomis macrochirus*). *Environ Sci Technol* **7**:1127–1130.

Calasans CF, Malm O. 1997. Elemental mercury contamination survey in a chlor-alkali plant by the use of transplanted Spanish moss, *Tillandsia usneoides* (L.). *Sci Total Environ* **208**:165–177.

Canesi L, Viarengo A, Leonzio C, Filippelli M, Gallo G. 1999. Heavy metals and glutathione metabolism in mussel tissues. *Aquat Toxicol* **46**:67–76.

Canuel R, de Grosbois SB, Atikesse L, Lucotte M, Arp P, Ritchie C, Mergler D, Chan HM, Amyot M, Anderson R. 2006. New evidence on variations of human body burden of methylmercury from fish consumption. *Environ Health Perspect* **114**:302–306.

Carlberg I, Mannervik B. 1985. Glutathione reductase. *Methods Enzymol* **113**:484–490.

Carrington CD, Bolger MP. 2002. An exposure assessment for methylmercury from seafood for consumers in the United States. *Risk Anal* **22**:689–699.

Cernichiari E, Brewer R, Myers GJ, Marsh DO, Lapham LW, Cox C, Shamlaye CF, Berlin M, Davidson PW, Clarkson TW. 1995a. Monitoring methylmercury during pregnancy: Maternal hair predicts fetal brain exposure. *Neurotoxicology* **16**:705–709.

Cernichiari E, Toribara TY, Liang L, Marsh DO, Berlin MW, Myers GJ, Cox C, Shamlaye CF, Choisy O, Davidson P, Clarkson TW. 1995b. The biological monitoring of mercury in the Seychelles study. *Neurotoxicology* **16**:613–627.

Clarkson TW, Magos L. 2006. The toxicology of mercury and its chemical compounds. *Crit Rev Toxicol* **36**:609–662.

Clarkson TW, Magos L, Myers GJ. 2003. Human exposure to mercury: The three modern dilemmas. *J Trace Elem Exp Med* **16**:321–343.

Council Directive 82/176/EEC of 22 March 1982 on limit values and quality objectives for mercury discharges by the chlor-alkali electrolysis industry.

Coyle P, Philcox JC, Carey LC, Rofe AM. 2002. Metallothionein: The multipurpose protein. *Cell Mol Life Sci* **59**:627–647.

Crump KS, Kjellstrom T, Shipp AM, Silvers A, Stewart A. 1998. Influence of prenatal mercury exposure upon scholastic and psychological test performance: Benchmark analysis of a New Zealand cohort. *Risk Anal* **18**:701–713.

De Faverney CR, Lafaurine M, Girard JP, Rahmani R. 2000. Effects of heavy metals and 3-methylcholanthrene on expression and induction of CYP1A1 and metallothionein levels in trout (*Oncorhynchus mykiss*) hepatocyte cultures. *Environ Toxicol Chem* **19**:2239–2248.

de Lafontaine Y, Gagne F, Blaise C, Costan G, Gagnon P, Chan HM. 2000. Biomarkers in zebra mussels (*Dreissena polymorpha*) for the assessment and monitoring of water quality of the St Lawrence River (Canada). *Aquat Toxicol* **50**:51–71.

Diez S, Bayona JM. 2002. Determination of methylmercury in human hair by ethylation followed by headspace solid-phase microextraction-gas chromatography-cold-vapor atomic fluorescence spectrometry. *J Chromatogr A* **963**:345–351.

Diez S, Montuori P, Querol X, Bayona JM. 2007. Total mercury in the hair of children by combustion atomic absorption spectrometry (Comb-AAS). *J Anal Toxicol* **31**:144–149.

Diez S, Montuori P, Pagano A, Sarnacchiaro P, Bayona JM, Triassi M. 2008. Hair mercury levels in an urban population of southern Italy: Fish consumption as a determinant of exposure. *Environ Int* **34**:162–167.

Di Giulio RT, Benson WH, Sanders BM, Van Veld PA. 1995. Biochemical mechanisms: metabolism, adaptation, and toxicity. In: Rand G, editor. *Fundamentals of Aquatic Toxicology, Effects, Environmental Fate, and Risk Assessment*. London: Taylor & Francis, pp. 523–561.

Doadrio I, Elvira B, Bernat Y. 1991. *Peces Continentales Españoles*. Madrid: Coleccion Técnica. ICONA-CSIC.

Dolbec J, Mergler D, Passos CJS, de Morais SS, Lebel J. 2000. Methylmercury exposure affects motor performance of a riverine population of the Tapajos river, Brazilian Amazon. *Int Arch Occup Environ Health* **73**:195–203.

Dufresne J, Cyr DG. 1999. Effects of short-term methylmercury exposure on metallothionein mRNA levels in the testis and epididymis of the rat. *J Androl* **20**:769–778.

Ebinghaus R, Kock HH, Temme C, Einax JW, Lowe AG, Richter A, Burrows JP, Schroeder WH. 2002. Antarctic springtime depletion of atmospheric mercury. *Environ Sci Technol* **36**:1238–1244.

EC. 2005. Commission Regulation (EC) No 78/2005 of 19 January 2005 amending Regulation (EC) No 466/2001 as regards heavy metals.

EC. 1999. Canadian Sediment Quality Guidelines for Mercury. Scientific Supporting Document. National Guidelines and Standard Office, Environmental Quality Branch, Environment Canada, Otawa, Ontario.

Eljarrat E, De la Cal A, Raldúa D, Durán C, Barceló D. 2004. Ocurrence of polybrominated diphenylethers and hexabromocyclododecane in sediments and fish from Cinca River (Spain). *Environ Sci Technol* **38**:2603–2608.

Eljarrat E, Martínez MA, Sanz P, Concejero MA, Piña B, Quirós L, Raldúa D, Barceló D. 2008. Distribution and biological impact of dioxin-like compounds in risk zones along the Ebro River basin (Spain). *Chemosphere* **71**:1156–1161.

EPA US. 1997. *Mercury Study Report to Congress*. Washington, DC: Office of Air Quality Planning and Standards and Office of research and Development, EPA 452/R-97-0003, USA, December 1997.

EPA US. 2001. *Water Quality Criterion for the Protection of Human Health: Methyl Mercury.* EPA 0823-R-01-001. Washington, DC: U.S. Environmental Protection Agency.

Faria M, Grimalt JO, Barata C. 2007. SETAC Europe 17th Annual Meeting 20-24 May 2007, Porto, Portugal, 2007, 217 pp.

Fernandez N, Beiras R. 2001. Combined toxicity of dissolved mercury with copper, lead and cadmium on embryogenesis and early larval growth of the *Paracentrotus lividus* sea-urchin. *Ecotoxicology* **10**:263–271.

Ferrari CP, Dommergue A, Boutron CF, Skov H, Goodsite M, Jensen B. 2004. Nighttime production of elemental gaseous mercury in interstitial air of snow at Station Nord, Greenland. *Atmos Environ* **38**:2727–2735.

Fitzgerald WF, Engstrom DR, Mason RP, Nater EA. 1998. The case for atmospheric mercury contamination in remote areas. *Environ Sci Technol* **32**:1–7.

Ghosh MC, Ghosh R, Ray AK. 2001. Impact of copper on biomonitoring enzyme ethoxyresor-ufin-o-deethylase in cultured catfish hepatocytes. *Environ Res Section A* **86**:167–173.

Goering PL, Galloway WD, Clarkson TW, Lorscheider FL, Berlin M, Rowland AS. 1992. Toxicity assessment of mercury-vapor from dental amalgams. *Fundam Appl Toxicol* **19**:319–329.

Grandjean P, Weihe P, White RF, Debes F, Araki S, Yokoyama K, Murata K, Sorensen N, Dahl R, Jorgensen PJ. 1997. Cognitive deficit in 7-year-old children with prenatal exposure to methylmercury. *Neurotoxicol Teratol* **19**:417–428.

Grandjean P, White RF, Nielsen A, Cleary D, Santos ECD. 1999. Methylmercury neurotoxicity in Amazonian children downstream from gold mining. *Environ Health Perspect* **107**:587–591.

Hakanson L. 1984. Metals in fish and sediments from the river Kolbacksan water-system, Sweden. *Arch Hydrobiol* **101**:373–400.

Halliwell B, Gutteridge JMC. 1999. *Free Radicals in Biology and Medicine*. Oxford: Oxford University Press.

Halsey NA. 1999. Limiting infant exposure to thimerosal in vaccines and other sources of mercury. *J Am Med Assoc* **282**:1763–1766.

Hansen G, Victor R, Engeldinger E, Schweitzer C. 2004. Evaluation of the mercury exposure of dental amalgam patients by the Mercury Triple Test. *Occup Environ Medi* **61**:535–540.

Harada M. 1995. Minamata Disease—Methylmercury Poisoning in Japan Caused by Environmental-Pollution. *Crit Rev Toxicol* **25**:1–24.

Harakeh S, Sabra N, Kassak K, Doughan B. 2002. Factors influencing total mercury levels among Lebanese dentists. *Sci Total Environ* **297**:153–160.

Hayes JD, Pulford DJ. 1995. The glutathione S-transferase supergene family: Regulation of GST and the contribution of the isozymes to cancer chemoprotection and drug resistance. *Crit Rev Biochem Mol Biol* **30**:445–600.

Hightower JM, Moore D. 2003. Mercury levels in high-end consumers of fish. *Environ Health Perspect* **111**:604–608.

Jaffe D, Prestbo E, Swartzendruber P, Weiss-Penzias P, Kato S, Takami A, Hatakeyama S, Kajii Y. 2005. Export of atmospheric mercury from Asia. *Atmos Environ* **39**:3029–3038.

JECFA. 2006. Joint FAO/World Health Organization (WHO) Expert Committee on Food Additives. Available at www.chem.unep.ch/mercury/Report/JECFA-PTWI.htm

Kamencic H. 2000. Monochlorobimane fluorometric method to measure tissue glutathione. *Anal Biochem* **286**:35–37.

Ketterer B, Coles B, Meyer DJ. 1983. The role of glutathione in detoxification. *Environ Health Perspect* **49**:59–69.

Kerper LE, Ballatori N, Clarkson TW. 1992. Methylmercury transport across the blood–brain barrier by an amino acid carrier. *Am J Physiol* **267**:R761–R765.

Lamborg CH, Fitzgerald WF, O'Donnell J, Torgersen T. 2002. A non-steady-state compartmental model of global-scale mercury biogeochemistry with interhemispheric atmospheric gradients. *Geochim Cosmochim Acta* **66**:1105–1118.

Lavado R, Ureña R, Martin-Skilton R, Torreblanca A, del Ramo J, Raldúa D, Porte C. 2006. The combined use of chemical and biochemical markers to assess water quality along the Ebro River. *Environ Pollut* **139**:330–339.

Lebel J, Mergler D, Branches F, Lucotte M, Amorim M, Larribe F, Dolbec J. 1998. Neurotoxic effects of low-level methylmercury contamination in the Amazonian Basin. *Environ Res* **79**:20–32.

Leibovitz BE, Siegel BV. 1980. Aspects of free radical reactions in biological sustems: Aging. *J Gerontol* **35**:45–56.

Li TY, Xiong ZT. 2004. A novel response of wild-type duckweed (*Lemna paucicostata* Hegelm.) to heavy metals. *Environ Toxicol* **19**:95–102.

Lindberg SE, Brooks S, Lin CJ, Scott KJ, Landis MS, Stevens RK, Goodsite M, Richter A. 2002. Dynamic oxidation of gaseous mercury in the Arctic troposphere at polar sunrise. *Environ Sci Technol* **36**:1245–1256.

Livingstone DR. 2001. Contaminated-stimulated reactive oxygen species production and oxidative damage in aquatic organisms. *Mar Pollut Bull* **42**:656–666.

Livingstone DR. 2003. Oxidative stress in aquatic organisms in relation to pollution and aquaculture. *Revue Med Vet* **154**:427–430.

Malm O. 1998. Gold mining as a source of mercury exposure in the Brazilian Amazon. *Environ Res* **77**:73–78.

Marchi B, Burlando B, Moore MN, Viarengo A. 2004. Mercury and copper-induced lysosomal membrane destabilisation depends on $[Ca^{2+}]_i$ dependent phospholipase A2 activation. *Aquat Toxicol* **66**:197–204.

Mason RP, Fitzgerald WF, Morel FMM. 1994. The biogeochemical cycling of elemental mercury—Anthropogenic influences. *Geochim Cosmochim Acta* **58**:3191–3198.

Meinelt T, Kruger R, Pietrock M, Osten R, Steinberg C. 1997. Mercury pollution and macrophage centres in pike (*Esox lucius*) tissues. *Environ Sci Pollut Res* **4**:32–36.

Montuori P, Jover E, Alzaga R, Diez S, Bayona JM. 2004. Improvements in the methylmercury extraction from human hair by headspace solid-phase microextraction followed by gas-chromatography cold-vapour atomic fluorescence spectrometry. *J Chromatogr A* **1025**:71–75.

Montuori P, Jover E, Diez S, Ribas-Fito N, Sunyer J, Triassi M, Bayona JM. 2006. Mercury speciation in the hair of pre-school children living near a chlor-alkali plant. *Sci Total Environ* **369**:51–58.

Moore JW, Ramamoorthy S. 1983. *Heavy Metals in Natural Waters. Applied Monitoring and Impact Assessment*. New York: Springer Verlag.

Morton J, Mason HJ, Ritchie KA, White M. 2004. Comparison of hair, nails and urine for biological monitoring of low level inorganic mercury exposure in dental workers. *Biomarkers* **9**:47–55.

National Listing of Fish and Wildlife Advisories. 2001. EPA-823-F-01-010; U.S. Environmental Protection Agency, Office of Water; U.S. Government Printing Office: Washington, DC, 2001.

Niimi AJ, Kissoon GP. 1994. Evaluation of the critical body burden concept based on inorganic and organic mercury toxicity to rainbow trout (*Oncorhynchus mykiss*). *Arch Environ Contam Toxicol* **26**:167–178.

NRC. 2000. NRC (National Research Council). *Committee on the Toxicological Effects of Methylmercury 2000. Toxicological Effects of Methylmercury.* Washington, DC: National Academy Press.

Ochi T, Ostuka F, Takahashi K, Ohsawa M. 1988. Glutathione and metalothioneins as cellular defense against cadmium toxicity in cultured chinese hamster cell. *Biol Interact* **65**:1–14.

Oliveira-Ribeiro CA, Bleger L, Pelletier E, Rouleau C. 2002. Histopathological evidence of inorganic mercury and methylmercury toxicity in the artic charr (*Salvelinus alpinus*). *Environ Res* **90**:217–225.

OSPAR Commission 2004. Mercury losses from the chlor-alkali industry (1982–2002).

Pichichero ME, Gentile A, Giglio N, Umido V, Clarkson T, Cernichiari E, Zareba G, Gotelli C, Gotelli M, Yan L, Treanor J. 2008. Mercury levels in newborns and infants after receipt of thimerosal-containing vaccines. *Pediatr* **121**:208–214.

Quiros L, Piña B, Solé M, Blasco J, López MA, Riva MC, Barceló D, Raldúa D. 2007. Environmental monitoring by gene expression biomarkers in Barbus graellsii: Laboratory and field studies. *Chemosphere* **67**:1144–1154.

Rabestein DL, Guevremont R, Evans CA. 1985. Glutathione and its metal-complexes. In: Siegel H, editor. *Metal Ions in Biological Systems.* New York: Marcel Dekker, pp. 104–141.

Rainbow PS. 1985. The biology of heavy-metals in the sea. *Int J Environ Stud* **25**:195–211.

Raldúa D, Pedrocchi C. 1996. Mercury concentration in three different species of freshwater fish from Gallego and Cinca rivers, Spain. *Bull Environ Contam Toxicol* **57**:597–602.

Raldúa D, Ferrando P, Durán C, Pedrocchi C. 1997. The influence of place of capture, sex, and season on the organochlorine pesticide content in barbel (*Barbus graellsi*) from northeastern Spain. *Chemosphere* **35**:2245–2254.

Raldúa D, Diez S, Bayona JM, Barceló D. 2007. Mercury levels and liver pathology in feral fish living in the vicinity of a mercury cell chlor-alkali factory. *Chemosphere* **66**:1217–1225.

Ravichandran M. 2004. Interactions between mercury and dissolved organic matter—A review. *Chemosphere* **55**:319–331.

Razagui IBA, Haswell SJ. 2001. Mercury and selenium concentrations in maternal and neonatal scalp hair—Relationship to amalgam-based dental treatment received during pregnancy. *Biol Trace Elem Res* **81**:1–19.

Regoli F, Nigro M, Orlando E. 1998. Lysosomal and antioxidant responses to metals in the Antartic scallop *Adamassium colbecki. Aquat Toxicol* **40**:375–392.

Regoli F. 1992. Lysosomal responses as a sensitive stress index in biomonitoring heavy-metal pollution. *Mar Ecol Prog Ser* **84**:63–69.

Rikans LE, Hornbrook KR. 1997. Lipid peroxidation, antioxidant protection and aging. *Biochim Biophys Acta Mol Basis Dis* **1362**:116–127.

Rising L, Viterella D, Kimelberg HK, Aschner M. 1995. Metallothioenin induction in neonatal primary astrocyte cultures protects against methylmercury toxicity. *J Neurochem* **65**:1562–1568.

Roesijadi G. 1992. Metallothionein in metal regulation and toxicity in aquatic animals. *Aquat Toxicol* **22**:81–114.

Rowland I, Davies M, Grasso P. 1977. Biosynthesis of methylmercury compounds by intestinal flora of rat. *Arch Environ Health* **32**:24–28.

Sandvik M, Beyer J, Goksoyr A, Hylland K, Egaas E, Skaare JU. 1997. Interaction of benzo[a]pyrene, 2,3,3',4,4',5-hexachlorobiphenyl (PCB-156) and cadmium on biomarker responses in flounder (Platichthys flesus L). *Biomarkers* **2**:153–160.

Schionning JD, Moller-Mandesen B, Dansher G. 1991. Mercury in the dorsal root ganglia of rats treated with inorganic and organic mercury. *Environ Res* **56**:48–56.

Schlenk D, Zhang YS, Nix J. 1995. Expression of hepatic metallothionein messenger RNA in feral and caged fish species correlates with muscle mercury levels. *Ecotox Environ Safe* **31**:282–286.

Schlenk D, Chelius M, Wolford L, Khan S, Chan KM. 1997. Characterization of hepatic metallothionein expression in channel catfish Ictalurus punctatus by reverse transcriptase polymerase chain reaction. *Biomarkers* **2**:161–167.

Schroeder WH, Anlauf KG, Barrie LA, Lu JY, Steffen A, Schneeberger DR, Berg T. 1998. Arctic springtime depletion of mercury. *Nature* **394**:331–332.

Sohal R. 1981. Metabolic rate, aging and lipofucsin accumulation. In: Sohal R, editor. *Age Pigments*. Amsterdam: Elsevier/North Holland Biochemical Press, pp. 303–316.

Stohs SJ, Bagchi D. 1995. Oxidative mechanisms in the toxicity of metal ions. *Free Radical Biol Med* **2**:321–336.

Teh SJ, Adams SM, Hinton DE. 1997. Histopathologic biomarkers in feral freshwater fish populations exposed to different types of contaminant stress. *Aquat Toxicol* **37**:51–70.

Thiele DJ. 1992. Metal-regulated transcription in eukaryotes. *Nucleic Acid Res* **20**:1183–1191.

Travnikov O. 2005. Contribution of the intercontinental atmospheric transport to mercury pollution in the Northern Hemisphere. *Atmos Environ* **39**:7541–7548.

Ullrich SM, Tanton TW, Abdrashitova SA. 2001. Mercury in the aquatic environment: A review of factors affecting methylation. *Crit Rev Environ Sci Technol* **31**:241–293.

Verslycke T, Vangheluwe M, Heijerick D, De Schamphelaere K, Van Sprang P, Janssen CR. 2003. The toxicity of metal mixtures to the estuarine mysid Neomysis integer (Crustacea: Mysidacea) under changing salinity. *Aquat Toxicol* **64**:307–315.

Viarengo A. 1994. Heavy metal cytotoxicity in marine organisms: Effects of Ca^{2+} homeostasis and possible alteration of signal transduction pathways. *Adv Comp Environ Physiol* **20**:85–110.

Viarengo A, Pertica M, Canesi L, Accomando R, Mancinelli G, Orunesu M. 1989. Lipid peroxidation and level of antioxidant compounds (GSH, vitamin E) in the digestive glands of mussels of three different age groups exposed to aneaerobic and aerobic conditions. *Mar Environ Res* **28**:291–295.

Viarengo A, Ponzano E, Dondero F. 1997. A simple spectophotometric method for metallothionein evaluation in marine organisms: An application to mediterranean and Antarctic molluscs. *Mar Environ Res* **44**:69–84.

Wang QR, Kim D, Dionysiou DD, Sorial GA, Timberlake D. 2004. Sources and remediation for mercury contamination in aquatic systems—A literature review. *Environ Pollut* **131**:323–336.

Wobeser G. 1975. Prolonged oral administration of methylmercury chloride to rainbow trout (*Salmo gairdneri*) fingerlings. *J Fish Res Board Can* **32**:2015–2023.

Wolfe DA, 1992. Selection of bioindicators of pollution for marine monitoring programmes. *Chem Ecol* **6**:149–167.

Wolfe MF, Schwarzbach S, Sulaiman RA. 1998. Effects of mercury on wildlife: A comprehensive review. *Environ Toxicol Chem* **17**:146–160.

Wong CSC, Duzgoren-Aydin NS, Aydin A, Wong MH. 2006. Sources and trends of environmental mercury emissions in Asia. *Sci Total Environ* **368**:649–662.

16 Cadmium as an Environmental Contaminant: Consequences to Plant and Human Health

SARITHA V. KURIAKOSE and M. N. V. PRASAD

Department of Plant Sciences, University of Hyderabad, Hyderabad 500 046, India

1 INTRODUCTION

Cadmium (Cd) is a naturally occurring metallic element; it is also one of the components of the earth's crust and is present everywhere in our environment. Its existence was revealed in 1817. It owes its name to "cadmia fornacum," the "zinc flowers" that formed on the walls of zinc distillation furnaces.

Its industrial applications were developed, particularly during the first half of the twentieth century, based on its unique chemical and physical properties. Its potential risks to human health have been well-studied and are now well-controlled. The principal use of Cd in industries constitutes Ni–Cd batteries (79%), rest comprising pigment formulations (11%), surface coatings and plating (7%), stabilizers for synthetics and plastics (2%), and nonferrous alloys (1%) [Minerals and Metals Sector, 2005]. Cadmium emissions into the environment resulting from its use have been continually decreasing since the 1960s. Today, these are insignificant and will approach zero in less than a decade time, when cadmium products will be almost totally recycled, thereby eliminating any perceived problems. Beginning in the 1950s, the scientific community has drawn its attention to the potential toxicity of cadmium and to the risks presented by its accumulation in man.

Trace Elements as Contaminants and Nutrients: Consequences in Ecosystems and Human Health, Edited by M. N. V. Prasad
Copyright © 2008 John Wiley & Sons, Inc.

2   CADMIUM IS NATURAL

Cadmium is a potentially toxic metal [WHO, 1992] that is classified as a Class I human carcinogen by the International Agency for Research on Cancer [IARC, 1993]. It is a naturally occurring group IIB element that has no biological significance and is present in the earth's crust at an average concentration of 0.2 ppm. Cadmium ores are rare and is usually found associated with zinc ores such as sphalerite (zinc sulfide, ZnS). Greenockite (CdS) is the only mineral of any consequence which contains Cd. As a metal, Cd is soft, malleable, and silver white in color with relatively low melting (321°C) and boiling (765°C) points and is mainly produced as a by-product of zinc, copper, and lead mining and refining. As a soft Lewis acid, Cd can form strong complexes with sulfur compounds (S^{2-}, HS^-, organic sulfides and thiols) and halide ions (e.g., Cl^-) [Traina, 1999].

The supply of Cd is more dependent on Zn production than on Cd demand. A trend analysis shows that production and usage of Cd would continue unless non-hazardous substitutes are found. Naturally a very large amount of Cd is released into the environment, about 25,000 tons a year. About half of this Cd is released into rivers through weathering of rocks, and some Cd is released into air through forest fires and volcanoes. Human activities, such as manufacturing, also contributes to a major amount. Because various anthropogenic activities have led to the addition of significant levels of Cd to the agricultural soils, environmental health concerns are being raised. Cadmium exhibits excellent resistance to corrosion, particularly in alkaline and seawater environments, and has high electrical and thermal conductivity. It also possesses outstanding resistance to high stresses and temperatures and deters UV degradation of certain plastics, giving it a lucrative prospect in pigment, electroplating, and coating and stabilizer industries, but its most widespread use is in rechargeable Ni–Cd battery production. Finally, cadmium pigments produce intense colorings such as yellow, orange, and red and are well-known pigments in artists' colors, plastics, glasses, ceramics, and enamels. Cadmium in the environment requires 40 years of steady decrease to present (after Boutron et al. [1995]) Pg = 10E-12 g. Cadmium has the ability to absorb neutrons, so it is used as a barrier to control nuclear fission. However, Cd also enters the ecosystem from recycled materials like nickel–Cd batteries and phosphate fertilizer application. The natural presence of cadmium in the environment results mainly from gradual phenomena, such as rock erosion and abrasion, and from singular occurrences, such as volcanic eruptions. Cadmium is therefore naturally present in air, water, soil, and foodstuffs. Cadmium is also a by-product of the primary nonferrous metal industry. Rather than disposing of it as a waste, engineers have been able to utilize its unique properties for many important industrial applications. Cadmium levels in the environment reached a peak in the 1960s. Since then, these levels have been constantly decreasing due to improved technology for the production, use, and disposal of cadmium and cadmium-containing products.

Industrial emissions are now tightly controlled due to the significant improvement in pollution control technology and strict regulation and legislation, particularly in the metals industry. The problems of incinerating waste containing cadmium can be solved using existing best available technology to capture more than 99% of incinerator fume emissions. With regard to end-of-life disposal of products containing cadmium, it should be emphasized that, in many of its applications, cadmium is embedded in a product matrix and hence not directly bioavailable. In the very long term, the limited traces of cadmium eventually released from waste products will transform to a stable chemical form (oxide or sulfide) and thus return to the original state found in nature.

Cadmium-containing batteries, coatings, and alloys are totally recyclable; the techniques have been mastered, and existing capacities are available throughout Europe. The costs involved are economically acceptable and hence viable. Increased recycling of cadmium products using modern technologies will further decrease its dispersion into the environment resulting from human activity.

Fate and Transport. Cadmium can enter the environment in several ways. It can enter the air from the burning of coal and household waste and from metal mining and refining processes. It can enter water from disposal of wastewater from households or industries. Fertilizers often have some cadmium in them, and fertilizer use causes cadmium to enter the soil. Spills and leaks from hazardous waste sites can also cause cadmium to enter soil or water. Cadmium attached to small particles may get into the air and travel a long way before coming down to earth as dust or in rain or snow. Cadmium does not break down in the environment but can change into different forms. Most cadmium stays where it enters the environment for a long time. Some of the cadmium that enters water will bind to soil, but some will remain in the water. Cadmium in soil can enter water or be taken up by plants. Fish, plants, and animals take up cadmium from the environment.

Exposure Pathways. Food and cigarette smoke are the largest potential sources of cadmium exposure for members of the general population. Average cadmium levels in U.S. foods range from 2 to 40 parts of cadmium per billion parts of food (ppb). Average cadmium levels in cigarettes range from 1000 to 3000 ppb. Air levels in U.S. cities are low, ranging from 5 to 40 ng per cubic meter. The level of cadmium in most drinking water supplies is less than 1 ppb. In the United States, the average person eats food with about 30 micrograms (μg) of cadmium in it each day. About $1-3$ μg per day of cadmium is absorbed from food, and smokers absorb an additional $1-3$ μg per day from cigarettes. Smoke from other people's cigarettes probably does not cause nonsmokers to take in much cadmium. Cadmium is found at hazardous waste sites at average concentrations of about 4 ppb in soil and 5 ppb in water. Workers can be exposed to cadmium in air from making cadmium products such as batteries or paints. Workers can also be exposed from working with metal by soldering or welding. Each year almost 90,000 workers are exposed to cadmium in the United States.

3 PAST AND PRESENT STATUS

The world cadmium production during the last five years is shown in Fig. 1. In the years 2002–2006, world Cd production dramatically increased due to demanding

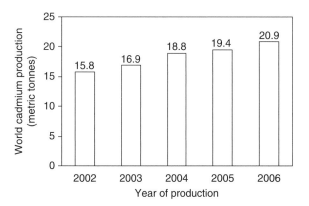

Figure 1. World Cadmium Production during the years 2002–2006.

industries like Ni–Cd batteries, coating for corrosion protection on ships and aerospace applications and in pigments. The average Cd production ranged from 17.9 to 20.2 mt during the decade 1991–2001 [Minerals and Metals Sector, 2003–2005; Kuck, 2007].

3.1   Natural Sources

Cadmium emissions occur in the three major compartments of the environment—air, water, and soil—but there may be considerable transfer between the three compartments after initial deposition. Emissions in air are considered more mobile than those to water, which in turn are considered more mobile than those in soils. Even though the average Cd concentration in the earth's crust is generally placed between 0.1 and 0.5 ppm, much higher levels may accumulate in sedimentary rocks, and marine phosphates and phosphorites have been reported to contain levels as high as 500 ppm [Cook and Morrow, 1995; WHO, 1992]. Weathering and erosion of parent rocks result in the transport (by rivers) of large quantities, recently estimated at 15,000 metric tonnes (mt) per annum, of Cd to the world's oceans [WHO, 1992; OECD, 1994]. Volcanic activity is also a major natural source of Cd release to the atmosphere, and estimates on the amount have been placed as high as 820 mt per year [Nriagu, 1980, 1989; WHO, 1992; OECD, 1994]. Forest fires have also been reported as a natural source of Cd air emissions, with estimates from 1 to 70 mt emitted to the atmosphere each year [Nriagu, 1980].

3.2   Technogenic Sources

Anthropogenic Cd emissions arise either from the manufacture, use, and disposal of products intentionally utilizing Cd (like Ni–Cd batteries, Cd-pigmented plastics, ceramics, glasses, paints and enamels, Cd-stabilized polyvinylchloride (PVC) products, Cd-coated ferrous and nonferrous products, Cd alloys, Cd electronic compounds) or from the presence of Cd as a natural impurity in non-Cd-containing products [for instance, nonferrous metals and alloys of zinc, lead, and copper, iron and steel products, fossil fuels (coal, oil, gas, peat, and wood), cement, phosphate fertilizers].

Examination of the many studies that attempt to present a comprehensive overview of anthropogenic Cd emissions to air, water, and soil and their specific sources [Nriagu and Pacyna, 1988; Nriagu, 1989; ERL, 1990; Jensen and Bro-Rasmussen, 1992; Van Assche and Ciarletta, 1992; Jackson and MacGillivray, 1995; Jones et al., 1993; WHO, 1992; OECD, 1994; Cook and Morrow, 1995] immediately indicates that there are two factors of primary importance in determining the levels of Cd emissions. First, Cd emission factors, which are the amounts of Cd emitted to the environment per unit of Cd processed, are generally much lower in the more technologically advanced regions of the world such as North America, Western Europe, and Japan than in other regions [Nriagu and Pacyna, 1988; WHO, 1992; Jackson and MacGillivray, 1995]. Second, many countries have only partial data and often do not include significant Cd emission sources, particularly those where Cd is not intentionally added.

Cadmium emissions may be considered as arising from either point sources such as large manufacturing or production facilities or from diffuse sources such as may occur from the use and disposal of products by many consumers over large areas.

One of the concerns expressed by some is that increasing incineration of Cd-containing products will eventually lead to increased Cd emissions to the environment and increased risk to human health and the environment.

Most of the earlier studies indicate that the vast majority of Cd emissions, approximately 80–90%, partition initially to soils. While some transfer does occur from soils back to the air or water compartments, the net flux into the soil is generally regarded as positive since there is deposition from both air and water onto soils. Thus, most Cd emissions eventually return to soil.

Cadmium emissions to air arise, in decreasing order of importance, from the combustion of fossil fuels, iron and steel production, nonferrous metals production, and municipal solid waste combustion [ERL, 1990; Jackson and MacGillivray, 1995; Jones et al., 1993; Van Assche and Ciarletta, 1992; Cook and Morrow, 1995]. Cadmium emissions to water arise, in decreasing order of importance, from phosphate fertilizers, nonferrous metals production, and the iron and steel industry [ERL, 1990; Van Assche and Ciarletta, 1992; OECD, 1994]. Cadmium emissions to soils must be considered in three distinct categories: as inputs to agricultural soils, as emissions to nonagricultural soils, and as depositions in controlled landfills. In the first case, the main inputs to agricultural soils which are of primary relevance to human exposure to Cd arise from atmospheric deposition, sewage sludge application, and phosphate fertilizer application [Jensen and Bro-Rasmussen, 1992; Van Assche and Ciarletta, 1992]. In the second case, inputs to nonagricultural soils arise mainly from the iron and steel industry, nonferrous metals production, fossil fuel combustion, and cement manufacture [ERL, 1990; Jackson and MacGillivray, 1995; OECD, 1994]. In the case of Cd present in controlled landfills, these amounts can arise from disposal of spent Cd-containing products, non-Cd-containing products that may contain Cd impurities, and naturally occurring wastes such as grass, food, and soil, which inherently contain trace levels of Cd [Chandler, 1996].

3.3 In Agricultural Soils: Cadmium from Phosphate Fertilizers

A significant contributor to Cd in the arable soils is phosphate fertilizers [McLaughlin et al., 1996]. Cadmium input to agricultural soils is of far greater relevance to human health than Cd input to nonagricultural soils, and input to controlled landfills is of even less importance because Cd is largely immobilized in controlled landfills. The amount contributed by a single application may be insignificant compared to the soil volume and may not be readily detectable, but repeated application leads to gradual buildup over time. In Europe, 74% of the total Cd load to arable soils is from phosphate fertilizers [EUROSTAT, 1995]. Of the total Cd input in areas that are not heavily polluted or industrialized, 50% was contributed by phosphatic fertilizers [de Meeüs et al., 2002]. Andrews et al. [1996] found phosphatic fertilizers to be the main source of Cd in Australia and New Zealand pasture soils. Loganathan et al. [1996] showed a clear relationship between P fertilizer use and Cd accumulation in cropland soils. Phosphatic fertilizers can contain high Cd levels due to the presence of the metal in the phosphate rock used for the manufacture of almost all phosphate fertilizers [McLaughlin et al., 1996]. The extent to which Cd originally present in the phosphate rock is transmitted to the final product may partly depend on the manufacturing process. The use of high-Cd-containing fertilizers may thus cause an accumulation of the metal in the soil, which may in turn be transferred to agricultural products. For instance, Semu and Singh [1996] reported a significant Cd enrichment of tobacco soils that received high-P fertilizers for many years.

3.4 Induction of Oxidative Stress as a Fall-Out of Cadmium Toxicity

One of the primary responses of metal stress in a biological system is the production of toxic reactive oxygen species (ROS) through various mechanisms involving electron transfer [Dietz et al., 1999]. The manifestations of oxidative damage are multifarious, running the gamut from altered membrane fluidity and permeability attributable to lipid peroxidation, through loss of conformation and enzyme activity to genomic damage arising from scission of DNA [Thompson et al., 1987; Davies, 2003]. The reactive oxygen species (ROS) capable of causing oxidative damage include the superoxide anion ($O_2^{\cdot-}$), perhydroxyl radical ($HO_2^{\cdot-}$, the protonated form of superoxide), hydrogen peroxide (H_2O_2), hydroxyl radical ($OH^{\cdot}$), alkoxyl radical ($RO^{\cdot}$), peroxyl radical ($ROO^{\cdot}$), organic hydroperoxide (ROOH), singlet oxygen ($^1O_2$), and excited carbonyl (RO^*) [Thompson et al., 1987; Fleschin et al., 2000]. These are more involved in chemical reactions than molecular oxygen because of the unpaired electrons [Halliwell and Gutteridge, 1990; Bergendi et al., 1999]. Superoxide can act either as an oxidant, where it can oxidize sulfur, ascorbate (AsA), or NADPH, or as a reductant reducing cytochrome C and metal ions; it can also be dismutated to H_2O_2 nonenzymatically proceeding through $HO_2^{\cdot-}$ or in an enzyme-catalyzed reaction [Gebicki and Bielski, 1981].

3.5 Oxidative Damage to Membranes

One of the most damaging effects of ROS and their products in cells is the peroxidation of membrane lipids, a process initiated by abstraction of hydrogen atom from the methylene group (—CH2—) of the polyunsaturated fatty acids of the membrane lipid by hydroxyl radical. The presence of a double bond in the fatty acid weakens the CH bonds on the carbon atom adjacent to the double bond and thus makes the abstraction of a hydrogen atom easier [Kappus, 1985], leaving behind a carbon-centered radical (CH·) that can stabilize by molecular rearrangement to form conjugated dienes, which then further react with oxygen molecule to form peroxy radical (ROO·) [Logani and Davies, 1980]. This, in turn, abstracts another hydrogen atom from an adjacent lipid molecule, propagating the chain reaction further and finally forming lipid hydroperoxides (ROOH). An alternative fate for ROO· is to form cyclic peroxides, which finally get fragmented to aldehydes (malondialdehyde-MDA) and various other polymerization products [Fridovich, 1986]. ROOH can decompose to form alkoxyl (RO·) and peroxyl (RO$_2$) radicals, which in turn can further propagate lipid peroxidation by chain branching [Tadolini et al., 1989].

Lipid peroxidation can also be induced enzymatically by phospholipases and lipoxygenases (LOX), the former by lipolysis releases unsaturated fatty acids, which subsequently acts as the substrate for LOX, a nonheme Fe (III) dioxygenase yielding cis–trans conjugated dienes as the main product [Lacan and Baccou, 1998; Brash, 1999; Oliw, 2002].

3.6 Oxidative Damage to Chloroplasts

The structure of the thylakoid membrane is directly affected by ROS through peroxidation and oxidative stress, altering the lipid composition of the thylakoid membranes [Mohanty and Mohanty, 1998] that leads to changes and disorganization [Stoyanova and Tchakalova, 1999] of the grana stacks with dilated thylakoid membranes observable as plastoglobules [Baszyński et al., 1980]. The levels of phosphatidylcholine and phosphatidylglycerol associated with the inner membrane of chloroplasts for the efficient PS II activity decrease [Baszyński et al., 1984; Maksymiec and Baszyński, 1988; Krupa et al., 1994], with a simultaneous increase in galactolipase activity and hence degradation of acyl lipids specifically monogalactosyl diacylglycerol [Skórzynska and Baszyński, 1993]. This ultimately leads to the inactivation of oxygen-evolving centers and impaired electron transport [Sanita di Toppi et al., 2003].

Metal ions specifically inhibit chlorophyll biosynthesis through δ-aminolevulinic acid dehydratase (ALA dehydratase) [Myśliwa-Kurdziel and Strzałka, 2002] and protochlorophyllide reductase [Baszyński et al., 1980; Gadallah, 1995; Ouzounidou, 1995; Myśliwa-Kurdziel et al., 2004] because of the oxidation prone —SH group [Prasad and Strzałka, 1995b] leading to the lower production of 5-aminolevulinic acid (ALA), the first common precursor for all the tetrapyrroles, thereby impairing chlorophyll biosynthesis.

3.7 Protein Oxidation

Oxidative attack of ROS on proteins results in site-specific amino acid modifications, fragmentation of the peptide chain, aggregation of cross-linked reaction products, altered electrical charge, and increased susceptibility to proteolysis [Davies, 1987]. Since the rate constants for reaction of superoxide anion with amino acid side chains are higher than those with most other cellular targets, proteins would be the major targets for ROS [Ho Kim et al., 2001; Davies, 2003]. ROS modify proteins directly or indirectly by reaching targets within protein through "secondary toxic messengers" such as malondialdehyde (MDA) and 4-hydroxynonenal (HNE) generated from fatty acid degradation [Esterbauer et al., 1991], which, unlike free radicals, are long-lived and can therefore attack targets quite distant from their site of production [Cabiscol et al., 2000]. Conversion of —SH groups to disulfides and other species (e.g., oxyacids—glycine to glyoxylic acid, alanine to acetaldehyde, acetic acid) is one of the earliest observable events during the radical mediated oxidation of proteins [Davies, 1987; Dean et al., 1997]. The oxidation of aliphatic amino acids to hydroxylated derivatives by hydroxyl radical (histidine to oxo-histidine, proline to hydroxyproline, glutamic semialdehyde, etc.) and aromatic residues to phenoxyl derivatives [tyrosine to dityrosine, chlorotyrosine, dihydroxyphenylalanine (DOPA)] in the absence of any reductants (thiols, vitamin E) to repair amino-acid-derived radicals is common [Wright et al., 2002; Winterbourn and Kettle, 2003]. This ultimately leads to peptide bond cleavage [Shacter, 2000a], cross-linking [Davies et al., 1987a; Stadtman and Levine, 2000] and increased susceptibility to proteolysis [Wolff et al., 1986; Davies et al., 1987b]. Normally in efficient systems, radical damaged proteins are rapidly removed in vivo due to enhanced susceptibility to proteolysis. Inhibition of proteolysis due to inactivation of proteolytic enzymes leads to the accumulation of oxidized proteins within cells completely impairing cellular function [Wolff et al., 1986; Davies et al., 1987b; Cabiscol et al., 2000]. The oxidative degradation of a protein is further enhanced by site-specific metal (Fe, Cu)-catalyzed oxidations, where the bound transition metal reacts with H_2O_2 in a Fenton reaction to form an amino acid side-chain-bound hydroxyl radical that is highly destructive to the protein [Stadtman and Oliver, 1991; Requena and Stadtman, 1999]. Extensive oxidation leads to unfolding of the protein and loss of native fluorescence as well as specific tertiary interactions of the aromatic amino acid residues [Anfinsen, 1973; Davies and Delsignore, 1987; Ali et al., 1999; Shacter, 2000a,b]. A strong correlation between increased hydrophobicity on the surface of protein and oxidatively modified proteins exists [Pacifici and Davies, 1990; Chao et al., 1997]. Oxidatively modified enzymes can have either mild or severe effects on cellular and systemic metabolism, depending on the percentage of molecules that are modified and the chronicity of modification [Shacter, 2000a].

3.8   Oxidative Damage to DNA

Cadmium induces a number of genome-related changes including chromosomal aberrations [Zhang and Xiao, 1998], decrease of mitotic index in root cells [Zhang and Yang, 1994], and abnormalities in nucleolar structure [Zhang and Yang, 1994]. Although cadmium is a genotoxic metal, the molecular basis of cadmium genotoxicity is not well-defined. The possible pathway(s) of cadmium-induced genotoxicity are still unknown, but may involve the interaction of this metal with DNA, either directly [Hossain and Huq, 2002a,b] or indirectly [Gichner et al., 2004]. Because both nucleobases and sugar moieties are targets of ROS, nucleic acids are highly susceptible to metal-catalyzed oxidations [Imalay and Linn, 1986; Jabs et al., 1996]. ROS induce numerous lesions in DNA that cause deletions, mutations, and other lethal genetic effect [Breen and Murphy, 1995]. The primary effect is the oxidation of sugar moiety by the hydroxyl radical in a metal (bound to DNA by chelation to phosphodiester linkage)-catalyzed reaction, thereby leading to the oxidation of the adjacent sugar or nitrogenous base which in turn provokes a broad spectrum of DNA lesions [Halliwell and Gutteridge, 1990].

The DNA lesions include DNA single- and double-strand breaks, apurinic/apyrimidinic sites, DNA-protein cross-links, and base modifications [Hartwig and Schwerdtle, 2002]. Cadmium is also reported to induce base modifications. Various products of oxidized bases are cytosine glycol, 5,6-dihydroxycytosine, 8-oxoguanine, 7,8-dihydro-8-oxoguanidine, 7,8-dihydro-8-oxoadenine, 5-hydroxymethyl uracil, thymine glycol, and so on [Dizgaroglu and Bergtold, 1986]. Cross-linking of protein to DNA is another consequence of hydroxyl attack on either the DNA or protein, generating covalent linkages such as thymine-cysteine adducts. This makes the DNA and protein inseparable, thereby being lethal to the system if replication or transcription precedes repair mechanism [Oleinick et al., 1986; Hartwig, 2001; Valverde et al., 2001; Hengstler et al., 2003]. DNA repair systems are also important targets for metals, leading to diminished removal of endogenous and exogenous DNA damage, which is extremely degenerative to the system [Hartwig and Schwerdtle, 2002; Fatur et al., 2003]. Chromosomal aberrations, micronucleus formation, and chromatin fragmentation in Cd-treated plants have also been documented [Koppen and Verschaeve, 1996; Fojtova and Kovarık, 2000]. Investigation of Cd-induced genotoxicity using PCR-based techniques like RAPD analysis [Liu et al., 2005] and ISSR analysis [Kuriakose and Prasad, 2007] revealed modifications of band intensity and lost bands which may be due to one or a combination of the following events:

1. Changes in oligonucleotide priming sites mainly due to genomic rearrangements.
2. Interactions of DNA polymerase with damaged DNA.

These events could act to block or reduce (bypass event) polymerization of DNA in the PCR reaction [Nelson et al., 1996]. The bypass event is a complicated process that depends on the enzymatic properties of the DNA polymerase, the structure of the lesion, and the sequence context of its location [Nelson et al., 1996; Atienzar et al., 2002]. The presence of diverse types of DNA lesions and mutations may also induce important structural changes that can significantly affect the kinetics of PCR events [Bowditch et al., 1993]. Appearance of new PCR products occurred because some oligonucleotide priming sites could become accessible to oligonucleotide primers after structural change or because of some changes in DNA sequence due to mutations (resulting in new annealing events), and/or large deletions (bringing two preexisting annealing sites closer), and/or homologous recombination (juxtaposing two sequences that match the sequence of the primer) [Atienzar et al., 1999]. New bands may also be the result of genomic template instability related to the level of DNA damage and the level of the efficiency of DNA repair and replication. Changes in PCR-based markers' banding patterns may be associated with heritable mutations, chromosomal rearrangements, or other DNA lesions [Theodorakis, 2001].

3.9   Antioxidant Defense Mechanisms in Response to Cadmium Toxicity

ROS detoxification reactions involve the right balance between the formation and detoxification of active oxygen species. The defense strategies range from prevention, through interception, to repair. Since hydroxyl radicals are far too reactive to be controlled easily, defense mechanisms are based on the elimination of its precursors, namely, O_2^- and H_2O_2. Removal of ROS and cellular homeostasis are regulated by antioxidant systems, which include the enzymatic as well as nonenzymatic components [Larson, 1988]. The enzymatic antioxidants include enzymes such as superoxide dismutase (SOD), catalase (CAT), and peroxidases (POD) including ascorbate peroxidase (APX) and guaiacol peroxidase (GPX) and a complex antioxidant system: the ascorbate–glutathione cycle (AGC) [Zhang and Kirkham, 1996] and the associated glutathione metabolism enzymes [Rennenberg and Brunold, 1994; Nagalakshmi and Prasad, 2001] comprising γ-glutamylcysteine synthetase (γ-GCS), glutathione-S-transferase (GST), and glutathione peroxidase (GSH-PX). A schematic representation of the antioxidant enzymes, ascorbate–glutathione cycle and glutathione metabolism, is shown in Fig. 2. The endogenous nonenzymatic

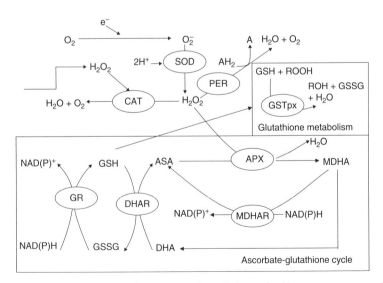

Figure 2. An integrated schematic representation of the antioxidant enzymes, ascorbate–glutathione cycle, and glutathione metabolism enzymes that are involved in the antioxidant defense mechanism.

antioxidants include carotenoids, α-tocopherol, flavonoids, phenolic acids, amino acids, polyamines, ascorbate (AsA) [Horemans et al., 2000, Pallanca and Smirnoff, 2000], thiols (—SH), and glutathione (GSH) [Foyer et al., 2001], which are effective free radical scavengers. Singlet oxygen is mainly quenched by carotenoids [Polykov et al., 2001], and the superoxide radical is dismutated to molecular oxygen and H_2O_2 by SOD [Scandalios et al., 1997]. Hydrogen peroxide detoxifying mechanisms become pivotal in the defense against active oxygen species since, H_2O_2 is potentially capable of reacting with O_2^- to form OH^- [Cakmak et al., 1993]. Subsequently H_2O_2 is scavenged by CAT and POD enzymes—APX as well as GPX—in cytosol and peroxisomes and in chloroplasts via coupling of reduction of H_2O_2 to the oxidation of GSH by GSH-PX [Asada, 1994]. The products of oxidative damage like 4-hydroxyalkenals, 4-hydroxynonenals (membrane lipid peroxides), and base propenals (products of oxidative DNA degradation) are highly cytotoxic. Glutathione-S-transferases (GSTs) detoxify such endogenously produced electrophiles by conjugation with GSH [Marrs, 1996].

A very important antioxidant in the cellular milieu responsible for maintenance of the antioxidative machinery of the cells intact under stress is glutathione [Rennenberg, 1982; Noctor and Foyer, 1998], and it functions as a stress indicator occurring in two distinct redox forms, promptly responding to oxidative stress [May et al., 1998; Devi and Prasad, 1998]. GSH is a particularly adequate electron donor/acceptor in physiological reactions due to the chemical reactivity of the thiol group, its relative stability, and high solubility in water [Potters et al., 2002]. Glutathione is synthesized from glutamate, cysteine, and glycine in two ATP-dependent reactions catalyzed by γ-GCS in the first step of glutamate–cysteine coupling and glutathione synthetase (GS) in the second step of glycine addition [Hell and Bergmann, 1990]. Glutathione also detoxifies metal ions by scavenging them through the formation of phytochelatins and thereby facilitate their transport to vacuoles [Cobbett, 1999; Rauser, 2000].

Another vital antioxidant that is an excellent electron donor participating in various reactions of the plant system even involved in a photoprotective xanthophylls cycle is ascorbate (AsA) [Smirnoff, 1996; Horemans et al., 2000], which also aids in the generation of the lipophilic chloroplastic antioxidant α-tocopherol (vitamin E) from the α-chromanoxyl radical [Asada, 1994; Arrigoni and De Tullio, 2000]. The ascorbate–glutathione cycle, a major H_2O_2 scavenging pathway where H_2O_2 is reduced to water by ascorbate peroxidase (APX) producing ascorbyl radical (monodehydroascorbate) and oxidized form of ascorbate (dehydroascorbate) [Hausladen and Kunert, 1990], operates both in chloroplast and in the cytosol [Zhang and Kirkham, 1996]. The regeneration of ascorbate from monodehydroascorbate and dehydroascorbate is catalyzed by NAD(P)H-dependent monodehydroascorbate reductase (MDHAR) [Hossain et al., 1984] and GSH-dependent dehydroascorbate reductase (DHAR) [De Tullio et al., 1998], coupled with glutathione reductase (GR) (refer to Fig. 4) [Smith et al., 1989]. Hence, operation of the AGC not only maintains the reduced active forms of GSH and AsA in cells on a suitable level (adjusting the cellular redox potential), but also participates in ROS detoxification [Kingston-Smith and Foyer, 2000; Ma and Cheng, 2003].

A number of organic acids, amino acids, and some members of mugineic acids that occur in plant tissues are possible ligands for metal complexation and may hence confer metal tolerance [Łobiński and Potin-Gautier, 1998; Hall, 2002]. Amino acids are present in living systems and are major candidates for metal ion binding ligands stabilizing various macromolecules structurally and aid in vital cellular functions [Pohlmeier, 2004]. The sulfur atoms of cysteine are responsible for the major covalent cross-links in protein structures, where the disulfide bridge formed between two cysteine molecules is important in stabilizing protein conformation [Komarnisky et al., 2003]. Protective thiol-group-containing amino acids like methionine and cysteine are reported to prevent methyl mercury toxicity by chelation [Peraza et al., 1998]. Similar reports exist for the Ni-histidine complex [Krämer et al., 1996; Kerkeb and Krämer, 2003] in the xylem sap of *Alyssum lesbiacum* and *Brassica juncea*. Amino acids cysteine and glutamate are the basic components of GSH [Noctor and Foyer, 1998], a crucial antioxidant of the plant cell, along with phytochelatins [PCs (poly(γ-glutamyl-cysteinyl)-glycines)], metal-binding peptides that chelate free metal ions and transport them to vacuoles [Rauser, 2000]. Therefore amino acids indirectly modulate detoxification of xenobiotic compounds and scavenging of free radicals and hence oxidative stress [Bray and Taylor, 1993]. Phytochelatins consist mainly of glutamic acid, cysteine, and glycine in molar ratios of 2:2:1 to 11:11:1 and are also referred to as class III metallothioneins [Reddy and Prasad, 1992; Rauser, 1995]. Extensive studies on glutathione synthesis in plants have indicated that GSH synthesis is regulated by cysteine and glutamic acid availability, feedback inhibition by γ-GCS, transcriptional control of γ-GCS, and translational regulation of γ-GCS by the ratio of reduced to oxidized glutathione [Xiang and Oliver, 1998; Foyer et al., 2001]. Low-molecular-weight organic acids—especially citric, oxalic, and malic acids—are capable of forming complexes with metals that can affect their fixation, mobility, and availability to plants. Metal–organic acids interactions in the soil–plant system are found to be important for solubilizing metals from highly insoluble phases [Cieslinski et al., 1998; Jones, 1998; Wu et al., 2003]. The toxicity of aluminum to plants is known to be handled efficiently by Al-citrate, Al-aconitate, Al-malate, and Al-oxalate complexes [Ma, 2000; Jonnarth et al., 2000]. Similar reports exist for Ni-citrate in *Sebertia acuminata* latex [Sagner et al., 1998]. Phytoremediation of Cd-contaminated soils with organic acid amendments [Elkhatib et al., 2001] also indicates the potential role of organic acids in detoxification of toxic metal ions.

3.10 Cadmium Availability and Toxicity in Plants

In line with other toxic metals like As, Cr, Co, Cu, Ni, Sr, and Zn, Cd is a borderline toxic metal and is phytotoxic either at all concentrations or above certain threshold levels [Nieboer and Richardson, 1980]. Lane and Morel [2000] provided evidence of a biological role for Cd in the marine diatom *Thalassiosira weissflogii* under conditions of low zinc, typical of the marine environment. Addition of Cd to Zn-limited cultures enhanced the growth rate as reflected in increased levels of cellular carbonic anhydrase (CA) activity. This is the only evidence for a "nutritive" role for cadmium which is most commonly referred to as a toxic metal. Cadmium in food is a potential health risk. According to Olsson et al. [2005], approximately 75% of the dietary intake originates from cereals and vegetables for nonsmokers. Since the diet is the main source of exposure to cadmium for the general population [Wagner, 1993; Tsukahara et al., 2003], intensive research has been done on the accumulation of cadmium in edible plant tissues [Stalikas et al., 1997; Kashem and Singh, 2001]. The effects of this toxic metal on plant and plant defense systems have been extensively investigated and reviewed by different authors [Prasad, 1995a; Das et al., 1997; Sanita di Toppi and Gabbrielli, 1999]. The different sources of Cd to plants ans the resulting plant responses are briefly summarized in Fig. 3.

Cadmium is a nonessential element for plants, and the most evident symptoms of its toxicity are chlorosis and stunting. While chlorosis is the result of Cd toxicity on the uptake, transport, and use of essential elements like Ca, Mg, Fe, Mn, Cu, Zn, P, and K, leading to reduction of Mn and Fe absorption and changes in Fe:Zn ratios [Das et al., 1997; Baryla et al., 2001], the reduction of plant development is due to the

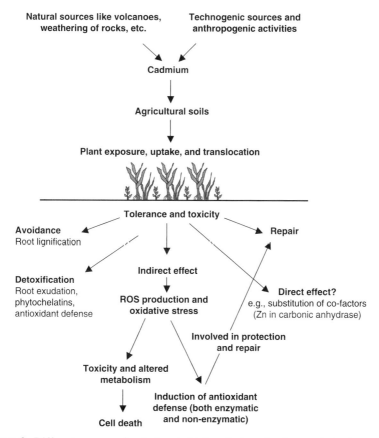

Figure 3. Different sources of cadmium to higher plants and subsequent plant response.

interference of Cd in several important physiological processes. In our experiments we have noticed a significant reduction in seed germination and seedling growth (especially root growth) in Cd-treated seeds. Cadmium alters the hormonal balance [Poschenrieder et al., 1989] and disturbs the plant water status through a decrease of water absorption reduction of root hydraulic conductivity into xylem vessels, decrease of transpiration rate, and increase of stomatal resistance [Barcelo and Poschenrieder, 1990; Vassilev and Yordanov, 1997]. At the cellular level, turgor loss and cell expansion inhibition by decrease of the cellwall elasticity [Barcelo and Poschenrieder, 1990; Prasad, 1995a] is a result of Cd toxicity. This may result in damage to the photosynthetic apparatus [Siedlecka and Krupa, 1996; Vassilev and Yordanov, 1997]. Cadmium can bind to the sulfhydryl groups of proteins—leading to activity inhibition or to structure disruption—or can even displace essential elements, resulting in deficiency effects [Van Assche and Clijsters, 1990; Hall, 2002]. Induction of oxidative stress is another very important aspect of Cd toxicity [Shah et al., 2001]. The free-radical reactions initiated by Cd directly affect the structure and function of macromolecules by oxidative reactions [Stadtman and Oliver, 1991; Szuster-Ciesielska et al., 2000], displayed as an increase in lipid peroxidation, disruption of membrane integrity, oxidative modifications of proteins, and DNA damage [Romero-Puertas et al., 2002]. Cadmium alters functionality of membranes by inducing changes in lipid composition and by affecting the enzymatic activities associated with membranes such as H^+-ATPase [Fodor et al., 1995]. An increase in peroxidase and hydrolytic enzyme activity representing signals of premature senescence has often been observed in Cd-treated plants and is due to the induced oxidative stress [Prasad, 1995a,b]. Cadmium ions directly affect the biosynthesis of photosynthetic pigments and the associated photosynthetic processes, leading to impaired carbon-utilization and respiratory processes [Krupa, 1988; Atal et al., 1991; Greger and Ogren, 1991; Tukendorf, 1993; Prasad, 1995a,b; Šeršeň and Králová, 2001; Sanita di Toppi et al., 2003; Prasad, 2004] in the plant. In different plant species, an increase in ethylene production and a reduction of soluble protein content in response to Cd has also been recorded [Fuhrer, 1982; Kevresan et al., 1998; Hsu and Kao, 2003].

Sunflower, durum wheat, and flax are naturally higher in Cd than other crops [Simon, 1998]. One of the strategies for Cd tolerance is the synthesis of phytochelatins. In some "Cd-shoot excluders" such as pea, an increased Cd supply leads to increased concentrations of soluble proteins in shoots and roots, where a putative phytochelatin appeared [Lozano-Rodriguez et al., 1997]. An interspecific difference in shoot Cd concentration has been reported for some economically important crops [John, 1973]. Intraspecific variation in Cd concentration has also been found in soybean [Bogess et al., 1978], maize [Hinesly et al., 1982], and lettuce [Thomas and Harrison, 1991]. McLaughlin et al. [1994] found that potato cultivars grown commercially in Australia exhibited significant difference in tuber Cd concentration. Genotypic differences in grain Cd concentration have been reported for durum wheat [Penner et al., 1995], common wheat [Oliver et al., 1995; Zhang et al., 2002], and sunflower [Li et al., 1995]. Oliver et al. [1995] found considerable variations in Cd concentration in grains of many of the Australian wheat cultivars studied. Li et al. [1995] reported that Cd concentration in sunflower kernels of nine major commercial non-oilseed hybrids grown at uncontaminated sites in North Dakota and Minnesota varied from 0.79 to 1.17 mg/kg. *Brassica juncea*, or Indian mustard, is known to accumulate high levels of heavy metals including Cd, Cr, Cu, Ni, Pb, and Zn under some conditions that particularly enhance the solubility of the metal cations [Liphadzi and Kirkham, 2005]. *Nicotiana tabacum* can accumulate relatively high levels of Cd [Lugon-Moulin et al., 2004]. In a recent survey of Swedish winter wheat, the average Cd concentration in grain was 0.044 mg kg^{-1} [Eriksson et al., 2000]. Wheat accumulates more Cd than the other commonly grown cereals; the order is rye < barley < oats < wheat [Jansson, 2002].

3.11 Metal–Metal Interactions

Unfavorable effects of toxic metals on plants are manifested, among others, by inhibiting the normal uptake and utilization of mineral nutrients [Burzyński, 1987; Trivedi and Erdei, 1992]. One of the crucial factors of metal influence on plant metabolism and physiological processes is their relationships with other mineral nutrients [Marschner, 1995; Siedlecka, 1995]. There is a dearth of information on the specifics of Cd toxicity on essential (Cu, Fe, Zn) and nonessential metals (Pb, Hg) [Rauser, 2000] and there have not been much studies designed specifically to address the effect of micronutrient status on toxicity from exposure to nonessential metals [Peraza et al., 1998]. Metalliferous environments are often contaminated by more than one metal in potentially toxic concentrations [Wallace, 1982; Siedlecka, 1995]. Earlier workers [Taylor, 1989; Symeonidis and Karataglis, 1992] have classified plant responses to combinations of metals in the growth medium into additive, antagonistic, and synergistic.

An increased concentration of the fertilizer cation (Ca^{2+}) can cause an increased concentration of Cd ions (Cd^{2+}) in the soil solution by ion exchange, and therefore also a higher Cd uptake and a higher Cd concentration in plant tissues [Brown et al., 1994; Lorenz et al., 1994]. The increased concentration of Ca^{2+} ions may also reduce the binding of Cd^{2+} at sites in the root cell walls.

Cadmium and Zn belong to group IIB transition elements, and hence both these elements have similar geochemical and environmental properties [Nan et al., 2002]. Cadmium has been described as an antimetabolite of Zn by scientists due to the observed Zn deficiency in most of the Cd-treated systems [Peraza et al., 1998]. Jalil et al. [1994] reported that Cd application decreased the concentration of K, Zn, and Mn in roots and shoots of durum wheat while Fe and Cu concentrations remained unaffected. In direct contrast, Yang et al. [1998] reported a decrease in Fe, Cu, Mn, Ca, and Mg in cabbage, ryegrass, maize, and white clover while P concentrations increased on Cd application.

Fertilizer application can affect Cd phytoavailability not only directly through addition of Cd as a contaminant in P fertilizers but also indirectly through alteration of the soil conditions and also influence Cd speciation and complexation, which affects the movement of Cd to plant roots and perhaps also its absorption into the roots. In addition, rhizosphere composition, root growth, and general crop growth are likely to be affected by the application of fertilizers [Grant et al., 1999]. An increased rate of N fertilizer tended to elevate the grain Cd concentration of winter wheat, irrespective of site and cultivar [Wångstrand et al., 2007]. Nitrogen fertilizers may increase Cd concentrations in plants, even if the fertilizers do not contain significant levels of Cd. Jönsson and Eriksson [2003] found a strong positive correlation between N and Cd concentrations in grain of winter and spring wheat. The increased Cd uptake may be caused by an increase in ionic strength of the soil solution and ensuing ion exchange reactions, or by soil acidification. The type of N fertilizer determines the effect of the fertilization. Ammonium fertilizers have been shown to cause higher Cd concentrations in crops than nitrate fertilizers as a result of the decrease in pH caused by nitrification or plant uptake of NH^{4+} [Grant et al., 1999]. The NH^{4+} in ammonium fertilizer may also have an exchange effect on Cd^{2+}, but it is probably less for a monovalent ion than for Ca^{2+}. However, any nitrification of the NH^{4+} has an acidifying effect, which may enhance Cd solubility [Eriksson, 1990]. However, the reported effects of N fertilizer on the Cd concentration in plants are not unambiguous.

3.12 Uptake and Transport of Cadmium by Plants

Cadmium primarily enters the plant through the roots. Smeyers-Verbeke et al. [1978] concluded from their experiments that cadmium absorption by wheat roots is similar in all respects to the uptake of other ions which are known to be absorbed by metabolically mediated processes. Therefore it is very probable that, besides involving an exchangeable fraction, the uptake also involves a metabolic one. The temperature dependence of the uptake and the effect of metabolic inhibitors strengthen this conclusion. Once in the roots, the cadmium can either be stored or exported to the shoots. Transport to the shoot primarily takes place through the xylem. Xylem loading is likely facilitated by yet-unidentified membrane-transport processes. Cadmium can reach the xylem *via* symplastic transport, but probably through apoplastic transport (extracellular spaces) under high-level exposure [Lugon-Moulin et al., 2004]. In the xylem, transpiration-driven mass flow may play an important role in the movement of Cd, probably in the form of noncationic, organic-acid complexes, such as Cd-citrate [Senden et al., 1995]. Gong et al. [2003] demonstrated that phytochelatins (PCs) can be transported from roots to shoots and are indeed involved in root-to-shoot transport of Cd in *Arabidopsis thaliana* despite reports to the contrary by Salt et al. [1998].

Herren and Feller [1994] suggested that once the cadmium is transported to the shoots, it is possible that some Cd may be redistributed in the plant via the phloem. The transport mechanisms of Cd from root to shoot remains to be elucidated in many plants and most likely differs with Cd exposure level. The pattern of Cd accumulation and root and shoot distribution may differ substantially from genus to genus and within genus also. Cadmium accumulation is also dependent on age of the plant as reported in *Nicotiana tabacum* [Wagner and Yeargan, 1986].

Cd can be found in various cell components like cell wall, cytoplasm, chloroplast, nucleus, vacuoles, and so on, within the plant cell [Ramos et al., 2002]. Distribution varies depending on the species, organ, tissue, and concentrations used in the study. Cd accumulation in the cell walls of higher plants serve to protect the cell from toxic effects of Cd, though it is not clear to what extent Cd can bind to the cell wall [Lugon-Moulin et al., 2004].

The uptake of Cd by plants is also influenced by (a) external factors like pH of the soil, cation exchange capacity, and Cd content of the soil, (b) other soil properties like clay content, organic matter content, and presence of other metals like Cu and Zn, and (c) agronomic practices like sewage sludge application, liming, fertilizers, and irrigation practices.

3.13 Consequences to Human Health

Cadmium can enter the human body from food, water, or breath. Very little cadmium enters through the skin. The human body rapidly takes in about one-quarter of the cadmium from breath and about one-twentieth of the cadmium from food. The rest of the cadmium is breathed out or excreted in feces. Iron deficiency leads to greater Cd uptake than usual. Cigarette smoke has cadmium in it, and so smokers breathe in Cd. Other people who breathe in cadmium are those who work with Cd, and people who live near hazardous waste sites or factories that release cadmium into the air. The general population and people living near hazardous waste sites may eat or drink cadmium in food, dust, or water. Cadmium that enters the body stays in liver and kidneys. Cadmium leaves the body slowly, in urine and feces, and has a biological half-life of 10–30 years. Cd can be easily bio-concentrated through the food chain [Dudka and Miller, 1999] as shown in Fig. 4. The major route for cadmium intake is ingestion and is largely due to the presence of trace levels of cadmium in foodstuffs of natural origin or of the use of phosphate fertilizers and sludge on agricultural soils.

Investigations conducted around the world show that, for the general population, the average daily cadmium intake, from all sources, is very low and at the lower end of the total range of 10–25 μg/day. The tolerable daily cadmium intake established by WHO is 60 and 70 μg/day, respectively, for adult women and men. In industrial environments, workers could be exposed to cadmium through inhalation. In modern enterprises, workers' exposure to cadmium is controlled by rigorous industrial hygiene practice and continuously monitored by medical tests and follow-up, thus preventing any health risk for the workers. The carcinogenic risk of cadmium by inhalation is increasingly being contested by newer studies [Sorahan et al., 1995; Sorahan and Lancashire, 1997], demonstrating the importance of confounding factors (e.g., arsenic) in older studies.

Cadmium mainly accumulates in one is specific organ: The kidneys and high dietary intake can lead to kidney damage with time. However, the most recent studies have shown that these effects are reversible, at least at low exposures, once exposure to cadmium is reduced. Cadmium is classified as a class I human carcinogen by the International Agency for Research on Cancer in 1993 [IARC, 1993]. Exposure to Cd can also cause brittle bones, that is, osteoporosis [Alfven et al., 2000]. For nonsmokers, approximately 75% of the dietary intake originates from cereals

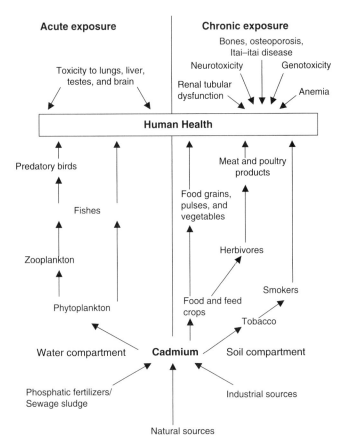

Figure 4. Bio-concentration of cadmium.

and vegetables [Olsson et al., 2005]. Effects on human kidney function have already been reported at current exposure levels in areas with moderate pollution [Järup et al., 2000; Olsson et al., 2005]. Cd content in many agricultural soils tends to be gradually increasing [Nicholson and Jones, 1994; Satarug et al., 2003; Martin and Ruby, 2004]. Thus adverse human health effects due to soil–plant and plant–human transfer of Cd have been enhanced [McLaughlin et al., 1999]. Excessive human intake of Cd is of concern because it not only impairs kidney function but also elevates the risk of osteoporosis, by inhibiting mineralization, vitamin D activation, and calcium uptake [Järup et al., 1998; McLaughlin et al., 1999]. Extreme cases of chronic Cd toxicity can result in osteomalacia and bone fractures, as characterized by the disease called Itai–Itai (meaning "ouch! ouch!") in Japan during the 1950s and 1960s, where local populations were exposed to Cd-contaminated rice. Eating food or drinking water with very high cadmium levels severely irritates the stomach, leading to vomiting and diarrhea. Animals ingesting cadmium sometimes show high blood pressure, iron poor blood, liver disease, and nerve or brain damage.

Cadmium is believed to cause damage even at very low concentrations [Järup et al., 1998]. Therefore, the World Health Organization [WHO, 1972] set the maximum permissible concentration (MPC) of 0.1 mg Cd/kg for cereal grains (Table 1). Cadmium principally occurs in human diet as a result of its uptake and accumulation from soil by crop plants, primarily of cereals [Stolt et al., 2003], although the whole pathway of Cd transfer into the food chain is also involved in input from atmosphere, water, and aquatic life. The European Community limit value for wheat grain is

TABLE 1. Limits Set by European Union (EU) and Australian and New Zealand Food Authority (ANFA) for Various Foodstuffs

Food	Australia/New Zealand (mg kg^{-1} fresh weight)	European Union (mg kg^{-1} fresh weight)
Plant-Based Foods		
Rice (including bran, germ)	0.10	0.20
Wheat (including bran, germ)	0.10	0.20
Other cereals (excluding wheat rice, bran, germ)	—	0.10
Soybean	—	0.20
Peanuts	0.10	—
Vegetables and fruits	—	0.05
Leafy vegetable (fresh herbs, celery, cultivated fungi)	0.10	0.20
Stem vegetables	—	0.10
Root and tuber vegetables	0.10	0.10 (root/tuber) 0.05 (other vegetables)
Animal-Based Foods		
Kidney (cattle, sheep, pig, and poultry)	2.5	1.0
Liver (cattle, sheep, pig, and poultry)	1.25	0.5
Meat	0.05	0.05 0.20 (horse)
Fish	—	0.05–0.10 (depending on species)
Seafood	2 (molluscs excluding dredge/buff oysters and queen scallops)	0.5–1.0 (depending on species)
Others		
Chocolate and cocoa products	0.50	—
Drinking water	—	5 μg L^{-1}

Source: Australia New Zealand Food Standards Code, European Union: Commission Regulation (EC) No 466/2001 of 8 March 2001.

0.2 mg/kg [EC, 2001]. However, if wheat and other food products were to reach the maximum permitted levels given by the European Union, according to Olsson et al. [2005], 10–25% of the Swedish population would be exposed to levels above the provisional tolerable weekly intake (7 g Cd/kg body weight). About 43% of the Cd ingested with food in Sweden comes from wheat products [Hellstrand and Landner, 1998].

In mammalian cells, cadmium enhances the mutagenicity of UV light, suggesting its interference with DNA repair processes and the activity of detoxifying enzymes (reviewed in Beyersmann and Hechtenberg [1977] and Hartwig et al. [1994]). Cadmium toxicity may represent a new mechanism by which the genomes may be destabilized [McMurray and Tainer, 2003]. Reported mutagenic effects of Cd include generation of reactive oxygen species, inhibition of several types of DNA repair, depletion of glutathione, and alteration of apoptosis. McMurray and Tainer, [2003] reviewed that the primary cause of Cd genotoxicity is the inactivation of an essential DNA repair activity showing specific effect targeting the mismatch repair system in particular, among the different repair systems.

Breathing air with very high levels of cadmium severely damages the lungs and can cause death. Breathing lower levels for years leads to a buildup of cadmium in the kidneys that can cause kidney disease. Other effects that may occur after breathing cadmium for a long time are lung damage and fragile bones. Workers who inhale cadmium for a long time may have an increased chance of getting lung cancer, though no proof has been found that mice or hamsters. However, some rats that breathe in cadmium do develop lung cancer. Female rats and mice that breathe high levels of cadmium have shown lower fecundity with high birth defects than usual. Breathing cadmium causes liver damage and changes in the immune system in rats and mice.

3.14   Options for Cadmium Minimization

The various approaches and options available to decrease plant availability of soil Cd can be grouped into three categories: (a) molecular and biochemical approaches, (b) breeding strategies, and (c) soil Cd remediation.

3.15   Molecular and Biochemical Approaches

Recent advances in molecular and biochemical approaches have helped in understanding better the mechanisms of ion uptake, extrusion, transport, and sequestration. Various options are available to decrease the plant Cd uptake or modify plant genome to facilitate hyperaccumulation, depending on the need of the hour.

Root Exudates. Root exudates contain low-molecular-weight organic acids (LMWOA) that can induce changes in the physiochemical properties of the surrounding soil affecting Cd availability. As LMWOA modify Cd solubilization and accumulation in plants, genes that facilitate their release could be introduced by genetic engineering to reduce or enhance Cd uptake. Phytosiderophores released by graminaceous plants in response to nutrient deficiencies can mobilize Cd from solid phase but does not necessarily increase Cd phytoextraction efficiency [Shenker et al., 2001]; however, it may be able to reduce Cd uptake.Vacuolar chelation may predominate over cytosol mechanisms because root cells are mostly mature cells with large vacuoles [Rauser, 1999].

Cadmium Transporters. Cadmium as well as other nonessential trace metals such as Cr, Hg, and As are most likely carried across plant membranes via transporters. This represents a point where Cd transport can be modified. No specific Cd transporters have been unambiguously identified to date, though various enzymes are known [Guerinot, 2000; Theodoulou, 2000; Williams et al., 2000; Clemens, 2001]. The regulation of genes encoding these transporters may be complex with regard to transcriptional and posttranslational regulation. The most commonly studied group of transporters include the *ZIP* family (zinc and iron regulated transporters) [Grotz et al., 1998], *ABC* transporters (ATP binding cassette) [Theodoulou, 2000], *Nramp* family (Natural resistance-associated macrophage protein) [Williams et al., 2000], *P*-type ATP*ase* (distinguished by formation of phosphorylated intermediate) [Williams et al., 2000], *CDF* family (cation diffusion facilitator) [Mäser et al., 2001], cation/proton antiporters (can exchange protons for metal ions in the vacuole sap leading to metal accumulation in vacuole) [Salt and Wagner, 1993], *LCT1* (low-affinity cation transporter from wheat) [Schachtman et al., 1997], *MATE* family (multidrug and toxic compound extrusion) [Brown et al., 1999].

Phytochelatins, Metallothioneins, and Glutathione. Phytochelatins (PCs), metallothioneins (MTs), and glutathione (GSH) are low-molecular-weight polypeptides that are rich in cysteine and can bind various metals. The role of PCs and MTs in metal tolerance and detoxification has been subject to various recent reviews and hence is not being detailed here (refer to Cobbett and Goldsbrough [2002] and Lugon-Moulin et al. [2004]. While MTs can bind to various metals like Cd in mammalian systems and are difficult to isolate due to the high sensitivity of thiol groups to oxygen, PCs are not gene-encoded but are instead the product of a biosynthetic pathway. Lefebvre et al. [1987] provided the first evidence of MTs in plants (wheat). PCs are synthesized from glutathione or its analogues by phytochelatin synthase (an enzyme activated by Cd or other metal ions) [Grill et al., 1989]. Despite the ability of Pcs to form complexes with transition metal ions and its apparent prominent role in Cd detoxification, there is no evidence that an elevated production of PCs is responsible for resistance to toxic metals in all plants. Glutathione is necessary not only for synthesis of PCs but also for Cd-GSH transport via YCF1-type ABC transporters. Overexpression of key enzymes in the GSH biosynthesis pathway like γ-glutamylcysteine synthetase and glutathione synthetase result in high GSH levels.

Other Approaches. Other transporters or compounds may be linked to Cd transport, sequestration, or detoxification. Several metal-binding proteins are as yet uncharacterized, and their genes are yet to be identified. Also the impact of metal ions on other nutrient assimilative pathways (for instance, Cd impact on sulfur assimilative pathway) needs to be taken into account.

3.16 Breeding Strategies

A principal step in breeding programs is to analyze quantitative variation and uncover its potential genetic basis since quantitative variation is a feature of any important traits like yield, quality, or disease resistance. Cadmium uptake and accumulation in plants must undoubtedly be under the control of multiple genes which contribute quantitatively in stage-specific, tissue-specific, and environment-specific manner to Cd transport, accumulation, and sequestration in plants [Lugon-Moulin et al., 2004]. The strong variability in Cd accumulation and tolerance observed between varieties/species, which for the most part is untapped, can be exploited using traditional breeding methods. The development of DNA molecular marker technology is of great assistance in breeding programs since the DNA markers near genes of interest can be used as tools of selection in breeding programs and hence can decrease the time involved compared to traditional methods.

3.17 Soil Cadmium Regulation

Potential strategies for reducing phytoavailable cadmium includes reducing availability of Cd to plants by application of amendments, using plants to extract Cd from soil (phytoremediation) or use of other biological organisms to reduce soil Cd.

Phytoremediation. The use of plants to clean up contaminated land and water either by extracting, volatilizing, stabilizing, or inactivating soil organic or inorganic pollutants is termed phytoremediation [Chaney et al., 1997]. This relatively low-cost alternative has received considerable attention in recent years and has been reviewed in detail by many authors [Salt et al., 1998; Pilon-Smits, 2005]. An ideal plant for Cd phytoremediation must be able to tolerate high levels of Cd, have high rates of Cd translocation from roots to shoots, accumulate Cd in their harvestable parts, rapidly uptake Cd from soils, show rapid growth, and have high biomass. At present, no natural plant has all the desirable characteristics. The use of conventional plants to decontaminate Cd-polluted soil may take decades, so, several mutually non exclusive means can be used to ameliorate Cd phytoextraction efficiency. For instance, agronomic practices can be modified to improve biomass and hence Cd uptake.

Soil Amendments. Various synthetic and natural nonbiological amendments can be used to remediate Cd-polluted soils either by enhancing Cd mobilization for subsequent phytoremediation or by decreasing phytoavailability. A list of some common amendments and their effect on Cd phytoavailability is given in Table 2. The mechanisms of metal sequestration by these amendments are not fully known. Traditionally, the use of unassisted or noninvasive natural processes as a part of

TABLE 2. Possible Strategies for Cadmium Minimization in Agroecosystems by Using Nonbiological Soil Amendments

Amendment	Reference
Immobilization	
Hydroxyapatite	Boisson et al. [1999], Jeanjean et al. [1995], Mandjiny et al. [1998], Seaman et al. [2001], Xu et al. [1994]
Rock phosphate	Basta et al. [2001], Chen et al. [1997]
K_2HPO_4	Bolan et al. [1999], Naidu et al. [1994], Pearson et al. [2000], Pierzynski and Schwab [1993]
$Ca(H_2PO_4)_2$	Bolan et al. [1999]
$CaCO_3$	Levi-Minzi and Petruzzelli [1984], McGowen et al. [2001], Pierzynski and Schwab [1993]
Adsorption	
$(NH_4)_2HPO_4$	Andersson and Siman [1991], Bingham et al. [1979], Brown et al. [1997], Chaney et al. [1977], Han and Lee [1996], He and Singh [1994], Hooda and Alloway [1996], John and Van Laerhoven [1976], John et al. [1972], Lehoczky et al. [2000], Maclean [1976], Oliver et al. [1995], Singh and Myhr [1998], Singh et al. [1995], Tyler and Olsson [2001]
Inorganic components	Basta et al. [2001], Basta and Sloan [1999], Keefer et al. [1984], Pietz et al. [1983], Soon [1981]
Sewage sludge	Hyun et al. [1998], John and Van Laerhoven [1976], Street et al. [1978]
Biosolid	Brown et al. [1998], Li et al. [1995]
Decreased Availability	
$Ca(OH)_2$	Basta and Sloan [1999], Chaney et al. [1977]
CaO	Brallier et al. [1996], Gray et al. [1999], Vasseur et al. [1998]
$MgCO_3$	Williams and David [1976]
$CaMgCO_3$	Kreutzer [1995]
Milorganite	John and Van Laerhoven [1976]
Anaerobic digested biosolid	del Castilho et al. [1993], Merrington and Madden [2000]
Cattle manure	Pearson et al. [2000]
Lime-stabilized biosolid	Pierzynski and Schwab [1993]
Poultry manure	Riffaldi et al. [1983]
Secondary digested sewage sludge	Riffaldi et al. [1983]
Differential Response (Increase/Decrease)	
Paper mill sludge and biosolids (decreased Cd adsorption but increased Cd in ryegrass)	Merrington and Madden [2000]

site remediation strategy is called natural attenuation. For widespread contamination by metals (e.g., fertilizer-derived Cd input in pasture soils), remediation options generally include amelioration of soils to minimize metal solubility (i.e., immobilize) and eventual bioavailability. Bioavailability can be regulated by using a range of inorganic materials and organic residues by immobilization and adsorption. A number of studies have examined the potential value of various soil amendments in immobilizing metals in soils, thereby reducing their bioavailability.

There exist three main routes of exposure to soil-borne metals: soluble contaminants migrating with soil water (ingestion via drinking); assimilation by plants and aquatic organisms (via food chain pathway); and, to a lower extent, volatilization (gaseous inhalation). The other pathways concern direct ingestion or inhalation of air-borne particles. Hence any thorough evaluation of sustainability of trace element immobilization should consider physicochemical and biological methods. Evaluation of efficacy of soil metal immobilization mandates supplementary observation *in situ*. Understanding of key processes and factors controlling the dynamics of trace elements in soils is essential to exploit the benefits of natural attenuation of these contaminants.

Biological Amendments. Fungi as well as root-colonizing bacteria can play an important role in determining bioavailability of Cd. Arbuscular mycorrhizal fungi form associations with roots of many angiosperms and can influence Cd uptake by plants. El-Kherbawy et al. [1989] suggest that original soil Cd concentration plays a role in plant responses to mycorrhizae. Bacteria may also affect Cd uptake by plants, and several strains are being used as biosorbents for bioremediation of Cd contaminated sites or wastewaters [Carlot et al., 2002]. Bacteria can be genetically engineered to be used in remediation of heavy metal-polluted sites. Bang et al. [2000a,b] used genetically modified bacteria to enhance Cd removal from soil by precipitating it as Cd-sulfide. Peptides or proteins can be engineered to enhance metal specificity producing a protein containing a heavy-metal binding motif [Pazirandeh et al., 1998].

4   CONCLUSIONS

Cadmium is a naturally occurring and potentially toxic carcinogen that has no biological significance and is released into the ecosystem through natural processes such as volcanoes and erosion and through anthropogenic activities. The industrial supply of Cd is more dependent on Zn production than on Cd demand. Industrial emissions are now tightly controlled due to the significant improvement in pollution control technology and strict regulation and legislation, particularly in the metals industry. Phosphatic fertilizers can contain high Cd levels.

Diet is the main source of Cd for nonsmoking human population and is naturally present in air, water, soil, and foodstuffs. It is easily available to plants due to its high solubility. Because various anthropogenic activities have led to the addition of significant levels of Cd to the agricultural soils, environmental health concerns are being raised.

One of the primary responses of Cd stress in a biological system is the production of toxic reactive oxygen species (ROS). Some of the most damaging effects of ROS and their products in cells is the peroxidation of membrane lipids, site-specific amino acid modifications in proteins, fragmentation of the peptide chain, aggregation of cross-linked reaction products, altered electrical charge, and increased susceptibility to proteolysis and a number of genome-related changes. ROS detoxification reactions involve right balance between the formation and detoxification of active oxygen species. The defense strategies range from prevention, through interception, to repair.

Molecular and biochemical approaches, breeding strategies, and soil Cd remediation are approaches being followed to minimize Cd phytoavailability. In depth analyis of the different methods being employed gives an idea of their short-comings and strategies to be employed to overcome them. Thus Cd phytoavailabilty can be minimized, in turn minimizing the risk to human health.

REFERENCES

Alfven T, Elinder CG, Carlsson MD, Grubb A, Hellström L, Persson B, Pettersson C, Spång G, Schűtz A, Järup L. 2000. Low level cadmium exposure and osteoporosis. *J Bone Miner Res* **15**:1579–1586.

Ali V, Prakash K, Kulkarni S, Ahmad A, Madhusudhan KP, Bhakuni V. 1999. 8-Anilino-1-naphthalene sulfonic acid (ANS) induces folding of acid unfolded cytochrome c to molten globule state as a result of electrostatic interactions. *Biochemistry* **38**:13635–13642.

Andersson A, Siman G. 1991. Levels of Cd and some other trace-elements in soils and crops as influenced by lime and fertilizer level. *Acta Agric Scand* **41**:3–11.

Andrews M, Raven JA, Sprent JI. 1996. Environmental effects on the partitioning of dry matter between root and shoot of higher plants: the importance of tissue protein content. *J Exp Bot* **47**:1317.

Anfinsen CB. 1973. Principles that govern the folding of protein chains. *Science* **181**: 223–230.

Arrigioni O, De Tullio MC. 2000. The role of ascorbic acid in cell metabolism: between gene-directed functions and unpredictable chemical reactions. *J Plant Physiol* **157**: 481–488.

Asada K. 1994. Production and action of active oxygen species in photosynthetic tissues. In: Foyer CH, Mullinaeux PM, editors. *Causes of Photooxidative Stress and Amelioration of Defense Systems in Plants*. Boca Raton FL: CRC Press, pp. 77–103.

Atal N, Pardhasaradhi P, Mohanty P. 1991. Inhibition of chloroplast photochemical reactions by treatment of wheat seedlings with low concentrations of cadmium: analysis of electron transport activities and changes in fluorescence yield. *Plant Cell Physiol* **32**:943–951.

Atienzar FA, Conradi M, Evenden AJ, Jha AN, Depledge MH. 1999. Qualitative assessment of genotoxicity using random amplified polymorphic DNA: Comparison of genomic template stability with key fitness parameters in *Daphnia magna* exposed to benzo[*a*]pyrene. *Environ Toxicol Chem* **18**:2275–2282.

Atienzar FA, Venier P, Jha AN, Depledge MH. 2002. Evaluation of the random amplified polymorphic DNA (RAPD) assay for the detection of DNA damage and mutations. *Mutat Res* **521**:151–163.

Australia New Zealand Food Standards Code http://www.foodstandards.gov.au/thecode/foodstandardscode.cfm

Bang SW, Clark DS, Keasling JO. 2000a. Cadmium, lead and zinc removal by expression of the thiosulfate reductase gene from *Salmonella typhimurium* in *Escherichia coli*. *Biotechnol Lett* **22**:1331–1335.

Bang SW, Clark DS, Keasling JO. 2000b. Engineering hydrogen sulfide production and cadmium removal by expression of the thiosulfate reductase gene (phsABC) from *Salmonella enterica* serovar *typhimurium* in *Escherichia coli*. *Appl Environ Microbiol* **66**:3939–3944.

Barcelo J, Poschenreider C. 1990. Plant water relations as affected by heavy metals: A review. *J Plant Nutr* **13**:1–37.

Baryla A, Carrier P, Franck F, Coulomb C, Sahut C, Havaux M. 2001. Leaf chlorosis in oilseed rape plants (*Brassica napus*) grown on cadmium polluted soil: Causes and consequences for photosynthesis and growth. *Planta* **212**:696–709.

Basta NT, Gradwohl R, Snethen KL, Schroder JL. 2001. Chemical immobilization of lead, zinc and cadmium in smelter contaminated sols using biosolids and rock phosphate. *J Environ Qual* **30**:1222–1230.

Basta NT, Sloan JJ. 1999. Application of alkaline biosolids to acid soils: Changes in solubility and bioavailability of heavy metals. *J Environ Qual* **28**:633–638.

Baszyński T, Tukendorf A, Lyszcz S. 1984. Copper binding proteins in spinach tolerant to excess copper. *J Plant Physiol* **11**:351–360.

Baszyński T, Krol WM, Wolinska D, Krupa Z, Tukendorf A. 1980. Photosynthetic activities of cadmium-treated plants. *Physiol Plantarum* **48**:365–370.

Bergendi L, Beneš L, Ďuračková Z, Fereneik M. 1999. Chemistry, physiology and pathology of free radicals. *Life Sci* **65**:1865–1874.

Beyersmann D, Hechtenberg. 1977.

Bingham FT, Page AL, Mitchell GA, Strong JE. 1979. Effects of liming an acid soil amended with sewage sludge enriched with Cd, Cu, Ni, and Zn on yield and Cd content of wheat grain. *J Environ Qual* **8**:202–207.

Bogess SF, Willavize S, Koeppe DE, 1978. Differential response of soybean cultivars to soil cadmium. *Agron J* **70**:756–760.

Boisson JA, Ruttens A, Mench M, Vangronsveld J. 1999. Evaluation of hydroxyapatite as a metal immobilizing additive for remediation of polluted soils: I. Influence of hydroxyapatite on metal exchangeability in soil, plant growth, and plant metal accumulation. *Environ Pollut* **104**:225–233.

Bolan NS, Naidu R, Khan MAR, Tillman RW, Syers JK. 1999. The effects of anion sorption on sorption and leaching of cadmium. *Aust J Soil Res* **37**:445–460.

Boutron CF. 1995. Historical reconstruction of the earth's past atmospheric environment from Greenland and Antarctic snow and ice cores. *Environ Rev* **3**:1–28.

Bowditch BM, Albright DG, Williams JGK. 1993. Use of randomly amplified polymorphic DNA markers in comparative genomic studies. *Methods Enzymol* **224**:294–309.

Brallier S, Harrison RB, Henry CL, Dongsen X. 1996. Liming effects on availability of Cd, Cu, Ni and Zn in a soil amended with sewage sludge 16 years previously. *Water Air Soil Pollut* **86**:195–206.

Brash A. 1999. Lipoxygenses: occurrence, functions, catalysis and acquisition of substrate. *J Biol Chem* **274**:23679–23682.

Bray TM, Taylor CG. 1993. Tissue glutathione, nutrition and oxidative stress. *Can J Physiol Pharmacol* **71**:746–752.

Breen AP, Murphy JA. 1995. Reactions of oxy radicals with DNA. *Free Rad Biol Med* **18**:1033–1077.

Brown KR, Grant CA, Racz GJ. 1994. The effect of nitrogen source, rate and placement on Cd bioavailability. In: Papers presented at the 37th Annual Manitoba Society of Soil Science Meeting,Winnipeg, 4 and 5 January 1994, pp. 167–175.

Brown MH, Paulsen IT, Skurray RS. 1999. The multidrug efflux protein NorM is a prototype of a new family of transporters. *Mol Microbiol* **31**:393–395.

Brown S, Chaney R, Angle JS. 1997. Subsurface liming and metal movement in soils amended with lime-stabilized biosolids. *J Environ Qual* **26**:724–732.

Brown S, Chaney R, Angle JS, Ryan JA. 1998. phytoavailability of cadmium to lettuce in long-term biosolid amended soil. *J Environ Qual* **27**:1071–1078.

Burzyński M. 1987. The influence of lead and cadmium on the absorption and distribution of potassium, calcium, magnesium and iron in cucumber seedlings. *Acta Soc Bot Pol* **53**:77–86.

Cabiscol E, Tamarit J, Ros J. 2000. Oxidative stress in bacteria and protein damage by reactive oxygen species. *Int Microbiol* **3**:3–8.

Cakmak I, Marschner H. 1993. Effect of zinc nutritional status on activities of superoxide radical and hydrogen peroxide scavenging enzymes in bean leaves. *Plant Soil* **155/156**:127–130.

Carlot M, Giacomini A, Casella S. 2002. Aspects of plant-microbe interactions in heavy metal polluted soil. *Acta Biotechnol* **22**:13–20.

Chandler AJ. 1996. Characterising Cadmium in Municipal Solid Waste. Sources of Cadmium in the Environment, Inter-Organisation Programme for the Sound Management of Chemicals (IOMC), Organisation for Economic Co-operation and Development (OECD), Paris, France.

Chaney WR, Strickland RC, Lamoreaux RJ. 1977. Phytotoxicity of cadmium inhibited by lime. *Plant Soil* **47**:275–278.

Chao CC, Ma YS, Stadtman ER. 1997. Modification of protein surface hydrophobicity and methionine oxidation by oxidative systems. *Proc Natl Acad Sci USA* **94**:2969–2974.

Chen X, Wright JV, Conca JL, Peurrung LM. 1997. Evaluation of heavy metal remediation using mineral apatite. *Water Air Soil Pollut* **98**:57–78.

Cieslinski G, Van Rees KCJ, Szmigieslka AM, Krishnamurti GSR, Huang PM. 1998. Low-molecular-weight organic acids in rhizosphere soils of durum wheat and their effect on cadmium bioaccumulation. *Plant Soil* **203**:109–117.

Clemens S. 2001. Molecular mechanisms of plant metal homeostasis and tolerance. *Planta* **21**:475–486.

Cobbett C. 1999. A family of phytochelatin synthase genes from plant, fungal and animal species. *Trends Plant Sci* **4**:335–337.

Cobbett CS, Goldsbrough PB. 2002. Phytochelatins and metallothioneins: roles in heavy metal detoxification and homeostasis. *Annu Rev Plant Biol* **53**:159–182.

Cook ME, Morrow H. 1995. Anthropogenic Sources of Cadmium in Canada. National Workshop on Cadmium Transport Into Plants, Canadian Network of Toxicology Centres, Ottawa, Ontario, Canada.

Das P, Samantaray S, Rout GR. 1997. Studies on cadmium toxicity in plants: A review. *Environ Pollut* **98**:29–36.

Davies KJA. 1987. Protein damage and degradation by oxygen radicals. I. General aspects. *J Biol Chem* **262**:9895–9901.

Davies KJA, Delsignore ME. 1987. Protein damage and degradation by oxygen radicals III. modification of secondary and tertiary structure. *J Biol Chem* **262**:9908–9913.

Davies KJA, Delsignore M, Sharon WL. 1987a. Protein damage and degradation by oxygen radicals II. modification of amino acids. *J Biol Chem* **262**:9902–9907.

Davies KJA, Lin SW, Pacifici RE. 1987b. Protein damage and degradation by oxygen radicals IV. Degradation of denatured protein. *J Biol Chem* **262**:9914–9920.

Davies MJ. 2003. Singlet oxygen-mediated damage to proteins and its consequences. *Biochem Biophys Res Commun* **305**:761–770.

de Meeüs C, Eduljee GH, Hutton M. 2002. Assessment and management of risks arising from exposure to cadmium in fertilizers I. *Sci Total Environ* **291**:167–187.

De Tullio MC, Gara LD, Paciolla C, Arrigoni O. 1998. Dehydroascorbate-reducing proteins in maize are induced by the ascorbate biosynthesis inhibitor lycorine. *Plant Physiol Biochem* **36**:433–440.

Dean RT, Fu S, Stocker R, Davies MJ. 1997. Biochemistry and pathology of radicalmediated protein oxidation. *Biochem J* **324**:1–18.

del Castilho P, Chandron WJ, Salomons W. 1993. Influence of cattle manure slurry application on the solubility of cadmium, copper, and zinc in a manured acidic loamy sand soil. *J Environ Qual* **22**:689–697.

Devi SR, Prasad MNV. 1998. Copper toxicity in *Ceratophyllum demersum* L. (coontail), a free floating macrophyte: Response of antioxidant enzymes and antioxidants. *Plant Sci* **138**:157–165.

Dietz KJ, Baier M, Krämer U. 1999. Free radicals and reactive oxygen species as mediators of heavy metal toxicity in plants. In: Prasad MNV, Hagemeyer J, editors. *Heavy metal stress in Plants—From Molecules to Ecosystems*, first edition. Berlin: Springer-Verlag, pp. 73–98.

Dizgaroglu M, Bergtold DS. 1986. Characterization of free radicals-induced base damage in DNA at biologically relevant levels. *Anal Biochem* **156**:182–188.

Dudka S, Miller WP. 1999. Accumulation of potentially toxic elements in plants and their transfer to human food chain. *J Environ Sci Health B* **34**:681–708.

Elkhatib EA, Thabet AG, Mahdy AM. 2001. Phytoremediation of cadmium contaminated soils: Role of organic complexing agents in cadmium phytoextraction. *Land Contam Reclam* **9**:359–366.

El-Kherbawy M, Angle JS, Heggo A, Chaney RL. 1989. Soil pH, rhizobia, and vesicular arbuscular mycorrhizae inoculation effects on growth and heavy metal uptake of alfalfa (*Medicago sativa* L.). *Biol Fertil Soil* **8**:61–65.

Environmental Resources Limited (ERL). 1990. Evaluation of the sources of human and environmental contamination by cadmium. Prepared for the Commission of the European

Community, Directorate General for Environment, Consumer Protection and Nuclear Safety, London.

Eriksson J. 1990. Effects of nitrogen-containing fertilizers on solubility and plant uptake of cadmium. *Water Air Soil Pollut* **49**:355–368.

Eriksson J, Stenberg B, Andersson A, Andersson R. 2000. Tillståndet isvensk åkermark och spannmålsgröda (Current Status of Swedish Arable Soils and Cereal Crops). Naturvårdsverket, rapport 5062 (in Swedish with English abstract).

Esterbauer H, Schaur RJ, Zollner H. 1991. Chemistry and biochemistry of 4-hydroxynonenal, malondialdehyde and related aldehydes. *Free Rad Biol Med* **11**:81–128.

European Commission (EC). 2001. Commission regulation (EC) No 466/2001 of 8 March 2001 setting maximum levels for certain contaminants in foodstuffs. *Official J Eur Communities* **L77**:1–13.

EUROSTAT 1995. *Europe's Environment: Statistical Compendium for the Dobris Assessment.* Luxembourg: Office for Official Publications of the European Communities, European Environment Agency.

Fatur T, Lah T, Filipič M. 2003. Cadmium inhibits repair of UV-, methyl methane sulfonate- and *N*-methyl-*N*-nitrosourea-induced DNA damage in chinese hamster ovary cells. *Mutat Res* **529**:109–116.

Fleschin S, Fleschin M, Nita S, Pavel E, Mageanu V. 2000 Free radicals mediated protein oxidation in biochemistry. *Roum Biotechnol Lett* **5**:479–495.

Fodor A, Szabó-Nagy A, Erdei L. 1995. The effects of cadmium on the fluidity and H⁺-ATPase activity of plasma membrane from sunflower and wheat roots. *J Plant Physiol* **14**:787–792.

Fojtova M, Kovarik A. 2000. Genotoxic effects of cadmium is associated with apoptotic changes in tobacco cells. *Plant Cell Environ* **23**:531–537.

Foyer CH, Theodoulou F, Delrot S. 2001. The functions of inter- and intracellular glutathione transport systems in plants. *Trends Plant Sci* **6**:486–492.

Fridovich I. 1986. Biological effects of the superoxide radical. *Arch Biochem Biophys* **247**:1–11.

Fuhrer J. 1982. Ethylene biosynthesis and cadmium toxicity in leaf tissue of beans (*Phaseolus vulgaris* L.). *Plant Physiol* **70**:162–167.

Gadallah MΛΛ. 1995. Effects of cadmium and kinetin on chlorophyll content, saccharides and dry matter accumulation in sunflower plants. *Biol Plant* **37**:233–240.

Gebicki JM, Bielski BHJ. 1981. Comparison of the capacities of the perhydroxyl and superoxide radicals to initiate chain oxidation of linoleic acid. *J Am Chem Soc* **103**:7020–7022.

Gichner T, Patková Z, Száková J, Demnerová K. 2004. Cadmium induces DNA damage in tobacco roots, but no DNA damage, somatic mutations or homologous recombination in tobacco leaves. *Mutat Res* 49–57.

Gong JM, Lee DA, Schroeder JI. 2003. Long-distance root-to-shoot transport of phytochelatins and cadmium in *Arabidopsis. Proc Natl Acad Sci USA* **100**:10118–10123.

Grant CA, Bailey LD, McLaughlin MJ, Singh BR. 1999. Management techniques to reduce cadmium transfer from soils to plants. A review. In: McLaughlin MJ, Singh BR, editors. *Cadmium in Soils and Plants*. Dordrecht, The Netherlands: Kluwer Academic Publishers, pp. 151–198.

Gray CW, McLaren RG, Roberts AHC, Condron LM. 1999. Effect of soil pH on cadmium phytoavailability in some New Zealand soils. *NZ J Crop Horti* **27**:169–179.

Greger M, Ogren E. 1991. Direct and indirect effects of Cd^{2+} on photosynthesis in sugarbeet (*Beta vulgaris*). *Physiol Plant* **83**:129–135.

Grill E, Löffler S, Winnacker E-L, Zenk MH. 1989. Phytochelatins, the heavy-metal-binding peptides of plants, are synthesized from glutathione by a specific glutamylcysteine dipeptidyl transpeptidase (phytochelatin synthase). *Proc Natl Acad Sci USA* **86**:6838–6842.

Grotz N, Fox T, Connolly E, Park W, Guerinot ML, Eide D. 1998. Identification of a family of zinc transporter genes from *Arabidopsis* that respond to zinc deficiency. *Proc Natl Acad Sci USA* **95**:7220–7225.

Guerinot ML. 2000. The ZIP family of metal transporters. *Biochim Biophys Acta* **1465**:190–198.

Hall JL. 2002. Cellular mechanisms for heavy metal detoxification and tolerance. *J Exp Bot* **53**:1–11.

Halliwell B, Gutteridge JMC. 1990. Role of free radicals and catalytic metal ions in human disease: an overview. In: Methods Enzymol, Part B, 186 London: Academic Press, pp. 1–85.

Han DH, Lee JH, 1996. Effects of liming on uptake of lead and cadmium by *Raphanus sativa*. *Arch Envion Contam Toxicol* **31**:488–493.

Hartwig A. 2001. Zinc finger proteins as potential targets for toxic metal ions: differential effects on structure and function. *Antioxid Redox Signal* **3**:625–634.

Hartwig A, Schwerdtle T. 2002. Interactions by carcinogenic metal compounds with DNA repair processes: toxicological implications. *Toxicol Lett* **127**:47–54.

Hartwig A, Kruger I, Beyersmann D. 1994. Mechanisms in nickel genotoxicity: The significance of interactions with DNA repair. *Toxicol Lett* **72**:353–358.

Hausladen A, Kunert KJ. 1990. Effects of artificially enhanced levels of ascorbate and glutathione on the enzymes monodehydroascorbate reductase, dehydroascorbate reductase, and glutathione reductase in spinach (*Spinacia oleracea*). *Physiol Plant* **79**:384–388.

He QB, Singh BR. 1994. Plant availability of cadmium in soils: 2. Factors related to the extractability and plant uptake of cadmium in cultivated soils. *Acta Agric Scand* **43**:142–150.

Hell R, Bergmann L. 1990. γ-Glutamylcysteine synthetase in higher plants: Catalytic properties and subcellular localization. *Planta* **180**:603–612.

Hellstrand S, Landner L. 1998. Cadmium in fertilizers, soil, crops and foods—The Swedish situations. In: Cadmium Exposure in the Swedish Environment. KEMI report no. 1/98.

Hengstler JG, Audorff UB, Faldum A, Janssen K, Reifenrath M, Gotte W, Jung D, Mayer-Popken O, Fuchs J, Gebhard S, Bienfait HG, Schlink K, Dietrich C, Faust D, EpeB, Oesch F. 2003. Occupational exposure to heavy metals: DNA damage induction and DNA repair inhibition prove co-exposures to cadmium, cobalt and lead as more dangerous than hitherto expected. *Carcinogenesis* **24**:63–73.

Herren T, Feller U. 1994. Transfer of zinc from xylem to phloem in the peduncle of wheat. *J Plant Nut* **17**:1587–1598.

Hinesly TD, Alexander DE, Redborg KE, Ziegie EL. 1982. Differential accumulations of cadmium and zinc by corn hybrids grown on soil amended with sewage sludge. *Agron J* **74**:469–474.

Ho Kim Y, Berry AH, Spencer DS, Stites WE. 2001. Comparing the effect on protein stability of methionine oxidation versus mutagenesis: Steps toward engineering oxidative resistance in proteins. *Protein Eng* **14**:343–347.

Hooda PS, Alloway BJ. 1996. The effect of liming on heavy metal concentrations in wheat, carrots and spinach grown on previously sludge-applied soils. *J Agric Sci* **127**:289–294.

Horemans N, Foyer CH, Potters G, Asard H. 2000. Ascorbate function and associated transport systems in plants. *Plant Physiol Biochem* **38**:531–540.

Hossain MA, Nakano Y, Asada K. 1984. Monodehydroascorbate reductase in spinach chloroplasts and its participation in regeneration of ascorbate for scavenging hydrogen peroxide. *Plant Cell Physiol* **25**:385–395.

Hossain Z, Huq F. 2002a. Studies on the interaction between Cd^{2+} ions and DNA. *J Inorg Biochem* **90**:85–96.

Hossain Z, Huq F. 2002b. Studies on the interaction between Cd^{2+} ions and nucleobases and nucleotides. *J Inorg Biochem* **90**:97–105.

Hsu YT, Kao CH. 2003. Role of abscisic acid in cadmium tolerance of rice (*Oryza sativa* L.) seedlings. *Plant Cell Environ* **26**:867–874.

Hyun H, Chang AC, Parker DR, Page AL. 1998. Cadmium solubility and phytoavailability in sludge-treated soil: effects of soil organic matter. *J Environ Qual* **27**:329–334.

IARC. 1993. Beryllium, cadmium, mercury and exposures in the glass manufacturing industry. In: *IARC Monographs on the Evaluation of Carcinogenic Risks to Humans*, Vol. 58. Lyon: International Agency for Research on Cancer, pp. 41–117.

Imaly JA, Linn S. 1986. DNA damage and oxygen radical toxicity. *Science* **240**:1302–1309.

Jabs T, Dietrich A, Dangl JL. 1996. Initiation of runaway cell death in an *Arabidopsis* mutant by extracellular superoxide. *Science* **273**:1853–1856.

Jackson T, MacGillivray A. 1995. Accounting for cadmium-tracking emissions of cadmium from the global economy, *Chem Ecol* **11**(3):44.

Jalil A, Selles F, Clarke JM. 1994. Effect of cadmium on growth and uptake of cadmium and other elements by durum wheat. *J Plant Nut* **17**:1839–1858.

Jansson G. 2002. Cadmium in Arable Crops, The Influence of Soil Factors and Liming. Ph.D. thesis, Department of Soil Sciences, The Swedish University of Agricultural Sciences, Uppsala, ISBN 91-576-6192-8.

Järup L, Berglund M, Elinder CG, Nordberg G, Vahter M. 1998. Health effects of cadmium exposure—A review of the literature and a risk estimate. *Scand J Work Environ Health* **24**:1–51.

Järup L, Hellström L, Alfven T, Carlsson MD, Grubb A, Persson B, Pettersson C, Spång G, Schütz G, Elinder C-G. 2000. Low level exposure to cadmium and early kidney damage: The OSCAR study. *Occup Environ Med* **57**:668–672.

Jeanjean J, Fedoroff M, Faverjon F, Vincent U, Corset J. 1995. Influence of pH on the sorption of cadmium ions on calcium hydroxyapatite. *J Mater Sci* **30**:6156–6160.

Jensen A, Bro-Rasmussen F. 1992. Environmental contamination in Europe. *Rev Environ Contam Toxicol* **125**:101–181.

John MK. 1973. Cadmium uptake by eight food crops as influenced by various soil levels of cadmium. *Environ Pollut* **4**:7–15.

John MK, Van Laerhoven CJ, Chuah H. 1972. Factors affecting plant uptake and phytoavailability of cadmium added to soils. *Environ Sci Technol* **6**:1005–1009.

John MK, Van Laerhoven CJ. 1976. Effects of sewage sludge composition, application rate, and lime regime on plant availability of heavy-metals. *J Environ Qual* **5**:246–251.

Jones DL. 1998. Organic acids in rhizosphere—A critical review. *Plant Soil* **205**:25–40.

Jones R, Lapp T, Wallace D. 1993. Locating and estimating air emissions from sources of cadmium and cadmium compounds. *Prepared by Midwest Research Institute for the U.S. Environmental Protection Agency, Office of Air and Radiation*, Report EPA-453/R-93-040.

Jonnarth UA, Van Hees PAW, Lundstrom U, Finlay RD. 2000. Organic acids produced by mycorrhizal *Pinus sylvestris* exposed to elevated aluminium and heavy metal concentrations. *New Phytol* **146**:557–567.

Jönsson JÖ, Eriksson J. 2003. The effect of fertilisation for higher protein content on Cd-level in winter wheat grain. In: *7th International Conference on the Biogeochemistry of Trace Elements, Uppsala, Conference Proceedings,* Vol. 1, pp. 242–243.

Kappus H. 1985. Lipid peroxidation: Mechanism, analysis, enzymology and biological relevance. In: Sies H, editor. *Oxidative Stress* New York: Academic Press, pp. 273–309.

Kashem MA, Singh BR. 2001. Metal availability in contaminated soils. II. Uptake of Cd, Ni and Zn in rice plants grown under flooded culture with organic matter addition. *Nutr Cycl Agroecosyst* **61**:257–266.

Keefer RE, Codling EE, Singh RN. 1984. Fractionation of metal–organic components extracted from sludge-amended soil. *Soil Sci Soc Am J* **48**:1054–1059.

Kerkeb L, Krämer U. 2003. The role of free histidine in xylem loading of nickel in *Alyssum lesbiacum* and *Brassica juncea*. *Plant Physiol* **1312**:716–724.

Kevresan S, Petrovic N, Popovic M, Kandrac J. 1998. Effect of heavy metals on nitrate and protein metabolism in sugar beet. *Biol Plant* **41**:235–240.

Kingston-Smith AH, Foyer CH. 2000. Over expression of manganese superoxide dismutase in maize leaves leads to increased monodehydroascorbate reductase, dehydroascorbate reductase and glutathione reductase activities. *J Exp Bot* **51**:1867–1877.

Komarnisky LA, Christopherson DVM, Basu TK. 2003. Sulfur: Its clinical and toxicological aspects. *Nutrition* **19**:54–61.

Koppen G, Verschaeve L. 1996. The alkaline comet test on plant cells: A new genotoxicity test for DNA strand breaks in *Vicia faba* root cells. *Mutat Res* **360**:193–200.

Krämer U, Cotter-Howells JD, Charnock JM, Baker AJM, Smith JAC. 1996. Free histidine as a metal chelator in plants that accumulate nickel. *Nature* **379**:635–638.

Kreutzer K. 1995. Effects of forest liming on soil processes. *Plant Soil* **168–169**:447–470.

Krupa Z. 1988. Cadmium induced changes in the composition and structure of light harvesting chlorophyll a/b protein complex II in radish cotyledons. *Physiol Plant* **73**:518–524.

Krupa Z, Öquist G, Huner NPA. 1994. The effects of cadmium on photosynthesis of *Phaseolus vulgaris*—A fluorescence analysis. *Physiol Plant* **88**:626–630.

Kuck PH. 2007. Cadmium, *US Geological Survey*, Mineral Commodity Summaries. pp. 38–39.

Kuriakose SV, Prasad MNV. 2007. Marker assisted cadmium toxicity investigations in sorghum seedlings. In: Zhu Y, Lepp N, Naidu R, editors. *Biogeochemistry of Trace Elements: Environmental Protection, Remediation and Human Health*, Beijing, China: Tsinghua University Press, pp. 734–735.

Lacan D, Baccou JC. 1998. High levels of antioxidant enzymes correlate with delayed senescence in non-netted muskmelon fruits. *Planta* **204**:377–382.

Lane TW, Morel FM. 2000. A biological function for cadmium in marine diatomeas. *Proc Natl Acad Sci USA* **97**:4627–4631.

Larson RA. 1988. The antioxidants of higher plants. *Phytochemistry* **27**:969–978.

Lefebvre DD, Miki BL, Laliberte J-F. 1987. Mammalian metallothionein functions in plants. *Biotechnology* **5**:1053–1056.

Lehoczky E, Marth P, Szabados I, Szomolanyi A. 2000. The cadmium uptake by lettuce on contaminated soils as influenced by liming. *Commun Soil Sci Plant Anal* **31**: 2433–2438.

Levi-Minzi R, Petruzzelli G. 1984. The influence of phosphate fertilizers on Cd solubility in soil. *Water Air Soil Pollut* **23**:423–429.

Li YM, Chaney RL, Schneiter AA, Miller JF. 1995. Combining ability and heterosis estimates for kernel cadmium level in sunflower. *Crop Sci* **35**:1015–1019.

Liphadzi MS, Kirkham MB. 2005. Phytoremediation of soil contaminated with heavy metals: A technology for rehabilitation of the environment. *S Afr J Bot* **71**:24–37.

Liu W, Li PJ, Qi XM, Zhou QX, Zheng L, Sun TH, Yang YS. 2005. DNA changes in barley (*Hordeum vulgare*) seedlings induced by cadmium pollution using RAPD analysis. *Chemosphere* **61**(2):158–167.

Łobiński R, Potin-Gautier M. 1998. Metals and biomolecules-bioinorganic analytical chemistry. *Analusis* **26**:21–24.

Loganathan P, Hedley MJ, Gregg PEH, Currie LD. 1996. Effect of phosphate fertiliser type on the accumulation and plant availability of cadmium in grassland soils. *Nutrient Cycling Agroecosystems* **46**:169–178.

Logani MK, Davies RE. 1980. Lipid oxidation: Biologic effects and antioxidants—A review. *Lipids* **15**:485–495.

Lorenz SE, Hamon RE, McGrath SP, Holm PE, Christensen TH. 1994. Applications of fertilizer cations affect cadmium and zinc concentrations in soil solutions and uptake by plants. *Eur J Soil Sci* **45**:159–165.

Lozano-Rodriguez E, Hernandez LE, Bonay P, Carpena-Ruiz RO. 1997. Distribution of cadmium in shoot and root tissues of maize and pea plants: Physiological disturbances. *J Exp Bot* **48**:123–128.

Lugon-Moulin N, Zhang M, Gadani F, Rossi L, Koller D, Krauss M, Wagner GJ. 2004. Critical review of the science and options for reducing cadmium in tobacco (*Nicotiana tabacum* L) and other plants. *Adv Agron* **83**:111–180.

Ma JF. 2000. Role of organic acids in detoxification of aluminum in higher plants. *Plant and Cell Physiol* **41**:383–390.

Ma F, Cheng L. 2003. The sun-exposed peel of apple fruit has higher xanthophylls cycle dependent thermal dissipation and antioxidants of the ascorbate–glutathione pathway than shaded peel. *Plant Sci* **165**:819–827.

Maclean AJ. 1976. Cadmium in different plant species and its availability in soils as influenced by organic-matter and additions of lime, P, Cd and Zn. *Cana J Soil Sci* **56**:129–138.

Maksymiec W, Baszyński T. 1988. The effect of Cd^{2+} on the release of proteins from thylakoid membranes of tomato leaves. *Acta Soc Bot Pol* **57**:465–474.

Mandjiny S, Matis KA, Fedoroff M, Jeanjean J, Rouchaud JC, Toulhoat N, Potocek V, Maireles-Torres P, Jones D. 1998. Calcium hydroxyapatites: Evaluation of sorption properties for cadmium ions in aqueous solution. *J Mater Sci* **33**:5433–5439.

Marrs KA. 1996. The functions and regulation of glutathione-*S*-transferase in plants. *Annu Rev Plant Physiol Plant Mol Biol* **47**:127–158.

Marschner H. 1995. *Mineral Nutrition of Higher Plants*, second edition. London: Academic Press.

Martin TA, Ruby MV. 2004. Review of *in situ* remediation technologies for lead, zinc, and cadmium in soil. *Remediation J* **14**(3):35–53.

Mäser P, Thomine S, Schroeder JI, Ward Jm, Hirschi K, Sze H, Talke IN, Amtmann A, Maathuis FJM, Sanders D, Harper JF, Tchieu J, Gribskov M, Persans MW, Salt DE, Kim SA, Gueirnot ML. 2001. Phylogenetic relationships within cation transporter families of *Arabidopsis*. *Plant Physiol* **126**:1646–1667.

May MJ, Vernoux T, Leaver C, Van Montagu M, Inze D. 1998. Glutathione homeostasis in plants: Implications for environmental sensing and plant development. *J Exp Bot* **49**:649–667.

McGowen SL, Basta NT, Brown GO. 2001. Use of diammonium phosphate to reduce heavy metal solubility and transport in smelter-contaminated soil. *J Environ Qual* **30**:493–500.

McLaughlin MJ, Williams CMJ, McKay A, Kirkman R, Gunton J, Jackson KJ, Thompson R, Dowling B, Partington D, Smart MK, Tiller KG. 1994. Effect of cultivar on uptake of cadmium by potato tubers. *Aust J Agric Res* **45**:1483–1495.

McLaughlin MJ, Tiller Kg, Naidu R, Stevens DP. 1996. Review: The behaviour and environmental impact of contaminants in fertilizers. *Aust J Soil Res* **34**:1–54.

McLaughlin MJ, Parkerb DR, Clarke JM. 1999. Metals and micronutrients—Food safety issues. *Field Crops Res* **60**:143–163.

McMurray CT, Tainer JA. 2003. Cancer, cadmium and genome integrity. *Nat Gen* **34**(3): 239–241.

Merrington G, Madden C. 2000. Changes in Cd and Zn phytoavailability in an agricultural soil after amendment with papermill sludge and biosolids. *Commun Soil Sci Plant Anal* **31**:759–776.

Minerals and Metals Sector. 2003. Cadmium. *Canadian Minerals Year Book*, pp. 13.1–13.6.

Minerals and Metals Sector. 2004. Cadmium. *Canadian Minerals Year Book*, pp. 13.1–13.3.

Minerals and Metals Sector. 2005. Cadmium. *Canadian Minerals Year Book*, pp. 13.1–13.7.

Mohanty N, Mohanty P. 1998. Cation effects on primary processes of photosynthesis. In: Singh R, Sawhney SK, editors. *Advances in Frontier Areas of Plant Biochemistry*, Delhi: Prentice Hall India, pp. 1–18.

Myśliwa-Kurdziel B, Prasad MNV, Strzałka K. 2004. Photosynthesis in metal stressed plants. In: Prasad MNV, editor. *Heavy Metal Stress in Plants: From Biomolecules to Ecosystems*, second edition. Heidelberg: Springer-Verlag, pp. 146–181.

Myśliwa-Kurdziel B, Strzałka K. 2002. Influence of metals on the biosynthesis of photosynthetic pigments. In: Prasad MNV, Strzałka K, editors. *Physiology and Biochemistry of Metal Toxicity and Tolerance in Plants*. Dordrecht, Netherlands: Kluwer Academic, pp. 201–228.

Nagalakshmi N, Prasad MNV. 2001. Responses of glutathione cycle enzymes and glutathione metabolism to copper stress in *Scenedesmus bijugatus*. *Plant Sci* **160**:291–299.

Naidu R, Bolan NS, Kookona RS, Tiller KG. 1994. Ionic-strength and pH effects on the sorption of cadmium and the surface charge of soils. *Eur J Soil Res* **45**:419–429.

Nan Z, Li J, Zhang J, Cheng G. 2002. Cadmium and zinc interactions and their transfer in soil–crop system under actual field conditions. *Sci Tot Environ* **285**:187–195.

Nelson JR, Lawrence CW, Hinkle DC. 1996. Thymine–thymine dimer bypass by yeast polymerase. *Science* **272**:1646–1649.

Nicholson FA, Jones KC. 1994. Effect of phosphate fertilizers and atmospheric deposition on long-term changes in the cadmium content of soils and crops. *Environ Sci Technol* **28**:2170–2175.

Nieboer E, Richardson DHS. 1980. The replacement of the non-descriptive term "heavy metals" by a biologically and chemically significant classification of metal ions. *Environ Pollut Ser B* **1**:3–26.

Noctor G, Foyer CH. 1998. Ascorbate and glutathione: Keeping active oxygen under control. *Annu Rev Plant Physiol Plant Mol Biol* **49**:249–279.

Nriagu JO, Pacyna JM. 1988. Quantitative assessment of world-wide contamination of air, water and soils by trace metals. *Nature* **333**:134–139.

Nriagu JO. 1980. Cadmium in the atmosphere and in precipitation. In: Nriagu JO, editor. *Cadmium in the Environment, Part 1, Ecological Cycling.* New York: John Wiley & Sons, pp. 71–114.

Nriagu JO. 1989. A global assessment of natural sources of atmospheric trace metals. *Nature* **338**:47–49.

Oleinick NL, Chiu S, Ramakrishnan N, Xue L. 1986. The formation, identification, and significance of DNA-protein cross links in mammalian cells. *Br J Cancer* **55**(8):135–140.

Oliver DP, Gartrell JW, Tiller KG, Correll R, Cozens GD, Youngberg BL. 1995. Differential responses of australian wheat cultivars to cadmium concentration in wheat grain. *Aust J Agric Res* **46**:873–886.

Oliw EH. 2002. Plant and fungal lipoxygenases. *Prostaglandins Other Lipid Mediat* **68–69**:313–323.

Olsson I-M, Eriksson J, Öborn I, Skerfving S, Oskarsson A. 2005. Cadmium in food production systems—a health risk for sensitive population groups. *Ambio* **34**:344–351.

Organisation for Economic Co-operation and Development (OECD) 1994. *Risk Reduction Monograph No. 5.* Paris, France: Cadmium OECD Environment Directorate.

Ouzounidou G. 1995. Cu-ions mediated changes in growth, chlorophyll and other ion contents in Cu-tolerant *Koeleria splendens. Biol Plant* **37**:71–79.

Pacifici RE, Davies KJA. 1990. Protein degradation as an index of oxidative stress. *Methods Enzymol* **186**:485–502.

Pallanca JE, Smirnoff N. 2000. The control of ascorbate synthesis and turnover in pea seedlings. *J Exp Bot* **51**:669–674.

Pazirandeh M, Wells BM, Ryan RL. 1998. Development of bacterium-based heavy metal biosorbents: Enhanced uptake of cadmium and mercury by *Escherichia coli* expressing a metal binding motif. *Appl Environ Microbiol* **64**:4068–4072.

Pearson MS, Maenpaa K, Pierzynski GM, Lydy MJ. 2000. Effects of soil amendments on the bioavailability of lead, zinc, and cadmium to earthworms. *J Environ Qual* **29**:1611–1617.

Penner GA, Clarke J, Bezte LJ, Leisle D. 1995. Identification of RAPD markers linked to a gene governing cadmium uptake in durum wheat. *Genome* **38**:543–547.

Peraza MA, Fierro FA, Barber DS, Casarez E, Rael LT. 1998. Effects of micronutrients on metal toxicity. *Environ Health Perspect* **106**(1): 203–216.

Pierzynski GM, Schwab AP. 1993. Bioavailability of zinc, cadmium and lead in a metal contaminated alluvial soil. *J Environ Qual* **22**:247–254.

Pietz RI, Peterson JR, Hinesly TD, Ziegler EL, Reborg KE, Lue-Hig C. 1983. Sewage sludge application to calcareous strip-mine spoil: Effect on spoil and corn Cd, Cu, Ni, and Zn. *J Environ Qual* **12**:463–467.

Pilon-Smits E. 2005. Phytoremediation. *Annu Rev Plant Biol* **56**:15–39.

Pohlmeier A. 2004. Metal speciation, chelation and complexing ligands in plants. In: Prasad MNV, editor. *Heavy Metal Stress in Plants: From Biomolecules to Ecosystems*, Second edition. Heidelberg, Narosa, New Delhi: Springer-Verlag, pp. 28–46.

Polykov NE, Kruppa AI, Leshina TV, Konovalova TA, Kispert LD. 2001. Carotenoids as antioxidants: spin trapping EPR and optical study. *Free Rad Biol Med* **31**:43–52.

Poschenrieder C, Gunse B, Barcelo J. 1989. Influence of cadmium on water relations, stomatal resistance and abscisic acid content in expanding bean leaves. *Plant Physiol* **90**:1365–1371.

Potters G, De Gara L, Asard H, Horemans N. 2002. Ascorbate and glutathione: Guardians of the cell cycle, partners in crime? *Plant Physiol Biochem* **40**:537–548.

Prasad MNV. 1995a. Cadmium toxicity and tolerance in vascular plants. *Environ Exp Bot* **35**:525–545.

Prasad MNV. 1995b. Inhibition of maize leaf chlorophylls, carotenoids and gas exchange functions by cadmium. *Photosynthetica* **31**:635–640.

Prasad MNV. 2004. *Heavy Metal Stress in Plants: From Biomolecules to Ecosystems*, second edition. Heidelberg: Springer-Verlag, 462 pp.

Prasad MNV, Strzałka K. 1999. Impact of heavy metals on photosynthesis. In: Prasad MNV, Hagemeyer J, editors. *Heavy Metal Stress in Plants: From Molecules to Ecosystems*. Berlin: Springer-Verlag, 117–138.

Ramos I, Esteban E, Lucena JJ, Garate A. 2002. Cadmium uptake and subcellular distribution in plants of *Lactuca* sp. Cd–Mn interaction. *Plant Sci* **162**:761–767.

Rauser WE. 1995. Phytochelatins and related peptides. *Structure, Biosynthesis, and Function. Plant Physiol* **109**:1141–1149.

Rauser WE. 1999. Structure and function of metal chelators produced by plants: The case for organic acids, amino acids, phytin and metallothioneins. *Cell Biochem Biophys* **31**:19–48.

Rauser WE. 2000. The role of thiols in plants under metal stress. In: Brunold C, Rennenberg H, De Kok LJ, Stulen I, Davidian JC, editors. *Sulfur Nutrition and Sulfur Assimilation in Higher Plants*. Bern, Switzerland: Paul Haupt, pp. 169–183.

Reddy GN, Prasad MNV. 1992. Characterization of cadmium binding protein from *Scenedesmus quadricauda* and cadmium toxicity reversal by phytochelatin constituting amino acids and citrate. *J Plant Physiol* **140**:156–162.

Rennenberg H. 1982. Glutathione and possible biological roles in higher plants. *Phytochemistry* **21**:2771–2780.

Renneberg H, Brunold C. 1994. Significance of glutathione metabolism in plants under stress. *Prog Bot* **55**:142–156.

Requena JR, Stadtman ER. 1999. Conversion of lysine to *N*-(carboxymethyl)lysine increases susceptibility of proteins to metal-catalyzed oxidation. *Biochem Biophys Res Commun* **264**:201–211.

Riffaldi R, Levi-Minzi R, Saviozzi A, Tropea M. 1983. Sorption and release of cadmium by some sewage sludges. *J Environ Qual* **12**:253–256.

Romero-Puertas MC, Palma JM, Gómez M, del Río LA, Sandalio LM. 2002. Cadmium causes the oxidative modification of proteins in pea plants. *Plant Cell Environ* **25**:677–686.

Sagner S, Kneer R, Wagner G, Cosson J-P, Deus-Neumann B, Zenk MH. 1998. Hyperaccumulation, complexation and distribution of nickel in *Sebertia acuminata*. *Phytochem* **47**:339–347.

Salt DE, Wagner GJ. 1993. Cadmium transport across tonoplast of vesicles from oat roots. *J Biol Chem* **268**:12297–12302.

Salt DE, Smith RD, Raskin I. 1998. Phytoremediation. *Annu Rev Plant Physiol Plant Mol Biol* **49**:643–668.

Sanita di Toppi L, Gabbrielli R. 1999. Response to cadmium in higher plants. *Environ Exp Bot* **41**:105–130.

Sanita di Toppi L, Gremigni P, Pawlik Skowronska B, Prasad MNV, Cobbett CS. 2003. Responses to heavy metals in plants—Molecular approach. In: Sanita di Toppi L, Pawlik Skowronska B, editors. *Abiotic Stresses in Plants*. Dordrecht, Netherlands: Kluwer Academic, pp. 33–156.

Satarug S, Baker JR, Urbenjapol S, Haswell-Elkins M, Reilly PEB, Williams DJ, Moore MR. 2003. A global perspective on cadmium pollution and toxicity in non-occupationally exposed population. *Toxicol Lett* **137**:65–83.

Scandalios JG, Guan LM, Polidoros A. 1997. Catalases in plants: Gene structure, properties, regulation and expression. In: Scandalios JG, editor, Oxidative Stress and Molecular Biology of Antioxidant Defense. Plainview, NY: Cold Spring Harbor Laboratory Press, pp. 343–406.

Schachtman DP, Kumar R, Schroeder JI, Marsh EL. 1997. Molecular and functional character-ization of a novel low affinity cation transporter (LCT1) in higher plants. *Proc Natl Acad Sci USA* **94**:11079–11084.

Seaman JC, Arey JS, Bertsch PM. 2001. Immobilization of nickel and other metals in contami-nated sediments by hydroxyapatite addition. *J Environ Qual.* **30**:460–469.

Semu E, Singh BR. 1996. Accumulation of heavy metals in soil and plants after long-term use of fertilisers and fungicides in Tanzania. *Fertiliser Res* **44**:241–248.

Senden MHMN, Van Der Meer AJGM, Wolterbeck HT. 1995. Effects of cadmium on the longitudinal and lateral xylem movement of citric acid through tomato stem internodes. *Acta Bot Neerl* **44**:129–138.

Šeršeň F, Králová K. 2001. New facts about $CdCl_2$ action on the photosynthetic apparatus of spinach chloroplasts and its comparison with $HgCl_2$ action. *Photosynthetica* **39**:575–580.

Shacter E. 2000a. Protein oxidative damage. *Methods Enzymol* **319**:428–436.

Shacter E. 2000b. Quantification and significance of protein oxidation in biological samples. *Drug Metabolism Rev* **32**:307–326.

Shah K, Kumar RG, Verma S, Dubey RS. 2001. Effect of cadmium on lipid peroxidation, superoxide anion generation and activities of antioxidant enzymes in growing rice seedlings. *Plant Sci* **161**:1135–1144.

Shenker M, Fan TWM, Crowley DE. 2001. Phytosiderophores influence on cadmium mobilization and uptake by wheat and barley plants. *J Environ Qual* **30**:2091–2098.

Siedlecka A. 1995. Some aspects of interactions between heavy metals and plant mineral nutrients. *Acta Soc Bot Pol* **3**:265–272.

Siedlecka A, Krupa Z. 1996. Interaction between cadmium and iron: Accumulation and distribution of metals and changes in growth parameters of *Phaseolus vulgaris* L. seedlings. *Acta Soc Bot Pol* **66**:1–6.

Simon L. 1998. Cadmium accumulation and distribution in sunflower plant. *J Plant Nut* **21**:341–352.

Singh BR, Myhr K. 1998. Cadmium uptake by barley as affected by Cd sources and pH levels. *Geoderma* **84**:185–194.

Singh BR, Narwal RP, Jeng AS, Almas A. 1995. Crop uptake and extractability of cadmium in soils naturally high in metals at different pH levels. *Commun Soil Sci Plant Anal* **26**:2123–2142.

Skórynzska E, Baszyński T. 1993. The changes in PSII complex polypeptides under cadmium treatment—Are they of direct or indirect nature? *Acta Physiol Plant* **15**:263–269.

Smeyers-Verbeke J, De Graeve M, Francois M, De Jaegere R, Massart DL. 1978. Cd uptake by intact wheat plants. *Plant Cell Environ* **1**:291–296.

Smirnoff N. 1996. The function and metabolism of ascorbic acid in plants. *Ann Bot* **78**:661–669.

Smith IK, Vierheller TL, Thorne CA. 1989. Properties and functions of glutathione reductase in plants. *Physiol Plant* **77**:449–456.

Soon YK. 1981. Solubility and sorption of cadmium in soils amended with sewage sludge. *Soil Sci* **32**:85–95.

Sorahan T, Lancashire RJ. 1997. Lung cancer mortality in a cohort of workers employed at a cadmium recovery plant in the United States: an analysis with detailed job histories. *Occup Environ Med* **4**(3):194–201.

Sorahan T, Lister M, Gilthorpe S, Harrington JM. 1995. Mortality of copper cadmium alloy workers with special reference to lung cancer and non-malignant diseases of the respiratory system, 1946–92. *Occup Environ Med* **52**(12):804–812.

Stadtman ER, Levine RL. 2000. Protein oxidation. *Ann NY Acad Sci* **899**:191–208.

Stadtman ER, Oliver CN. 1991. Metal-catalyzed oxidation of proteins. *J Biol Chem* **266**:2005–2008.

Stalikas CD, Mantalovas AC, Pilidis GA. 1997. Multi-element concentrations in vegetable species grown in two typical agricultural areas of Greece. *Sci Tot Environ* **206**:17–24.

Stolt JP, Sneller FEC, Bryngelsson T, Lundborg T, Schat H. 2003. Phytochelatin and cadmium accumulation in wheat. *Environ Exp Bot* **49**:21–28.

Stoyanova D, Tchakalova T. 1999. Cadmium induced ultrastructural changes in shoot apical meristem of *Elodea Canadensis* Rich. *Photosynthetica* **37**:47–52.

Street JJ, Sabey BR, Lindsay WL. 1978. Influence of pH, phosphorus, cadmium, sewage sludge, and incubation time on the solubility and plant uptake of cadmium. *J Environ Qual* **7**:286–290.

Symeonidis L, Karataglis S. 1992. Interactive effects of cadmium, lead and zinc on root growth of two metal tolerant genotypes of *Holcus lanatus* L. *Biometals* **5**:173–178.

Szuster-Ciesieslska A, Stachura A, Słotwinska M, Kamińska T, Snieżka R, Paduch R, Abramczyk D, Filar J, Kandefer-Szerszeń M. 2000. The inhibitory effect of zinc on cadmium-induced cell apoptosis and reactive oxygen species (ROS) production in cell cultures. *Toxicology* **145**:159–171.

Tadolini B, Fiorentini D, Landi L, Cabrini L. 1989. Lipid peroxidation. Definition of experimental conditions for selective study of the propagation and termination phases. *Free Rad Res Commun* **5**:245–252.

Taylor GJ. 1989. Multiple metal stress in *Triticum aestivum* L., differentiation between additive, antagonistic, and synergistic effects. *Can J Bot* **67**:2272–2276.

Theodorakis CW. 2001. Integration of genotoxic and population genetic endpoints in biomonitoring and risk assessment. *Ecotoxicology* **10**:245–256.

Theodoulou FL. 2000. Plant ABC transporters. *Biochim Biophys Acta* **1465**:79–103.

Thomas GM, Harrison HC. 1991. Genetic line effects on parameters influencing cadmium concentration in lettuce. *J Plant Nutr* **14**:953–962.

Thompson JE, Legge RL, Barber RF. 1987. The role of free radicals in senescence and wounding. *New Phytol* **105**:317–344.

Traina SJ. 1999. The environmental chemistry of cadmium. In: McLaughlin MJ, Singh BR, editors. *Cadmium in Soils and Plants.* Dordrecht: Kluwer Academic Publishers, pp. 11–37.

Trivedi S, Erdei L. 1992. Effects of cadmium and lead on the accumulation of Ca^{2+} and K^{+} and on the influx and translocation of K^{+} in wheat of low and high K^{+} status. *Physiol Plant* **84**:94–100.

Tsukahara T, Ezaki T, Moriguchi J, Furuki K, Shimbo S, Matsuda-Inoguchi N, Ikeda M. 2003. Rice as the most influential source of cadmium intake among general Japanese population. *Sci Tot Environ* **305**:41–51.

Tukendorf A. 1993. The response of spinach plants to excess of copper and cadmium. *Photosynthetica* **28**:573–575.

Tyler G, Olsson T. 2001. Plant uptake of major and minor mineral elements as influenced by soil acidity and liming. *Plant Soil* **230**:307–321.

Valverde M, Trejo C, Rojas E. 2001. Is the capacity of lead acetate and cadmium chloride to induce genotoxic damage due to direct DNA–metal interaction? *Mutagenesis* **16**:265–270.

Van Assche F, Clijsters H. 1990. Effects of metals on enzyme activity in plants. *Plant Cell Environ* **13**:195–206.

Van Assche FJ, Ciarletta P. 1992. Cadmium in the environment: Levels, trends and critical pathways. In: *Edited Proceedings Seventh International Cadmium Conference—New Orleans, Cadmium Association*, London, Cadmium Council, Reston VA, International Lead Zinc Research Organisation, Research Triangle Park, NC.

Vasseur L, Fortin MJ, Cyr J. 1998. Clover and cress as indicator species of impacts from limed sewage sludge and landfill wastewater land application. *Sci Tot Environ* **217**:231–239.

Vassilev A, Yordanov I. 1997. Reductive analysis of factors limiting growth of cadmium-treated plants: A review. *Bulg J Plant Physiol* **23**:114–133.

Wagner GJ. 1993. Accumulation of cadmium in crop plants and its consequences to human health. *Adv Agron* **51**:173–212.

Wagner GJ, Yeargan R. 1986. Variation in cadmium accumulation potential and tissue distribution of cadmium in tobacco. *Plant Physiol* **82**:274–279.

Wallace A. 1982. Additive, protective and synergistic effects on plants with excess trace elements. *Soil Sci* **133**:319–323.

Wångstrand H, Eriksson J, Öborn I. 2007. Cadmium concentration in winter wheat as affected by nitrogen fertilization. *Eur J Agrono* **26**:209–214.

Williams CH, David DJ. 1976. The accumulation in soil of cadmium residues from phosphate fertilizers and their effect on the cadmium content of plants. *Soil Sci* **121**:86–93.

Williams LE, Pittman JK, Hall JL. 2000. Emerging mechanisms for heavy metal transport in plants. *Biochim Biophys Acta* **1465**:104–126.

Winterbourn CC, Kettle AJ. 2003. Radical-radical reactions of superoxide: A potential route to toxicity. *Biochem Biophys Res Commun* **305**:729–736.

Wolff SP, Garner A, Dean R. 1986. Free radicals, lipids and protein degradation. *Trends Biochem Sci* **11**:27–31.

World Health Organization (WHO). 1972. Evaluation of certain food additives and of the contaminants mercury, lead and cadmium. FAO Nutrition Meetings Report Series No. 51. WHO Technical Report Series 505.

World Health Organization, (WHO). 1992. *Cadmium* (Environmental Health Criteria 134), Geneva.

Wright A, Bubb WA, Hawkins CL, Davies MJ. 2002. Singlet oxygen-mediated protein oxidation: Evidence for the formation of reactive side chain peroxides on tyrosine residues. *Photochem Photobiol* **76**:35–46.

Wu LJ, Luo YM, Christie P, Wong MH. 2003. Effects of EDTA and low molecular weight organic acids on soil solution properties of a heavy metal polluted soil. *Chemosphere* **50**:819–822.

Xiang CB, Oliver DJ. 1998. Glutathione metabolic genes coordinately respond to heavy metals and jasmonic acid in *Arabidopsis. Plant Cell* **10**:1539–1550.

Xu Y, Schwartz FW. 1994. Lead immobilization by hydroxyapatite in aqueous solutions. *J Contam Hydrol* **15**:187–206.

Yang MG, Lin XY, Yang XE. 1998. Impact of cadmium on growth and nutrient accumulation of different plant species. *Chinese J Appl Ecol* **9**:89–94.

Zhang GP, Fukami M, Sekimoto H. 2002. Influence of cadmium on mineral concentrations and yield components in wheat genotypes differing in Cd tolerance at seedling stage. *Field Crops Res* **4079**:1–7.

Zhang J, Kirkham MB. 1996. Enzymatic responses of the ascorbate-glutathione cycle to drought in sorghum and sunflower plants. *Plant Sci* **113**:139–147.

Zhang Y, Xiao M. 1998. Antagonistic effect of calcium, zinc and selenium against Cd induced chromosomal aberration and micronuclei in root cells of *Hordeum vulgare. Mutat Res* **420**:1–6.

Zhang Y, Yang X. 1994. The toxic effects of cadmium on cell division and chromosomal morphology of *Hordeum vulgare. Mutat Res* **312**:121–126.

17 Trace Element Transport in Plants

DANUTA MARIA ANTOSIEWICZ

Department of Ecotoxicology, Faculty of Biology, The University of Warsaw, 02-096 Warsaw, Poland

AGNIESZKA SIRKO

Institute of Biochemistry and Biophysics, Polish Academy of Sciences, 02-106 Warsaw, Poland

PAWEŁ SOWIŃSKI

Department of Plant Physiology, Institute of Botany, Faculty of Biology, University of Warsaw, 02-096 Warsaw, Poland; and Plant Biochemistry and Physiology Department, Plant Breeding and Acclimatization Institute, 05-870 Błonie, Radzików, Poland

1 INTRODUCTION

A number of transition metals, such as Cu, Zn, Mn, Fe, Ni, and Co, are trace elements necessary in small amounts for the proper growth and development of organisms, including plants. Because plants constitute the main source of food for humans, their mineral composition determines their nutritive value for humans, meaning that not only must trace elements be present in proper amounts, but also nonessential heavy metals such as Cd^{2+}, Pb^{2+}, Hg^{2+} must be absent in edible parts. Metal nutrients are components of cellular structures and act as cofactors in a wide range of physiological processes [Marschner et al., 1996]. However, when present in excess, they become extremely toxic. The concentration of free metal ions in a cell must, therefore, be under strict control and all mechanisms maintaining metal ion homeostasis are crucial to the normal development of plants, both under conditions of adequate nutrient supply and their excess as well as in the presence of high concentrations of nonessential heavy metals. Transmembrane transport is one of the basic elements of both mechanisms determining metal homeostasis and the level of tolerance to their excess.

Trace Elements as Contaminants and Nutrients: Consequences in Ecosystems and Human Health, Edited by M. N. V. Prasad

413

TRACE ELEMENT TRANSPORT IN PLANTS

TABLE 1. Internet Resources

TOPPRED	http://bioweb.pasteur.fr/seqanal/interfaces/toppred.html
PFAM	http://www.sanger.ac.uk/Software/Pfam/
PHYLIP	http://evolution.genetics.washington.edu/phylip.html
PlantsT database	http://plantst.genomics.purdue.edu/
	http://plants.sdsc.edu
Arabidopsis thaliana YSL family	http://www.cbs.dtu.dk/services/TMHMM-2.0

The use of modern molecular biology techniques has enabled identification of new families of genes involved in the transport of metals through biological membranes. Sequencing the entire genome of *Arabidopsis thaliana* in turn enables the identification and functional analysis of all members of a given transporter family of one species. Information about advances in work on transporters is available on a website called the PlantsT database (Table 1). Understanding the mechanisms of

TABLE 2. The Complete List of *A. thaliana* Genes Encoding Members of the Transporter Families Described in this Work

Family	Protein Name	Accession Number
ZIP	AtIRT1	At4g19690
	AtIRT2	At4g19680
	AtIRT3	At1g60960
	AtZIP1	At3g12750
	AtZIP2	At5g59520
	AtZIP3	At2g32270
	AtZIP4	At1g10970
	AtZIP5	At1g05300
	AtZIP6	At2g30080
	AtZIP7	At2g04032
	AtZIP8	At5g45105
	AtZIP9	At4g33020
	AtZIP10	At1g31260
	AtZIP11	At1g55910
	AtZIP12	At5g62160
	NP_187477	At3g08650
	AtIAR1	At1g68100
NRAMP	AtNRAMP1	At1g80830
	AtNRAMP2	At1g47240
	AtNRAMP3	At2g23150
	AtNRAMP4	At5g67330
	AtNRAMP5	At4g18790
	AtNRAM6	At1g15960
	EIN2-NRAMP	At5g03280

TABLE 2. *Continued*

Family	Protein Name	Accession Number
CDF	AtCDF5	At2g04620
	AtMTP1	At2g46800
	AtMTPa1	At3g61940
	AtMTPa2	At3g58810
	AtMTPb1	At2g47830
	AtMTPc1	At3g12100
	AtMTPc2	At1g51610
	AtMTPd1	At3g58060
	AtMTPd2	At1g79520
	AtMTPd3	At1g16310
	AtMTPd4	At2g39450
CAX/CaCA	AtCAX1	At2g38170
	AtCAX2	A3t13320g
	AtCAX3/HCX1	At3g51860
	AtCAX4	At5g01490
	AtCAX5	At1g55370
	AtCAX6	At1g55720
	AtCAX7	At5g17860
	AtCAX8	At5g17850
	AtCAX9	At3g14070
	AtCAX10	At1g54110
	AtCAX11	At1g08960
P_{1B}-ATPase	AtHMA1	At4g37270
	AtHMA2	At4g30110
	AtHMA3	At4g30120
	AtHMA4	At2g19110
	AtHMA5	At1g63440
	AtHMA6/AtPAA1	At4g33520
	AtHMA7/AtRAN1	At5g44790
	AtHMA8/AtPAA2	At5g21930
YSL	AtYSL1	At4g24120
	AtYSL2	At5g24380
	AtYSL3	At5g53550
	AtYSL4	At5g41000
	AtYSL5	At3g17650
	AtYSL6	At4g24120
	AtYSL7	At1g65730
	AtYSL8	At1g48370
COPT	AtCOPT1	At5g59030
	AtCOPT2	At3g46900
	AtCOPT3	At5g59040
	AtCOPT4	At2g37920
	AtCOPT5	At5g20650

uptake and transport of metals in a plant, as well as their transport in its tissues and organs, will enable engineering by a transgenic approach to crop plants having the desired mineral composition of edible parts ("biofortification") or selection of appropriate varieties. No proteins specific for transmembrane transport of toxic, nonessential metals like Cd, Pb, or Hg have yet been detected. Their similarity to essential elements allows them to cross membranes via transporters for chemically similar cations [Clemens, 2006]. To be able to engineer crop plants able to take up, accumulate, and distribute essential micronutrients and to eliminate the toxic, nonessential ones, the understanding of the biochemistry of substrate specificity is one of the most important problems that needs to be solved.

This chapter presents a broad overview of a range of transition metal transport proteins differing in their structure and substrate specificity, able to transfer ions against their electrochemical gradient. The presented families of transport proteins have been divided into two classes according to the direction of transport: (i) metal uptake protein families, including Nramp, ZIP, COPT, YSL; (ii) metal efflux proteins: P_{1B}ATPase, CDF, CAX (MRP from the ABC family is discussed in another chapter of this book). Membrane transport proteins—channel proteins, facilitating ion transfer across membranes along the concentration gradient—are a separate class and are not covered by this overview. Table 2 presents the list of *A. thaliana* genes encoding members of the transporters families described in this work.

2   SHORT-DISTANCE TRANSPORT

2.1   Metal Uptake Proteins

2.1.1   *Nramp Family*

2.1.1.1   General Characteristics. The family is conserved in eukaryotes and bacteria. The first member, mammalian *nramp1*, was cloned as a gene responsible for mouse resistance to infection by mycobacteria [Vidal et al., 1993]. Its product, NRAMP1 (natural resistance-associated macrophage protein) functions in metal transport across the phagosomal membrane of macrophages, and defective NRAMP1 causes sensitivity to several intracellular pathogens. Members of NRAMP family from yeasts, plants, and insects share a significant sequence identity with the mammalian proteins, and they all contain the NRAMP signature (PFAM:01566). In mammals the polymorphism of genes encoding NRAMPs is associated with iron disorders and immune diseases. For an extensive general description of this family, with emphasis put on mammalian members, the reader is referred to the excellent recent reviews [Courville et al., 2006; Nevo and Nelson, 2006].

2.1.1.2   Structure and Regulation. The NRAMP transporters have at least 10 predicted transmembrane domains, and sometimes 11 or even 12 transmembrane domains can be identified (Fig. 1). The conserved transmembrane core containing the NRAMP signature enables metal and proton cotransport.

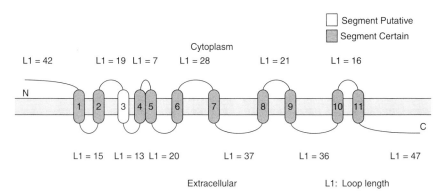

Figure 1. Predicted topology of AtNRAMP1, a member of NRAMP family. The protein consists of 532 amino acid residues and is predicted to cross the membrane 11 times (TOPPRED [Claros and von Heijne, 1994; von Heijne, 1992]; see also Table 1).

A. thaliana genome contains seven genes encoding proteins with homology to NRAMPs (Table 2); these genes include EIN2-NRAMP, which is phylogenetically more distinct from the others. EIN2-NRAMP is an integral membrane protein; however, it also possesses a long intracellular C-terminal tail. It has been recently identified as a cross-link node in ethylene, abscisic acid, and stress signaling pathways [Wang et al., 2007]; however, its direct link to metal transport is uncertain. The genes encoding AtNRAMP1, AtNRAMP3, and AtNRAMP4 are up-regulated under Fe-deficiency [Curie et al., 2000; Thomine et al., 2000]. On the other hand, the rice genes encoding NRAMP were shown to respond to infectious stimuli [Zhou and Yang, 2004].

2.1.1.3   Functional Properties and Physiological Role.   The mammalian NRAMPs have been studied very intensively; therefore, most information about the mechanisms of transport comes from the studies on DCT1 (product of human gene *nramp2*), where it appeared that the first N-terminal external loop and the transmembrane domains 3 and 4 are important for the transport activity [Nevo and Nelson, 2006]. The nature of the driving force for transport is not apparent however, the transport is energy- and temperature-dependent [Chaloupka et al., 2005]. Protons are cotransported with metal ions thus metal transport is dependent on proton concentration in the external site of the membrane [Tandy et al., 2000]. Here it is worth mentioning an interesting phenomenon observed in mammalian and yeast NRAMPs (and also other proton cotransporters) that is called "proton slippage" [Nelson et al., 2002]. This evolutionarily conserved mechanism results in low metal-to-proton transport, and thus it is energetically disadvantageous; however, it can serve as a protective mechanism against overloading the cell with metal ions.

The function of AtNRAMP proteins have been studied both in the heterologous yeast expression system and *in planta*. Although heterologous expression failed to detect metal transport activity of EIL2-NRAMP (see above and [Alonso et al., 1999]), the other tested AtNRAMPs appeared to be able to transport not only Fe but also Cd and Mn [Curie et al., 2000; Thomine et al., 2000]. In summary, different members of the family can perform different physiological functions, and at least some of them are involved in Fe and Cd transport and homeostasis. Nevertheless, the primary biological function of *A. thaliana* NRAMPs seem to be Fe homeostasis. It has been recently shown that AtNRAMP3 and AtNRAMP4 (both proteins reside in tonoplast) are involved in remobilization of Fe stored in vacuoles during Fe deficiency [Thomine et al., 2003; Lanquar et al., 2005]. It is yet unknown whether they play any role in release of Mn and Zn.

2.1.2   ZIP Family

2.1.2.1   General Characteristics. The ZIP family consists of zinc transport proteins, iron transport proteins, and many putative metal transporters containing conserved ZIP domain (PFAM:02535). Members of this family are present in organisms of all phylogenic levels including bacteria, fungi, protozoa, plants, and mammals [Guerinot, 2000; Gaither and Eide, 2001; Eide, 2004]. The family name (ZIP = zinc-regulated transporter/iron-regulated transporter-like protein) comes from the first three identified members: ScZRT1 and ScZRT2, which are, respectively, the high- and low-affinity Zn transporters in *Saccharomyces cerevisiae* and AtIRT1, which is a major Fe transporter in *Arabidopsis thaliana* and was the first ZIP protein identified in plants [Eide et al., 1996].

2.1.2.2   ZIP Genes and Their Regulation. Analysis of *A. thaliana* genomic sequence reveals that there are at least 17 genes encoding ZIP family members. Evidently, the expression of several genes from this family is increased under Zn (*AtZIP1* to *AtZIP5, AtZIP9* to *AtZIP12, AtIRT3*) or under Fe (*AtIRT1, AtIRT2*) limiting conditions (for recent reviews and references see Hall and Williams [2003], Krämer et al. [2007], and Maser et al. [2001]. However, the responses of numerous other *A. thaliana ZIP* genes to changes in trace metals supplies need to be explored. A bunch of data concerning expression of *ZIP* genes comes from analysis of the metal hyperaccumulating species, such as *Thlaspi caerulescens* and *A. halleri*, where the expression of the closest homologues of *ZIP4, ZIP10*, and *IRT3* remain responsive to changes in Zn supply, while *ZIP9* and *ZIP6* do not. The comparative microarray studies lead to the conclusion that *ZIP6* is the possible candidate gene for a role in Zn hyperaccumulation since this is consistently highly expressed in the leaves of all hyperaccumulating species [Filatov et al., 2006].

2.1.2.3   Protein Structure. ZIP proteins consist of variable number of amino acid residues (for example, in *Arabidopsis thaliana*—from 315 to 595). They usually have eight transmembrane domains and contain a region of variable length located between transmembrane domains 3 and 4 with variable number of His residues depending on the transporter. The topological predictions suggest that in most cases this loop is located in the cytoplasm (see Fig. 2).

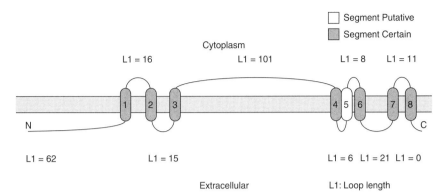

Figure 2. Predicted topology of AtZIP4, a typical representative of the ZIP family. The protein consists of 408 amino acid residues (including potential chloroplast-targeting N-terminal peptide) and is predicted to cross the membrane eight times (TOPPRED [Claros and von Heijne, 1994; von Heijne, 1992]; see also Table 1). The loop located between transmembrane domains 3 and 4 contains an His-rich region.

The function of the loop between transmembrane domains 3 and 4 is still not clear; however, it has been recently demonstrated that the His-motif present in the region spanning this loop is involved in the Zn uptake activity of human hZIP1 [Milon et al., 2006]. Moreover, these authors suggested that in hZIP1 this region might have a topology of a reentrant loop, which resides in the membrane but does not traverse from one side to the other; instead, it enters and leaves the same side of the membrane. This hypothesis, although requiring verification, seems to be consistent with the model proposing that in ZIP proteins the His-rich loop is located in the proximity of His residues present in transmembrane domains 3 and 4 to allow for initial binding and transport of the metal across the membrane.

The most conserved portion of ZIP proteins occurs in their transmembrane domain 4, suggesting that this might be an important protein part. Accordingly, it has been shown for AtIRT1 that the His residues present in the transmembrane regions 4 and 5 are essential for the function of the transporter [Rogers et al., 2000].

2.1.2.4 Function. The ZIP proteins are able to transport a variety of cations including Zn, Fe, Cd, Mn, Cu, and Ni [Guerinot, 2000]. The proposed significant role of ZIP transporters in Zn transport in plants is supported by characterization of their homologues from other species. For example, in *S. cerevisiae*, ScZRT1 and ScZRT2 function in uptake of Zn across the plasma membrane, while ScZRT3 functions in mobilization of Zn from the vacuole under conditions of Zn deficiency [MacDiarmid et al., 2000]. It is presumed that the plant ZIP proteins can be also localized in different cellular membranes. The presence of chloroplast targeting sequence in AtZIP4 suggests that it is most probably located in plastids [Grotz et al., 1998]; however, other plant ZIP proteins located in organellar membranes have not yet been identified. Nevertheless, the key feature of ZIP proteins seems to be that they transport metal ions from the extracellular space or organellar lumen (such as Golgi membrane-located mammalian ZIP7 [Huang et al., 2005] or tonoplast-located ScZRT3 [MacDiarmid et al., 2000]) into the cytoplasm. In contrast to the above statement is the recent suggestion that TcZNT5, which in *T. caerulescens* is expressed in roots only, might mediate the Cd efflux from cytoplasm to apoplast [Plaza et al., 2007].

The presence of multiple *ZIP* genes in almost each species allows predicting that some of their functions are redundant. For example, several ZIP proteins out of 14 known members encoded by the human genome can participate in dietary Zn absorption [Eide, 2006]. Thus, the overexpression studies should be more informative in terms of assigning the role of the uncharacterized members of ZIP family than the loss-of-function studies.

The metal selectivity of several plant transporters, including AtZIP1, AtZIP2, AtZIP3 [Grotz et al., 1998], AtIRT1, and AtIRT2 [Korshunova et al., 1999; Rogers et al., 2000; Vert et al., 2001] from *Arabidopsis*, LeIRT1 and LeIRT2 from tomato [Eckhardt et al., 2001], and GmZIP1 from soybean [Moreau et al., 2002], was characterized by functional complementation studies in the respective yeast mutant cells. The results of such studies allow concluding that the ZIP family members significantly differ in terms of metal specificity and affinity. For example, despite the fact that AtIRT1 and AtIRT2 are both expressed in the roots of Fe-deficient plants, they have different substrate specificities. Namely, AtIRT1 seems to mediate uptake of Fe, Zn, and Mn [Eide et al., 1996; Korshunova et al., 1999], while AtIRT2 transports mostly Fe and Zn [Vert et al., 2001], also Cd and Mn ions are transported by AtIRT1 but not by AtIRT2.

Phylogenic analysis including both characterized and uncharacterized family members can be also used as a particular approach to establish functions of novel proteins. The phylogenic tree of ZIP family members from *A. thaliana* and *T. caerulescens* is shown in Fig. 3.

At least six clusters containing the following *A. thaliana* ZIP transporters can be distinguished: (i) AtZIP4, AtZIP9, and AtIRT3; (ii) AtZIP6; (iii) AtZIP3 and AtZIP5; (iv) AtZIP1 and AtZIP12; (v) AtIRT1, AtIRT2, AtZIP7, AtZIP8, and AtZIP10; (vi) AtZIP2 and AtZIP11. Two *A. thaliana* proteins (AtIAR1 and NP_187477) are apparently more distant phylogenetically from the other proteins. Location of the uncharacterized *T. caerulescens* ZIP proteins to the respective clusters might help to identify their functional counterparts in *A. thaliana*.

The mechanism of transport used by ZIP proteins is unclear and might differ among the members. For example, in the case of ScZRT1 it is energy-dependent, but in the case of human hZIP2 it is energy-independent, possibly driven by the pH gradient across the membrane [Zhao and Eide, 1996; Gaither and Eide, 2000].

2.1.3   COPTs. COPT is a small *Arabidopsis* family of putative plasma membrane proteins mediating Cu uptake with high affinity. These are Ctr-related proteins, eukaryotic high-affinity copper transporters, containing three transmembrane domains (TMDs) and methionine, histidine, and serine residues within the N-terminus [Kampfenkel et al., 1995; Sancenon et al., 2004]. Five members of

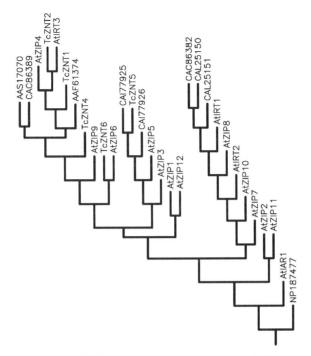

Figure 3. Phylogenetic tree of ZIP transporters from *Arabidopsis thaliana* and *Thlaspi caerulescens*. All *A. thaliana* proteins, except NP_187477, are labeled with "At" letters. The *T. caerulescens* proteins are labeled with either "Tc" or the respective accession numbers. The tree was constructed using full-length proteins by the parsimony method and 100 bootstrap replicates (SEQBOOT, PROTPARS, and CONSENSE of the Phylip v.3.63 program package).

this family have been identified in *Arabidopsis* (COPT1-5; see Table 2) [Sancenon et al., 2003] and are able to functionally complement the ctr1Δctr3Δ yeast mutant, which points to their ability to transport copper [Sancenon et al., 2004]. They are constitutively expressed in roots, leaves, stems, and flowers with different intensity, and their expression either is down-regulated in the presence of excess copper (COPT1, COPT2) or remains unaffected (COPT3, COPT4, COPT5) [Sancenon et al., 2003]. *AtCOPT1* is the one best characterized so far. Specific expression in the root tips together with the reduced uptake and accumulation of Cu in *COPT1* antisense plants strongly supports their suggested putative role in copper uptake. Developmental disturbances of antisense plants, restored under the conditions of copper supplementation, have, however, indicated a role of COPT1 in the copper delivery necessary for normal development of certain organs [Sancenon et al., 2004].

2.1.4 YSL Family. Metal cations are taken up by plants not only in their ionic form but also complexed by organic compounds. It is known that graminaceous plants release phytosiderophores (PS) to chelate Fe(III) in the soil and metal–PS complexes are taken up from the environment [Takagi et al., 1984]. The precursor for the synthesis of PS, nicotianamine (NA), however, is produced in all plants but is not released by roots. Instead, it plays a role as a chelator in metal homeostasis and detoxification. It also participates in the transport of metals within a plant body, since metal–NA complexes are substrates for transmembrane transport by specific YSL family transporting proteins belonging to oligopeptide transporters (OPT), possessing 12–15 TMDs [Stephan et al., 1996; von Wiren et al., 1999; Yen et al., 2001; Colangelo and Guerinot, 2006].

Studies on YS1 from *Zea mays* provided the first description of the transport mechanism employed by OPT-related transporters. Schaaf et al. [2004] demonstrated a proton-coupled symport of Fe–PS complexes. ZnYS1 has broad substrate specificity because it mediates the transport of Fe^{III}–, Zn–, Cu–, Ni–, Mn–, and Cd–PS complexes, and unexpectedly also Fe^{II}–, Fe^{III}–, and Ni–NA complexes, and likely is responsible for their uptake by roots [Curie et al., 2001; Roberts et al., 2004; Schaaf et al., 2004]. Its expression under Fe-sufficient conditions has been detected only in roots; however, its up-regulation was detected when Fe was withdrawn from the medium. In addition, under such conditions, expression in leaves was also detected [Roberts et al., 2004].

Eight YSL homologues of the metal-PS/NA transporter *ZmYSA1* have been identified in *A. thaliana* (Table 2). They encode transmembrane proteins possessing 15 TMDs, in general with the longer hydrophilic loops between TMD II-VIII [DiDonato et al., 2004] (Table 1; internet resource). The major variable region occurs in the extracellular linker between TMDVII-VIII and the amino termini, although the amino acid composition of this tail is highly similar in all eight YSLs of *Arabidopsis*. In addition, the characteristic feature is the presence of numerous acidic glutamate and aspartate residues [DiDonato et al., 2004]. While the functional analysis by Schaaf et al. [2005] with the use of yeast mutants and *Xenopus* oocytes did not confirm that *AtYSL2* was involved in the mediation of the transport of metals chelated by nicotianamine, the study by DiDonato et al. [2004] employing yeast complementation analysis confirmed its role in the transport of Fe or Cu chelates. The expression of the *AtYSL2* promoter:GUS was detected in the vasculature consistently by both authors [DiDonato et al., 2004; Schaaf et al., 2005]. In addition, transcriptional regulation by Fe and Zn was demonstrated, and therefore it was believed that this protein plays a role in the homeostasis of both metals. Less is known about the role of *AtYSL1*. The insertional loss-of-function mutant analysis led to the conclusion that the protein regulates the Fe and NA level in shoots and seeds [Le Jean et al., 2005]. The gene is constitutively expressed in shoots, and its level goes up under increased Fe concentration. Similarly, AtYSL1 and AtYSL3 were implicated in the delivery of metal micronutrients (mainly Fe but also Mn, Zn, and Cu) to and from vascular tissues [Waters et al., 2006].

In another monoct plant, *Oryza sativa*, 18 putative YSL genes have been identified [Koike et al., 2004]; however, transcripts were found only for six of them. The constitutive expression detected in shoots and/or in roots was regulated for the majority of genes by Fe-deficiency (induced or suppressed). The best characterized is *OsYSL*. It has 14 TMDs. Surprisingly, its expression in *Xenopus* oocytes did not demonstrate involvement in the transport of Fe^{3+}–PS; instead, transport of

Fe^{2+}–NA and Mn^{2+}–NA complexes was shown. Its tissue-specific expression investigated by promoter:GUS analysis showed its presence in the phloem of roots, leaves, and reproductive organs, suggesting the role of metal–NA complexes in long-distance Fe and Mn transport. The use of the OsYSL2:GFP protein for the study points to its plasma membrane localization [Koike et al., 2004].

Five members of YSL family have been also identified in the metal-hyperaccumulator Thlaspi caerulescens [Gendre et al., 2007] of which TcYSL3 appeared to be Fe/Ni–NA influx transporter.

2.2 Metal Efflux Proteins

2.2.1 *P$_{1B}$ATPases*

2.2.1.1 General Information. The P$_{1B}$-type ATPases, efflux pumps involved in ATP-driven transport of heavy metals across membranes out of the cytoplasm, belong to the P-type ATPase family of proteins found in a wide range of organisms belonging to prokaryotes and eucaryotes, including plants. P-type ATPases have been divided into five subfamilies based on sequence similarities, topological arrangements, and ion specificity. They include Type I ATPases (heavy metal pumps), Type II ATPases (Ca^{2+}-ATPases, Na^+/K^+-ATPases, and H^+/K^+-ATPases), Type III ATPases (H^+ and Mg^{2+}-ATPases); Type IV ATPases (phospholipid pumps) and Type V ATPases (transporters with still undetermined substrate specificity) [Axelsen and Palmgren, 1998, 2001; Williams and Mills, 2005].

Heavy-metal transporting Type 1$_B$-ATPases, also known as HMAs (heavy-metal ATPases) or CPx-ATPases (possessing the conserved CPx motif: Cysteine-proline-cysteine/histidine/serine), pump cations of transition metals (such as Cu^{2+}, Zn^{2+}, Cd^{2+}, Pb^{2+}, Ag^+, and Co^{2+}) against their electrochemical gradient out of the cytosol. ATP-driven cation transmembrane transport, similar to all ATPases, is coupled to ATP hydrolysis with the formation of a phosphorylated intermediate [Moller et al., 1996]. These enzymes have been divided into two groups based on sequence similarity and substrate specificity: (i) Cu/Ag transporters and (ii) Zn/Cd/Co/Pb transporters [Axelsen and Palmgren, 2001; Cobbett and Day, 2003]. Most of the information on the structure and function of these proteins has come from studies on *Arabidopsis thaliana*. In it, eight genes belonging to this subfamily have been identified (Table 2), of which AtHMA1 to AtHMA4 belong to Zn/Cd/Co/Pb pumps, and AtHMA5, AtHMA6 (named initially PAA1), AtHMA7 (known as RAN1), and AtHMA8 (known also as PAA2) belong to Cu/Ag pumps. The number of identified 1$_B$-ATPases is growing. They were also cloned from *Thlaspi caerulescens* [Bernard et al., 2004; Papoyan and Kochian, 2004], *Oryza sativa* and *Hordeum vulgare* [Baxter et al., 2003; Williams and Mills, 2005], and *Brassica napus* [Southron et al., 2004], as well as from algae like *Chlamydomonas reinhardtii* and *Cyanidioschizon merole* [Rosakis and Koster, 2004; Hanikenne et al., 2005]. Initial characterization pointed to their involvement in a variety of physiological functions, of which the most important are regulation of metal ion homeostasis and resistance to toxic metals.

2.2.1.2   Structure and Regulation of Activity. In contrast with other P-type ATPases, the P_{1B} type is a membrane protein characterized by the presence of eight transmembrane domains (TM) [Axelsen and Palmgren, 1998, 2001]. Another distinctive feature is the presence of CPx motifs (cysteine-proline-cysteine/ histidine/serine) in the TM6, which is proposed to be essential for translocating metal ions through the membrane. An exception is the HMA1 (AtHMA1 and OsHMA1), in which the SPC motif (serine/proline/cysteine) is found instead of CPx [Rutherford et al., 1999]. Studies on AtHMA4 show that mutations in the CPx motif destroy its transport function, pointing to a key role of this sequence in the translocation of cations [Mills et al., 2005]. Determination of ion selectivity, however, likely resides in TM6, TM7, and TM8 as well [Arguello, 2003; Mandal and Arguello, 2003].

P_{1B}-type ATPase's possess C- and N-terminal extensions on the cytoplasmic side. They are different in length but share a common feature, which is the presence of one or more putative metal-binding domains (MBDs), characterized by the presence of cystein and/or histidine residues in various sequences and repetitions [Moller et al., 1996; Williams and Mills, 2005]. Thus, the N-terminus of P_{1B}-type ATPases possesses domains with the core signature sequence CxxC (which in most cases appears within a conserved GMxCxxC sequence) or CCxx motif; however, variability has been detected in their presence and location [Verret et al., 2005; Williams and Mills, 2005]. In most cases, the N-terminus of the Cu/Ag class of proteins possess one or two such putative binding sites (in AtHMA5, AtHMA6/PAA1; AtHMA7/RAN1, whereas within Zn/Cd/Pb/Co pumps in AtHMA2-4, the CxxC pattern is replaced by a cysteine pair in the sequence GICC(T/S)SE [Tabata et al., 1997; Eren and Arguello, 2004; Verret et al., 2005; Williams and Mills, 2005]. In contrast, however, at the N-terminal extension, AtHMA1 and OsHMA1 possess repeats of His residues instead of any variant of a Cys-containing motif.

There are also MBDs within C-terminal extensions. Variant motifs containing repeats of histidine and/or cysteine as CysCys, CysHis, or HisCys dipeptide sequences, as well as consecutive His residues, with variable contents and locations, have been identified. The C-terminal tails of Zn/Cd/Pb/Co proteins differ from that in Cu/Ag proteins possessing an abundance of the above-mentioned MBDs. For example, in the Zn/Cd/Pb/Co group, CysCys pairs were detected throughout the long C-terminal extension of AtHMA4 (13 pairs) and TcHMA4 (16 pairs). Both proteins also contain consecutive His residues (11 and 9, respectively) [Mills et al., 2003; Papoyan and Kochian, 2004]. In turn, the string of His residues is lacking in AtHMA2, where only Cys repeats have been identified [Eren and Arguello, 2004]. The C-terminal extension of the Cu/Ag group of pumps is short and, in the majority of cases, lack MBDs [Williams and Mills, 2005]. The role of the C-terminal tail is not clear. A mutant of AtHMA4 lacking a C-terminal extension remained functional [Mills et al., 2005]. Interestingly, yeast transformed with the C-terminus of TcHMA4 alone containing numerous potentially metal-binding sites were more tolerant to cadmium relative to yeast expressing the full-length gene [Papoyan and Kochian, 2004].

In general, the role of MBDs present in C- and N-terminal tails in the regulation of transport activity of P_{1B}-type ATPases has not been precisely defined; thus their involvement in the binding and/or sensing of a range of metals and in the regulation of various processes, including determination of substrate selectivity and enzyme activity, is considered [Papoyan and Kochian, 2004; Williams and Mills, 2005]. It was demonstrated for some ATPases that the mutant lacking MBDs in the N-displayed altered catalytic activity [Mana-Capelli et al., 2003]. It has been suggested that MBDs may participate in regulating the affinity of the intramembrane metal-binding sites and the transport rate by influencing cation release and that they may influence the binding of nucleotides as a result of interaction with the ATP-binding cytoplasmic domain [Mitra and Sharma, 2001; Tsivkovskii et al., 2001; Bal et al., 2003; Huster and Lutsenko, 2003; Mandal and Arguello, 2003; Liu et al., 2005; Verret et al., 2005; Williams and Mills, 2005]. In addition, with the use of a yeast two-hybrid system, it was demonstrated that MBDs of AtHMA5 interact with Atx1-like Cu chaperones, pointing to their involvement in metal trafficking throughout the cytosol [Andres-Colas et al., 2006].

Topological analysis revealed two characteristic cytoplasmic loops between transmembrane segments (TM). The first, a small loop, is located between helices 4 and 5, the second, a large loop, is located between helices 6 and 7. Within them are three putative functional domains. The first is the A-domain, present within the small loop. The other two are within the large loop and have been named P-domain and N-domain [Williams and Mills, 2005]. Most of the information on the participation of the amino acid residues of these domains in the regulation of transport comes from studies on other than plant material. These domains contain certain conserved motifs, which is why it is believed that P_{1B}-type ATPases from plants may have functions similar to those of the well-characterized Ca-ATPase from sarcoplasmic reticulum (SERCA) [Hirayama et al., 1999; Toyoshima et al., 2000; Toyoshima and Inesi, 2004; Gravot et al., 2004; Verret et al., 2005]. It can, therefore, be supposed that the role of the A-domain is to control the gating mechanism, that the P-domain contains the phosphorylation site, and that the N-domain contains the ATP-binding site. Within the nucleotide-binding domains (N-domain) in P_{1B}-type ATPases, there is a characteristic HP motif containing histidine and a nearby glutamate, with a putative role in nucleotide coordination [Efremov et al., 2004; Morgan et al., 2004].

2.2.1.3 Physiological Function. Various methods have been applied to elucidate the physiological function of this important family of proteins, leading to the identification of their localization, substrate specificity, and regulation mechanisms. The varied experimental approaches have included studies on (a) mutants in which the gene of interest was disrupted or with a deletion in its sequence and (b) homologous/heterologous expression. However, we are still far from fully understanding their role in plants.

In general, P_{1B}-type ATPases are involved in (a) mineral nutrition, transport of nutrients, and their distribution throughout the plant body, (b) storage, and (c) delivery of metals to target proteins. On the other hand, it is proposed that they contribute to the detoxification of the excess of micronutrients and of metal ions with no known biological function (like Pb^{2+}, Cd^{2+}, Hg^{2+}). The transport activity of these proteins is related, at least in some cases, to the activity of metallochaperones, involved in metal trafficking throughout the cytoplasm [Harrison et al., 2000; Andres-Colas et al., 2006]. Current knowledge about the function of P_{1B}-type ATPases comes mainly from studies on *Arabidopsis thaliana* and *Thlaspi caerulescens*. The physiological function of these proteins is largely determined by their localization, and the results of studies to date suggest a role of a given pump in several different metabolic processes, probably due to the frequent broad tissue and organ specificity of these pumps, among others.

The Physiological Function of the Zn/Cd/Co/Pb Group. P_{1B}-type ATPases localized to the outer plasma membrane pump cations out of the cytosol to the apoplast. As suggested, the major role of proteins in such sites could be involvement in the transport of micronutrients throughout the plant body. By transporting metal cations out of a cell, they might contribute, among others, to xylem/phloem loading/unloading and to increasing ion concentrations in the apoplast and thus contribute to their availability for intracellular transport including the long-distance root-to-shoot transport.

Among the proteins belonging to the Zn/Cd/Co/Pb group, plasma membrane localization has been demonstrated for *AtHMA4* and *AtHMA2* [Hussain et al., 2004; Verret et al., 2004]. Both mediate Zn and Cd efflux [Hussain et al., 2004; Mills et al., 2003, 2005]. The evidence suggests that they play a role in both mineral nutrition, mainly zinc, but also in detoxification under an excess of metal ions. *AtHMA2* and *AtHMA4* expression in cells in the stele surrounding the vascular tissue of roots, stems, and leaves points to their involvement in Zn and Cd loading in the xylem and in root-to-shoot translocation [Eren and Arguello, 2004, Hussain et al., 2004; Verret et al., 2004]. For example, the *AtHMA4* knockout mutant accumulated significantly less Zn in the shoots relative to the wild type [Hussain et al., 2004]. Its heterologous expression in yeast increased tolerance to Zn, Cd, and Pb [Mills et al., 2003; Verret et al., 2005]. The role of both genes in metal transport is not limited, however, to the xylem, since the presence of these proteins was demonstrated in phloem as well, suggesting a function in phloem loading or unloading. In addition, they have been found in developing anthers (tapetum cells mainly) and at the base of the developing silique, likely supplying Zn to the appropriate tissues of various organs [Mills et al., 2003; Verret et al., 2004]. An *hma2* and *hma4* double mutant was characterized by increased accumulation of Zn in roots accompanied by a reduction in shoots [Hussain et al., 2004]. AtHMA2 is ubiquitously expressed in all plant organs, and similar levels were observed upon exposure to metals [Eren and Arguello, 2004].

Being active in the efflux of cations to the cell wall, AtHMA4 may also play an important role in metal detoxification [Mills et al., 2005]. An *Arabidopsis hma4* mutant appeared to be more sensitive to both Zn and Cd supplied at elevated concentrations. The mechanism underlying AtHMA4-dependent detoxification processes is not known, however. It is supposed that the protein may also be involved in Co transport. The overexpression of this protein in *Arabidopsis* led to increased Co tolerance, whereas expression in yeast conferred moderate sensitivity to Co [Mills et al., 2003; Verret et al., 2004]. AtHMA4 expressed in all plant parts was up-regulated in roots in the presence of Zn and Mn, but down-regulated by Cd [Mills et al., 2003]. For comparison, the expression of *HMA4* from *Thlaspi caerulescens*, a well-known Zn and Cd hyperaccumulator, is also up-regulated under exposure to Zn and Cd, and the gene is expressed almost exclusively in roots. Expression was also up-regulated under Zn-deficiency, which is considered a strategy for maintaining shoot Zn status under this type of nutrient stress. *TcHMA4* expression in yeast conferred tolerance to Zn, Cd, and Pb by their active efflux out of cells. Its overexpression in *Arabidopsis* led to increased accumulation of Cd and Zn in leaves. It was concluded that it may be involved in xylem loading of metals and, by so doing, plays a role in heavy-metal hyperaccumulation [Papoyan and Kochian, 2004]. Similarly, HMA4 from Zn/Cd hyperaccumulator *Arabidopsis halleri* was also localized in the plasma membrane and mediated Zn and Cd transport, contributing to the high tolerance to both metals [Courbot et al., 2007].

A functional *AtHMA3* yeast expression study demonstrated the involvement of this ATPase in the detoxification of Cd and Pb, likely by transporting metals into the vacuole, because the fusion protein AtHMA3::GFP was localized at the vacuole. Its constitutive expression was detected mainly in roots and leaves and was not induced by Cd or Zn [Gravot et al., 2004]. So far, the function of AtHMA1 has remained uncertain. Based on the SPC motif characteristic of cobalt transporter CoaT of cyanobacterium, a role in cobalt translocation was ascribed by Cobbett and Day [2003].

The Physiological Function of the Cu/Ag Group.   Although Cu is a trace element, it is toxic to plants when it occurs in the soil in high concentrations; therefore a precise mechanism for the regulation of homeostasis of this element, as well as detoxification of its excess, is crucial for a plant growth and development.

Relatively much is known about the function of three *Arabidopsis* Cu-pumps, AtHMA6-8, supposed to play a role in metal delivery into intercellular compartments to Cu-requiring proteins. Both AtHMA6/PAA1 and AtHMA8/PAA2 are part of Cu transport in chloroplasts [Shikannai et al., 2003; Abdel-Ghany et al., 2005]. It was shown that AtHMA6/PAA1 transports Cu across the plastid envelope, and metal in the stroma binds into Cu/Zn superoxide dismutase (AtCSD2). There is another pump, AtHMA8/PAA2, which transports the metal from the stroma to the thylakoid lumen to supply plastosyanin (a protein involved in photosynthetic electron transport). AtHMA8/PAA2 has been found only in shoots where it likely plays a unique role in chloroplasts. In contrast, AtHMA6/PAA1 was detected also in roots, indicating a broader spectrum of possible functions in plants, not restricted to green plant parts only. It is believed, however, that both genes probably participate directly or indirectly in other yet unknown functions, which is suggested by the seedling lethal phenotype of a double mutant *paa1 paa2* [Abdel-Ghany et al., 2005].

The role of another gene, AtHMA7/RAN1, in copper transmembrane transport was demonstrated by complementation of the yeast *ccc2* mutant. An attempt to determine the physiological role by phenotypic analysis of the *ran1-3* mutant indicated that this constitutive ethylene-signaling mutant had altered substrate binding specificity of the ethylene receptor. Because copper is required for high-affinity ethylene binding [Rodriguez et al., 1999], it was proposed that AtHMA7/RAN1 delivers Cu into a post-Golgi compartment to the ETR1 apoprotein (the membrane-targeted ethylene receptor), to make it functional [Hirayama et al., 1999; Woeste and Kieber, 2000]. The study of the *ran1-3* mutant also suggested the contribution of AtHMA7/RAN1 to physiological processes not dependent on ethylene. Detecting a rosette-lethal phenotype led to the hypothesis that it could be due to the malfunction of an unidentified cuproenzyme due to lack of Cu delivery [Woeste and Kieber, 2000].

A recent study by Andres-Colas et al. [2006] indicated that AtHMA5, a member of the Cu/Ag group, mediates Cu compartmentalization and detoxification in *A. thaliana*, presumably by pumping Cu out of the cytosol. A T-DNA insertion mutant appeared to be Cu-hypersensitive.

2.2.2   CDFs

2.2.2.1   General Information. The proteins of a ubiquitous CDF family (cation diffusion facilitator) occur in bacteria, yeast, animals, and plants [Paulsen and Saier, 1997; Maser et al., 2001; Haney et al., 2005]. They are divided into three groups: CDF I (mostly eubacterial and archaeal proteins), CDF II, and CDF III, that encompass both prokaryotic and eukaryotic proteins [Gaither and Eide, 2001]. They contain transmembrane proteins that transport a wide range of divalent transition metal cations (primarily Zn^{2+} but also Mn^{2+}, Cd^{2+}, Co^{2+}, Fe^{2+}, and Ni^{2+}) against their electrochemical gradient. The majority of the data, mostly from studies conducted on prokaryotes and mammals, suggest that the translocation of substrates is driven by proton antiports, however, it is not excluded that they may have another energy source for the transport [Chao and Fu, 2004]. CDFs have six transmembrane domains (TMD) localized on both intracellular membranes and in the plasma membrane. They mediate the transport of cations out of the cytosol either to cellular organelles/compartments or out of a cell [Gaither and Eide, 2001].

In general, functional analysis has shown the involvement of CDF proteins in the regulation of metal homeostasis and tolerance to the presence of excess transition metals in the environment. It has also been suggested that they constitute one element of the intracellular signal transduction pathway. Data obtained from experiments performed on bacterial and mammal CDFs indicate that they may receive substrates as a result of interaction with other cation-binding molecules—for example from chaperones [Paulsen and Saier, 1997; Maser et al., 2001; Nies et al., 2003; Haney et al., 2005; Martinoia et al., 2007].

Information about the structure and function of plant members of the CDF family comes from recent studies on genes cloned from several organisms such as *Arabidopsis thaliana*, *Arabidopsis halleri*, *Thlaspi caerulescens*, *T. goesingense*, and *Stylosanthes hamata*. Eight genes were identified in *Arabidopsis thaliana* as having homology to the CDF family [Maser et al., 2001]. The first gene cloned from this plant species was called ZAT (zinc transporter of *Arabidopsis thaliana* [van der Zaal et al., 1999]; however, later on it was renamed MTP1 (metal tolerance proteins [Maser et al., 2001]), and the term MTP has since been used to designate other plant members of this family [Krämer, 2005; Martinoia et al., 2007]. Its topology is shown in Fig. 4.

2.2.2.2 Structure and Regulation of Activity. The characteristic topological elements common among CDFs are six TMDs in most of the members, C- and N-terminal tails on the cytoplasmic side, and a long loop region located between TMDs IV and V [Anton et al., 1999; Kobae et al., 2004; Haney et al., 2005; Krämer, 2005]. One of the most important signature sequences is located within TMDs I, II, and V. Conserved sequences have also been identified in most of the cytoplasmic N- and C-terminal tails and in the cytoplasmic loop [Paulsen and Saier, 1997; van der Zaal et al., 1999; Gaither and Eide, 2001; Dräger et al., 2004; Kobae et al., 2004; Haney et al., 2005]. The size of the CDF proteins varies considerably, ranging from about 280 to 740 residues [Paulsen and Saier, 1997; Haney et al., 2005]. Out of eight *Arabidopsis* members of the CDF family (Table 2), four possess all of the common structural features, whereas the other two (previously named *AtMTPc1* to *AtMTPc4*) display differences, such as four to five TMDs, the presence of only part of the N-terminal signal sequence, and missing His-rich domain [Maser et al., 2001].

In eukaryota, the cytoplasmic loop between TMDs IV and V frequently has a histidine-rich region (motif $[HX]_n$, where $n = 3-6$ and $X = $ often S or G]) [Paulsen and Saier, 1997; van der Zaal et al., 1999; Gaither and Eide, 2001]. It is believed that histidine residues may be involved in metal-binding, although it is not clear whether they contribute directly to the translocation of the substrate or are involved in the regulation processes—for example, in conformational changes of the protein leading to exposing the substrate-binding site [Huang and Gitschier, 1997; Williams et al., 2000; Nies, 2003]. Plant CDF–AtZAT1 (MTP1) contains

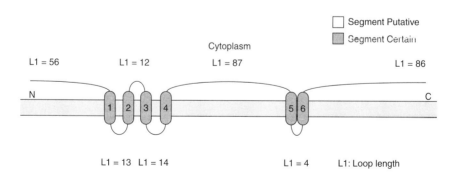

Figure 4. Predicted topology of AtMTP1/AtZAT1, a member of CDF family. The protein consists of 398 amino acid residues and is predicted to cross the membrane 6 times (TOPPRED [Claros and von Heijne, 1994; von Heijne, 1992]; see also Table 1).

21 His residues there [van der Zaal et al., 1999]. Based on the results of mutational analysis of *EcZitB* [Lee et al., 2002; Anton et al., 2004], it is believed that in ZitB and probably in other CDFs, a combination of histidine, glutamate, and/or aspartate take part in the transport of zinc ions, whereas probably glutamate and aspartate contribute to proton translocation. In plants, the replacement of aspartate from the second TMD of AhMTP1-3, which is conserved in all CDFs, by alanine led to a non-functional protein [Dräger et al., 2004], confirming that it is necessary in the transport process. Similarly, in *Populus*, site-directed mutagenesis of Asp-93 from TMD II of PtdMTP1 also rendered the protein inactive [Blaudez et al., 2003]. The results of studies on bacterial ZitB suggest that the transport of the substrate by CDF is two-staged. The first stage is the binding of the substrate on the cis side, and the second stage involves conformational changes of the protein leading to release of the substrate on the opposite, trans side of the membrane. The binding site is then protonized, leading to the closure of the protein structure on the trans side and its reopening on the cis side. Cysteine and histidine participate in binding the substrate [Chao and Fu, 2004]. Recent experiments indicate that these transporters may function as dimers or oligomers [Blaudez et al., 2003].

It is believed that in substrate translocation by CDF proteins, an important role is played by the C-terminus, which contains the characteristic cation-binding motif (H-D/E-X-H-X-W-X-L-T-X8-H) [Rosakis and Koster, 2004]. The studies by Jirakulaporn and Muslin [2004] suggest that in bacterial CDFs the role of the C-terminus may be either the binding and accumulating of the substrate or the binding of the chaperone that supplies metal ions for transport (chaperone-docking site). Its importance for protein function in plants is suggested by the results of experiments conducted on the *AtZAT1* (*MTP1*) mutant lacking the C-terminal tail [van der Zaal et al., 1999]. The presence of the conserved putative Zn-binding site at the C-terminus has been shown in AhMTP1-3, AtMTP1, and AlMTP1 [Dräger et al., 2004].

2.2.2.3   Physiological Function.   It is generally believed that primarily, CDFs play a role in the homeostasis of a variety of cations of transition metals and determine the given organism's tolerance to their excess in the environment, mainly by contribution to metal sequestration within cellular compartments and extrusion out of a cell, but also to tissue and organ partitioning.

ZAT1 from *Arabidopsis thaliana* (*AtMTP1*) is most closely related to the mammalian CDFs designated as ZnTs (Zn transporter) and shares 35–40% amino acid identity with ZnT 2–4, as well as 40–50% identity at the nucleotide level. It is constitutively expressed throughout a plant, and the lack of induction at the mRNA level in response to increases in Zn concentration indicates that it could be a housekeeping gene [van der Zaal et al., 1999] involved in the regulation of Zn homeostasis and also in contributing to Zn tolerance of a plant, likely by mediating metal detoxification [Krämer, 2005]. *Arabidopsis* overexpressing *AtZAT* (*AtMTP1*) had both greater Zn accumulation in roots and Zn-tolerance to the excess of this metal than untransformed plants. The above-mentioned putative function is supported by experiments demonstrating that a T-DNA insertion mutant was more sensitive to high Zn concentrations [van der Zaal et al., 1999; Kobae et al., 2004], and RNAi-mediated silencing rendered plants hypersensitive to this metal [Desbrosses-Fonrouge et al., 2005]. It has been proposed that *AtMTP1* plays its physiological function by mediating vesicular/vacuolar metal sequestration. More data in support of this conclusion were provided by the study of Bloss et al. [2002]. He demonstrated Zn accumulation in proteoliposomes with reconstituted AtMTP1, and direct evidence for the tonoplast localization of MTP1 was provided later on by Kobae et al. [2004] and Desbrosses-Fonrouge et al. [2004]. Another CDF plant protein, MTP1 from the metal hyper-accumulator *Arabidopsis halleri*, was also localized to the vacuolar membrane [Dräger et al., 2004]. It shares 91% identity with *A. thaliana* MTP1 and 97% identity with *A. lyrata*. The protein is also constitutively expressed; up-regulation was, however, detected in roots upon exposure to high Zn concentrations. Its function as a protein mediating Zn detoxification is supported, among others, by the fact that in an interspecies cross between *A. halleri* and *A. lyrata*, *MTP1* loci co-segregate with zinc tolerance [Dräger, 2004]. Studies by Arrivault et al. [2006] led to the characterization of another CDF member from *Arabidopsis thaliana*, *AtMTP3*. Functional analysis by heterologous expression implicated its role in Zn tolerance and metal-partitioning in the plant body. Their results suggest that MTP3 contributes to Zn exclusion from the shoot under conditions of high metal concentration. Also localized to the tonoplast, it likely contributes to Zn and Co sequestration in the vacuole. Interestingly, it is up-regulated not only by high nontoxic concentrations of Zn and Co, but also under Fe-deficiency conditions. A role in Zn vacuolar seques-tration has also been ascribed to *MTP1* from the hybrid poplar *Populus trichocarpa × Populus deltoides* [Blaudez et al., 2003]. *PtdMTP1* heterologously expressed in a mutant strain of yeast complemented hypersensitivity to Zn, but not to Cd, Co, Mn, or Ni. Overexpression in *Arabidopsis* conferred Zn tolerance. The protein is targeted to the vacuolar membrane of yeast and plant cells, suggesting its role in Zn vacuolar sequestration.

In an attempt to find the molecular basis for the Zn-hyperaccumulation phenotype of *Thlaspi caerulescens*, another member of the CDF family ZAT (MTP) gene was identified. It shares 90% of nucleotide identity and 75% amino acid identity with *Arabidopsis* ZAT1 (MTP1) and was called *ZTP1* (Zn transporter 1) [Assunção et al., 2001]. The highest expression was detected in leaves, whereas a lower level was detected in roots. Comparison of its expression level in plants from different populations demonstrated the highest expression in Zn-tolerant plants inhabiting soil rich in Zn, Cd, and Pb. Lower levels were found in plants from uncontaminated as well as Ni-enriched sites. The suggestion is that TcZTP1 may be involved in high Zn compartmentation and tolerance of Zn-hyperaccumulating plants.

A role of CDFs in vacuolar sequestration of numerous transition metals has been ascribed to MTP1 from an Ni-hyperaccumulator, *Thlaspi goesingense* [Persans et al., 2001]. A *TgMTP1* genomic sequence gives rise to two types of transcripts with different substrate specificity. The first, an unspliced *TgMTP1t1*, when expressed in yeast, confers tolerance to Zn and also to Cd and Co, but the second, a spliced tran-script (*TgMTP1t2*), confers Ni- and Co-tolerance to yeast. Their diverse substrate specificity probably arises from the difference within a central cytoplasmic histidine-rich putative metal-binding domain. The lack of this region almost entirely in *TgMTP1t2* was proposed to confer Ni specificity. The gene is constitutively expressed in plants with a higher level in shoots, and its expression is not regulated by the presence of high Ni concentrations [Persans et al., 2001]. It was demonstrated by the immunolocalization of TgMTP1::HA that the protein was targeted in yeast *S. cerevisiae* to both vacuolar and plasma membranes. In turn, transient expression of TgMTP1::GFP in protoplasts of *Arabidopsis* showed plasma membrane localization [Kim et al., 2004]. This result is in agreement with the general knowledge regarding the CDFs' ability to extrude cations out of a cell and to the internal cellular compartments [Gaither and Eide, 2001].

The involvement of CDFs in the translocation of another cation, Mn^{2+}, was suggested for *MTP1* from *Stylosanthes hamata*, a tropical legume able to grow on acid soil and to accumulate high Mn concentrations [Delhaize et al., 2003]. The *Sh*MTP1::GFP fusion protein was located on the vacuolar membrane, and the gene, when heterologously expressed in both yeast and *Arabidopsis*, conferred increased Mn tolerance.

In yeast, *S. cerevisiae*, two members of the CDF family, Zrc1p and Cot1p, are both localized on the vacuolar membrane, and they confer resistance to zinc when over-expressed but confer sensitivity when deleted [Regalla and Lyons, 2006]. Zrt3p is closely related to the Zn^{2+} uptake transporter ZupT from *E. coli*, and it is involved in the release of Zn^{2+} from vacuolar stores.

The role of the CDF family of proteins is likely not to be limited only to involvement in the regulation of metal tolerance and homeostasis, but also, as shown in the study on prokaryota and eukaryota other than plants, in the oncogene pathway and in drug resistance [Jirakulaporn and Muslin, 2004]. It has also been suggested that the role played by CDFs in the cell may stem from their participation in the activation or deactivation of proteins like kinases, for example, whose function is dependent upon the presence/absence of a bound substrate like Zn [Michaud et al., 1995; Hoyos et al., 2002].

2.2.3 CAXs. The CAX family is one of the five families constituting the large family of Ca^{2+}/cation antiporters (CaCA) [Cai and Lytton, 2004]. The signature of the family of cation exchangers (PFAM: 01699) is present in at least 11 *A. thaliana* proteins (see Table 2); however, AtCAX7–AtCAX11 are more closely related to K^+-dependent Na^+/Ca^{2+} exchangers than to CAX [Shigaki and Hirschi, 2006]. The six remaining AtCAX can be classified into two distinct phylogenetic groups that most probably serve different functions: type IA (AtCAX1, AtCAX3 and AtCAX4) and type IB (AtCAX2, AtCAX5, and AtCAX6) [Shigaki and Hirschi, 2006]. The CAX proteins are present in plants, fungi, and bacteria and have not been found in the genomes of human, mouse, fruitfly, or *Caenorhabtitis elegans*. The first transporters from *A. thaliana* (AtCAX1 and AtCAX2) were cloned in yeast as suppressors of a Ca^{2+} transport defective mutant [Hirschi et al., 1996]. Interestingly, it appeared later that these cDNA clones contained truncated open reading frames with deleted first 36 amino acids and the full-length AtCAX1 and AtCAX2 [Pittman et al., 2002, 2004]. It is now recognized that the

N-terminus present in various CAX has an autoinhibitory function and is capable of binding the regulatory proteins [Cheng et al., 2004].

The CAX proteins belong to the best-characterized tonoplast proteins [Cai and Lytton, 2004; Martinoia et al., 2007]. It is predicted that these proteins have 11 transmembrane domains and a central hydrophilic region with an acidic motif between domains 6 and 7 [Hall and Williams, 2003; Shigaki and Hirschi, 2006].

Although the physiological functions of AtCAX1 in plants suggested its major role in homeostasis of calcium ions and Ca^{2+} signaling [Hirschi, 1999], the AtCAX2 is probably predominantely a heavy metal transporter [Hirschi et al., 2000]. The subsequent studies led to the identification of His-338 as an amino acid crucial for the metal specificity of AtCAX1 [Shigaki et al., 2005]. Therefore, the suggestion that some CAX variants can be used for Cd or Zn transport in order to obtain an elevated accumulation of these metals in plants remains justified [Shigaki and Hirschi, 2006].

2.3   Alternative Plant Metal Transporter

Based on K^+-transport-deficient yeast mutant complementation, LCT1 (low-affinity cation transporter) from wheat roots was cloned [Schachtman et al., 1997]. Although its cellular localization remained undiscovered, a yeast expression study suggested that LCT1 mediated the uptake of cadmium in addition to sodium, rubidium, and calcium [Clemens et al., 1998]. Investigation of transgenic tobacco expressing *TaLCT1* under the 35S CaMV promoter demonstrated its involvement in the uptake of calcium by plants; however, it did not indicate that this protein could be an important entry route for cadmium and lead, unless calcium was present in very low concentrations. Instead, the results suggested that TaLCT1 contributes to the regulation of the well-known phenomenon, "Ca-dependent Cd/Pb-tolerance" [Antosiewicz and Hennig, 2004; Wojas et al., 2007].

3   INTERCELLULAR AND LONG-DISTANCE TRANSPORT

Heavy metals are natural constituents of many soils, being in some cases and in small quantities micronutrients, necessary for normal plant growth and development. Heavy metals are transported in plants through the same transport compartments and by the same transport mechanism as other ions. Therefore, before considering the particular transport features of heavy metals, especially in hyperaccumulating plants, the short description of the intercellular and long-distance transport mechanisms in higher plants is warranted.

The plant body consists of two compartments, the symplast and the apoplast. The symplast is the continuum of cells linked by plasmodesmata. The apoplast is the continuum of cell walls and intercellular spaces. Both these compartments are closely linked, since one is located inside the another; however, they are separated by cell membranes. Transport between neighboring cells may be apoplasmic, if a molecule leaves the donor cell, moves in the apoplast, and enters the acceptor cell, or it may be symplasmic (i.e., through plasmodesmata) without leaving the cell. Both types of transport occur in the same plant tissues, but in different proportions, depending on the tissue, developmental status, and environmental demand. However, the symplasmic transport is more efficient (approximately two orders of magnitude) than the apoplasmic one [Patrick and Offler, 1995] and seems to be predominant in many tissues.

In apoplasmic transport the transported molecule crosses the cell membrane barrier twice, when it leaves the donor cell and when it is taken up by the acceptor cell. The transport through the cell membrane could be passive and involve channels, or it could be active with the participation of carriers. Active transport is driven by a proton gradient across the membrane, created by a proton pump, H^+-ATPase [Palmgren, 2001]. Channels allow the passage rather small molecules, such as simple ions. Carriers transport molecules of various masses, from ions to oligopeptides.

Symplasmic transport depends on the features of plasmodesmata. These structures represent narrow ($\sim$40–60 nm in diameter) tubes crossing cell walls [Roberts and Oparka, 2003]. In the center of the channel a desmotubule is located connecting the ER systems of the neighboring cells. The desmotubule is bound to plasmalemma lining cell wall sleeve by proteinaceous, possibly contractile the filaments forming a sort of spokes. Plasmodesmata function as molecular sieves. They allow low-molecular-mass molecules to pass—that is, those of molecular mass less than $\sim$1 kDa [Roberts and Oparka, 2003]. This value is called the size exclusion limit (SEL). Transport of small molecules is believed to be driven by diffusion. Plasmodesmata may also allow the passage of high-molecular-weight molecules—that is, proteins. In the case of some proteins, the SEL may increase much above the value of 10 kDa, allowing the trafficking of protein together with their mRNAs, in some cases. This very important phenomenon is outside the scope focus of this chapter, and the interested reader will find it described in many excellent reviews (e.g. Chen and Kim [2006]).

Two tissues specialize in long-distance transport of solutes in plants: xylem and phloem. In xylem, the conducting elements are either vessels or tracheids. They are the secondary walls of tracheary elements which underwent cell death to remove their cellular contents at an early stage of xylem differentiation. The ends of each vessel element are perforated, so the vessels form long tubes with relatively low resistance to the flow of water, allowing xylem sap to reach the above-ground part of the plant. The conducting elements in phloem are sieve elements. These are live cells; however, in higher plants they have no nucleus. Also, almost all other cellular substructures are absent. Pores in the sieve plates at the ends of sieve elements are modified plasmodesmata strongly increased in diameter, thus allowing very fast and efficient transport of phloem sap.

In both conductive systems, the transport proceeds by mass flow—that is, the movement of solutes together with the solvent (i.e., water). Also in both systems, three stages of the transport may be identified: loading, long-distance translocation, and unloading. Xylem loading takes place in roots and concerns the movement of water and mineral nutrients from the surrounding tissue. It may be powered by

H^+-ATPase localized in the vascular parenchyma [Jahn et al., 1998] as well as by ion carriers [Sondergaard et al., 2004]. The driving force for xylem sap flow is a pressure gradient. It is negative with respect to the atmospheric pressure under transpiration conditions. It becomes a positive, when the root pressure is higher than transpiration. Xylem unloading takes place in distinct, generally above-ground organs, with a participation of bundle sheath cell and/or vascular parenchyma cells [De Boer and Volkov, 2003]. Phloem loading takes place in mature leaves. Carbohydrates and other products of photosynthesis (i.e., amino acids) are transported into the companion cell/sieve element (cc/se) complex, which decreases the osmotic potential of phloem sap. In consequence, water transport from xylem is promoted, increasing the hydrostatic pressure. Water moves to phloem by osmosis and/or through aquaporins which are strongly expressed along the sieve tubes [Patrick et al., 2001]. There are several types of phloem loading mechanisms, including both apoplasmic and symplasmic ones (for a review, see Turgeon [2006]). The driving force for phloem sap flow is the hydrostatic pressure gradient between the phloem loading zone and the phloem unloading zone, where sucrose and other constituents of phloem sap leave the sieve tubes and move to acceptor tissues by means of post-sieve transport, which has either symplasmic or apoplasmic character or is a combination of both [Patrick and Offler, 1995].

Every molecule that is taken up from the soil and moves along the plant uses the above transport routes on its way. Nutrients from the soil need to enter the symplasm of the root cortex to cross the endodermis, which is a barrier for apoplasmic transport of solutes to xylem. Heavy metal ions enter the root symplasm either as free ions or in the form of chelates, as do other micronutrients. The uptake of Fe^{3+} seems to be the best-known mechanism of metal ion uptake [Grusak et al., 1999] and is believed to be a good model for the uptake of other ions (e.g., Cu^{2+} and Mn^{3+}). There are two strategies for iron ion uptake: (a) Strategy I in all dicots and monocots except grasses and (b) Strategy II in grasses. In the first case, the soil pH is lowered by protons released from root cells. Before uptake, Fe^{3+} (the main form of iron in most soils) must be reduced to Fe^{2+} by a membrane-localized, Fe(III) reductase. The reduction requires iron to be chelated by organic acids and/or other chelators released from roots and thus solubilized. Following reduction, Fe^{2+} is transported to the root cells by carriers in the free or chelated form. Strategy II has evolved in grasses. Here, Fe^{3+} is not reduced before being taken up by a root cell. Instead, Fe^{3+} ions are chelated with phytosiderophores released from the roots. The Fe chelate is then transported by a carrier into root cells. It is not clear if phytosiderophores participate in the transport of other heavy metal ions into the root cells. The ways and forms of heavy metal transport in the root are not clear. It is assumed that once the ions enter the root symplasm, they move inside this compartment up to the pericycle. The presence of numerous plasmodesmata linking cortex cells is in line with such a hypothesis [Giaquinta et al., 1983; Ma and Peterson, 2001]. However, some experiments have shown that symplasmic continuity concerns only undifferentiated cells very closely to the root tip [Duckett et al., 1994; Hukin et al., 2002]. It seems therefore that the way of transport of solutes in roots depends on plant species, with exception of crossing the endodermis, which is a barrier to apoplasmic transport.

The definition of a transport form of metal ions in the root epidermis and cortex is not clear. All metal ions, including micronutrients, are toxic for cells, so once they enter a cell, they are chelated and eventually transported to the vacuole for storage and detoxification. Several types of chelators are discussed in the literature [Clemens, 2001]. Some of them, like citrate or malate, could easily move through plasmodesmata because of their low molecular mass. The same concerns amino acids; however, histidine, which is to be assumed a good Ni chelator [Kramer et al., 1996], has been found to be excluded from symplasmic transport [Tucker and Tucker, 1993]. The same limitation concerns several other amino acids [Ding et al., 1999]; however, there is no information on symplasmic transport of nicotinamine (NA), a nonproteinaceous amino acid known as a chelator of many heavy metals. Also, phytochelatins (PCS) and metallthioneins (MT), known metal ion protein chelators [Clemens, 2001; Gong et al., 2003], have not been tested with respect to their mobility in the symplast. It is known, however, that PCS and MT are involved in detoxification rather than transport of metals in cells [Clemens, 2001; Salt and Rauser, 1995]. In general, the symplasmic transport of chelators of molecular mass as higher than the standard SEL (1 kDa) might be hampered. Additionally, hydrophilic molecules with a charge of -4 to -2 were found not to diffuse through plasmodesmata [Tucker and Tucker, 1993]. Generally, both the mass and the charge of chelates might determine the ability of a given molecule to move along the symplast.

The transport of an ion from root to shoot begins with xylem loading, when the ion crosses the barrier between the symplast and the apoplast; that is, it is released from the pericycle and xylem parenchyma into the dead tracheary vessels. This process involves many ion channels and transporters. In the case of some ions (e.g., K^+), xylem loading goes along an electrochemical gradient, while others like Zn^{2+} require active transport [Sondergaard et al., 2004]. It seems that xylem loading of heavy metal ions is preceded by a change whereby the chelates bind a given ion. For example, Cd ions are chelated by PCS in Indian mustard roots; however, the transport of Cd ions in xylem was transported chelated with oxygen or nitrogen ligands [Salt et al., 1995]. In rice, transcripts of nicotinamine synthase *OsNAS1* and *OsNAS2* were detected specifically in companion cells and pericycle cells in Fe-sufficient plants [Inoue et al., 2003]. In the metal hyperaccumulator *Thlaspi caerulescences*, expression of nicotinamine synthase, *TcNAS*, was detected only in leaves [Mari et al., 2006]. The authors postulated that NA involved in the root-to-shoot transport of nickel is transported from leaves. This finding underlines the role of chelators in the long-distance transport of metal ions as well. Among them, apart from nicotinamine, the histidine and citrate [Hell and Stephan, 2003] have been postulated to participate in transport of heavy metal ions from root to shoot. It was also found that phytochelatins, in addition to their role in transport of heavy metal ions into the vacuole [Salt and Rauser, 1995], might participate in the long-distance transport of Cd as well [Gong et al., 2003]. Apparently, metal ions might be transported in the xylem in the form of free ions, too. The equilibrium between free and chelated metal ions depends on xylem sap pH as well on the binding of ions by cell walls [Clemens et al., 2002].

In above-ground parts of the plant, water and solutes should reach tissues distant from the tracheary vessels (e.g., palisade cells or spongy mesophyll). Two routes of water movement are possible. First, water could be taken up by bundle sheath cells and then flow from cell to cell with the participation of aquaporins. Second, water may exit the xylem through cell walls (i.e., the apoplast) [Sack and Holbrook, 2006]. In similar ways, solutes, including heavy metal ions, should move outside the xylem. They are taken up actively—that is, with the participation of proton-linked carriers, possibly in bundle sheath cells—and trafficked to other cells. Here, heavy-metal ions might be sequestered in the vacuole. Alternatively, ions might reach distinct tissues through the apoplast. If not deposited in the apoplast, as—as, for example, in the case of zinc deposits in leaves of *Armeria maritima* [Heumann, 2002]—heavy-metal ions enter the symplasm, where they probably move on the distance of several cells. It concerns especially the trichomes, were heavy metals are often deposited [Salt et al., 1995; Garcia-Hernandez et al., 1998]. Trichomes are particularly adapted to an intensive transport of a wide spectrum of substances, including toxic ones [Echeverria, 2000; Waigmann and Zambryski, 2000]. The adaptation concerns particular properties of plasmodesmata with SEL > 1 kDa and open desmotubules, which apparently allows communication between the ER of neighboring cells [Waigmann and Zambryski, 2000]. The detailed mechanism of heavy-metal ion transport and accumulation in trichomes remains to be elucidated.

When they reach the shoot, nutrients and other ions are further distributed along the plant through phloem. Mineral nutrients undergo cycling [Marschner et al., 1996] to maintain the homeostasis of particular ions. Additionally, some organs may be supplied with mineral nutrients almost exclusively by phloem, since the transpiration stream is too low for xylem transport to be effective (e.g., in tubers and fruits). The mobility of distinct heavy-metal ions in xylem and phloem differs strongly. For example, Page and Feller [2005] found in wheat that ^{63}Ni was highly mobilized in the phloem, while ^{54}Mn was mobile only in the xylem. Loading of metal ions into the cc/se complex generally proceeds through the apoplast; however, the symplasmic mechanism is also possible [Schmidke et al., 1999]. Phloem loading of iron was suggested to be preceded by the binding of Fe by a chelator [Grusak et al., 1999], possibly NA [Stephan et al., 1995]. It was later found [Schmidke et al., 1999] that iron, zinc, copper, and manganese might be loaded to phloem in a more complicated manner, with a participation of NA only as a temporary chelator, before phloem loading and inside the companion cell. According to the authors, the transport chelator of metal ions in sieve tubes is a high-molecular-weight molecule. This concept is supported by the finding that one of phloem sap proteins is a copper chaperone (CCH) known to be involved in Cu homeostasis in the cell [Mira et al., 2001]. The final step of phloem transport is phloem unloading, which, as a rule, goes symplasmically. There are no particular data on this process in respect to heavy metal ion transport, one may, however, suppose that all aspects of symplasmic transport discussed earlier for roots could be applicable.

To summarize, the intercellular and long-distance transport in plants is a chain of processes proceeding in both apoplast and symplast. There are several sieve-like barriers for a transported molecule to cross. These barriers are located in the plasma-lemma (selective transport), but also in the apoplast (Casparian bands, bundle sheath cell wall, binding of ions in the cell wall) and in the symplasm (molecular mass and physical properties of molecules passed by plasmodesmata). In general, elucidating the role of different chelators in the transport of heavy metal ion in plants seems to be crucial for a better understanding the mechanisms of heavy-metal hyperaccumulation and phytoremediation.

4 THE IMPORTANCE OF PLANT MINERAL STATUS FOR HUMAN HEALTH

Heavy-metal contamination in soils has markedly increased in the past few decades due to the industrial and agricultural processes that contribute to the elevation of trace metals in soils. For example, most fertilizers, manures, and chemicals used for plant protection contain trace amounts of trace elements [He et al., 2005]. In view of the fact that some heavy metals (e.g., Cd [Martelli et al., 2006]) enter the human and animal body mostly through the water and food chain after being accumulated in plant tissues, the knowledge of mechanisms of their transport in plants is important for an effective protection of human and animal populations from exposure to these metals. There is a need to intensify the studies on metal transporting activities in various plants, especially in the real field conditions. The genes responsible for transport and distribution capabilities of trace elements in plants and/or the genes responsible for regulatory mechanisms controlling these processes might become a target for the future plant breeding or plant engineering programs. Particularly, characteristics of genes (and gene expression) in metal hyperaccumulating and hypertolerant plants, such as *T. caerulescens* or *A. halleri*, should be an important element of such programs.

Finally, we would like to emphasize that both the trace elements transport capabilities and the degrees of crop tolerance to the toxic concentrations of metals are of particular significance for human and animal nutrition. On one hand, accumulation of some toxic metals (e.g., Cd, Pb, Cr) can occur in plants to the level that is potentially health-threatening for consumers without causing symptoms of phytotoxicity; on the other hand, crops biofortified with some trace elements (e.g., Fe, Zn, Ni, Se, Co) can be used as an important supplement of a diet in food or feed.

ACKNOWLEDGMENTS

The authors Danuta Maria Antosiewicz and Agnieszka Sirko participate in the network of coordinated national research projects—COST Action 859 [http://www.gre.ac.uk/cost859/]. The work in Agnieszka Sirko's laboratory is supported by the grants from the Polish Ministry of Education and Science (PBZ-KBN-110/P04/2004 and SPB/COST/112/2005). The work in Danuta Maria Antosiewicz's laboratory is funded by FP6 EU project PHIME.

REFERENCES

Abdel-Ghany SE, Muller-Moule P, Niyogi KK, Pilon P, Shikanai T. 2005. Two P-type ATPases are required for copper delivery in *Arabidopsis thaliana* chloroplasts. *Plant Cell* **17**:1233–1251.

Alonso JM, Hirayama T, Roman G, Nourizadeh S, Ecker JR. 1999. EIN2, a bifunctional transducer of ethylene and stress responses in *Arabidopsis*. *Science* **284**:2148–2152.

Andres-Colas N, Sancenon V, Rodriguez-Navarro S, Mayo S, Thiele DJ, Ecker JR, Puig S, Penarrubia L. 2006. The Arabidopsis heavy metal P-type ATPase HMA5 interacts with metallochaperones and functions in copper detoxification of roots. *Plant J* **45**:225–236.

Anton A, Weltrowski A, Haney CJ, Franke S, Grass G, Rensing C, Nies DH. 2004. Characteristics of zinc transport by two bacterial cation diffusion facilitators from *Ralstonia metallidurans* CH34 and *Escherichia coli*. *J Bacteriol* **186**:7499–7507.

Antosiewicz DM, Hennig J. 2004. Overexpression of *LCT1* in tobacco enhances the protective action of calcium against cadmium toxicity. *Environ Pollut* **129**:237–245.

Arguello JM. 2003. Identification of ion-selectivity determinants in heavy-metal transport P1B-type ATPases. *J Membr Biol* **195**:93–108.

Arrivault S, Senger T, Krämer U. 2006. The Arabidopsis metal tolerance protein AtMTP3 maintains metal homeostasis by mediating Zn exclusion from the shoot under Fe deficiency and Zn oversupply. *Plant J* **46**:861–879.

Assunção AGL, Costa Martins PDA, De Folter S, Vooijs R, Schat H, Aarts MGM. 2001. Elevated expression of metal transporter genes in three accessions of the metal hyperaccumulator *Thlaspi caerulescens*. *Plant Cell Envrion* **24**:217–226.

Axelsen KB, Palmgren MG. 1998. Evolution of substrate specificities in the P-type ATPase superfamily. *J Mol Evol* **46**:84–101.

Axelsen KB, Palmgren MG. 2001. Inventory of the superfamily of P-type ion pumps in Arabidopsis. *Plant Physiol* **126**:696–706.

Bal N, Wu CC, Catty P, Guillain F, Mintz E. 2003. Cd^{2+} and the N-terminal metal-binding domain protect the putative membranous CPC motif of the Cd^{2+}-ATPase of *Listeria monocytogenes*. *Biochem J* **369**:681–685.

Baxter I, Tchieu J, Sussman MR, Boutry M, Palmgren MG, Gribskov M, Harper JF, Axelsen KB. 2003. Genomic comparison of P-Type ATPase ion pumps in *Arabidopsis* and Rice. *Plant Phys* **155**:201–210.

Bernard C, Roosens N, Czernic P, Lebrun M, Verbruggen N. 2004. A novel CPx-ATPase from the cadmium hyperaccumulator *Thlaspi caerulescens*. *FEBS Lett* **31**:489–498.

Blaudez D, Kohler A, Martin F, Sanders D, Chalot M. 2003. Poplar metal tolerance protein 1 confers zinc tolerance and is an oligomeric vacuolar zinc transporter with an essential leucine zipper motif. *Plant Cell* **15**:2911–2928.

Bloss T, Clemens S, Nies DH. 2002. Characterization of the ZAT1p zinc transporter from *Arabidopsis thaliana* in microbial model organisms and reconstituted proteoliposomes. *Planta* **214**:783–791.

Cai X, Lytton J. 2004. The cation/Ca(2+) exchanger superfamily: Phylogenetic analysis and structural implications. *Mol Biol Evol* **21**:1692–1703.

Chao Y, Fu D. 2004. Kinetic study of the antiport mechanism of an *Escherichia coli* zinc transporter, ZitB. *J Biol Chem* **279**:12043–12050.

Chaloupka R, Courville P, Veyrier F, Knudsen B, Tompkins TA, Cellier MF. 2005. Identification of functional amino acids in the Nramp family by a combination of evolutionary analysis and biophysical studies of metal and proton cotransport *in vivo*. *Biochemistry* **44**:726–733.

Chen X-Y, Kim J-Y. 2006. Transport of macromolecules through plasmodesmata and the phloem. *Plant Physiol* **126**:560–571.

Cheng NH, Pittman JK, Zhu JK, Hirschi KD. 2004. The protein kinase SOS2 activates the Arabidopsis H(+)/Ca(2+) antiporter CAX1 to integrate calcium transport and salt tolerance. *J Biol Chem* **279**:2922–2926.

Claros MG, von Heijne G. 1994. TopPred II: An improved software for membrane protein structure predictions. *CABIOS* **10**:685–686.

Clemens S. 2001. Molecular mechanisms of plant metal tolerance and homeostasis. *Planta* **212**:475–486.

Clemens S. 2006. Toxic metal accumulation, responses to exposure and mechanisms of tolerance in plants. *Biochimie* **88**:1707–1719.

Clemens S, Antosiewicz DM, Ward JM, Schachtman DP, Schroeder JI. 1998. The plant cDNA *LCT1* mediates the uptake of calcium and cadmium in yeast. *Proc Natl Acad Sci USA* **95**:1243–1248.

Clemens S, Palmgren MG, Kramer U. 2002. A long way ahead: Understanding and engineering plant metal accumulation. *Trends Plant Sci* **7**:309–315.

Cobbett P, Day TA. 2003. Functional voltage-gated Ca^{2+} channels in muscle fibers of the platyhelminth *Dugesia tigrina*. *Comp Biochem Physiol A Mol Integr Physiol* **134**:593–605.

Colangelo EP, Guerinot ML. 2006. Put the metal to the petal: metal uptake and transport throughout plants. *Curr Opin Plant Biol* **9**:322–330.

Courbot M, Willems G, Motte P, Arvidsson S, Roosens N, Saumitou-Laprade and Verbruggen N. 2007. A major quantitative trait locus for cadmium tolerance in *Arabidopsis halleri* colocalizes with *HMA4*, a gene encoding a heavy metal ATPase. *Plant Phys* **144**:1052–1065.

Courville P, Chaloupka R, Cellier MF. 2006. Recent progress in structure-function analyses of Nramp proton-dependent metal-ion transporters. *Biochem Cell Biol* **84**:960–978.

Curie C, Alonso JM, Le Jean M, Ecker JR, Briat JF. 2000. Involvement of NRAMP1 from *Arabidopsis thaliana* in iron transport. *Biochem J* **347**(Pt 3):749–755.

Curie C, Panaviene Z, Loulergue C, Dellaporta SL, Briat JF, Walker EL. 2001. Maize *yellow stripe1* encodes a membrane protein directly involved in Fe(III) uptake. *Nature* **409**:346–349.

De Boer AH, Volkov V. 2003. Logistics of water and salt transport through the plant: Structure and functioning of the xylem. *Plant Cell Environ* **26**:87–101.

Delhaize E, Kataoka T, Hebb DM, White RG, Ryan PR. 2003. Genes encoding proteins of the cation diffusion facilitator family that confer manganese tolerance. *Plant Cell* **15**:1131–1142.

Desbrosses-Fonrouge A-G, Voigt K, Schröder A, Arrivault S, Thomine S, Krämer U. 2005. *Arabidopsis thaliana* MTP1 is a Zn transporter in the vacuole membrane which mediates Zn detoxification and drives leaf Zn accumulation. *FEBS Lett* **579**:4165–4174.

DiDonato RJ, Jr., Roberts LA, Sanderson T, Eisley RB, Walker EL. 2004. Arabidopsis Yellow Stripe-Like2 (YSL2): A metal-regulated gene encoding a plasma membrane transporter of nicotianamine–metal complexes. *Plant J* **39**:403–414.

Ding B, UItaya A, Woo Y.-M. 1999. Plasmodesmata and cell-to-cell communication in plants. *Int Rev Cyt* **190**:251–316.

Dräger DB, Desbrosses-Fonrouge A-G, Krach C, Chardonnens N, Meyer RC, Saumitou-Laprade P, Krämer U. 2004. Two genes encoding *Arabidopsis halleri* MTP1 metal transport proteins co-segragate with zinc tolerance and account for high *MTP1* transcript levels. *Plant J* **39**:425–439.

Duckett CM, Oparka KJ, Prior DAM, Dolan L, Roberts K. 1994. Dye-coupling in the root epidermis of *Arabidopsis* is progressively reduced during development. *Development* **120**:3247–3255.

Echeverria E. 2000. Vesicle-mediated solute transport between the vacuole and the plasma membrane. *Plant Physiol* **123**:1217–1226.

Eckhardt U, Mas Marques A, Buckhout TJ. 2001. Two iron-regulated cation transporters from tomato complement metal uptake-deficient yeast mutants. *Plant Mol Biol* **45**:437–448.

Efremov RG, Kosinsky YA, Nolde DE, Tsivkovskii R, Arseniev AS, Lutsenko S. 2004. Molecular modelling of the nucleotide-binding domain of Wilson's disease protein: Location of the ATP-binding site, domain dynamics and potential effects of the major disease mutations. *Biochem J* **382**:293–305.

Eide D, Broderius M, Fett J, Guerinot ML. 1996. A novel iron-regulated metal transporter from plants identified by functional expression in yeast. *Proc Natl Acad Sci USA* **93**:5624–5628.

Eide DJ. 2004. The SLC39 family of metal ion transporters. *Pflugers Arch* **447**:796–800.

Eide DJ. 2006. Zinc transporters and the cellular trafficking of zinc. *Biochim Biophys Acta* **1763**:711–722.

Eren E, Arguello JM. 2004. Arabidopsis HMA2, a divalent heavy metal-transporting P(IB)-type ATPase, is involved in cytoplasmic Zn^{2+} homeostasis. *Plant Physiol* **136**:3712–3723.

Filatov V, Dowdle J, Smirnoff N, Ford-Lloyd B, Newbury HJ, Macnair MR. 2006. Comparison of gene expression in segregating families identifies genes and genomic regions involved in a novel adaptation, zinc hyperaccumulation. *Mol Ecol* **15**:3045–3059.

Gaither LA, Eide DJ. 2000. Functional expression of the human hZIP2 zinc transporter. *J Biol Chem* **275**:5560–5564.

Gaither LA, Eide DJ. 2001. Eukaryotic zinc transporters and their regulation. *Biometals* **14**:251–270.

Garcia-Hernandez M, Murphy A, Taiz L. 1998. Metallothioneins 1 and 2 have distinct but overlapping expression patterns in *Arabidopsis*. *Plant Physiol* **118**:387–397.

Gendre D, Czernic P, Conejero G, Pianelli K, Briat JF, Lebrun M, Mari S. 2007. TcYSL3, a member of the YSL gene family from the hyper-accumulator *Thlaspi caerulescens*, encodes a nicotianamine-Ni/Fe transporter. *Plant J* **49**:1–15.

Giaquinta RT, Lin W, Sadler NL, Franceschi VR. 1983. Pathway of phloem unloading of sucrose in corn roots. *Plant Physiol* **72**:362–367.

Gong JM, Lee DA, Schroeder JI. 2003. Long-distance root-to-shoot transport of phytochelatins and cadmium in Arabidopsis. *Proc Natl Acad Sci USA* **100**:10118–10123.

Gravot A, Lieutaud A, Verret F, Auroy P, Vavasseur A, Richaud P. 2004. AtHMA3, a plant P1B-ATPase, functions as a Cd/Pb transporter in yeast. *FEBS Lett* **561**:22–28.

Grotz N, Fox T, Connolly E, Park W, Guerinot ML, Eide D. 1998. Identification of a family of zinc transporter genes from *Arabidopsis* that respond to zinc deficiency. *Proc Natl Acad Sci USA* **95**:7220–7224.

Grusak MA, Pearson JN, Marentes E. 1999. The physiology of micronutrient homeostasis in field crops. *Field Crops Res* **60**:41–56.

Guerinot ML. 2000. The ZIP family of metal transporters. *Biochim Biophys Acta* **1465**:190–198.

Hall JL, Williams LE. 2003. Transition metal transporters in plants. *J Exp Bot* **54**:2601–2613.

Haney CJ, Grass G, Franke S, Rensing C. 2005. New developments in the understanding of the cation diffusion facilitator family. *J Ind Microbiol Biotechnol* **32**:215–226.

Hanikenne M, Kramer U, Demoulin V, Baurain D. 2005. A comparative inventory of metal transporters in the green alga *Chlamydomonas reinhardtii* and the red alga *Cyanidioschizon merolae*. *Plant Physiol* **137**:428–446.

Harrison MD, Jones CE, Solioz M, Dameron CT. 2000. Intracellular copper routing: the role of copper chaperones. *Trends Biochem Sci* **25**:29–32.

He ZL, Yang XE, Stoffella PJ. 2005. Trace elements in agroecosystems and impacts on the environment. *J Trace Elem Med Biol* **19**:125–140.

Hell R, Stephan UW. 2003. Iron uptake, trafficking and homeostasis in plants. *Planta* **216**:541–551.

Heumann H-G. 2002. Ultrastructural localization of zinc in zinc-tolerant *Armeria maritima* ssp. *halleri* by autometallography. *J Plant Physiol* **159**:191–203.

Hirayama T, Kieber JJ, Hirayama N, Kogan M, Guzman P, Nourizadeh S, Alonso JM, Dailey WP, Dancis A, Ecker JR. 1999. RESPONSIVE-TO-ANTAGONIST1, a Menkes/Wilson disease-related copper transporter, is required for ethylene signaling in *Arabidopsis*. *Cell* **97**:383–393.

Hirschi KD. 1999. Expression of Arabidopsis CAX1 in tobacco: altered calcium homeostasis and increased stress sensitivity. *Plant Cell* **11**:2113–2122.

Hirschi KD, Zhen RG, Cunningham KW, Rea PA, Fink GR. 1996. CAX1, an H^+/Ca^{2+} anti-porter from *Arabidopsis*. *Proc Natl Acad Sci USA* **93**:8782–8786.

Hirschi KD, Korenkov VD, Wilganowski NL, Wagner GJ. 2000. Expression of arabidopsis CAX2 in tobacco. Altered metal accumulation and increased manganese tolerance. *Plant Physiol* **124**:125–133.

Hoyos B, Imam A, Korichneva I, Levi E, Chua R, Hammerling U. 2002. Activation of c-Raf kinase by ultraviolet light. *Regulation by retinoids*. *J Biol Chem* **277**:23949–23957.

Huang L, Gitschier J. 1997. A novel gene involved in zinc transport is deficient in the lethal milk mouse. *Nat Gene* **17**:292.

Huang L, Kirschke CP, Zhang Y, Yu YY. 2005. The ZIP7 gene (Slc39a7) encodes a zinc transporter involved in zinc homeostasis of the Golgi apparatus. *J Biol Chem* **280**:15456–15463.

Hukin D, Doering-Saad C, Thomas CR, Pritchard J. 2002. Sensitivity of cell hydraulic conductivity to mercury is coincident with symplasmic isolation and expression of plasmalemma aquaporin genes in growing maize roots. *Planta* **215**:1047–1056.

Hussain D, Haydon MJ, Wang Y, Sherson SM, Young J, Camakaris J, Harper JF, Cobbett C. 2004. P-type ATPases heavy metal transporters with roles in essential zinc homeostasis in *Arabidopsis*. *Plant Cell* **16**:1327–1339.

Huster D, Lutsenko S. 2003. The distinct roles of the N-terminal copper-binding sites in regulation of catalytic activity of the Wilson's disease protein. *J Biol Chem* **278**:32212–32218.

Inoue H, Higuchi K, Takahashi M, Nakanishi H, Mori S, Nishizawa NK. 2003. Three rice nicotianamine synthase genes, OsNAS1, OsNAS2, and OsNAS3 are expressed in cells involved in long-distance transport of iron and differentially regulated by iron. *Plant J* **36**:366–381.

Jahn T, Baluska F, Michalke W, Harper JF, Volkmann D. 1998. Plasma membrane H+ ATPase in the root apex: evidence for strong expression in xylem parenchyma and asymmetric localization within cortical and epidermal cells. *Physiol Plant* **104**:311–316.

Jirakulaporn T, Muslin AJ. 2004. Cation diffusion facilitator proteins modulate Raf-1 activity. *J Biol Chem* **279**:27807–27815.

Kampfenkel K, Kushnir S, Babiychuk E, Inze D, Van Montagu M. 1995. Molecular characterization of a putative *Arabidopsis thaliana* copper transporter and its yeast homologue. *J Biol Chem* **270**:28479–28486.

Kim D, Gustin JL, Lahner B, Persans MW, Baek D, Yun D-J, Salt D. 2004. The plant CDF family member TgMTP1 from the Ni/Zn hyperaccumulator *Thlaspi goesiongense* acts to enhance efflux of Zn at the plasma membrane when expressed in *Saccharomyces cerevisiae*. *Plant J* **39**:237–251.

Kobae Y, Uemura T, Sato MH, Ohnishi M, Mimura T, Maeshima M. 2004. Zinc transporter of *Arabidopsis thaliana* AtMTP1 is localized to vacuolar membranes and implicated in zinc homeostasis. *Plant Cell Phys* **45**:1749–1758.

Koike S, Inoue H, Mizuno D, Takahashi M, Nakanishi H, Mori S, Nishizawa NK. 2004. OsYSL2 is a rice metal-nicotianamine transporter that is regulated by iron and expressed in the phloem. *Plant J* **39**:415–424.

Korshunova YO, Eide D, Clark WG, Guerinot ML, Pakrasi HB. 1999. The IRT1 protein from *Arabidopsis thaliana* is a metal transporter with a broad substrate range. *Plant Mol Biol* **40**:37–44.

Krämer U. 2005. MTP1 mops up excess zinc in *Arabidopsis* cells. *Trends Plant Sci* **10**:313–315.

Krämer U, Cotter-Howells JD, Charnock JM, Baker AJM, Smith JAC. 1996. Free histidine as a metal chelator in plants that accumulate nickel. *Nature* **379**:635–638.

Krämer U, Talke IN, Hanikenne M. 2007. Transition metal transport. *FEBS Lett* **581**:2263–2272.

Lanquar V, Lelievre F, Bolte S, Hames C, Alcon C, Neumann D, Vansuyt G, Curie C, Schröder A, Kramer U, Barbier-Brygoo H, Thomine S. 2005. Mobilization of vacuolar iron by AtNRAMP3 and AtNRAMP4 is essential for seed germination on low iron. *EMBO J* **24**:4041–4051.

Le Jean M, Schikora A, Mari S, Briat JF, Curie C. 2005. A loss-of-function mutation in AtYSL1 reveals its role in iron and nicotianamine seed loading. *Plant J* **44**:769–782.

Lee SM, Grass G, Haney CJ, Fan B, Rosen BP, Anton A, Nies DH, Rensing C. 2002. Functional analysis of the *Escherichia coli* zinc transporter ZitB. *FEMS Microbiol Lett* **215**:273–278.

Liu J, Stemmler AJ, Fatima J, Mitra B. 2005. Metal-binding characteristics of the amino-terminal domain of ZntA: Binding of lead is different compared to cadmium and zinc. *Biochemistry* **44**:5159–5167.

Ma F, Peterson CA. 2001. Frequencies of plasmodesmata in *Allium cepa* L. roots: Implications for solute transport pathways. *J Exp Bot* **52**:1051–1061.

MacDiarmid CW, Gaither LA, Eide D. 2000. Zinc transporters that regulate vacuolar zinc storage in *Saccharomyces cerevisiae*. *EMBO J* **19**:2845–2855.

Mana-Capelli S, Mandal AK, Arguello JM. 2003. *Archaeoglobus fulgidus* CopB is a thermophilic Cu^{2+}-ATPase: functional role of its histidine-rich-N-terminal metal binding domain. *J Biol Chem* **278**:40534–40541.

Mandal AK, Arguello JM. 2003. Functional roles of metal binding domains of the *Archaeoglobus fulgidus* Cu(+)-ATPase CopA. *Biochemistry* **42**:11040–11047.

Mari S, Gendre D, Pianelli K, Ouerdane L, Lobinski R, Briat JF, Lebrun M, Czernic P. 2006. Root-to-shoot long-distance circulation of nicotianamine and nicotianamine-nickel chelates in the metal hyperaccumulator *Thlaspi caerulescens*. *J Exp Bot* **57**:4111–4122.

Marschner H, Kirkby EA, Cakmak I. 1996. Effect of mineral nutrition status on shoot-root partitioning of photoassimilates and cycling of mineral nutrients. *J Exp Bot* **47**:1255–1263.

Martelli A, Rousselet E, Dycke C, Bouron A, Moulis JM. 2006. Cadmium toxicity in animal cells by interference with essential metals. *Biochimie* **88**:1807–1814.

Martinoia E, Maeshima M, Neuhaus HE. 2007. Vacuolar transporters and their essential role in plant metabolism. *J Exp Bot* **58**:83–102.

Maser P, Thomine S, Schroeder JI, Ward JM, Hirschi K, Sze H, Talke IN, Amtmann A, Maathuis FJ, Sanders D, Harper JF, Tchieu J, Gribskov M, Persans MW, Salt DE, Kim SA, Guerinot ML. 2001. Phylogenetic relationships within cation transporter families of *Arabidopsis*. *Plant Physiol* **126**:1646–1667.

Michaud NR, Fabian JR, Mathes KD, Morrison DK. 1995. 14-3-3 is not essential for Raf-1 function: Identification of Raf-1 proteins that are biologically activated in a 14-3-3-independent and Ras-independent manner. *Mol Cell Biol* **15**:3390–3397.

Mills RF, Krijger GC, Baccarini PJ, Hall JL, Williams LE. 2003. Functional expression of AtHMA4, a P1B-type ATPase of the Zn/Co/Cd/Pb subclass. *Plant J* **35**:164–176.

Mills RF, Francini A, Ferreira da Rocha PS, Baccarini PJ, Aylett M, Krijger GC, Williams LE. 2005. The plant P1B-type ATPase AtHMA4 transports Zn and Cd and plays a role in detoxification of transition metals supplied at elevated levels. *FEBS Lett* **579**:783–791.

Milon B, Wu Q, Zou J, Costello LC, Franklin RB. 2006. Histidine residues in the region between transmembrane domains III and IV of hZip1 are required for zinc transport across the plasma membrane in PC-3 cells. *Biochim Biophys Acta* **1758**:1696–1701.

Mira H, Martinez-Garcia F, Penarrubia L. 2001. Evidence for the plant-specific intercellular transport of the Arabidopsis copper chaperone CCH. *Plant J* **25**:521–528.

Mitra B, Sharma R. 2001. The cysteine-rich amino-terminal domain of ZntA, a Pb(II)/Zn(II)/Cd(II)-translocating ATPase from Escherichia coli, is not essential for its function. *Biochemistry* **40**:7694–7699.

Moller JV, Juul B, Lemaire M. 1996. Structural organisation, ion transport, and energy transduction of P-type ATPases. *Biochim Biophys Acta* **1286**:1–51.

Moreau S, Thomson RM, Kaiser BN, Trevaskis B, Guerinot ML, Udvardi MK, Puppo A, Day DA. 2002. GmZIP1 encodes a symbiosis-specific zinc transporter in soybean. *J Biol Chem* **277**:4738–4746.

Morgan CT, Tsivkovskii R, Kosinsky YA, Efremov RG, Lutsenko S. 2004. The distinct functional properties of the nucleotide-binding domain of ATP7B, the human copper-transporting ATPase: Analysis of the Wilson disease mutations E1064A, H1069Q, R1151H, and C1104F. *J Biol Chem* **279**:36363–36371.

Nelson N, Sacher A, Nelson H. 2002. The significance of molecular slips in transport systems. *Nat Rev Mol Cell Biol* **3**:876–881.

Nevo Y, Nelson N. 2006. The NRAMP family of metal-ion transporters. *Biochim Biophys Acta* **1763**:609–620.

Nies DH. 2003. Efflux-mediated heavy metal resistance in procaryotes. *FEMS Microbiol Rev* **27**:313–339.

Page V, Feller U. 2005. Selective transport of zinc, manganese, nickel, cobalt and cadmium in the root system and transfer to the leaves in young wheat plants. *Ann Bot (Lond)* **96**:425–434.

Palmgren MG. 2001. Plant plasma membrane H^+-ATPases: Powerhouses for nutrient uptake. *Annu Rev Plant Physiol Plant Mol Biol* **52**:817–845.

Papoyan A, Kochian LV. 2004. Identification of *Thlaspi caerulescens* genes that may be involved in heavy metal hyperaccumulation and tolerance. Characterization of a novel heavy metal transporting ATPase. *Plant Physiol* **136**:3814–3823.

Patrick J, Offler C. 1995. Post-sieve element transport of sucrose in developing seeds. *Functional Plant Biol* **22**:681–702.

Patrick JW, Zhang W, Tyerman SD, Offler CE, Walker NA. 2001. Role of membrane transport in phloem translocation of assimilates and water. *Functional Plant Biol* **28**:697–709.

Paulsen IT, Saier MH, Jr. 1997. A novel family of ubiquitous heavy metal ion transport proteins. *J Membr Biol* **156**:99–103.

Persans MW, Nieman K, Salt DE. 2001. Functional activity and role of cation-efflux family members in Ni hyperaccumulation in *Thlaspi goesingense*. *Proc Nat Acad Sci USA* **98**:9995–10000.

Pittman JK, Shigaki T, Cheng NH, Hirschi KD. 2002. Mechanism of N-terminal autoinhibition in the Arabidopsis $Ca(2+)/H(+)$ antiporter CAX1. *J Biol Chem* **277**:26452–26459.

Pittman JK, Shigaki T, Marshall JL, Morris JL, Cheng NH, Hirschi KD. 2004. Functional and regulatory analysis of the *Arabidopsis thaliana* CAX2 cation transporter. *Plant Mol Biol* **56**:959–971.

Plaza S, Tearall KL, Zhao FJ, Buchner P, McGrath SP, Hawkesford MJ. 2007. Expression and functional analysis of metal transporter genes in two contrasting ecotypes of the hyperaccumulator *Thlaspi caerulescens*. *J Exp Bot* **58**:1717–1728.

Regalla LM, Lyons TD. 2006. Zinc in yeast: mechanisms involved in homeostasis. In: Tamás MJ, Martinoia E, editors. *Molecular Biology of Metal Homeostasis and Detoxification. From Microbes to Man*. Berlin: Springer, pp. 37–58.

Roberts AG, Oparka KJ. 2003. Plasmodesmata and the control of symplastic transport. *Plant Cell Environ* **26**:103–124.

Roberts LA, Pierson AJ, Panaviene Z, Walker EL. 2004. Yellow stripe1. Expanded roles for the maize iron-phytosiderophore transporter. *Plant Physiol* **135**:112–120.

Rodriguez FI, Esch JJ, Hall AE, Binder BM, Schaller E. Bleecker AB. 1999. A copper cofactor for the ethylene receptor ETR1 from *Arabidopsis*. *Science* **283**:996–998.

Rogers EE, Eide DJ, Guerinot ML. 2000. Altered selectivity in an *Arabidopsis* metal transporter. *Proc Natl Acad Sci USA* **97**:12356–12360.

Rosakis A, Koster W. 2004. Transition metal transport in the green microalga *Chlamydomonas reinhardtii*–Genomic sequence analysis. *Res Microbiol* **155**:201–210.

Rutherford JC, Cavet JS, Robinson NJ. 1999. Cobalt-dependent transcriptional switching by a dual-effector MerR-like protein regulates a cobalt-exporting variant CPx-type ATPase. *J Biol Chem* **274**:25827–25832.

Sack L, Holbrook NM. 2006. Leaf hydraulics. *Annu Rev Plant Biol* **57**:361–381.

Salt DE, Rauser WE. 1995. MgATP-dependent transport of phytochelatins across the tonoplast of oat roots. *Plant Physiol* **107**:1293–1301.

Salt DE, Prince RC, Pickering IJ, Raskin I. 1995. Mechanisms of cadmium mobility and accumulation in Indian mustard. *Plant Physiol* **109**:1427–1433.

Sancenon V, Puig S, Mira H, Thiele DJ, Penarrubia L. 2003. Identification of a copper transporter family in *Arabidopsis thaliana*. *Plant Mol Biol* **51**:577–587.

Sancenon V, Puig S, Mateu-Andres I, Dorcey E, Thiele DJ, Penarrubia L. 2004. The Arabidopsis copper transporter COPT1 functions in root elongation and pollen development. *J Biol Chem* **279**:15348–15355.

Schaaf G, Ludewig U, Erenoglu BE, Mori S, Kitahara T, von Wiren N. 2004. ZmYS1 functions as a proton-coupled symporter for phytosiderophore- and nicotianamine-chelated metals. *J Biol Chem* **279**:9091–9096.

Schaaf G, Schikora A, Haberle J, Vert G, Ludewig U, Briat JF, Curie C, von Wiren N. 2005. A putative function for the *Arabidopsis* Fe-Phytosiderophore transporter homolog AtYSL2 in Fe and Zn homeostasis. *Plant Cell Physiol* **46**:762–774.

Schachtman DP, Kumar R, Schroeder JI, Marsh EL. 1997. Molecular and functional characterization of a novel low-affinity cation transporter (LCT1) in higher plants. *Proc Natl Acad Sci USA* **94**:11079–11084.

Schmidke I, Kruger C, Frommichen R, Scholz G, Stephan UW. 1999. Phloem loading and transport characteristics of iron in interaction with plant-endogenous ligands in castor bean seedlings. *Physiol Plant* **106**:82–89.

Shigaki T, Barkla BJ, Miranda-Vergara MC, Zhao J, Pantoja O, Hirschi KD. 2005. Identification of a crucial histidine involved in metal transport activity in the *Arabidopsis* cation/H$^+$ exchanger CAX1. *J Biol Chem* **280**:30136–30142.

Shigaki T, Hirschi KD. 2006. Diverse functions and molecular properties emerging for CAX cation/H$^+$ exchangers in plants. *Plant Biol (Stuttg)* **8**:419–429.

Shikanai T, Müller-Moulé P, Munekage Y, Niyogi KK, Pilon M. 2003. PAA1, a P-type ATPase of *Arabidopsis* functions in copper transport. *Plant Cell* **15**:1333–1346.

Sondergaard TE, Schulz A, Palmgren MG. 2004. Energization of transport processes in plants. Roles of the plasma membrane H$^+$-ATPase. *Plant Physiol* **136**:2475–2482.

Southron JL, Basu U, Taylor GJ. 2004. Complementation of *Saccharomyces cerevisiae* ccc2 mutant by a putative P1B-ATPase from *Brassica napus* supports a copper-transporting function. *FEBS Lett* **566**:218–222.

Stephan UW, Schmidke I, Pich A. 1995. Phloem translocation of Fe, Cu, Mn and Zn in *Ricinus* seedlings in relation to the concentration of nicotianaminem an endogenous chelator of divalent metal ionsm in different seedlings parts, In Abadia J. editor. *Iron Nutrition in Soils and Plants*. The Netherlands: Kluwer Academic, pp. 43–50.

Stephan UW, Schmidke I, Stephan VW, Scholz G. 1996. The nicotianamine molecule is made-to-measure for complexation of metal micronutrients in plants. *BioMetals* **9**:84–90.

Tabata K, Kashiwagi S, Mori H, Ueguchi C, Mizuno T. 1997. Cloning of a cDNA encoding a putative metal-transporting P-type ATPase from *Arabidopsis thaliana*. *Biochim Biophys Acta* **1326**:1–6.

Takagi S, Nomoto K, Takemoto T. 1984. Physiological aspect of mugineic acid, a possible phytosiderophore of graminaceous plants. *J Plant Nutr* **7**:469–477.

Tandy S, Williams M, Leggett A, Lopez-Jimenez M, Dedes M, Ramesh B, Srai SK, Sharp P. 2000. Nramp2 expression is associated with pH-dependent iron uptake across the apical membrane of human intestinal Caco-2 cells. *J Biol Chem* **275**:1023–1029.

Thomine S, Wang R, Ward JM, Crawford NM, Schroeder JI. 2000. Cadmium and iron transport by members of a plant metal transporter family in *Arabidopsis* with homology to *Nramp* genes. *Proc Natl Acad Sci USA* **97**:4991–4996.

Thomine S, Lelievre F, Debarbieux E, Schroeder JI, Barbier-Brygoo H. 2003. AtNRAMP3, a multispecific vacuolar metal transporter involved in plant responses to iron deficiency. *Plant J* **34**:685–695.

Toyoshima C, Inesi G. 2004. Structural basis of ion pumping by Ca^{2+}-ATPase of the sarcoplasmic reticulum *Annu Rev Biochem* **73**:269–292.

Toyoshima C, Nakasako M, Nomura H, Ogawa H. 2000. Crystal structure of the calcium pump of sarcoplasmic reticulum at 2.6 A resolution. *Nature* **405**:647–655.

Tsivkovskii R, MacArthur BC, Lutsenko S. 2001. The Lys1010-Lys1325 fragment of the Wilson's disease protein binds nucleotides and interacts with the N-terminal domain of this protein in a copper-dependent manner. *J Biol Chem* **276**:2234–2242.

Tucker EB, Tucker JE. 1993. Cell-to-cell diffusion selectivity in staminal hairs of *Setcreasea purpurea*. *Protoplasma* **174**:36–44.

Turgeon R. 2006. Phloem loading: how leaves gain their independence. *Bioscience* **56**:15–24.

van der Zaal BJ, Neuteboom LW, Pinas JE, Chardonnens AN, Schat H, Verkleij JA, Hooykaas PJ. 1999. Overexpression of a novel Arabidopsis gene related to putative zinc-transporter genes from animals can lead to enhanced zinc resistance and accumulation. *Plant Physiol* **119**:1047–1055.

Verret F, Gravot A, Auroy P, Leohhardt N, David P, Nussaume L, Vavasseur A, Richaud P. 2004. Overexpression of AtHMA4 enhances root-to-shoot translocation of zinc and cadmium and plant metal tolerance. *FEBS Lett* **576**:306–312.

Verret F, Gravot A, Auroy P, Preveral S, Forestier C, Vavasseur A, Richaud P. 2005. Heavy metal transport by AtHMA4 involves the N-terminal degenerated metal binding domain and the C-terminal His11 stretch. *FEBS Lett* **579**:1515–1522.

Vert G, Briat JF, Curie C. 2001. *Arabidopsis* IRT2 gene encodes a root-periphery iron transporter. *Plant J* **26**:181–189.

Vidal SM, Malo D, Vogan K, Skamene E, Gros P. 1993. Natural resistance to infection with intracellular parasites: isolation of a candidate for Bcg. *Cell* **73**:469–485.

van der Zaal BJ, Neuteboom LW, Pinas JE, Chardonnens AN, Schat H, Verkleij JAC, Hooykaas PJJ. 1999. Over-expression of a novel *Arabidopsis* gene related to putative zinc-transporter genes from animals can lead to enhanced zinc resistance and accumulation. *Plant Phys* **119**:1047–1055.

von Heijne G. 1992. Membrane protein structure prediction: Hydrophobicity analysis and the 'positive inside' rule. *J Mol Biol* **225**:487–494.

von Wiren N, Klair S, Bansal S, Briat JF, Khodr H, Shioiri T, Leigh RA, Hider RC. 1999. Nicotianamine chelates both FeIII and FeII. Implications for metal transport in plants. *Plant Physiol* **119**:1107–1114.

Waigmann E, Zambryski P. 2000. Trichome plasmodesmata: a model system for cell-to-cell movement. *Adv Bot Res* **31**:261–283.

Wang Y, Liu C, Li K, Sun F, Hu H, Li X, Zhao Y, Han C, Zhang W, Duan Y, Liu M, Li X. 2007. *Arabidopsis* EIN2 modulates stress response through abscisic acid response pathway. *Plant Mol Biol* **64**:633–644.

Waters BM, Chu H-H, DiDonato RJ, Roberts LA, Eisley RB, Lahner B, Salt DE, Walker EL. 2006. Mutations in *Arabidopsis Yellow Stripe-Like1* and *Yellow Stripe-Like3* reveal their roles in metal ion homeostasis and loading of metal ions in seeds. *Plant Phys* **141**:1446–1458.

Williams LE, Mills RF. 2005. P_{1B}-ATPases – An ancient family of transition metal pumps with diverse functions in plants. *Trends Plant Sci* **10**:491–502.

Williams LE, Pittman JK, Hall JL. 2000. Emerging mechanisms for heavy metal transport in plants. *Biochim Biophys Acta* **1465**:104–126.

Woeste KE, Kieber JJ. 2000. A strong loss-of-function mutation in RAN1 results in constitutive activation of the ethylene responsive pathway as well as a rosette-lethal phenotype. *Plant Cell* **12**:443–455.

Wojas S, Ruszczynska A, Bulska E, Wojciechowski M, Antosiewicz DM. 2007. Ca(2+)-dependent plant response to Pb(2+) is regulated by LCT1. *Environ Pollut* **147**:584–592.

Yen MR, Tseng YH, Saier MH, Jr. 2001. Maize Yellow Stripe1, an iron-phytosiderophore uptake transporter, is a member of the oligopeptide transporter (OPT) family. *Microbiology* **147**:2881–2883.

Zhao H, Eide D. 1996. The ZRT2 gene encodes the low affinity zinc transporter in *Saccharomyces cerevisiae*. *J Biol Chem* **271**:23203–23210.

Zhou X, Yang Y. 2004. Differential expression of rice *Nramp* genes in response to pathogen infection, defense signal molecules and metal ions. *Physiol Mol Plant Pathol* **65**:235–243.

18 Cadmium Detoxification in Plants: Involvement of ABC Transporters

SONIA PLAZA

Plant Biology, University of Fribourg, 1700 Fribourg, Switzerland

LUCIEN BOVET

Philip Morris International R&D, Philip Morris Products SA, 2000 Neuchâtel, Switzerland

1 CADMIUM IN PLANTS

1.1 Cadmium Effects in Plants

Cadmium is a strongly phytotoxic heavy metal, which is widely recognized to be an important environmental pollutant [McLaughlin and Singh, 1999]. Cd is a nonessential element, except for a marine diatom *Thalassiosira weissflogii* in which active carbonic anhydrase may contain Cd as cofactor instead of zinc [Lane et al., 2005]. In plants and at elevated concentrations, Cd inhibits growth and can even lead to death [di Toppi and Gabbrielli, 1999; Deckert, 2005]. The main symptoms are stunting and chlorosis. Molecular and biochemical studies have suggested that Cd^{2+} causes the perturbation of intracellular calcium level and interferes with Ca^{2+} signaling by substituting it in calmodulin [Ghelis et al., 2000; Perfus-Barbeoch et al., 2002]. The mitogen-activated protein kinase (MAPK) pathway has been shown to be activated by Cd^{2+} in alfalfa [Jonak et al., 2004] and in rice [Yeh et al., 2004]. In plants, high level of Cd^{2+} ($100-500\ \mu M$) decreased the activity of RNase, resulting in an increase of RNA content [Shah and Dubey, 1995] and induced DNA damage in tobacco [Fojtová and Kovarik, 2000; Gichner et al., 2004], broad bean [Koppen and Verschaeve, 1996], and soybean [Sobkowiak and Deckert, 2004]. Proteins are also subject to Cd-induced damages, mainly resulting from high Cd affinity to cysteine, glutamate, aspartate, and histidine and competition

Trace Elements as Contaminants and Nutrients: Consequences in Ecosystems and Human Health, Edited by M. N. V. Prasad
Copyright © 2008 John Wiley & Sons, Inc.

with zinc for a variety of binding sites in cells including those which are involved in gene regulation or enzyme activity [Waalkes, 2000]. Therefore, cellular toxicity can result from various effects of Cd-linked damages in the photosynthetic apparatus [Siedlecka and Krupa, 1996], the disturbance of carbohydrate metabolism [di Toppi and Gabbrielli, 1999], changes in nitrate absorption and reduction [Hernandez et al., 1996], interferences with water balance [Perfus-Barbeoch et al., 2002], and inhibition of several enzyme activities [di Toppi and Gabbrielli, 1999]. In addition, reactive oxygen species (ROS) may accumulate in cells in response to high Cd exposure [Sandalio et al., 2001; Romero-Puertas et al., 2004]. However, in contrast to other heavy metals (like Cu and Fe), Cd is apparently not able to induce alone the production of ROS through a Fenton-like reaction since it is not an electron acceptor or donor.

1.2   Genes Regulated by Cd Stress

Cd is thought to enter the plant mainly via divalent cation transporters exhibiting broad substrate specificity like members of the ZIP family (zinc-regulated transporter (ZRT)-iron-regulated transporter (IRT)-like protein), notably IRT1 [Plaza et al., 2007]. In addition, Cd^{2+} uptake through plant plasma membranes may also be performed by Nramp (natural resistance-associated macrophage protein), LCT1 (low-affinity cation transporter) in monocots, and Ca^{2+} channels [Korshunova et al., 1999; Pence et al., 2000; Thomine et al., 2000; Williams et al., 2000; Clemens et al., 2002; Perfus-Barbeoch et al., 2002; Hall and Williams, 2003; Lindberg et al., 2004]. In the cytosol, at equilibrium, most of the Cd^{2+} is bound to chelators, such as phytochelatins (PCs), metallothioneins (MTs), chaperones, organic acids, and amino acids [Krämer et al., 1996; Clemens, 2001; Clemens et al., 2002; Hall, 2002]. It is subsequently sequestrated in vacuoles and the cell wall [Carrier et al., 2003; Van Belleghem et al., 2007]. Cd transport through cell membranes involved P-type heavy-metal ATPases (HMA), cation diffusion facilitator (CDF), Ca^{2+}/H^+ antiport (CAX), and ABC-binding cassette (ABC) proteins [Rauser 1995; Clemens, 2001; Persans et al., 2001; Bovet et al., 2003; Mills et al., 2003; Bernard et al., 2004; Gravot et al., 2004; Hussain et al., 2004; Koren'hov et al., 2007]. As epidermal structure, leaf glandular trichomes of tobacco [Choi et al., 2001] and leaf trichomes of Arabidopsis [Ager et al., 2003] accumulate Cd^{2+}, when plants are exposed to this heavy metal. In *Arabidopsis*, such an accumulation also occurs at very low Cd concentrations—for instance, following labeling with $^{109}CdCl_2$ (Bovet, unpublished data).

However, in most plants, Cd^{2+} tolerance is likely dependent on sulfur metabolism, antioxidative response, and the Cd transport into internal stores or outside the cell. The expression of "sulfur" enzymes varies upon Cd^{2+} exposure, notably ATP sulfurylase (APS) and adenosine 5′-phosphosulphate reductase (APR), which may help to maintain the S^{2-} supply for cysteine synthesis [Heiss et al., 1999; Herbette et al., 2006]. Furthermore, in some plants treated with Cd^{2+}, up-regulation of the expression of Ser acetyltransferase (SAT) and *O*-acetyl-ser (thiol)-lyase (OASTL) was also observed [Dominguez-Solis et al., 2001, 2004; Howarth et al., 2003], attesting the importance of producing thiol compounds to

maintain Cd tolerance. Reduced glutathione (GSH), which is an important antioxidant, is also the precursor of PCs, which is known to be involved in Cd detoxification.
Therefore, overexpression of γ-glutamylcysteine synthetase (γ-ECS, GSHI) and
glutathione synthetase (GS, GSHII), which catalyze GSH synthesis from Cys, can
improve Cd tolerance in plants [Cobbett et al., 1998; Zhu et al., 1999]. Finally,
several reports attest to the main role of phytochelatins in Cd tolerance via Cd
complex formation and transport in some plants [Clemens et al., 1999; Gong et al.,
2003; Heiss et al., 2003; Li et al., 2004; Pomponi et al., 2006; Roth et al., 2006;
Lindberg et al., 2007; Gasic and Korban, 2007]. In this contribution, we will
discuss the involvement of ABC transporters in Cd transport and tolerance.

2 ABC TRANSPORTERS

2.1 Functions of ABC transporters in plants

The ATP-binding cassette (ABC) proteins form one of the largest known protein
family with more than 120 open reading frames (ORFs) in *Arabidopsis thaliana*
and *Oryza sativa* [Sanchez-Fernandez et al., 2001, Jasinski et al., 2003; Bovet
et al., 2005]. Most of the ABC proteins, but not all, are transmembrane proteins
forming pores allowing active transport of molecules from one compartment
to another.

ABC transporters are known to be implicated in many different processes in plants
like auxin polar transport [Geisler et al., 2005], alkaloid import [Yasaki et al., 2001;
Shitan et al., 2003], tissue pigmentation [Goodman et al., 2004], vacuolar xenobiotic
sequestration [Martinoia et al., 2000, 2003; Glombitza et al., 2004], stomatal regulation [Klein et al., 2003, 2004], disease resistance [Stukkens et al., 2005], lipid
catabolism [Footitt et al., 2002; Pighin et al., 2004], cuticular wax deposition [Pighin
et al., 2004], antibiotic resistance [Mentewab and Stewart, 2005; Rea, 2007], assembly
of redox-active cytosolic Fe/S proteins [Kushnir et al., 2001; Hanikenne et al., 2005;
Kim et al., 2006] and heavy-metal transport (see below for references).

2.2 Characteristics of ATP-Binding Cassette Transporters

There are three basic characteristics of the ABC protein-mediated transport [Rea
et al., 1998]. First, the transport is directly energized by MgATP, but not by free
ATP or nonhydrolyzable ATP analogues. Depending on the transporter, UTP or
GTP can partially substitute for ATP. Second, the transport is insensitive to the transmembrane H^+ electrochemical potential difference. Dissipation of pH gradient and
electrical potential, which are established across membrane by primary H^+ pumps,
does not usually inhibit transport. Thirdly, the transport is sensitive to vanadate, a
metastable analogue of orthophosphate. It arrests the catalytic cycle by substituting
for the released phosphate and trapping ADP (hydrolysis product of ATP) in the
nucleotide-binding site.

ABC proteins and (in particular) ABC transporters (full and half size) are
composed of two core structural domains: transmembrane domains (TMDs),

containing multiple membrane-spanning α-helices (usually 4–6) and nucleotide-binding folds (NBFs also named nucleotide-binding domains, NBDs; Higgins [1992]). The TMDs are hydrophobic, allowing the movement of solutes across the membrane lipid bilayer by forming a pore and determining the specificity or selectivity of the transporter [Martinoia et al., 2000]. Classification is designated on the basis of NBFs, which share 30–40% sequence identity between family members [Higgins, 1992]. Each NBF encompasses around 200 amino acid residues and contains three defined sequence motives. A Walker A box ($GX_4GK[ST]$) and a Walker B box [(hydrophobic)$_4$[DE]] flanked by approximately 120 amino acid residues containing the ABC signature (alias C) motif ([LIVMFY]S[GX$_3$[RKA][LIVMYA]X[LIVDM][AG]), commonly cited as LSSG, is responsible for the ATP hydrolysis and ADP release [Walker et al., 1982; Altenberg, 2003]. Most functional ABC transporters (one full-size polypeptide or two half-size ABC proteins) consist of a combination of two NBFs and two TMDs assembled as an internal twofold or pseudo-twofold geometry. The core domain of ABC transporters may be expressed either as two separate polypeptides or as a multidomain protein. For instance, in *Schizosaccharomyces pombe* one TMD is fused with one NBF to form a "half molecule" called the heavy metal tolerance factor 1 (HMT1).

2.3   Subfamilies of ATP-Binding Cassette Proteins

Plant ABC proteins are differentiated into 13 subfamilies on the basis of the protein size (full, half, or soluble molecules), orientation (forward (TMD1-NBF1-TMD2-NBF2) or reverse (NBF1-TMD1-NBF2-TMD2)), the presence or absence of idiotypic transmembrane and/or linker domains, and overall sequence similarity [Sanchez-Fernandez et al., 2001]. The full molecules include multidrug resistance homologues (MDRs, equivalent to PGPs, P-glycoproteins), multidrug resistance-associated protein homologues (MRPs), pleiotropic drug resistance homologues (PDRs), peroxisomal membrane protein homologues (PMPs), and ABC1 homologues (AOH). The half-molecule transporters include the white-brown complex homologues (WBCs), ABCs homologues (ATHMs), ABC transporters of the mitochondrion homologues (ATMs), and transporter associated with antigenic processing homologues (TAPs). Finally, soluble proteins include the 2′,5′-oligoadenylate-activated RNase inhibitor homologues (RLIs), general control nonrepressible homologues, like the yeast GCN20 (GCNs), and structural maintenance of chromosome homologues (SMCs). The only category of plant ABC proteins that cannot be properly classified is the subfamily of nonintrinsic ABC proteins (NAPs). For a complete survey of the function and localization of major families please see Rea [2007]. In the following part, we will focus on the ABC protein members that are able to transport heavy metals and particularly Cd in plants.

2.4   Involvement of ABC Transporters in Cadmium Detoxification in Plants

The inventory of genes, resulting from the complete sequencing of the entire genome of the plant model *A. thaliana* [AGI, 2000; Blanc et al., 2000], helps to understand the relation between protein structure and function. Based on that, new methods of investigation were developed and used to elucidate the role of transporters in heavy-metal transport. Data came first from transcriptomic studies using cDNA-AFLP [Suzuki et al., 2001] and then microarray analyses [Becher et al., 2004; Weber et al., 2004; Bovet et al., 2005; Herbette et al., 2006]. However, cloning and functional analyses of ABC members implicated in Cd transport is still scarce. The functional link between ABC proteins and heavy metal(s) came first from a human MRP which was found to transport glutathione (GSH) S-conjugates through membranes [Ishikawa, 1992]. Such GS-X pumps were shown later to be involved in detoxification of Cd and other heavy metals. Indeed, the first full-length ABC transporter YCF1 (yeast cadmium factor 1) implicated in the detoxification of Cd by complex formation with GSH was found in yeast by Li et al. [1996]. Interestingly, Tommasini et al. [1996] found that the human MRP1 was able to complement YCF1 in yeast. Thus it was rapidly postulated that some plant MRPs (YCF1 belongs to the MRP subfamily) might also be key players in Cd detoxification [Rea, 1999]. But the first attempts were not successful, for instance Sanchez-Fernandez et al. [1998] tried to correlate the expression of *AtMRP4* with the presence of Cd in cell suspension of *Arabidopsis thaliana* without positive results. Historically, another half-size ABC transporter HMT1 (heavy-metal transporter 1) from *Schizosaccharomyces pombe* was shown to be able to transport phytochelatin Cd complexes into vacuoles [Ortiz et al., 1995]. Based on sequence comparison, no *Sp*HMT1 homologues were found in plants so far, thereby suggesting that another(other) ABC protein(s) or transporter(s) is(are) responsible for the transport of PC-Cd in plants. The sole homology found is with NBF domains of *At*ATMs, but not with the hydrophobic domain forming the pore. An HMT1 homologue (*Ce*HMT-1) was found in *Caenorhabtitis elegans* [Vatamaniuk et al., 2005], confirming that PCs transporters play a major role in Cd detoxification and tolerance.

The first study using a microarray approach to identify putative candidates of the MRP subfamily that might be involved in Cd detoxification was performed by Bovet et al. [2003]. The same group, using a cDNA sub-microarray, extended their search to all the genes coding for putative ABC proteins to find candidates responding to Cd [Bovet et al., 2005]. They identified a total of 17 ABC genes, which were differentially regulated in *Arabidopsis thaliana* (as a Cd-nonadapted plant) after Cd treatment in liquid culture. The results showed that the 17 respective and putative ABC proteins were equally distributed between full-size (6), half-size (6), and soluble (5). ABC members belonging to the following subfamilies were up-regulated by Cd: MRP, PDR, PGP (MDR), ATH, ATM, WBC, GCN, and NAP, but most of them have still unknown functions in plants. More recently, early Cd responsive genes of *A. thaliana* were identified by using a genome-wide transcriptome profiling [Herbette et al., 2006]. In this study, five members of the MRPs subfamily (*AtMRP2*, *AtMRP4*, *AtMRP5*, *AtMRP6*, and *AtMRP12*) were found to be up-regulated in roots treated with 50 μM Cd for 30 h.

Figure 1 shows the possible involvement of ABC transporters in cell detoxification in plants. Removal of Cd from the cytosol requires either transport in the vacuolar compartment or export outside the cell for subsequent Cd immobilization. Therefore four possible transport ways have to be taken into consideration: (1) transport of

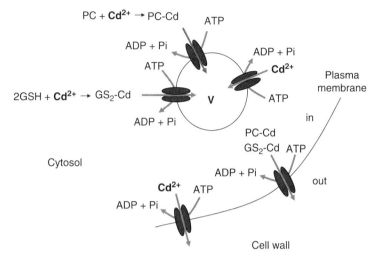

Figure 1. Involvement of ABC transporters in Cd detoxification of plant cell. Elimination of Cd triggered by ABC transporters likely occurs by Cd sequestration in the tonoplast compartment or by Cd^{2+} bonding within the cell wall. Both the import of Cd^{2+} and/or Cd-complexes in the vacuole (V) and the export of Cd^{2+} and/or Cd-complexes in the apoplastic compartment may exist and might be driven by ABC transporters.

Cd-complexes (likely with GSH or PC) into the vacuolar compartment, (2) transport of Cd^{2+} into the vacuole "as free ion," (3) transport of Cd-complexes in the apoplast, and (4) transport of Cd^{2+} "as a free ion" in the apoplast.

2.4.1 The MRP Subfamily. This subfamily of ABC proteins has been shown to be involved in the transport of a broad range of substrates, comprising GS-conjugates, endogenous or exogenous toxins, linearized tetrapyroles, sterol glucuronides, folates, and heavy metals (see Rea [2007] and references herein). They have also been suggested to be able to modulate the activity of other channels [Klein et al., 2003].

cDNA spotted arrays and RT-PCR approaches showed that Cd treatments did not modify AtMRP transcript levels in leaves of 4-week-old *A. thaliana*. In roots, induction of *AtMRP3* and *AtMRP6*, and *AtMRP7* and *AtMRP14* to a lower extent, were found; the highest induction was observed for *AtMRP3* (about 2.5-fold compared to control conditions; Bovet et al. [2003]). Therefore from the 15 members of the MRP subfamily in *A. thaliana*, AtMRP3 appeared to be the best candidate implicated in Cd transport in *Arabidopsis* [Bovet et al., 2003]. The fact that this induction was only observed in roots suggests that AtMRP3 is involved in the sequestration of Cd in roots but not in the shoot. To test whether Cd might enhance the expression of *AtMRP3* in leaves, roots were removed from 4-week-old plants and leaves directly treated with Cd. Under these conditions, an increase of *AtMRP3* transcript levels was observed. In contrast, in young plantlets, *AtMRP3* transcript levels increased in both the root and shoot in presence of Cd and a far higher portion of Cd^{2+} was translocated to the aerial part compared to adult plants [Bovet et al., 2005]. The stronger root retention of Cd in adult plants compared to plantlets, as judged by experiments using ^{109}Cd labeling, suggests that Cd translocation is more active in young plants and that AtMRP3 might be involved directly or indirectly in such a transport. Interestingly, in young plantlets, *AtMRP3* transcript levels were not strongly affected by oxidative stress (H_2O_2 and menadione), thereby indicating that the induction of *AtMRP3* in these plants is mainly due to the presence of Cd^{2+} or Cd-complexes rather than to oxidative stress possibly generated by Cd excess. In addition, Cd induction of *AtMRP3* was not strongly affected in mutants exhibiting reduced level of glutathione or phytochelatin (*cad2* and *cad1-3*; Howden et al., 1995a, 1995b and Bovet et al., 2003) and following a BSO treatment (buthionine sulfoximine, Fig. 2a). BSO is known to inhibit GSHI activity [Griffith, 1982]. This suggests that Cd itself is responsible for *AtMRP3* induction. In comparison, BSO and Cd have no effect on the transcript levels of *AtMRP1* and 2, as well as on phytochelatin synthase I (PCSI) and GSHI. This shows that the maintenance of the GSH pool in *Arabidopsis*, which is crucial for plant viability (Fig. 2b), does not affect the *AtMRP3* Cd up-regulation process, unlike for *AtATM3* (see Kim et al. [2006] and the end of the present review), to possibly support vacuolar transport of Cd. AtMRP3 is likely to be localized in vacuolar membranes, according to recent proteomic data [Jacquinot et al., 2007].

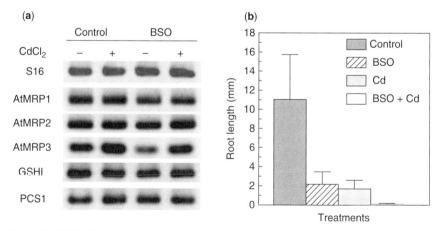

Figure 2. BSO effect on the expression of *AtMRP3*. Arabidopsis seedlings were grown for one week on 1/2 MS bactoagar plates in the presence or absence of 10 μM $CdCl_2$ and/or 0.5 mM BSO. (**a**) RNA was isolated from seedlings and subjected to two-step semiquantitative RT-PCR analyses using specific primers for *S16* (housekeeping gene), *AtMRP1*, *AtMRP2*, *AtMRP3*, *GSHI*, and *PCS1*. The data showed that *AtMRP3*, compared to the five other genes, is induced by Cd in the absence or presence of BSO. It can be noted that in the presence of BSO the level of *AtMRP3* transcripts is globally reduced. (**b**) Under the same experimental conditions, Cd and BSO strongly affected root elongation. The combination of Cd and BSO blocked almost completely the growth of the roots. Root elongation was measured between two time points during root growth.

Nevertheless, the exact function of AtMRP3 in Cd transport remains unclear. However, as AtMRP3 (i) is able to partially complement the Cd sensitivity of the ΔYCF1 yeast mutant [Tommasini et al., 1998; T. Eggmann, unpublished data], (ii) is localized in the tonoplast, and (iii) is up-regulated by Cd, AtMRP3 is likely a good candidate to be a Cd detoxifier in *Arabidopsis*. Furthermore, *AtMRP3* has also been described to be up-regulated by xenobiotics, like sulfonylurea herbicides [Tommasini et al., 1996; Glombitza et al., 2004], thereby suggesting low substrate specificity not restricted to Cd.

The role of MRPs in Zn/Cd hyperaccumulator plants that are able to store high concentration of Zn and Cd in leaves has been recently investigated. Indeed, in the Zn/Cd hyperaccumulator *Thlaspi caerulescens*, *TcMRP10*, highly homologous to *AtMRP4* and *AtMRP10*, was up-regulated by high Zn concentrations in the roots and shoots. Expression of this gene in yeast was able to complement Cd and Cu to some degrees. Because the yeast Cd content did not change when *TcMRP10* was expressed, it was suggested that TcMRP10 was not able to sequester Cd in the vacuole [Hassinen et al., 2007].

To study the involvement of MRPs in plant Cd detoxification, not only plants MRPs have been studied, but also MRPs from yeast (YCF1), human (MRP1), and *Chlamydomonas reinhardtii*, mostly via heterologous expression.

YCF1, which has been shown to pump Cd^{2+} conjugated to glutathione into yeast vacuoles, was expressed in transgenic *Arabidopsis thaliana* in order to investigate the accumulation of Cd and lead [Song et al., 2003]. The plant expressing YCF1 gene under the control of four copies of the cauliflower mosaic virus (CaMV) 35S were grown onto heavy metal-containing media for 3 weeks. An enhanced resistance to Cd was observed in transgenic lines, which also contained a higher amount of Cd. In order to determine YCF1-specific transport of Cd, intact vacuole transport activity of labelled glutathione in the absence and presence of Cd was performed. The transport of glutathione was about four times higher in *YCF1*-expressing plants than in the wild-type vacuoles, indicating that the expression of *YCF1* strongly increases the transport of glutathione-conjugated Cd^{2+} in the vacuoles. In the absence of Cd, there was very little transport of glutathione into isolated vacuoles. These data suggest that YCF1 sequesters Cd^{2+} into the vacuoles, thereby contributing to heavy metal resistance in the transgenic *Arabidopsis*. The slightly greater concentration of heavy metal in the shoot of the transgenic plants might be interpreted as a more active translocation from root to shoot followed by an increase in shoot sink of cadmium through YCF1-mediated vacuolar sequestration. Such a Cd storage should allow transgenic plants to grow better and produce higher biomass than control plants. Tobacco plants were also engineered with the same YCF1 construct designed by Song et al. [2003]; see above. The authors particularly thank Won Young Song for providing them with the construct. Seeds were collected from more than 30 independent lines. After antibiotic selection (2 weeks), 10 positive lines were cultivated for 10 weeks in agricultural soil containing Cd (0.75 ppm total soil Cd). Bottom leaves were then collected and subjected to Cd and Zn determination. When compared to wild-type, YCF1 transgenic lines showed significantly higher concentrations of Cd and Zn in the leaves, correlated with an increase in the fresh biomass of the leaves (Figs. 3a, 3b, and 3c, respectively), thus making a total Zn and Cd accumulation from 1.2 to 2.2 times more than the control, depending on the lines. This confirms the data published by the group of Prof. Dr. Youngsook Lee in *Arabidopsis* [Song et al., 2003]. In conclusion *ScYCF1* is a promising candidate to engineer plants exhibiting high biomass, like tobacco, for phytoremediation purposes.

Another non-plant MRP has been expressed in plant and its application for phytoremediation investigated. Yazaki et al. [2006] chose to express the human ABC transporter hMRP1, which is involved in the multidrug resistance of cancer cells, in tobacco plant [Muller et al., 1994; Jedlitschky et al., 1996]. When expressed in tobacco, this protein was localized in the vacuolar membrane. Transgenic tobacco had an increased tolerance to Cd as observed by higher fresh cell mass and root length of seedlings compared to wild-type. This tolerance was also true for daunorubicin. A time-course experiment on Cd uptake by cultured cells was performed over a period of 15 days. A decrease of Cd in the media was observed for the wild-type and the hMRP1-expressing tobacco cells, but this decrease was faster in the transgenics than in the wild-type, particularly at the beginning of the time course. In order to determine in which part (root, leaf) Cd accumulated in these plants, Cd

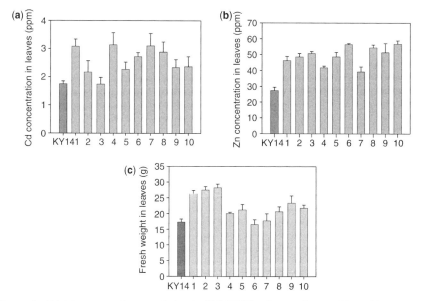

Figure 3. Cd tolerance and accumulation in 35S-YCF1 tobacco plants cultivated in a greenhouse. 35S-YCF1 tobacco plants (background KY14) were grown in a pot assay on cadmium contaminated soil for 10 weeks. Amount of Cd (**a**) and Zn (**b**) in 10 different lines were determined in the four bottom leaves of each plant by ICP-MS. (**c**) Fresh weight (in gram) of the four bottom leaves used for ICP-MS analyses was determined just after harvest. An increase in fresh weight of transformed tobacco was observed in the majority of the lines. The box represents the mean ± standard error (for each line, 6 plants).

concentration was measured by atomic absorption spectroscopy analyses in the aerial and underground parts. No differences were detected in root or leaf tissues between the wild-type and the hMRP1-expressing tobacco plants, suggesting that the absorption of Cd was not influenced by this gene. However, Western blot analyses showed that the expression of hMRP1 in roots was considerably lower than in leaves. Therefore, these data altogether suggested that hMRP1, when expressed in tobacco, contributed to a more active sequestration of Cd in the vacuoles compared to the control plants. This may reduce the rapid cell damage due to the presence of elevated concentrations of Cd^{2+} in the cytosol, thereby increasing Cd tolerance.

In *Chlamydomonas reinhardtii*, CrMRP2 shares a high similarity with HsMRP1, AtMRP3, and ScYCF1 (all involved in Cd tolerance). Interestingly, CrMRP2 was able to restore the function of ScYCF1 in yeast mutant lacking this gene. The data also showed that *CrMRP2* was implicated in the formation/accumulation of a stable high-molecular-weight (HMW) PC–Cd complex [Wang and Wu, 2006], making it a candidate for Cd transport in this unicellular green alga.

Several *MRPs* are suitable candidates to engineer plants for phytoremediation or food safety programs. However, the use of specific promoters to drive heavy-metal contaminants to the right cells (instead of using a blind approach with 35S or nonspecific promoters) and the use of appropriated transporters (functioning at the right place) or silencing have to be considered in future studies.

2.4.2 The PDR Subfamily. The PDR family is usually associated with general plant defense response [Jasinski et al., 2001; Van den Brule and Smart, 2002], but some members seem also to be involved in heavy metal transport, such as AtPDR8 and OsPDR9 for Cd, and AtPDR8 and AtPDR12 for Pb [Moons, 2003; Lee et al., 2005; Kim et al., 2007]. In contrast to YCF1 conferring resistance to cadmium and lead [Song et al., 2003], AtPDR12 only provides apparent tolerance to Pb and not to Cd in *A. thaliana* [Lee et al., 2005].

AtPDR8 has a dual function in *A. thaliana*. It was first shown to contribute to pathogen resistance by possibly facilitating export to plasma membrane of substances related to defense against pathogens [Kobae et al., 2006; Stein et al., 2006], but also more recently to confer resistance to Cd [Kim et al., 2007]. Therefore AtPDR8 is a perfect example of a multifunctional ABC transporter [Martinoia et al., 2002]. The expression of this gene was up-regulated in the root and the shoot of Cd- or Pb-treated *Arabidopsis* plants. By a fusion between its promoter and the GUS reporter gene, GUS expression was found in shoots, stems, and roots as well as in trichomes, mesophyll, and leaf guard cells of 2-week-old *Arabidopsis* plants. In roots, GUS activity was very high at the tip, absent in the elongation zone, and again high in root hairs. AtPDR8 was localized mainly in the plasma membrane of protoplasts. To test if AtPDR8 might play a role in heavy-metal detoxification, RNAi, knocked-out, and overexpressing plants were tested for heavy-metal (Cd, Cu, and Pb) sensitivity, content, and transport. Interestingly, AtPDR8 RNAi and KO plants were more sensitive to Cd (similar behaviors between them) than the wild-type *Arabidopsis*. On the other hand, overexpressing plants were more resistant than wild-type when shoot FW and chlorophyll content were compared. In addition,

RNAi plants exhibited higher Cd contents, while inversely overexpressing plants exhibited lower Cd contents, than did wild-type, thereby strongly suggesting that AtPDR8 contributes to Cd resistance by excluding Cd from the cell. To obtain the direct evidence that the mechanism leading to Cd resistance is driven by cell Cd extrusion via AtPDR8, flux assays using radioactive Cd^{2+} and protoplasts, isolated either from RNAi plants or overexpressing plants and wild-type, were performed. The data show that after 1 h of $^{109}Cd^{2+}$ loading, the radioactivity in the protoplasts isolated from AtPDR8 silenced plants was higher than in the wild-type protoplasts, whereas the protoplasts isolated from plants overexpressing AtPDR8 had a lower amount of radiolabelled Cd inside the cell. This indicated that AtPDR8 might pump Cd out across the plasma membrane, thereby reducing intracellular Cd levels. This difference in Cd efflux was confirmed by using different inhibitors of various types of transporters.

In rice, *PDR9* was also found to be induced by Cd as well as by zinc, polyethylene glycol, and hypoxic and salt stress in the roots of rice seedlings [Moons et al., 2003]. Moreover, the plant growth regulator jasmonic acid, the auxin analogue α-naphthalene acetic acid, and the cytokinin 6-benzylaminopurine also triggered *OsPDR9* expression. The expression of *OsPDR9* under Cd stress was root-specific, suggesting a possible function in the root-to-shoot metal transport.

2.4.3 The ATM Subfamily.

ATMs belong to the half-size ABC transporter subfamily. There are three ATM family members in *Arabidopsis thaliana*, namely ATM1, ATM2, and ATM3 [Sanchez-Fernandez et al., 2001]. The closest homologue of AtATM3 is a yeast (*Saccharomyces cerevisiae*) ABC transporter, Atm1p, which has been shown to be localized in the mitochondrial matrix [Leighton and Schatz, 1995]. Atm1p has been suggested to function as an exporter of iron–sulfur clusters (Kispal et al., 1997, 1999; Chloupkova et al., 2004). Interestingly, such a function also appeared to be driven by AtATM3, alias STA1. AtATM3 is a mitochondrial ABC transporter involved in the biogenesis of iron–sulfur clusters, the maturation of iron–sulfur cluster proteins, and iron homeostasis [Kushnir et al., 2001]. More recently, Kim et al. [2006] found that this gene was up-regulated in roots of Arabidopsis plants treated with Cd or Pb. When overexpressed in *A. thaliana*, transgenic plants had enhanced resistance to Cd, whereas *atatm3* mutants were more sensitive than wild-type. The overexpressing plants also had a higher Cd content in the shoot than the wild-type whereas no significant differences were observed in the roots.

To test if the increased sensitivity of the *atatm3* mutant to Cd may reflect a reduced glutathione level in the cytosol, levels of nonprotein thiols (NPSH) were measured. However, *atatm3* was surprisingly found to contain more NPSH than the wild-type, with NPSH often being correlated with Cd resistance [Zhu et al., 1999; Freeman et al., 2004]. In plants overexpressing AtATM3, no differences were found in comparison to the wild-type under control conditions. These results suggested that AtATM3 controlled intracellular NPSH (mostly GSH) levels and that either lack or excess of AtATM3 in the mitochondrial membrane causes abnormal compartmentalization of NPSH. Moreover, genetic or pharmacological inhibition of glutathione biosynthesis using BSO (L-buthionine-(S,R)-sulfoximine) led to an elevated expression of *AtATM3*, whereas the expression of the glutathione synthase gene (*GSH1*) was increased under Cd stress and in the *atatm3* mutants. So, in conclusion, AtATM3 possibly functions in the mitochondrial membranes as a transporter of GS-conjugated Cd(II) and may interfere with maturation of Fe–S cluster proteins. Hanikenne et al. [2005] have previously shown that a protein closely related to AtATM3 was also able to confer Cd tolerance in *Chlamydomonas reinhardtii*.

2.4.4   The GCN Subfamily. GCN20 from yeast has a marked homology with the five GCN putative protein sequences of *A. thaliana* (GCN20 > AtGCN3 > AtGCN4 > AtGCN1 > AtGCN5 > AtGCN2). The yeast protein GCN20 is a soluble ABC protein whose function is connected to the binding of tRNAs to ribosomes. In yeast, GCN20 is a positive regulator of GCN2, a protein kinase that regulates protein synthesis initiation in amino acid starved cells by phosphorylation of eukaryotic translation initiation factor 2 [Dong et al., 2004]. Therefore a function of translational controller is highly suspected for AtGCNs, but has not been demonstrated yet.

Based on experiments aimed at finding ABC transporters involved in Cd response, we run cDNA microarrays with TOM1 tomato chips, using mRNA collected from tobacco leaves treated with 50 μM CdCl$_2$ for 24 hr. The experimental procedure is detailed in Bovet et al. [2006]. The results showed that one tobacco sequence hybridized a TOM1 cDNA that is homologous to the ABC protein AtGCN5. Having not previously found a tobacco sequence closely related to *AtGCN5*, we run preliminary experiments with *Arabidopsis*, with T-DNA mutants being available in SALK lines. By RT-PCR, we first confirmed the data obtained in tobacco showing that *AtGCN5* is induced by Cd in both roots and leaves (Fig. 4a). *AtGCN5* was weakly expressed in roots compared to leaves, in accordance with Genevestigator expression data (gene Atlas; Zimmermann et al. [2004]). Apparently, *AtGCN5* was not found to markedly respond to abiotic stress, such as light, wounding, osmotic, oxidative, and drought stress (stress-specific expression profile in Genevestigator). Using Target-P prediction (www.expasy.ch), a possible location for AtGCN5 might be the chloroplast, which would make sense with its constitutive expression in green tissues. The gene (2737 bp) is based on chromosome 5 and consists of seven exons, having after slicing a putative cds of 2079 bp. BLAST analyses show that *AtGCN5* has strong homologies with *AtGCN2* and other plant *GCN* sequences, but also with Cyanobacteria like Synechococcus, Nostoc, Prochlorococcus (the chloroplast ancestor), and the blue green algae Anabaena, which often contain photosynthetic pigments to perform photosynthesis. This suggests that the location of AtGCN5 in the chloroplast is probably correct, but needs to be further demonstrated in plants. We have selected a KO mutant (homozygote line identified by PCR and kanamycin selection, data not shown) in SALK lines (SALK 048287). In this mutant, the T-DNA is inserted in the last exon (Fig. 4b). We analyzed the growth of the mutant on 1/2 MS bactoagar plates supplemented with 10 μM CdCl$_2$ or 200 μM ZnCl$_2$. Interestingly, when there is a lack of *AtGCN5* transcripts, we observed an increase tolerance to both heavy metals, Cd and Zn (Fig. 4c, d). Indeed the growth of roots

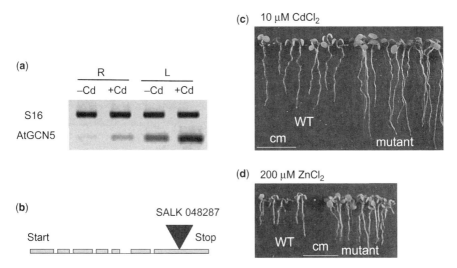

Figure 4. (**a**) Involvement of AtGCN5 in cadmium response. Three-week-old *Arabidopsis* (*A. thaliana*, Columbia-0) plants cultivated under hydroponic conditions were treated with 10 μM CdCl$_2$ for 24 h. RNA was isolated and subjected to RT-PCR analyses using specific primers for *S16* and *AtGCN5*. The data showed that *AtGCN5* is more expressed in leaves (L) than in the roots (R) and is induced by Cd in both leaves and roots. (**b**) Homozygote lines of a T-DNA inserted mutant (SALK 048287) were selected. AtGCN5 is located in chromosome 5 and has 7 exons. In this mutant, T-DNA is inserted in the last exon. The mutants were grown vertically on 1/2 MS bactoagar for 14 days in the presence of 10 μM CdCl$_2$ (**c**) or 200 μM ZnCl$_2$ (**d**). In both case, roots of homozygote lines (mutant) grew better than wild-type (WT), and cotyledons were larger in the mutant compared to WT.

and cotyledons were significantly less affected by these two heavy metals in the mutant, compared to WT (Fig. 4c, d; statistical analyses not shown). These data suggest that AtGCN5 is a possible regulator or switch that can be altered by Cd and Zn. However, as long as the function of this gene is unknown, the implication of AtGCN5 in the chloroplast translation apparatus with high sensitivity to metals or ion homeostasis remains pure speculation.

2.4.5 *Expression of AtMRP3, AtMRP4, AtPDR8 and AtATM3 Homologues in Tobacco after Cd Treatment.*

In Fig. 5, the expression of tobacco putative ABC members possibly involved in cadmium detoxification in *Arabidopsis* were analyzed. Tobacco plants were germinated in vertical plates (1/2 MS) for two weeks before being transferred in hydroponics (1/2 MS) for one month. Afterwards, different concentrations of Cd were added to the media (1 and 5 μM CdCl$_2$) for one week. Root and leaf samples were collected separately. RNA was extracted and subjected to RT-PCR using specific primers amplifying coding fragments of the tobacco ABC genes homologous to *AtMRP3* ("*NtMRPa*"), *AtMRP4* ("*NtMRPb*"), *AtPDR8* ("*NtPDRa*"), and *AtATM3* ("*NtATMa*"). Elongation factor A ("*NtEFa*") was used as the housekeeping gene. Most of the tobacco

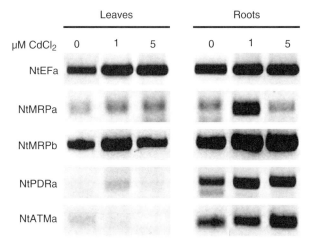

Figure 5. Effect of Cd on the expression of selected genes coding for ABC proteins in leaves and roots of tobacco plants. Plants were grown under hydroponics (1/2 MS) for 4 weeks before treatment with two different concentrations of cadmium (1 and 5 μM CdCl₂) for 1 week. RNAs were isolated and subjected to two-step semiquantitative RT-PCR analyses, using "NtEFa" as internal control. PCR fragments of tobacco genes homologous to AtMRP3 ("NtMRPa"), AtMRP4 ("NtMRPb"), AtPDR8 ("NtPDRa") and AtATM3 ("NtATMa") were separated in agarose gel. An induction of the four genes in their respective expression compartment was observed after Cd treatments, confirming the data obtained in *Arabidopsis thaliana*.

sequences were found in TGI (Tobacco Genome Initiative). All the genes showed different expression levels in roots and leaves. Under control condition, higher expression in roots was observed for the whole set of genes tested. When Cd was added to the culture medium, the expression of all the selected genes increased to a different extent. The up-regulation was more pronounced in the roots, confirming the possible involvement of these proteins in cadmium detoxification (see above). "*NtMRPb*" also appeared to be very sensitive to the presence of Cd in the leaves. The expression data in tobacco were similar to those observed in *A. thaliana*.

3   CONCLUSION

To date only a few plant ABC transporters have been characterized, but the function of more members is sure to be investigated in the near future, particularly in the field of the maintenance of metal homeostasis in plant tissues or the detoxification of heavy metals. The fact that some ABC proteins (i.e., AtPDR8) have been localized in the plasma membrane will considerably modify the basic concept of detoxification involving ABC transporters. Indeed, ABC proteins were exclusively viewed as vacuolar transporters importing toxic compounds (heavy metals, herbicides, etc.) or complex/conjugates into the vacuoles. At this point, efflux transport driven by ABC proteins has also to be considered.

ACKNOWLEDGMENTS

The authors greatly thank the following persons for making this research feasible: Paolo Donini, Ferruccio Gadani, and Irfan Gunduz; Nikolai Ivanov, Nicolas Lugon-Moulin, Florian Martin, and Luca Rossi at Philip Morris International R&D, Philip Morris Products SA, Neuchâtel (Switzerland); Prof. Enrico Martinoia at the University of Zürich (Switzerland); Prof. Jean-Pierre Métraux, Amélie Fragnière, Adrien Pairraud, and Régis Mark at the University of Fribourg (Switzerland), Prof. Youngsook Lee and Do-Young Kim at the Postech Institute in Pohang (South Korea).

REFERENCES

Ager FJ, Ynsa MD, Dominguez-Solis JR, López-Martin MC, Gotor C, and Romero LC. 2003. Nuclear micro-probe analysis of *Arabidopsis thaliana* leaves. *Nucl Instrum Methods Phy Res Sect B: Beam Interact Mater Atoms* **210**:401–406.

AGI, Arabidopsis Genome Initiative. 2000. Analysis of the genome sequence of the flowering plant *Arabidopsis thaliana*. *Nature* **408**:796–815.

Altenberg GA. 2003. The engine of ABC proteins. *News Physiol Sci* **18**:191–195.

Becher M, Talke IN, Krall L, Kramer U. 2004. Cross-species microarray transcript profiling reveals high constitutive expression of metal homeostasis genes in shoots of the zinc hyper-accumulator *Arabidopsis halleri*. *Plant J* **37**:251–268.

Bernard C, Roosens N, Czernic P, Lebrun M, Verbruggen N. 2004. A novel CPx-ATPase from the cadmium hyperaccumulator *Thlaspi caerulescens*. *FEBS Lett* **569**:140–148.

Blanc G, Barakat A, Guyot R, Cooke R, Delseny M. 2000. Extensive duplication and reshuffling in the *Arabidopsis* genome. *Plant Cell* **12**:1093–1102.

Bovet L, Eggmann T, Meylan-Bettex M, Polier J, Kammer P, Marin E, Feller U, Martinoia E. 2003. Transcript levels of AtMRPs after cadmium treatment: Induction of AtMRP3. *Plant Cell Environ* **26**:371–381.

Bovet L, Feller U, Martinoia E. 2005. Possible involvement of plant ABC transporters in cadmium detoxification: A cDNA sub-microarray approach. *Environ Inter* **31**:263–267.

Bovet L, Rossi L, Lugon-Moulin N. 2006. Cadmium partitioning and gene expression studies in *Nicotiana tabacum* and *Nicotiana rustica*. *Physiol Plant* **128**:466–475.

Carrier P, Baryla A, Havaux M. 2003. Cadmium distribution and microlocalization in oilseed rape (*Brassica napus*) after long-term growth on cadmium-contaminated soil. *Planta* **216**:939–950.

Choi YE, Harada E, Wada M, Tsuboi H, Morita Y, Kusano T, Sano H. 2001. Detoxification of cadmium in tobacco plants: Formation and active excretion of crystals containing cadmium and calcium through trichomes. *Planta* **213**:45–50.

Chloupkova M, Reaves SK, LeBard LM, Koeller DM. 2004. The mitochondrial ABC transporter Atm1p functions as a homodimer. *FEBS Lett* **569**:65–69.

Clemens S. 2001. Molecular mechanisms of plant metal tolerance and homeostasis. *Planta* **212**:475–486.

Clemens S, Kim EJ, Neumann D, Schroeder JI. 1999. Tolerance to toxic metals by a gene family of phytochelatin synthases from plants and yeast. *EMBO J* **18**:3325–3333.

Clemens S, Palmgren MG, Krämer U. 2002. A long way ahead: Understanding and engineering plant metal accumulation. *Trends Plant Sci* **7**:309–315.

Cobbett C, May M, Howden R, Rolls B. 1998. The glutathione-deficient, cadmium-sensitive mutant, cad2-1, of *Arabidopsis thaliana* is deficient in gamma-glutamylcysteine synthetase. *Plant J* **16**:73–78.

Deckert J. 2005. Cadmium toxicity in plants: Is there any analogy to its carcinogenic effect in mammalian cells? *BioMetals* **18**:475–481.

di Toppi LS, Gabbrielli R. 1999. Response to cadmium in higher plants. *Environ Exp Bot* **41**:105–130.

Dominguez-Solis JR, Gutierrez-Alcala G, Vega JM, Romero LC, Gotor C. 2001. The cytosolic *O*-acetylserine(thiol)lyase gene is regulated by heavy metals and can function in cadmium tolerance. *J Biol Chem* **276**:9297–9302.

Dominguez-Solis JR, Lopez-Martin MC, Ager FJ, Ynsa MD, Romero LC, Gotor C. 2004. Increased cysteine availability is essential for cadmium tolerance and accumulation in *Arabidopsis thaliana*. *Plant Biotechnol J* **2**:469–476.

Dong J, Lai R, Nielsen K, Fekete CA, Qiu H, Hinnebusch AG. 2004. The essential ATP-binding cassette protein RLI1 functions in translation by promoting preinitiation complex assembly. *J Biol Chem* **279**:42157–42168.

Footitt S, Slocombe SP, Larner V, Kurup S, Wu YS, Larson T, Graham I, Baker A, Holdsworth M. 2002. Control of germination and lipid mobilization by COMATOSE, the *Arabidopsis* homolog human ALDP. *EMBO J* **21**:2912–2922.

Fojtová M, Kovarik A. 2000. Genotoxic effect of cadmium is associated with apoptic changes in tobacco cells. *Plant Cell Environ* **23**:531–537.

Freeman JL, Persans MW, Nieman K, Albrecht C, Peer W, Pickering IJ, Salt DE. 2004. Increased glutathione biosynthesis plays a role in nickel tolerance in Thlaspi nickel hyper-accumulators. *Plant Cell* **16**:2176–2191.

Gasic K, Korban SS. 2007. Transgenic Indian mustard (*Brassica juncea*) plants expressing an *Arabidopsis* phytochelatin synthase (AtPCS1) exhibit enhanced As and Cd tolerance. *Plant Mol Biol* **64**:361–369.

Geisler M, Blakeslee JJ, Bouchard R, Lee OR, Vincenzetti V, Bandyopadhyay A, Titapiwatanakun B, Peer WA, Bailly A, Richards EL, Ejendal KF, Smith AP, Baroux C, Grossniklaus U, Muller A, Hrycyna CA, Dudler R, Murphy AS, Martinoia E. 2005. Cellular efflux of auxin catalyzed by the *Arabidopsis* MDR/PGP transporter AtPGP1. *Plant J* **44**:179–194.

Ghelis T, Dellis O, Jeanette E, Bardat F, Miginiac E, Sott B. 2000. Abscisic acid plasmalemma perception triggers a calcium influx essential for *RAB18* gene expression in *Arabidopsis thaliana* suspension cells. *FEBS Lett* **483**:67–70.

Gichner T, Patkova Z, Szkova J, Dmnerova K. 2004. Cadmium induces DNA damage in tobacco roots, but not DNA damage, somatic mutations or homologous recombination in tobacco leaves. *Mutat Res* **559**:49–57.

Glombitza S, Dubuis PH, Thulke O, Welzl G, Bovet L, Gotz M, Affenzeller M, Geist B, Hehn A, Asnaghi C, Ernst D, Seidlitz HK, Gundlach H, Mayer KF, Martinoia E, Werck-Reichhart D, Mauch F, Schaffner AR. 2004. Crosstalk and differential response to abiotic and biotic stressors reflected at the transcriptional level of effector genes from secondary metabolism. *Plant Mol Biol* **54**:817–835.

Gong JM, Lee DA, Schroeder JI. 2003. Long-distance root-to-shoot transport of phytochelatins and cadmium in *Arabidopsis*. *PNAS* **100**:10118–10123.

Goodman CD, Casati P, Walbot V. 2004. A multidrug resistance-associated protein involved in anthocyanin transport in *Zea mays*. *Plant Cell* **16**:1812–1826.

Gravot A, Lieutaud A, Verret F, Auroy P, Vavasseur A, Richaud P. 2004. AtHMA3, a plant P-1B-ATPase, functions as a Cd/Pb transporter in yeast. *FEBS Lett* **561**:22–28.

Griffith OW. 1982. Mechanism of action, metabolism, and toxicity of buthionine sulfoximine and its higher homologs, potent inhibitors of glutathione synthesis. *J Biol Chem* **257**:13704–13712.

Hall JL. 2002. Cellular mechanisms for heavy metal detoxification and tolerance. *J Exp Bot* **53**:1–11.

Hall JL, Williams LE. 2003. Transition metal transporters in plants. *J Exp Bot* **54**:2601–2613.

Hanikenne M, Motte P, Wu MCS, Wang T, Loppes R, Matagne RF. 2005. A mitochondrial half-size ABC transporter is involved in cadmium tolerance in *Chlamydomonas reinhardtii*. *Plant Cell Environ* **28**:863–873.

Hassinen VH, Tervahauta AI, Halimaa P, Plessl M, Peräniemi S, Schat H, Aarts MGM, Servomaa K, Kärenlampi SO. 2007. Isolation of Zn-responsive genes from two accessions of the hyperaccumulator plant *Thlaspi caerulescens*. *Planta* **225**:977–998.

Heiss S, Shäfer H, Haag-Kerwer A, Rausch T. 1999. Cloning sulfur assimilation genes of *Brassica juncea* L.: Cadmium differentially affects the expression of a putative low-affinity sulfate transporter and isoforms of ATP sulfurylase and APS reductase. *Plant Mol. Biol* **39**:847–857.

Heiss S, Wachter A, Bogs J, Cobbett C, Rausch T. 2003. Phytochelatin synthase (PCS) protein is induced in Brassica juncea leaves after prolonged Cd exposure. *J Exp Bot* **54**:1833–1839.

Herbette S, Taconnat L, Hugouvieux V, Piette L, Magniette M-LM, Cuine S, Auroy P, Richaud P, Forestier C, Bourguignon J, Renou JP, Vavasseur A, Leonhardt N. 2006. Genome-wide transcriptiome profiling of the early cadmium response of *Arabidopsis* roots and shoots. *Biochimie* **88**:1751–1765.

Hernandez LE, Carpena-Ruiz R, Garate A. 1996. Alterations in the mineral nutrition of pea seedlings exposed to cadmium. *J Plant Nutri* **19**:1581–1598.

Higgins CF. 1992. ABC transporters: From microorganisms to man. *Annu Rev Cell Biol* **8**:67–113.

Howarth JR, Dominguez-Solıs JR, Gutierrez-Alcal G, Wray JL, Romero LC, Gotor C. 2003. The serine acetyltransferase gene family in *Arabidopsis thaliana* and the regulation of its expression by cadmium. *Plant Mol Biol* **51**:689–598.

Howden R, Andersen CR, Goldsbrough PB, Cobbett CS. 1995a. A cadmium-sensitive, glutathione-deficient mutant of *Arabidopsis thaliana*. *Plant Physiology* **107**:1067–1073.

Howden R, Goldsbrough PB, Andersen CR, Cobbett CS. 1995b. Cadmium-sensitive, cad1 mutants of *Arabidopsis thaliana* are phytochelatin deficient. *Plant Physiol* **107**:1059–1066.

Hussain D, Haydon MJ, Wang Y, Sherson SM, Young J, Camakaris J, Harper JF, Cobbett CS. 2004. P-Type ATPase heavy metal transporters with roles in essential zinc homeostasis in *Arabidopsis*. *Plant Cell* **16**:1327–1339.

Ishikawa T. 1992. The ATP-dependent glutathione *S*-conjugate export pump. *Trends Biochem Sci* **17**:433–438.

Jacquinot M, Villiers F, Kieffer-Jacquinot S, Hugouvieux V, Bruley C. Garin J, Bourguignon J. 2007. A proteomics dissection of *Arabidopsis thaliana* vacuoles isolated from cell cultures. *Mol Cell Proteomics* **6**:394–412.

Jasinski M, Stukkens Y, Degand H, Purnelle B, Marchand-Brynaert J, Boutry M. 2001. A plant plasma membrane ATP binding cassette-type transporter is involved in antifungal terpenoid secretion. *Plant Cell* **13**:1095–1107.

Jasinski M, Ducos E, Martinoia E, Bourtry M. 2003. The ATP-binding cassette transporters: Structure, function, and gene family comparision between rice and *Arabidopsis*. *Plant Physiol* **131**:1169–1177.

Jedlitschky G, Leier I, Buchholz U, Barnouin K, Kurz GN, Keppler D. 1996. Transport of glutathione, glucuronate, and sulphate conjugates by the MRP gene-encoded conjugate export pump. *Cancer Res* **56**:988–994.

Jonak C, Nakagami H, Hirt H. 2004. Heavy metal stress. Activation of distinct mitogen-activated protein kinase pathways by copper and cadmium. *Plant Physiol* **136**:3276–3283.

Kim DY, Bovet L, Kushnir S, Noh EW, Martinoia E, Lee Y. 2006. AtATM3 is involved in heavy metal resistance in *Arabidopsis*. *Plant Physiol* **140**:1–11.

Kim DY, Bovet L, Meashima M, Martinoia E, Lee Y. 2007. The ABC transporter AtPDR8 is a cadmium extrusion pump conferring heavy metal resistance. *Plant J* **50**:207–218.

Kispal G, Csere P, Guiard B, Lill R. 1997. The ABC transporter Atm1p is required for mitochondrial iron homeostasis. *FEBS Lett* **418**:346–350.

Kispal G, Csere P, Prohl C, Hill R. 1999. The mitochondrial proteins Atm1p and Nf1p are essential for biogenesis of cytosolic Fe/S proteins. *EMBO J* **18**:3981–3989.

Klein M, Perfus-Barbeoch L, Frelet A, Gaedeke N, Reinhardt D, Mueller-Roeber B, Martinoia E, Forestier C. 2003. The plant multidrug resistance ABC transporter AtMRP5 is involved in guard cell hormonal signalling and water use. *Plant J* **33**:119–129.

Klein M, Geisler M, Suh SJ, Kolukisaoglu Ü, Azevedo L, Plaza S, Curtis MD, Richter A, Weder B, Shulz B, Martinoia E. 2004. Disruption of *AtMRP4*, a guard cell plasma membrane ABCC-type ABC transporter, leads to deregulation of stomatal opening and increased drought susceptibility. *Plant J* **39**:219–236.

Kobae Y, Sekino T, Yoshioka H, Nakagawa T, Martinoia E, Maeshima M. 2006. Loss of AtPDR8, a plasma membrane ABC transporter of *Arabidopsis* thaliana, causes hypersensitivity cell death upon pathogen infection. *Plant Cell Physiol* **47**:309–318.

Koppen G, Verschaeve L. 1996. The alkaline comet test on plant cell: A new genotoxicity test for DNA strand breaks in *Vicia faba* root cells. *Mutat Res* **360**:193–200.

Koren'hov V, Park S, Cheng NH, Sreevidya C, Lachmansingh J, Morris J, Hirschi K, Wagner GJ. 2007. Enhanced Cd(2+)-selective root-tonoplast-transport in tobacco expressing *Arabidopsis* cation exchangers. *Planta* **225**:403–411.

Korschunova YO, Eide D, Clark WG, Guerinot ML, Pakrasi HB. 1999. The IRT1 protein from Arabidopsis thaliana is a metal transporter with a broad substrate range. *Plant Mol Biol* **40**:37–44.

Krämer U, Cotter-Howells JD, Charnock JM, Baker AJM, Smith JAC. 1996. Free histidine as a metal chelator in plants that accumulate nickel. *Nature* **379**:635–638.

Kushnir S, Babiychuk E, Storozhenko S, Davey MW, Papenbrock J, De Rycke R, Engler G, Stephan UW, Lange H, Kispal G, Lill R, Van Montagu M. 2001. A mutation of the mitochondrial ABC transporter Sta1 leads to dwarfism and chlorosis in the *Arabidopsis* mutant starik. *Plant Cell* **13**:89–100.

Lane TW, Saito MA, George GN, Pickering IJ, Prince RC, Morell FMM. 2005. A cadmium enzyme from marine diatom. *Nature* **435**:42.

Lee M, Lee K, Lee J, Noh EW, Lee Y. 2005. AtPDR12 contributes to lead resistance in *Arabidopsis*. *Plant Physiol* **138**:827–836.

Leighton J, Schatz G. 1995. An ABC transporter in the mitochondrial inner membrane is required for normal growth of yeast. *EMBO J* **14**:188–195.

Li Y, Dhankher OP, Carreira L, Lee D, Chen A, Schroeder JI, Balish RS, Meagher RB. 2004. Overexpression of phytochelatin synthase in *Arabidopsis* leads to enhanced arsenic tolerance and cadmium hypersensitivity. *Plant Cell Physiol* **45**:1787–1797.

Li ZS, Szczypka M, Lu YP, Thiele DJ, Rea PA. 1996. The yeast cadmium factor protein (YCF1) is a vacuolar glutathione S-conjugate pump. *J Biol Chem* **271**:6509–6517.

Lindberg S, Landberg T, Greger M. 2004. A new method to detect cadmium uptake in protoplasts. *Planta* **219**:526–532.

Lindberg S, Landberg T, Greger M. 2007. Cadmium uptake and interaction with phytochelatins in wheat protoplasts. *Plant Physiol Biochem* **45**:47–53.

Martinoia E, Massonneau A, Frangne N. 2000. Transport processes of solutes across the vacuolar membrane of higher plants. *Plant Cell Physiol* **41**:1175–1186.

Martinoia E, Klein M, Geisler M, Bovet L, Forestier C, Kolukisaoglu HU, Mueller-Roeber B, Schulz B. 2002. Multifunctionality of plant ABC transporters: More than just detoxifiers. *Planta* **214**:345–355.

McLaughlin MJ, Singh BR. 1999. Cadmium in soils and plants. In: McLaughlin MJ, Singh BR, editors. *Developments in Plant and Soil Sciences*. Dordrecht, The Netherlands: Kluwer Academic Publishers, pp. 1–7.

Mentewab N, Stewart CN. 2005. Overexpression of an *Arabidopsis thaliana* ABC transporter confers kanamycin resistance to transgenic plants. *Nat Biotechnol* **23**:1177–1180.

Mills RF, Krijger GC, Baccarini PJ, Hall JL, Williams LE. 2003. Functional expression of AtHMA4, a P_{1B}-type ATPase in the $Zn/Co/Cd/Pb$ subclass. The *Plant J* **35**:164–175.

Moons A. 2003. *Ospdr9*, which encodes a PDR-type ABC transporter, is induced by heavy metals, hypoxic stress and redox perturbations in rice roots. *FEBS Lett* **553**:370–376.

Muller M, Meijer C, Zaman GJ, Borst P, Scheper RJ, Mulder NH, de Vries EG, Jansen PL. 1994. Overexpression of the gene encoding the multidrug resistance-associated protein results in increased ATP-dependent glutathione S-conjugate transport. *PNAS* **91**:13033–13037.

Ortiz DF, Ruscitti T, Mccue KF, Ow DW. 1995. Transport of metal-binding peptides by Hmt1, a fission yeast ABC-Type vacuolar membrane-protein. *J Biol Chem* **270**:4721–4728.

Pence NS, Larsen PB, Ebbs SD, Letham DLD, Lasat MM, Garvin DF, Eide D, Kochian LV. 2000. The molecular physiology of heavy metal transport in the Zn/Cd hyperaccumulator *Thlaspi caerulescens*. *PNAS* **97**:4956–4960.

Perfus-Barbeoch L, Leonhardt N, Vavasseur A, Forestier C. 2002. Heavy metal toxicity: Cadmium permeates through calcium channels and disturbs the plant water status. *Plant J* **32**:539–548.

Persans M, Nieman K, Salt D. 2001. Functional activity and role of cation-efflux family members in Ni hyperaccumulation in *Thlaspi goesingense*. *PNAS* **98**:9995–10000.

Pighin JA, Zhen H, Balakshin LJ, Goodman IP, Western TL, Jetter R, Kunst L, Samuels AL. 2004. Plant cuticular lipid export requires an ABC transporter. *Science* **306**:702–704.

Plaza S, Tearall KL, Zhao FJ, Buchner P, McGrath SP, Hawkesford MJ. 2007. Expression and functional analysis of metal transporter genes in two contrasting ecotypes of the hyperaccumulator *Thlaspi caerulescens*. *J Exp Bot* **58**:1717–1728.

Pomponi M, Censi V, Di G, De Paolis A, di Toppi L.S, Aromolo R, Constantino P, Cardarelli M. 2006. Overexpression of Arabidopsis phytochelatin synthase in tobacco plants enhances Cd(2+) tolerance and accumulation but not translocation to the shoot. *Planta* **223**:180–190.

Rauser WE. 1995. Phytochelatins and related peptides—Structure, biosynthesis and function. *Plant Physiol* **109**:1141–1149.

Rea PM. 1999. MRP subfamily ABC transporters from plants and yeast. *J Exp Bot* **50**:895–913.

Rea PM. 2007. Plant ATP-binding cassette transporters. *Annu Rev Plant Biol* **58**:347–375.

Rea PA, Li ZS, Lu YP, Drozdowicz YM, Martinoia E. 1998. From vacuolar GS-X pumps to multispecific ABC transporters. *Annu Rev Plant Physiol Plant Mol Biol* **49**:727–760.

Romero-Puertas MC, Rodríguez-Serrano M, Corpas FJ, Gómez M, del Río LA, Sanalio LM. 2004. Cadmium-induced subcellular accumulation of O_2 and H_2O_2 in pea leaves. *Plant Cell Environ* **27**:1122–1134.

Roth U, von Roepenack-Lahaye E, Clemens S. 2006. Proteome changes in *Arabidopsis thaliana* roots upon exposure to Cd^{2+}. *J Exp Bot* **57**:4003–4013.

Sanchez-Fernandez R, Ardiles-Diaz W, Van Montagu M, Inzé D, May MJ. 1998. Cloning and expression analyses of *AtMRP4*, a novel *MRP*-like gene from *Arabidopsis thaliana*. *Mol Gen Genet* **258**:655–662.

Sanchez-Fernandez R, Davies TGE, Coleman JOD. 2001. Do plants have more genes than humans? Yes, when it comes to ABC proteins. *Trends Plant Sci* **6**:347–348.

Sandalio LM, Dalurzo, HC, Gomez, M, Romero-Puertas, MC, del Rio, LA. 2001. Cadmium-induced changes in the growth and oxidative metabolism of pea plants. *J Exp Bot* **52**:2115–2126.

Shah K, Dubey RS. 1995. Effect of cadmium on RNA level as well as acitivity and molecular forms of ribonuclease in growing rice seedlings. *Plant Physiol Biochem* **33**:577–584.

Shitan N, Bazin I, Dan K, Obata K, Kigawarm K, Ueda K, Sato F, Forestier C, Yazaki K. 2003. Involvement of CjMDR1, a plant multidrug-resistance-type ATP-binding cassette protein, in alkaloid transport in *Coptis japonica*. *PNAS* **100**:751–756.

Siedlecka A, Krupa Z. 1996. Interaction between cadmium and iron and its effects on photosynthetic capacity of primary leaves of *Phaseolus vulgaris*. *Plant Physiol Biochem* **34**:833–841.

Sobkowiak R, Deckert J. 2004. The effect of cadmium on cell cycle control in suspension culture cells of soybean. *Acta Physiol Plant* **26**:335–344.

Song W-Y, Sohn EJ, Martinoia E, Lee YJ, Yang Y-Y, Jasinski M, Forestier C, Hwang I, Lee Y. 2003. Engineering tolerance and accumulation of lead and cadmium in transgenic plants. *Nat Biotechnol* **21**:914–919.

Stein M, Dittgen J, Sanchez-Rodriguez C, Hou BH, Molina A, Schulze-Lefert P, Lipka V, Somerville S. 2006. *Arabidopsis* PEN3/PDR8, an ATP binding cassette transporter, contributes to nonhost resistance to inappropriate pathogens that enter by direct penetration. *Plant Cell* **18**:731–746.

Stukkens Y, Bultreys A, Grec S, Trombik T, Vanham D, Boutry M. 2005. NpPDR1, a pleistropic drug resistance-type ATP-binding cassette transporter from *Nicotiana plumbaginifolia*, plays a major role in plant pathogen defense. *Plant Physiol* **139**:341–352.

Suzuki N, Koizumi N, Sano H. 2001. Screening of cadmium-responsive genes in *Arabidopsis* thaliana. *Plant Cell Environ* **24**:1177–1188.

Thomine S, Wang R.C, Ward JM, Crawford NM, Schroeder JI. 2000. Cadmium and iron transport by members of a plant metal transporter family in *Arabidopsis* with homology to Nramp genes. *PNAS* **97**:4991–4996.

Tommasini R, Evers R, Vogt E, Mornet C, Zaman GJR, Schinkel, AH, Borst P, Martinoia E. 1996. The human multidrug resistance-associated protein functionally complements the yeast cadmium resistance factor 1. *PNAS* **93**:6743–6748.

Tommasini R, Vogt E, Fromenteau M, Hortensteiner S, Matile P, Amrhein N, Martinoia E. 1998. An ABC-transporter of *Arabidopsis thaliana* has both glutathione-conjugate and chlorophyll catabolite transport activity. *Plant J* **13**:773–780.

Van Belleghem F, Cuypers A, Brahim S, Smeets K, Vangronsveld J, d'Haen J, Valcke R. 2007. Subcellular localization of cadmium in roots and leaves of *Arabidopsis thaliana*. *New Phytol* **173**:495–508.

Van den Brûle S, Smart CC. 2002. The plant PDR family of ABC transporters. *Planta* **216**:95–106.

Vatamaniuk OK, Bucher EA, Sundaram MV, Rea PA. 2005. CeHMT-1, a putative phytochelatin transporter, is required for cadmium tolerance in *Caenorhabdtis elegans*. *J Biol Chem* **280**:23684–23690.

Waalkes MP. 2000. Cadmium carcinogenesis in review. *J Inorg Chem* **79**:241–244.

Walker JE, Saraste M, Runswick MJ, Gay NJ. 1982. Distantly related sequences in α- and β-subunits of ATP synthase, myosin, kinases and other ATP-requiring enzymes and a common nucleotide binding fold. *EMBO J* **1**:945–951.

Wang T, Wu M. 2006. An ATP-binding cassette transporter related to yeast vacuolar ScYCF1 is important for Cd sequestration in *Chlamydomonas reinhardtii*. *Plant Cell Environ* **29**:1901–1912.

Weber M, Harada E, Vess C, Roepenack-Lahaye E, Clemens S. 2004. Comparative microarray analysis of *Arabidopsis thaliana* and *Arabidopsis halleri* roots identifies nicotianamine synthase, a ZIP transporter and other genes as potential metal hyperaccumulation factors. *Plant J* **37**:269–281.

Williams LE, Pittman JK, Hall JL. 2000. Emerging mechanisms for heavy metal transport in plants. *Biochim Biophys Acta—Biomembr* **1465**:104–126.

Yasaki N, Shitan N, Takamatsu H, Ueda K, Sato F. 2001. A novel *Coptis japonica* multidrug resistant protein preferentially expressed in the alkaloid accumulating rhizome. *J Exp Bot* **52**:877–879.

Yasaki K, Yamanaka N, Masuno T, Konagai S, Kaneko S, Ueda K, Sato F. 2006. Heterologous expression of a mammalian ABC transporter in plant and its application to phytoremediation. *Plant Mole Biolo* **61**:491–503.

Yeh CM, Hsiao LJ, Huang HJ. 2004. Cadmium activates a mitogen-activated protein kinase gene and MBP kinase in plant. *Plant Cell Physiol* **45**:1306–1312.

Zhu Y, Pilon-Smits E, Tarun A, Weber S, Jouanin L, Terry N. 1999. Cadmium tolerance and accumulation in Indian mustard is enhanced by overexpressing gamma-glutamylcysteine synthetase. *Plant Physiol* **121**:1169–1177.

Zimmermann P, Hirsch-Hoffmann M, Hennig L, Gruissem W. 2004. GENEVESTIGATOR. *Arabidopsis* microarray database and analysis toolbox. *Plant Physiol* **136**:2621–2632.

19 Iron: A Major Disease Modifier in Thalassemia

BPS LAB-Centre for Diagnostic Hematology, Sankatmochan, Varanasi-221005 (UP), India

1 INTRODUCTION

Thalassemias are the most widely occurring group of inherited disorders of hemoglobin caused by mutations in the globin genes, leading to a reduced production of one or more globin chains. An imbalance of the globin chains results in deficient production of hemoglobin due to unavailability of one of the three essential components of hemoglobin: globin, protoporphyrin, and iron. There are essentially two main types of thalassemias—alpha thalassemia and beta thalassemia—depending on whether the mutation is in the alpha globin gene or the beta globin gene. The pathophysiology of the resultant chronic anemia is complex and involves three major mechanisms: decreased overall hemoglobin production, death of immature erythroid precursors in the marrow (ineffective erythropoiesis), and early destruction of red blood cells in the peripheral circulation (hemolysis). The disease has two main clinical phenotypes: *thalassemia major* and *thalassemia intermedia*. Disease phenotypes of this essentially globin chain disorder has been the subject of various reviews, articles, and monographs. Variation in phenotypic expression of these monogenic disorders has been a subject of intrigue and research for the past few decades for geneticists and clinical researchers. Natural history of disease in the two phenotypic categories have been studied and well-documented. Thus *thalassemia major* can be described as a chronic microcytic hypochromic anemia, characterized by evidence of ineffective erythropoiesis and peripheral hemolysis manifesting as growth failure, short stature, splenomegaly, hepatomegaly, bone marrow expansion and fragility of bones, ultimately leading to death due to cardiac failure. Severity of the aforesaid manifestations is reduced in *thal intermedia*. The defect responsible for

Trace Elements as Contaminants and Nutrients: Consequences in Ecosystems and Human Health, Edited by M. N. V. Prasad
Copyright © 2008 John Wiley & Sons, Inc.

anemia is such that it is not amenable to correction by drugs or biochemical supplementation or therapy. Supportive management of this chronic anemia by regular red blood cell transfusion regimens has led to an increased lifespan and survival of the affected patients. Hypertransfusion regimens aimed at maintaining average hemoglobin (Hb) levels around near normal levels of $11-14 \, \text{gm/dl}$ are able to counter many of the effects of chronic anemia with adequate tissue oxygen delivery. However, the resultant increased lifespan provides an opportunity for the effects of iron overload on different organs and organ systems to manifest. This is further complicated by a variable amount of iron chelation provided or taken by the patients. Genetic modifiers of globin genes have dominated the spectre of modifiers of the disease. Much of the variability in phenotypic expression of the disorders has been attributed to the variable severity of the large number of mutations (>200 alone in beta globin gene) associated with the disease. In this chapter we shall focus on the role of iron in pathophysiology, phenotypic expression, complications through the natural course of the disease and as a result of therapeutic intervention and management of the disease, leaving the genetics of globin chains behind.

1.1 Hemoglobin: The Tetramer Molecule

Hemoglobin is a tetramer molecule with four globin chains combining with four molecules of heme. HbA or the adult hemoglobin is the predominant human hemoglobin comprising $97-98\%$ of the total hemoglobin present in humans. HbA has two α and two β globin chains forming the tetramer $\alpha_2\beta_2$. It is the heme molecule which contains iron (Fe) in its ferrous form. The ferrous iron (Fe^{2+}) is present in heme as a prosthetic group and serves to function as a transporter of oxygen to tissues. It is important to keep balance between levels of globin chains and heme groups, because excess of any of these is toxic to erythroid cells and precursors. Heme iron accounts for majority of the total iron content of the body in a healthy adult [Gardenghi et al., 2007]. Hemoglobin (Hb) is the most abundant of the heme proteins—the metalloprotoporphyrins containing iron—and comprises 90% of the mature red blood cell (RBC) weight.

1.2 Erythropoiesis and Erythroid Differentiation

Pronormoblast, the earliest erythroid precursor, does not contain any detectable hemoglobin; but during its course of differentiation and maturation into an enucleated mature RBC, it gets packed with Hb. The cells of the erythron undergo division and multiplication and in due course acquire a certain level of Hb per cell when they cease to divide. The cell division ceases at the stage of polychromatophilic erythroblast while hemoglobinization continues for a while to the stage of reticulocyte (see Fig. 1). Thus cell division and hemoglobinization of the erythroid cells are closely correlated. The parameters of the red cells which can be measured are (i) red cell size expressed as mean corpuscular volume (MCV), (ii) hemoglobin content or the mean corpuscular hemoglobin (MCH), and (iii) concentration of hemoglobin in the red cells, mean corpuscular hemoglobin concentration (MCHC). It is the concentration of hemoglobin

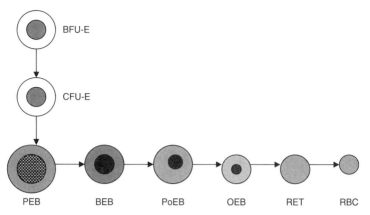

Figure 1. Schematic diagram of erythron showing stages during maturation. Proerythroblast (PEB) is the earliest identifiable erythroid precursor which is derived by division of erythroid colony-forming unit cells (CFU-E). First traces of hemoglobin are seen in basophilic erythroblast (BEB). Cell division does not occur beyond polychromatic erythroblast (PoEB) because the nucleus of orthochromatic erythroblast (OEB) is not capable of division. Hemoglobinization continues up to reticulocyte (RET) stage, which is slightly larger than the mature red blood cell (RBC). Hb enters the nucleus of OEB and inactivates it. BFU-E, erythroid bursa forming unit.

in red cells, the MCHC, which is nearly similar in all mammalian red blood cells [Dessipyris and Sawyer, 2004]. MCHC of 32–36% is easily accommodated in the biconcave shape of the red cell. The maximum surface-to-volume ratio achieved by this concentration facilitates gas transfer as well as deformability during cell movement through microcirculation [Telen and Kaufman, 2004]. Thus biochemical and rheological properties of the red cells are governed to an extent by the MCHC. Resistance to change in MCHC, by attaining maximum possible Hb, is perhaps the adaptive response of the body during disorders and diseases.

Hence, the microcytic cells, due to increased number of cell divisions, are a result of the effort on part of the erythron to acquire as much of Hb as possible to reach the optimum required concentration. These cells are microcytic and hypochromic as indicated by a decreased MCV and MCH values. The reverse is the case in erythropoietin-induced acceleration of hemoglobin synthesis that leads to earlier nuclear degeneration and cessation of cell division, and the mature erythrocyte is a macrocyte with enough hemoglobin. The classic example is the macrocytic anemia due to deficiency of folic acid and cobalamin leading to "megaloblastic" erythropoiesis as a result of hampered nuclear maturation and division. From the above discussion, it is obvious that the size of the RBC is dependent on the Hb content. Deficient supply of any of the three essential components of the Hb molecule-globin chains, iron and protoporphyrin IX (PPIX), leads to production of a hypochromic and microcytic red cell. This microcytic hypochromic red cell is the morphologic hallmark of thalassemia (Thal), erythrocytic protoporphyria (EPP), and iron deficiency anemia (IDA), examples of disorders caused by deficiency of globin chains, PPIX, and iron, respectively.

1.3 Pathophysiology of Thalassemia

Ever since thalassemias were identified to be disorders caused by globin chain imbalance, their pathophysiology has been explained primarily on the basis of this primary defect. Decrease in one of the globin chains leads to decrease in availability of globin tetramer for hemoglobin synthesis, directly lowering the Hb levels due to deficient production. Precipitation of excess free globin chains on cell membranes of erythroid precursors, causing damage and cell death, leads to ineffective erythropoiesis and dyserythropoiesis. Hypoxia-induced erythroid hyperplasia of the marrow and extramedullary erythropoiesis leads to bone expansion and hepatosplenomegaly. Further decrease in hemoglobin level occurs with peripheral hemolysis due to precipitates of globin chains, splenic sequestration, pooling, and hemolysis. Role of excess unutilized iron as a result of excessive absorption and iron released from destruction of immature precursors was mainly thought to produce long-term damage to organs by siderosis affecting functioning of organs, functional atrophy followed by fibrosis, and finally organ failure. Moreover, increased iron absorption from the gut contributing to iron overload was mainly thought to be a feature of thalassemia intermedia, the milder phenotype attributed to less severe beta or alpha thalassemia mutations.

Tremendous progress made in the field of iron metabolism with discovery of iron-related genes and the study of the proteins expressed by them has brought about a turnover in understanding of all iron-related disorders whether of deficiency or overload. Understanding of thalassemic disorders, especially beta thalassemia, has also undergone a dimensional change by a decade of iron research. In the following sections, an effort will be made to elucidate them.

2 IRON METABOLISM: CURRENT CONCEPTS AND ALTERATIONS IN THALASSEMIA

Iron is required by the cells for cell proliferation and respiratory functions. Total body iron varies with age and sex. In adult men and in postmenopausal women, total iron content is 50 mg/kg body weight [Andrews, 2004]. Due to menstrual losses, iron content in postpubertal women averages around 35 mg/kg body weight. Iron is absorbed in duodenum, the proximal most part of the intestine, and is taken up and transported to liver, where it is stored. It is taken up from the liver for biosynthesis of various functional forms like heme and iron–sulfur (Fe4–S4) cluster proteins. During these processes it is shunted between its divalent ferrous (Fe^{2+}) and trivalent ferric (Fe^{3+}) forms. Though essential, iron is highly toxic in its free form as well as in heme form. Hydrogen peroxide undergoes the Fenton reaction in the presence of ferrous ions to generate the hydroxyl radical, an extremely reactive radical [Harrison and Arosio, 1996]. The resultant ferric ions are again reduced by reductants to produce ferrous ions. Hence, ferric ions are sequestered from ions in biological fluids by ligands such as transferrin and ferritin. The hydroxyl radical attacks lipids, proteins and DNA, sometimes leading to cell death or cancer [Taketani, 2005].

Erythropoietic cells require a large amount of iron for synthesis of hemoglobin. Heme is synthesized in mitochondria of the erythroid cells by an enzyme ferrochelatase (Fch), which facilitates the insertion of Fe in PPIX, the last step in the biosynthesis of heme. Heme has a high reactivity toward oxygen, to which it binds reversibly, and it is this property which makes it a transporter of oxygen to tissues. However, the same property can render it toxic to membranes and lipids if the porphyrin ring is destroyed by hydroxyl radicals. Heme oxygenase (HO) is a rate-limiting enzyme of heme catabolism which regulates the cellular level of heme. The iron formed by HO is reutilized [Taketani, 2005].

The high utility and high toxicity of iron and heme render it necessary for the body to have a tight regulatory mechanism which regulates absorption, uptake, and utilization of iron. In recent years, studies on iron-related genes and their expressions have had a remarkable impact on understanding the mechanisms of iron overload (see Table 1). The important genes associated with iron metabolism are *TfR1* (gene encoding for transferrin receptor 1: TFR1), *TfR2* (gene encoding for transferrin receptor 2: TFR2), *Hfe* (hemochromatosis gene), *Hjv* (juvenile hemochromatosis gene), *Hamp1* (hepcidin: HAMP) *Fpn1* (ferroportin), *NGAL*, *Nramp2*, *FTH* and *FTL* (ferritin), and genes for iron regulatory proteins. Discovery of these genes and

TABLE 1. Important Iron-Related Genes Involved in Absorption and Uptake of Iron

Gene	Gene Product	Phenotypic Expression
Hamp/Hamp1	Hepcidin antimicrobial peptide1	Triggers degradation of ferroportin by its internalization
Fpn1	Ferroportin	Facilitates iron export to plasma from basolateral surface of enterocytes
Nramp	Nramp (DMT1/DCT1)	An IRE containing transmembrane divalent metal ion transporter protein
NGAL	Neutral gelatinase Associated lipocalin	Non transferrin dependent iron uptake protein
Tf	Transferrin	Responsible for plasma transport of iron
TfR1	Transferrin receptor 1	Major protein involved in plasma iron uptake
Hfe	Hemochromatosis	Controls Hamp expression
Hjv/Hfe2	Hemochromatosis type 2	Juvenile hemochromatosis-related gene
Cebpa	CCAAT/enhancer-binding protein-α	Controls Hamp expression
FTH1	Ferritin heavy chain1	Heavy-chain subunit of storage protein ferritin
FTL1	Ferritin light chain1	Light-chain subunit of storage protein ferritin

their role in normal iron metabolism led to sequential studies in mouse models of thalassemia and hemochromatosis and other disorders related to iron overload. In the sections ensuing we shall discuss some of these studies and their implications on understanding of the disease pathophysiology in humans.

2.1 Iron Absorption and Uptake

Level of iron in the body is regulated by absorption. Intestinal iron absorption has two distinct steps: uptake by mucosal cells and export to plasma through lamina propria. Dietary iron is absorbed from the duodenum, which has an acidic pH as a result of its proximity to gastric outlet. It is in this portion of the intestine that the proteins involved in absorption are maximally expressed. Dietary iron is available in nonheme and heme form. Heme iron forms only 10–15% of the total dietary iron in a Western meat-rich diet, and its absorption is generally unaffected by other dietary factors. Vegetarian diets contain negligible amounts of heme iron. Heme is absorbed intact by enterocytes. Heme carrier protein 1 (HCP1) present in proximal intestine mediates heme absorption. In the enterocyte it is cleaved by HO, and released iron joins the nonheme iron cycle. A small amount of heme passes into plasma, where it circulates bound to hemopexin and albumin. Heme circulating in plasma is uptaken by various cells including hepatocytes. Presence of heme in diet enhances absorption of nonheme dietary iron [Andrews, 2004].

Nonheme iron is available in ferric form (Fe^{3+}) and has to be converted to ferrous form (Fe^{2+}) for uptake by mucosal cells of the intestine. The absorption of nonheme iron is greatly affected by other dietary constituents and can vary up to tenfold [Andrews, 2004]. The discovery of the sequence of iron responsive element (IRE) in the $5'$ noncoding region of ferritin mRNA [Hentze et al., 1987] and in five sites of $3'$ noncoding region of transferrin receptor 1 [Klausner et al., 1993] and identification of IRE binding iron regulatory proteins (IRP) has greatly enhanced the understanding of iron absorption.

The transferrin receptor 1(TfR1)-mediated pathway is the classic pathway of cellular iron uptake mechanism but is most important in the erythroid iron uptake. TfR2 has also been identified, is expressed mainly on hepatocytes, and lacks IREs but has a higher affinity for transferrin than does TfR1. However, the small intestinal cells, which lack transferrin receptors, take up iron by an Nramp2-mediated mechanism. Nramp2 is a transmembrane transporter protein (also known as DCT1 or DMT1, a divalent metal ion transporter). Ferric iron is insoluble and has to be reduced to ferrous form by a cytochrome b-like hemoprotein Dcytb at the plasma membrane. Nramp2 mRNA contains IREs, and its expression is regulated posttranscriptionally by the iron level in the body [Andrews, 2004; McKie et al., 2002]. Another transferrin-independent mechanism is mediated by a siderophore, a lipocalin-neutral gelatinase-associated lipocalin (NGAL/24p3) discovered by Yang et al. [2002]. The mechanism is similar to that of bacterial iron uptake (the iron chelator drug deferrioxamine is also a siderophore). Iron internalized thus can regulate the IREs. This is an important mechanism in early stages of embryo when Tf/TfR1 mechanism is not well-developed.

After the intestinal uptake, export of iron from intestinal cells involves transport of iron across the basolateral membrane of the enterocyte into the portal vein and is mediated by ferroportin1 (FPN1), an iron-regulated transporter, whose expression increases on increase in iron absorption. Fpn1 mRNA contains an IRE in 5′ noncoding region. In nonintestinal cells, ceruloplasmin is required in addition to FPN, for converting Fe^{2+} to Fe^{3+} so as to facilitate binding of iron to Tf present in the plasma. Hephaestin is probably the ceruloplasmin homologue in intestinal enterocytes [Anderson, 1999; Lee Gelbart et al., 2001]. Ferroportin is a key player in regulating dietary iron absorption in the duodenum, whereas serum hepcidin levels control the level of ferroportin on basolateral surface of the enterocytes [Nemeth et al., 2004]. Hepcidin targets ferroportin and triggers its degradation by internalizing it.

Hepcidin (HAMP), a peptide hormone, was simultaneously discovered in human urine for its microbicidal effects [Park et al., 2001] and in murine liver, where it was shown to be preferentially expressed in conditions of iron overload [Bridle et al., 2003; Nicolas et al., 2004]. Hepcidin is synthesized in the liver and secreted in the bloodstream and inhibits iron absorption from duodenum, inhibits release of iron from reticuloendothelial cells, and inhibits iron transport across placenta [Ganz, 2003]. Murine *Hamp1* mRNA levels are homeostatically regulated by body iron stores, erythroid iron needs, hypoxia, and inflammation [Nicolas et al., 2001, 2002; Fleming and Sly, 2001; Park et al., 2006; Wrighting and Andrews, 2006]. Studies in patients with hereditary hemochromatosis (HH) have helped considerably in understanding iron metabolism. HH is known to be associated with iron overload and subsequent organ failure due to iron deposition. Findings of inappropriately low hepcidin expression or presence of mutant forms of FPN1 in HH due to the presence of mutations in different genes—*Hfe* (hemochromatosis gene), *TfR2* (transferrin receptor 2 gene), and *Hjv* (Hemochromatosis type 2 gene)—have helped in establishing the key role of HAMP and FPN1 in iron absorption [Nemeth et al., 2004; Camaschella and Merlini, 2005]. Hepcidin expression in relation to iron absorption was further elucidated by studies in GSD1a patients with hepatic adenomas where its overexpression was found to be associated with severe iron deficiency anemia unresponsive to iron supplementation [Weinstein et al., 2002]. Increased hepcidin expression has also been observed in response to erythropoietic stimuli like induced hemolysis, bleeding, hypoxia, and erythropoietin administration. In these conditions associated with erythroid hyperplasia, the positive effect of iron on hepcidin was overridden by their negative effect leading to decreased hepcidin expression [Nicolas et al., 2002].

Thus, iron absorption is mostly mediated by the balance between the levels of HAMP made in the liver and Fpn1 produced in the duodenum [Taketani, 2005].

2.2   Regulation of Expression of Transferrin Receptors

Iron circulates in plasma mostly bound to transferrin, a plasma transport protein, which has two binding sites for the trivalent ferric ion. Transferrin-mediated uptake of iron is the major cellular uptake mechanism. Cellular uptake of transferrin is mediated by transferrin receptors (TfR). The delivery of iron to erythroid precursors is an important event in the iron cycle and is mediated by TfR1. Another type of transferrin receptor, TfR2 has been found to express on hepatocytes, although its role in cellular iron uptake is not yet clear.

Regulation of expression of transferrin receptors occurs at three levels [Taketani, 2005]. First, the number of receptors expressing on the cell surface differs in different cells, and only those receptors present on the cell surface are able to take up iron [Richardson and Ponka, 1997; Ponka and Lok, 1999]. Second, at the transcription level in growing cells as the receptors are synthesized during G2S phase of cell division. The proximal region in the 5′ promoter region of the transferrin receptor gene and AP1 and SP1 binding sites control cell division. Expression of transferrin receptors, along with other genes related to hemoglobin synthesis, might be influenced by several other erythroid-related elements in the promoter region of the TfR gene. The third level is at the posttranscriptional level. Discovery of the iron response element (IRE), a novel sequence in the 5′-uncoding region of ferritin mRNA and at five sites in the 3′-uncoding region of TfR1mRNA, has shed much light on the mechanism of cellular iron uptake [Klausner et al., 1993]. Binding of IRE-binding protein (IRP) to IRE stabilizes TfR1 mRNA, resulting in increased synthesis of receptor protein in iron-deficient cells while release of IRP from IRE results in decrease of TfR mRNA. A reverse effect is exerted by IRP binding to IRE of ferritin mRNA. Binding leads to blocking of translation of ferritin mRNA, resulting in decreased synthesis of ferritin, and its release causes increased synthesis with increased sequestering of iron in cells overloaded with iron. Two iron regulatory proteins are identified. IRP1 exhibits aconitase activity both in the mitochondria [Rouult et al., 1990] and the cytoplasm [Ponka and Lok, 1999] and is regulated mainly by heme and nitric oxide [Kim and Ponka, 2002]. IRP2 has an iron degradation domain consisting of 73 amino acids. The increase in intracellular level of iron leads to degradation of IRP2. Thus IRP2 functions mainly in the regulation of cellular homeostasis of iron [Iwai et al., 1995] (Fig. 2).

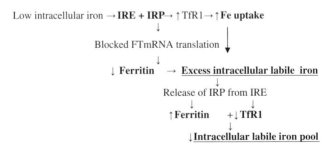

Figure 2. IRP–IRE-mediated regulation of intracellular iron. Binding of iron regulatory proteins (IRP1 and IRP2) to iron responsive element (IRE) on mRNAs of transferrin receptor1 (TfR1) and ferritin (FT) plays a major role in regulating intracellular iron. Many other proteins and factors, involved in the IRP-IRE mechanism of iron regulation, have neither been depicted in the figure nor discussed in detail in the text.

2.3 Alterations in Iron Absorption and Uptake in Thalassemia

Thalassemia presents a unique situation where anemia coexists with iron overload. While anemia is associated with decreased hepcidin expression, the latter, iron overload is known to decrease hepcidin expression [Weizer-Stern et al., 2006]. Weizer-Stern et al. studied hepcidin expression in murine models of thalassemia intermedia (th3/+) and thalassemia major (th3/th3). The thal major mouse models (Hbb$^{th3/th3}$) were created for the purpose by bone marrow transplantation of beta globin null th3/th3 fetal liver cells into Hbb mice. These mice rapidly developed a severe phenotype of thalassemia major and succumbed to ineffective erythropoiesis in 60 days. It was ascertained that the transplant itself did not lead to increased iron absorption. In their subsequent studies they were able to demonstrate down-regulation of hepatic hepcidin mRNA in thal intermedia and thal major mice expressing disease phenotype similar to that of humans. Moreover, decrease in hepcidin was much greater: 16-fold in th3/th3mice as compared to 0.75-fold in th3/+ thal intermedia mice. In the absence of any significant correlation of hepcidin with other iron regulatory genes or hemoglobin levels, the researchers suggested the possibility of existence of a putative "erythropoietic regulator" produced in response to ineffective erythropoiesis.

Untill these studies were reported, little was known regarding factors controlling Hamp1 expression. Mixed results with thal intermedia mouse models th1/th1 and th3/+, having slightly different levels of Hb, regarding hepcidin levels and its expression at different ages prompted longitudinal studies on expression of iron regulatory genes, taking into account ineffective erythropoiesis and organ iron content [Gardenghi et al., 2007].

The explicitly designed study [Gardenghi et al., 2007] on transplanted and nontransplanted and transfused and nontransfused mouse models including wild-type (+/+), thal intermedia (th1/th1 & th3/+), and thal major(th3/th3) spread over a year with analyses at 2 months, 5 months, and 12 months provided some very valuable results and insights into iron metabolism of thalassemia. The major conclusions of the study were that organ distribution of iron is dictated by ineffective erythropoiesis (IE). Thus in severe thal major the parenchymal liver cells are the primary site of iron deposition, while in thal intermedia the spleen followed by Kupffer cells in liver show iron deposit. This also indicates that iron overload occurs primarily through intestinal absorption and not through destruction of erythroid cells. The increased amount of iron absorbed remained unutilized by erythron and deposited in parenchymal cells. Extreme IE overrides the positive effect of iron overload on HAMP levels resulting in continued increase in iron absorption. Non-transferrin-bound iron (NTBI) further contributes to this parenchymal iron overload. The study also shows that in thal intermedia the net increase in organ iron content is governed by balance between IE and iron overload. Thus in young mice, *Hamp* was down-regulated, whereas in old mice the *Hamp1* levels were similar to wt, though still low as compared to their iron stores. Although net increase in iron content of the body continues due to a persistent increase in *Fpn1* mRNA expression or the protein levels, factors mediating this persistent up-regulation of Fpn1 in thal intermedia need to be studied. A possible role of *Hfe* and *Cebpa* (CCAAT/enhancer binding protein-α) in influencing *Hamp1* expression is also indicated by their down-regulation in thal intermedia mice expressing low *Hamp1* levels.

Similar results have been observed by other group of researchers [DeFranceschi et al., 2006]. However, a decrease in Hamp1 has not been reported by them in Th3/+ mice at 4, 8, and 10 months of age. The reduced Hamp1 levels observed in thal intermedia mice at 2 months of age may be explained on the basis of total iron content at that early age which may be too low to counter the negative effect of anemia on Hamp levels.TfR1 and Fpn1 were not altered in the liver and duodenum of both thal intermedia mouse models but were highly up-regulated in thal major transplanted mice, (th3/th3)tp. Furthermore, this up-regulation was shown to be due to extramedullary hematopoiesis by estimation of alpha globulins and Tfr1 expression [Gardenghi et al., 2007].

The results of the above studies done on mouse models may not be exactly reproducible in humans but certainly provide close corroborations. Decreased urinary hepcidin levels in thal intermedia patients do provide a basis for the postulations derived from the above-mentioned studies.

Tfr2 expression was also found to be decreased in both thal intermedia (Hbb$^{th3/+}$) and thal major (Hbb$^{th3/th3}$) mouse models [Weizer-Stern et al., 2006]. Precise mechanisms affecting TfR2 expression are not known. Non-transferrin-bound iron (NTBI) is known to be elevated in thalassemia patients [al-Refaie et al., 1992]. The finding of increased expression of NGAL, involved in transport of NTBI, in thalassemia mouse models is supportive of the role of NTBI proteins in abnormal regulation of iron in thalassemia [Weizer-Stern et al., 2006].

3 HEME SYNTHESIS AND ITS ROLE IN REGULATION OF ERYTHROPOIESIS

Heme is synthesized in mitochondria of erythropoietic tissue in bone marrow and hepatocytes. Erythropoietic cells require a large amount of iron for synthesis of hemoglobin. Increase in number of transferrin receptor expression occurs on the surface of cells during erythroid differentiation. While 60–70% of iron taken up by hepatocytes is stored as ferritin, the rest is transported to mitochondria for synthesis of heme. Of the mitochondrial iron in hepatocytes, 30–50% is heme and 50–70% is nonheme iron. Sixty percent of the nonheme iron pool of hepatic mitochondria consists of labile iron pool utilized for heme synthesis, and the rest is present in iron-containing proteins including Fe–S cluster proteins.

In erythroid cells, mitochondria play a major role, because 90% of iron taken up by erythroid cells is used for heme biosynthesis. The first and last three steps in the biosynthesis of heme take place in mitochondria. δ-Aminolaevulinic acid synthase (ALAS) catalyzes the first step of condensation of succcinyl-CoA and glycine to form ALA. The final step of heme synthesis of insertion of iron in the protoporphyrin IX ring is catalyzed by the enzyme ferrochelatase. Absence of ferrochelatase leads to accumulation of PPIX in cells causing erythrocytic protoporphyria (EPP), a metabolic disorder in humans characterized by liver dysfunction and cutaneous photosensitivity.

There are major differences in heme synthesis in hepatic and erythroid mitochondria, that is, there are differences in activity of the enzymes involved, regulation and utilization of intracellular iron pool, and storage and transport of heme to cytosol. The studies so far indicate that biosynthesis of heme is tightly coupled with homeostasis of iron in erythroblasts. Intracellular transport mechanisms, cellular mobilization, and distribution of iron and its regulation are not well understood. Although a number of factors and proteins related to iron have been identified, their biological function and significance needs to be studied.

3.1 Role of Heme in Globin Regulation and Erythroid Differentiation

Heme regulates transcription of globin genes through its binding to the transcriptional factor Bach1 during erythroid differentiation [Taketani, 2005; Igrashi and Sun, 2006]. Heme controls the translation in erythroid precursors by modulating eIF2 α kinase activity through heme-regulated eIF2 α kinase, HRI [Chen, 2000]. This control of translational function of globin genes by HRI prevents excessive synthesis of globin proteins or chains. HRI was discovered in reticulocytes produced under conditions of iron or heme deficiency. HRI activity causes inhibition of protein synthesis at initiation by disaggregation of polysomes [Grayzel, 1966; Legon et al., 1973]. HRI acts by phosphorylating the α subunit of eIF2 (erythroid initiation factor 2), a key translation initiation factor. Heme binding affinity of HRI is of the same order as that of Bach1 and heme oxygenase (HO). HRI activity is initiated by multiple phosphorylation in heme deficiency. Whereas presence of hemin inhibits binding of ATP to HRI in a concentration-dependent manner, thus inhibiting kinase activity of HRI [Chen et al., 1991]. HRI has two distinct types of heme-binding sites [Chefalo et al., 1998]. One type is stable with incorporation of endogenous heme, while the other type is available for reversible binding of exogenous heme. It is this second type of site which is likely to be involved in down-regulation of HRI activity by heme [Chen, 2007]. HRI belongs to a family of four eIF2 α kinases. The other eIF2 α kinases expressed in erythroid precursors are GCN2 protein kinase and PKR, the RNA-dependent eIF2 α kinase. The former gets activated by nutrient starvation stress, and the latter responds to viral infection [Kaufman, 2000]. The fourth kinase is the resident endoplasmic reticulum (ER) kinase, PERK, activated by ER stress [Ron and Harding, 2000]. Only HRI gets activated by heat shock, osmotic shock, and arsenite shock in the presence of sufficient heme.

3.2 Pivotal Role of HRI in Microcytic Hypochromic Anemia

Heme-regulated eIF2 α kinase activity controls the translation in erythroid precursors through the heme-regulated inhibitor eIF2 α kinase (HRI) [Chen, 2000]. HRI is essential to reduce excessive synthesis of globin proteins and phenotypic severity of anemias in IDA, EPP, and Thal. HRI is responsible for the morphologic phenotype of hypochromic microcytosis in these three anemias studied by experimental mouse models.

Action of HRI in controlling globin chain synthesis by sensing cellular heme levels has been proven by experiments involving $Hri-/-$ mice [Chen, 2007]. In absence of HRI, the globin chain synthesis would continue, leading to precipitation of these chains and damage and death of cells. This would be responsible for the hemolysis of the Hb-deficient cells in IDA. HRI-deficient mice also show marked erythroid hyperplasia in marrow and spleen and an accentuated decrease in red cell counts, due to death of late erythroid precursors by precipitated globin chains, seen as variably sized inclusions in reticulocytes and to a lesser extent, in mature red cells. Furthermore, the phenotype of the red cell was slightly hyperchromic and normocytic. Thus, HRI is critical and solely responsible for the adaptation of microcytic hypochromic phenotype in IDA [Chen, 2007]. Similar results were observed in $Hri-/- fech^{m1pas/m1pas}$ mice deficient in the enzyme ferrochelatase required for incorporation of iron in PPIX, the last step in heme biosynthesis. This leads to erythrocyte protoporphyria due to accumulation of PPIX in liver as in humans.

3.3   Role of HRI in Beta Thalassemia Intermedia

In both of the previous microcytic hypochromic anemias, the stress was heme deficiency. However, HRI has been known to be activated by nonheme stress like oxidative stress, osmotic shock, and heat shock but is not activated by ER stress, nutrient starvation, and viral infection, which are the stress activators for other eIF2α kinases, namely, PERK, GCN2 and PRK. That HRI plays a protective role in conditions of nonheme stress also was corroborated by studies in thal intermedia mouse models [Chen, 2007]. The $Hbb-/-$ mice lacking both copies of β-globin major gene showed activation of HRI and exhibited thal intermedia phenotype with severe microcytic hypochromic anemia simlar to the thal intermedia phenotype seen in humans. That activation of HRI was essential for survival of $Hbb-/-$ mice was demonstrated in $Hri/Hbb-/-$ mice, which had a much more severe phenotype (being much smaller and paler at E 15.5) and died uniformly by E18.5. The red blood cells from these mice showed increase in inclusions composed entirely of α-globin chains despite the presence of $β^{minor}$-globin in soluble hemoglobin. The increased phenotypic severity of RBCs in mice lacking one copy of Hri gene showed that both copies of Hri were required to mitigate the severity of thalassemic phenotype by reducing production and denaturation of α-globin chains. These studies also indicate that HRI might turn out to be a strong nonglobin gene genetic modifier of thalassemia phenotype in humans [Chen, 2007].

3.4   Iron and Pathobiology of Thalassemia

Accumulation of unpaired globin chains is a cardinal feature of the pathophysiology of thalassemia, and their precipitation and denaturation results in liberation of both heme and nonheme iron [Shinar and Rachmilewitz, 1990; Scott et al., 1993]. This leads to destruction of cell membranes leading to premature cell death. Deposition of free iron (i.e., nonheme, nonferritin iron), on the cytoplasmic surface of the red blood cells in thalassemia was also demonstrated by Repka et al. [1993]. That this free cellular iron redistribution leads to membrane injury and ultimately cell death through oxidative cytochemistry of iron by production of reactive iron species and hydroxyl radicals was elucidated in further studies in mice models of severe beta thalassemia [Advani et al., 1992] and by Olivieri et al. [1994a]. To corroborate the hypothesis suggested by these studies that iron-mediated cytochemical reactions on the cytoplasmic surface of red cell membranes is of fundamental significance in the pathophysiology of thalassemias, Browne et al. [1997] carried out studies on mice homozygous for murine beta globin gene deletion having a thalassemia intermedia phenotype. They subjected these mice to iron chelation by administration of oral iron chelator deferiprone (L1), which is known to permeate cell membranes and successfully removed iron deposits on cytoplasmic surface of the red blood cells. The study, which provided experimental evidence of reduced membrane iron deposits and reduced oxidative membrane damage, provided evidence of amelioration of hematologic parameters with a significant increase in hematocrit (Hct) and mean cell volume (MCV) and a decrease in mean corpuscular hemoglobin (MCH). The most important result was the increased RBC survival in L1-treated mice ($t_{1/2}$ of 13.2 ± 1.2 days) as compared to saline-treated control mice ($t_{1/2}$ of 11.8 ± 1.6 days) with a p value of 0.007. The evidence and postulations suggested by the study were very encouraging. The researchers suggested the possibility of improvement in human pathobiology of thalassemia, by use of iron chelators which can permeate cell membrane, by decreasing not only mature red cell lysis but also intramedullary hemolysis.

3.5 Iron Storage and Its Effects on Parenchymal Tissues and Organs

Ferritin is the ubiquitous iron storage protein that sequesters the toxic unutilized iron in its core. The multimeric protein, with 24 heavy (H) and light (L) chain subunits, can accommodate 4500 atoms of iron in its ferric hydroxide core [Harrison and Arosio, 1996]. These H- and L-chain mRNAs of ferritin contain IREs that are sensitive not only to the availability of iron but also to the oxidative status of the cell. Thus ferritin responds not only to iron levels but also to oxidative stress and inflammation. The IRP-IRE regulatory mechanism for ferritin is iron-dependent. Binding of IRP to IREs of ferritin- and erythroid-specific δALAS mRNAs blocks the translation of mRNAs into proteins, leading to their decrease in iron-deficient conditions. Release of IRPs under iron-loaded conditions triggers synthesis of these proteins to sequester excess intracellular iron. The H subunit is overexpressed and is the main regulator of ferritin activity. Overexpression of H subunit in MEL (murine erythroleukemia) cells led to production of an iron-deficient phenotype [Picard et al., 1996]. The cells overexpressing H chains have also shown to be resistant to oxidative stress probably due to removal of potentially toxic ferrous ions [Epsztejn et al., 1999].

Increased iron accumulation in the parenchymal tissues of different organs like heart, endocrine glands, and liver in the form of ferritin leads to siderosis of the organs, affecting their function and ultimately leading to organ failure. Iron accumulates initially in the reticuloendothelial system and then in heart and the endocrine glands [Rund and Rachmilewitz, 2005; Forget, 1993]. Although ferritin is the major storage form of iron, its serum levels do not correlate very well with overall iron burden in condition of iron overload. Serum ferritin levels are known to increase in conditions of stress and inflammation. Moreover, iron storage may be organ-specific, thus requiring evaluation of tissue siderosis of specific organs.

4   EFFECT OF TRANSFUSIONAL IRON OVERLOAD ON IRON HOMEOSTASIS AND MORBIDITY AND MORTALITY

4.1   Iron Homeostasis in Transfusional Iron Overload

Regular chronic red cell transfusions are a part of the standard management strategy for thalassemia major and severe cases of thalassemia intermedia. The regimen significantly overcomes the anemia and ineffective erythropoiesis-related effects on growth retardation and hypoxic cardiac failure and pulmonary hypertension, increasing the lifespan of the thalassemic patients. However, the resultant iron overload leads to significant morbidity and mortality, even in presence of chelation therapy. In order to manage or avoid the attendant complications, it is necessary to understand the changes in iron homeostasis in these multiply transfused patients.

To establish clinical correlates observed in animal models, several studies have been undertaken in thalassemia and sickle cell patients on hypertransfusion regimens.

It has been observed that despite expression of IRP/IRE-mediated iron regulatory mechanisms in hepatic tissue, both iron uptake and storage continue to occur in liver in hypertransfusion iron overload. Jenkins and colleagues studied five iron homeostatic genes—genes for iron regulatory proteins (*IRP1 and IRP2*), ferritin (*FTH* and *FTL*), and hepcidin (*Hamp*)—in liver tissue of thal and sickle cell disease patients on hypertransfusion regimen [Jenkins et al., 2007].

Liver biopsies were taken within 3 days of the last transfusion. This fact needs to be taken into account while analyzing the results because these patients essentially had their anemia corrected, which is one of the major factors affecting iron homeostasis. *Hamp* expression was increased five- to eight-fold, reflecting most likely the domination of iron overload signals over anemia or hypoxia signals. A significant increase was also observed in IRP1 and IRP2 mRNA expression. The responses again appear to be different from predicted responses in cells with increased iron content. Expression of *FTH* and *FTL* showed no significant alteration, though a marked decrease in H:L subunit mRNA ratio was observed in comparison to the H:L subunit proteins [Jenkins et al., 2007].

The above observations from the solitary study need to be substantiated by further studies in humans before drawing any postulations. However, they do provide an indication that regulatory mechanisms at the transcriptional and posttranscriptional level undergo adaptive changes in response to transfusion-corrected anemia with concomitant appropriate chelation therapy as compared to uncorrected anemia or partially corrected anemia with no or inappropriate iron chelation.

4.2 Transfusion Iron Overload-Associated Morbidity and Mortality

Iron overload is the major complication of chronic regular transfusion therapy in thalassemia patients. Iron-related growth failure, hypogonadism, diabetes, hypothyroidism, and cardiac disease are the complications contributing to morbidity and quality of life. Hospitalization is required for treatment of cardiac failure and hepatic failure. Mortality is considerably high as compared to normal adult population, with cardiomyopathy being the leading cause of death in transfused thal subjects (55). Endocrine dysfunction, liver disease, and heart disease are the major complications of long-term transfusion therapy.

4.3 Endocrinopathy in Thalassemia

Endocrine organs are the earliest to show iron-related dysfunction. A multicenter study of a cohort of thal patients with comparable groups of patients of sickle cell disease (SCD) showed that thal subjects had a greater risk of developing iron-related endocrinopathy [Fung et al., 2007]. Hypogonadism was the most common of endocrinopathies followed by growth failure. That gonadotrophs have a special requirement for iron was shown by Atkin and colleagues [Atkin et al., 1996], and the observations were confirmed by histologic studies [Bergeron and Kovacs, 1978]. They were found to be the earliest cells of the pituitary to be affected by transfusional siderosis [Bergeron and Kovacs, 1978]. Hypogonadotrophic hypogonadism due to transfusional iron overload is partially responsive to GnRH therapy but not when organ damage is severe [Chatterjee and Katz, 2000]. Diabetes and thyroid dysfunction have been found to be the least prevalent according to several studies [Fung et al., 2007; Cunningham et al., 2004]. Overall incidence of endocrinopathy of 56% in thal patients was significantly higher than the 13% incidence in transfused SCD patients [Anderson, 1999]. Cellular damage to the siderotic tissues takes place through the Fenton reaction leading to generation of hydroxyl radicals. In adult patients, where assessment of iron-induced damage was made by noninvasive method of T2 relaxometry, small pituitary size with low T2 values was explained by shrinkage of the gland by iron deposition and cell death. In pancreas, where the damage to beta cells is selective, the high T2 values reflected siderosis followed by cell death and fatty degeneration of the pancreatic tissue. Both pituitary and pancreatic siderosis show an increase with age [Argyropoulou, 2007].

4.4 Liver Disease

The liver is the earliest organ to show iron overload with increased parenchymal deposition apparent after 2 years of transfusion therapy [Cohen, 1987]. Although hepatomegaly is one of the most common presentations of thalassemia, multiple causes lead to its occurrence including extramedullary hematopoiesis, iron deposition, and viral infections. Hepatocytes are the primary site of deposition of iron resulting from increased absorption secondary to ineffective erythropoiesis. Fibrosis and cirrhosis ultimately leads to decompensated liver failure. However, in most cases, cardiac failure precedes liver failure.

4.5 Heart Disease

Heart disease is the leading cause of mortality in all groups of thalassemia patients. In untreated thal major patients, iron-induced ventricular dysfunction is the major cause, whereas in thal intermedia patients belonging to a higher age group, pulmonary hypertension and LV remodeling dominate the manifestations of cardiac disease. There has been a marked improvement in cardiac outcome with chronic regular transfusion and appropriate iron chelation, with one of the studies reporting heart failure in <4% of patients at the mean age of 28 years [Aessopos et al., 2005]. Iron-induced cardiomyopathy is characterized by restrictive left ventricular (LV) filling affecting echocardiographic diastolic indexes. This occurrence of restrictive filling, as indicated by echocardiography, was noted in both fit thal major patients and those with evident heart disease and probably reflects the effect of iron overload [Aessopos et al., 2005].

5   EVALUATION AND MANAGEMENT OF IRON OVERLOAD

5.1   Evaluation of Iron Overload

For effective management of iron overload, an accurate assessment is a precondition to avoid toxic side effects of the chelators and to avoid excessive chelation. Paucity of such markers or evaluation methods has also been a part of the difficult management of thalassemic disorders. A brief discussion of available markers and tools with their limitations follows.

Serum ferritin (SF) is the most widely available and most commonly used assay for measuring iron content of the body. One microgram of ferritin corresponds to 8–10 mg or 120 μ of storage iron/kg body weight. Despite its proven competency as a marker for iron deficiency, its usefulness in assessing iron in conditions characterized by iron excess such as hemochromatosis—and more so in thalassemia—is limited. Part of it can be attributed to its response to inflammation. But serum ferritin levels have not proved to be a consistent predictor of liver iron or cardiac disease. An SF level of 1000 μ/liter is the recommended value to be maintained in thalassemics on a regular transfusion regimen [Capellini et al., 2000]. However, in other studies SF levels of 2500 μg/liter and 1500 μg/liter have been found to be associated with significantly less heart disease. On the other hand significant cardiac complications have been found at SF levels below 1000 μg/liter. Moreover, the fall in SF appears to be more rapid as compared to actual liver tissue iron on effective chelation therapy. The difference in serum ferritin levels and tissue iron levels in liver may also be due to serum ferritin reflecting reticuloendothelial iron rather than parenchymal iron [Olivieri et al., 1994b; Telfer et al., 2000; Fischer et al., 2003]. These observations limit the use of SF as a predictor of cardiac and liver disease in thalassemic patients on iron chelation therapy. Recently, a systematic evaluation of serum ferritin (SF) for LIC against magnetic susceptometry (SQUID) showed that SF underestimated liver iron concentration (LIC) more in non transfused Thal patients than in transfused patients of thalassemia [Pakbaz et al., 2007].

Liver biopsy has remained the gold standard for measurement of LIC for years. However, it can be inaccurate because of fibrosis, cirrhosis or uneven distribution of iron. An LIC of >15 mg/g dry wt was considered as an index of high risk of death from cardiac disease [Brittenham et al., 1994] and for liver cirrhosis and fibrosis [Angelucci et al., 2002], as 59 patients >7 years of age who died had LIC greater than this. $7-15$ mg/g dry wt was suggested as high risk for HCV-positive patients and <7 no risk. LIC as predictor of cardiac risk is questioned on the basis of $T2^*$ MRI studies which showed no correlation. Differential action of chelators led to decreased myocardial iron and improved cardiac function in patients treated with deferiprone.

Endomyocardial biopsy of the right atrium for estimation of myocardial iron, besides being a high risk procedure, is inappropriate as iron locates mainly in the ventricles. MRI-based method $T2^*$MRI is now considered to be the most appropriate technique for assessment of cardiac iron, and a value of <20 msec is found to correlate with cardiac dysfunction detected by echocardiography [Anderson et al., 2001a,b].

Magnetic resonance imaging (MRI)-based noninvasive techniques have come to be regarded as a reasonably good noninvasive method for assessment of organ-specific iron overload.

SQUID (superconducting quantum interface device) biomagnetic liver susceptometry is a precise, reproducible, and noninvasive method for estimating liver iron concentration but is limited by its availability [Beutler et al., 2003].

5.2   Basis of Iron Chelation Therapy and Iron Chelator Drugs

In TM, about $100-200$ ml of pure red cells/kg/year are transfused. In TI, iron absorption is around 0.1 mg/kg /day, $5-10$ times the normal amount [Beutler et al., 2003]. Iron chelation therapy has to be given to prevent organ damage and morbidity and mortality due to their dysfunction. An effective chelator should be able to attain negative iron balance to prevent, control, and, wherever possible, reverse the damage caused by excess iron deposition. Monitoring for toxic effects and assessment of damage to tissues and organ functions forms an important part of the iron chelation therapy.

Deferrioxamine (DFO), a hexadentate chelator binding to iron atoms in 1:1 ratio, does not permeate cell membranes and thus is not absorbed orally. It is given as continuous intravenous infusion or subcutaneous infusion over $8-12$ hr. Deferiprone (DFP) is an orally active bidentate chelator with three molecules binding one atom of iron. The drug has low efficacy with a short half-life and easy degradation. However, its ability to penetrate membranes and remove iron from intact red blood cells and from RES as well from transferrin, ferritin, and hemosiderin is being taken as an advantage and is likely to have an increased role in iron chelation therapy. The latest addition to the group of chelators available for use is Deferasirox, an orally active tridentate chelator with two molecules binding with one atom of iron and a longer half-life of $8-12$ hr, which permits once daily administration. Besides the prohibitive cost, long-term results on safety and efficacy are yet to be available. Some other key issues regarding its efficacy and dosage remain to be resolved for its appropriate use alone or in combination.

According to a review published in 2003 [Beutler et al., 2003], 72,000 patients of Thal major and Thal intermedia received regular blood transfusions worldwide and about 42,000 received no chelation. Twenty-five thousand patients were prescribed desferal and about 5000 received deferiprone (mainly in India). Some 2000–4000 thalassemia patients died each year of iron overload. Lack of a safe, cheap, and orally active chelator was found to be the main reason for no chelation therapy to these patients. In many of the developing and poor countries, the option of transfusion was not considered at all because transfusion without chelation would inevitably lead to death. Hence, only 3500 of the 27,000 children born with transfusion-dependent disease were receiving transfusions.

However, presently with availability of three chelators and the possibility of development of newer chelators, the clinician managing iron overload has an option of combination therapies. Some other compounds and modulations of the existing ones are under evaluation in various laboratories worldwide. Detailed discussion is beyond the scope of this chapter.

5.3 Potential Role of Iron Chelation Therapy in Improving Basic Pathophysiology of Beta Thalassemia

Besides reducing morbidity and mortality due to transfusion siderosis, with better understanding of properties of chelator drugs, iron chelation therapy has a possible role in amelioration of the basic pathophysiology of thalassemia.

As earlier mentioned, DFP, already shown to remove red cell membrane iron in murine thalassemia, has a potential role in reducing peripheral hemolysis and marrow ineffective erythropoiesis. Another potential role of chelators is in their ability to remove NTBI, the most toxic of iron species present in most patients of thalassemia major and intermedia.

6 SUMMARY

Iron plays a significant role in modifying thalassemia, which is essentially a globin chain disorder. Tight regulatory mechanisms involved in absorption and uptake of iron contribute to fundamental pathobiology of disease affecting its phenotypic severity. Ineffective erythropoiesis with its attendant increased iron absorption is one of the major pathways altering the ferrokinetics of anemia in thalassemia Heme-regulated globin synthesis by way of heme-regulated inhibitor of globin synthesis plays a major role in phenotypic manifestations and their severity. The biochemical properties of iron, related to its toxicity to membranes and organelles, also contribute significantly to organ tissue damage and complications in natural history of disease. Highly secure iron sequestration mechanisms involving ligands like ferritin greatly affect management regimens and their outcomes. The unique properties of iron also render itself to difficulties in its evaluation, assessment, and chelation. Thus in the globin chain disorder, thalassemia, iron plays a major role in modifying and affecting all aspects of the disease. Deeper understanding of the complex mechanisms of iron homeostasis at genetic and cellular level and its correlation with clinical aspects can transform the clinical outcome of the disease.

REFERENCES

Advani R, Rubin E, Mohandas N, Schrier SL. 1992. Oxidative red blood cell membrane injury in the pathophysiology of severe mouse β-thalassemia. *Blood* **79**:1064–1067.

Aessopos A, Farmakis D, Deftereos S, Tsironi M, Tassiopoulos S, Moyssakis I, Karagiorga M. 2005. Thalassemia heart disease. *Chest* **127**:1523–1530.

al-Refaie F, Wickens D, Wonke B, Kontoghiorghes G, Hoffbrand A. 1992. Serum non-transferrin-bound iron in beta-thalassemia major patients treated with desferrioxamine and L1. *Br J Haematol* **82**:431–436.

Anderson GJ. 1999. Hephaestin a ceruloplasmin homologue implicated in intestinal iron transport is defective in the sla mouse. *Nat Genet* **21**:195–199.

Anderson L, Bunce N, Davis B, et al. 2001a. Reversal of siderotic cardiomyopathy: A prospective study with cardiac magnetic resonance (CMR) [Abstract]. *Heart* **85**(suppl 1):33.

Anderson L, Holden S, Boris B, et al. 2001b. Cardiovascular T2-star (T2*) magnetic resonance for the early diagnosis of myocardial iron overload. *Eur Heart J* **22**:2171–2179.

Andrews N. 2004. Iron deficiency and related disorders. In: Greer JP, Foerster J, Leukens NJ, Rodgers GM, Paraskevas F, Glader B, editors *Wintrobe's Clinical Hematology* Philadelphia: Lippincott Williams & Wilkins, pp. 979–1009.

Angelucci E, Muretto P, Nicolucci A, et al. 2002. Effects of iron overload and hepatitis C virus positivity in determining progression of liver fibrosis in thalassemia following bone marrow transplantation. *Blood* **100**:17–21.

Argyropoulou MI, Kiortsis DN, Astrakas L, Metafratzi Z, Chalissos N, Efremidis SC. 2007. Liver, bone marrow, pancreas and pituitary gland iron overload in young and adult thalassemic patients: A T2 relaxometry study. *Eur Radiol* **17**(12):3025–3030.

Atkin SL, Burnett HE, Green VL, et al. 1996. Expression of the transferrin receptor in human anterior pituitary adenomas is confined to gonadotrophinomas. *Clin Endocrinol (Oxford)* **44**:467–471.

Bergeron C, Kovacs K. 1978. Pituitary siderosis: A histologic immunocytologic and ultrastructural study. *Am J Pathol* **93**:295–309.

Beutler E, Hoffbrand VA, Cook JD. 2003. Iron deficiency and overload. In: *Hematology*, Vol. 1., American Society of Hematology Education Program Book, 40.

Bridle KR, Frazer DM, Wilkins SJ, Dixon JL, Purdie DM, Crawford DH, Subramaniam VN, Powell LW, Anderson GJ, Ramm GA. 2003. Disrupted hepcidin regulation in HFE-associated haemochromatosis and the liver as a regulator of body iron homoeostasis. *Lancet* **361**:669–673.

Browne PV, Shalev O, Kuypers FA, Brugnara C, Solovey A, Mohandas N, Schrier SL, Hebbel RP. 1997. Removal of erythrocyte membrane iron in vivo ameliorates the pathobiology of murine thalassemia. *J Clin Invest* **100**:1459–1464.

Camaschella C, Merlini R. 2005. Inherited hemochromatosis from genetics to clinics. *Minnerva Med* **96**:207–222.

Capellini N, Cohen A, Eleftheriou A, Piga A, Porter J, editors. 2000. In: *Guidelines for Clinical Management of Thalassemia Nicosia.* Cyprus: Thalassemia International Federation.

Chatterjee R, Katz M. 2000. Reversible hypogonadotrophic hypogonadism in sexually infantile male thalassaemic patients with transfusional iron overload. *Clin Endocrinol* **53**:33–42.

Chefalo P, Oh J, Rafie-Kolpin M, Chen J-J. 1998. Heme-regulated eIF-2a kinase purifies as a hemoprotein. *Eur J Biochem* **258**:820–830.

Chen J-J. 2000. Heme-regulated eIF-2α kinase. In: Sonenberg N, Hershey JWB, Matthews MB, editors. *Translational Control of Gene Expression.* Cold Spring Harbor, NY: Cold Spring Harbor Laboratory Press, pp. 529–546.

Chen J-J. 2007. Regulation of protein synthesis by the heme regulated eIF2α kinase: Relevance to anemias. *Blood* **109**:2693–2700.

Chen J-J, Pal JK, Petryshyn R, et al. 1991 Amino acid microsequencing of the internal tryptic peptides of heme-regulated eukaryotic initiation factor 2α subunit kinase: Homology to protein kinases. *Proc Nat Acad Sci USA* **88**:315–319.

Cohen A. 1987. Management of iron overload in the pediatric patient. *Hematol Oncol Clin North Am* **1**:522–544.

Cunningham MJ, Macklin EA, Neufield EJ, Cohen AR. Thalassemia Clinical Research Network. 2004. Complications of beta thalassemia in North America. *Blood* **104**:34–39.

DeFranceschi L, Daraio F, Filippini A, et al. 2006. Liver expression of hepcidin and other iron genes in two mouse models of beta-thalassemia. *Haematologica* **91**:1336–1342.

Dessipyris EN, Sawyer ST. 2004. Erythropoiesis. In: Greer JP, Foerster J, Leukens NJ, Rodgers GM, Paraskevas F, Glader B, editors *Wintrobe's Clinical Hematology*, Philadelphia: Lippincott Williams & Wilkins, pp. 195–216.

Epsztejn S, Glickstein H, Picard V, Slotki LN, Breuer W, Beaumont C, Cabantchik ZI. 1999. H-ferritin subunit overexpression in erythroid cells reduces the oxidative stress response and induces multidrug resistance properties. *Blood* **94**:3593–3603.

Fischer R, Longo F, Nielsen P, et al. 2003. Monitoring long term efficacy of iron chelation therapy by deferiprone and desferrioxamine in patients with thalassemia major: Application of SQUID biomagnetic liver susceptometry. *Br J Haematol* **121**:938–948.

Fleming RE, Sly WS. 2001. Hepcidin: A putative iron regulatory hormone relevant to hereditary hemochromatosis and the anemia of chronic disease. *Proc Natl Acad Sci USA* **98**:8160–8162.

Forget BG . 1993. The pathophysiology and molecular genetics of beta thalassemia. *Mt Sinai J Med* **60**:95–103.

Fung EB, Harmatz P, Milet M, Ballas SK, DeCastro L, Hagar W, Owen W, Olivieri N, Smith-Whitley K, Darbari D, Wang W, Vichinsky E, Multi-Center Study of Iron Overload Research Group. 2007. Morbidity and mortality in chronically transfused subjects with thalassemia and sickle cell disease: A report from the multi-center study of iron overload. *Am J Hematol* **82**:255–265.

Ganz T. 2003. Hepcidin a key regulator of iron metabolism and mediator of anemia of inflammation. *Blood* **102**:783–788.

Gardenghi S, Marongiu MF, Ramos P, et al. 2007. Ineffective erythropoiesis in β-thalassemia is characterized by increased iron absorption mediated by down-regulation of hepcidin and up-regulation of ferroportin. *Blood* **109**(11):5027–5035.

Grayzel AI, Horchner P, London IM. 1966. The stimulation of globin synthesis by heme. *Proc Natl Acad Sci USA* **55**:650–655.

Harrison PM, Arosio P. 1996. The ferritins: molecular properties iron storage function and cellular regulation. *Biochim Biophys Acta* **1275**:161–203.

Hentze MW, Caughman SW, Rouault TA, Barriocanal JG, Dancis A, Harford JB, Klausner RD. 1987. Identification of the iron-responsive element for the translational regulation of human ferritin mRNA. *Science* **238**:1570–1573.

Igarashi K, Sun J. 2006. The heme-Bach1 pathway in the regulation of oxidative stress response and erythroid differentiation. *Antioxid Redox Signal* **8**:107–118.

Iwai K, Klausner RD, Roualt TA. 1995. Requirements for iron-regulated degradation of the RNA binding protein iron regulatory protein-2. *EMBO J* **14**:5350–5357.

Jenkins ZA, Hagar W, Bowlus CL, Johansson HE, Harmatz P, Vichinsky EP, Theil EC. 2007. Iron homeostasis during transfusional iron overload in β-thalassemia and sickle cell disease: Changes in iron regulatory protein, hepcidin and ferritin expression. *Pediatr Hematol Oncol* **24**(4):237–243.

Kaufman RJ. 2000. Double-stranded RNA-activated protein kinase PKR. In: Sonenberg N, Hershey JWB, Matthews M, editors. *Translational Control of Gene Expression*. Cold Spring Harbor, NY: Cold Spring Harbor Laboratory Press, pp. 503–528.

Kim S, Ponka P. 2002. Nitrogen monoxide-mediated control of ferritin synthesis: Implications for macrophage iron homeostasis. *Proc Natl Acad Sci USA* **99**:12214–12219.

Klausner RD, Rouault TA, Harford JB. 1993. Regulating the fate of mRNA: The control of cellular iron metabolism. *Cell* **72**:19–28.

Lee Gelbart T, West C, Halloran C, Beutler E. 2001. Seeking candidate mutations that affect iron homeostasis. *Blood Cells Mol Dis* **29**:471–487.

Legon S, Jackson RJ, Hunt T. 1973. Control of protein synthesis in reticulocyte lysates by haemin. *Nat New Biol* **241**:150–152.

Mckie AT, Latunde-Dada GO, Miret S, McGregor JA, Anderson GJ, Vulpe CD, Wrigglesworth JM, Simpson RJ. 2002. Molecular evidence for the role of a ferric reductase in iron transport. *Biochem Soc Trans* **30**:722–724.

Nemeth E, Tuttle MS, Powelson J, Vaughn MB, Donovan A, Ward DM, Ganz T, Kaplan J. 2004. Hepcidin regulates cellular iron efflux by binding to ferroportin and inducing its internalization. *Science* **306**:2090–2093.

Nicolas G, Bennoun M, Devaux I, et al. 2001. Lack of hepcidin gene expression and severe tissue iron overload in upstream stimulatory factor 2 (USF2) knockout mice. *Proc Natl Acad Sci USA* **98**:8780–8785.

Nicolas G, Chauvet C, Viatte L, et al. 2002. The gene encoding the iron regulatory peptide hepcidin is regulated by anemia, hypoxia, and inflammation. *J Clin Invest* **110**:1037–1044.

Nicolas G, Andrews NC, Kahn A, Vaulont S. 2004. Hepcidin, a candidate modifier of the hemochromatosis phenotype in mice. *Blood* **103**:2841–2843.

Olivieri NF, Nathan DG, MacMillan JH, et al. 1994a. Survival in medically treated patients with homozygous β-thalassemia. *N Engl J Med* **331**:574–578.

Olivieri O, De Franceschi L, Capellini MD, Girelli O, Corrocher R, Brugnara C. 1994b. Oxidative damage and erythrocyte membrane transport abnormalities in thalassemias. *Blood* **84**:315–320.

Pakbaz Z, Fischer R, Fung E, Nielsen P, Harmatz P, Vichinsky E. 2007. Serum ferritin underestimates liver iron concentration in transfusion independent thalassemia patients as compared to regularly transfused thalassemia and sickle cell patients. *Pediatr Blood Cancer* **49**(3):329–332.

Park CH, Valore EV, Waring AJ, Glanz T. 2001. Hepcidin, a urinary antimicrobial peptide synthesized in liver. *J Biol Chem* **276**:7806–7810.

Park M, Lopez MA, Gabayan V, Ganz T, Rivera S. 2006. Suppression of Hepcidin during anemia requires erythropoietic activity. *Blood* **108**:3730–3735.

Picard V, Renaudie F, Porcher C, Hentze MW, Grandchamp B, Beaumont C. 1996. Overexpression of the ferritin H subunit in cultured erythroid cells changes the intracellular iron distribution. *Blood* **87**:2057–2064.

Ponka PN, Lok CN. 1999. The transferrin receptor: Role in health and disease. *Int J Biochem Cell Biol* **31**:1111–1137.

Repka TO, Shalev R, Reddy R, Yuan J, Abrahamov A, Rachmilewitz EA, Low PS, Hebbel RP. 1993. Nonrandom association of free iron with membranes of sickle and β-thalassemic erythrocytes. *Blood* **82**:3204–3210.

Richardson DR, Ponka P. 1997. The Molecular mechanisms of the metabolism and transport of iron in normal and neoplastic cells. *Biochim Biophys Acta* **1331**.1–40.

Ron D, Harding HP. 2000. PERK and translational control by stress in endoplasmic reticulum. In: Sonenberg N, Hershey JWB, Matthews M, editors. *Translational Control of Gene Expression*. Cold Spring Harbor, NY: Cold Spring Harbor Laboratory Press, pp. 547–560.

Rouault TA, Tang CK, Kaptain S, Burgess WH, Haile DJ, Samaniego F, McBride OW, Harford JB, Klausner RD. 1990. Cloning of cDNA encoding an RNA regulatory protein— The human iron responsive element binding protein. *Proc Natl Acad Sci USA* **87**:7958–7962.

Rund D, Rachmilewitz E. 2005. Beta-thalassemia. *N Engl J Med* **353**:1135–1146.

Scott MD, van den Berg JJM, Repka T, Rouyer-Fessard P, Hebbel RP, Beuzard Y, Lubin BH. 1993. Effect of excess α-hemoglobin chains on cellular and membrane oxidation in model β-thalassemic erythrocytes. *J Clin Invest* **91**:1706–1712.

Shinar E, Rachmilewitz EA. 1990. Oxidative denaturation of red blood cells in thalassemia. *Semin. Hematol.* **27**:70–82.

Taketani S. 2005. Acquisition, mobilization and utilization of cellular iron and heme: Endless findings and growing evidence of tight regulation. *Tohuku J Exp Med* **207**:297–318.

Telen MJ, Kaufman RE. 2004. The mature erythrocyte. In: Greer JP, Foerster J, Leukens NJ, Rodgers GM, Paraskevas F, Glader B, editors. *Wintrobe's Clinical Hematology*. Philadelphia: Lippincott Williams & Wilkins, pp. 217–247.

Telfer PT, Prescott E, Holden S, et al. 2000. Hepatic iron concentration combined with long term monitoring of serum ferritin to predict complications of iron overload in thalassemia major. *Br J Haematol* **110**:971–977.

Weinstein DA, Roy CN, Fleming MD, Loda MF, Wolfsdorf JI, Andrews NC. 2002. Inappropriate expression of hepcidin is associated with iron refractory anemia: Implications for the anemia of chronic disease. *Blood* **100**:3776–3781.

Weizer-Stern O, Adamsky K, Amariglio N, Rachmilewitz E, Breda L, Rivella S, Rechavi G. 2006. mRNA expression of iron regulatory genes in β-thalassemia intermedia and β-thalassemia major mouse models. *Am J Hematol* **81**:479–483.

Wrighting DM, Andrews NC. 2006. Interleukin-6 induces hepcidin expression through STAT3. *Blood* **108**:3204–3209.

Yang J, Goetze D, Li JY, Wang W, Mori K, Setlik D, Du T, Erdjument-Bromage H, Tempst P, Strong R, Barasch J. 2002. An iron delivery pathway mediated by a lipocalin. *Mol Cell* **10**:1045–1056.

20 Health Implications: Trace Elements in Cancer

RAFAEL BORRÁS AVIÑÓ and JOSÉ RAFAEL LÓPEZ-MOYA

ABBA Chlorobia S.L., Citriculture Department, School of Agronomists, Polytechnic University of Valencia, 46022 Valencia, Spain

JUAN PEDRO NAVARRO-AVIÑO

ABBA Chlorobia S.L., Citriculture Department, School of Agronomists, Polytechnic University of Valencia, 46022 Valencia, Spain; and Department of Agrarian Sciences and of the Natural Environment, School of Technology and Experimental Sciences, University "Jaume I," 12071 Castellón, Spain

1 INTRODUCTION

"Heavy metals" is an inexact term used to describe a group of elements that are metals or metalloids (elements that have both metal and non-metal characteristics). Generally, heavy metals have densities above 5 g/cm^3. Unlike organic contamination such as the originated by fuel, or pesticides, the contamination generated by heavy metals is persistent in all parts of the environment because they cannot be degraded or destroyed. Human activity affects the natural geological and biological redistribution of heavy metals through pollution of the air, water, and soil. The primary anthropogenic sources of heavy metals are point sources such as mines, foundries, smelters, and coal-burning power plants, as well as diffuse sources such as combustion by-products and vehicle emissions. Humans also affect the natural geological and biological redistribution of heavy metals by altering the chemical form of heavy metals released to the environment.

Heavy metals are associated with numerous adverse health effects, including allergic reactions (e.g., chromium), neurotoxicity (e.g., lead), nephrotoxicity (e.g., mercuric chloride, cadmium chloride), and cancer (e.g., arsenic, hexavalent Cr). Humans are often exposed to heavy metals in various ways, mainly through the inhalation of metals in the workplace or polluted neighborhoods, or through the ingestion of food (particularly fish and seafood) that contains high levels of heavy metals.

There are five heavy metals that are commonly cited as being of the greatest public health concern: lead, cadmium arsenic, mercury, and hexavalent chromium. Apparently in general there is no biological need for any of these heavy metals, but they are very useful for men. Cadmium, for instance, has many commercial applications, including electroplating and the manufacture of batteries. However, exposure to cadmium (as for most of heavy metals) can occur in the workplace or from contaminated foodstuffs and can result in emphysema, renal failure, cardiovascular disease, and perhaps cancer.

1.1   General Nutritional and Medical Benefits

In small quantities, certain heavy metals are nutritionally essential for a healthy life. Some of these are referred to as trace elements (e.g., iron, copper, manganese, zinc, etc.). These elements, or some form of them, are commonly found naturally in foodstuffs, in fruits and vegetables, and in commercially available multivitamin products. Diagnostic medical applications include direct injection of gallium during radiological procedures, dosing with Cr in parenteral nutrition mixtures, and the use of lead as a radiation shield around X-ray equipment.

2   TOXIC HEAVY METALS

Heavy metals become toxic when they enter in excess and when they are not metabolized by the body and accumulate in the soft or hard tissues (e.g., Pb can accumulate in bones). Heavy metals may enter the human body through food, water, air, or absorption through the skin when they take place in contact with humans—for example, in agriculture and in manufacturing, pharmaceutical, industrial, or residential settings. Industrial exposure accounts for a common route of exposure for adults. Ingestion is the most common route of exposure in children [Roberts, 1999]. Children may develop toxic levels from the normal hand-to-mouth activity of small children who come in contact with contaminated soil or by eating objects that are not food (for instances paint chips). Less common routes of exposure are medical applications, as during a radiological procedure, from inappropriate dosing or monitoring during intravenous (parenteral) nutrition, or from a broken thermometer [Smith et al., 1997].

As a rule, acute poisoning is more likely to result from inhalation or skin contact to dust, fumes or vapors, or materials in the workplace. However, lesser levels of contamination may occur in residential settings, particularly in older homes with lead paint or old plumbing. The Agency for Toxic Substances and Disease Registry (ATSDR) located in USA, Atlanta, Georgia (a part of the U.S. Department of Health and Human Services) was established by congressional mandate to perform specific functions concerning adverse human health effects and diminished quality of life associated with exposure to hazardous substances. The ATSDR is responsible for assessment of waste sites and providing health information concerning hazardous substances, response to emergency release situations, and education and training concerning hazardous substances [ATSDR Mission Statement, November 7, 2001]. In cooperation with the U.S. Environmental Protection Agency, the ATSDR has compiled a Priority List for 2001 called the "Top 20 Hazardous Substances." The heavy metals arsenic (1), lead (2), mercury (3), and cadmium (7) appear on this list.

2.1 Mercury

2.1.1 Characteristics of Mercury. The name mercury comes from the Greek word "hydrargyrum" (which is a compound word meaning "water" and "silver," i.e., liquid silver). Hg is a transition metal, and that is an important characteristic that explains much of the physical, chemical, and biological behavior and toxicity (for all the heavy metals in general). As mentioned above, Hg can be included inside the group of the most toxic heavy metals. Hg is found in air, water, and soil and exists in several forms: elemental or metallic mercury, inorganic mercury compounds, and organic mercury compounds.

Elemental mercury (Hg^0) or metallic mercury is exceptional in one aspect: It is liquid at room temperature, with a state of oxidation (reduced element) that is rarely found in Nature. In fact, Hg is the only common metal that is in liquid state at ordinary temperatures. In this state, Hg is a shiny, silver-white, odorless liquid that slowly evaporates, forming a vapor. If elemental metallic mercury is heated, then it is a colorless and odorless gas, which makes it extraordinarily dangerous. Humans have the ability to convert this inorganic Hg to an organic form once it has become absorbed into the bloodstream. However, in Nature a chemical form normally much more present in contact with human beings is an organic compound: methyl mercury ($CH_3 \cdot HgCl$). This substance is primarily a product of the activity of microscopic organisms that live in the water and soil which convert elemental mercury to methyl mercury and dimethyl mercury, escaping into the atmosphere. Furthermore, methyl mercury is the chemical form that habitually metabolizes also the microorganisms and many living organisms, including human beings.

2.1.2 Localization. In Nature, to produce inorganic Hg compounds, Hg combines with other elements such as chlorine, sulfur, and oxygen to form Hg salts that habitually have the aspect of white powder or of crystals, with the exception of mercuric sulfide (HgS, present in cinnabar), which is red. Other examples of salts are: fulminate $Hg(ONC)_2$, used in explosives: Hg_2Cl_2, used in medicine and mercuric chloride: and $HgCl_2$, which is considered a poison and is used to disinfect wounds.

Hg is an extremely rare element in the earth's crust, having an average crustal abundance by mass of all forms of Hg in the earth's crust of only 0.08 parts per million. Because it does not blend geochemically with those elements that comprise the majority of the crustal mass, mercury ores can be extraordinarily concentrated considering the element's abundance in ordinary rock. As mentioned above, it is found either as a native metal (rare) or in cinnabar (principal ore in Nature) also in corderoite, livingstonite, and other minerals. Inorganic mercury (metallic mercury and inorganic mercury compounds) enters the air via mining ore deposits, coal combustion, and manufacturing plants. It enters in water or soil via natural deposits, disposal of wastes, and volcanic activity.

Both inorganic and organic mercury compounds are absorbed through the gastrointestinal tract and affect other systems via this route. However, organic mercury compounds are more readily absorbed via ingestion than inorganic mercury compounds. Elemental (metallic) mercury, however, primarily causes health effects when it is breathed as a vapor, where it can be absorbed through the lungs. Hg can also be absorbed through the unbroken skin.

2.1.3 Contamination by Hg.

The contact with Hg occurs mainly when breathing contaminated air, drinking contaminated water, and eating contaminated food, as well as from medical and dental procedures and from skin contact while working. The central nervous system is very susceptible to all the forms of Hg.

Hg is also released into the environment during hot weather, because it is retained in soils and is volatile. Another source is the Hg liberated from the working and smelting of ores such as of copper, gold, lead, silver, and zinc, which normally contain traces of Hg. However, one of the mayor sources of exposure is chlor-alkali plants. In general, the sources of Hg are industrial and domestic products, such as thermometers, batteries, and electrical switches, which account for a significant loss of Hg to the environment, ultimately become solid waste in major urban areas. Anthropogenic sources of airborne Hg may also arise from the operation of metal smelters or cement manufacture. It is also used in pesticides. For instance, elemental Hg has been employed in India to fumigate grain in closed containers and protected insect infestation. Twenty thousand tons of Hg are released into the environment each year by human activities such as combustion of fossil fuels and other industrial release.

Three characteristics make Hg a unique heavy metal toxic for living organisms, including human beings. These are: volatility, biotransformation, and bioconcentration or biomagnification. For example, regarding volatilization, in areas containing 10 ppm of Hg in soil, the concentration in near atmosphere is between 4 and 40 times higher. This problem of volatilization is also in both directions. The Hg that is present in the air subsequently is positioned in the land and in the water. This process of deposition and revolatilization takes place many times, with a residence time in the atmosphere of at least a few days. In the volatile phase it can be transported hundreds of kilometers.

Once deposited, certain microorganisms can change it into methyl mercury, which builds up in fish, shellfish, and animals that eat fish. That is, enters the food chain. Therefore there is a biotransformation (a chemical transformation made by an organism). This organic administration of Hg builds up more in some types of fish and shellfish than in others. The levels of methyl mercury in fish and shellfish depend on what they eat, how long they live, and how high they are in the scale of the food chain. In any case, fish can accumulate Hg to very high levels because accumulation is rapid and elimination is slow. Bioaccumulation therefore occurs when an organism absorbs a toxic substance at a rate greater than that at which the substance is lost. Consequently, predators (living organisms that are at a higher level in the food chain) achieve higher concentration than fish. Consequently, fish and shellfish are the main sources of methyl mercury exposure to humans. In the human body, methyl mercury and, for example, tetraethyl lead are organic compounds that can be accumulate as lipids and are stored in the body's fat; therefore when the fatty tissues are used for energy, the compounds are released and cause acute poisoning (Table 1).

The consequence is that human risk for accumulating Hg in the terrestrial and aquatic food chains mainly results through the consumption of fish from contaminated waters—especially predator species, such as tuna fish, swordfish, and other large oceanic fish, even if caught a considerable distance off shore. Mercury also comes from the following: other seafoods, including mussels and crayfish; fish-eating birds and mammals; and eggs of fish-eating birds. The EPA considers that the safe level of Hg provided by food in the human body has to be 0.1 μg/kg of weight.

2.1.4 Health in Humans. People consume Hg compounds of the environment by means of the inhalation of atmospheric dust, drinking groundwater, inhaling particles of dust from the floor or suspended in the air, and through the skin by direct contact with the soil or with the surface of contaminated water. Hg exposure at high levels can harm the brain, heart, kidneys, lungs, and immune system of people of all ages. It has also been demonstrated that high levels of methyl mercury in the bloodstream of unborn babies and young children may harm the developing nervous system, making the child less able to think and learn. Furthermore, researchers have found a positive correlation between complications seen in autistics and complications seen in Hg-poisoned individuals [Bernard et al., 2001].

2.1.5 Mercury and Cancer. There are few available data about the risk of cancer of the exposition to Hg [Boffetta et al., 1993]. However, EPA has determined that mercuric chloride ($HgCl_2$) and methyl mercury ($CH_3.HgCl$) are possible human carcinogens. Mercuric chloride has been found to cause increases in several types of tumors in rats and mice, and methyl mercury has been found to cause kidney tumors in male mice. Different groups of workers professionally exposed have

TABLE 1. Hg Biomagnification Found in Some Organisms[a]

Living Organism	Biomagnification
Freshwater fish	63,000
Marine water fish	10,000
Marine Invertebrates	100,000
Freshwater plants	1,000
Freshwater Invertebrates	100,000
Freshwater fish	1,000
Mycorrhizal fungi	63

[a]Biomagnification causes that the presence of legal Hg levels in water can be converted in high concern or danger.

been studied. The main conclusion that researchers deduced is that Hg did not produce cancer in the cases evaluated. Although, for instance, in three cases reported [Barregard et al., 1990; Ellingsen et al., 1993; Boffeta et al., 1998], different explanations are argued, the consistent fact is that an increment (although small) of the lung and liver cancer is produced.

2.2 Arsenic

2.2.1 Characteristics of As.
Arsenic is a metalloid. Elemental arsenic occurs in two solid modifications: yellow (molecular nonmetallic) and gray or metallic, with specific gravities of 1.97, and 5.73, respectively. The element is a steel gray, very fragile, crystalline, semimetallic (metalloid) solid. It is odorless and tasteless. It tarnishes in air and, when heated rapidly, oxidizes to arsenous oxide (arsenic trioxide As_2O_3), which has a garlic odor. It is commonly found as arsenite and arsenate compounds. Upon heating, arsenic and some minerals contain arsenic sublime.

2.2.2 Localization.
Arsenic is found as a free element occasionally, but is found more normally in a number of minerals. Mispickel (arsenopyrite, FeSAs) is the most common mineral. Locations include France, Germany, Italy, Romania, Siberia, and North America. It enters drinking water supplies from natural deposits in the earth or from agricultural and industrial practices. Arsenic also forms compounds mainly with halogens, oxygen, and hydrogen.

2.2.3 Contamination by Arsenic.
Arsenic occurs naturally in rocks, soil, water, air, plants, and animals. It can be further released into the environment through natural activities such as volcanic action, erosion of rocks, and forest fires, as well as through human actions. Arsenic compounds are used in making special types of glass, as a wood preservative, and, lately, in the semiconductor gallium arsenate, which has the ability to convert electric current to laser light. Arsine gas, AsH_3, has become an important dopant gas in the microchip industry, although it requires strict guidelines regarding its use because it is extremely toxic. Microorganisms release volatile methylarsines to the extent of 20,000 tonnes per year, but human activity is responsible for much more: 80,000 tonnes of arsenic per year are released by the burning of fossil fuels. However, for a long time the most frequent application of As has been that of insecticides, predominantly as lead arsenate and less often as arsenate and arsenite of calcium, sodium arsenite, arsenite of copper, and copper acetoarsenate ($C_4H_6As_6Cu_4O_{16}$). The last compound is a coloring agent, commonly called Paris Green, which, despite its toxicity, is used to produce some of the best blue colors when it is combined with potassium perchlorate. Arsenic derivates have also been used as herbicides for weeds. Arsenic is known to promote the growth in animals and was used as arsenilic acid in the diet of hogs and poultry. Also, it has been used in the glass industry, in electronic applications, in colors for digital clocks, in the textile industry, in pigments, in cosmetics and

plucking tools, in the industry of ceramics (AsO_5), in the firework industry (As_4S_4), in the metallurgical industry for cars radiators, and so on.

2.2.4 Arsenic and Health in Humans.

Arsenic is one of the most toxic elements that can be found in the environment. The long-term consequences of the exposition with regard to the inorganic forms of arsenic can be critical since these compounds are documented as carcinogenic and especially affect the liver. There are also some countries where the drinking water contaminated by natural sources has been related to cancer of skin.

Expositions prolonged for a long time are extremely harmful to human health, producing (as mentioned) diverse types of cancer, cardiovascular pathologies, diabetes, and anemia, as well as alterations in neurological, immune, and reproductive functions and in the development process. Arsenic is found in natural water as a dissolved species, which exists normally as oxyanions with arsenic in two states of oxidation: trivalent arsenic, arsenite [As(III)] and arsenic pentavalent, arsenate [As(V)]. The arsenic state of oxidation, and therefore its mobility, is essentially controlled by the redox conditions in the specific environment (redox potential) and by pH. The contents of arsenic in the subterranean, superficial, and natural water are not normally high.

There are anomalous cases, above all aquifers, that present exceptional values, as occurs in the case of Bangladesh. Inhabitants of Bangladesh and surrounding areas have been exposed to water that is naturally and heavily contaminated with As, causing what the World Health Organization has described as the worst mass poisoning in history [Jones, 2007]. The phenomenon of the existence of high contents of arsenic of natural origin in the water is controlled by three factors: first the primary source of arsenic (geosphere and atmosphere); second, the arsenic mobilization and retention processes in the interaction between the solid phase and the liquid phase; and finally the transportation of arsenic as watery species in the mass of the water.

Most of the aquifers with high contents of arsenic have an origin connected with natural geochemical processes. As opposed to the anthropogenic contamination, which generates a more local incidence, the occurrence of high concentrations of arsenic of natural origin affects to large areas. The abundant cases of natural "contamination" of groundwater by arsenic that exist in the world are related to very different geological environments: metasediments (sediment that shows evidence of having been subjected to metamorphism) with veins mineralized; volcanic formations; breakdown sedimentary formations; and tertiary and quaternary alluvial basins.

Arsenic is an extremely toxic element for humans—not only in high concentrations, where the exposition causes sharp effects that can become lethal, but also the exposition during a long period to low concentrations (e.g., by consumption of water). Therefore, As in superficial water (rivers, lakes, reservoirs) and subterranean water (aquifers), which may be utilized for consumption, constitutes a great threat for human health. This situation has motivated agencies such as the World Health Organization (WHO), the European Union (Managing 98/83), or the Agency of

American Environmental Protection (USEPA) to establish the reduction of the limit of the content of As in drinking water from 50 to 10 µg liter^{-1} [Jiménez Valverde, 1999].

2.2.5 Arsenic in the Subterranean Continental Water. In general, the As basal levels in groundwater are, for most cases, as low as 10 µg liter^{-1}. Nevertheless, the values cited in the literature for water in natural conditions define a very extensive rank between 0.5 and 5000 µg liter^{-1}. There are a great number of areas with groundwater that present contents of arsenic over 50 µg liter^{-1} in different places of the planet. The most important problems cited in the literature occur in Argentina, Bangladesh (Fig. 1), Chile, China, Ghana, Hungary, India, Mexico, Taiwan, Thailand, United Kingdom, and United States; the United States has

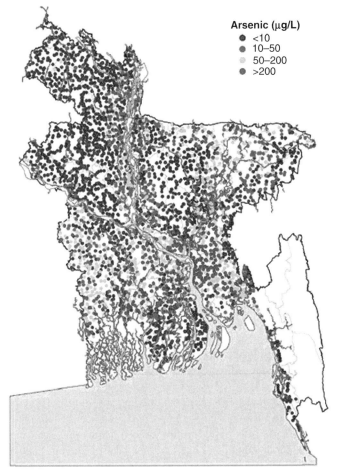

Figure 1. Distribution of arsenic in the water of Bangladesh. Arsenic concentration in groundwater (tubewells). Reported by British Geological Survey in 2001.

TABLE 2. Risk of Cancer[a]

Metal/Tissue	Hg	As	Cr	Cd	Pb
Lung	Possibly	Risk	Risk	Possibly	Possibly
Brain	Possibly				Concern
Stomach/exophage			Risk		Concern
Kidney	Risk	Risk			Concern
Colon					Risk
Rectal					No
Skin		Risk			
Prostate				Possibly	
Liver		Risk	Risk		
Pancreas				Risk	

[a]Different tissue-cancer produced by heavy metals from the reports consulted. The heavy metal form is considered in all of its different chemical possibilities.

received the least attention, whereas Bangladesh has been the object of deeper study. In all these places, human drinking water has been extracted from aquifers with high concentrations of As originated in natural sources (Table 3).

Areas with problems of As related to mineral deposits and the mining industry are known to exist in many parts of the world, with the most outstanding cases being those of Ghana, Greece, Thailand, Chile, and the United States. One of the most notable peculiarities of the problem of As of natural origin in groundwater is that there is not always a direct relation between the high content of As in the water and a high content of As in the materials that constitute the aquifer. There is no model (geological/hydrogeological) common for all the identified occurrences being found in water, with As occuring in various situations: reducing conditions as in oxidant conditions, or in exploited aquifers, in dry zones as in wetlands, or in superficial free aquifers as in deep confined aquifers. The presence of As in each case is the consequence of a geochemical environment and some specific hydrogeologycal conditions [Sarkar, 2002].

The pentavalent As is absorbed more easily than the trivalent, and the inorganic compounds are better than the organic ones. The absorption by the skin is faster for the trivalent As. After the absorption occurs in the lungs and through the gastrointestinal tract, 95–99% of the As is located in the erythrocytes being combined with the hemoglobin. The pentavalent As is excreted more quickly than the trivalent As, and the organic compounds are excreted more than the inorganic ones. Resembling Hg bioaccumulation, As bioaccumulation is also a problem. The bioaccumulation of As is very high in the zooplankton and in algae.

2.2.6 Arsenic and Cancer. Arsenic and phosphorus are right next in the same group of the periodic system. The compounds of As seem to be replacing the groups phosphate of DNA and can also react with enzymes by means of sulfhydryl groups. Arseniate interferes in the synthesis of ATP, forming unstable esters of arseniate. In some types of cells the As(V) can be reduced to As(III).

TABLE 3. As in Water of Different Countries

Country	Problem
Argentina	The most affected areas are extensive zones of the Pampa, some places of the plain of Chaco and some small areas of the Andes where the water of the wells contains from 50 to 2000 μg liter^{-1}.
Bangladesh	Near 50 million people drink water of wells with a content of arsenic over the acceptable limits. Damages related to the chronic exposition to As have been declared such as arsenical dermatosis, hiperkeratosis (excess of keratin production), hiperpigmentation and skin cancers. The demonstration of the first toxic effects was identified in the West of Bengala in 1978 but the agent that was poisoning to the population was not known until 1982. Near 200,000 people suffered arsenicosis. The aquifers that contained arsenic were confined to depths that went from the 20 to the 80 meters. The presence of arsenic cannot be attributed to anthropogenic activities developed in the region.
Chile	Approximately 400,000 north residents of Chile drink water originated in rivers that proceed of the mountains of the Andes. The high content of arsenic in the water of beverage has been associated with an increment of the mortality in cancers of bladder, kidney, lung and skin. The public system of water in Antofagasta has approximately $800-1000$ μg liter^{-1} of Arsenic. A population of near 200,000 inhabitants supplies themselves of treated water and that present less than 40 μg liter^{-1}.
China	People from extensive areas of the provinces of Xinjiang and Mongolia have drunk water of wells with high concentrations of arsenic ($50-1860$ μg liter^{-1}). The source of arsenic of the two provinces is geogenic and has given rise to endemic arsenicosis.
Ghana	Approximately 1600 km^2 of Obuasi have rivers contaminated with more than 7900 μg liter^{-1}. The concentration of arsenic in wells of this region is higher than 175 μg liter^{-1}. The main cause of contamination is the oxidation of arsenopyrite of gold mines.
Hungry	456,500 people distributed in 5 cities and 54 towns drink water originated in aquifers with a concentration of arsenic under the 50 μg liter^{-1}. The aquifers are formed by sediments placed by the river Danube. The concentration of arsenic in this region varies between 25 and 150 μg liter^{-1}. It has come to the conclusion that the concentration in the groundwater is controlled by the humic acid since humic substances may chelate multivalent cations.
Mexico	Groundwater is the only source of water of 10,000 inhabitants of the region of Zimapán. The levels of Arsenic were around 1100 μg liter^{-1}. Cases of poisoning by arsenic have been cited that have been related to cancers of skin, kidney and lung.

(Continued)

TABLE 3. *Continued*

Country	Problem
Taiwan	20,000 inhabitants were exposed to arsenic contained in water of wells that enclosed around 1820 µg liter^{-1}. In 1975 they were diagnosed 1141 cases of chronic exposition to arsenic. The exposition to arsenic associated to cancers of skin, kidney, ureter, urethra, lung.
Thailand	In 1997 were diagnosed 824 cases of poisoning by arsenic. The source of this contamination was the oxidation of the arsenopyrite during the operations of mining industry.
UK	The old mines of Cornwalll and Devon have soils with more than 1.000 mg of arsenic/kg of soil. The treated water is used to drink. The concentrations of arsenic in treated and not treated water is of 10–50 µg liter^{-1} and 10 µg liter^{-1} respectively.

Unlike nickel and Chromium, whose carcinogenic properties have been considered to affect the fidelity of polymerases involved in DNA synthesis, As does not affect polymerases. However, some arsenical compounds can produce chromosomal aberrations due to the replacement of phosphorus since As induces aberrant gene expressions. Arsenic can intervene also at the signal transduction level and therefore can modify gene expression [Abernathy et al., 1999].

There is a clear association among keratosis, epidermal carcinoma, lung cancer, and the exposition of humans to water with inorganic As through the water of beverage, beer, and wine. The therapies in which inorganic arsenicals are used are also associated with the development of precancerous skin wounds, multiple epiteliomatosis (specific alterations of the skin with incidence of light), and bronchial carcinoma. The inorganic As compounds have been described as carcinogenic of skin and liver.

Critical Participation of As in Cancer. Ingestion of folic acid (vitamin B$_9$) in food has been related to a general decrease in cancer risk. Nevertheless, it is still a matter of research and debate whether enhanced folate intake from foods or folic acid supplements may reduce the risk of cancer. Folate helps to synthesize the DNA required for rapidly growing cells and also participates in the prevention of DNA changes that may produce cancer. However, since folate is important in cells and tissues that rapidly divide and cancer cells divide rapidly, methotrexate (an inhibitor of tetrahydrofolic acid, the active form of folate) is used as a drug to treat cancer (Fig. 2). Tetrahydrofolic acid is produced from dihydrofolic acid by dihydrofolate reductase. Two As salts (sodium arsenite and sodium arsenate) induced a high frequency of methotrexate-resistant 3T6 cells, which were shown to have amplified copies of the dihydrofolate reductase gene. The ability of As to induce gene amplification may relate to its carcinogenic effects in humans since amplification of oncogenes is observed in many human tumors [Lee et al., 1988].

Most mammals are capable of methylating the inorganic compounds of As and are able to form monomethylarsonic acid (MMA) and dimethylarsinic acid (DMA),

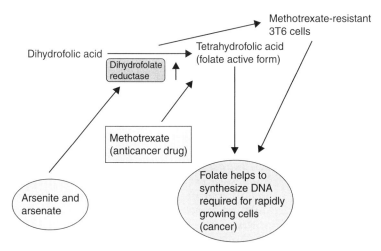

Figure 2. Arsenite and arseniate increase the production of dihydrofolate reductase in methotrexate-resistant cells, thereby overcoming the inhibition of cancer production carried out by methotrexate (inhibits the production of the active form of tetrahydrofolate) in nonresistant cells.

chemical forms that are frequently present in aquatic systems. Both are the primary metabolites of inorganic As. The methylated compounds have low toxicity and low affinity by the tissues; however, recent studies in vitro show that DMA is a powerful clastogenic agent (an agent that can cause structural changes as breaks in chromosomes) and induces chromosomal aberrations [Inoue et al., 1996]. Bladder tumors in patients with higher levels of As exposure showed higher levels of chromosomal instability. Most of the chromosomal alterations associated with As exposure were also associated with tumor stage and grade, raising the possibility that bladder tumors from As-exposed patients may behave more aggressively than tumors from unexposed patients [Moore et al., 2002]. The reduction As (V) may use glutathione (GSH) as a source of reducing equivalents or may be catalyzed by As (V) reductases. The oxidative methylation of trivalent arsenicals is catalyzed by a methyltransferase that uses S-adenosylmethionine as a methyl group donor [Inoue et al., 1996].

Methylated arsenicals have unique biological effects. Exposure to DMAV is teratogenic (causes congenital malformations) in the rat and mouse and causes renal injury in rats [Inoue et al., 1996]. DMA significantly enhanced the tumor induction in the urinary bladder, kidney, liver, and thyroid gland, with respective incidences at 400 ppm DMA. Induction of preneoplastic lesions (glutathione S-transferase placental form-positive foci in the liver and atypical tubules in the kidney) was also significantly increased in DMA-treated groups. Ornithine decarboxylase activity in the kidneys of rats treated with 100 ppm DMA was significantly increased compared with control values ($P < 0.001$). In conclusion, DMA is acting as a promoter of urinary bladder, kidney, liver, and thyroid gland carcinogenesis in rats, and this may be related to cancer induction by As in humans [Yamamoto et al., 1995].

The carcinogenic effect of DMA was marked by three different observations: significant early induction of tumors in both treated p53+/− knockout and wild-type mice, significant increase of the tumor multiplicity in 200 ppm-treated p53+/− knockout mice, and significant increase in the incidence and multiplicity of tumors (malignant lymphomas) in the treated wild-type mice. By the end of 80 weeks, tumor induction, particularly malignant lymphomas and sarcomas, were similar in treated and control p53+/− knockout mice. No evidence for organ-tumor specificity of DMA was obtained. Molecular analysis using PCR−SSCP techniques revealed no p53 mutations in lymphomas from either p53+/− knockout or wild-type mice. In conclusion, DMA primarily exerted its carcinogenic effect on spontaneous development of tumors with both of the animal genotypes investigated [Elsayed et al., 2003]. For instance, DMA and MMA cause DNA damage after interaction with ascorbic acid. Methylated As species and endogenous ascorbic acid can cause the release of iron from ferritin, the iron-dependent formation of reactive oxygen species, and DNA damage. This reactive oxygen species pathway could be a mechanism of action of As carcinogenesis in humans [Ahmad et al., 2000].

The administration of sodium 2,3-dimercapto-1-propane sulfonate (DMPS) to humans chronically exposed to inorganic As in their drinking water resulted in the increased urinary excretion of As, in the appearance and identification of mono-methylarsenous acid (MMA(III)) in their urine, and in a large decrease in the concentration and percentage of urinary dimethylarsinic acid (DMA). The experimental results support the hypothesis that DMPS competes with endogenous ligands for MMA(III), forming a DMPS−MMA complex that is readily excreted in the urine. These results indicate the need to study the biochemical toxicology of MMA(III) [Aposhian et al., 2000].

Arsenicals may also interfere with DNA methyltransferases, resulting in inactivation of tumor suppressor genes through DNA hypermethylation. Methylation within the promoter region of p53 was altered in human lung A549 cells exposed to arsenite over a 2-week period in culture. In cancer, methylation is very important. Adding a chemical group (methyl group) to DNA the way the gene behaves is changed. Genes may become *hypomethylated*, meaning that their "directions" tell them to produce proteins that can cause cancer. Genes may also become *hypermethylated*, meaning that "directions" that normally tell genes to stop cancers from growing become inactive and allow those cancers to grow. Methylation takes place in tumor cells in stretches of DNA that act as "control centers," called *CpG islands*. Genes in these islands tend to hypermythelate in cancer cells from certain tissues but not from others, making the cell behavior specific to the type of cancer. The methylation status of the 5′ control region of the tumor suppressor gene VHL, a gene known to be silenced transcriptionally by CpG methylation [von Hippel−Lindau syndrome (VHL)], was assessed. Six fragments were hypermethylated, and two were hypo-methylated, relative to untreated controls. Sequence analysis revealed the following: Two DNA fragments contained repeat sequences of mammalian-apparent LTR retrotransposons, five contained promoter-like sequences, and 13 CpG islands were identified. For cells exposed to arsenite, these results support the existence of a state of DNA methylation imbalance that could possibly disrupt appropriate gene expression [Zhong and Mass, 2001].

Long-term administration of DMA induced urinary bladder carcinomas in male F344 rats. Therefore, the results indicate that DMA is carcinogenic for the rat urinary bladder, which may be related to the human carcinogenicity of arsenicals [Wei et al., 1999].

Thioredoxin reductase (TR), an NADPH-dependent flavoenzyme that catalyzes the reduction of many disulfide-containing substrates, plays an important role in the cellular response to oxidative stress. Trivalent arsenicals, especially methyl As that contains trivalent As (MAs(III)), are potent noncompetitive inhibitors of TR purified from mouse liver. Because MAs(III) is produced in the biomethylation of As, it was postulated that the extent of inhibition of TR in cultured rat hepatocytes would correlate with the intracellular concentration of methyl As. Following exposure to As(III) or MAs(III), the extent of inhibition of TR activity correlated strongly with the intracellular concentration of MAs. These results suggest that arsenicals formed in the course of cellular metabolism of As are potent inhibitors of TR activity—in particular, MAs(III) [Lin et al., 2001].

Recently, the action mechanism of arsenite in lung cells was elucidated. Arsenite is a well-known metalloid human carcinogen, and epidemiological evidence has demonstrated its association with the increased incidence of lung cancer. However, the mechanism involved in its lung carcinogenic effect remains obscure. Exposure of human bronchial epithelial cells (Beas-2B) to arsenite resulted in a marked induction of cyclooxygenase (COX)-2, an important mediator for inflammation and tumor promotion. Exposure of the Beas-2B cells to arsenite also led to significant transactivation of nuclear factor of activated T-cells (NFAT), but not to activator protein-1 (AP-1) and NFB, suggesting that NFAT, rather than AP-1 or NFB, is implicated in the responses of Beas-2B cells to arsenite exposure. Inhibition of the NFAT pathway by either chemical inhibitors, dominant negative mutants of NFAT, or NFAT3 small interference RNA resulted in the impairment of COX-2 induction and caused cell apoptosis in Beas-2B cells exposed to arsenite. Site-directed mutation of two putative NFAT-binding sites between -111 and $+65$ in the COX-2 promoter region eliminated the COX-2 transcriptional activity induced by arsenite, confirming that those two NFAT-binding sites in the COX-2 promoter region are critical for COX-2 induction by arsenite. COX-2 induction by arsenite is through NFAT3-dependent and AP-1- or NFB-independent pathways and plays a crucial role in antagonizing arsenite-induced cell apoptosis in human bronchial epithelial Beas-2B cells [Ding et al., 2006].

2.3   Chromium

2.3.1   Characteristics of Cr.   Chromium is a lustrous, fragile, hard metal. Its color is silver-gray and it can be highly polished. It does not tarnish in air; when heated, it burns and forms the green chromic oxide. Cr is unstable in oxygen, and it immediately produces a thin oxide layer that is impermeable to oxygen and protects the metal below.

2.3.2 Localization. Chromium is found in Nature in rocks, animal, plants, soils, dust, and volcanic gases, almost always in mineral form, combined in the form of oxide (commonly as an oxide of iron). It is found as chromite ore—as, for example, Siberian red lead (crocoite, $PrCrO_4$), which is a chromium ore valued as a red pigment for oil paints. The most common forms of Cr are Cr VI and Cr 0, which are produced generally in industrial processes. Cr 0 or metallic Cr is used to manufacture steel and itself is not often presented in Nature. Most of the free Cr found in Nature exists as Cr III and Cr VI. Cr III and Cr VI are used in dyes and pigments, in tanning, and as a protector of wood. The natural compounds of Cr have neither flavor nor smell.

2.3.3 Contamination by Chromium. The main sources of exposition to toxic forms of Cr are dyes, air pollution, and dental fillings. The workers most exposed are those involved in the production of steel and chrome, those working in foundries for Cr and ferrochrome, and those involved in the production of paintings, where Cr is used as a pigment.

In air the compounds of Cr exist primarily as particles of fine dust that eventually are placed on land or water. When these particles adhere to soil, only a small quantity can be dissolved in water and can pass at a deeper level inside the soil and into groundwater. Fish accumulate Cr of the water in very low proportion. The main users of the compounds of Cr are the metallurgical and chemical industries. The commercial applications include alloys of steel, catalyzers, pigments, fireproof products, and the leather industry.

2.3.4 Chromium and Health in Humans. Chromium can be incorporated from the environment by means of inhalation of atmospheric dust, inhaling particles of dust coming from soil or suspended in air, or drinking groundwater, as will as through the skin by direct contact with the soil or with the surface of contaminated water. The maximum level of Cr in drinking water should be always less than 100 μg/liter. Chromic acid and hexavalent Cr contained in the air of workplaces should be less than $100\,\mu g/m^3$. Approximately 64% of the atmospheric Cr emerges from the liberation of hexavalent Cr released in the combustion of fuel oil and in the production of steel, while approximately 30% results from the liberation of trivalent Cr from nickel-plated items, chemical processing with Cr, production of metals, and cooling towers. The Cr present in soil comes mainly from insoluble salt of trivalent Cr and from oxides and carbonates present in the storage of commercial products that contain Cr, while most of the Cr in aquatic systems is derived from poured domestic water. Breathing Cr in high levels cause irritation of the nose, nosebleeds, ulcers and drillings in the nasal septum, stomach discomfort and ulcers, convulsions, damage in kidney and liver, and even death. Skin exposure to compounds of Cr, above all chromate and dichromate of sodium and potassium, can cause skin irritation and ulcers. The quantity of Cr that is incorporated into the body depends on the type of particles absorbed, its solubility in the organic fluids, the reaction with the respiratory mucous membrane, and on the state of oxidation. The hexavalent form of Cr is more easily absorbed for the lungs that the trivalent form. The soluble chromates are incorporated quickly through the epithelia of the alveolae to the circulatory torrent. Both forms penetrate in limited quantity through the skin. As mentioned above, hexavalent Cr can be absorbed with the drinking water, with any another type of beverage, and with food. The absorption of the ingested Cr in the digestive process is minute (less than 2%).

The inhalation of Cr VI takes place mostly in the handling of metallic Cr in mines, handling ferrochrome refined, nickel-plated Cr in metallic surfaces, welding with Cr, and working in the production of cement. The acute poisoning of Cr is produced generally by oral consumption. It can be found dissolved in both forms, but the trivalent Cr is rarely present in the water used in beverages. The consumption of hexavalent Cr causes gastrointestinal irritation, ulceration and corrosion, epigastric pain, queasiness, vomiting, diarrhea, vertigo, fever, muscular cramps, hemorrhagic diastesis, toxic nephritis, and renal failure, until even the occurrence of death. The severity of the symptomology depends exclusively of the dose consumed. Besides, the severe skin exposure to chromic acid causes local inflammation, ulceration of the skin, dermatitis, and even systemic toxicity.

The detection of Cr by means of analytic methods can take place from samples of hair, urine, serum, red blood cells, and complete blood: Cr is found in minuscule proportion in the biological tissues, for which we should avoid contamination during the simple handling or in the analytic processes of the tissues. It is not simple to distinguish by means of procedures of simple analysis the trivalent Cr of the hexavalent.

After being absorbed, Cr joins with the transferrin (a plasma glycoprotein, which binds iron very tightly but reversibly) and is distributed all over the body, being incorporated in low quantity into the erythrocytes (Fig. 3). The hexavalent Cr penetrates quickly in the blood cells, where it joins with the beta chains of the hemoglobin. In the interior of the erythrocyte the hexavalent Cr is reduced to trivalent Cr. The

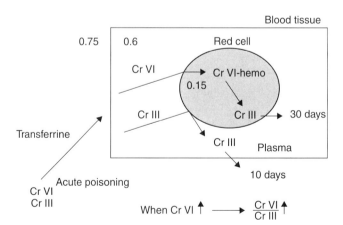

Figure 3. Metabolization and excretion of Cr(VI) and Cr(III). Cr(VI) causes acute poisoning, and Cr(III) does not. Cr(VI) is transported to red cells. It is bound to hemo groups and reduced to Cr(III). Recent exposure to Cr(VI) can be detected measuring internal chromium concentration in red cells and comparing the chromium ratio: red cell/plasma.

elimination of the Cr from the body takes place in several phases. It is excreted above all by urine in three half-lives: 7 hr, from 15 to 30 days, and from 3 to 5 years.

If chromium is found intracellularly, this would indicate a contamination of hexavalent chromium since only this can penetrate in red blood cells. The exposure to hexavalent Cr increases the proportion of Cr in red cells with respect to that of Cr in plasma. The erythrocytes free Cr very slowly. The level of Cr remains high in the red cells for a month or more, while the level of Cr plasma returns to its normal level in 10 days. Therefore the presence of Cr in blood would indicate recent exposure to any of the two forms of Cr. The content of Cr in blood should be less than 0.75 μg/liter, and in serum it should be less than 0.6 μg/liter. The apparition of high levels of chromium in urine signifies that there has been absorption of chromium particularly in the last 1 or 2 days. The content of Cr in urine for a normal exposure to hexavalent Cr should be less than 5 μg/g of creatinine (breakdown product of creatine phosphate in muscle, which is usually produced at a fairly constant rate by the body, depending on muscle mass). The determination of Cr in the urine of workers when carrying out the turn change is utilized as an indicator of combined exposure. Finally, the detection of Cr in hair is difficult to elucidate because of the difficulty to discern among those born with Cr in their hair after a systemic contamination and those with Cr in their hair as a result of external contamination.

The chemical determination of hexavalent chromium is carried out with the method of 1,5-diphenylcarbohydrazide. By means of this method, the hexavalent chromium is determined utilizing a formula of dry dust of reagent of chromium that contains magnesium sulfate and potassium pirosulfate, along with an acid buffer combined with 1,5-diphenylcarbohydrazide. The reaction causes a purple color that is proportional to the quantity of hexavalent chromium present. Iron, vanadium, mercury, and mercuric ions can interfere.

2.3.5 Chromium and Cancer. There is sufficient evidence showing that Cr(VI) is carcinogenic for humans and for animals. In particular, Cr(VI) increases the risk to develop cancer of lung, esophagus, and liver. A rare demonstration of the exposure to Cr is the cancer of the paranasal sinuses. The increase of the incidence of lung cancer is connected with the repeated inhalation of Cr hexavalent and is directly proportional to the time of exposition. The latency period expands from 13 to 30 years. The inhalation by means of aerosols of Cr used to paint can cause bronchial asthma and pulmonary edema. In some cases, it has been difficult to establish a direct correlation between cancer and Cr, particularly in the industrial manipulators. The difficulty to confirm this relation for cases of cancer in workers of industries such as that of leather or coal is due to the interaction of the factor Cr with other carcinogenic factors like the diet, smoking, and way of life. This also true for the rest of heavy metals.

2.4 Cadmium

2.4.1 Characteristics of Cd. Cadmium is a transition metal and is shiny, silver-white, bluish-white, and ductile. It is an extremely malleable metal and is available in many forms including foil, granules, powder, pellets, sheet, rod, shot, sticks, wire, and "mossy cadmium." Its surface has a bluish tinge and the metal is soft enough to be cut with a knife, but it tarnishes in air. It is soluble in acids but not in alkalis. In many respects it is similar to Zn, but it forms more complex compounds.

2.4.2 Localization. Cadmium can mainly be found in the earth's crust. No cadmium ore is mined for the metal, because more than enough is produced as a byproduct of the smelting of Zn from its ore, sphelerite (ZnS), in which CdS is a significant impurity, making up as much as 3%.

2.4.3 Contamination by Cadmium. Cadmium is used in nickel–cadmium batteries, PVC plastics, and paint pigments. It can be found in soils because insecticides, fungicides, sludge, and commercial fertilizers that use cadmium are used in agriculture. Cadmium may be found in reservoirs containing shellfish. Cigarettes also contain cadmium. Lesser-known sources of exposure are dental alloys, electroplating, motor oil, and exhaust. Inhalation accounts for 15–50% of absorption through the respiratory system; 2–7% of ingested cadmium is absorbed in the gastrointestinal system. Target organs are the liver, placenta, kidneys, lungs, brain, and bones.

2.4.4 Cadmium and Health in Humans. Acute exposure to cadmium generally occurs in the workplace, particularly in the manufacturing processes of batteries and color pigments used in paint and plastics, as well as in electroplating and galvanizing processes. Symptoms of acute cadmium exposure are nausea, vomiting, abdominal pain, and breathing difficulty. Chronic exposure to cadmium can result in chronic obstructive lung disease, renal disease, and fragile bones. Children must be protected by carefully storing products containing cadmium, especially nickel–cadmium batteries. Symptoms of chronic exposure could include alopecia, anemia, arthritis, learning disorders, migraines, growth impairment, emphysema, osteoporosis, loss of taste and smell, poor appetite, and cardiovascular disease.

The monograph program of the International Agency for Research of on Cancer has evaluated many trace elements for their carcinogenicity to humans. For this agency, five groups of compounds were considered human carcinogens: As and As compounds, beryllium and beryllium compounds, cadmium and cadmium compounds, hexavalent Cr compounds, and nickel compounds. Antimony trioxide, cobalt and cobalt compounds, lead and inorganic lead compounds, methyl mercury compounds, and metallic nickel were considered possibly carcinogenic to humans. Antimony trisulfide, trivalent Cr compounds, metallic Cr, ferric oxide, organolead compounds, metallic mercury, inorganic mercury compounds, selenium and selenium compounds, and titanium dioxide were not classifiable. Trace elements studied to a limited extent include copper, manganese, tin, vanadium, and Zn. Among the problems are the lack of relevant data, the definition of active species, the extrapolation of the results of experimental studies to humans, the methodological problems of epidemiologic studies, and the possible anticarcinogenic activity of some trace elements [Boffetta et al., 1993].

2.4.5 Cadmium and Cancer. Pancreatic cancer is one of the most deadly forms of cancer. Although it is the fourth leading cause of cancer deaths in the United States. It accounts for only 2% of all newly diagnosed cancers each year. Five-year survival rates for patients diagnosed with pancreatic cancer in the United States average only 4.4%. In developing countries, pancreatic cancer appears to be extremely rare. Studies in Egypt, Algeria, and Iraq suggest a low incidence of the disease in the Middle East. Risk factors for this disease have generally been grouped into four categories: (a) cigarette smoking; (b) chronic pancreatitis (largely from alcohol consumption) and genetic predisposition; (c) diabetes mellitus and macro- or micronutrients; and (d) occupational and environmental contamination from exposure to pesticides and fertilizers, manufacturing paints and pigments, metalworking, and soldering [Kriegel et al., 2006].

Little is known about the etiology of pancreatic cancer, which is an important cause of cancer mortality in developed countries. Schwartz and Reis [2000] hypothesize that exposure to cadmium is a cause of pancreatic cancer. Cadmium is a nonessential metal that is known to accumulate in the human pancreas. The major risk factors for pancreatic cancer (increasing age, cigarette smoking, and occupations involving exposure to metalworking and pesticides) are all associated with increased exposure to cadmium. Cadmium can cause the transdifferentiation of pancreatic cells, increases in the synthesis of pancreatic DNA, and increases in oncogene activation. Thus, cadmium is a plausible pancreatic carcinogen. The cadmium hypothesis provides a coherent explanation for much of the descriptive epidemiology of pancreatic cancer [Schwartz and Reis, 2000].

Two main risk factors for pancreatic cancer, age, and cigarette smoking are also associated with cadmium exposure. Cadmium accumulates in the body over time because there are no specific mechanisms for its removal. The half-life of this metal in the body ranges from 10 to 30 years, with an average of 15 years [Jin et al., 1998]. Urinary cadmium levels, commonly used as an indication of life-long exposure, are significantly higher in smokers than in nonsmokers [Schwartz et al., 2003].

Cadmium is a potent carcinogen in rodents. $CdCl_2$ injection causes a significant elevation of incidence of the pancreatic islet cell tumors (8.5% versus 2.2%) regardless of any other experimental treatment. These results provide further evidence that the divalent carcinogenic metals may exert their activity through an antagonism with the physiologically essential divalent metals [Poirier et al., 1983]. In experiments designed to determine the carcinogenic effects of repeated exposure to $CdCl_2$, Fischer and Wistar rats showed a very high incidence of dose-related pancreatic metaplasia, as reflected by pancreatic hepatocyte formation [Konishi et al., 1990]. These pancreatic hepatocytes are thought to arise from the ductal and interstitial cells of the pancreas, which resemble oval cells of the liver [Rao et al., 1986].

Cadmimum has been shown to transform cells in vitro. However, the carcinogenic mechanisms of Cd as a mutagen and/or promoter are not well-clarified. $CdCl_2$ in a range of 1.5–360 ng/ml enhanced transformation of Balb/3T3 A31 cells induced by *N*-methyl-*N*′-nitro-*N*-nitrosoguanidine (MNNG, 0.1 g/ml). Cadmium stimulated cell proliferation on MNNG-initiated cells through inactivation of p53 and p27 and increase of proliferating cell nuclear antigen (PCNA) expression after 24 hr treatment. During the transformation process of MNNG-treated cells, Cd may activate oncogenes such as c-myc, mdm2, and cellular tumor antigen p53, inhibiting tumor suppressor genes such as wild-type p53 and p27 and consequently accelerating the proliferation of initiated cells. Thus Cd affects the genes involved in growth regulation on initiated cells during the promotion stage of in vitro cell transformation [Fang et al., 2002].

Regarding human chelatin agents, basal or induced metallothionein (MT) perturbs Cd-induced apoptotic cell death in various cell lines, and a strong negative correlation exists between cellular MT content and the rate of apoptosis induced by Cd [Shimoda et al., 2001].

A possible association between exposure to CdO and cancer was first raised in the United Kingdom by the death certificates of eight Ni–Cd battery workers with 14–38 years of exposure to cadmium. Five of the eight workers died of cancer, three of them from prostate cancer. This finding initiated the epidemiological survey of 248 workers with at least 1 year of exposure. The first study in a Cd-smelter in the United States seemed to confirm the role of cadmium in prostate cancer. The vital status of 292 cadmium production workers with at least 2 years of employment was recorded from the first day of 1940 to the last day of 1969. There were four prostatic epidemiological studies that failed to support the role of cadmium in the causation of prostate cancer. Epidemiological studies on cadmium-exposed workers have not given information on the contribution of smoking to the induction of lung cancer and did not differentiate between the different chemical and physical forms of cadmium. The lack of smoking data is always a cause for concern, but the problem is amplified when the target is the lung and there is a concomitant exposure to As.

In the first experiments on the carcinogenic potential of cadmium, the route of administration was either intramuscular or subcutaneous. The injection of suspensions of 25 mg finely divided cadmium, cadmium sulfide, or oxide regularly gave rise to rhabdomyosarcomas or fibrosarcomas in rats. The tumors were metastasizing and pleomorphic with high yield. It is well known that the injection of cadmium at this dose level has a necrotic effect at the site of injections, and necrosis may have contributed to the development of nonmetastasizing testicular tumors in rats and mice treated subcutaneously with soluble cadmium salts (chloride or sulfate) [Magos, 1991].

2.4.6   Chelation Therapy.   Chelation is a chemical process that has applications in many areas, including medical treatment, environmental site rehabilitation, water purification, and so forth. In the medical environment, chelation is used to treat cardiovascular disease and heavy-metal toxicity, as well as to remove metals that accumulate in body tissues because of genetic disorders (hemochromatosis). Chelation therapy, simply defined, is the process by which a molecule encircles and binds (attaches) to the metal and removes it from tissue. Depending on the drug used, chelating agents specific to the heavy metal involved are given orally, intramuscularly, or intravenously. Once the bound metal leaves the tissue, it enters the bloodstream, is filtered from the blood in the kidneys, and then is eliminated in the urine.

Chelation is effective in treating arsenic, lead, iron, mercury, and aluminum poisoning. However, there is no known medical chelating method that is effective for the treatment of cadmium toxicity; however, it may be used in cases of acute oral cadmium poisoning to help prevent additional absorption of cadmium in the gastrointestinal tract. Prevention or elimination of exposure is all that is available at this time for cadmium toxicity (ATSDR ToxFAQs for cadmium).

An agent frequently used in chelation therapy is dimercaprol (also known as BAL or British anti-lewisite). Oral chelating agents used as alternatives to BAL are 2,3-demercaptosuccinic acid (DMSA), dimercaptopropanesulfonate (DMPS), and D-penicillamine (ATSDR MMG). Another agent, deferoxamine, is often used to chelate iron. Ethylenediamintetraacetic acid (ETDA) also has an affinity for lead and was one of the first chelators developed.

2.5 Lead

2.5.1 Characteristics of Pb. The word "plumbing" derives from the Latin word for lead, *plumbum*. Pb (atomic number 82 and atomic weight 207.19) is a heavy metal unlike other trace elements such as As or Se. It is inelastic, malleable, and easily fusible, and it melts ($327.4°C$) and boils ($1740°C$) at high temperature, unlike Hg. Lead can be found in small amounts in the earth's crust, existing in different chemical forms: metallic (pure metal); inorganic compounds, such as lead oxide, lead sulfate, lead chromates, lead silicates, lead arsenates, and lead chloride; and organic compounds, such as tetraethyl lead. These last compounds are especially toxic as mentioned above for methyl mercury since they enter fatty tissues.

As a pure metal, Pb is very soft, bluish-white, ductile, and relatively poor conductor of electricity. It is very resistant to corrosion but tarnishes upon exposure to air. Because of the high resistance to corrosion, it is used to contain corrosive liquids (e.g., sulfuric acid). It is relatively resistant to attack by sulfuric and hydrochloric acids but dissolves slowly in nitric acid. Lead forms many salts, oxides, and organometallic compounds. It has an amphoteric behavior.

2.5.2 Localization. Lead rarely occurs in its elemental state. The most common ore is the sulfide, galena (PbS). The other minerals of commercial importance are the carbonate, cerussite ($PbCO_3$), and the sulfate, anglesite ($PbSO_4$), which are much more rare. Lead also occurs in various uranium and thorium minerals, arising directly from radioactive decay. Commercial lead ores may contain as little as 3% lead, but a lead content of about 10% is most common. The ores are concentrated to 40% or greater lead content before smelting.

2.5.3 Contamination by Lead. Lead compounds are toxic; therefore they have to be handled carefully since they have resulted in poisoning of workers from misuse and overexposure.

The most important lead compounds in the industry are the lead oxides and tetraethyl lead. Lead is used in industry in the form of alloys since it forms alloys with many metals (arsenic, cadmium, tin, antimony, bismuth, copper) and since it is also very often employed in the form of alloys in most applications. The primary source of lead in drinking water is from lead-based plumbing materials. The corrosion of such materials will continue to increase concentrations of lead in municipal drinking water. As in the case of Hg, an important hazard arises from the inhalation of vapor or dust. Cigarette smoke is also a significant source of lead exposure. Acute exposure to lead is also more likely to occur in the workplace, particularly in manufacturing processes that include the use of lead (e.g., where batteries are manufactured or lead is recycled). Lead is in batteries, paints, ceramics, and soldering and building materials. It is even used in printing ink. Tetraethyl lead (PbEt$_4$) is still used in some grades of petrol (gasoline) but is being phased out on environmental grounds. Some fertilizers also contain lead. Therefore until recently, lead was added to automobile fuel. For many years this has been the major source of exposure. Automobile exhaust contained lead and people inhaled it in contaminated air. Lead from automobile exhaust also contaminated soil near roadways. More recently, since lead has been taken out of gasoline, the major source of exposure has been old lead-containing paint, which can be found in older homes and on structures such as bridges. However, the largest single use of lead is for the manufacture of storage batteries followed by cable covering, construction (because its excellent resistance to corrosion), pigments, solder, and ammunition.

2.5.4 Lead and Health in Humans. The Agency for Toxic Substances and Disease Registry (ATSDR) estimates that more than one million workers in 100 occupations are exposed to lead. It is now recognized that even low levels of lead exposure are associated with adverse health effects. The U.S. Centers for Disease Control and Prevention (CDC) has identified a lead concentration of 10 µg/dl of blood as the level of concern above which significant health risks occur.

Symptoms include abdominal pain, convulsions, hypertension, renal dysfunction, loss of appetite, fatigue, and insomnia. Other symptoms are hallucinations, headache, numbness, arthritis, and vertigo. Chronic exposure to lead may result in birth defects, mental retardation, autism, psychosis, allergies, dyslexia, hyperactivity, weight loss, shaky hands, muscular weakness, paralysis (beginning in the forearms), and problems in the central nervous system, blood pressure, the kidneys, the male reproductive system, and vitamin D metabolism. Children are particularly sensitive to lead (absorbing as much as 50% of the ingested dose) and are prone to ingesting lead because they chew on painted surfaces and eat products not intended for human consumption (e.g., hobby paints, cosmetics, hair colorings with lead-based pigments, and even playground dirt). However, another important potential cause of Pb accumulation in children is the substitution of Ca by Pb. Since children undergo a rapid and strong development, Pb may be included in tissues that demand important amounts of Ca—for instance, bones.

Lead appears to affect not only cognitive development of young children but also other areas of neuropsychological function. Young children exposed to lead may exhibit mental retardation, learning difficulties, shortened attention spans (ADHD), increased behavioral problems (aggressive behaviours), and reduced physical growth. Lead has been determined by many health experts to be the first threat to developing children in our industrial societies. Despite this impressive decrease in blood lead levels, more than one million children in the United States have blood lead levels above 10 µg/dl and are at risk of permanent neurological impairment. Because of its many uses, lead is now a common contaminant in waste sites, appearing in over half of the US Superfund sites [ATSDR, 1990]. Superfund is the program of the US federal government to evaluate and clean up uncontrolled or abandoned sites where dangerous waste is located, potentially posing a risk to local ecosystems or people.

2.5.5   *Lead and Cancer.* In the United States, there are at least three different types of classifications for the exposures that may be carcinogenic. They are established by the National Toxicology Program (included in the Reports on Carcinogens, published every 2 years), the International Agency for Research on Cancer (IARC), and the Environmental Protection Agency (EPA), through its Integrated Risk Information System. The EPA classified lead and lead compounds (inorganic) as probable human carcinogens.

It is suggested that lead exposure can interfere with repair of DNA damage caused by other chemicals [Hayes, 1997]. Lead acetate and lead phosphate cause kidney tumors in mice and rats [Hayes, 1997]. Lead inhibits α-aminolevulinate (ALA) dehydratase and ferrochelatase, preventing both porphobilinogen formation and the incorporation of iron into protoporphyrin IX, the final step in heme synthesis. Inhibition of both of these steps results in ineffective heme synthesis and subsequent microcytic (hemoglobin-poor) anemia.

As in the case of other heavy metals, one of the essential problems that is produced is the competence of the heavy metals with the nutrients, especially with the ones that share a certain chemical similarity. In this case, Pb and Ca have shown competence. This, for instance, has been shown very recently when lead toxicity was reduced by supplementation with calcium in mice [Prasanthi and Reddy, 2006].

2.6   Benefits in Cancer

2.6.1   *Zinc.* Zinc is the most abundant trace element in cells. It plays an important role in both genetic stability and function. Zinc is a component of chromatin, and it also plays a role in DNA replication, transcription, and repair. Zinc is found in over 300 enzymes, including copper/zinc superoxide dismutase, which is an important antioxidant enzyme, and in several proteins involved in DNA repair. Zn also assists to protect cellular components from oxidation and damage.

Zinc deficiency is a problem not only in developing countries but also in rich countries. It can occur in populations with low dietary Zn intake and high intake of phytate, a reservoir substance found in seeds and cereal grains that binds strongly to Zn, making it biologically unavailable. Populations that are at high risk include infants and young children, whose requirements for Zn are high. In addition, the elderly have an increased risk of Zn deficiency because Zn absorption may become impaired with age. Elderly also tends to consume low-Zn diets. Foods rich in Zn include red meat, seafood, and several plant sources, such as whole grains and legumes. However, the Zn present in plant foods is much less bioavailable.

The link between Zn deficiency and cancer has now been established by human, animal, and cell culture studies. It is known that Zn level is compromised in cancer patients compared to healthy people. Oxidative DNA damage and chromosome breaks have been reported in animals fed a Zn-deficient diet. In rats, dietary Zn deficiency causes an increased susceptibility to tumor development when the rats are exposed to carcinogens. Zinc also appears to play an important role in maintaining prostate health, but the precise function of Zn in the prostate is unknown. Prostate cancer is the second leading cause of cancer deaths in American men, and most elderly men have some abnormal prostate cells. Still, the cause of prostate cancer is unclear. As an antioxidant and a component of many DNA repair proteins, Zn plays an important role in protecting DNA from damage. Zinc also functions as an anti-inflammatory agent and can promote programmed cell death, or apoptosis. As with most therapeutics, higher doses do not always equate with an increase in efficacy. The efficacy of Zn supplements in preventing prostate cancer is controversial. Zinc supplementation may be more helpful in the early stages of cancer development rather than as cancer treatment.

2.6.2   Selenium. Selenium is a mineral that is found in small quantities in meat, cereals, shellfish, and some dry fruits. Initial evidence suggests that the supplements of Se reduce the prostate cancer risk in men having a referential line of normal level of specific antigen of the prostate and low levels of Se in the blood. Selenium can protect not only against the colon cancer, but also against the prostate and lung cancer. One of these mechanisms is the role played by Se in activating genes that prevent cancer.

For many researchers the deficiency of Se is a vast problem that currently faces humanity. Selenium protects against the toxic effects of Cd and behaves as an antagonist of As, Hg, and Cu. Selenium is also involved in the synthesis of prostaglandins, increases the effectiveness of the vitamin E, and works as an antioxidant that helps to prevent the break of chromosomes. It has been related to development of colorectal cancer. Selenium activates glutathione peroxidase, which is the only known enzyme that contains this trace element. In patients whose blood has a low level of glutathione peroxidase, the incident of cancer is higher. Higher levels of Se in the soil (and therefore higher levels of Se in the products grown in that soil) reduce the risk of cancer. Serum levels of Se are low in patients with certain cancers. Selenium also inhibits cancer in animals. In vitro evidence shows that Se reduces tumoral growth and stimulates the apoptosis, therefore suggesting that Se inhibits both the progress and the promotion of cancer in late stages [Morris et al., 2005].

2.6.3   Arsenium. Despite its notoriety as a deadly poison, As is an essential trace element for some animals, and maybe even for humans, although the necessary intake may be as low as 0.01 mg/day. Arsenic has been used with therapeutic objective for 2000 years. It stimulates the production of hemoglobin and is related to the metabolism of arginine, Zinc, and manganese. This explains why As was prescribed for the cure of the anemia. It has also been effective to treat rheumatism, arthritis, asthma, malaria, infections by trypanosoma, tuberculosis, and diabetes. Some arsenicals (melarsoporol) still remain in use in some African countries to treat African sleeping sickness. The trioxide of As is used in leukemia in patients that do not respond to chemotherapies.

2.6.4 Chromium. There are a great number of studies in human beings and animals on the effects of Cr. Chromium is not an essential element for plants, but it is an essential nutrient for human beings. Shortages may cause heart conditions, disruptions of metabolism, and diabetes. More important, trivalent Cr is considered as a nutrient, a trace element that participates in the formation of the factor of tolerance to glucose and in the metabolism of insulin. On the contrary, hexavalent Cr is a toxic substance, a strong oxidant agent. The excess of uptake of Cr(III) can cause health effects as well—for instance, skin rashes.

2.6.5 Heavy Metals in General. The links between the trace elements and free radicals are extremely strong and complex. For example, certain mechanisms of enzymatic protection against free radicals cause the participation of enzymes that contain metallic cofactors such as Zn, Cu, Se, Fe, or Mn. Most of the time, metal exposure involves a mixture of metals and almost never a single metal. Heavy metals are an important class of carcinogens. Cadmium, Chromium, and Nickel have been proven to be carcinogenic in animal models. Also, carcinogenics are As and As compounds, Be and Be compounds, Cd and Cd compounds, Cr^{6+} and Cr^{6+} compounds, and Ni and Ni compounds.

3 GENERAL CONCLUSIONS

The literature published to date allow us to make the following conclusions: First, most of the studies establish a cause–effect relationship between heavy metals (analyzed in this review) and cancer. Although in most cases possible alternative explanations are argued, or failures in the studies are reported, the results are persistent. Second, attending to the results published and keeping in mind that all the heavy metals (especially those analyzed in this review) produce oxidative stress, this common factor would explain the general tendency of this group to be carcinogenic. Third, besides this oxidative effect produced by heavy metals, it should be emphasized that normally they compete with other nutrients; therefore they can produce an effect of interference and even of deposition in some tissues, depending of the type of heavy metal analyzed.

REFERENCES

Abernathy CO, Liu Y-P, Longfellow D, Aposhian HV, Beck B, Fowler B, Goyer R, Menzer R, Rossman T, Thompson C, Waalkes M. 1999. Arsenic: Health effects, mechanisms of actions, and research issues. *Environ Health Perspect* **107**:593–597.

Ahmad S, Kitchin KT, Cullen WR. 2000. Arsenic species that cause release of iron from ferritin and generation of activated oxygen. *Arch Biochem Biophys* **382**:195–202.

Aposhian HV, Zheng B, Aposhian MM. 2000. DMPS-arsenic challenge test. II. Modulation of arsenic species, including monomethylarsonous acid (MMA (III)), excreted in human urine. *Toxicol Appl Pharmacol* **165**:74–83.

ATSDR. Agency for Toxic Substances and Disease Registry, 1825 Century Blvd, Atlanta, GA 30345. Tel: (404) 498-0110/Public Inquiries: (888) 422–8737.

Barregard L, Sallsten G, Jarvholm B. 1990. Mortality and cancer incidence in chloralkali workers exposed to inorganic mercury. *Br J Ind Med* **47**:99–104.

Bernard S, Enayati A, Redwood L, Roger H. 2001. Autism: A novel form of mercury poisoning. *T Med Hypotheses* **56**:462–471.

Boffetta P, Merler E, Vainio H. 1993. Carcinogenicity of mercury and mercury compounds. *Scand J Work Environ Health* **19**:1–7.

Boffetta P, Garcia-Gomez M, Pompe-Kirn V. 1998. Cancer occurrence among European mercury miners. *Cancer Causes & Control* **9**:591–599.

Ding J, Li J, Xue C, Wu K, Ouyang W, Zhang D, Yan Y, Huang C. 2006. Cyclooxygenase-2 induction by arsenite is through a nuclear factor of activated T-cell-dependent pathway and plays an antiapoptotic role in Beas-2B cells. *J Biol Chem* **281**:24405–24413.

Ellingsen DG, Andersen A, Nordhagen HP, Efskind J, Kjuus H. 1993. Incidence of cancer and mortality among workers exposed to mercury vapour in the Norwegian chloralkali industry. *Br J Ind Med* **50**.875–880.

Elsayed S., Wanibuchi H, Morimura K, Wei M, Mitsuhashi M, Yoshida K, Endo G, Fukushima S. 2003. European mercury miners. *Carcinogenesis* **24**:335–342.

Fang MZ, Mar W, Cho MH. 2002. Cadmium affects genes involved in growth regulation during two-stage transformation of Balb/3T3 cells. *Toxicology* **177**:253–265.

Hayes RB. 1997. The carcinogenicity of metals in humans. *Cancer Causes & Control* **8**:371–385.

Inoue Y, Date Y, Yoshida K, Chen SH, Endo G. 1996. Speciation of arsenic compounds in the urine of rats orally exposed to dimethylarsinic acid ion chromatography with ICP-MS as an element-selective detector. *Appl Organometallic Chem* **10**:707–711.

Jiménez-Valverde G. 1999. Arsenic speciation of in soils contaminated after the mining accident of Aznalcóllar. Experimental Master's in thesis chemistry (analytical chemistry). University of Barcelona.

Jin T, Lu J, Nordberg M. 1998. Toxicokinetics and biochemistry of cadmium with special emphasis on the role of metallothionein. *Neurotoxicol* **19**:529–535.

Jones FT. 2007. A broad view of arsenic. *Poultry Sci* **86**:2–14.

Konishi N, Ward JM, Waalkes MP. 1990. Pancreatic hepatocytes in Fischer and Wistar rats induced by repeated injection of cadmium chloride. *Toxicol Appl Pharmacol* **104**:149–156.

Kriegel AM, Soliman AS, Zhang O, El-Ghawalby N, Ezzat F, Soultan A, Abdel-Wahab M, Fathy O, Ebidi G, Bassiouni N, Hamilton SR, Abbruzzese JL, Lacey MR, Blake DA. 2006. Serum cadmium levels in pancreatic cancer patients from the East Nile Delta Region of Egypt. *Environ Health Perspect* **114**:113–119.

Lee TC, Tanaka N, Lamb PW, Glimer TM, Barrett JC. 1988. Induction of gene amplification by arsenic. *Science* **241**:79–81.

Lin S, Del Razo LM, Styblo M, Wang C, Cullen WR, Thomas, DJ. 2001. Arsenicals inhibit thioredoxin reductase in cultured rat hepatocytes. *Chem Res Toxicol* **14**:305–311.

Magos, L. 1991. Epidemiological and experimental aspects of metal carcinogenesis: Physicochemical properties, kinetics, and the active species. *Environ Health Perspect* **95**:157–189.

Moore LE, Smith AH, Eng C, Kalman D, DeVries S, Bhargava V, Chew K, Moore II D, Ferreccio C, Rey OA, Waldman FM. 2002. Arsenic-related chromosomal alterations in bladder cancer. *J Natl Cancer Inst* **94**:1688–1696.

Morris J, Pramanik R, Zhang X, Carey A, Ragavan N, Martin F, Muir G. 2005. Selenium- or quercetin-induced retardation of DNA synthesis in primary prostate cells occurs in the presence of a concomitant reduction in androgen-receptor activity. *Cancer Lett* **239**:111–122.

Occupational Safety and Health Information Centre, Geneva, International Labour Office, 1999.

Poirier LA, Kasprzak KS, Hoover KL, Weak ML. 1983. Effects of calcium and magnesium acetates on the carcinogenicity of cadmium chloride in Wistar rats. *Cancer Res* **43**:4575–4581.

Prasanthi R, Reddy G. 2006. Calcium or zinc supplementation reduces lead toxicity: Assessment of behavioral dysfunction in young and adult mice. *Nutr Res* **26**:537–545.

Rao MS, Scarpelli DS, Reddy JK. 1986. Transdifferentiated hepatocytes in rat pancreas. *Curr Top Dev Biol* **20**:63–78.

Roberts JR. 1999. Metal toxicity in children. In: *Training Manual on Pediatric Environmental Health: Putting It into Practice*. Emeryville, CA: Children's Environmental Health Network.

Sarkar B, editor. 2002. *Heavy Metals in the Environment*. Toronto: The Hospital for Sick Children and University of Toronto.

Schwartz GG, Reis IM. 2000. Is cadmium a cause of human pancreatic cancer? *Cancer Epidemiol Biomarkers Prev* **9**:139–145.

Schwartz GG, Il'yasova D, Ivanova A. 2003. Urinary cadmium, impaired fasting glucose, and diabetes in the NHANES III. *Diabetes Care* **26**:468–470.

Shimoda R, Nagamine T, Takagi H, Mori M, Waalkes MP. 2001. Induction of apoptosis in cells by cadmium: Quantitative negative correlation between basal or induced metallothionein concentration and apoptotic rate. *Toxicol Sci* **64**:208–215.

Smith SR, Jaffe DM, Skinner MA. 1997. Case report of metallic mercury injury. *Pediatr Emerg Care* **13**:114–116.

Wei M, Wanibuchi H, Yamamoto S, Li W, Fukushima S. 1999. Urinary bladder carcinogenicity of dimethylarsenic in male F344 rats. *Carcinogenesis* **20**:1873–1876.

Yamamoto S, Konishi Y, Matsuda T, Murai T, Shibata M-A, Matsui-Yuasa I, Otani S, Kurada K, Endo G, Fukushima S. 1995. Cancer induction by an organic arsenic compound, dimethylarsinic acid (cacodylic acid) in F3444/DuCrj rats after pretreatment with five carcinogens. *Cancer Res* **55**:1271–1276.

Zhong CX, Mass MJ. 2001. Both hypomethylation and hypermethylation of DNA associated with arsenite exposure in cultures of human cells identified by methylation-sensitive arbitrarily-primed PCR. *Toxicol Lett* **122**:223–234.

21 Mode of Action and Toxicity of Trace Elements

ARUN K. SHANKER

Central Research Institute for Dryland Agriculture (CRIDA), Indian Council of Agricultural Research (ICAR), Santoshnagar, Hyderabad, 500 059, India

1   INTRODUCTION

Trace elements have made their entry into plant and animal systems due to their distinct chemical properties such as reduction and oxidation reactions under physiological conditions. On the other hand, the very same chemical reactions that make some of these trace metal ions obligatory for life are also the primary cause of their toxicity when present in surfeit. In analytical chemistry, a trace element is an element in a sample that has an average concentration of less than 100 parts per million (ppm) atoms, or less than 100 micrograms per gram. In biochemistry, a trace element is a chemical element that is needed in minute quantities for the proper growth, development, and physiology of the organism. In biochemistry, a trace element sometimes is also referred to as a micronutrient. In geochemistry, a trace element is a chemical element whose concentration is less than 1000 ppm or 0.1% of a parent rock's composition. The major difference between metals and other toxic substances and xenobiotics are that they are not created by humans, although several anthropomorphic activities contribute to the prevalence of these metals in proximity of human and plant ecosphere which facilitates their entry into the biological systems [Beijer and Jernelov, 1986]. The alteration of elemental stoichiometry by human activity is also another important way by which the trace elements acquire toxic properties. Historically, *The Ebers papyrus* (1500 B.C.E.) contains information pertaining to inquiries into the poisonous nature of metals such as lead, copper, and antimony. Metal toxicity and cure is documented from as early as 370 B.C.E. when Hippocrates described abdominal colic in a man who worked with metal extraction. From then on, there are several other references of studies in

Trace Elements as Contaminants and Nutrients: Consequences in Ecosystems and Human Health, Edited by M. N. V. Prasad

toxic interaction of metals—mainly lead, arsenic, and mercury—in humans. It must be emphasized here that most of the concerns until the mid-20th century among scientists were mainly on the acute and visible clinical effects mostly confined to humans.

The landmark inquiry into mechanism of action of heavy metals was the work by Voegtlin et al. [1923], wherein they reported the mechanism of action of arsenic on protoplasm. Diagnosis and treatment at a clinical level was more prevalent than inquiries into the mode of action and toxicity of trace elements. Later the logical evolution of the science moved toward the more subtle, chronic, or long-term effects of numerous trace elements which paved the way for the modern science of trace element toxicology at cellular and molecular levels. If one traces the history of the toxicology of metals over the past half-century, the role of the Pharmacology Department of the University of Chicago is paramount. The story commences with the use of uranium for the "bomb" and continues today with research on the role of metals in their relations with DNA, RNA, and growth factors. In fact, the Manhattan Project created a productive atmosphere that resulted in the commencement of quantitative biology, radiotracer technology, and inhalation toxicology. These innovations have revolutionized modern biology, chemistry, therapeutics, and toxicology. The behavior of dissolved metals has been studied for over 25 years. In the early 1970s, much research concerned the influence of pH and water hardness on metal toxicity to algae and other aquatic organisms. This work led to the development of a model to predict metal toxicity based on pH and water hardness [US EPA, 1986]. Currently, more interest is shown in trace element interaction with plants, microorganism, and animals due to the established link of theses biological systems with human life.

The list of trace elements essential for both animals and plants is large: It comprises vanadium to zinc in the first row series of the periodic table, plus molybdenum in the second row series. The physiological range for essentiality of these elements between deficiency, sufficiency, and toxicity is consequently exceedingly narrow, and there exists a controlled metal homeostasis network to adjust to fluctuations starting from nonavailability to toxicity [Beijer and Jernelov, 1986]. At a larger scale the effects of trace elements on the ecosphere is very detrimental as proved by the classical study of Watson et al. [1976], wherein emissions of heavy metals from a lead ore-processing complex caused perturbations to the litter–arthropod food chain in a forest ecosystem. Elevated concentrations of lead (Pb), zinc (Zn), copper (Cu), and cadmium (Cd) caused reduced arthropod density and microbial activity, resulting in a lowered rate of decomposition and a disturbance of forest nutrient dynamics.

An understanding of the mechanisms of toxicity is of practical importance. Knowledge on the mode of action provides a coherent basis for (a) interpretation of toxicity data, (b) estimation of the probability and the extent of harmful effects, (c) establishment of protocols to prevent toxic effects, and (d) discovering drugs that can counteract these effects. This chapter reviews the cellular and molecular mechanisms that contribute to the expression of toxicities. Complete mechanism of action information is rarely available and is not required for toxicity assessment. Mode of action is defined as the sequence of key cellular and biochemical events that result in a toxic effect, while mechanism of action implies a more detailed understanding of the molecular basis of the toxic effect. Hence in this chapter it is important to distinguish between actions of metals and their effects. Actions of metals are the biochemical and/or physiological mechanisms by which the chemical produces a response in living organisms. The effect is the observable consequence of a toxic action. Although such mechanisms may be dealt with elsewhere in this book, an attempt is made to discuss this process in detail in this chapter in an integrated and complete manner with more emphasis on the mode rather than on the effect. This chapter further focuses integration of these mechanisms that have been identified definitively in animals and plants.

2 MODE OF ACTION AND TOXICITY OF TRACE ELEMENTS IN GENERAL

Hypothetically, the strength of a toxic effect of all trace metals depends principally on the absorption, concentration, and persistence of the eventual toxicant at its location of action. The final toxicant is the metal species that reacts with the endogenous target molecule such as receptors, enzymes, DNA, protein, or lipid or critically alters the biological environment, producing structural and functional changes that result in toxic damage. More often than not, the principal toxicant is the metal species to which the organism is exposed. However, in some cases the toxicant may be a metabolite of the parent compound or reactive oxygen or reactive nitrogen species (ROS or RNS) generated during the in vivo transformations of the toxicant. In few cases an endogenous molecule or compound synthesized as a response to primary toxicant exposure may be the principal toxicant. The buildup of the definitive toxicant at its target is facilitated by various biological processes that involve absorption, distribution to the site of action, reabsorption, and metabolic activation [Langman and Kapur, 2006]. Contrastingly, the predominant processes involved in counteracting toxic exposure contributing to tolerance are presystemic elimination, distribution away from the site of action, excretion, and detoxication which work against the accumulation of the ultimate toxicant at the target molecule. The logical steps by which toxicity results are typically as follows: delivery of toxicants > reaction of toxicants with target molecules > manifestation of dysfunction > counter reaction (repair)/failure of counter reaction (disrepair) > toxicity. Mode of action typically starts with the reaction of metals with target molecules and ends with toxic manifestations.

After mediation of toxicity by chemical reactions with a target molecule, a succession of secondary, tertiary, and quaternary biochemical events take place; this leads to dysfunction or injury that is apparent at diverse levels of biological organization, which may include the target molecule itself, cell organelles, cells, tissues and organs, the organism, and finally the community as a whole. Hence, mechanism of action and toxicity is typically characterized by the target molecules attributes, the reactions type's trace elements and target molecules, and the effects of trace elements on the target molecules. In addition, manifestation of toxicity may also be due to the alteration of the biological environment in which molecules, cell organelles, cells, and organs operate in the absence of contact chemical reactions initiated by the toxicant.

The properties that make a molecule or substance qualify as a toxicant are (1) it reacts with the target and adversely affects its function, (2) it reaches an effective concentration at the target site, and (3) it alters the target in a way that is mechanistically related to the observed toxicity. Progression toward toxicity involves (1) the attributes of target molecules, (2) the types of reactions between ultimate toxicants and target molecules, and (3) the effects of toxicants on the target molecules [Gregus and Klaassen, 2001]. The reaction types after contact of the trace element and the target molecule would involve (i) noncovalent binding by the formation of hydrogen and ionic bonds involving the interaction of some trace metals with targets such as membrane receptors, intracellular receptors, and ion channels and (ii) covalent bonding by most traces metals that react with nucleophilic atoms that are plentiful in biological macromolecules, such as amino acids, proteins, and nucleic acids. Soft electrophilic metals such as silver and mercury react with (a) soft nucleophiles like sulfur in thiols, cysteinyl residues in proteins, and glutathione and sulfur in methionine (thiols and thiolates) and (b) hard electrophiles such as lithium, calcium, and barium, which react preferentially with hard nucleophiles (oxygen of purines and pyrimidines in nucleic acids phosphate oxygen in nucleic acids). On the other hand, chromium, zinc, and lead fall between these two extremes and exhibit universal reactivity with all nucleophiles including nitrogen in primary and secondary amino groups of proteins, nitrogen in amino groups, and in purine bases in nucleic acids. The reactivity of an electrophile determines which endogenous nucleophiles can react with it and become a target.

A typical target molecule having the functions of regulation, maintenance, and signaling when effectively attacked will initiate a cascade of effects as diversified as (a) changes in gene expression causing improper cell division apoptosis, (b) impaired protein synthesis, (c) changes in internal and external maintenance causing impaired ATP synthesis, and (d) altered membrane function leading to cell injury in plants and animals. Furthermore, in animals these changes may cause unbalanced homeostasis, bleeding, inappropriate neuromuscular activity-like tremors, convulsion, and paralysis. After the trace element has acted on the target, the last process would be the process of repair which, if absent, may cause damage at higher levels of the biological hierarchy in the organism, which in effect would be irreversible. The mechanisms of repair take place at molecular, cellular, and tissue levels. It is noteworthy that even repair mechanisms can contribute to the toxicity of trace elements; the best examples of this are (1) if excessive amounts of biomolecules are cleaved by the enzymes that assist in repairing broken DNA strands and (2) when too much reducing power is consumed for the repair of oxidized proteins and endogenous reductants.

Perturbations in signaling is one of the principal modes of action resulting in toxicity which is common to most trace elements (see Fig. 1). Oxidative processes enhancement resulting in the increase of superoxide anion radical ($O^{2.-}$), H_2O_2, and hydrogen peroxide radical ($OH^{\bullet}$) is the base for other connections with signaling response. Oxidative stress affects numerous cellular components, such as DNA, lipids, and proteins, through oxidation reactions. These alterations in structure

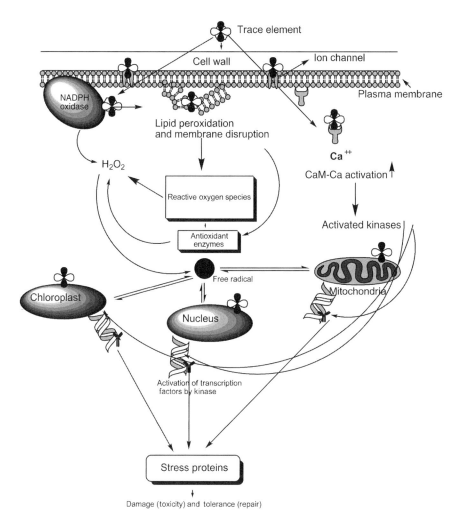

Figure 1. Generalized scheme of signaling induced by trace elements in living systems. In a very simplistic model of signaling systems leading toward toxicity or tolerance, the following would be the train of events. Trace element's entry in the system is facilitated by transport protein or diffusion. At the topical level there is increased production of H_2O_2 by direct action on NADPH oxidase; at the systemic level there is disruption of phospholipid bilayer due to lipid peroxidation, leading to production of ROS inducing synergistic action of SOD, CAT, and APx and increasing H_2O_2 levels especially by SOD. In due time course, excess metals enter cellular organelles like mitochondria and chloroplasts (plants), act as a sink in the electron flow, or misdirect the electron flow (depending on the redox status of the metal), which causes production of free radicals. Free radicals, in turn, initiate the antioxidant systems (Halliwel Asada Pathway) to quench H_2O_2. Unquenched H_2O_2 in addition to other free radicals gives rise to singlet oxygen. In addition, receptor metal complex in the plasma membrane causes excess calcium ion concentration which initiates the calmodulin-Ca^{2+} system activating various kinases. These reactive molecules and kinases act as signals on the transcription factors present in the nuclear as well as the organelle DNA, leading to the production of stress proteins and secondary metabolites that can act as either damage-causing agents or stress-countering agents.

produce significant changes in the function and may result in pathogenesis. Most trace elements produce reactive oxygen species, resulting in an increased lipid peroxidation (LPO), depletion of sulfhydryls, altered calcium homeostasis, and finally DNA damage. Oxidative stress refers to the cytopathological consequences of a disproportion between the production of free radicals and the capability of the cell to protect against them. Oxidative lipid injuries, named lipid peroxidation (LPO), create a progressive loss of membrane fluidity, thus reducing membrane potential and increasing its permeability to ions such as Ca^{2+}. Active oxygen species are continuously produced in tissues by (a) the action of the mitochondrial electron transport system and reduced nicotinamide adenine dinucleotide phosphate (NADPH) oxidase and (b) photophosphyration in chloroplasts of plants, among other sources. There are several antioxidant defense systems, including the action of some enzymes, to antagonize oxidative damage; these include catalase (CAT) and superoxide dismutase (SOD) for hydrogen peroxide. This system is the primary line of defense by living organisms against trace elements.

A growing body of verification demonstrates that most trace metals mediate gene expression by activation of signal transduction pathways. Signal transduction is a progression by which information from an extracellular indication is transmitted into the cell through the plasma membrane and along an intracellular sequence of signaling molecules to excite a cellular response. The response may be the commencement of a gene transcription that occurs through various regulatory proteins that bind to specific DNA sequences in a gene. The net result of this binding is usually transcription of that gene and is referred to as transcriptional activation. Cell transformation is a complex process involving a variety of transcription factors and signaling pathways. Transcription factors, such as AP-1, NFκB, and the classical mitogen-activated protein kinase (MAPK) cascade, play important roles in cell proliferation, differentiation, and transformation [Jonak et al., 2004]. Apoptosis or cell death induced by most trace elements has been best studied. Apoptotic cascade pathways induced by metals have mitochondrial dysfunction as its originator. Mitochondrial dysfunction initiated by the opening of the mitochondrial transition pore leads to mitochondrial depolarization, release of cytochrome C, activation of a variety of caspases, and cleavage of downstream death proteins and ultimately results in apoptotic cell death.

3 SPECIFIC MODE OF ACTION OF MAJOR TRACE ELEMENTS

3.1 Arsenic

Arsenic is one of the few elements that is yet to be characterized as a single element. This is mainly due to its highly complex chemistry and also due to the existence of a plethora of compounds. It is trivalent or pentavalent and has a wide distribution. The most common inorganic trivalent arsenic compounds are arsenic trioxide, sodium arsenite, and arsenic trichloride. Pentavalent inorganic compounds are arsenic pentoxide, arsenic acid, and arsenates, such as lead arsenate and calcium arsenate.

Organic compounds may also be trivalent or pentavalent, such as arsanilic acid, or may even occur in methylated forms as a consequence of biomethylation by organisms in soil, fresh water, and seawater. Inhalation exposure to arsenic occurs in industrial settings such as lead, copper, and zinc smelting, from fossil fuel combustion in power plants, in the semiconductor industry, as well as in pesticide production. Nonoccupational exposure to arsenic can take place through ingestion of contaminated food and water. Contamination of well water with arsenic leached from underground sediments has occurred in large areas of India and Bangladesh, and hundreds of thousands of people have developed precancerous skin lesions due to arsenic ingestion [National Research Council Report, 1999]. High concentrations of arsenic in drinking water are also found within the United States in the western and southwestern states and in Alaska. Exposure to high concentrations of arsenic is associated with skin, lung, liver, and bladder cancers [Morales et al., 2000].

3.1.1   Mode of Toxic Action.   The toxicity of arsenic is highly dependent on its oxidation state and chemical composition. Arsenite is taken into cells by passive diffusion, while arsenate competes with phosphate for uptake. Arsenite is extremely thiol-reactive. It can affect enzyme activities by binding to critical vicinal cysteinyl residues, such as those in the lipoamide of pyruvate dehydrogenase, tyrosine phosphatases, and enzymes involved in protein ubiquination. It is thought that arsenite is a sulfhydryl reagent having a high affinity mainly for vicinyl dithiols and also thiols located near hydroxyls. In contrast, arsenate is similar to phosphate in structure and may interfere with oxidative phosphorylation by forming an unstable arsenate ester. Thus, arsenate affects phosphotransfer reactions, which are required for ATP generation. Furthermore, arsenate is excreted more rapidly than arsenite from the body. Arsenite is therefore considerably more toxic and carcinogenic than arsenate. It is also believed that inorganic arsenic was more toxic than organic arsenic, and the methylation of inorganic arsenic was thought to be a detoxification process [Huang et al., 2004].

It is well established that the trivalent compounds are the principal toxic form of As and that pentavalent arsenic compounds have little or no effect. Sulfhydryl proteins and enzymes are extensively altered by exposure to arsenic. Reversal of alterations is possible in the presence of glutathoine in reduced or in oxidized form, and hence the Halliwell Asada pathway forms an important site for action and counteraction in As toxicity. The cellular toxicity of As is primarily because of impairment of mitochondrial enzymes and resultant blockage of oxidiative phosphorylation. The accumulation of As in mitochondria resulting in uncoupling of NAD linked substrates in the electron transport chain. Reactions between dihyrolipoic acid and arsenite ion prevents the oxidation of substrate. In addition, inhibition of succinic dehydrogenase by arsenite also contributes to increase in the activity of mitochondrial ATPase as a result of uncoupled oxidative phosphorylation. Arsenic inhibits energy-linked functions of mitochondria in two ways: (1) competition with phosphate during oxidative phosphorylation and (2) inhibition of energy-linked reduction of NAD. Inhibition of mitochondrial respiration results in decreased cellular production of ATP and increased production of hydrogen peroxide, which might cause oxidative stress, and production of reactive oxygen species (ROS) (see Fig. 2 for a generalized scheme). Intracellular production of ROS results in observed induction of major stress protein families [Shi et al., 2004]. Arsenic compounds induce metallothionein in vivo. Potency is dependent on the chemical form of arsenic. As(III) is most potent, followed by As(V), monomethylarsenate, and dimethylarsenate [Kreppel et al., 1993].

Metallothionein is thought to have a protective effect against arsenic toxicity and may be responsible, at least in part, for its self-induced tolerance. Metallothionein-null mice are more sensitive than wild-type mice to the hepatotoxic and nephrotoxic effects of chronic or injected inorganic arsenicals [Liu et al., 2001]. In addition, in plants As clearly increases homophytochelatin (hPC) and phytochelatins (PC). It has been demonstrated that the formation of As–PC complexes is in accordance with a detoxifying role for the peptides [Gupta et al., 2004]. The role of asenical-induced oxidative stress and ROS may play a role in mediating DNA damage and initiating the carcinogenic process [Kligerman et al., 2005]. Although a well-recognized human carcinogen, arsenic itself is not a potent mutagen and has been thought to act through epigenetic mechanisms that modify DNA methylation

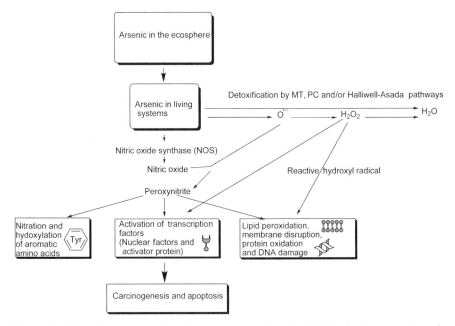

Figure 2. Schematic representation of arsenic mode of action in biological systems. Arsenic enters in vivo from the ecosphere and is either subjected to detoxification or proceeds to cause toxicity that is dose-dependent. Detoxification is mainly by antioxidant enzymes and metal-binding peptides which finally reduce the toxic free radicals to water. Unquenched and unscavenged ions generated by excess H_2O_2 and peroxynitrite cause nitration of aromatic amino acids (tyrosine) and activation of transcription factors that promote carcinogenesis and apoptosis.

patterns, perhaps in conjunction with DNA damaging agents. Hypomethylation of DNA is thought to be an early epigenetic mechanism that coincides with malignant transformation and a mechanism that transmits inappropriate gene expression patterns transgenerationally. Exposure to physiologically relevant concentrations of arsenic mediates hypomethylation of chromatin by competition for methyl donors and inhibition of DNA methyltransferase reactions that utilize S-adenosylmethionine as a cofactor [Reicharda et al., 2007].

The arsenite (+3) and arsenate (+5) forms have different modes of action. Arsenite binds to sulfhydryl groups and has been reported to inhibit over 100 different enzymes, while the arsenate can substitute for phosphate in various high-energy intermediates, resulting in arsenolysis. In addition, when arsenate is reduced to arsenite in vivo, it can also cause toxicity as that species. Recently, it has been demonstrated that chromosome fragments that are not incorporated into the nucleus at cell division (micronuclei formation) is the one of the principal manifestations of arsenic in plants and animals [Yi et al., 2007]. There are several possible hypotheses on the pathways of MN formation. The first pathway depends upon arsenic-induced oxidative stress. The induction of reactive oxygen species (ROS) was found in arsenic-exposed mammalian and plant cells [Requejo and Tena, 2005]. The attacks of ROS on purine and pyrimidine bases and on deoxyribose in DNA can cause DNA strand breakage, which can increase the probability of chromosome/chromatid fragmentation, thereby leading to the observed increases of MN formation. The second pathway involves the changes of protein sulfhydryl groups (−SH). Binding to sulfhydryl groups, arsenic can inactivate some important enzymes involved in DNA repair and expression, alter DNA repair mechanism, and cause an increase in MN frequency. There is some evidence to suggest that the cytoskeleton is an important cellular target of arsenic toxicity. Arsenic by interfering with microtubule assembly and spindle formation causes chromosomal lagging and leads to a higher frequency of micronuclated cells [Binet et al., 2006]. It has been reported that arsenite is a potent stimulator of proto-oncogene c-fos and c-jun expression and AP-1 transactivational activity. A DNA-binding protein composed of the Jun and Fos proteins, AP-1 regulates the transcription of various genes governing cellular processes, such as inflammation, proliferation, and apoptosis. Arsenic has been found to induce activation of JNKs, ERKs, and several other kinases. ERKs activation may contribute to the carcinogenic effects of arsenic, whereas JNKs activation is associated with apoptosis and results in the anti-carcinogenic effects of arsenic [Huang et al., 1999]. Although the mechanism by which JNKs mediates cell apoptosis in response to arsenite is not clear, several recent studies have shown that JNKs translocation to mitochondria is an important step in the resultant toxicity. JNKs play a central role in regulation of apoptosis and the release of cytochrome c and apoptotic proteases. Very recently, it has been found that JNK perturbation occurred during arsenite-induced malignant transformation, which is resistant to apoptosis as compared with passage-matched control cells, suggesting that apoptotic control mechanisms are disrupted as cells become transformed through arsenic exposure. This apoptotic disruption may allow damaged cells to inappropriately escape apoptosis and potentially proliferate, thereby providing initiating events in carcinogenic development. Thus, this may lead to accumulation of genetically damaged cells that have a potential to become malignant [Huang et al., 2004]. Considering the fact that carcinogenesis is the principal toxic action of As, at present the mechanisms by which arsenic causes human cancers are not well understood. Arsenic is an atypical carcinogen because it is classified in neither the initiator nor the promoter categories of carcinogenic agents. Thus, arsenic probably does not act as a classical carcinogen, but rather enhances the carcinogenic action of other carcinogens.

3.2   Cadmium

Cadmium is a modern toxic metal. It was discovered as an element in 1817, and its industrial use was minor until about 50 years ago. But now it is a very important metal with many applications. Because of its noncorrosive properties, its main use is in electroplating or galvanizing. It is also used as a color pigment for paints and plastics and as a cathode material for nickel–cadmium batteries. Cadmium is a by-product of zinc and lead mining and smelting, which are important sources of environmental pollution. During the past century, Cd and its compounds have been used extensively in the smelting and electroplating industries and in the manufacturing of batteries, dyes, paints and plastics. Cadmium pollution was shown to have severe consequences on human, animal, and plant systems. Large amounts of Cd have also been released into the environment through the burning of refuse materials that contain Cd and through the use of Cd-contaminated sludge and phosphate salts as fertilizers. Tobacco contains significant amounts of Cd, and smoking is one of the primary sources of Cd exposure in the general population [Satarug and Moore, 2004]. Exposure to Cd can result in a variety of adverse effects in humans and animals. Depending on the dose, route, and duration of exposure, Cd can damage various organs including the lung, liver, kidney, bone, testis, and placenta. The most important effects are renal injuries, immune deficiencies, apathies, bone injuries, femoral pain, lumbago, and skeleton deformations. Cadmium exposure has detrimental effects on the CNS, with symptoms including headache and vertigo, parkinsonian-like symptoms, slowing of visuomotor functioning, peripheral neuropathy, decreased equilibrium, and decreased ability to concentrate. In addition, Cd has been shown to have teratogenic and carcinogenic activities.

Exposure to Cd has been associated with a wide variety of cardiovascular pathologies including hemorrhagic injury, atherosclerosis, hypertension, and cardiomyopathy [Navas-Acien et al., 2005]. The exact mechanism(s) through which cadmium produces its neurotoxic effects is still unresolved. Even though Cd represents a major environmental health problem, the specific mechanisms by which it produces its adverse effects have yet to be fully elucidated. Studies to address this issue have shown that Cd has a variety of biochemical, metabolic, and cytotoxic mechanism of action. However, in most cases, the relationships between these effects and the specific toxic actions of Cd in various target organs have not been firmly established. In plants, cadmium is a nonessential element, and the most evident symptoms of its toxicity are chlorosis and stunting. Chlorosis seems to be the result of the effects of Cd on the uptake, transport, and use of several elements (Ca, Mg, Fe, Mn, Cu, Zn, P, and K), with the consequent reduction of Mn and Fe absorption and changes in Fe:Zn ratios. On the other hand, reduction of plant development seems to be the result of Cd interference with several important physiological processes: Cd alters the hormonal balance and disturbs the plant water status through a decrease of water absorption, reduction of root hydraulic conductivity into xylem vessels, decrease of transpiration rate, and increase of stomatal resistance [Mishra et al., 2006; Aina et al., 2007].

3.2.1   Mode of Toxic Action. Cadmium has various effects on molecules, cells, and organelles. Oxidative stress has often been discussed as a primary effect of Cd^{2+} exposure even though Cd is not a redox-active metal and will not take part in Fenton and Haber−Weiss reactions. Rather, symptoms of oxidative stress such as lipid peroxidation are a consequence of GSH depletion due to binding of Cd^{2+} to GSH and formation of GSH-derived PCs [Schützendübel and Polle, 2002]. Cell cycle progression, DNA replication and repair differentiation, proliferation, and apoptotic pathways are altered and affected by cadmium (Fig. 3). By activating

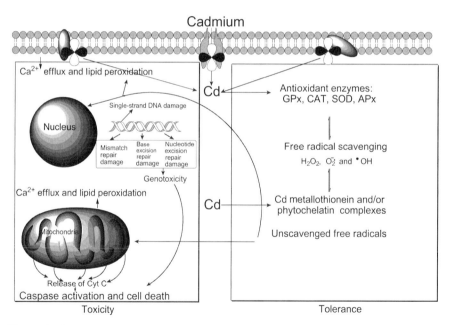

Figure 3. Scheme depicting cadmium-induced neurotoxicity and genotoxicity in cell. Cd ion enters cells mediated by channels or transport proteins. After entering, Cd induces a decrease or increase (in a dose-dependent manor) in the activities of scavenging enzymes. Metallothionein (MT) and/or phytochelatins are activated as scavengers of the free radicals. Excess free radical production increases the levels of lipid peroxidation and disruption of membranes [cellular, mitochondrial, and chloroplast (plants)]. Furthermore, Cd causes a decrease in intracellular ATP levels, leading to mitochondrial membrane breaking. This causes Ca^{2+} and cytochrome C to escape into the cellular matrix and activate various caspases by binding to Apaf-1 (apoptosis protease activating factor 1) and induce apoptosis and/or necrosis.

cellular signals, it regulates cell cycle progression and inhibition of DNA methylation and/or interference with cell adhesion. Almost all the effects including the effects on DNA synthesis and cell proliferation are clearly dose-dependent. Although it enhances DNA synthesis and cell proliferation at lower concentration than 1 μM, cadmium exposure above this concentration inhibits DNA synthesis and cell division [Yang et al., 2004; Dong et al., 2001]. RAS proteins are small enzymes, which serve as master regulators of a myriad of signaling cascades involved in highly diverse cellular processes. The RAS signaling pathway involved in cell proliferation and differentiation is one of the primary sites of cadmium action as a consequence of which many genes involved in cell cycle regulation are overexpressed and many proteins are up-regulated.

Cadmium also increases the level of several kinases of the RAS pathway, with MAPK being the one of the chief ones. Furthermore, cadmium treatment induces translocation of adhesion molecules and chaperones from the cytosolic side to the membrane side, thereby increasing DNA-binding activity of these molecules. Overexpression of the proto-oncogenes—namely, *c-fos*, *c-myc*, and *c-jun*—are induced by cadmium. These genes actively promote proliferation of cells and development of tumors after cadmium exposure. In mammalian cells, Cd^{2+} induces cyclin D and cyclin E expression by activation of a Myb-type transcription factor; in plants, Cd^{2+} causes inhibition of cyclin B [Deckert, 2005]. Several stress-inducible proteins such as hemeoxygenase-1(HO-1), regulating interleukin-10 (IL-10), and (tumor necrosis factor alpha (TNF-α) production are induced by cadmium. Cadmium also affects several genes involved in the stress response to pollutants or toxic agents.

Zn-dependent proteins and Zn-binding molecules are "candidate" targets of Cd^{2+} toxicity. The chemical similarity of these two ions makes it a possibility that Cd^{2+} ions can replace Zn and, in so doing, interfere with some of the many Zn reliant processes. Indirect confirmation for such consequences is the transcriptional up-regulation of recognized Zn^{2+} uptake systems, an observation that has been made in several biological systems [Weber et al., 2006]. It implies that a Zn^{2+}-sensing molecule can be engaged by Cd^{2+} ions as a result of Cd^{2+} exposure. Similarly, Ca^{2+}-binding proteins such as calmodulin might well be prime intracellular binding sites of Cd^{2+}, and such binding will most likely be detrimental to cellular signaling cascades. Still, in spite of a large body of work on metal toxicity, primary target sites of toxic metal effects in general, or Cd^{2+} in particular, remain to be identified [Clemens, 2006].

Metallothioneins are the most-studied proteins in relation to Cd and its mechanism of toxicity because it is readily induced in high quantities after cadmium exposure. Mammalian metallothioneins (MT) are cysteine-rich heavy-metal-binding proteins that can protect against cadmium toxicity and oxidative stress. They contain about 30% of cysteine residues, which are known for their ability to chelate free-cadmium. Intoxication of cells or animals with cadmium results in an increase of the production of metallothioneins, which belong to the principal pathway of detoxification of this heavy metal. The importance of metallothioneins is shown by the fact that administration of exogenous metallothionein to rats exposed to cadmium results in a decrease of oxidative stress. Metallothionein (MT) gene transcription is induced by cadmium and oxidative stress. Metal response elements (MRE) present in the proximal promoters of MT genes and metal-responsive six zinc finger transcription factor MTF-1 are instrumental in the transactivation of MTs. The next important mode of action relates to the effects of Cd on heat shock proteins (HSPs). Cadmium intoxication alters the expression levels of HSPs, which are active stress-responsive proteins. After Cd^{2+} exposure, overexpression of the genes encoding HSPs has been widely reported. HSPs are cellular chaperone proteins, which can be induced by various environmental stresses including toxic exposure to cadmium. HSP induction is generally considered as an adaptive response of cells to stress, closely linked with cell survival. HSP proteins participate not only in a principal role of chaperoning folding but also in the degradation of proteins. In addition, apoptosis pathways are also activated or inhibited by HSPs. Cd^{2+}-mediated induction of HSPs is largely due to its effects on proteins. Denaturation and oxidation of proteins by cadmium are responsible for the overexpression of HSP chaperones [Gaubin et al., 2000]. HSPs induction after Cd^{2+} exposure is also linked to the increase of oxidative stress. Cadmium intoxication induces overexpression of at least 10–15 families of heat shock chaperone proteins.

Apoptosis is one of the well-studied effects of Cd in living tissues other than plants. Cell death resulting from cadmium (Cd) intoxication has been confirmed to occur through apoptosis by morphological and biochemical studies. However, it is still not clear whether Cd itself or metallothionein (MT) induced by Cd is the major factor responsible for the apoptosis. CdMT may play a paradoxical role, providing protection against the cadmium ion in the intracellular milieu, but promoting cadmium toxicity when it is present in sufficient amounts in the parts of living systems where absorption is fast. Thus, the function of MT in relation to the effect of Cd is rather debatable. As discussed above, oxidative damage due to Cd is a well-proven mechanism of action. Taking these together, it can be said that Cd causes apoptosis through its indirect oxidative effects routed through overexpression of MT genes. Three apoptotic pathways have been described: (1) mitochondria-dependent pathway, (2) death-receptor-dependent pathway, and (3) endoplasmic reticulum (ER) pathway. There is an individual initiator caspase in each apoptosis pathway: caspase-9, in the mitochondrial pathway; caspase-8, in the death-receptor-dependent pathway; and caspase-12, in the ER pathway. Cadmium probably induces apoptosis through the mitochondrial pathway; Lopez et al. [2006] showed that the decreases in the ATP intracellular levels at the highest concentration of cadmium were accompanied by ATP release, indicating mitochondrial and cytosolic membrane breaking. This mechanism, produced by mitochondrial toxicity with fall in ATP, breakdown of mitochondrial membrane potential, and ROS formation, induces apoptosis with disruption of cellular membranes and necrosis.

Nerurotoxicity is also induced by Cd, and it is known that oxidative stress caused by Cd in nerve cells completely blocks CNTF family neurokinase and IFN-γ-mediated Jak/STAT signaling; this seems to be a novel mechanism for mediating cadmium neurotoxicity [Monroe and Halvorsen, 2006]. It seems that the administration of cadmium initially affects the integrity and permeability of the vascular endothelium, and necrotic changes in nerve cells occur secondarily to this effect.

Cadmium inhibits all of the known pathways of cellular Ca^{2+} influx and acts as a competitive ion to Ca^{2+} at the voltage-dependent Ca^{2+} channels, and it is a potent blocker of the Ca^{2+}-dependent neurotransmitter release. This effect on Ca^{2+} influx is due to the interaction of the heavy-metal ion with thiol groups of proteins involved in intracellular Ca^{2+} sequestration. On the other hand, Cd has been reported to elevate the intracellular Ca^{2+} concentrations. This sustained increase is believed to be the main cause for cellular death. In addition, an increase of free radicals and Ca^{2+} levels associated with Cd exposure induces mitochondrial disruption and release cytochrome C into the cytosol. The change of the membrane potential of the mito-chondria membrane opens the transition pore, stimulating again the release of cyto-chrome c. Cytochrome c is a component of the electron transport chain and is involved in the production of ATP; cytochrome c activates caspases by binding to specific proteins, inducing it to associate with procaspase-9 (holoenzyme), thereby triggering caspase-9 activation and initiating the proteolytic cascade and eventually cell death. In addition to causing cell death, neurotoxicty, and genotoxicity, cadmium targets the DNA repair mechanism itself, thus posing as a great threat among all the known toxic trace elements.

DNA repair is major protection machinery against DNA injury caused by regular metabolic activities and environmental factors. It includes a multiplicity of biochemi-cal mechanisms that add to the preservation of genetic sequence, minimizing cell killing, mutations, duplication errors, persistence of DNA harm, and genomic vola-tility. Cadmium interferes with multiple DNA repair process, and at low biological relevant concentrations it enhances the mutagenicity induced by other DNA-damaging agents. Cadmium (a) interferes with base excision repair (BER) by redu-cing the repair capacity of formamidopyrimidine DNA glycosylase, (b) interferes with nucleotide excision repair (NER) by disturbing in a dose-dependent manner, DNA-protein interactions essential for the initiation of NER, and (c) interferes with mismatch repair (MMR) by inhibiting the capacity of the MMR process to correct base–base and insertion/deletion mismatches, which had escaped the proofreading function of replicative polymerases. It can thus be postulated that this interferences with DNA repair mechanisms is the cause of many other effects of Cd on living systems. As Giaginis et al. [2006] have suggested, an ample amount of evidence shows that Cd might interfere with DNA repair procedure, leading to increased buildup of damaged DNA bases. The failure of DNA repair systems to correct affected bases can lead to harmful mutations, genomic instability, or cell death. In higher eukaryotes, the damage that occurs in DNA stability genes accountable for DNA repair and cell cycle control can result in tumor creation. Although Cd is mainly implicated in the initial steps of DNA repair process, additional indications suggest that further steps could also be inhibited. Specifically, it has been established that many proteins, involved in these DNA repair systems, are possible targets of Cd toxicity. Furthermore, it is well known that Cd is able to complex with DNA repair proteins either by substitution of Zn from their Zn finger motif or by binding to nega-tively charged surface residues on their structure, leading to conformational changes and disturbing the DNA–protein interactions, essential for DNA repair process and maintenance of the genome integrity. The complete, detailed, step-by-step mechanism by which Cd inhibits DNA repair is yet to be elucidated, and this essentially contributes to the lack of explicit understanding of Cd-induced mutagenicity and carcinogenicity.

3.3 Chromium

Chromium (Cr) was first discovered in the Siberian red lead ore (crocoite) in 1798 by the French chemist Vauquelin. It is a transition element located in group VIB of the periodic table with a ground-state electronic configuration of $Ar3d^54s^1$. The stable forms of Cr are the trivalent Cr(III) and the hexavalent Cr(VI) species, although there are various other valence states that are unstable and short-lived in biological systems. Cr(VI) is considered the most toxic form of Cr, which usually occurs associated with oxygen as chromate (CrO_4^{2-}) or dichromate ($Cr_2O_7^{2-}$) oxyanions. Cr(III) is less mobile and less toxic and is mainly found bound to organic matter in soil and aquatic environments. Cr(III) is a micronutrient important in the biological activity (receptor binding) of insulin and, accordingly, can be found in many dietary supplements. Contamination of soil and groundwater due to the use of Cr in various anthropomorphic activities has become a serious source of concern to plant and animal health over the past decade. It is estimated that several hundred thousand workers are potentially exposed to high levels of Cr(VI). Occupational exposure to Cr (Cr(III) and Cr(VI)) by inhalation depends upon the job function and industry, but can reach several hundred micrograms per cubic meter. These estimated exposures have been significantly lowered in the past few decades as industrial hygiene practices and worker controls have been implemented. Nonoccupational exposure to Cr occurs from automobile emissions and cigarette smoke.

It has been estimated by IARC that, on average, cigarettes produced in the United States contain 0.24–6.3 mg Cr/kg. In the environment, elevated levels of Cr have been reported in areas near landfills, hazardous waste disposal sites, chromate industries, and highways. Chromium, in contrast to other toxic trace metals like cadmium, lead, mercury, and aluminum, has received lesser attention from scientists. Its complex electronic chemistry has been a major hurdle in unraveling its toxicity mechanism in general. The impact of Cr contamination in biological systems depends on the metal speciation, which is responsible for its mobilization, subsequent uptake, and resultant toxicity in the biological systems. Chromium toxicity in plants is observed at multiple levels, from reduced yield, through effects on leaf and root growth, to inhibition on enzymatic activities and mutagenesis [Shanker et al., 2005; O'Brien et al., 2003].

3.3.1 Mode of Toxic Action. Chromium(VI) has long been recognized as a carcinogen in human and mammalian systems. Since hexavalent chromium does not damage DNA in vitro without a reducing agent, it is believed that the actual mutagenic species is one or more of the reactive intermediates produced in the reduction of Cr(VI) to Cr(III) (see Fig. 4). Of the many reducing agents available in the cellular environment, glutathione (GSH) is suspected to be a prime reductant due to its fairly high concentration in the cytosol, its favorable reduction potential,

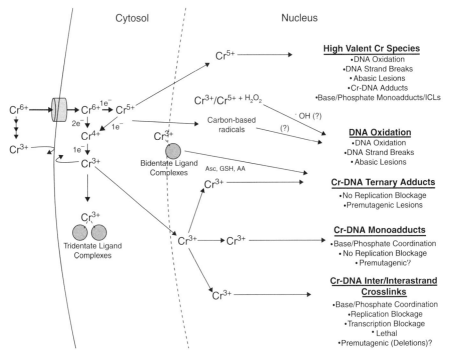

Figure 4. Major pathways involved in the formation of genetic lesions by Cr. This scheme illustrates the interrelationships between Cr metabolism and genotoxicity. Cr^{6+} enters the cell via anionic transporters and is rapidly reduced by a one- or two-electron mechanism to Cr^{3+} which cannot cross the cell membrane. An initial one-electron reduction can lead to the generation of high valent Cr species (Cr^{5+}/Cr^{4+}). Both Cr(III) and Cr(V) display an appreciable affinity for both DNA bases and the phosphate backbone leading to the formation of Cr–DNA monoadducts. Cr^{5+} can directly oxidize DNA bases and sugars and produce DNA strand breaks. Cr^{3+} is the ultimate DNA reactive species and is critical for the formation of DNA adducts. Both Cr^{3+} and Cr^{5+} may generate the hydroxyl radical in the presence of elevated levels of H_2O_2 and lead to "oxidative" DNA damage. Note that the relative location of some reactions/species is not intended to imply that these occur exclusively within that intracellular compartment. (Adapted from O'Brien et al. [2003] with kind permission from Elsevier.)

and its ability to produce long-lived Cr(V/IV) intermediates during the reduction of Cr(VI). Glutathione Cr interactions in plants have been fairly well elucidated. Chromium stress can induce three possible types of metabolic modification in plants: (i) alteration in the production of pigments that are involved in the life sustenance of plants (e.g., chlorophyll, anthocyanin), (ii) increased production of metabolites [malondialdehyde (MDA), H_2O_2 (see Fig. 5) glutathione, ascorbic acid] as a direct response to Cr stress which may cause damage to the plants [Shanker, 2003], and (iii) alterations in the metabolic pool to channelize the production of new biochemically related metabolites which may confer resistance or tolerance to Cr stress (e.g., phytochelatins, histidine).

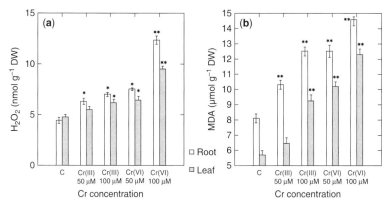

Figure 5. Levels of H_2O_2 (**a**) and lipid peroxidation expressed as malondialdehyde (MDA) (**b**) in roots and leaves of sorghum treated with different concentrations of Cr(III) and Cr(VI). Data represent mean $\pm$ SE of five separate experiments. *Significant at $P < 0.05$. **Significant at $P < 0.01$.

Glutathione pool dynamics of plants, in terms of GSH and GSSG and the GSH/GSSG ratio, is affected by Cr speciation stress (Fig. 6), indicating the role of this pathway in the mechanism of action of Cr [Shanker and Pathmanabhan, 2004]. There is a marked decline in the GSH pool under Cr speciation stress more severely in roots (Fig. 5). Chromium-induced oxidation has been observed in different cellular thiols such as GSH and cysteine by Cr(VI) in in vitro studies. Dichromate reacts with GSH at the sulfhydryl group forming an unstable glutathione–CrO_3^- complex. Thiolate complexes of Cr(VI) with g-glutamylcysteine, N-acetylcysteine, and cysteine have also been described. The interconversion of reduced and oxidized forms of glutathione to maintain redox status of the cell as well as to scavenge free

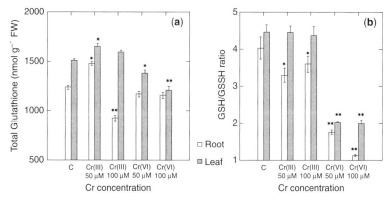

Figure 6. Levels of total glutathione (**a**) and GSH/GSSG ratio (**b**) in roots and leaves of sorghum treated with different concentrations of Cr(III) and Cr(VI). Data represent mean $\pm$ SE of three separate experiments. *Significant at $P < 0.05$. **Significant at $P < 0.01$.

radicals could be one of the roles of GSH under Cr stress. Metal-binding peptides like metallothionein have been reported to have increased under Cr(VI) stress [Shanker et al., 2004b].

The role ROS scavenging enzymes have been clearly elucidated by Shanker et al. [2004a] in plant parts under both Cr(III) and Cr(VI) stress in plants. Scavenging enzymes are unaffected by lower concentration of Cr(III) because there is insignificant ROS production. The combined action of SOD and CAT is critical in mitigating the effects of oxidative stress, since the former merely acts on the superoxide anion converting it to another reactive intermediate (H_2O_2) and the latter acts on H_2O_2 by converting it to water and oxygen. SOD, in contrast to CAT, is more active in scavenging chromium-induced ROS. APX, which essentiality catalyzes the same reaction as CAT, compensates for the reduced CAT activity in plant roots. This could be because unlike CAT, which is present only in the peroxisome and has low substrate affinities since it requires simultaneous access of two molecules of H_2O_2, APX is present throughout the cell and has higher substrate affinity in the presence of ascorbic acid as a reductant. Cr(VI), apart from generating ROS which consequently sets off a signaling response for active scavenging, could actually inhibit the scavenging enzyme activities at acute concentrations. The high contents of dihydro ascorbate (DHA) in combination with an absence of active scavenging and blockage of normal cell cycle progression by DHA is one of the main mechanisms for inducing toxicity in plants by Cr. The role of GSH is more of a signal intermediate rather than direct participation in detoxification of ROS. This mechanism is more likely because of the reported absence of phytochelatin production under Cr stress and because glutathione serves as a precursor for phytochelatin production. The cellular redox status is maintained by the ratio of GSH/GSSG.

The GSH–GSSG redox pair can function effectively only when there is an adequate supply of NADPH and because GSH itself can serve as a cellular sensor to maintain the NADPH pool. Cr(VI) can function as a hill reagent and can inhibit electron transport both in the photosynthetic and mitochondrial apparatus, thus accounting for reduced NADPH pool. The critical balance between the available NADPH pool and ROS production by Cr would decide the redox status of the cell in both plants and animals. Chromium–DNA interactions are the one of the well-explained mechanisms of action of Cr in causing cell death and cancer. There are three theories that account for Cr-induced DNA injury, all of which involves ROS. First, Cr(VI), or reactive intermediate species (Cr(VI)/Cr(III)), may participate in a "Fenton-like" reaction with H_2O_2 leading to hydroxyl radical production. Second, reaction of Cr(VI) with hydrogen peroxide produces the superoxide anion and hydroxyl radical and leads to ROS-mediated DNA strand breaks. Third, oxidized DNA bases (and ROS) generated by the metabolic reduction of Cr(VI) has been suggested to be related to the production of metal-mediated mechanisms DNA injury.

Cr(VI)-containing compounds are genotoxic and can induce gene mutations, sister chromatid exchanges, and chromosomal aberrations. Cr(VI) alone does not react with isolated DNA, because the cellular constituents of reductive metabolism must be present for Cr to damage macromolecules. Chromium associates with both DNA bases and the phosphodiester backbone and the binding occurs through both coordinate covalent binding or electrostatic/ionic interactions. The base-specific binding of Cr has revealed a general, but not absolute, preference toward the formation of Cr(III)–guanine DNA adducts (a DNA adduct is an abnormal piece of DNA covalently bonded to a carcinogenic chemical) and polyriboguanylic acid (poly(G)) in the case of RNA [O'Brien et al., 2003]. The major mechanism involved in the formation of genetic lesions by Cr is explained in Fig. 5. DNA protein cross-links (DPCs) are also one the modes by which Cr induces cytotoxic effects. Cr(VI), and not Cr(III), salts cross-link proteins to DNA in mammalian cells. Unlike nickel, which catalyzes DPC formation largely through oxidative mechanisms, Cr(VI) directly couple proteins to DNA. Cr–DPCs have been reported to extensively develop respectively between DNA and nonhistone proteins and RNA and cytoplasmic proteins in many animal systems [Reem et al., 2007].

Consistent with their ability to generate DNA strand breaks, Cr(VI)-containing compounds are well-documented clastogens. Clastogens are chemical agents that increase the rate of genetic mutation by interfering with the function of nucleic acids. Clastogens are usually specific mutagens that cause breaks in chromosomes. In most cases, damaged metaphases have been observed for both water-soluble and insoluble chromates. Evidence also indicates that chromosomal abnormalities (micronuclei) and genomic instability (microsatellite instability) are possibly involved in the induction of cancer by Cr(VI). DNA interstrand cross-links (ICLs) are caused by Cr(III) interacting with reaction centers on the complementary strand of DNA. DNA adducts are covalent adducts between chemical mutagens and DNA. Such couplings activate DNA repair processes and, unless repaired prior to DNA replication, may lead to nucleotide substitutions, deletions, and chromosome

TABLE 1. Apoptosis Induced by Ar, Cd, and Cr and Their Possible Mechanism of Action

Metal	Compound	Apoptosis target or Possible Mechanism
Arsenic	$NaAsO_2$	Apoptosis by activation of caspase-3, apoptosis by increase in MAPK, JNK, and p38 phosphorylation in T lymphocytes.
	As_2O_3	Apoptosis by disruption of mitochondrial membrane in acute promyelocytic leukemia cells.
Cadmium	$CdCl_2$	Apoptosis by breakdown of mitochondrial membrane and activation of caspase-9 and rapid phosphorylation of MAPKm-induced activation of JNK.
	$Cd(C_2H_3O_2)_2$	Apoptosis in rat primary epithelial lung cells by increase in protein expression in rat primary epithelial lung cells.
Chromium	Na_2CrO_4	Apoptosis by increased release of cytochrome c, depressed oxygen consumption, and decrease in mitochondrial NADH level.
	$K_2Cr_2O_7$	Apoptosis in human lung epithelial cells by reduced mitochondrial membrane potential and increase in p53 protein level and ROS in general.

rearrangements. Chromium complexation with glutathione and other metabolites of ROS induce the formation of DNA adducts, leading to genotoxicity. The relative role of reactive oxygen species (ROS) to the mutagenicity and carcinogenicity of Cr(VI) is not implicit and is still a topic of research by many groups. A notion that has received much attention is that intracellular Cr(VI) mediates Fenton-like reaction and produces ROS that are responsible for nearly all of the toxicity and genotoxicity caused by Cr(VI) [Shanker et al., 2005]. This has led some to assume that cells respond to Cr(VI)—and subsequent production of high valent Cr(V), ROS, and organic radicals—with a classic oxidative stress response. Table 1 gives an overview of the apoptosis induced by Ar, Cd, and Cr and their possible mechanism of action.

4   SPECIFIC MODE OF ACTION OF OTHER METALS

4.1   Nickel

Nickel was discovered by Cronstedt in 1751, and it is the 24th element in order of natural abundance in the earth's crust. It is widely distributed in the environment. Natural sources of atmospheric nickel include dusts from volcanic emissions and the weathering of rocks and soils. Natural sources of aqueous nickel are derived from biological cycles and solubilization of nickel compounds from soils. Global input of nickel into the human environment is approximately 150,000 metric tonnes per year from natural sources and 180,000 metric tonnes per year from anthropogenic sources, including emissions from fossil fuel consumption, and the industrial production, use, and disposal of nickel compounds and alloys [Kasprzak et al., 2003]. The consumption of nickel-containing products leads to environmental pollution by nickel and its by-products at various stages of production, recycling, and disposal. Human exposure to nickel occurs primarily via inhalation and ingestion. Wearing or handling of jewelry, coins, or utensils that are fabricated from nickel alloys or that have nickel-plated coatings may result in cutaneous nickel absorption. Occupational exposure to nickel occurs predominantly in mining, refining, alloy production, electroplating, and welding [Sunderman, 2008]. In 1990, the International Committee on Nickel Carcinogenesis in Man suggested that respiratory cancer risks are primarily related to exposure to soluble nickel concentrations above $1 \, mg/m^3$ and to exposure to less soluble forms at concentrations above $10 \, mg/m^3$. Significant amounts of nickel in different forms may be deposited in the human body through occupational exposure and diet over a lifetime. Since nickel has not been recognized as an essential element in humans, it is not clear how nickel compounds are metabolized.

4.1.1   Mode of Toxic Action. Very similar to other metals, nickel too induces oxidative stress that depletes glutathione and activates Ap1, NF-kB, and other oxidatively sensitive transcription factors. The molecular basis of nickel carcinogenesis has proved vague since carcinogenic nickel compounds are inadequately mutagenic in most assay systems even though they produce oxidative DNA damage and inhibit DNA repair activity. Various compounds of nickel vary in their activity to produce carcinogenic effects due to differences in their uptake, transport, distribution, retention, and the capacity to deliver Ni ions to specific cells and target molecules. Nickel compounds generate specific morphologic chromosomal damage. However, this may not be sufficient to produce mutations.

Hypoxic signaling pathway activation by nickel has emerged as a new theory to explain toxic manifestation of nickel. Nickel substitutes for iron in a hypothetical oxygen sensor, thus switching a cell's metabolism to a state that mimics permanent hypoxia. Since hypoxia is common in solid tumors and selects for more malignant phenotype, this information provides new prospects for understanding the molecular mechanisms of nickel carcinogenesis [Denkhaus and Salnikow, 2002]. The exposure of cells to nickel triggers cellular reactions typical of hypoxia, including the expression of genes involved in glucose transport and intracellular metabolism. The GeneChip microarray technique revealed that genes coding for glycolytic enzymes, glucose transporters, and other hypoxia-inducible genes regulated by hypoxia-inducible factor 1 (HIF-1) are induced by nickel [Salnikow et al., 2003]. Additionally, cellular responses to hypoxic stress include inhibition of cell proliferation and, when cell damage is irreversible, apoptosis [Salnikow et al., 2002]. Further progress in understanding molecular mechanisms of nickel carcinogenicity has been achieved in a study showing that nickel compounds increase the extent of DNA methylation that leads to the inactivation of gene expression.

Nickel interacts with DNA by replacement of Zn^{2+} with Ni^{2+} on the Zn^{2+}-binding sites of DNA-binding proteins. Ni^{2+} has an ionic radius similar to that of Zn^{2+}. DNA-binding proteins or "Zn finger loops" have been identified on some proto-oncogenes and are thought to be likely targets for metal toxicity. There are possibly two mechanisms by which cell transformations are caused by nickel although being weakly mutagenic. One method is by DNA damage and the other method is by epigenetic changes exerted by nickel compounds and the induction of cytosine methylation and histone deacetylation which leads to the inactivation of the senescence/tumor suppressor gene(s). Synergistic action of nickel with many mutagenic carcinogens in enhancing cell transformation both in vitro and in vivo is also known. Nickel-induced carcinogenesis is known to be both tissue- and species-dependent. All these theories of mechanism of action of nickel suggests that genetic predispositions, including variations in metabolism and antioxidant capacities of different species and strains of animals, may also play an important role in nickel carcinogenesis. The possible role of oxidative pathogenesis caused by nickel also has been explained. As compared with copper, iron, cobalt, and other redox-active metals, nickel produces relatively low, but measurable, levels of ROS in cells The oxidative effects of nickel depend on its ability to form the $Ni(III)/Ni(II)$ redox couple at around pH 7.4 when $Ni(II)$ is complexed by some natural ligands, including peptides and proteins. A number of proteins with high affinity to nickel have been identified in recent years. They are mainly involved in nickel transport, detoxification, and excretion.

The capacity of nickel ions to interact with a number of proteins raises the likelihood that nickel may appreciably change intracellular homeostasis by altering protein functions and producing stress comparable to unfolded protein response. Unlike most other metals, the metal-binding protein metallothionein does not appears to constitute a major nickel-binding component in different tissues. Change of the oxidation status of redox-dependent regulatory proteins by nickel complexation may disturb the timely and orderly generation of cellular messengers like oncogenes, tumor suppressor genes, and many others, eventually resulting in improper progression of the cell cycle and/or apoptosis. In addition, DNA–protein cross-links, oxidative damage, and the inhibition of nucleotide excision repair which have been implicated in metals described above are also believed to be some of the mechanisms of nickel-induced cellular toxicity. Molecular mechanisms of nickel toxicity are still in its nascent stage as compared to the degree of progress with respect to other, more toxic metals.

4.2   Lead

Lead is an omnipresent toxic metal and is detectable in practically all phases of the inert environment and in all biological systems. Lead is one of the oldest-established poisons. Knowledge of its general toxic effects stretches back three millennia, and knowledge of its effects in children spans over 100 years. The major issue regarding lead is determining the dose at which it becomes toxic. Since it is toxic to both plants and animals at high exposures, there is no proven biological need for it. In human health, lead affects your children most. Exposure to lead at concentrations lower than recognized as threshold by the Center for Disease Control and Prevention adversely affects cognitive, neurobehavioral, and neurophysiological development of young children. The main route of contact for humans and animals is food, and sources that produce excess exposure and toxic effects are usually environmental and presumably controllable. These sources include lead-based indoor paint in old dwellings, lead in dust from environmental sources, lead in contaminated drinking water, lead in air from combustion of lead containing industrial emissions, hand-to-mouth activities of young children living in polluted environments, and lead-glazed artifacts.

4.2.1 Mode of Toxic Action. Possible mechanisms for lead toxicity include competition with and substitution for calcium, disruption of calcium homeostasis, stimulation of release of calcium from mitochondria, and opening of mitochondrial transition pore. In addition, direct damage to mitochondria and mitochondrial membranes by generation of ROS is also seen. Disruption of tissue oxidant/antioxidant balance, alteration of lipid metabolism, and substitution for zinc in various zinc-mediated processes are some of the metabolic repercussions of lead toxicity [Ahamed et al., 2007]. Lead toxicity development due to calcium interaction has gained considerable amount of attention by researchers. Calcium blocks the uptake of lead through the intestine because lead is a strong blocker of calcium channels.

Lead and calcium compete for the same binding sites on a large family of ion-binding proteins composed of calmodulin and related proteins. Calmodulin serves as a sensor for the concentration of calcium within cells. Lead acts by displacing calcium ions bound to calmodulin. Lead impairs normal calcium homeostasis and uptake by calcium membrane channels and substitutes for calcium in calcium sodium ATP pumps. Impact of lead on brain activity is explained by the fact that lead also blocks access of calcium into nerve terminals, thereby blocking calcium movement into the mitochondria of brain cells, resulting in a decrease in energy production to perform brain functions. Lead also blocks heme synthesis, thereby increasing levels of the precursor daminolevulinic acid (ALA). ALA suppresses GABA-mediated neurotransmission by inhibiting its release and also possibly by competing with GABA at receptors. Lead also can produce anemia, both by interfering with heme synthesis and by decreasing iron absorption from the gut. Lead's ability to substitute for zinc, mentioned previously, affords another avenue by which lead can act as a neurodevelopmental toxicant. By displacing zinc, lead can alter the regulation of genetic transcription through sequence-specific DNA-binding zinc finger protein or zinc-binding sites in receptor channels [Lidsky and Schneider, 2003].

4.3 Mercury

Mercury is a metal that is a liquid at room temperature. The chemical symbol, Hg, is derived from "hydrargyros," the Greek word for "water silver." Mercury is generally found in three oxidation states, each with a specific toxic profile. In the zero oxidation state, Hg^0, elemental mercury exists as the liquid metallic form or as vapor. The two other inorganic forms of mercury are the mercurous (Hg^+) or the mercuric (Hg^{2+}). The mercuric state also forms organic compounds such as methyl, ethyl, phenyl, or dimethyl mercury. Mercury is used as a component of barometers, thermometers, dental products (amalgam), and electrical equipment, as well as in pest control chemicals and in fungicides. It is also used in the gold industry. Mercurous chloride is one of the oldest known pharmaceuticals and is continuously used for its antiseptic properties. It prevents fungus contamination in seeds, and it is good to amalgamate other metals. Cinnabar (mercuric sulfide), the natural form of mercury, was used for the red coloring in cave drawings thousands of years ago. Tombs adorned with elemental mercury have been noted in Egypt dating back over 3500 years.

Mercury's medicinal use probably originated in China and India nearly 2000 years ago. Medications containing this element have been used as antibacterial (syphilis), antiseptics, dermatologic ointments, teething compounds, laxatives, and diuretics. In the United States in the 1800s, the etiology of an epidemic within the hat industry referred to as "hatters' shakes" or "Danbury shakes" resulted from mercury exposure that occurred during production of felt. Exposure to Hg occurs via inhalation, ingestion, and parenteral or subcutaneous administration. Exposure to the vapor may come from sources including the burning of fossil fuels, emissions from volcanic activity, smelting processes in mining activities, the industrial electrolytic production of HCl and NaOH, industrial and medical waste incineration, degassing from the natural erosion of the earth's crust, evaporation from water, vaporization from dental amalgams, and crematoriums [US EPA, 1997]. Table 2 gives a summary of mode of action of Lead, Mercury and Nickel in brief.

4.3.1 Mode of Toxic Action. Inorganic mercurials and organomercury compounds have different modes of action due to the difference in their chemical structure and reactivity. The principal feature of organomercurials is the presence

TABLE 2. Summary of Mode of Action of Lead, Mercury, and Nickel

Metal	Position in Periodic Table	Atomic Number and Mass	Electronic Configuration and Valence State	Available Forms	Principal Use	Mode of Action on Biological Compounds
Lead (Pb)	IVA	82 and 207.19	$+2, +4$ [Xe] $4f^{14}5d^{10}6s^26p^2$	Oxides, sulfides, acetates, chlorates, and chlorides	Used in solder, shielding against radiation and in batteries.	Lead binds strongly to a large number of molecules like amino acids, several enzymes, DNA, and RNA; thus it disrupts many metabolic pathways. Common ligands attacked are (γGlu–Cys)2Gly, (γGlu–Cys)3Gly, (γGlu–Cys), cysteine, acid-soluble thiol, and glutathione. Direct damage to mitochondria. Calcium channel blocks. Displace Zn in Zn finger proteins.

Element	Group	Atomic no. and weight	Electron configuration / oxidation states	Forms	Uses	Mechanism
Mercury (Hg)	IIB	80 and 200.6	$+1, +2$ [Xe]$4f^{14}5d^{10}6s^2$	Organometallic compounds as methyl, ethyl, phenyl, or dimethyl mercury; inorganic salts mostly chlorides	Used in thermometers, barometers, and batteries; Also used in electrical switches and mercury-vapor lighting products.	Attacks thiol groups, $CONH_2$ and NH_2 of proteins, and amino acids, phosphate group in DNA, cysteine, glutathione, and sulfhydryl (—SH) of GABA. Impairs the activity of glutathione peroxidase by binding to Se. Induces protein precipitation. Mercury appears to modify nuclear antigens that are capable of eliciting reactions from T-cells.
Nickel (Ni)	VII (iron cobalt transition group)	28 and 58.5	[Ar]$3d^84s^2$ $+2 +3$		Used in electroplating and metal alloys because of its resistance to corrosion; also in nickel–cadmium batteries, as a catalyst and for coins.	Oxidative DNA damage and inhibits DNA repair activity. Hypoxic signaling pathway activation. Induction of cytosine methylation and histone deacetylation. Inhibition of nucleotide excision repair.

of both the metal center (Hg) and the organic moieties covalently bonded to Hg atom in their molecules. The coordination ability of mercury atom and the cleavage of C–Hg bond is associated with the various pathways that are involved in biochemical processes that cause toxicity by multiple mechanisms. Mercury is one of the few metals due to which there is has been an acute outbreak of toxicity in human populations. In the early 1950s, massive methyl mercury (MeHg) poisoning of residents living around Minamata Bay, a small inlet located on the southwestern coast of Kyushu island, Japan, first raised awareness of the resulting severe neurological disease [Shiraki and Takeuchi, 1971]. Acute MeHg poisoning caused various toxic symptoms such as visual and hearing impairment, ataxia, and psychological disturbances [Kaur et al., 2006]. This led to increased research on the toxic mechanism of action of this metal.

The presence of MeHg in food is linked to its high toxicity. In general, mercury and its properties assist excessive free radical formation, and mercury is implicated as one of major causative factors in the toxic cell damage associated with organo-mercury compounds. It is known to exert major toxic effects on the central nervous system (CNS). The developing brain, in particular, is vulnerable to methyl mercury toxicity, leading to several neurodevelopmental disorders. The role of glutathione (GSH) and reactive oxygen species (ROS) in MeHg-induced neurotoxicity has been proven. Depletion of GSH increases MeHg accumulation and enhances MeHg-induced oxidative stress; conversely, supplementation with GSH precursor protects against MeHg exposure in vitro. Similarly, Hg-induced modulation of GABA currents is also considered a primary mechanism of action of mercury in which Hg is capable of interacting with the sulfhydryl (—SH) groups on GABA receptors, resulting in increased neurocurrent. Another possible mechanism of action involves altering the phosphorylation of the GABA receptor complex [Fitsanakis and Aschner, 2005]. In summary, neurotoxicity of Hg occurs as a result of a modulation of amino acid metabolism, catabolism, and its regulation. Organic Hg and its implications in cardiovascular diseases (CVD) has received considerable attention recently. This stems predominantly from initial epidemiological findings from Finland that high mercury content in hair was associated with an increased progression of atherosclerosis and risk of CVD. These adverse effects on CVD have been observed at methyl mercury levels lower than those associated with neurotoxicity. The mechanisms by which mercury exerts its effects on CVD are not fully understood. High affinity of Hg for thiol groups and its ability to bind selenium into an insoluble complex reduces antioxidative defenses and promote free radical stress and lipid peroxidation.

Mercury has a high affinity for selenium, and it readily binds selenium to form insoluble mercury selenide complexes. This interaction between mercury and selenium may represent one mechanism through which mercury increases the risk of CVD. Mercury reduces the bioavailability of selenium and impairs the activity of glutathione peroxidase, thus promoting lipid peroxidation and, subsequently, atherosclerosis [Virtanena et al., 2007]. There are at least four different glutathione peroxidases, and all contain selenium in their active site; these enzymes are actively involved in the defense against ROS. Binding to selenium, mercury can reduce the bioavailability of selenium for incorporation into glutathione peroxidase, thereby inactivating the enzyme. Similar to other trace metals, Hg can act as a catalyst in Fenton-type reactions, which result in the formation of highly reactive hydroxyl free radicals. It has been demonstrated that mercury alters the structural integrity of the mitochondrial inner membrane, resulting in loss of normal cation selectivity. Mercury induces autoimmunity, which is the failure of an organism to recognize its own submolecular levels. The ultimate mechanisms of mercury-induced autoimmunity are not yet understood clearly, although some insights have been gained in relation to T-cell involvement. Mercury appears to modify nuclear antigens that are capable of eliciting reactions from T-cells. There is accumulating research pointing to various mechanisms by which Hg causes toxicity. A variety of studies oriented toward the assessment of biochemical mechanisms responsible for the effects of organomercury compounds (mostly methylmercury) on living organisms are underway. However, the toxicity mechanism is still under debate.

5 MODE OF ACTION: WHAT IS THE FUTURE?

The mode of action (MOA) of trace elements has attracted the attention of researchers and regulators for many years. The literature is abounding with appropriate studies, and MOA-based approaches are becoming important in risk assessments for individual trace elements and in generic risk appraisal guidance documents. In summary, it is seen that most trace metals have the highest common mechanism as DNA perturbations, induction of chromosomal aberrations, and synthesis of new proteins by activation of various kinases and release and activation of capsases, leading to carcinogenicity and cell death as being important aspects of the MOA. ROS and free radical generation and lipid peroxidation is also a common phenomenon seen in all metals either directly or indirectly. In the laboratory, researchers are uncovering new information on the biochemical and cellular changes underlying toxic effects. In the regulatory context, scientists and policymakers are working to harmonize some of the principles and practices that underline toxic effects. All this information will prove its worth only when it is useful in the context of alleviating toxicity after its development apart from prophylactic measures. Microarray analyses reveal global changes in gene expression in response to environmental changes and, thus, are well-suited to provide a detailed picture of trace element MOA [Kawata et al., 2007]. Specifically, these responses are represented by patterns of gene expression signatures, which provide insight into the MOA as well as the general physiological responses of living systems to metal stresses.

DNA microarray technology has helped in uncovering a large amount of MOA information in minimal time and, in addition, has proved to be phenomenal in the area of drug discovery. This is because microarray technology enables high-throughput testing of gene expression to explore a variety of toxicological questions. This, in turn, creates an opportunity to use this information to discover and test novel pathways and therapeutic applications. There is much optimism concerning the use of functional cells derived from stem cell cultures in drug discovery and toxicology.

On the other hand, nanotechnology has helped in drug delivery; elucidation of MOA has helped in developing nanoscale delivery vehicles capable of controlling the release of chemotherapeutic agents directly inside cancer cells targeting the trace metal or the target molecule. Active targeting can be achieved by the functionalization of potential drugs with targets such as antibodies, peptides, nucleic acid, carbohydrates, and small molecules. Transitional structural chemogenomics [Chan et al., 2006] is one of the possible future methods that could prove to be highly useful in trace element therapeutics. This is a method by which one can regulate gene expression, employing ultrasensitive small-molecule drugs targeted toward nucleic acids. Gene expression can be regulated by using chemicals to target transitional changes in the helical conformations of single and double-stranded DNA.

MOA studies have now almost reached the limits of "reductionist" approach. The field has taken a turn to interdisciplinary approach with a systems biology perspective that looks for strength in biology, metabolic engineering, and idiotypic networks. Recently, attention is being focused on harmonizing risk assessment approaches for all toxic endpoints. Harmonization refers to using a biologically dependable approach to risk assessment for all endpoints. As research reveals more MOA information, biological linkages become increasingly clear. MOA throws light on key events in the toxicity pathway specific for an organ, tissue, or cell. These should be conceptualized for toxic effects in various organisms culminating in the ecosphere level. Multiscale models of MOA studies integrate information at different length and time scales, with the horizontal two-way link from the DNA to ecosphere, while amalgamating both the continuum processes (e.g., activation of a specific kinase) and stochastic processes (e.g., the eventual effect on the ecosphere). Multiscale models of MOA will also be taking the human relevance factor, which is a key aspect in the progress of MOA science. Human relevance factor will specifically determine if there is a remedy in the future. It is possible that detailed molecular study of MOA will proceed within the framework of these multiscale models that sum up knowledge of the expansive biological context in an accurate and quantitative manner.

REFERENCES

Ahamed M, Kaleem M, Siddiqui J. 2007. Environmental lead toxicity and nutritional factors. *Clin Nutr* **26**:400–408.

Aina R, Labra M, Fumagalli P, Vannini C, Marsoni M, Cucchi U, Marcella Bracale, Sergio Sgorbati, Sandra Citterio. 2007. Thiol-peptide level and proteomic changes in response to cadmium toxicity in *Oryza sativa* L. roots. *Environ Exp Bot* **59**:381–392.

Beijer K, Jernelov M. 1986. Sources, transport and transformation of metals in the environment. In: Friberg L, Nordberg GF, Vouk VB, editors. *Handbook on the Toxicoloy of Metals*, second edition. *General Aspects.* Amsterdam: Elsevier, pp. 68–74.

Binet F, Cavalli H, Moisan E, and. Girard D. 2006. Arsenic trioxide (AT) is a novel human neutrophil pro-apoptotic agent: Effects of catalase on AT-induced apoptosis, degradation of cytoskeletal proteins and de novo protein synthesis. *Brasilian J Haematolo* **132**:349–358.

Chan NJ, Gagna CE, Yam T, Lambert WC. 2006. Transitional structural chemogenomics. *Microsc Microanal* **12**:414–415.

Clemens S. 2006. Toxic metal accumulation, responses to exposure and mechanisms of tolerance in plants. *Biochim* **88**:1707–1719.

Deckert J. 2005. Cadmium toxicity in plants: Is there any analogy to its carcinogenic effect in mammalian cells. *Biometals* **18**:475–481.

Denkhaus E, Salnikow K. 2002. Nickel essentiality, toxicity, and carcinogenicity. *Crit Rev Oncol Hematol* **42**:35–56.

Dong S, Shen HM, Ong CN. 2001. Cadmium-induced apoptosis and phenotypic changes in mouse thymocytes. *Mol Cell Biochem* **222**:11–20.

Fitsanakis VA, Aschner M. 2005. The importance of glutamate, glycine, and γ-aminobutyric acid transport and regulation in manganese, mercury and lead neurotoxicity. *Toxicol Appl Pharmacol* **204**:343–354.

Gaubin YF, Vaissade F, Croute B, Beau J, Soleilhavoup J. 2000. Implication of free radicals and glutathione in the mechanism of cadmium-induced expression of stress proteins in the A549 human lung cell-line. *Biochim Biophys Acta* **1495**:4–13.

Giaginis C, Gatzidou E, Theocharis S. 2006. DNA repair systems as targets of cadmium toxicity. *Toxicol Appl Pharmacol* **213**:282–290.

Gregus Z, Klaassen CD. 2001. Mechanisms of toxicity. In: Klaassen CD, editor. *Casarett and Doull's Toxicology: The Basic Science of Poisons*, sixth edition. New York: McGraw-Hill, pp. 35–81.

Gupta DK, Tohoyama H, Joho M, Inouhe M. 2004. Changes in the levels of phytochelatins and related metal-binding peptides in chickpea seedlings exposed to arsenic and different heavy metal ions. *J Plant Res* **117**:253–256.

Huang C, Ma WY, Li J, Goranson A, Dong Z. 1999. Requirement of Erk, but not JNK, for arsenite-induced cell transformation. *J Biol Chem* **274**:14595–14601.

Huang C, Ke Q, Costa M, Shi X. 2004. Molecular mechanisms of arsenic carcinogenesis. *Mol Cell Biochem* **255**:57–66.

Jonak C, Nakagami N, Hirt H. 2004. Heavy metal stress. Activation of distinct mitogen-activated protein kinase pathways by copper and cadmium. *Plant Physiol* **136**:3276–3283.

Kasprzak KS, Sunderman FW, Jr. Salnikowa K. 2003. Nickel carcinogenesis. *Mutat Res* **533**:67–97.

Kaur P, Aschner M, Syversen T. 2006. Glutathione modulation influences methyl mercury induced neurotoxicity in primary cell cultures of neurons and astrocytes. *Neurotoxicol* **27**:492–500.

Kawata K, Yokoo H, Shimazaki R, Okabe S. 2007. Classification of heavy-metal toxicity by human DNA microarray analysis. *Environ Sci Technol* **41**:3769–3774.

Kligerman AD, Doerr CL, Tennant AH. 2005. Oxidation and methylation status determine the effects of arsenic on the mitotic apparatus. *J Mol Cell Biochem* **279**:113–121.

Kreppel H, Liu J, Reichl FX. 1993. Zinc-induced arsenite tolerance in mice. *Fundam Appl Toxicol* **23**:32–37.

Langman LJ, Kapur BM. 2006. Toxicology: Then and now. *Clin Biochem* **39**:498–510.

Lidsky TI, Jay S. Schneider. 2003. Lead neurotoxicity in children: Basic mechanisms and clinical correlates. *Brain* **126**:5–19.

Liu SX, Athar M, Lippai I, Waldren C, Hei TK. 2001. Induction of oxyradicals by arsenic: Implication for mechanism of genotoxicity. *Proc Natl Acad Sci USA* **98**:1643–1648.

Lopez E, Figueroa S, Oset-Gasque MJ, Gonzalez MP. 2003. Apoptosis and necrosis: two distinct events induced by cadmium in cortical neurons in culture. *Br J Pharmacol* **138**:901–911.

Mishra S, Srivastava S, Tripathi RD, Govindarajan R, Kuriakose SV, Prasad MNV. 2006. Phytochelatin synthesis and response of antioxidants during cadmium stress in *Bacopa monnieri* L. *Plant Physiol Biochem* **44**:25–37.

Monroe RK, Halvorsen SW. 2006. Cadmium blocks receptor-mediated Jak/STAT signaling in neurons by oxidative stress. *Free Rad Biol Med* **41**:493–502.

Morales KH, Ryan L, Kuo TL, Wu MM, Chen CJ. 2000. Risk of internal cancers from arsenic in drinking water. *Environ Health Perspect* **108**:655–661.

National Research Council Report. 1999. *Arsenic in the Drinking Water*. Washington, DC: National Academy Press.

Navas-Acien A, Silbergeld EK, Sharrett R, Calderon-Aranda E, Selvin E, Guallar E. 2005. Metals in urine and peripheral arterial disease. *Environ Health Perspect* **113**:164–169.

O'Brien TJ, Ceryak S, Patierno SR. 2003. Complexities of chromium carcinogenesis: Role of cellular response, repair and recovery mechanisms. *Mutat Res* **533**:3–36.

Reem HE, Ayman OS, El-Kadi XY. 2007. Transcriptional activation and posttranscriptional modification of Cyp1a1 by arsenite, cadmium, and chromium. *Toxicol Lett* **172**:106–119.

Reicharda JF, Schnekenburgera M, Puga A. 2007. Long term low-dose arsenic exposure induces loss of DNA methylation. *Biochem Biophys Res Commun* **352**:188–192.

Requejo R, Tena M. 2005. Proteome analysis of maize roots reveals that oxidative stress is a main contributing factor to plant arsenic toxicity. *Phytochemisty* **66**:1519–1528.

Salnikow K, Davidson T, Costa M. 2002. The role of hypoxiainducible signaling pathway in nickel carcinogenesis. *Environ Health Perspect* **110**(Suppl 5):831–834.

Salnikow K, Davidson T, Kluz T, Chen H, Zhou D, Costa M. 2003. GeneChip analysis of signaling pathways effected by nickel. *J Environ Monitor* **5**:1–5.

Satarug S, Moore MR. 2004. Adverse health effects of chronic exposure to low-level cadmium in foodstuffs and cigarette smoke. *Environ Health Perspect* **112**:1099–1103.

Schützendübel A, Polle A. 2002. Plant responses to abiotic stresses: Heavy metal-induced oxidative stress and protection by mycorrhization, *J Exp Bot* **53**:1351–1365.

Shanker AK, Pathmanabhan G. 2004. Speciation dependant antioxidative response in roots and leaves of Sorghum (*Sorghum bicolor* (L) Moench cv CO 27) under Cr(III) and Cr(VI) stress. *Plant Soil* **265**:141–151.

Shanker AK, Djanaguiraman M, Sudhagar R, Chandrashekar CN, Pathmanabhan G. 2004a. Differential antioxidative response of ascorbate glutathione pathway enzymes and metabolites to chromium speciation stress in green gram (*Vigna radiata* (L) R Wilczek, cv CO 4) roots. *Plant Sci* **166**:1035–1043.

Shanker AK, Djanaguiraman M, Sudhagar R, Jayaram R, Pathmanabhan G. 2004b. Expression of metallothionein 3 (MT3) like protein mRNA in Sorghum cultivars under chromium(VI) stress. *Curr Sci* **86**:901–902.

Shanker AK, Cervantes C, Loza-Tavera H, Avudainayagam S. 2005. Chromium toxicity in plants. *Environ Int* **31**:739–753.

Shi H, Shi X, Liu KJ. 2004. Oxidative mechanism of arsenic toxicity and carcinogenesis. *Mol Cell Biochem* **255**:67–78.

Shiraki H, Takeuchi T. 1971. Minamata disease. In: Minckler J, editor. *Pathology of the Nervous System (II)*. New York: McGraw-Hill, pp. 1651–1665.

Sunderman FW, Jr. 2008. Nickel. In: Anke M, Ihnat M, Stoeppler M, editors. *Elements and Their Compounds in the Environment*. Weinheim: Wiley/VCH, in press.

US Environmental Protection Agency. (US EPA). 1986. *Quality Criteria for Water*. Appendix A. EPA 440/5–86-0001, Washington, DC: US EPA Office of Water Regulation and Standards.

US Environmental Protection Agency (US EPA). 1997. *Mercury study report to congress*, Vol. IV: *An Assessment of Exposure to Mercury in the United States*.

Virtanena JK, Rissanena TH, Voutilainena S, Tuomainen T-P. 2007. Mercury as a risk factor for cardiovascular diseases. *J Nutr Biochem* **18**:75–85.

Voegtlin C, Dyer HA, Leonard CS. 1923. On the mechanism of the action of arsenic upon protoplasm. *Public Health Rep* **38**:1882–1912.

Watson AP, Van Hook RI, Jackson DR, Reichle DE. 1976. Impact of a lead mining-smelting complex on a forest-floor fitter arthropod fauna in the new lead belt region of southeast Missouri. ORNWNSF/EATC-30. Oak Ridge, TN: Oak Ridge National Laboratory.

Weber M, Trampczynska Λ, Clemens S. 2006. Comparative transcriptome analysis of toxic metal responses in *Arabidopsis thaliana* and the Cd^{2+}-hypertolerant facultative metallophyte *Arabidopsis halleri*. *Plant Cell Environ* **29**:950–963.

Yang PM, Chiu SJ, Lin KA, Lin LY. 2004. Effect of cadmium on cell cycle progression in Chinese hamster ovary cells. *Chem Biol Interact* **149**:125–136.

Yi H, Wua L, Jianga L. 2007. Genotoxity of arsenic evaluated by Allium-root micronucleus assay. *Sci Total Environ* **383**:232–236.

Zheng W, Aschner M, Ghersi-Egea J-F. 2003. Brain barrier systems: A new frontier in metal neurotoxicological research. *Toxicol Appl Pharmacol* **192**:1–11.

22 Input and Transfer of Trace Metals from Food via Mothermilk to the Child: Bioindicative Aspects to Human Health[*]

SIMONE WUENSCHMANN

Fliederweg 17, D-49733 Haren, Germany

STEFAN FRÄNZLE and BERND MARKERT[†]

International Graduate School (IHI) Zittau, Department of Environmental High Technology, D-02763 Zittau, Germany

HARALD ZECHMEISTER

University of Vienna, Faculty of Life Sciences, Department of Conservation Biology, Vegetation, and Landscape Ecology, A-1090, Vienna, Austria

1 INTRODUCTION

An increased transfer of chemical elements into the different environmental compartments may bring about damages to health which, on the level of an entire organism, can be changes of their respective morphological, histological, or cell structures yet may also be changes in biochemistry or metabolic physiology.

For several decades now, several scientific disciplines such as geochemistry, geomedicine, environmental sciences, and ecotoxicology are concerned with

---

[*]The main parts of this work correspond to the dissertation of Dr. Simone Wuenschmann, Bestimmung chemischer Elemente im System Nahrung/Muttermilch—ein Beitrag zur Bioindikation? Dissertationsschrift, Hochschule Vechta, Germany, 2007.
[†]*Bernd Markert's present address*: Fliederweg 17, D-49733 Haren-Erika, Germany.

---

investigating this problem in order to identify and quantify possible risks, yet focusing on different aspects of this problem. Workgroups who adhere to traditional lines of geochemistry have addressed the effects of elements or compounds thereof on biological systems since early in the 19th century. Some corresponding phenomena were already known at that time—and used in prospection of ore lagerstätten—like "witch springs" containing larger amounts of dissolved heavy metals; somewhat later, occupational diseases of workers processing certain metals were attributed to the latter exposition ("hut fever" due to Mn or Zn vapors/dusts or the mad hatter syndrome [owing to Hg exposure]). In this very 19th century the extent of various chemical elements are used for most diverse purposes increased rapidly. Accordingly, the ratio (technophilic index) between release by anthropogenic activities following the minery (ore processing) technical application of metal or alloy sequence and erosion to be described by geochemistry does increase in the same way. Once again, first occupational health hazards turned up; accordingly, geochemical aspects were replaced or at least augmented by health considerations. Thus geochemistry or (speaking in a modern manner) environmental chemistry turned into environmental medicine. As both anthropogenic emissions and their health effects occur somewhere on the surface of earth, they can in addition be spatially described, producing the new branch of geographical medicine. It is possible to survey amounts of emission also by considering their biological/toxicological effects. So health damages in man caused by exposure to some chemical and bioindication are linked to each other, including aspects of spatial distribution pattern and causality likewise. Eventually this gave way to ecotoxicology. According to Markert et al. [1997], bioindicators are organisms or communities of organisms that provide pieces of information on quality and changes of the environment by the organisms contents of certain elements or compounds and/or their morphological, histological, or cellular structures, processes in metabolism and physiology, and their behavior or population structures, each including possible changes of these parameters [Markert et al., 1997].

With environmental sciences being the more inclined toward medicine, they are especially concerned with human exposure and burdening. As compared to both animals and plants, retrieval of human samples is more difficult, except for various excretion and structures attached to skin (hair, fingernails, toenails). The different functions of the excretion correspond to the extent to which inorganic and organic species pass over: The latter is the main task of urine, whereas sweat does contain, for example, Fe and Zn, yet this is but a minor aspect of secretion functions, whereas mother's milk is meant to be a nutrient rather than a source of possible pollutants. Yet investigation of mother's milk might provide information on pollutions the mother is exposed to, while there are also implications for both supply and exposure of a breast-fed child.

2 AIMS AND SCOPES

The meaning of bioindication is analyzing the state of the environment by means of analyses of biogenic material. With statements on a possible exposure of man being most relevant to toxicology and epidemiology, human-sample-based biomonitoring is most important.

This work is focused at investigating contents of chemical elements in mother's milk, concerning a possible application in human biomonitoring. Using experimental data on the (maternal) food/milk system, one can address distributions of selected chemical elements but also address their physiological requirement (if any). For this purpose, some transfer factor (TF) is determined for essential elements iron, iodine, copper, manganese, and zinc and nonessential ones barium, lead, cadmium, cesium, rubidium, and strontium. This TF gives extent of transfer (pass-over) of chemical elements from maternal food into mother's milk, being defined as the quotient of element concentration in fresh (undried) mother's milk versus its daily supply given some equilibrium state.

Mother's milk has a peculiar significance for being the best-quality and physiologically best-suited source of food for nutrition of small children. For that reason, there is a history of several decades of extensive research on quantification of hazardous substances in mother's milk. As important as these works were, they could not attribute detected burdens to any path of uptake, nor could they attribute them to that via maternal food. Acknowledging the human organs system to be highly complex in all biological, physiological, and biochemical aspects, the development of such methods is required which allow for pinpointing the origins and behavior of invading materials, giving clues to certain environmental influences on mother's milk quality and composition.

The main aspects of this work are as follows:

(a) Comparative analyses of mother's milk samples from some highly polluted (mining) area in Southern Poland (Olkusz/Woivodship Małopolska) and milk samples from a region that is known to be lowly to moderately polluted (Euroregion Neisse).

(b) Calculations of element-specific transfer factors in the system (maternal) food/mother's milk to estimate the possible exposition of the child via breastfeeding, additionally providing valuable hints regarding which properties of some element control the extent of transfer into mother's milk.

The transfer factor TF—defined and reproduced as a simple number so far—may depend on the environmental status of the mother's surroundings, too, and this possibility must be checked using samples from an area that is more polluted than Euroregion Neisse. However, there is a difficulty: Nowhere else in the world have multielement studies on mother's milk been done so far using the duplicate[1]

[1]In the duplicate approach, duplicates of the food really eaten by the mothers during the time they provided the milk samples are analyzed like the contemporary milk samples that are covering the 24 hr/day intake over the entire duration of the experiment ranging from 2 to 8 weeks individually. The selection and amount of food were up to the nursing mothers; there were no dietary recommendations whatsoever. This method (only) allows for an (approximate) determination of the diurnal intake and possible trends in element uptake over all the time of the investigation. The other methods used to estimate individual supply by elements, such as retrospective interrogation (dietary history), provide much less reliable data on diurnal element uptake since effects brought about by, for example, the individual ways of preparing foods cannot be measured in this manner.

method. By means of duplicate analyses, it is at least feasible to find out how the transfer responds to an increased supply; this is done by making use of the variations in supply amounts due to different kinds of diet. There are two different possible reasons for corresponding variations, not only reactions to some larger intake quantity but also interactions among different elements in the maternal diet. For one specific area, it is likely that the range of intake amounts is smaller than the differences between lightly and highly heavy-metal burdened areas, yet the former up to now gives the only way to modeling corresponding second-order effects. Information derived in this way yet provide some preliminary basis for extrapolation on highly burdened areas. Man is omnivorous; accordingly, his/her guts are considerably shorter than those of herbivorous animal models (cow, goat, rabbit, etc.) and adapted to other kinds of food; hence it can be anticipated that resorption of certain chemical elements does differ also. The data for TF in the system food/milk determined in this work are meant to become the basis of more advanced medicinal and analytical investigations.

3   PRINCIPLES

3.1   Transfer of Chemical Elements

Living beings are exposed to chemicals in the environment throughout. Elements may be resorbed on each dermal, inhalative, or oral pathway depending on their physicochemical properties, getting into cells of whatever kind/differentiation, causing, given corresponding properties, an active exposure of the target organ to an element just resorbed. Limited by their active or passive elimination in different ways (urine, feces, other body liquids, etc.), they are often capable of accumulating in organ systems and tissues, sometimes augmented by a high affinity toward the corresponding target organ or tissue. Taking histological and biochemical effects into consideration, Fraenzle [2007] derived typical accumulation and resorption properties by correlating the electrochemical ligand parameter and the corresponding complex dissociation constants. Like this approach covers ligand properties of biological matter throughout a species [Fraenzle et al., 2004], also transfer/relocation processes within some organism will be understood better when considering rules of coordination chemistry. Even without discussing this reasoning in much detail, it is feasible to describe the pathway of chemical elements by a so-called transfer factor. This also gives information on exposure of the breast-fed child. Figure 1 explains what this transfer factor refers to.

After quantification of chemical elements in both maternal food and mother's milk, the element status in both analytical matrices is linked and some transfer factor is calculated. It is feasible to derive the amount of some element the breast-fed child may take up either by multiplying concentrations in milk and the diurnal consumption of milk or by calculating directly from the transfer factor pertinent to the specified element.

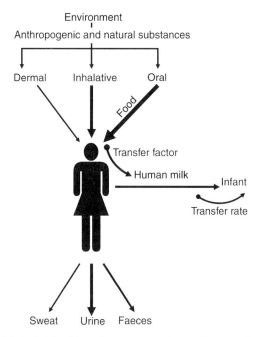

Figure 1. A simplified sketch of how humans can take up (by dermal, inhalative, and oral paths) and excrete (via sweat, urine, feces, and mother's milk, respectively) anthropogenic or natural substances from the environment. For estimating transfer factors in the food/milk system, analytical data of food samples are related to those of milk from the same mothers. This approach has the drawback of ignoring both other (i.e., dermal or inhalative) pathways of uptake and a possible mobilization from inbody depots (e.g., bones). The diurnal transfer *rate* of certain elements into the breast-fed child (which means supply but not resorption amounts!) can be calculated using the daily rate of consumption of mother's milk.

3.2 Physiology of Lactation

With onset of puberty, female breast glands start to grow due to the action of estro-genes. During pregnancy, estrogen and gestagene levels do increase even further, causing formation of distal alveolis and lobuli of the breast gland. To accomplish this, in addition the placental hormone lactogen (equivalent to prolactine hormone) and other hormones insulin, thyroxine, and cortisol are required.

During breast-feeding, three different kinds of milk are produced which differt with respect to their contents in both energy and nutrients [Kees-Aigner, 2002]. The first kind that starts to be produced already during the fourth week of preg-nancy and terminates just a few days after delivery is called colostrum. This "pre-liminary" milk consists of fat droplets produced by phagocytosis, being relatively concentrated. Colostrum has a yellowish color owing to its carotine content and in addition contains vitamins A, C, and E, together with immunoglobulin A. Between the fifth and 15th day after delivery transitory milk is produced. Eventually, starting at 15th day after childbirth, mature milk is formed. Being

TABLE 1. The Composition of Nutrients in the Various Lactation Phases of Mother's Milk (According to Martius [1983], Deutsche Forschungsgemeinschaft [1984], Morris et al. [1986], Bates Prentice [1994])

Phases of Lactation	Age (days)	Fat (%)	Protein (%)	Carbohydrate (%)	pH Value	Production of Milk[a] (g/d)
Colostrum	0–4	44	17	39	7.45	50[b]–350
Transitory milk	5–15	48	9	43	7.04	410–590
Mature milk	>15	50	7	43	7–7.4	600–850

[a]When breast-feeding only.
[b]First day after childbirth; this increases to 170 g at day 2.

white in colour, its protein and salt levels are lower than that of colostrum but it contains more lactose and high concentrations of essential fatty acids and of lipase, diastase, and vitamins A and C. Noteworthy for their contribution to the child's immune status are contents of maternal immune globulines and blood group antibodies. Though protein and mineral contents do substantially decrease when changing from colostrum to mature milk, the volume produced per day does increase substantially (Table 1).

Human milk differs from that of other mammals with respect to both its contents of proteins and minerals being rather low. Hambraeus et al. [1984] give the following average concentrations (g/100 g milk): protein 0.9, fat 3.8, lactose (milk sugar) 7.0, ash 0.2. Generally speaking, milk contains water, sugar(s), fat, proteins, trace elements, vitamins, immunoactive compounds, and enzymes.

3.3   Transfer of Chemical Elements into Human Milk

Elements get along different pathways and transport mechanisms from maternal blood into mother's milk. According to Jödicke and Neubert [2004], most elements are transported through the membrane of breast glands simply by diffusion, while an active chemical transport appears likely for just a few chemical compounds. Active transport was implied, for example, for iodine, which exists in milk as both iodide and iodate. Both the breast tissues and milk contain much fat, causing some affinity to lipophilic compounds, yet alkalinic compounds might likewise accumulate because the pH of milk (about 6.9) is lower than that of blood.

Chemical elements use to be transported into milk after being bound to proteins, with the mechanisms responsible for this effect not yet clearly elucidated. Moreover, information is lacking to *which* proteins elements (ions) are bound when transported into milk. Yet it is established that heavy metals such as lead or cadmium are more highly enriched in colostrum, owing to the higher protein content of the latter as compared to later (mature) lactation phases wherein concentrations of these elements tend to remain fairly constant.

4   MATERIALS AND METHODS

4.1   A Comparison of the Two Experimental Regions Euroregion Neisse and Woivodship Małopolska with Respect to Factors that Cause Environmental Burdens

The mother's milk samples investigated in this study were taken in Euroregion Neisse and Woivodship Małopolska and tested for their elemental composition. For comparing environmental effects on element contents of mother's milk from either lowly to moderately polluted areas to that of highly polluted ones, atmospheric deposition in either region as known from earlier work is to be considered.

For instance, a comparison concerning Pb levels in the Netherlands, Germany, and Poland by Herpin et al. [1996] is given in Fig. 2. Concerning Poland, Woivodships Śląnskie (Silesia, Katowice district) and Małopolska ("Little Poland") are distinguished by high lead concentrations in mosses ($>45\,\mu g/g$). The situation is similar with zinc ($>110\,\mu g/g$) and also iron ($>4000\,\mu g/g$).

Taking average concentrations of heavy metals and other chemical elements in mosses [Markert, 1996], the environment must be considered moderately to heavily polluted in the corresponding parts of Poland. Using analytic data from 1990–1992, levels of Zn, Pb, and Fe are 3, 4, and 10 times higher than on average in those Polish areas, respectively. "Basic" or "average"/minumum

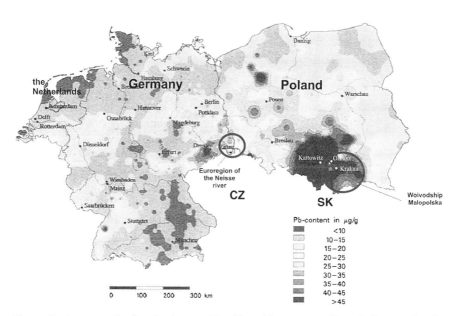

Figure 2. An example for lead exposition/deposition across Central Europe in the Netherlands, Germany, and Poland (after Herpin et al. [1996], modified). The data were obtained via measuring metal accumulation in mosses in 1990–1992. The circles mark those regions where the mother's milk samples for this work were taken.

concentration levels were derived from the works by Ellison et al. [1976] (background level), Bowen [1979] (minimum contents), and Markert [1993] (limiting values) [Markert, 1996].

The UNECE moss survey gives proof of an elevated atmospheric input occurring in Eastern Europe in 2000/2001 [Buse et al., 2003]. Levels of Cd, Fe, Zn, and Pb in Southern Poland were 4 to 6 times those given as basic or background concentrations. Among the 27 states participating in this study, Bulgaria and Poland still had considerably higher environmental burdens than did the Western European states [Buse et al., 2003], with data from the 2005 moss survey not yet being available when this work was finished.

Schröder and Pesch [2004a,b] report metal bioaccumulation in mosses to have decreased in Germany by up to 60% during the decade 1990–2000. For the German part of Euroregion Neisse, a similar trend is discerned when comparing data for 1990 and 2005 concerning many different metals [Schroeder et al., 2007].

Table 2 gives a comparison of numeric data for selected metal concentrations found in mosses obtained in German part of Euroregion Neisse and those from Southern Poland in different sampling years. According to the data in Table 2, metal concentrations in mosses were decreasing since the early 1990s in all the regions discussed here. For moss plants taken from the German part of Euroregion Neisse recently, the "basic" concentration was reached and even lower levels were detected, whereas at Katowice the levels were considerably above "basic" as late as 2000/2001.

In the recent "Environmental Performance Index" (EPI) report by Esty et al. [2006] giving a global comparison and ranking of environmental situations of the states, Poland is rated as one of the most highly polluted countries in Europe. The EPI covers 16 indicators from environmental health, air quality, states of water, biodiversity, natural resources, and energy.

4.1.1 Euroregion Neisse. The Euroregion Neisse comprises an area around the border triangle of Germany, Poland and the Czech Republic, more exactly speaking Upper Lusatia and Lower Silesia, Northern Bohemia and the Polish part around Jelenia Gora in Western Poland. Until the early 1990ies, this area was referred to as "black triangle," with emission amounts being as high as 32% of dust, 13% of SO_2 and 3.3% of NO_x emitted all over the territories of the then (12) EEC member states [Staatsministerium für Umwelt und Landesentwicklung, 1994]. This situation urged the ministers of environmental protection of the three states to create a trilateral workgroup that focused on surveillance and sanitation of the lignite power plant grid in the Northern Bohemian lowlands in particular. Life expectancy there was 9 years shorter than Czech average values. Sanitation was accomplished by partly disconnecting these power plants, partly equipping them with flue gas cleaning gear, adhering to emission limit values, which since the later 1990s became identical to European (EEC/EU) standards in the Czech Republic.

Concerning SW Poland, the main interest of this trilateral workgroup was with cross-border effects of lignite open pits and of the Turow power plant. The latter,

TABLE 2. Metal Concentrations (μg/g DW) in Mosses Obtained in German Part of Euroregion Neisse and Those from Southern Poland in Different Sampling Years

Euroregion of the Neisse River	Fe	Zn	Cu	Cd	Pb	
German part 1990	4800	46.2	13.9	N.D.	30.1	Schroeder et al. [2007]
German part 1995	684	50.7	9.36	0.29	10.6	Schroeder et al. [2007]
German part 2000	424	41.1	6.38	0.23	5.75	Schroeder et al. [2007]
German part 2005	438	45.2	6.80	0.26	6.10	Schroeder et al. [2007]
Basic concentration	*400*	*40*	*12*	*0.2*	*10*	Markert [1996]
1990–1992	>4000	>110	N.D.	N.D.	>45	Herpin et al. [1996]
1996	2000	500	18	8	100	Wappelhorst [1999]
2000/2001	1000–1500	80–>160	8–16	>0.8	20–>60	Buse et al. [2003]

N.D. No data given. Data concerning Southern Poland were deduced from the data published by the mentioned authors.

too, was equipped with flue gas cleaning gear, one power plant unit was shut down, to be replaced by six modern agitated-bed combustion systems. Turow power plant is scheduled to continue electricity production until 2035.

According to Wappelhorst et al. [2000a], atmospheric deposition measurements by mosses here show low- to intermediate-level emission levels as compared to other Central European countries, except for Fe and Ti emisions which are above average.

4.1.2 Woivodship Małopolska.

4.1.2 Woivodship Małopolska. Śląnskie and Małopolska regions are considered to be the mostly polluted areas in Central Europe, mainly due to the following kinds of industrial activities: mining and processing of metals such as iron, zinc, lead, and others, thermal power plants, heavy industries and electroindustries, crude oil refining, construction of cars; and locomotives and railway cabs.

Polish mother's milk samples were taken in NW parts of Woivodship Małopolska, from cities of Olkusz, Wolbrom, Bukowno, and Bolesław which are located about 40 km (Wolbrom some 70 km) east of Katowice. Between 1986 and 1989, epidemiological studies were done around Katowice [Norska-Borówka et al., 1996] in order to account for the enhanced infant mortality and to relate morbidity of small children there to their further development up to the second year of life. In addition, ecological burdens were determined by analyzing air and soil for the following substances: fluorine, mobile dust, SO_2, NO_2, tarry compounds, lead, cadmium, copper, formaldehyde, and phenol.

The main reasons for child mortality were congenital deformations, perturbations acquired during or shortly after birth, and lung inflammations. Child mortality was highest in the most polluted areas. Preterm births had a level of 20% as compared to a Polish average value of only 6%.

A biomonitoring study conducted by Dmowski [2000] showed Śląnskie and Małopolska regions (the latter next to Olkusz) to be more polluted by As, Pb, Cd, Zn, Ge, and Tl than Polish average (Fig. 3). In Śląnskie Woivodship, this study demonstrated a substantial decrease of Se, Pb, Cd, Zn, Cu, Ge, Sn, Sb, Hg, and Ba contents which Dmowki [2000] attributed to introduction of various flue gas filtering devices. Figure 3 gives the element contents found in feathers of magpies. In the Małopolska region, effects of environmental pollution can still be seen with animals, plants, and also open waters. Metals are released by mining to the environment, to be distributed by wind particularly during dry summer months. As a result, metals get into humans not only by way of inhalation but also by consumption of fruits and vegetables grown in their own gardens (for more details, see, e.g., Norska-Borowka et al. [1993] and Dmuchowski and Badurek [2003]).

4.2 Origins and Sampling of Food and Milk Samples

4.2.1 Food and Milk Samples from Mothers from Euroregion Neisse. During the period 1998–2001, food and mother's milk samples from a total of 23 mothers, most of whom were living in Euroregion Neisse (16 Germans, six Poles, one Czech woman) were analyzed for element contents [Wappelhorst et al., 2000b;

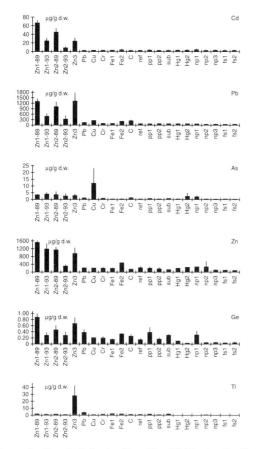

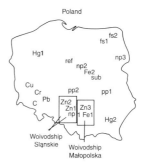

Zn^1: zinc smelter at Katowice

Zn^2: Poland's largest zinc smelter "Miasteczko Śląnskie"

Zn^3: Boleslaw zinc smelter near Olkusz

Figure 3. Sites in Poland polluted by Cd, Pb, As, Zn, Ge and Tl as determined by biomonitoring (magpie [*Pica pica*] feathers) [from Dmowski [2000]). Sites Zn^1 and Zn^2 are close to Katowice (Woivodship Śląnskie), site 3 is at Olkusz (Woivodship Małopolska). The other sites on the *x*-axis refer to other regions of Poland.

Wuenschmann et al., 2003]. Sampling was done by Wappelhorst et al. [2000b]. All participants of this study were nonsmokers; their average age was 29. Pregnancies and deliveries were reported to be without difficulties. During the period of investigation, the youngest breast-fed child was 2 weeks old while the oldest still nurtured had an age of 71.2 weeks; this is to say that the mothers were producing mature milk already (starting with day 15 after delivery). Because no mothers could be spotted ready to participate in so early a stage of motherhood after childbirth, there were no samples to represent colostrum (<5 days after birth) or transitory milk (6th–14th day after birth) production.

Analyses of food consumed by mothers were done according to the duplicate method. This means analyzing duplicates (samples of all foods and drinks set aside by consumption, covering 24 hr a day) of the food really consumed for a defined longer period of time (2–8 weeks). Food was selected by mothers̓ prefer-ences; there were no dietary recommendations.

This method allows for an estimate of daily uptake and possible trends in element supply throughout the period of investigation. Comparable approaches that, for example, rely upon retrospective inquiries (dietary history) of food supply and example analyses provide some far less reliable estimate of daily element supplies only for individual changes brought about by, for example, preparation of foods that cannot be recognized.

For sampling, the mothers received balances (for weighing their foods) and sampling containers (made of polythene). Subsequent to a precise teaching, they made protocols of food intake including date, time, amount, and kind of both solid and liquid foods of whatever kind. Some aliquot of each kind of food was bypassed to a sampling container, given a letter code, closed, and frozen immediately to −18°C in a common household deep freezer.

For taking milk samples the mothers received milk pumps that they could use if necessary. They attempted to obtain have a sample of at least 30 ml for two reasons:

(a)   There must be a minimum amount for elemental analysis.

(b)   In some cases, a repeated analysis might prove necessary.

Like with food samples, the mothers made a protocol of sampling time (meant to be the same throughout the experiment) and date. Recognizing the situation of nursing mothers, we considered taking only one sample per day (from both breasts) sufficient. When one sampling provided less than 30 ml, sampling was to be repeated several times a day, pooling the samples. Once again, the milk samples were frozen to −18°C immediately.

Top priority in sampling was given—besides getting a precise sampling proto-col—to avoidance of contamination that might arise, for example, from storage of unclosed samples or by washing containers with water or other liquids. Mothers were instructed how to avoid contamination.

4.2.2   Milk Samples by Mothers from Woivodship Małopolska. Mother's milk samples from Woivodship Małopolska (Southern Poland) were obtained by support of the director of Olkusz County Hospital (Szpital Powiatowy im. Franciszka Kryszta owicza) during the second half of 2003. The practical way of sampling was settled during a discussion with director Dr. W. Głuch and the leading gynecologist of the hospital, Dr. Ziobro-Kruckowiecka. Methods and protocol of sampling were identical as with Euroregion Neisse sampling campaign. This time, 22 mothers from cities Olkusz, Wolbrom, Bukowno, and Bolesław (Fig. 4) took part, with their children all being younger than 1 week. Unlike the mature milk from Euroregion Neisse, we now deal with colostrum in Woivodship Małopolska. Sampling was done on two consecutive days between 8:00 a.m. and 8:00 p.m. All the mothers had given birth to at least one child before.

Figure 4. Living places of the Polish mothers (bracketed). As can be seen from this map segment, cities Bukowno, Bolesław, Olkusz, and Wolbrom are located in between larger cities Katowice (to the West) and Cracow (to the East) in Southern Poland.

Milk was transferred to polythene containers and frozen to $-18°C$ at the site. During car transfer of samples to Germany (some 10 h), these were in a mobile refrigerator at $-18°C$.

4.3 Analytical Methods

Prior to analytical element quantification, the frozen food and milk samples were gently molten at ambient temperature under air. Owing to diversity of food samples, providing unlike sample matrices, sample preparation was done for both sample kinds in different ways.

Solid foods were homogenized by means of a titanium-covered "Ultra-Turrax" vibrating stirrer to get an even distribution of contents. Some part of each sample was kept—re-frozen—as a backup sample if repetition of analysis became necessary or for additional measurements of yet other elements.

Complete digestion of the matrix was accomplished in a closed system by pressure-supported microwave liquid digestion. Matter [1994] states that the advantages of this method include avoidance of evaporation losses (closed system) but also that mistakes due to adsorption are small given the relationship between reaction vessel surface and sample size/weight. Nowadays, microwave digestion devices are generally equipped with PTFE vessels which show little adsorption due to purity and chemical inertness. Figure 5 gives the analytical procedure which is explained in the text.

Digestions of food samples were done using a microwave oven MLS ETHOS plus II 1000/10s (Microwave Lab System GmbH, Leutkirch/Germany), fitted with a type HPR 1000/6 rotor with PTFE reaction vessels. Depending on kind of food, between 0.4 and 2.5 g (fresh weight) of food were placed in a reaction vessel, then 4 ml 65% nitric acid and 2 ml 30% hydrogen peroxide were added under a ventilating hood

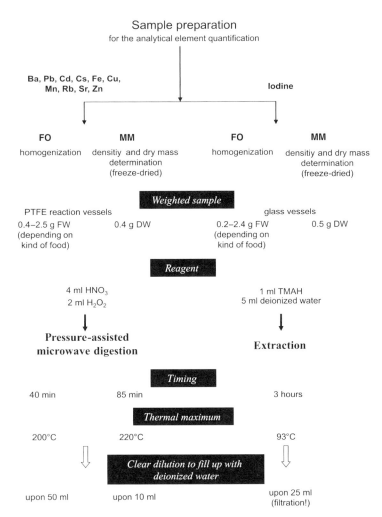

Figure 5. Sample preparation for food and milk samples, going to be digested by microwave-assisted liquid digestion except for determination of iodine. Iodine is extracted by TMA hydroxide (see text): FO, food, MM, mother's milk; FW, fresh weight, DW, dry weight.

(Merck Suprapur). Digestion of food samples took about 40 min according to the schedule, including cooling-down; the maximum temperature was 200°C. When the samples were cooled entirely (clear, not turbid solution), they were opened under the hood again and diluted to 50 ml by deionized water. Measurement was done by ICP-MS or ICP-OES, respectively.

Unlike food samples, milk had to be freeze-dried before digestion, owing to its relatively low element contents. For determination of element contents in fresh milk on a per-volume basis, dry weight must be known, too.

Though milk samples were digested by pressure-assisted microwave digestion, too, the exact procedure was somewhat different: 400 mg of freeze-dried milk were transferred to a 20-ml glass vessel, and 4 ml 65% nitric acid and 2 ml 30% hydrogen peroxide (Merck Suprapur again) were added and eventually poured into a PTFE vessel already containing 8 ml of deionized water, 1 ml hydrogen peroxide, and one drop (about 0.05 ml) of nitric acid. Digestion was done in the same MLS ETHOS plus II 1000/10s device but following another protocol, taking 85 min including cool-down after a thermal maximum of 220°C. When the samples were cooled entirely (clear, not turbid solution), they were opened again and diluted to 10 ml by deionized water. Measurement was done by ICP-MS or ICP-OES, respectively.

According to Fecher et al. [1999], iodine should not be determined after microwave digestion. They noted substantial losses after heating to about 300°C this caused iodide oxidation to iodine, which is volatile. Therefore, in both milk and food samples, iodine was determined according to §35 in Lebensmittel-und Bedarfsgegenständegesetz (LMBG [1998]; German Regulations on Food and Domestic Items) by extraction with tetramethyl ammonium hydroxide TMAH.

Depending on kind of food, 0.2–2.4 g was filled into a 100 ml glass vessel, because freeze-dried milk sample size was 0.5 g. Thereafter, 1 ml TMAH and 5 ml deionized water were added and the glass vessels were tightly closed by a screw cap. Extraction of samples took place for 3 h at $90 \pm 3°C$. After cooling down to ambient temperature, these were diluted to 25 ml by deionized water. Particles that might influence analytical data were removed by filtration through cellulose acetate filters of pore sizes 0.45–5 μm [LMBG, 1998, §35; Fecher et al., 1999].

Digested food or milk samples are clear solutions then to be diluted with deionized water after digestion or extraction, respectively. Calibration of ICP-MS measurements was done for elements barium, lead, cadmium, rubidium, strontium, copper, manganese, and zinc by using Merck ICP Multielement Standard VI, for iron Merck ICP Multielement Standard IV, and for cesium Kraft multielement standard solution, respectively.

Except for iron, quantification was done by ICP-MS (Perkin–Elmer ELAN 6000) making use of atomization and ionization of sample molecules and subsequent separation according to their specific mass/weight ratios [Fecher, 1994; Markert, 1996]. Because detection thresholds are in the ng/l range, ICP-MS is best suited for determinations of trace elements.

Iron was determined by ICP-OES (atomic emission spectroscopy using excitation by inductively coupled plasma) because of its relatively higher concentration in both food and milk. Here, element quantification is affected by emission of quanta of electromagnetic radiation when electrons change between different energetic levels. For this work a Perkin–Elmer ICP-OES Optima 3000 was used.

4.4 Quality Control Measures for Analytic Data

Concentrations of elements measured in this work were in the ppb range for both food and milk samples. In order to check quality and correctness of analytic work, both blind samples and certified standard reference materials (CRM) were used in every analytical process (microwave digestion or alkalinic extraction), namely peach leaves, pine needles, tea, apple leaves, tomato leaves, hay powder, skimmed milk powder, and skimmed milk powder with lower element levels. Blind samples do not contain any sample material but just the reagents required and used for digestion or extraction of samples, respectively (here, nitric acid, hydrogen peroxide, and TMAH). They are used to check the purity of reagents concerning the absence of elements to be analyzed.

4.5   Calculation of Transfer Factors in the System Food/Mother's Milk

The transfer factor is a measure for the extent of transfer of chemical elements from maternal food to mother's milk; it is defined as the ratio of concentration of a certain element in fresh (un-dried) milk to the daily intake by the (single) mother; equilibrium conditions are assumed. The transfer factor (TF) is given by unit [d/kg] according to the following reasoning:

After having determined the element concentration in food actually consumed and multiplication of this value with the total of consumed foodstuffs,[2] one gets the element intake of each mother (μg/day), which then is averaged over the duration of the experiment (2–8 weeks). As the element concentration in milk is known, too, TF now can be calculated for each element and mother [Eq. (1)]:

$$TF_d = c(mm_d)_{FW}/EI_d \qquad (1)$$

where TF_d means the TF for the corresponding day, $c(mm_d)_{FW}$ is the element concentration in fresh (FW = fresh weight) milk [μg/kg] for this day, and EI_d denotes the element intake (oral only) by the mother [μg/day]. When the milk consumption of the breast-fed child is known, the share (in%) of the element E_p which is partitioned into mother's milk is given by the product of the transfer factor and amount of milk I(mm) the baby drinks per day [Eq. (2)]:

$$E_p = TF_d \times I(mm) \times 100[\%] \qquad (2)$$

5   RESULTS

5.1   A Comparison of Element Concentrations Detected in Colostrum and Mature Milk Sampled in Different Countries

Krachler et al. [1998] did a study of element contents in milk of nursing mothers who were in different phases of lactation in Styria (Austria). This set of data enabled us to calculate factors typical for a given element for concentration differences between Factor colostrum and mature milk: Cd 6.5; Zn, Ba, Mo 6; La 5.5; Cu, Sb 3; Ti 2.5; Mn 2.2; Co 1.5; Pb 1.4; Cs, Rb, Sr 1.

---

[2]This implies to take the individual uptake amounts into account. The average total of all solid and liquid foodstuffs was 3.0 kg/day (1.0–7.4 kg/day).

These factors then were used to compare colostrum element levels in milk samples from Woivodship Małopolska to those of mature milk obtained in Euroregion Neisse and milk from both phases in Styria. The product values then are (extrapolated) concentrations in colostrum.

Likewise, colostrum concentrations measured in Małopolska samples were transformed into those for mature milk. Because the latter concentrations are usually lower or constant (Rb, Cs, Sr) at best, the measured values were divided by the corresponding factor.

When comparing data thus treated for the three sampling regions, there is no significant difference of element concentrations. Values are very similar, except for Cd, where $2.0 \mu g/kg$ were detected in presumably strongly pollution-influenced Małopolska colostrum, whereas values for Styria and Euroregion Neisse were 35% lower each. Because these values are in a low-level trace range, this exceptional result must be regarded with caution.

Among the elements listed in Table 3 for the three sampling areas, those in the lower half cannot be "converted" into corresponding values for other lactation phases because no factors are available; nevertheless, the values are given for sake of information and completeness.

Generally speaking, Table 3 shows that there are no such elevated concentrations in milk samples from the most highly polluted Central European region Małopolska, which might cause concern, particularly regarding toxic elements like Ba, Cd, Pb, Sb, or Tl. Given the number and size of zinc smelters and other heavy-metal processing plants in this district, it is noteworthy that essential metals Co, Mn, Mo, or Zn are neither elevated nor too low. The level of Cu, only, is about 40% decreased in milk from Woivodship Małopolska as compared to levels in both Styria and Euroregion Neisse.

Concentrations of the physiologically less significant metals—Cs, La, Sr, and Rb—were similar in all three areas.

During ongoing lactation, cobalt contents, unlike those of other elements, do increase, which is attributed by Krachler et al. [1998] to an increasing content of vitamin B_{12} during mature lactation phase.

It must be stressed that concentrations of elements listed in Table 3 hardly differ among the three regions. Except for Cu, there are no influences on those elements that occur in larger amounts, regardless whether they are essential like Zn or more or less insignificant such as Rb. Thus, there is no plausible correlation to geochemical factors or to anthropogenic pollution of some site. Yet there may be interaction effects among different elements which depend on changes in element intake and are discussed in due detail in Section 6.

Table 4 gives the calculated element intakes of babies in regions Styria, Euroregion Neisse, and Woivodship Małopolska, respectively, assuming a daily intake of 0.85 kg/day in mature lactation phase and of 0.25 kg/day during colostrum production.

Like with the results for mother's milk contents in the three regions (Table 3), element supply rates of children in the highly polluted Małopolska do not get significantly higher amounts, especially of the toxicologically relevant elements Ba, Cd, Pb, and Tl via milk.

TABLE 3. Element Concentrations in Colostrum and Mature Milk from Styria (A), Euroregion Neisse (D, Pl, Cz) and Woivodship Małopolska (Pl)

	Phase of Lactation, Colostrum Mature					
	Styria [Krachler et al., 1998]		Euroregion Neisse		Woivodship Małopolska	
			Calculated[a]	Measured	Measured	Calculated
Zn	6040	1013	4656	776	5170	861
Rb	660	716	440	440	555	555
Cu	570	193	585	195	350	117
Sr	35	34	40	40	46	46
Ba	19	3.3	31	5.2	12	2.0
Mn	9.4	4.3	10	4.6	12	5.5
Mo[b]	8.9	1.5	3.6	0.6	9	1.5
Cs	2.7	2.9	2.2	2.2	2.2	2.2
Pb	2.3	1.6	3.5	2.5	1.9	1.3
Co[b]	1.5	2.3	0.52	0.78	0.9	1.35
Cd	1.3	0.2	1.3	0.2	2.0	0.3
Sb[b]	1.1	0.36	0.4	0.13	0.3	0.1
La[b]	0.6	0.11	0.4	0.07	0.4	0.07
Tl	0.2	0.08			0.2[c]	0.08
Ti[b]				282	173	
Fe				230	524	
Ni				14	9	
Cr[b]				10	33	
Zr				1.28	8.3	
Ag[b]				0.65	1.4	
Ga[b]				0.48	1.8	
Y				0.15	0.2	
Ce[b]				0.12	2.0	
Th[b]				0.02	1.7	

[a]See text.
[b]Data taken from Wappelhorst et al. [2000b], corresponding to mature milk obtained from mothers living in Euroregion Neisse.
[c]Values that were measured were below detection limit. Empty spaces denote that no data are available. All values are given in µg/kg.

Because the concentrations in milk do not correspond to the environmental burden of metals, there is no obvious effect of pollution on it. There are quite a number of essential elements, the demands of which of older children cannot be met by milk alone any longer: Mn, Mo, Fe, I, Cu, and Zn (the Co demand of small children is not specified). Accordingly, in these cases there is no "benefit" of enhanced heavy-metal amounts in the environment with respect to milk status and breast-fed child nutrition. Conversely, a lower degree of pollution with "obviously toxic" elements such as Cd does not improve, or at best hardly improves, milk quality and thus the situation of nursed children. Notably, these latter statements refer to the oral path of uptake only. Both positive and negative environmental effects are balanced and eliminated here, yet—considering elements like Cd and the higher intensity of

TABLE 4. Element Supplies to Nursed Babies in Euroregion Neisse and in Woivodship Małopolska, Assuming a Daily Milk Consumption of 0.25 kg in Colostrum and 0.85 kg in Mature Phase[a]

	Styria		Euroregion Neisse		Woivodship Małopolska	
	Phase of Lactation		Phase of Lactation		Phase of Lactation	
	Colostrum	Mature	Colostrum	Mature	Colostrum	Mature
Zn	1510	860	1164	660	1293	732
Rb	165	608	110	374	139	472
Cu	142	164	146	166	88	99
Sr	8.7	29	10	34	12	39
Ba	4.8	2.8	7.8	4.4	3	1.7
Mn	2.4	3.6	2.5	3.9	3	4.7
Mo	2.2	1.3	0.9	0.5	2.2	1.3
Cs	0.7	2.5	0.55	1.87	0.55	1.87
Pb	0.6	1.4	0.88	2.1	0.48	1.1
Co	0.4	1.9	0.13	0.7	0.2	1.1
Cd	0.3	0.17	0.33	0.17	0.50	0.26
Sb	0.27	0.3	0.1	0.1	0.08	0.09
La	0.15	0.09	0.1	0.06	0.1	0.06
Tl	0.05	0.07			0.05	0.07
Ti			240	43		
Fe			195	131		
Ni			12	2.3		
Cr			8.5	8		
Zr			1	2		
Ag			0.55	0.4		
Ga			0.4	0.5		
Y			0.12	0.05		
Ce			0.1	0.5		
Th			0.017	0.43		

[a]Following suggestions on recommended intakes for children younger and older than months filed by DGE (Deutsche Gesellschaft für Ernährung, German Nutrition Society); data for mature lactation phases were split into these two categories.
All values given in µg/day.

breathing and thus inhalation in small children compared to adults—maternal homoeostatic regulation of milk composition cannot influence those additional pathways of element intake, neither to the better nor to the worse of the baby's situation.

A comparison was made between element supplies via mother's milk and the estimated demands for children between birth and 4 months of age (according to DGE); most demands will be met, but for older children the metal contents of milk are far short of demands. Here additional food must supply the elements. For children <4 months nursed by mature milk, estimated demands are higher than oral uptake for Fe and Mn, while amounts did match for Cu, I, and Zn, using DGE recommendations again.

In the early colostrum, phase element supplies were lower than recommended. Yet this is not meant to denote some deficit in supply because depots were created in the foetus and his/her organs, which can be tapped in case of demand. For example, for iron an additional supply will not be needed before the 4th month because Fe (as hemoglobin) is "donated" to the child via the placenta.

5.2   Transfer Factors for All the Investigated Elements (Specific Ones) in the Food/Milk System and Extent of Partition of Elements into Mother's Milk

Equation (1) provides element-specific transfer factors (TF) from food to milk which hold for mothers in Euroregion Neisse. To obtain these data, a relation between the concentration of the element in fresh milk and its daily supply by maternal food is constructed.

Generally speaking, human food is not exclusively derived from local sources; moreover, the mothers who participated could not be selected along any regionalized sampling grid, but we had to accept them wherever they lived in the corresponding region. Thus the problem of regional representation is different, in a more urgent way, than (for example) with moss biomonitoring (the latter, for example, done according to sampling guidelines and grids from the ICP vegetation monitoring campaigns); in addition, a "region" in this sense is larger than in studies of moss monitoring. This is because these samples are taken from humans rather than plants because the remain at their sites. The time scale (extent) of the experiment (≥ 14 days) will efficiently "smooth down" effects from single episodes of extreme administration of some element. This agrees with the fact that, although there were such episodes in the food composition data, no reproducible extreme (maximal or minimal) values of the transfer factors followed after either one or several days.

Table 5 gives the average element-specific uptake amounts via food for the mothers from Euroregio Neisse, combined with the average concentrations in milk. The TF were calculated on a daily schedule from the corresponding raw data.

The data in Table 5 mostly agree with both values from the literature and recommended intakes of essential elements via maternal food except for Fe and I; here average intakes were 50% (Fe) and 17% or 37% (I) below the amounts recommended by DGE. For nonessential elements, uptakes of Pb and Cs were found to be about 70% lower than data reported by WHO and Emsley [2001], respectively.

Comparing to literature values, amounts of nonessential elements Rb and Cs in mother's milk were also considerably lower than reported before (Rb 40%, Cs 30% less).

When element-specific uptakes via maternal food are correlated with concentrations in mother's milk, there is a significant correlation.

The calculated TF values for the elements investigated in this study are collected in Table 6 (see Fig. 6). Additional TF data were taken from Wappelhorst et al. [2000b] and Wuenschmann et al. [2003, 2004a,b] and the corresponding partitioning rates as defined by Eq. (2) in order to get information about physiological interactions of

TABLE 5. Average Element-Specific Uptakes via Maternal Food and Concentrations in Mature Milk from Euroregion Neisse as Compared to Literature Data

	Intake via Food (μg/day)	Literature Data (μg/day)	Concentration in Mother's Milk (μg/kg)	Literature Data (μg/kg or μg/liter)
Zn	11.500	11.000[a]	776	700–2000[f]
Fe	10.000		230	200–1.700[i]
Mn	3.500	2.000–3.800[c]	4.6	3–4[f]
Cu	1.200	1.000–1.500[a]	195	180–751[i]
Rb	2.100	1.500–4.500[d]	440	720[j]
Sr	1.400	800–2.000[d]	40	33[j]
Ba	718	<1.000[e]	5.2	2.1–4,6[j]
I	166[l]	200[f] 260[a]	81[l]	110[k]
Pb	77	200–300[f]	2.5	2–5[f]
Cd	14	10–12[g]	0.2	<0,18–0,2[j]
Cs	8	30[h]	2.2	3[j]

[a]Required quantitiy after DGE.   [b]Illing et al. [1993].   [c]Anke et al. [1991].   [d]Bertram [1992].   [e]Binder [1999].   [f]WHO [1989].   [g]Müller et al. [1993].   [h]Emsley [2001].   [i]Caroli et al. [1994]. [j]Krachler et al. [1998].   [k]Hamosh [1999].   [l]Wuenschmann et al. [2002].

elements and their biochemicals reasons, eventually forming one focus of interest in the discussion section of this chapter.

Generally speaking, only a few percent of the oral uptake of most elements get partitioned into milk. Even without considering the shares of gastrointestinal resorption, which are poorly—if at all—known for most of the elements discussed here, this contrasts with the fact that milk corresponds to one-fourth of the total liquids secreted by a nursing woman. Accordingly, there is nothing like enrichment of metals in milk, not either of essential elements.

Because mothers in Euroregion Neisse consume a total of 3 liters of liquid daily on average,[3] any metal cation that does not undergo any significant chemical interactions (complexation) would be partitioned to about 25% into the 0.8 liter/day of milk because such an ion will be distributed at equal concentrations into milk, urine, sweat, and so on. The heavy alkali metals Rb and Cs are close to this situation, and their partition values into milk are close to 25% indeed. Given an efficient gastrointestinal resorption, values above (iodine) or below this value give proof for some (bio-)chemically caused fractionation among the various sinks.

It is likely that certain elements with low TF are mainly deposited in the skeleton of the mother (Sr, Ba, Mn). Most other elements take part in transport processes that are mainly due to their coordination[4] chemistry; thus a relationship between TF

---

[3]In addition, about half a liter of water per day is produced by metabolic oxidation of H-containing compounds such as sugars and fats.

[4]This is because, except for I and the "semi-metal" Sb, metals only were investigated which interact with biogenic materials mostly by forming coordination complexes.

TABLE 6. Calculated Transfer Factors (TF) for Food/Mother's Milk and Partition of Various Elements into Milk

	TF (d/kg)	± RSD[a] (%)	Partitioning into Milk (%)		TF (d/kg)	± RSD (%)	Partitioning into Milk (%)
Mn	0.001	67	0.1	Ba	0.007	57	0.6
Mo[b]	0.005	127	0.5	Cd	0.014	82	1.2
Fe	0.023	114	2.0	Y[b]	0.021	80	2.0
Co[b]	0.046	80	4.0	Ga[b]	0.022	74	2.0
Zn	0.077	119	6.5	Th[b]	0.022	76	2.0
Cu	0.156	70	13.2	Ce[b]	0.025	82	2.1
I	0.560	41	47.6	Sr	0.027	44	2.3
				La[b]	0.030	204	2.5
				Au[b]	0.032	140	2.7
				Pb	0.033	60	2.8
				Sb[b]	0.035	64	3.0
				Ni	0.054	51	4.6
				Cr[b]	0.065	60	5.5
				Zr[b]	0.065	98	5.5
				Ti[b]	0.067	76	5.7
				Ag[b]	0.152	253	13.0
				Rb	0.200	36	17.0
				Cs	0.271	35	23.0

[a]RSD, relative standard deviation.
[b]Data taken from Wappelhorst et al. [2000b] and Wuenschmann et al. [2003, 2004a,b].

values reported in this work and complex formation properties can be anticipated. Yet no accumulation of metals in mother's milk relative to the other liquid secretions is observed—quite the contrary. This holds for both essential and nonessential elements—for example, Ti, Zr, or REEs Y, La, Ce. Accordingly, the assumption that low TFs for toxic metals Ba, Cd, and Pb would be due to some kind of retention mechanism, provided that protection is not at all substantiated.

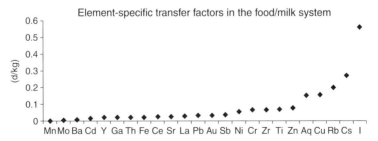

Figure 6. Element-specific transfer factors in the food/milk system (visual representation of data collected in Table 6).

6 DISCUSSION

6.1 Physiological and Dynamic Features of Chemical Elements in the Food/Milk System

Transfer factor (TF) is a measure for transfer of chemical elements from maternal food into mother's milk. TF values are used to identify aspects of coordination behaviour of specific elements including possible mechanisms of protection and regulation in the maternal organism. Thus we attempt to understand why there are no dangerously elevated concentration levels of elements such as Cd and Pb, but also why other elements that are more abundant in a polluted area do not get into milk up to a proportional scale with maternal intake. In fact, mother's milk from the highly polluted Małopolska region contains amounts of metals very similar to those detected in samples taken at less burdened Euroregion Neisse and Styria sites.

Though these results are hard to explain even when taking interactions between metals into account, they do imply an advantage of breast-feeding just in polluted regions, except if there are high levels of organic contaminants such as PCBs or DDT due to special nutritional habits. Up to now, the latter was observed only with indigenous peoples of the Upper Arctic, like Inuits and Chukhots [Muckle et al., 2001; Czub, 2004] except for sites of accidents with corresponding chemicals (e.g., at Seveso in 1976). For Germany, levels of organics like PCBs or DDT in mother's milk have been constantly decreasing for many years, being below acutely toxic threshold levels [Böse-O'Reilly et al., 1999].

Now we will try to interpret these data and the fact that different levels of regional pollution are not represented in/by milk composition, given variations of TF. There are two topics of interest:

- How does the milk level of an element and thus its TF respond to an increase or decrease in maternal supply? Is there regulation or (even) homoeostasis?
- Do elements that are chemically similar or dissimilar influence each other on their biochemical transportation ways until they get into milk, be it synergetistic or antagonistic?

TF values determined in Euroregion Neisse imply that just 0.1% and 0.5%, respectively, of Mn and Mo are partitioned into milk ($TF_{Mn} = 0.001 \, d/kg$; $TF_{Mo} = 0.005 \, d/kg$). Their TF values thus are far lower than those of the other elements. On the other hand, both essential I (highest extent of transfer—about 50% or $TF_I = 0.56 \, day/kg$) and physiologically less significant elements Rb and Cs display the highest TF values ($TF_{Rb} = 0.20$; $TF_{Cs} = 0.27 \, d/kg$). This large extent of elimination via milk might have been anticipated, given the similarity of these heavy alkali ions with the colloquial electrolyte ions Na and K, with Rb likely undergoing electrolyte fractionation at membrane interfaces in the same way and manner as K [Anke and Angelov, 2004]. Cs presumably will distribute among the sinks without any preceding interaction with other binding partner because it hardly forms complexes (see Kaim and Schwederski [1993]). With mother's milk representing about one-fourth of secreted liquid volumes, it can be expected that one-fourth of Cs resorbed in the gastrointestine will show up in milk eventually. Given the simplicity of this model, the agreement with the actual extent of partitioning of cesium (23%) is outright excellent.

A more general problem is why TF value and partition for, for example, Zn are much higher than those for Mn. Possible reasons include interactions of the elements with one another but also effects on the transport of other elements—for example, if they compete for carrier binding sites. Given the latter, TF values should be a complex function of different chemical factors. The only practical way to address this problem is to correlate (nonaveraged, individual) TF values of one element with the amounts consumed of another (or this very) element consumed by the mother. If so, that is, if the TF of one element significantly responds to changing uptake amounts of another one, there is an influence—either synergistic or antagonistic—to be detected in TF variations. Correlation diagrams for the corresponding data are summarized in Table 7. There are no significant effects of maternal intake amounts of other elements on TF values of Pb or Cd.

The relatively low TF of Cd (0.014 day/kg) is interpreted as a result of its considerably lower stability of complexes as compared to those of, for example, Ni, Zn, or Cu (0.054, 0.077, and 0.156 day/kg) (cf. data in Irving and Williams [1953]); therefore biochemical transport of Cd and some other metals during production of milk will take place mainly as simple aquaions, much like with alkali metals and alkaline earths [Wuenschmann et al., 2004a, p. 107]. Accordingly, Cd (or Sr or Ba) cannot effectively compete with other metals—neither those supplied in far larger amounts like Fe, Zn, Mn, or Cu nor those that are similarly rare (10–20 µg/day $\approx$ 100–200 nmol/day like Y, Cs, Co, or Zr)—for carriers. The latter carrier may be both transport proteins and ligands, the amounts of which in milk far more outweigh the metal contents (chloride, citrate, oxalate, some amino acids like glutamate [Neville, 1991]). Thus there is no effect of other elements on the TF of Cd and its partition into milk. Although there is no corresponding effects for Pb in these data, the fragmentary data on complex stability between Pb^{2+} and "milk ligands" preclude similar statements like those referring to Cd.

Likewise, TF values of essential elements Cu, Mo, and Zn and nonessential elements Ag, Au, La, Ba, and Cs are not influenced by supplies of other elements. Transportation of Cu cannot be compared to those of other elements [daSilva and Williams, 2001] because it usually involves special, Cu-specific proteins (metallochaperons) in order to avoid toxic effects from the very strong binding of Cu^{2+} to a multitude of ligands. This strong coordination is the probable reason for the highest TF of any di- or higher valent ion observed with Cu ($TF_{Cu} = 0.156$ d/kg). Mo is taken up and resorbed as molybdate ion by means of the sulfate carrier [Anke et al., 2004]; hence Mo is unlikely to compete with "genuine" cations for their carriers.

The interaction of some element with itself is particularly interesting because it bears some information on both homoeostasis and deposition within in-body depots (e.g., bones). Table 7 shows that several elements absorbed in larger amounts (Rb, Sr, Ti) which are neither toxicologically relevant nor essential

TABLE 7. Analyses of Correlation Between Element-Specific Transfer Factors ((d/kg), Vertical Columns) vs. Element-Specific Maternal Uptake Amounts (µg/d) via Food to Determine Synergistic or Antagonistic Effects. Dark Grey Shadowed Values: Significant Correlations ($r \geq 0.5$), Light-Grey: Self-Correlation Effects

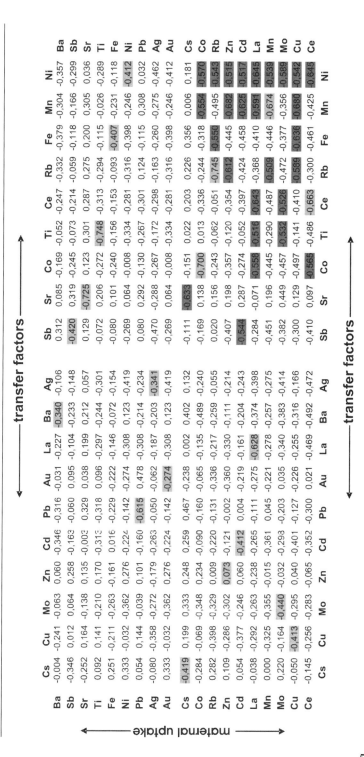

transfer factors →

maternal uptake	Cs	Cu	Mo	Zn	Cd	Pb	Au	La	Ba	Ag	Sb	Sr	Co	Ti	Ce	Rb	Fe	Mn	Ni
Ba	-0,004	-0,241	-0,063	0,060	-0,346	-0,316	-0,031	-0,227	0,340	-0,106	0,312	0,085	-0,169	-0,052	-0,247	-0,332	-0,379	-0,304	-0,357
Sb	-0,346	0,012	0,064	0,258	-0,163	-0,060	0,095	-0,104	-0,233	-0,148	-0,420	0,319	-0,245	-0,073	-0,214	-0,059	-0,118	-0,166	-0,299
Sr	-0,252	0,164	-0,138	0,135	-0,003	0,329	0,038	0,199	0,212	0,057	0,129	-0,725	0,123	0,301	0,287	0,275	0,200	0,305	0,036
Ti	0,092	0,141	-0,210	0,170	-0,313	-0,318	0,096	-0,297	-0,244	-0,301	-0,072	0,206	-0,272	-0,748	-0,313	-0,294	-0,115	-0,026	-0,289
Fe	0,251	-0,211	-0,263	-0,161	0,016	-0,229	-0,222	-0,146	-0,072	-0,154	-0,080	0,101	-0,240	-0,156	-0,153	-0,093	-0,407	-0,231	-0,118
Ni	0,333	-0,032	-0,362	0,276	-0,224	-0,142	-0,274	-0,308	0,123	-0,419	-0,269	0,064	-0,008	-0,334	-0,281	-0,316	-0,398	-0,246	-0,412
Pb	0,054	0,144	-0,039	0,101	-0,160	0,615	0,478	-0,308	0,123	-0,234	0,080	0,292	-0,130	-0,267	-0,301	0,124	-0,115	0,308	0,032
Ag	-0,080	-0,358	-0,272	-0,179	-0,263	-0,050	-0,062	-0,187	-0,214	-0,341	-0,470	-0,288	-0,267	-0,172	-0,298	-0,163	-0,260	-0,275	-0,462
Au	0,333	-0,032	-0,362	0,276	-0,224	-0,142	-0,274	-0,308	0,123	-0,419	-0,269	0,064	-0,008	-0,334	-0,281	-0,316	-0,398	-0,246	-0,412
Cs	0,419	0,199	0,333	0,248	0,259	0,467	-0,238	0,002	0,402	0,132	-0,111	0,633	-0,151	0,022	0,203	0,226	0,356	0,006	0,181
Co	-0,284	-0,069	-0,348	0,234	-0,090	-0,160	-0,065	-0,135	-0,489	-0,240	-0,169	0,138	0,700	0,013	-0,336	-0,244	-0,318	0,554	0,570
Rb	0,282	-0,398	-0,329	0,009	-0,220	-0,131	-0,336	-0,217	-0,259	-0,055	0,020	0,156	-0,243	-0,062	-0,051	0,745	0,556	-0,495	0,543
Zn	0,109	-0,286	-0,302	0,073	-0,121	-0,002	-0,360	-0,330	-0,111	-0,214	-0,407	0,198	-0,357	-0,120	-0,354	0,612	-0,445	0,682	0,515
Cd	0,054	-0,377	-0,246	0,060	0,412	0,004	-0,219	-0,161	-0,204	-0,243	0,544	0,287	-0,274	-0,052	-0,397	-0,424	-0,458	0,625	0,517
La	-0,038	-0,292	-0,263	-0,238	-0,265	-0,111	-0,275	0,628	-0,374	-0,398	-0,284	-0,071	0,558	0,516	0,643	-0,368	-0,410	0,591	0,645
Mn	0,000	-0,325	-0,355	-0,015	-0,361	0,045	-0,221	-0,278	-0,257	-0,275	-0,451	0,196	-0,445	-0,290	-0,487	-0,509	-0,446	-0,674	0,535
Mo	0,220	-0,164	-0,440	-0,032	-0,293	-0,203	0,035	-0,340	-0,383	-0,414	-0,382	0,449	-0,457	0,532	0,526	-0,472	-0,377	-0,356	0,589
Cu	-0,050	-0,413	-0,295	0,040	-0,401	-0,127	-0,226	-0,255	-0,316	-0,166	-0,300	0,129	-0,497	-0,141	-0,410	-0,589	0,636	0,680	0,542
Ce	-0,145	-0,256	-0,283	-0,065	-0,352	-0,300	0,021	-0,469	-0,492	-0,472	-0,410	0,097	0,665	-0,486	-0,663	-0,300	-0,461	-0,425	0,645

← transfer factors →

display a pronounced negative self-correlation (green colored), while, though other authors postulate homoeostatic regulation for this element, there is no relationship between supply and TF for zinc. Up to now, this can only be noted as a kind of phenomenon that warrants further explanation, as do the many different interelemental interactions that influence TFs of Ni and Mn. TF values of Ni and Mn depend on the uptake rates of following elements (cf. the last two columns of Table 7):

TF_{Mn} depends on supply of elements: Cu Zn Co Cd La

TF_{Ni} depends on supply of elements: Cu Zn Co Cd La Ce Mn Mo Rb

Mn is affected by increasing supplies of Zn, Cu, or Cd (pronounced negative correlation in each case); possibly these metals which are known to induce oxidative stress by radical chain reactions [Huang et al., 1994] induce oxidation of manganese to Mn(III), causing stronger retention to transferrin. Co, with more than 90% of its total not being absorbed as cobalamin (vitamin B_{12})[5] but as Co^{2+} aquaion or some complexes far less stable than cobalamine, does correlate negatively with uptakes of La and Ce. La and Ce in turn are coordinated to transferrin, like some heavier REEs (Nd, Sm, Yb [Hirano and Suzuki, 1996]), yet not as strongly as trivalent Fe, Mn, or Y do. This strongly suggests a competition for (protein) carrier binding sites. Figure 7 symbolizes the modes of interaction for the two metals distinguished by lowest TF values, namely, Mn and Mo. Whereas Mn responds to increased allowances of many other elements with a decreased TF (see above), Mo does not because it is resorbed as an oxoanion by means of the sulfate carrier [Williams and daSilva, 1996] and thus is not perturbed by cation carrier competition.

Mother's milk does contain conspicuously low amounts of Mn and Mo. Because both are essential, one is caused to ask for depots built up in the fetus before delivery. Rossipal et al. [2000] analyzed blood samples from both arteries and vein of the umbilical cord for contents of Zn, Mo, Mn, Sn, Ca, Mg, Co, Se, and Cu. Taking differences of concentrations of Mn in arteries and vein, the irreversible part of Mn transfer from maternal blood into the fetus-placenta amounts to 0.4 mg/liter blood serum (16.2% of a primary 2.4 µg/liter content); the corresponding value for Mo is 0.12 µg/liter (16.7% of 0.7 µg/liter). Considering the fact that these values were measured at the very end of pregnancy, that period of time when weight gain of the fetus per day is largest, one would anticipate the largest possible uptake of metals into whatever depot, too. Moreover, metabolic activity of the going-to-be mother is increased by some 30% as compared to a nonpregnant woman of similar activity patterns. At low physical activity levels, some 8–10 liters of blood per minute are circulated through the body; accordingly, the state of pregnancy will

---

[5]That >90% of cobalt uptake are not bound to cobalamine is shown by the following consideration: The molar mass of B_{12} is 1350 g/mol [Kaim and Schwederski, 1993], containing just one (58.9 g/mol) cobalt ion per molecule or 4.4% by weight. Man requires some 1.5 µg Co/day [Emsley, 2001], at an average intake of about 5 µg/day [Emsley, 2001]. Because the human B_{12} demand is about 3 µg/day, this corresponds to a diurnal allowance of about 130 ng (!) Co as cobalamine, less than 9% of recommended or 2.6% of actual average intake. Thus 91–97% of cobalt intake is not cobalamine.

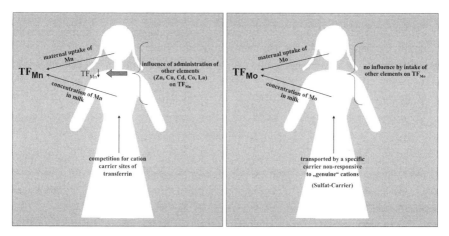

Figure 7. Influences of administration of other elements on TF values of Mn and Mo, which have the lowest TF values of all the metals investigated. The reduction of TF_{Mn} may be due to competition for cation carrier sites also employed by Zn, Cu, Cd, Co, or La. TF_{Mo}, on the other hand, is not influenced by intakes of other elements, presumably because it is transported by a specific carrier nonresponsive to "genuine" cations.

add some 2.5–3 liter/min of blood circulation, that is, about 4.000 liter/day. Because Mn is not given away from the placenta by the umbilical cord vein at all, some 1.6 mg/day[6] of manganese will remain in the fetus. For Mo, Rossipal et al. [2000] detected a chance of back-transfer, rendering the calculated uptake of some 0.4–0.5 mg/day just a conservative upper limit. Given the very small demand even of adults for this ultratrace element (≤ 0.1mg/day), this value would be an extremely high uptake. This is to say that depots may be constructed prenatally for both Mn and Mo. Given that such depots exist indeed, a supply of either element via milk thereafter would not be required by physiology.

Unlike many other mammals (e.g., ungulates), humans are physiologically very premature when just born and thus depend on milk nutrition entirely for a longer period of time. On the other hand, birth weight just about doubles in humans until breast-feeding is terminated. One can compare this situation with that of marsupial mammals or monotremes (kangaroo, platypus, etc.), whose youngsters are very small when born or hatching at less than 1 g of weight [Grzimek, 1965]; thus they cannot have acquired any relevant amounts of depots before delivery. When milk nutrition ends in these mammals, they did grow to (at least) several hundred times their weight at birth or hatching. Compared to both human and cow's milk, milks of both platypus [Griffiths, 1988] and some nonspecified marsupial (most likely gray giant kangaroo] contain 50–100 times as much iron (20–30 mg/liter). Obviously, milk has to cover all the metal demands of young platypuses directly from delivery onward. Data for Mn, Zn, Mo, or other metals are lacking. While

---

[6]4.000 liter/day additional blood circulation times a carryover of 0.4 mg/liter correspond to 1.6 mg/day, the transfer being irreversible (see above).

the body weights of humans and giant kangaroos (gray or red) are similar when exclusive milk nutrition comes to an end (of order 5 kg), the Fe depot in a human newborn of about 1 g is identical to the entire weight of a newborn kangaroo, with material balances being correspondingly much different. African summoprimates other than humans can also be raised/nursed with infant milk formulae based on cow's milk, implying that the situation (existence of metal depots) is similar in the great apes—even including gorillas, which feed on plants only.

This agrees with the DGE stating that there is no Fe demand whatsoever for the "pure" (human) phase of breast-feeding (<4 months).

Ni also interacts with numerous other elements; though it is coordinated for example, to albumin, this complex is so labile in both kinetic and thermodynamic terms that Ni^{2+} will be transferred to the latter already upon addition of simple amino acids [Tabata and Sarkar, 1992], suggesting intense interactions. Unlike Mn, Ni will not undergo redox reactions in aerobic organisms, rendering effects by other heavy metals by generation of oxidizing radicals unlikely. Nevertheless, increased maternal administrations of Cu, Cd, Co, and Zn here lower TF_{Ni} once again. Mn exerts an unsymmetrical effect: Mn lowers transfer of Ni but not the reverse. Such kind of asymmetry is to be anticipated in exactly those cases where transport of some pair of elements is accomplished using different pathways and/or carriers. There are similar asymmetries with obviously the same reasons in the element couple Cu/Fe.

6.2   Lack of an Effect of Regional Pollution on Chemical Element Composition in Mother's Milk

The differences in regional environmental burdens caused by heavy metals in the regions where the milk samples were taken were already discussed in Section 4.1. In comparison to other parts/countries of Western Europe, Euroregion Neisse is considered lowly to intermediately burdened by airborne metal particles [Wappelhorst et al., 2000a]. These results, which draw upon moss monitoring surveys done in the 1990s, were recently corroborated by results from the 2005 moss monitoring survey [Schröder et al., 2007].

However, according to Dmowski [2000], the NW of Małopolska region (next to Olkusz) is distinguished by enhanced levels of Cd, Pb, As, Zn, Ge, and Tl due to presence and activity of smelters that process zinc-containing ores. In that area used for mining, there are layers of relocated soils and ponds in which density differences among Zn minerals, sand, lime, and so on, are used to separate ZnO from the other materials after suspending all fractions in water. The ZnO-enriched fraction thus obtained then is subjected to hutting. When these ponds or pools partly become dry in summer, Zn-rich dust can be spread by the wind, contaminating the surroundings. Eventually, metal hutting in the zinc smelters causes volatilization of Zn, as well as other heavy elements such as that of the corresponding oxides (Zn, Cd, Hg, Pb, Tl), mobilizing the latter at least as long as there are no filter facilities. As a result, there will be multiple ways by which nonferrous metal mining produce emissions to the atmosphere [Nriagu and Pacyna, 1988] unless there are very contrived countermeasures. In Southern Poland, these were not implemented until recently.

TABLE 8. Element Contents of a Highly polluted Relocated Soil at Olkusz (Southern Poland) Compared to Element Contents of Unpolluted Soils

	Relocated Soil at Olkusz (Southern Poland) (mg/kg)	Unpolluted Soils [Markert, 1993] (mg/kg)
Cd	170	0.01–3
Pb	8,500	0.1–200
Tl	94	0.01–0.5
Zn	33,200	3–300
Ca	121,000	1,000–12,000
Mg	43,000	500–5,000

In 2003, we did studies of our own of the relocated soils around Olkusz which provided elevated concentrations of Cd, Pb, Tl, and Zn but also Ca and Mg (Table 8) as expected. Compared to nonpolluted soils (standard concentrations in Markert [1993]), levels of Tl were 200 times higher, for Zn 100 times, for Cd and Pb 50 times each, and for Ca and Mg 10 times the average values (neither As nor Ge were quantified). Levels of Ni, Mn, or Fe match those of nonpolluted soils.

Mother's milk samples from this highly burdened region were analyzed for their element contents and compared to corresponding data from Euroregion Neisse and those from Styria (Austria; Krachler et al. [1998]). The results are reported in Fig. 8. Transformation of the data—which hardly display relevant differences in concentrations among the regions except for Cu and Cd—referring to colostrum or mature milk, respectively, is done using the interaction data of the transfer into milk reported in Table 7.

Table 7 includes the self-correlations between maternal uptake amount and TF for a multitude of elements including Cd, Zn, and Pb; usually self-correlation is distinctly negative except for Zn. As a result of this negative self-correlation with respect to Cd or Pb, TF values are decreased so much that the milk contents of these metals essentially are identical to those from Euroregion Neisse or Styria even though there is a manifest increased level of heavy metal pollution in Małopolska. This fact suggests that the correlation data collected in Table 7 also pertain to subtoxic heavy-metal uptake levels even though most of them were measured originally in Euroregion Neisse. Assuming a competition for carrier sites at proteins such as transferrin, the blood/serum level of which is fairly low, this can be anticipated. Though comparative data are missing, this phenomenon should also be quite pronounced with thallium (Tl^+), which, given the fact that it behaves partly similar to Ag and partly emulates the heavier alkali metals K or Rb [Kolditz, 1990], should display a rather high TF value. Rather, although this has to be inferred in an indirect manner by combining several sets of data, TF_{Tl} appears to be extremely low. Besides those self-correlations, interactions between uptakes of several other elements and the TF of Mn merit consideration; the corresponding correlation coefficients are somewhat higher (i.e., less strongly negative) than those for the self-correlations. Accordingly, in Małopolska there is no effect of the larger inputs of Cd or Pb on the manganese contents of milk, nor there is any enrichment of milk in Cd or other elements. Mn here is a

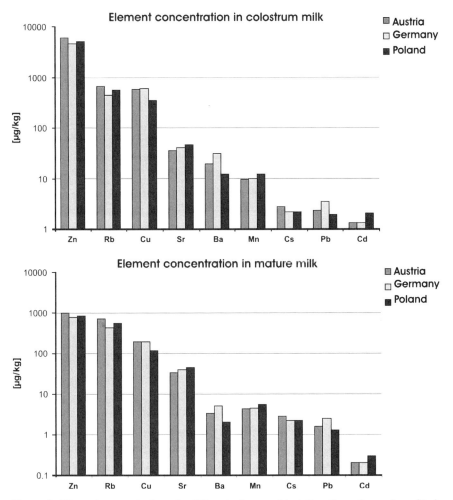

Figure 8. Element concentrations in different phases of lactation from the regions Styria (Austria), Malopolska (Poland) and the Euroregion Neisse (Germany).

suitable tracer since on one hand—like with Ni, which was omitted in Figure 8—Mn displays pronounced crosswise interactions with other elements, yet one need not expect increased levels of Mn from either mining/metal production technologies or from combustion of fossil fuels (lignite, etc.).

7 CONCLUSION: IS THERE A ROLE FOR HUMAN MILK IN METAL BIOINDICATION?

Trying to do bioindication by means of mother's milk does imply that concentration differences in oral uptakes or other administration pathways (inhalation, etc.) are not canceled by a decreased transfer factor eventually providing the same levels of the metals in milk. This can be checked by determination of both sign and extent of the self-correlations between uptake and TF of the same element. When there is "regulation"—brought about by whatever mechanism (e.g., carrier saturation)—sign and extent of self-correlation are strongly negative. Table 7 shows this to be the usual situation, holding for all the investigated elements except of Zn, Ag, and Au.

The contrary behavior—that is, a strongly positive self-correlation coefficient—would be "ideal" in terms of bioindication. This would cause some more than proportional enrichment of the element in milk, providing most sensitive bioindication. There is no element displaying such behavior in the set investigated. The model that attributes the negative uptake/TF-correlations to carrier saturation (see Guo and Sadler [1999]), is corroborated by the data obtained for nonessential elements which, like Ti($r = -0.748$), are taken up in substantial amounts. It should be mentioned that uptake amounts of such elements sometimes vary more strongly than the relative variation/increase brought about by a purposeful supplementation with essential elements like Fe, Zn, or I.

Two of the investigated elements respond to increased allowance of metals (especially toxic heavy metals) by substantial decreases of their TF values: Mn and Ni. Therefore we suggest to do indirect bioindication, employing the (changes of) Mn and Ni TF values in much the same manner as, for example, induction of metallothioneins [Braun et al., 1991]. In the latter case, organic pollutants or irritating agents cause the same reaction; future investigations on Mn and Ni in milk thus would have to show whether there would be a similar "blurring" of the bioindication signal by quite different chemical entities. It is a striking physiological fact that in both woman and animal (cow) the milk-producing gland is a site of high Mn enrichment in the corresponding organism [Emsley, 2001], an apparent contradiction to its very low TF value. However, transport proteins are not element-specific but bind and transport a considerable number of different metals which thereby compete for ligation sites. Probably the observed interaction effects are also due to this carrier-nonspecifity. Presumably, Mn is bound to transferrin or something else in the breast gland [Schäfer, 2004] and thus accumulates there; but, because Mn(III)transferrin is very stable [Williams and daSilva, 1996], it will be retained there rather than donated to milk (milk itself is an oxidizing liquid as evidenced by its containing iodine [also] as iodate). Other elements that bind to transferrin more weakly or that do not form similar depots will readily (high TF values) pass over into milk. With Ni there are similar sources of influence, and milk concentration levels of both metals are similar, too (Ni 14, Mn 5 μg/kg), but other elements with comparable concentrations do not display similar sensitivities or cross-correlations. This leads us to presume that limited transport capacities at or through the membrane are not the key problem, but different chemical properties of these two elements cause the observed peculiarity.

In this work a certain TF value is determined, though it is acknowledged that there is self-correlation with the extent of maternal oral uptake (for some metals there are also crosswise interactions), and TF is subject to significant variations (requiring us to use and report average values).

One might tackle both phenomena by replacing this simple number by some polynomial that includes these effects as higher-order correcting terms. As a zeroth-order approximation, the first term of this series expansion only might then be used. To test for numerical mathematical art-facts, self- and crosswise correlations with TF were also calculated by nonlinear methods; however, there is neither "improvement" nor a sign change in any correlation coefficient. Thus, expanding the TF into some second- or third-order polynomial would probably produce just a less tractable formula, though it might expose some few dominant influencing effects (intermetal interactions). Eventually the influences of all other elements investigated here might be recognized by one polynomial term each, providing an expression with >10 terms. However, it is very doubtful whether there is any point in this, given the limited amount of available data. TF values given as simple numbers are thus considered to be tractable semiquantitative data for the purpose of orientation at least; in addition, it is feasible to correlate this simplified representation to coordination chemistry properties of the corresponding metal ion directly. A polynomial of intermediate complexity that includes both self-correlation and the elements that influence transfers of Mn and Ni might also allow for a quantification of effects brought about by other elements. Possibly one could say whether $10\,\mu g/day$ of additional cadmium would bias transfers of certain other metals in the same way as $3\,mg/day$ more of iron in the mother's diet. For now, quantitative statements from the above approach provide just a description of effects and interactions which does not correspond to toxicological properties or denominators in any obvious way.

Both elements, the transfer of which is particularly sensitive toward other elements, also display peculiarities with respect to their own uptake: Mn is readily taken up by inhalation, causing severe neurological effects when administered in large doses [Schäfer, 2004]. Nickel is readily taken up dermally. Even including the available data on atmospheric deposition, it is hard to estimate the possible sizes of any contributions along these pathways to the results obtained at Małopolska. Yet two facts can be considered established:

(a) The TF value of Mn, though sensitive, cannot be substantially decreased anymore.

(b) Inhalation provides Mn in a form that can pass the blood–brain barrier.

However, the blood–brain barrier differs in structure from a "colloquial" membrane such as that in the breast gland. Though the amount of material detected in milk samples is linked to oral uptake only by this calculation, this very approach in turn implies an overestimation of TF when there are inhalative or dermal contributions to maternal uptake. Thus the TF value as calculated and defined here represents a conservative upper limit that includes some kind of safety factor, furthermore, inhalative administration, which is likely to occur in Małopolska, does not cause any increased metal transfer into milk.

TF and partition both give a measure for those parts of maternal element uptake that ultimately find their way into milk. Accordingly, it is tempting to obtain a more precise description by dividing TF values by the corresponding gastrointestinal resorption rates. However, this is tantamount to reducing results obtained from "genuine" oral consumption of metals to only those effects that occur during blood/milk transfers and thus correspond to some hypothetical intravenous administration of the element. It should be mentioned that the older, particularly radiobiochemical, literature contains numerous data that were obtained by measuring partition into milk after intravenous injection of radionuclides in both humans and animals (e.g., Nürnberger and Lipscomb [1952]; Jarvis et al. [1963]; Simon et al. [2002]). Conversely, such data permit an estimate of oral TF, given that the gastrointestinal resorption quota is known. This will be only an estimate because the compound that was injected and thus primary speciation in blood will hardly match that formed at the outer gut wall or in the liver or elsewhere, obviously affecting the TF values.

Colostrum is distinguished from mature milk by higher pH and also higher contents of both macromolecular and "simple" ligands (e.g., albumin, citrate [Morris et al., 1986]). From both factors and the membrane properties, TF values should be higher in colostrum than in mature milk. Taking into account the corresponding data by Krachler et al. [1998], the TF values in colostrum cannot be taken to be some simple (and generally holding) multiple of those obtained by us for mature milk. However, this enhances the consequences of the above statement that mothers from Olkusz and its outskirts produce milk (colostrum) with similar metal levels like those in Euroregion Neisse or Styria even though they are exposed to larger amounts of (probably mainly airborne) heavy metals.

This comparison of metal content data obtained in various environments shows that unlike for organics—especially halogenated aromatic compounds—bioindication using human milk is not feasible with respect to inorganic components. Unlike data from other human-derived biogenic matrices (e.g., urine), milk concentrations of metals thus cannot be introduced into some multiindicator strategy for estimating human health hazards caused by chemical species. Such a multiindicator strategy, namely the MMBC, was proposed by some of these authors before [Markert et al., 2003a,b, 2006]. This is not to put the information from human milk analyses concerning organic pollutants such as PCBs into any doubt.

Apart from this, it should be pointed that now so many different organismic and suborganismic test systems are employed in bioindication that there is no problem whatsoever caused by some matrix or test system proving useless for determining human health risks. As a rule, reliable statetments can be extracted from a well-designed combination of test systems which, though, are differently expensive, abundant, and easy to deal with; yet their ecological and ecochemical relationships with humans may sometimes be feeble. Another—mostly unsettled—issue is to which extent data from different matrices correspond to humans and his/her biochemistry, given either taxonomical relationship (mammals) or sympatry (animals or plants living next to man). Intraspecific effects seen in mother's milk (self-correlation and interactions among elements) suggest that we do not yet understand biochemical reasons for such effects in a way that would allow any realistic transfer of results. Likewise, we cannot produce statements on implications of biochemical differences among different kinds of mammals. Yet data on cow's and goat's milk [Krełowska-Kułas et al., 1999; Simsek et al., 2000] show that such differences exist and bear bioindicative implications. Unlike with humans, the general measures of food quality surveillance show that increased heavy-metal pollution (detected as atmospheric deposition on grass) also cause increased heavy-metal levels in ungulate milks. It is not yet known whether this is due to, for example, a diet differing from that of human, a different structure of the digestive tract, or other factors that might influence biochemistry.

REFERENCES

Anke M, Angelov L. 2004. Rubidium. In: Merian E, Anke M, Ihnat M, Stoeppler M, editors. *Elements and Their Compounds in the Environment*. Weinheim: VCH-Wiley, 547–563.

Anke M, Groppel B, Krause U, Arnhold W, Langer M. 1991. Trace element intake (zinc, manganese, copper, molybdenum, iodine, and nickel) of humans in Thuringia and Brandenburg of the FRG. *J Trace Electrolytes Health Dis* **5**:69–74.

Bates CJ, Prentice A. 1994. Breast milk as a source of vitamins, essential minerals and trace elements. *Pharmacol Ther* **62**.193–220.

Bertram HP. 1992. *Spurenelemente: Analytik, ökotoxikologische und medizinisch-klinische Bedeutung*. München: Urban & Schwarzenberg.

Binder H, editor. 1999. *Lexikon der chemischen Elemente*. Stuttgart: S. Hirzel Verlag.

Böse-O'Reilly S, Helbich HM, Mersch-Sundermann V. 1999. Frauenmilch (Stillen). In: Mersch-Sundermann V, editors. *Umweltmedizin. Grundlagen der Umweltmedizin, klinische Umweltmedizin, ökologische Medizin*. Stuttgart: Thieme-Verlag, pp. 359–372.

Bowen HJM. 1979. *Environmental Chemistry of the Elements*. London: Academic Press.

Braun R, Fuhrmann G, Legrum W, Steffen C. 1999. *Spezielle Toxikologie für Chemiker* Teubner: Eine Auswahl toxischer Substanzen.

Buse A, Norris D, Harmens H, Büker P, Ashenden T, Mills G. 2003. Heavy metals in European mosses: 2000/2001 survey. UNECE ICP Vegetation. Centre for Ecology & Hydrology.

Caroli S, Alimonti A, Coni F, Petrucci F, Senofonte O, Violante N. 1994. The assessment of reference values for elements in human biological tissues and fluids: A systematic review. *Crit Rev Anal Chem* **24**:363–398.

Czub G. 2004. Modellierung der Bioakkumulation persistenter organischer Umweltchemikalien im Menschen. Dissertation, Leibnitz-Institut für Ostseeforschung, Warnemünde.

daSilva JRF, Williams RJP. 2001. The biological chemistry of the elements. *The Inorganic Chemistry of Life*. New York: Oxford University Press.

Deutsche Forschungsgemeinschaft. 1984. Rückstände und Verunreinigungen in Frauenmilch. Mitteilung XII der Kommission zur Prüfung von Rückständen in Lebensmitteln. Weinheim: Verlag Chemie GmbH.

Deutsche Gesellschaft für Ernährung (DGE). 2000. Referenzwerte für die Nährstoffzufuhr. www.dge.de

Dmowski K. 2000. Environmental monitoring of heavy metals with magpie (Pica pica) feathers—An example of Polish polluted and control areas. In: Markert B, Friese K, editors. *Trace Elements. Their Distribution and Effects in the Environment*. Amsterdam: Elsevier, pp. 455–477.

Dmuchowski W, Badurek M. 2003. Biomonitoring of the Environmental Pollution in the Region of Olkusz (Southern Poland) Using Four Different Methods. Biomap workshop, Slovenia.

Emsley J, editor. 2001. *Nature's Building Blocks. An A–Z Guide to the Elements*. New York: Oxford University Press.

Esty D, Levy M, Srebotnjak T, de Sherbinin A, Kim C, Anderson B. 2006. Pilot 2006. Environmental Performance Index. New Haven: Yale Center for Environmental Law and Policy. www.yale.edu/epi/2006EPI_REPORT_full.pdf

Fecher P. 1994. Atomemissions-und Massenspektrometrie mit induktiv gekoppelten Plasma (ICP). In: Matter Le, editor. 1994. *Lebensmittel- und Umweltanalytik anorganischer Spurenbestandteile. Tips, Tricks und Beispiele für die Praxis*. Weinheim: VCH, pp. 11–43.

Fecher P, Walther C, Sondermann J. 1999. *Bestimmung von Iod in diätetischen Lebensmitteln mit der ICP-Massenspektrometrie*. Deutsche Lebensmittel-Rundschau, 95. Jahrgang, Heft 4, pp. 133–142.

Fraenzle S. 2007. *Prinzipien und Mechanismen der Verteilung und Essentialität von chemischen Elementen in pflanzlicher Biomasse—Ableitungen aus dem Biologischen System der Elemente*. Deutschland: Habilitationsschrift, Hochschule Vechta.

Fraenzle S, Markert B, Wuenschmann S. 2004. Quantitative aspect of interaction between metal ions and living matter: A novel method and applications for biomonitoring and phytoremediation. Abstract der 12. Tagung EcOPOLE, Duzhniki Zdroj, 21.-23.10.2004.

Griffiths M. 1988. Das Schnabeltier. *Spektrum Wissenschaft* **7**:76–83.

Grzimek B. 1965. *Mit Grzimek durch Australien. Vierfüßige Australier* Stuttgart: Bertelsmann-Mohn.

Guo Z, Sadler PJ. 1999. Metals in medicine. *Angew Chem* **111**:1610–1630.

Hambraeus L, Fransson GB, Lönnerdal B. 1984. Nutritional availability of breast protein. *Lancet* **ii**:167–168.

Hamosh M. 1999. Breastfeeding: Unraveling the mysteries of mother's milk. *Medscape Gen Med* **1**(1). http://www.medscape.com/viewarticle/408813.

Herpin U, Berlekamp J, Markert B, Wolterbeek B, Grodzinska K, Siewers U, Lieth H, Weckert V. 1996. The distribution of heavy metals in a transect of the three states the Netherlands, Germany and Poland, determined with the aid of moss monitoring. *Sci Total Environ* **187**:185–198.

Hirano S, Suzuki T. 1996. Environmental health issues. *Environ Health Perspect Supple* **104**(S1):1–19.

Huang X, Zhuang Z, Frenkel K, Klein CB, Costa M. 1994. The role of nickel and nickel-mediated reactive oxygen species in the mechanism of nickel carcinogenesis. *Environ Health Perspect* **102**:281–284.

Illing H, Anke M, Kräuter U. 1993. Iron intake in Germany. In: Anke M, Meissner CF, Mills CF, editors. *Trace Elements in Man and Animals* TEMA 8. Gersdorf: Verlag Media Touristik, pp. 243–246.

Irving H, Williams RJP. 1953. The stability series for complexes of divalent ions. *J Chem Soc* 3192–3210.

Jarvis AA, Brown JR, Tiefenbach B. 1963. Strontium-89 and strontium-90 levels in breast milk and in mineral-supplement preparations. *Can Med Assoc J* **88**:136–139.

Jödicke B, Neubert D. 2004. Reproduktion und Entwicklung. In: Marquardt H, Schäfer S, editors. *Lehrbuch der Toxikologie*. Stuttgart: Wissenschaftliche Verlagsgesellschaft mbH, pp. 491–544.

Kaim W, Schwederski B. 1993. *Bioanorganische Chemie*. Stuttgart: Teubner.

Kees-Aigner S. 2002. Untersuchung der Quecksilberbelastung von Muttermilch in Abhängigkeit von Amalgamfüllungen und weiteren Faktoren mit Berücksichtigung des Quecksilbergehaltes in Säuglingsnahrung. Dissertationsschrift, LMU München.

Kolditz L, editor. 1990. *Anorganische Chemie*. Berlin: Deutscher Verlag der Wissenschaften.

Krachler M, Shi Li F, Rossipal E, Irgolic J. 1998. Changes in the concentrations of trace elements in human milk during lactation. *Trace Elements Med Biol* **12**:159–176.

Krełowska-Kułas M, Kĕdzior W, Popek S. 1999. Content of some metals in goat's milk from southern Poland. *Nahrung/Food* **43**:317–319.

Lebensmittel- und Bedarfsgegenständegesetz (LMBG) §35. 1998. Untersuchungen von Lebensmitteln mit der ICP-MS. L. 49.00-6, Amtliche Sammlung von Untersuchungsverfahren.

Markert B. 1993. *Instrumentelle Multielementanalyse von Pflanzenproben*. Weinheim: VCH-Verlagsgesellschaft mbH, 266 S.

Markert B. 1996. *Instrumental Element and Multi-Element Analysis of Plant Samples— Methods and Applications*. New York: John Wiley & Sons.

Markert B, Oehlmann J, Roth M. 1997. Biomonitoring von Schwermetallen—eine kritische Bestandsaufnahme. *Z. Ökologie Naturschutz* **6**:1–8.

Markert B, Breure T, Zechmeister H, editors. 2003a. *Bioindicators & Biomonitors. Principles, Concepts and Applications*. Amsterdam: Elsevier.

Markert B, Breure T, Zechmeister H. 2003b. Definitions, strategies and priciples for bioindication/biomonitoring of the environment. In: Markert B, Breure T, Zechmeister H, editors. *Bioindicators & Biomonitors. Principles, Concepts and Applications*. Amsterdam: Elsevier, pp. 3–39.

Markert B, Wuenschmann S, Fraenzle S, Breulmann G, Djingowa R, Herpin U, Schröder W, Siewers U, Steinnes E, Wappelhorst O, Weckert V, Wolterbeek B, Zechmeister H. 2006. On the road from environmental biomonitoring to human health aspects of atmospheric deposition heavy metals by epiphytic plants. Present status and future needs. *Int J Environ Pollut* (Special Edition) Accepted 2006 (in press).

Martius H. 1983. *Lehrbuch der Geburtshilfe einschließlich der geburtshilflichen Operationen*. Stuttgart: Thieme-Verlag.

Matter L. 1994. Spurenelementanalytik in biologischen Matrizes. In: Matter L, editor. *Lebensmittel- und Umweltanalytik anorganischer Spurenbestandteile. Tips, Tricks und Beispiel für die Praxis*. Weinheim: VCH, pp. 1–9.

Morris FH, Brewer ED, Spedale SB, Riddle L, Temple DM, Caprioli RM, West MS. 1986. Relationship of human milk pH during course of lactation to concentration of citrate and fatty acids. *Am Acad Pediatr* **78**(3):458–464.

Muckle G, Ayotte P, Dewailly E, Jacobson S, Jacobson J. 2001. Prenatal exposure of the Northern Québec Inuit infants to environmental contaminants. *Environ Health Perspect* **109**(12):1291–1299.

Müller M, Anke M, Thiel C, Hartmann E. 1993. Zur Cadmiumaufnahme Erwachsener in den neuen Bundesländern. *Ernährungs-Umschau* **40**(6):240–243.

Neville MC. 1991. Appendix C: Summary of composition data for macronutrients of human milk. In: *Nutrition During Lactation* National Academic Press, pp. 279–286.

Norska-Borówka I, Kasznia-Kocot J, Bursa J. 1993. Umweltbelastung im Raum Kattowitz und ihre Auswirkung auf Säuglingssterblichkeit und Kindermorbidität. *Sozialpädiatri Praxis Klin* **15**(2):119–121.

Norska-Borówka I, Bursa J, Kasznia-Kocot J. 1996. Low Birthweight as the best Indicator of Environmental Disaster. *Polish J Environ Studies* **5**(4):29–31.

Nriagu JO, Pacyna JM. 1988. Quantitative assessment of worldwide contamination of air, water and soil by trace metals. *Nature* **333**:134–139.

Nurnberger CE, Lipscomb A. 1952. Transmission of radioiodine (131I) to infants through human maternal milk. *J Am Med Assoc* **150**:1398–1400.

Rossipal E, Krachler M, Li F, Micetic-Turk D. 2000. Investigation of the transport of trace elements across barriers in humans: Studies of placental and mammary transfer. *Acta Paediatr* **89**:1190–1195.

Schäfer U. 2004. Manganese. In: Merian E, Anke M, Ihnat M, Stoeppler M, editors. *Elements and Their Compounds in the Environment*. Weinheim: VCH-Wiley, pp. 901–930.

Schroeder W, Pesch R. 2004a. The 1990, 1995 and 2000 moss monitoring data in Germany and other European countries. Trends and statistical aggregation of metal accumulation indicators. Gate to Environ Health Sci [DOI: dx.doi.org/10.1065/ehs2004.06.011), pp. 1–25.

Schroeder W, Pesch R. 2004b. Spatial and temporal trends of metal accumulation in mosses. *J Atmos Chem* **49**:23–38.

Schroeder W, Hornsmann I, Pesch R, Schmidt G, Fraenzle S, Wuenschmann S, Heidenreich H, Markert B. 2007. Stickstoff- und Metallakkumulation in Moosen zweier Regionen Mitteleuropas als Spiegel ihrer Landnutzung? UWSF-Z., ecomed, online first.

Simon SL, Luckyanov N, Bouville A, VanMiddlesworth L, Weinstock RM. 2002. Transfer of 131I into human breast milk and transfer coefficients for radiological dose assessments. *Health Phys* **82**:796–806.

Simsek O, Gültekin R, Öksüz O, Kurultay S. 2000. The effect of environmental pollution on the heavy metal content of raw milk. *Nahrung/Food* **44**:360–363.

Staatsministerium für Umwelt und Landesentwicklung des Freistaates Sachsen. 1994. Umweltbericht, S. 61–64.

Tabata M, Sarkar B. 1992. Specific nickel(II) transfer process between the native sequence peptide representing the nickel(II)-transport site of human serum albumin and l-histidine. *J Inorg Biochem* **45**:93–104.

Wappelhorst O. 1999. Charakterisierung atmosphärischer Deposition durch ein terrestrisches Biomonitoring in der Euroregion Neiße. PhD Thesis, Internationales Hochschulinstitut Zittau.

Wappelhorst O, Korhammer S, Leffler US, Markert B. 2000a. Ein Moosmonitoring zur Ermittlung atmosphärischer Elementeinträge in die Euroregion Neiße (D, PL, CZ). UWSF-Z. *Umweltchemie Ökotoxikol* **12**(4):191–200.

Wappelhorst O, Kuehn I, Heidenreich H, Markert B. 2000b. Transfer von Elementen in die Muttermilch. Abschlussbericht des Forschungsvorhabens StSch 4155 für das Bundesamt für Strahlenschutz, Neuherberg.

Williams RJP, daSilva JRF. 1996. *The Natural Selection of the Chemical Elements*. Oxford: Clarendon.

World Health Organization (WHO). 1989. *Minor and Trace Elements in Breast Milk*. Geneva: World Health Organization.

Wuenschmann S. 2007. Bestimmung chemischer Elemente im System Nahrung/ Muttermilch—ein Beitrag zur Bioindikation? Dissertationsschrift, Hochschule Vechta Germany.

Wuenschmann S, Fraenzle S, Kuehn I, Heidenreich H, Markert B. 2002. Verteilung chemischer Elemente in der Nahrung und Milch stillender Mütter. Teil I: Iod. UWSF-Z *Umweltchem Ökotox* **14**(4):221–227.

Wuenschmann S, Kuehn I, Heidenreich H, Fraenzle S, Wappelhorst O, Markert B. 2003. Transfer von Elementen in die Muttermilch. Abschlussbericht der Forschungsvorhaben StSch 4155 und 4258 für das Bundesamt für Strahlenschutz, Neuherberg.

Wuenschmann S, Fraenzle S, Markert B. 2004a. Transfer von Elementen in die Muttermilch. Methoden, Modellierungen, Empfehlungen. Ecomed-Medizin-Verlagsgesell-schaft, Landsberg.

Wuenschmann S, Fraenzle S, Kuehn I, Heidenreich H, Wappelhorst O, Markert B. 2004b. Verteilung chemischer Elemente in der Nahrung und Milch stillender Mütter. Teil II. *UWSF-Z Umweltchem Ökotox* **15**(3):168–174.

23 Selenium: A Versatile Trace Element in Life and Environment

SIMONA DI GREGORIO

Department of Biology, University of Pisa, 56126 Pisa, Italy

1 WHAT IS SELENIUM?

Selenium, the element with properties of metalloid, belonging to Group VI A of the periodic table with atomic number 34, was discovered by the Swede Jöns Jacob Berzelius in 1817. The chemist reported an impurity in the sulfuric acid produced by a Swedish plant. The element was subsequently identified as a new one, and it was called selenium from the Greek word "*selene*," meaning "*moon*."

1.1 Selenium Industrial Applications

Nowadays, selenium is extracted from selenide minerals in many sulfide ores, such as those of copper, silver, or lead. It finds application as a catalyst in many chemical reactions and in various industrial and laboratory synthesis. It is also widely used in structure determination of proteins and nucleic acids by X-ray crystallography for the incorporation of one or more Se atoms that help with MAD phasing. The use of selenium is principally associated to glass and ceramic productions. In glass manufacturing it is specifically used to remove the natural green color of the material due to ferrous impurities, while cadmium sulfo-selenide red pigments were used to decorate glass and ceramics for their good heat stability. Moreover, selenium is used to reduce solar heat transmission in architectural plate glass. The metalloid is also added to copper and steel alloys to improve their mechanical properties, to replace the more toxic lead in plumbing applications, and to improve the abrasion resistance in vulcanized rubbers.

Having photovoltaic and photoconductive properties, selenium is used in photocopying, photocells, light meters, and solar cells. Sheets of amorphous selenium

Trace Elements as Contaminants and Nutrients: Consequences in Ecosystems and Human Health, Edited by M. N. V. Prasad
Copyright © 2008 John Wiley & Sons, Inc.

convert X-ray images to patterns of charge in xeroradiography and in solid-state flat panel X-ray cameras. Moreover, selenium is added to the toning of photographic prints because it intensifies and extends the tonal range of black and white photographic images, as well as improves the permanence of prints. On the other hand, selenium is used as dietary supplement in human and livestock nutrition because selenium is an essential trace element. Actually, dietary supplementation for livestock consists of the largest agricultural use. Finally, selenium is used as an anti-dandruff agent.

Growth in selenium industrial application was historically related by steady development of new uses, even though during the 1980s the photoconductor application declined as more and more copiers using organic photoconductors were produced. As rectifiers, selenium has been progressively replaced by silicon. Also electronic use, despite a number of applications, is declining. Presently, the largest use of selenium is related to glass manufacturing, followed by uses in chemicals and pigments production.

1.2   Selenium in the Environment

The naturally occurring element Se is rarely recovered in a free state. Its chemistry is related to sulfur and tellurium and occurs in the environment, especially as inorganic forms. Selenium oxidation states are $+6$ (VI), $+4$ (IV), 0, and -2. Selenate Se(VI) and selenite Se(IV) are the oxidized water-soluble forms recovered in soil solutions and in natural waters. The highly stable elemental or metallic selenium (Se^0) is also recovered in soil, but not in water solution because it is insoluble. Metallic selenium exists in several allotropic forms, and some of them have been identified: trigonal gray selenium (containing Se_n helical chain polymers), rhombohedral selenium (containing Se_6 molecules), three deep-red monoclinic forms, α, β-, and γ-selenium (containing Se_8 molecules), amorphous red selenium, and black vitreous selenium. The most thermodynamically stable form is the gray (trigonal) selenium, which contains countless helical chains of selenium atoms. Grey selenium is the only allotropic form that conducts electricity. The red and black amorphous allotropes are the forms that are most likely to occur in soils. At temperature greater than $30°C$, red amorphous selenium gradually reverts to the black amorphous form. The latter form is then slowly transformed into the much more stable gray hexagonal allotrope or, depending on the redox conditions and the pH of the soils, it is reoxidized.

Selenium is recovered in rocks combined with sulfide minerals with silver, copper, lead, and nickel. The metalloid is present also in coal and actually in significant amounts; in fact, a coal refinery is a severe source of selenium environmental contamination [Lawson and Macy, 1995]. Examples of selenide minerals are as follows: (i) crookesite, a copper selenide mineral containing also thallium and silver; it is formed by precipitation from hydrothermal fluids; (ii) tilleite, a zinc selenide found only as microscopic gray crystals associated with other selenides; (iii) tiemannite, a mercury selenide that occurs in hydrothermal veins associated with other selenides, or other mercury minerals such as cinnabar, and often with calcite; (iv) umangite, a copper selenide that occurs in small grains or fine granular aggregates with other copper minerals of the sulfide group [Kabayta-Pendias, 1998]. Selenium is distributed in the environment by processes such as volcanic activity and hot springs, combustion of fossil fuels, weathering of rocks and soils, soil leaching, sea salt spray, forest wildfires, groundwater movements, soil adsorption and desorption, chemical and biological redox reactions, and mineral formation, but also incineration of municipal waste, copper/nickel production, lead and zinc smelting, iron and steel production, crop growth and irrigation practices, and plant and animal uptake and release [McNeal and Balilestri, 1989; Mayland et al., 1989; Nriagu, 1989]. Estimated Se fluxes indicate that the natural sources of Se emission are as important as anthropogenic emission [Barrow and Whelan, 1989; McNeal and Balilestri, 1989; Masscheleyn et al., 1990; Nriagu, 1989]. Most soils contain between 0.1 and 2 mg Se/kg [Elrashidi et al., 1989; Mayland et al., 1989]. However, elevated concentrations of this metalloid are associated with various environments, notably those of marine sedimentary parent material and soils impacted by industrial activity [Haygarth, 1994; Presser et al., 1994; Kauffman et al., 1986; Weres et al., 1989]. Thus, as a general rule, Se concentration in soil or ground and fresh water depends upon the parent material, climate, topography, age of the soil, and agricultural or industrial utilization. Under acidic, reducing conditions in soils that may be waterlogged and rich in organic matter, elemental Se and selenides (Se^{2-}) are the predominant species [McNeal and Balilestri, 1989]. From pH 4 to 8, stable adsorption complexes or co-precipitates with sesquioxides are prevalent [Ullrey, 1981]. At moderate redox potentials in soil solution, Se(IV) is the predominant form, while at high redox potential in well-aerated, alkaline soils the predominant form is Se(VI) [Elrashidi et al., 1989], which does not form stable adsorption complexes or co-precipitates with sesqioxides [Ullrey, 1981]. Most natural waters have low concentrations of Se, ranging from 0.1 to 100 µg/liter. However, some evaporation ponds in the California San Joaquin Region, USA, have reached levels in excess of 1000 µg/liter [Thompson-Eagle et al., 1989]. Soils in this region were derived from seleniferous cretaceous sediments and contain high levels of Se [Presser et al., 1994] that is mobilized by infiltrating irrigation water that is then discharged to surface water by subsurface drainage or to groundwater by deep percolation. As in soils, in aquatic environment, under most pH and redox conditions, the two oxyanions are dominant with several forms of Se^2 also being present [Cutter and Bruland, 1984]. In these ecosystems, selenium can be absorbed by organisms, can form stable adsorption complexes with particulate/colloidal matter and sediments, or can be dissolved in solution. Most transport processes are governed by movement into and out of the top layer of sediment through biogeochemical processes. In contaminated aquifers showing high biological activity, a locally oxidative environment may occur. These environmental conditions oxidize and solubilize the reduced selenium forms that enter in the food web, and Se levels in biota can remain high for years after inputs have ceased [Lemly, 1997]. Also selenium accumulation in plants and algae from water can be considered a mobilization process, because the Se species are concentrated in a potentially biologically available form that can accumulate through the food chain [Denina et al., 2005; Fan et al., 2002]. On the contrary, processes that immobilize/sequester Se include chemical and microbial reduction of oxidized forms to Se^0 [Schlekat et al., 2000] as well as adsorption of selenate and selenite to clay, minerals, particularly iron, and dissolved organic carbon [Belzile et al., 2000].

2   BIOLOGICAL REACTIONS IN SELENIUM CYCLING

Selenium is predominantly cycled via biological pathways [Shrift, 1973]. Microorganisms and plants activate various reactions of oxidation and reduction that change the chemical properties, the distribution, and the bioavailability of the metalloid in the environment (Fig. 1). Microbial transformations are worth mentioning as among the main reactions involved in selenium cycling. Most of the work to date has focused on the reductive reactions in the Se cycle, whereas much less is known about Se oxidation. Actually, four biological transformations of Se are known to occur: (i) reduction reactions: assimilatory and dissimilatory reduction, (ii) oxidation, (iii) methylation, and (iv) demethylation.

In soil, sediment, and water, Se(IV) and Se(VI) oxyanions, which do not undergo chemical reductions under physiological conditions of pH and temperature, can be reduced to Se^0 by microbial catalyzed reactions [Jajaweera and Biggar, 1996; Oremland et al., 1989]. In fact, selenium is a trace essential element for micro-organisms; however, it can exert a toxic effect and selenium oxyanions reduction to elemental selenium [Lovely, 1993], and selenium methylation and volatilization

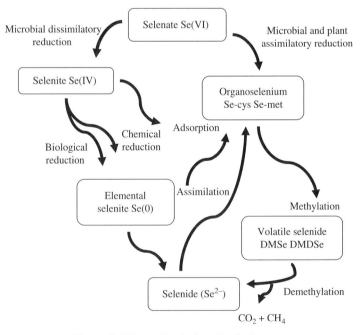

Figure 1. Biogeochemical cycle of selenium.

[Lortie et al., 1992] are two mechanisms of detoxification of the near microbial environment. On the other hand, phototrophic bacteria reduce selenium oxyanions to control reducing equivalents.

2.1 Microbial Assimilatory Reduction

Assimilatory selenium reduction is linked to the importance of selenium as an essential trace element and the importance of selenocysteine (Se-Cys) in the catalytic site of microbial enzymes [Böck et al., 1991]. In fact in bacteria, at least three enzymes containing Se as Se-Cys have been identified: a formate dehydrogenase in *Escherichia coli*, a hydrogenase found in both *Methanococcus vannielii* and *Desulfovibrio baculatus*, and finally a glycine reductase in *Clostridium sticklandii*. All these bacterial enzymes catalyze oxidation/reduction reactions [Stadtman, 1990]. The sulfidryl group of cysteine is mostly protonated at physiological pH, whereas the analogous groups of Se-Cys is dissociated, facilitating the catalytic role of Se in the selenoproteins as a redox center [Combs and Combs, 1984]. The incorporation of Se-Cys in the amino acidic sequence is a co-translational process directed by the UGA codon [Böck et al., 1991]. The UGA nucleotide triplet normally is a universal termination codon. In order for UGA to be a Se-Cys codon, both specific secondary structural elements in the mRNA and a unique Se-Cys-charged tRNA that contains the UGA anticodon are required [Stadtman, 1996]. Although it is recognized that trace amounts of Se are essential for some microbial proteins, when excess amounts of Se are present, the cell begins to indiscriminately substitute Se for its analogous S in cellular components. Because Se compounds are more reactive and less stable than S compounds, the organism begins to experience the toxic effect of excessive Se concentration in its environment.

In order to assimilate selenium, its bioavailable forms enter in the microbial cell. Selenate is believed to enter the cell through the sulfate transport system [Heider and Böck, 1993]. In fact, selenate microbial resistance occurs, mutagenizing the genes responsible for the sulfur transport inside the cell. Selenite enters the cell through at least a further mechanism because repression of sulfate permease expression does not completely inhibit its uptake [Turner et al., 1998].

2.2 Microbial Dissimilatory Reduction

Dissimilatory reduction of selenium oxyanions is actually a broadly distributed capacity among microorganism, comprising the major sink for Se oxyanions in anoxic sediments [Oremland et al., 1989]. Some example follows. *Sulfurospirillum barnesii* strain SES-3 [Oremland et al., 1994], attains respiratory growth with a variety of electron acceptors including As(V), NO_3^-, $S_2O_3^{2-}$, S^0, and Fe^{3+} in addition to Se(VI). However, the strain does not use sulfate as an electron acceptor, indicating that, regardless the similarity between S and Se, the microbial capacity to reduce the two oxyanions is not interdependent [Zehr and Oremland, 1987]. The characterization of SES-3 and other selenate respiring bacteria has clearly shown that the dissimilatory reduction of SO_4^{2-} to SO_3^{2-} and Se(VI) to Se(IV) and Se^0 are achieved by very different microbes, using different biochemical pathways and obtaining much different cellular energy yields. Actually in *Thauera selenatis*, a Se(VI), NO_3^-, and NO_2^- respiring bacterium isolated from seleniferous sediments of the San Joaquin Valley (California, USA) [Macy et al., 1994], the reduction of Se(VI) to Se(IV) and NO_3^- to NO_2^- occurs through the use of separate terminal reductases, the Se(VI) and NO_3^- reductase, respectively. The selenate reductase is a periplasmic protein, a component of the respiring electron transport chain [Krafft et al., 2000] that contains Se in a reduced form, probably organic, although the sequence does not indicate the presence of Se-Cys [Maher et al., 2004]. However, the complete reduction of Se(VI) to Se^0 occurs only when *Thauera selenatis* grows in presence of both Se(VI) and NO_3^- ions. In fact, selenite produced during the concomitant respiration of Se(VI) and NO_3^- is reduced by a periplasmic NO_2^- reductase [DeMoll-Decker and Macy, 1993].

Also, *Enterobacter cloacae* SLD1a-1, a facultative anaerobe [Losi and Frankenberger, 1997], and *Tauera selenatis* use Se(VI) and NO_3^- as terminal electron acceptors during anaerobic growth and reduce Se(VI) to Se^0 in microaerophilic conditions. Although *E. cloacae* respires Se(VI) anaerobically, the complete reduction of Se(VI) to Se^0 did not occur unless NO_3^- was present, suggesting that NO_3^- is necessary for the reduction of Se(IV) to Se^0 and that selenite reduction occurs only when denitrification occurs.

Bacillus sp. strain SF-1 [Fujita et al., 1997] uses Se(VI) as an electron acceptor and lactate as an electron donor in anaerobiosis, showing a stoichiometric balance between cell growth, lactate consumption, and selenate reduction. Strain SF-1 completely reduces up to 1 mM Se(VI) to amorphous Se^0 with transient accumulation of Se(IV). The presence of NO_3^- inhibits the Se(VI) reduction, suggesting a common enzyme for both reactions.

An *Enterobacter taylorae* strain was isolated from a rice straw bioreactor channel system tested to remove Se(VI) from agricultural drainage water [Zahir et al., 2003]. The strain is capable to reduce Se(VI) to Se^0. However, NO_3^- negatively affects Se(VI) reduction efficiency, confirming that NO_3^- may be a competitive electron acceptor in Se(VI) reduction to Se^0 in anaerobic environment [Zhang et al., 2003].

Wolinella succinogenes has been described to reduced Se(VI) and Se(IV) to Se^0 in solution culture after reaching the stationary growth phase [Tomei et al., 1992], but the two oxyanions, although enzimatically reduced, are not the terminal electron acceptors. A similar situation has been observed for *Desulfovibrio desulfuricans*, suggesting the involvement of a detoxification mechanism for either *Desulfovibrio* or *Wolinella* [Tomei et al., 1995].

Dissimilatory selenate reduction is distributed also among extremophilic microorganisms that also reduce selenium Se(VI), such as *Pyrobaculum arsenaticum*, a hyperthermophilic archaeon [Huber et al., 2000], the halophile *Selenihalanaerobacter shriftii* [Blum et al., 2001], *Bacillus arsenicoselenatis* strain E-1H, and *Bacillus selenitireducens* strain MLS-10, isolated from Mono Lake, California, an alkaline and hypersaline soda lake [Switzer-Blum et al., 1998]. Thus, the dissimilatory selenate reduction sustaining growth in anaerobic environments is actually a metabolic strategy fairly distributed among microbial different genera.

The metabolic strategy is usually associated with the use of short-chain fatty acids or H_2 as electron donor. Noteworthy is the capacity of *Sedimenticola selenatireducens* strain AK4OH1, isolated from estuarine sediment, that uses 4-hydroxybenzoate as the carbon and energy source [Narasingarao and Haeggblom, 2006]. On the other hand, the dissimilatory selenite reduction sustaining growth in anaerobiosis seems at present to be a metabolic strategy restricted to a narrower microbial biodiversity. In addition to the here reported examples of these process, microorganisms isolated from a terrestrial solfatara has been ascribed to be capable to sustain growth by the dissimilative reduction of selenite [Yurkov and Csotonyi, 2003]. However, any clear evidence of the presence of a selenite reductase has been collected to date, and the reduction of Se(IV) is hardly dissociated from reduction of NO_3^- [Kessi, 2006].

Recent results shows that lagoon sediments at mine sites harboring autochthonous Se(VI) and Se(IV) reducing bacteria are important sinks for soluble Se. Some microbial strains will certainly find application in removing Se from industrial effluents [Siddique et al., 2006, 2007].

2.3 Detoxification of Se Oxyanions by Reduction Reactions in Aerobiosis

The aerobic reduction of Se oxyanions to Se^0, not supporting microbial growth, is actually a mechanism of detoxification of the microbial growth milieu [Lovely, 1993]. In the last two decades, several bacteria capable of reducing selenium oxyanions to Se^0 in aerobic growth conditions have been isolated from seleniferous soil, sediments, and waters. Some examples are reported in Table 1 [Di Gregorio

TABLE 1. Selenium Reducing Bacteria in Aerobic Growth Condition

Organism	Catalyzed Reaction	Reference
Pseudomonas stutzeri	Reduces Se(IV) and Se(VI) to Se^0 under aerobic growth conditions	Lortie et al. [1992]
Stenotrophomonas maltophilia	Reduces Se(IV) and Se(VI) to Se^0 under aerobic and microaerophilic growth conditions	Dungan et al. [2003]
Ralstonia metallidurans	Reduces Se(IV) and Se(VI) to Se^0 and transforms Se(VI) to organic selenium compounds under aerobic growth conditions	Roux et al. [2001], Sarret et al. [2005]
Stenotrophomonas maltophilia SeITE02	Reduces Se(IV) to Se^0 showing the highest minimal inhibitory concentration to date (50 mm)	Di Gregorio et al. [2005]

et al., 2005; Dungan et al., 2003; Lortie et al., 1992; Rathgeber et al., 2002; Roux et al., 2001; Sarret et al., 2005]. Although microorganisms capable of reducing Se(VI) to Se^0 have been isolated, Se(IV) is more easily reduced than Se(VI) [Doran, 1982]. In both cases, aeration is necessary. Among the different microbial genera associated with the described detoxification mechanisms, *Ralstonia metallidurans* merits a particular interest. It has been initially isolated from a metal-rich sediment of a zinc factory in Belgium and has been demonstrated to prevail in industrial anthropogenic biotypes such as metallurgic wastes [Mergeay et al., 1985]. It harbors plasmid-borne multiple resistance to several heavy metals and oxyanions, and it is able to reduce both Se oxyanions to elemental selenium in aerobic growth conditions [Roux et al., 2001; Sarret et al., 2005]. However, kinetic reduction studies demonstrated that the reduction reaction to Se^0 occurs preferentially when selenite was present in the medium. Furthermore, selenate is reduced by an assimilative reduction process with the production of Se-alkyl compounds rather than elemental selenium forms [Sarret et al., 2005].

The need to reduce selenium oxyanions to Se^0 is related to the toxicity of the oxidized forms. Much has been done to elucidate mechanisms of selenium toxicity, and much has to be still elucidated. However, it has been proposed that selenium toxicity is in part due to its interaction with vicinal thiols in proteins and consequently production of reactive oxygen species (ROS), such as O^{2-} and H_2O_2. In fact, in vitro studies [Seko et al., 1989] have shown that the reduction of selenite involves reactions with sulfidryl groups of thiol-containing molecules such as glutathione (GSH), leading to the production of intermediate metabolites selenodiglutathione (GS-Se-SG), glutathioselenol (GS-SeH), hydrogen selenide (HSe-), and finally Se^0 (Fig. 2). The described pathway produces ROS that damage cell membranes and DNA [Seko et al., 1989]. Actually the oxyanion that is involved in the process is strictly selenite. Selenate is toxic only after being reduced to selenite. Morover, a number of different selenium compounds deriving from the entry of Se in sulfur cellular metabolism have been tested in order to determine their reaction with GSH to produce ROS [Spallhollz, 1994]. The hypothesis gained from this study was that (i) selenium compounds (i.e., selenite and selenium dioxide) can react with GSH and other thiols to form selenotrisulfides that will ultimately react to produce superoxide and hydrogen peroxide, resulting in the toxicity of the mentioned selenocompounds; (ii) diselenides (i.e., selenocystine and selenocystamine), in the presence of GSH and other thiols, are reduced to selenols (RSeH), highly reactive and toxic; (iii) selenium compounds that do not react with thiols [i.e., selenate and all tested selenoethers (RSeR)] do not produce superoxide or hydrogen peroxide in vitro and

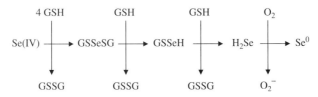

Figure 2. Reaction proposed by Seko et al. [1989] determining selenium toxicity.

are not toxic per se; (iv) selenate and selenoethers are toxic in tissue cultures or in vivo only after being reduced to selenite or selenol; (v). Se toxicity manifests itself acutely or chronically when oxidation damages excess antioxidant defenses or the ability of microbes, plants, or animals to form selenoproteins, selenoethers, or elemental selenium.

Actually, glutathione has been demonstrated to be involved in the reduction of selenite also in *E. coli* [Rabenstein and Tan, 1988]. Glutathione is present at millimolar levels in cyanobacteria and the alpha-, beta-, and gammaproteobacteria [Newton and Fahey, 1989]. The high reactivity of glutathione with selenite [Ganther, 1971; Kessi and Hanselmann, 2004] makes glutathione an obvious candidate for participating in the dissimilatory reduction of selenite in organisms that are able to synthesize this metabolite. Supporting this hypothesis, it has been reported that the synthesis of the glutathione reductase is induced in cultures of *Rhodobacter sphaeroides* and *E. coli* amended with selenite [Bebien et al., 2001, 2002]. Morerover, GS–Se–SG has also been shown to be a highly efficient oxidant of reduced bacterial thioredoxin [Bjoernstedt et al., 1992]. Thioredoxin and thioredoxin reductase may also be involved in the reduction of selenite in bacteria that are able to synthesize glutathione. Actually, a large induction of the synthesis of thioredoxin and thioredoxin reductase in *E. coli* amended with selenite has been reported by Bebien et al. [2002]. According to their results, glutathione and glutathione reductase and/or thioredoxin and thioredoxin reductase may be involved in the dissimilatory reduction of selenite in organisms containing high levels of glutathione.

However, it must be pointed out that the involvement of glutathione in the selenite reduction pathway in bacteria is restricted to the alpha-, beta-, and gammaproteobacteria and to the cyanobacteria, which are the only bacteria able to synthesize glutathione [Newton and Fahey, 1989]. In all other bacterial species, which lack the glutathione-dependent reduction system, the pathway for the dissimilatory reduction of selenite must be different from that of the alphaproteobacteria. High levels of other thiols, including coenzyme A, cysteine, lipoamide, and pantetheine, with their corresponding thiol/disulfide oxidoreductases [Newton and Fahey, 1989] could be involved.

2.4 Regulation of Reducing Equivalents

Phototrophic bacteria maintain the intracellular redox potential by the continuous removal of electrons [Moore and Kaplan, 1992, 1994]. During anaerobic growth in the light, the buildup of the pool of reducing equivalents occurs, definitely affecting photosynthesis. One mechanism to minimize the damage consists in removing electrons in excess by the reduction of oxidized forms, among other metal oxyanions. A side effect of this process is that the cells are resistant to high levels of heavy metal oxides like $Cr_2O_7^{2-}$; Rh_2O_3; Eu_2O_5; TeO_4^{2-}; TeO_3^{2-}; AsO_4^{3-} and Se(VI) [Moore and Kaplan, 1994]. The process definitely determines the reduction of Se oxyanions to elemental selenium. Although the exact mechanism has not been worked out, Moore and Kaplan [1992] claim the intervention of a membrane-bound, $FADH_2$-dependent metal reductase.

2.5 Oxidation of Reduced Se Forms

On the other hand, the oxidation of reduced forms of Se to Se oxyanions also occurs in natural environments. The process has been studied in several soil and sediment [Dowdle and Oremland, 1998; Losi and Frankenberger, 1998; Masscheleyn et al., 1990; Tokunaga et al., 1996]. However, little information is available on the oxidation mechanisms of the reduced Se compounds and on organisms that carry out the transformation. Losi and Frankenberger [1998] demonstrated that the oxidation of Se^0 to Se(VI) and Se(IV) was largely biotic in nature and occurs at relatively slow rates. The oxidation of Se^0 is carried out by both heterotrophic and autotrophic organisms. Dowdle and Oremland [1998] studied the oxidation of Se^0 in oxic soil slurries and found that Se(IV) was the main product, with lesser quantity of Se(VI) being produced. The microbial oxidation of Se^0 to Se(IV), by a group of unidentified autotrophic bacteria, was first reported by Lipman and Waksman [1923]. Torma and Habashi [1972] described the oxidation of copper (II) selenide to Se^0 by *Thiobacillus ferrooxidans*. Saratchandra and Watkinson [1981] reported the oxidation of Se^0 to Se(IV) and trace amounts of Se(VI) by the heterotrophic bacterium, *Bacillus megaterium*. The oxidation of Se^0 and other Se reduced forms has not been fully addressed and it is still an area of vast research opportunities.

2.6 Selenium Volatilization, Se Methylation and Demethylation

The formation of volatile methylated Se compounds from Se oxyanions and organo-Se compounds is known to occur in seleniferous soil, sediment, and water [Doran, 1982]. The methylation of Se has been shown to be mainly a biotic process and is primarily thought to be a protective mechanism to detoxify the surrounding environment. The predominant groups of Se-methylating organisms isolated from soils and sediments are bacteria and fungi [Doran, 1982; Karlson and Frankenberger, 1988]. Bacteria are thought to be the major Se methylating organisms in water [Thomson-Eagle and Frankenberger, 1991]. The main biotic volatile Se form is dimethylselenide (DMSe) [Doran, 1982; Karlson and Frankenberger, 1988; Thomson-Eagle and Frankenberger, 1991]. Other volatile Se compounds produced in much smaller amounts are dimethyldiselenide (DMDSe), dimethylselenylsulfide (DMSeS), methaneselenone, and methane selenol. Although the biological significance of Se methylation is not clearly understood, once volatile Se compounds are released to the atmosphere the metalloid is diluted and definitely loses its hazardous potential.

In soil and water, DMSe undergoes biological demethylation that can be defined as the removal of a methyl group from the central atom of a methylated compound. Several soil microorganisms have been isolated that are capable of demethylating volatile Se compounds. Doran and Alexander [1977] isolated from seleniferous clay a *Pseudomonas* strain able to demethylate tetramethylselenide (TMSe), two strains capable to demethylate DMSe, and strains of *Xanthomonas* and

Corynebacterium that were able to grow on DMDSe as the sole carbon source. Oremland et al. [1989] proposed that methylotrophic bacteria carry out demethylation, while certain hydrogen-oxidizing methanogens may be involved in reductive methylation. The magnitude of demethylation reactions in seleniferous environments has yet to be determined and may be of significant interest if Se volatilization is to be implemented as a remediation technique.

3 SELENIUM IN HUMANS AND ANIMALS

Selenium is an essential trace element also for humans and animals, however, even apparently low concentrations of selenium, in the order of few ppm, can provoke health disturbances [Vinceti et al., 2001]. Humans and animals require selenium for the function of a number of selenium-dependent enzymes. Also in humans and animals Se-Cys is incorporated into a very specific location in the amino acid sequence of selenoproteins, and in humans and animals the uncontrolled S substitution of Se may cause toxic effect. At least 11 selenoproteins have been characterized in animal systems. However, there is evidence that additional selenoproteins may be recovered. The first to be characterized was the glutathione peroxidases; more precisely, four selenium-containing glutathione peroxidases (GPx) have been identified: cellular or classical GPx, plasma or extracellular GPx, phospholipid hydroperoxide GPx, and gastrointestinal GPx [Holben and Smith, 1999]. These enzymes reduce damaging reactive oxygen species (ROS) oxidizing glutathione. A well-characterized example of this enzyme family is the sperm mitochondrial capsule selenoprotein, a phospholipid hydroperoxide GPx [Ursini et al., 1999]; this is an antioxidant enzyme that protects developing sperm from oxidative damage and is responsble for a structural protein required by mature sperm.

Se-Cys is present also in the active site of the thioredoxin reductase that, in conjunction with the compound thioredoxin, participates in the regeneration of several antioxidant systems in animal cells, possibly including vitamin C [Mustacich and Powis, 2000]. Moreover, maintenance of thioredoxin in a reduced form by thioredoxin reductase is important for regulating cell growth and viability [Holben et al., 1999; Mustacich and Powis, 2000].

Selenium-dependent iodothyronine deiodinase enzymes remove one iodine atom from thyroxine or T_4 to obtain the biologically active triiodothyronine or T_3. Three different selenium-dependent iodothyronine deiodinases (types I, II, and III) has been characterized that through their actions on T_3 and T_4 can both activate and inactivate thyroid hormone, thus making selenium an essential element for normal development, growth, and basal metabolism [Holben et al., 1999; Larsen et al., 1998].

Incorporation of selenocysteine into selenoproteins in humans and animals requires the enzyme selenophosphate synthetase. A selenoprotein itself, selenophosphate synthetase catalyzes the synthesis of monoselenium phosphate, a precursor of selenocysteine which is required for the synthesis of selenoproteins [Rayman, 2000; Holben et al., 1999].

Furthermore, selenoprotein P has been identified in plasma; its function has not been clearly delineated, and it has been suggested to function as a transport protein and also as an antioxidant capable of protecting endothelial cells from damage by a reactive nitrogen species (RNS) [Holben et al., 1999; Arteel et al., 1999].

On the other hand, a selenoprotein W has been found in muscle, whose function is presently unknown but thought to be related to muscle metabolism [Holben et al., 1999].

Deficiency in selenium intake in animals and humans principally results in decreased activity of glutathione peroxidases, and selenium-deficient individuals appear to be more susceptible to various physiological stresses [Burk et al., 1999]. Selenium deficiency is associated to muscular weakness and cardiomyopathy. These symptoms might be present at high frequency in chronically ill patients who were receiving total parenteral nutrition without added selenium for prolonged periods of time. People with large portion of the small intestine surgically removed or those with severe gastrointestinal problems, such as Crohn's disease, are also at risk for selenium deficiency due to impaired absorption of the trace element [Burk et al., 1999].

In selenium-deficient regions of the planet such as part of China, Keshan disease, a cardiomyopathy, affects young women and children. The acute form of the disease is characterized by the onset of cardiac insufficiency, while the chronic form results in moderate to severe heart enlargement with varying degrees of cardiac insufficiency. The incidence of Keshan disease is closely associated with the very poor selenium nutritional status. Selenium supplementation has been found to protect people from developing Keshan disease but cannot reverse heart muscle damage once it occurs [Burk et al., 1999; Foster and Sumar, 1997]. Instead in selenium-deficient areas like North Korea and eastern Siberia, Kashin–Beck disease has a high incidence. The disease is characterized by osteoarthritis that affects children. Severe forms of the disease may result in joint deformities and dwarfism. Unlike Keshan disease, there is little evidence that improving selenium nutritional status prevents Kashin–Beck disease [Burk et al., 1999; Foster and Sumar, 1997].

On the other hand, selenium appears to stimulate the immune response [Roy et al., 1994; Kiremidjian-Schumacher et al., 1994] in humans, playing a role in regulating the expression of cell signaling molecules called cytokines, which orchestrate the immune response [Baum et al., 2000].

Many studies suggest that selenium supplementation at high levels reduces the incidence of cancer in animals and that the methylated forms of selenium are the active species against tumors [Combs and Gray, 1998]. Results of epidemiological studies of cancer incidence in groups of humans with variable selenium intakes show a trend for individuals with lower blood selenium levels to have a higher incidence of several different types of cancer. The trend is less pronounced in women. Several mechanisms have been proposed for the cancer prevention effects of selenium: (1) maximizing the activity of antioxidant selenoenzymes and improving antioxidant status, (2) improving immune system function, (3) affecting the metabolism of carcinogens, and (4) increasing the levels of selenium metabolites that inhibit tumor cell growth, an interesting developing aspect of research in selenium reviewed in the next paragraph.

A two-stage model has been proposed to explain the different anticarcinogenic activities of selenium at different doses. At nutritional or physiologic doses ($\sim$40–100 μg/day in adults), selenium maximizes antioxidant selenoenzyme activity and probably enhances immune system function and carcinogen metabolism. At supranutritional or pharmacologic levels ($\sim$200–300 μg/day in adults) the formation of selenium metabolites, especially methylated forms of selenium, may also exert anticarcinogenic effects [Fleming et al., 2001].

Selenium recommended dietary allowance changes with human life stage. In fact, although selenium is required for health, high doses can be toxic. Acute and fatal toxicities have occurred with accidental or suicidal ingestion of gram quantities of selenium. Chronic selenium toxicity (selenosis) may occur with smaller doses of selenium over long periods of time. The most frequently reported symptoms of selenosis are hair and nail brittleness and loss, gastrointestinal disturbances, skin rashes, a garlic breath odor, fatigue, irritability, and nervous system abnormalities.

4 SELENIUM IN PLANTS

Nowadays, it has been demonstrated that also higher plants metabolize Se via the sulfur assimilation pathway [Zayed et al., 2000], synthesizing Se analogues of various S metabolites, as reviewed by Ellis and Salt [2003]. This involves the fact that the nonspecific incorporation of Se into selenoamino acids and proteins as well as Se volatilization occurs, when the metalloid is supplied to plants in excess of any potential, but not demonstrated, requirement. Plants differ in their ability to accumulate Se in their tissues. Actually, certain native plants are able to hyperaccumulate Se in their shoots when they grow on seleniferous soil. These species are called Se hyperaccumulators and include a number of species of *Astragalus*, *Stanleya*, *Morinda*, *Nepturia*, *Oonopsis*, and *Xylorhiza* [Beath et al., 1934; Brown and Shrift, 1982; Trelease and Trelease, 1939]. On the other hand, most forage and crop plants, as well as grasses, contain less than 25 mg Se/kg dry weight and do not accumulate Se much above 100 mg Se/kg dry weight when grown on seleniferous soils. These plants are referred to as Se nonaccumulators [Brown and Shrift, 1982; White et al., 2004]. A third category of plants, known as secondary Se accumulators [Brown and Shrift, 1982], grow on soil of low-to-medium Se content and accumulate up to 1000 mg Se/kg dry weight. Examples of plants in this group are species of *Aster*, *Astragalus*, *Atriplex*, *Castilleja*, *Comandra*, *Grayia*, *Grindelia*, *Gutierrezia*, and *Machaeran hera* [Parker and Page, 1994]. Indian mustard (*Brassica juncea*) and canola (*Brassica napus*), are also secondary Se accumulator plant species with a typical Se concentration of several hundred micrograms of Se/g dry weight in their shoots when grown on soils containing moderate levels of Se [Banuelos et al., 1997].

Although there are evidences that Se is required for the growth of algae [Price et al., 1987; Yokota et al., 1988], the question of the essentiality of Se as a micronutrient in higher plants is controversial. While there are indications that Se may be required for Se-accumulating plants [Broyer et al., 1972a,b], which are actually endemic on seleniferous soils, there is no evidence for a Se requirement in nonaccumulators [Shrift, 1969]. In order to investigate the essentiality of Se in higher plants, attempts have been made to establish whether plants contain essential seleno-proteins, such as those discovered for bacteria and animals. However, on the basis of all the available information published to date, no essential higher plant selenoprotein has been clearly identified by either protein or DNA sequences analysis [Terry et al., 2000]. Obscure is also the ecological significance of selenium hyperaccumulators except the hypothesis that protection from insect herbivore is a major ecological advantage of Se hyperaccumulation. The hypothesis is corroborated by the evidence that accumulation in leaves of two very different forms of Se, inorganic selenate in nonaccumulator plants and organic Se-methylcysteine in hyperaccumulators, had similar protective effects against herbivores. Additionally, an intermediate Se concentration caused an intermediate degree of herbivore protection. [Freeman et al., 2007]. These results shed light on the possible selection pressures that have driven the evolution of Se hyperaccumulation.

The rate and form of Se uptake by plants depend on the concentration and chemical form of Se in soil solution, as well as rhizosphere conditions such as pH. Sulfate and phosphate in the rhizosphere compete with Se for plant uptake [Bell et al., 1992; Blaylock and James, 1994; Dhillon and Dhillon, 2003]. Se is taken up in the following order: Se(VI) > organic Se compounds > Se(IV) [Asher et al., 1977; Martin et al., 1971; Shrift and Ulrich, 1969; White et al., 2004]. Both Se(VI) and organic Se compounds absorption in plants from the soil solution are active processes, whereas Se(IV) seems to be accumulated through passive diffusion and can be inhibited by phosphate [Abrams et al., 1990; Shrift and Ulrich, 1969]. On the other hand, the translocation of Se from root to shoot is also dependent on the form of Se supplied. Se(VI) is transported much more easily than Se(IV), or organic Se, such as SeMet. Zayed et al. [1998] showed that the shoot Se/root Se ratio ranged from 1.4 to 17.2 when Se(VI) was supplied, but was only 0.6 to 1 for plants supplied with SeMet and less than 0.5 for plants supplied with Se(IV). Arvy [1993] demonstrated that within 3 hr 50% of the Se(VI) taken up by bean plant roots moved to shoots, whereas in the case of Se(IV), most of the Se remained in the root and only a small fraction was found in the shoot. Time-dependent kinetics of Se uptake by *Brassica juncea* showed that only 10% of the Se(IV) taken up was transported from root to shoot, whereas Se(VI), which was taken up twofold faster than Se(IV), was rapidly transported into shoots [de Souza et al., 1998]. Thus, plants transport and accumulate substantial amounts of Se(VI) in leaves but much less Se(IV) or SeMet. The reason why Se(IV) is poorly translocated to shoots may be because it is rapidly converted to organic forms of Se such as SeMet [Zayed et al., 1998], which are retained in the roots. After Se(VI) is absorbed into the root via its transporter, it is translocated without chemical modification through the xylem to leaves [de Souza et al., 1998; Zayed et al., 1998]. Once inside the leaf, Se(VI) enters chloroplasts where it is metabolized by the enzymes of SO_4^{2-} assimilation. The first step in the reduction of Se(VI) is its activation by ATP sulfurylase to APSe (adenosine phosphoselenate). The ATP sulfurylase is rate-limiting for both Se(VI) reduction and Se accumulation [Pilon-Smits et al., 1999].

Subsequently, APSe can be reduced either nonenzymatically to GSH-conjugated selenide (GS-Se-) or enzymatically to selenide (Se^{2-}) via APS reductase and sulfite reductase. At the same time, Se(IV) is reduced to selenide via a nonenzymatic reaction to selenodiglutathione (GS-Se-SG), which is reduced to the selenol (GS-SeH). GSH-conjugated selenide is a substrate for Cys-synthase that synthesize Se-Cys [Ng and Anderson, 1978], which is the precursor of Se-Met [Droux et al., 2000]. Successively selenium is readily incorporated into proteins. The incorporation into proteins occurs through the nonspecific substitution of Se-Cys and Se-Met in place of Cys and Met, respectively [Brown and Shrift, 1982], which also in plants actually determines the toxic effect of the metalloid. However, SeMet may be also methylated to Se-methylSeMet and subsequently originate DMSe, the volatile Se organic compound. The most likely enzyme responsible for this reaction is S-methylMet hydrolase, which produces DMS from S-methylMet in higher plants [Gessler and Bezzubov, 1988]. Another possible pathway for DMSe production is via the intermediate dimethylselenoniopropionate (DMSeP) [Ansede and Yoch, 1997].

Regardless of the common traits of Se metabolism in plants, the existence of their different capacities to accumulate the metalloid is evidently related to different metabolic strategies. Strategies adopted by Se secondary accumulators and hyperaccumulators are consistent with two hypothesis. The first hypothesis reports the capacity of a plant to sequester Se in peripheral cells to exclude toxic Se from other, more sensitive tissues like the parenchyma and to allow for long-term storage of high concentrations of Se inside specialized cells. This strategy has been observed in the case of plant exposure at Ni and Zn [Kuepper et al., 2000, 2001]. The second hypothesis is related to the specific conversion of potentially toxic selenoamino acids into nonprotein derivatives such as Se-methylselenocysteine MeSeCys [Viropaksa and shrift, 1965; Wang et al., 1999]. Some *Brassica* and *Allium* species, when grown in Se-enriched medium, can accumulate $0.1–2.8\,\mu mol/g$ dry weight MeSeCys or its functional equivalent γ-glutamylmethylselenocysteine (γGluMeSeCys) [Dong et al., 2001; Ip et al., 2000; Medina et al., 2001; Kotrebai et al., 2000; Whanger et al., 2000; Whanger, 2002]. However, certain specialized Se-accumulating plants, such as *A. bisulcatus*, accumulate up to $68\,\mu mol/g$ dry weight Se ($6000\,\mu g/g$ dry weight), of which $90–95\%$ is MeSeCys in young leaves [Orser et al., 1999; Pickering et al., 2001, 2003]. Numerous studies have demonstrated the efficacy of MeSeCys in preventing mammary cancer in rat model systems [Dong et al., 2001; Finley et al., 2001; Ip and Ganter, 1992a,b; Ip et al., 2000; Lu et al., 1996; Medina et al., 2001]. These studies are at the base of the hypothesis that cancer prevention effects of selenium might be related to the biological production of selenium metabolites that inhibit tumor cell growth. Thus understanding mechanisms of selenium hyperaccumulation should lead to the development of anticarcinogenic Se-fortified crops.

5 SELENIUM OF ENVIRONMENTAL CONCERN: EXPLOITATION OF BIOLOGICAL PROCESSES FOR TREATMENT OF SELENIUM POLLUTED MATRICES

It is a matter of the fact that, besides the positive effect of the metalloid on biological systems, uncontrolled release of selenium in the environment may be troublesome to living organisms. Because Se undergoes microbial and plant transformations, their application may be potentially useful to the development of bioremediation strategies. Some example of real scale plants, based on Se bio-cycling, dedicated to Se-laden matrices will be described in the next paragraphs. Several different approaches have been developed, which include a variety of bioreactors utilizing bacteria with the ability to reduce the toxic, soluble Se oxyanions to insoluble Se^0. These systems are designed to remove Se from contaminated wastewater before their release into the environment. Another means to remove Se from contaminated soil and water involves stimulation of the autochthonous microbial community to volatilize Se. This process has proven to be effective as an in situ treatment for seleniferous soils in the San Joaquin Valley, California [Flury et al., 2001; Frankenberger and Karlson, 1995]. On the other hand, also plants have been shown to be highly effective in the remediation of Se contaminated soils and waters for their capacity to translocate Se from contaminated matrices to shoots, which can be harvested and removed from the site. Alternatively, once Se is in the shoots, it may be volatilized. Constructed wetlands are the additional technique for treating Se-laden wastewaters. The technology is based on the auto-depurative capacity of the complex ecosystems associated to wetlands. Reproducing and engineering wetlands in artificial environment end up with 90% of removal of Se from oil refinery effluents [Hansen et al., 1998].

5.1   Microbe-Induced Bioremediation

Oremland [1991] described a patented plant, based on a two-stage reduction process that uses algae in the aerobic process to deplete NO_3^- concentrations in contaminated water to <1 mM. The water is then fed to an anoxic bioreactor containing Se(VI) respiring bacteria where Se(VI) is reduced to insoluble Se^0 and deposited within the biomass. The Oswald Process [Gerhardt et al., 1991] uses aerobic algal growth to take up NO_3^-. The biomass suspension is then transferred to an anoxic unit where denitrifying and Se(VI) respiring bacteria carry out reduction of Se(VI) to Se(IV) rather than to Se^0. After the anoxic reduction step, $FeCl_3$ is added to precipitate Se(IV). The use of *Thauera selenatis* as Se(VI) respiring bacterium in a reactor system to remediate both selenium oxyanions from contaminated water using acetate as the electron donor has been developed by different groups such as Macy et al. [1993], Lawson and Macy [1995], and Cantafio et al. [1996]. Significant results were obtained with a pilot scale system, which consisted of a series of four medium-packed tanks, used to treat seleniferous agricultural drainage water [Cantafio et al., 1996]. NO_3^- was present in the system for Se(IV) completely reduction to Se^0, because of the evidence that nitrite reductase is capable to reduce selenite but the reaction takes place only when denetrification occurs.

Volatilization of Se is also a potentially important microbial process in relation to bioremediation. The process removes Se from soil and water permanently. Microbial volatilization of Se has been tested as a bioremediation approach to remove toxic levels of Se in soils at Kesterson Reservoir, California. Field investigations were conducted with the goal of identifying the most effective practices for accelerating volatilization and to obtain information necessary for the determination of time and factors affecting this technology [Flury et al., 2001; Frankenberger and Karlson, 1995]. At Kesterson Reservoir, 68–88% of the total Se inventory (0–15 cm) was dissipated over 100 months [Flury et al., 2001]. The highest Se removal rates were observed in soils amended with casein. At the Sumner Peck Ranch (California), 32% of the Se was removed over 22 months from the de-watered seleniferous sediment with the application of moisture and tillage, and this was increased to 59% with the addition of a carbon source [Frankenberger and Karlson, 1995]. Recently, Banuelos and Lin [2007] proposed milk as an organic amendment to improve selenium volatilization.

5.2 Selenium Plant-Assisted Bioremediation (Phytoremediation)

Phytoremediation is defined as the use of green plants to remove pollutants from the environment or to make them harmless [Kramer, 2005; Salt et al., 1998].

There are two distinct strategies in soil phytoremediation: phytostabilization and phytoextraction [Salt et al., 1998]. The former is used to provide a cover of vegetation for a moderately to heavily contaminated site, thus preventing wind and water erosion. Plants suitable for phytostabilization develop an extensive root system, provide good soil cover, show tolerance to the contaminants, and ideally immobilize them in the rhizosphere. Phytostabilization is often performed using species from plant communities occurring on local contaminated sites.

On the other hand, phytoextraction involves the cultivation of tolerant plants to the contaminants that concentrate them in their above-ground tissue that will be disposed. Bioconcentration capacity of phytoextracting plant species is the key factor for successful exploitation. Many species have been evaluated for their efficacy in selenium phytoremediation [Banuelos and Meek, 1990; Bell et al., 1992; Mayland et al., 1989; Wu et al., 1988]. Se hyperaccumulator plants are naturally capable of accumulating Se in their above-ground tissues, without developing any symptoms of toxicity [Baker and Brooks, 1989]. The concentration of Se in dry leaf biomass is usually up to 100-fold higher than the concentration in the polluted matrix [McGrath and Zhao, 2003], and the Se in shoot/root ratio is above one [Zayed et al., 1998]. However, Se hyperacumulators are generally slow-growing plants, and no agronomical practice is standardized yet. Moreover, the Se form accumulated in shoots is mostly soluble and can easily be leached from plant tissue back to the soil by rainfall [Cowgill, 1990]. *Brassica juncea*, which tipically contains 350 mg Se/kg dry weight, finds application in Se phytoremediation because of its capacity to accumulate and volatilize large amounts of Se. Moreover, this plant species grows rapidly and produces a large biomass, tolerates stressful growth conditions, and may provide a safe source of forage for Se-deficient livestock [Banuelos and Meek, 1990; Zayed et al., 2000]. In order to increase the bioconcentration factor of *B. juncea*, the plant species has been genetically modified to overexpress selenocysteine lyase (cpSL) and selenocysteine methyltransferase (SMT) and field trailed. The cpSL transgenic plants accumulated 2-fold more Se in the shoots and grew better on contaminated soil than did wild-type plants. At the same time, the SMT transgenic plants accumulated 1.6-fold more Se in their shoots than did wild-type plants. [Banuelos et al., 2007].

In the context of plant specie intervention in bio-based protocol to treat seleniferous matrices, the metalloid removal by constructed wetlands merit a particular interest. Constructed wetlands constitute a complex ecosystem, the biological and physical components of which interact to provide a mechanical and biogeochemical filter capable of removing many different types of contaminants from water. In the United States [Brown and Reed, 1994] and Europe [Haberl et al., 1995], constructed wetlands have been used as a low-cost treatment to remove a wide range of contaminants from polluted waters such as municipal wastewater and effluents from electricity-generating facilities and oil refineries. The anoxic/oxic environment and organic matter production in wetlands promote biological and chemical processes that transform contaminants to immobile or less toxic forms [Gao et al., 2003]. Plants support microbial-mediated transformations of contaminants by supplying fixed carbon as an energy source for bacteria and by altering the chemical environment in their rhizosphere [Oremland et al., 1990; Terry et al., 1992].

Plants also take up and accumulate metals and metalloids in their tissues [Zhu et al., 1999]. Once entering plant tissue, some metals and metalloids can be metabolized to nontoxic and/or volatile forms, which may escape the local ground ecosystem by release to the atmosphere [Hansen et al., 1998]. This is specifically the case for selenium. The first indication that wetlands might be useful in the removal of Se from wastewaters came from a study of 36-ha constructed wetland located adjacent to San Francisco Bay, California. Analysis of the wetland inlet and outlet waters showed that the constructed wetland was successful in removing at least 70% of the Se from the wastewater passing through it [Duda, 1992]. In 1996, an experimental wetland was constructed at the Tulare Lake Drainage District in the San Joaquin Valley. Its purpose was to evaluate the potential of constructed wetlands for the removal of Se from agricultural irrigation drainage waters. Ten individual cells were tested, either unvegetated or vegetated singly or with a combination of sturdy bulrush [*Schoenoplectus robustus* (Pursh) M.T. Strong], Baltic rush (*Juncus balticus* Willd.), smooth cordgrass (*Spartina alterniflora* Loisel.), rabbitfoot grass [*Polypogon monspeliensis* (L.) Desf.], saltgrass [*Distichlis spicata* (L.) Greene], cattail (*Typha latifolia* L.), tule [*Schoenoplectus acutus* (Muhl. ex Bigelow)], and widgeon grass (*Ruppia maritima* L.) [Gao et al., 2003]. On average, the wetland cells removed 69% of the total Se mass from the inflow. Vegetated wetland cells removed Se more efficiently than the unvegetated cell, without significant differences among vegetated cells [Lin and Terry, 2003]. Moreover, microcosm experiments has been used to evaluate the potential of constructed wetlands to remediate effluent containing highly toxic selenocyanate (SeCN), As, and boron (B) generated by a coal gasification plant [Ye et al., 2003]. The microcosms removed 79% Se, 67% As, 57% B, and 54% CN mass, significantly reducing the toxicity of the effluent. Because cattail (*Typhia latifolia* L.), *Thalia dealbata Fraser ex Roscoe*, and rabbit foot grass (*P. monspeliensis L. Desf.*) showed no growth retardation when supplied with the contaminated wastewater, constructed wetlands planted with these species are particular promising for remediating this highly toxic effluent. The potential of further plant species, *Phragmites australis* (common reed) and *Typha latifolia* (cattail), in the phytoremediation process of selenium was studied in subsurface-flow constructed wetland (SSF). The results obtained indicated that common reed is a very good species for Se phytoextraction and phytostabilization and that cattail is only a phytostabilization species [Azaizeh et al., 2003]. However, although constructed wetlands offer a less expensive alternative to other water-treatment methods, the approach needs to be optimized to enhance efficiency and reproducibility and reduce ecotoxic risk. Most of the contaminants removed from the waste stream are immobilized in the sediment. As an example, microcosm experiments previously described showed that the sediment contained 63% of the Se, 51% of As, and 36% of B, while only 2–4% was accumulated in plant tissue [Ye et al., 2003], the only part of the system that can be easily disposed of. In the Tulare Lake Drainage District wetlands, 41% of the supplied Se left the wetlands; the remaining 59% was retained in the wetland cells, partitioned between the surface sediment (0–20 cm; 33%), organic detrital layer (18%), fallen litter (2%), standing plants (<1%), and standing water (<1%) [Gao et al., 2003]. The Se in the agricultural drainage water entering the Tulare Lake Drainage District wetland was predominantly in the form of selenate (95%); it was reduced in sediment to a mixture of elemental Se (45%), organic Se (40%), and selenite (15%) [Lin and Terry, 2003]. Thus, there is a certain concern that, since Se concentrations in the organically rich surface sediments increased over time, Se could eventually enter the aquatic food chain and exert eco-toxic effects.

More recently, Banuelos and Lin [2005] conducted a greenhouse study where Se-laden drainage sediment from the San Luis Drain of Central California was mixed with uncontaminated soil to different ratios and vegetated with salt-tolerant plant species consisting of canola (*Brassica napus* var. Hyola 420), tall fescue (*Festuca arundinacea* var. Au Triumph), salado grass (*Sporobulus airoides*), and cordgrass (*Spartina patens* var. Flageo). Results obtained indicated that total Se concentrations in the soil were at least 20% lower at post harvest compared to pre-plant concentrations for all plant species at each ratio of sediment:soil used.

5.3 Plant-Microbe Interaction: Selenium Phytoremediation Processes

Rhizobacteria have been shown to have an significant effects on metal phytoremediation processes [Di Gregorio et al., 2006a] also in relation to Se phytoremediation [Di Gregorio et al, 2006b]. On the other hand, the plant has a positive effect on the metabolic activity of rhizobacteria. Actually, exudation of nutrients by plant roots creates a nutrient-rich environment in which microbial activity is stimulated [Vancura and Hovadik, 1965]. Moreover, the natural decay of the root apparatus provides nutrients for microbes [Lugtenberg and de Weger, 1992; Lynch and Whipps, 1990]. Conversely, these latter have a large impact on plants, showing root growth-stimulating or growth-inhibiting properties [Campbell and Greaves, 1990]. Rhizosphere bacteria can stimulate plant growth by producing phytohormones [Fallik et al., 1994], enhancing mineral and water uptake [Lin et al., 1983], producing antibiotics to inhibit pathogens [Lesinger and Margraff, 1979], and altering root morphology

[Kapulnik, 1996]. Moreover, several lines of evidence suggest that soil microorganisms possess mechanisms capable of altering environmental mobility of metal contaminants with subsequent effects on the potential for root uptake [Lasat, 2000; Lasat et al., 2002]. Chemolithotrophic bacteria have been shown to enhance environmental mobility of metal contaminants via soil acidification or, in contrast, to decrease their solubility due to precipitation as sulfides [Kelley and Tuovinen, 1988]. In addition, soil microorganisms have been shown to exude organic compounds that stimulate bioavailability and facilitate root absorption of a variety of metal ions including Fe^{2+}, Mn^{2+}, and Cd^{2+} [Bural et al., 2000; Salt et al., 1995].

Concerning Se phytoremediation process, rhizobacteria increase plant potential for Se phytoremediation because they facilitate Se accumulation and volatilization [de Souza et al., 1999]. In relation to Se volatilization, it is tempting to speculate that the synergistic interaction between microorganisms and plants increases the kinetic of the process, enhancing both the plant and microbial capacity to detoxify the surrounding environment through Se volatilization. However, the occurring of microbial dissimilative reduction of Se oxyanions decrease the bioavailability of the metalloid in the rhizosphere because of the reduction of Se(VI) and Se (IV) to metallic selenium (Di Gregorio et al., 2006b). This evidence is not necessarily detrimental but has to be considered in the context of a phytostabilization process. The application of the synergic effect on Se bioavailabilty of rhizobacteria and plants could be considered as an alternative to rhizofiltration, a phytoremediative technique designed for the removal of metals in aquatic environments [Dushenkov et al., 1995; Zhu et al., 1999]. In fact, continuous open bioreactors may be designed by exploiting the synergic action of plants such as *Astragalus bisulcatus* and *Brassica juncea*, which extract the metalloid and eventually concentrate Se in the shoot and/or volatilize it, and rhizobacteria eventually augmented the system. While plants extract Se from effluents flowing through sequencing hydroponic ponds, the dissimilative reduction operated by rhizobacteria precipitate the metalloid. In this context, the metabolic versatility of rhizobacteria can be exploited in the direction of selenium precipitation and volatilization, but also in the direction of increasing plant performance in terms of plant biomass production, response to the stressful growth condition, and, last but not least, capacity to volatilize the contaminant [Vallini et al., 2005].

In conclusion, implementation in Se phytoremediation is feasible and goes in the direction of confirmation that the plant-based technology in remediation of Se-laden matrices can be considered The low-cost environmentally friendly approach for managing soluble Se in the soil and water environment. Further improvements are provided by the fact that obtaining products with economic value (Se-fortified crops, production of anticarcinogenic molecules) from plants used in the cleanup of soil would certainly be an additional benefit to phytoremediation, which could help sustain its long-term use.

REFERENCES

Abrams MM, Burau RG, Zasoski RJ. 1990. Organic selenium distribution in selected California soils. *Soil Sci Soc Am J* **54**:979–982.

Ansede JH, Yoch DC. 1997. Comparison of selenium and sulfur volatilization by dimethylsulfoniopropionate lyase (DMSP) in two marine bacteria and estuarine sediments. *FEMS Microbiol Ecol* **23**:315–324.

Arteel GE, Briviba K, Sies H. 1999. Protection against peroxynitrite. *FEBS Lett* **445**:226–230.

Arvy MP. 1993. Selenate and selenite uptake and translocation in bean plants (*Phaseolus vulgaris*). *J Exp Bot* **44**:1083–1087.

Asher CJ, Butler GW, Peterson PJ. 1977. Selenium transport in root systems of tomato. *J Exp Bot* **28**:279–291.

Azaizeh HA, Salhani N, Sebesvari Z, Emons H. 2003. The potential of rhizosphere microbes isolated from a constructed wetland to biomethylate selenium. *J Environ Qual* **32**:55–62.

Baker AJM, Brooks RR. 1989. Terrestrial higher plants which hyperaccumulate metal elements—A review of their distribution, ecology, and phytochemistry. *Biorecovery* **1**:81–126.

Banuelos GS, Lin ZQ. 2005. Phytoremediation management of selenium-laden drainage sediments in the San Luis Drain: A greenhouse feasibility study. *Ecotoxicol Environ Safety* **62**:309–316.

Banuelos GS, Lin ZQ. 2007. Acceleration of selenium volatilization in seleniferous agricultural drainage sediments amended with methionine and casein *Environ Pollut* doi:10.1016/j.envpol.2007.02.009.

Banuelos GS, Meek DW. 1990. Accumulation of selenium in plants grown on selenium-treated soil. *J Environ Qual* **19**:772–777.

Banuelos GS, Ajwa HA, Mackey B, Wu L, Cook C, Akohoue S, Zambruzuski S. 1997. Evaluation of different plant species used for phytoremediation of high soil selenium. *J Environ Qual* **26**:639–646.

Banuelos G, LeDuc DL, Pilon-Smits EA, Terry N. 2007. Transgenic Indian mustard overexpressing selenocysteine lyase or selenocysteine methyltransferase exhibit enhanced potential for selenium phytoremediation under field conditions. *Environ Sci Technol* **41**:599–605.

Barrow NJ, Whelan BR. 1989. Testing a mechanistic model. VII. The effects of pH and of electrolyte on the reaction of selenite and selenate with a soil. *J Soil Sci* **40**:17–28.

Baum MK, Miguez-Burbano MJ, Campa A, Shor-Posner G. 2000. Selenium and interleukins in persons infected with human immunodeficiency virus type 1. *J Infect Dis* **182**(suppl 1): S69–S73.

Beath OA, Draize JH, Eppson HF, Gilbert CS, McCreary OC. 1934. Certain poisonous plants of Wyoming activated by selenium and their association with respect to soil types. *J Am Pharm Soc* **23**:94.

Bebien M, Chauvin JP, Adriano JM, Grosse S, Vermeglio A. 2001. Effect of selenite on growth and protein synthesis in the phototrophic bacterium *Rhodobacter sphaeroides*. *Appl Environ Microbiol* **67**:4440–4447.

Bebien M, Lagniel G, Garin J, Touati D, Vermeglio A, Labarre J. 2002. Involvement of superoxide dismutases in the response of *Escherichia coli* to selenium oxides. *J Bacteriol* **184**:1556–1564.

Bell PF, Parker DR, Page AL. 1992. Contrasting selenate sulfate interaction in selenium accumulating and non accumulating plant species. *Soil Sci Soc Am J* **56**:1818–1824.

Belzile N, Chen YW, Xu RR. 2000. Early diagenetic behavior of selenium in freshwater sediments. *Appl Geochem* **15**:1439–1454.

Bjoernstedt M, Kumar S, Holmgren A. 1992. Selenodiglutathione is a highly efficient oxidant of reduced thioredoxin and a substrate for mammalian thioredoxin reductase. *J Biol Chem* **267**:8030–8034.

Blaylock MJ, James BR. 1994. Redox transformations and plant uptake of selenium resulting from root–soil interactions. *Plant Soil* **158**:1–12.

Blum JS, Stolz JF, Oren A, Oremland RS. 2001. *Selenihalanaerobacter shriftii* gen. nov., sp. nov., a halophilic anaerobe from Dead Sea sediments that respires selenate. *Arch Microbiol* **175**:208–219.

Böck A, Forchhammer K, Heider J, Leinfelder W, Sawers G, Veprek B, Linoni F. 1991. Selenocysteine: The 21st amino acid. *Mol Microb* **5**:515–520.

Brown DS, Reed SC. 1994. Inventory of constructed wetlands in the United States. *Water Sci Tech* **29**:309–318.

Brown TA, Shrift A. 1982. Selenium—toxicity and tolerance in higher-plants. *Biol Rev* **57**:59–84.

Broyer TC, Huston RP, Johnson CM. 1972a. Selenium and nutrition of *Astragalus* 1. Effects of selenite or selenate supply on growth and selenium content. *Plant Soil* **36**:635–649.

Broyer TC, Johnson CM, Huston RP. 1972b. Selenium and nutrition of *Astragalus* 2. Ionic sorption interactions among selenium, phosphate, and macronutrient and micronutrient cations. *Plant Soil* **36**:651–669.

Bural GI, Dixon DG, Glick BR. 2000. Plant growth-promoting bacteria that decrease heavy metal toxicity in plants. *Can J Microbiol* **46**:237–245.

Burk RF, Levander OA. Selenium. 1999. *Nutrition in Health and Disease*, ninth edition. In: Shils M, Olson JA, Shike M, Ross AC, editors. Baltimore: Williams & Wilkins, pp. 265–276.

Campbell R, Greaves MP. 1990. Anatomy and community structure of the rhizosphere. In: Lynch JM, editor. *The Rhizosphere*. Chichester: John Wiley & Sons, pp. 11–34.

Cantafio AW, Hagen KD, Lewis GE, Bledsoe TL, Nunan KM, Macy JM. 1996. Pilot-scale selenium bioremediation of San Joaquin drainage water with *Thauera selenatis*. *Appl Environ Microbiol* **62**:3298–3303.

Combs GF, Combs SB. Jr. 1984. The nutritional biochemestry of selenium. *Annu Rev Nutr* **4**:257–280.

Combs GF, Gray WP. 1998. Chemopreventive agents: selenium. *Pharmacol Ther* **79**:179–192.

Cowgill UM. 1990. The selenium cycle in three species of *Astragalus*. *J Plant Nutr* **13**:1309–1318.

Cutter GA, Bruland KW. 1984. The marine biogeochemistry of selenium: A reevaluation. *Limnol Oceanogr* **29**:1179–1192.

DeMoll-Decker H, Macy JM. 1993. The periplasmic nitrite reductase of *Thauera selenatis* may catalyze the reduction of selenite to elemental selenium. *Arch Microbiol* **160**:241–247.

Denina B, Simmons D, Wallascheeger D. 2005. A critical review of the biogeochemistry and ecotoxicology of selenium in lotic and lentic environments. *Environ Toxicol Chem* **24**:1331–1343.

de Souza MP, Pilon-Smits EAH, Lytle CM, Hwang S, Tai J, Honma TSU, Yeh L, Terry N. 1998. Rate-limiting steps in selenium assimilation and volatilization by Indian mustard. *Plant Physiol* **117**:1487–1494.

de Souza MP, Huang CP, Chee N, Terry N. 1999. Rhizosphere bacteria enhance the accumulation of selenium and mercury in wetland plants. *Planta* **209**:259–263.

Dhillon KS, Dhillon SK. 2003. Distribution and management of seleniferous soils. *Adv Agron* **79**:119–184.

Di Gregorio S, Lampis S, Vallni G. 2005. Selenite precipitation by a rhizospheric strain of *Stenotrophomonas* sp. isolated from the root system of *Astragalus bisulcatus*: A biotechnological perspective. *Environ Int* **31**:233–241.

Di Gregorio S, Barbafieri M, Lampis S, Sanangelantoni AM, Tassi E, Vallini G. 2006a. Combined application of Triton X-100 and *Sinorhizobium* sp. Pb002 inoculum for the improvement of lead phytoextraction by *Brassica juncea* in EDTA amended soil. *Chemosphere* **63**:293–299.

Di Gregorio S, Lampis S, Malorgio F, Petruzzelli G, Pezzarossa B, Vallini G. 2006b. *Brassica juncea* can improve selenite and selenate abatement in selenium contaminated soils through the aid of its rhizospheric bacterial population. *Plant Soil* **285**:233–244.

Dong Y, Lisk D, Block E, Ip C. 2001. Characterization of the biological activity of gamma-glutamyl-Se-methylselenocysteine: A novel, naturally occurring anticancer agent from garlic. *Cancer Res* **61**:2923–2928.

Doran JW. 1982. Microorganisms and the biological cycling of selenium. In Marshall KC, editor. *Advances in Microbial Ecology*. New York: Plenum Press, pp. 1–32.

Doran W, Alexander M. 1977. Microbial formation of volatile selenium compounds in soil. *Soil Sci Soc Am J* **41**:70–73.

Dowdle PR, Oremland RS. 1998. Microbial oxidation of elemental selenium in soil slurries and bacterial cultures. *Environ Sci Technol* **32**:3749–3755.

Droux M, Gakiere B, Denis L, Ravanel S, Tabe L, Lappartient AG, Job D. 2000. Methionine biosynthesis in plants: Biochemical and regulatory aspects. In: Brunold C, editor. *Sulfur Nutrition and Sulfur Assimilation in Higher Plants*. Bern, Switzerland: Paul Haupt pp. 73–92.

Duda PJ. 1992. Chevron's Richmond Refinery Water Enhancement Wetland. Technical report submitted to the Regional Water Quality Control Board, Oakland, CA.

Dungan RS, Yates SR, Frankenberger WT Jr. 2003. Transformations of selenate and selenite by *Stenotrophomonas maltophilia* isolated from a seleniferous agricultural drainage pond sediment. *Environ Microbiol* **5**:287–295.

Dushenkov V, Kumar PBAN, Motto H, Raskin I. 1995. Rhizofiltration: The use of plants to remove heavy metals from aqueous steams. *Environ Sci Technol* **29**:1239–1245.

Ellis DR, Salt DE. 2003. Plants, selenium and human health. *Curr Opin Plant Biol* **6**:1–7.

Elrashidi MA, Adriano DC, Lindsay WL. 1989. Solubility, speciation, and transformations of selenium in soils. In: Jacobs LW, editor. *Selenium in Agriculture and the Environment*. SSSA Special Publication 23. Madison, WI: SSSA, pp. 51–63.

Fallik E, Sarig, S, Okon Y. 1994. Morphology and physiology of plant roots associated with *Azospirillum*. In: Okon Y, editor. Boca Raton, FL: *Azospirillum*/Plant Association, pp. 77–86.

Fan TWM, Teh SJ, Hinton DE, Higashi RM. 2002. Selenium biotransformations into proteinaceous forms by food web organisms of selenium-laden drainage waters in California. *Aquat Toxicol* **57**:65–84.

Finley JW, Ip C, Lisk DJ, Davis CD, Hiuntze KJ, Whanger PD. 2001. Cancer-protective properties of high-selenium broccoli. *J Agric Food Chem* **49**:2679–2683.

Fleming J, Ghose A, Harrison P. 2001. Molecular mechanisms for cancer prevention by selenium compounds. *Nutr Cancer* **40**:42–48.

Flury M, Frankenberger WT Jr, Jury WA. 1997. Long-term depletion of selenium from Kesterson dewatered sediments. *Sci Total Environ* **198**:259–270.

Foster LH, Sumar S. 1997. Selenium in health and disease: A review. *Crit Rev Food Sci Nutr* **37**:211–228.

Frankenberger WT, Karlson U. 1995. Volatilization of selenium from a dewatered seleniferous sediment: A field study. *Ind Microbiol J* **14**:226–232.

Freeman JL, Lindblom SD, Quinn CF, Fakra S Marcus MA, Pilon-Smits EAH. 2007. Selenium accumulation protects plants from herbivory by Orthoptera via toxicity and deterrence. *New Phytologist* **175**:490–500.

Fujita M, Ike M, Nishimoto S, Takahashi K, Kashiwa M. 1997. Isolation and characterization of a novel selenate-reducing bacterium *Bacillus* sp. SF-1. *J Fermen Bioeng* **83**:517–522.

Ganther HE. 1971. Reduction of the selenotrisulfide derivative of glutathione to a persulfide analog by glutathione reductase. *Biochemistry* **10**:4089–4098.

Gao S, Tanji KK, Lin ZQ, Terry N, Peters DW. 2003. Selenium removal and mass balance in a constructed flow-through wetland system. *J Environ Qual* **32**:1557–1570.

Gerhardt M, Green F, Newman R, Lundquist T, Tresan R, Oswald W. 1991. Removal of Se using a novel algal-bacterial process. *J Water Pollut Control Fed* **63**:799–805.

Gessler NN, Bezzubov AA. 1988. Study of the activity of *S*-methylmethionine sulfonium salt hydrolase in plant and animal tissues. *Prikl Biokhim Mikrobiol* **24**:240–246.

Haberl R, Perfler R, Mayer H. 1995. Constructed wetlands in Europe. *Water Sci Technol* **32**:305–315.

Hansen D, Duda PJ, Zayed A, Terry N. 1998. Selenium removal by constructed wetlands: Role of biological volatilization. *Environ Sci Technol* **32**:591–597.

Haygarth PM. 1994. Global importance and global cycling of selenium. In: Frankenberger WT Jr, Benson S, editors. *Selenium in the Environment*. New York: Marcel Dekker, pp. 1–27.

Heider J, Böck A. 1993. Selenium metabolism in micro-organisms. *Adv Microb Physiol* **35**:71–109.

Holben DH, Smith AM. 1999. The diverse role of selenium within selenoproteins: A review. *J Am Diet Assoc* **99**:836–843.

Huber R, Sacher M, Vollmann A, Huber H, Rose D. 2000. Respiration of arsenate and selenate by hyperthermophilic archaea. *Syst Appl Microbiol* **23**:305–314.

Ip C, Ganther HE. 1992a. Comparison of selenium and sulfur analogs in cancer prevention. *Carcinogenesis* **13**:1167–1170.

Ip C, Ganther HE. 1992b. Relationship between the chemical form of selenium and anticarcinogenic activity. In: Wattenberg I, Lipkin M, Boon CW, Kellott GJ, editors. *Cancer Chemoprevention*. Boca Raton, FL: CRC Press, pp. 479–488.

Ip C, Birringer M, Block E, Kotrebai M, Tyson J, Uden PC, Lisk D. 2000. Chemical speciation influences comparative activity of selenium-enriched garlic and yeast in mammary cancer prevention. *J Agric Food Chem* **48**:2062–2070.

Jajaweera GR, Biggar JW. 1996. Role of redox potential in chemical transformations of selenium in soils. *Soil Sci Soc Am J* **60**:1056–1063.

Kabayta-Pendias A. 1998. Geochemistry of selenim. *J Environ Toxicol Oncol* **17**:173–177.

Kapulnik Y. 1996. Plant growth promotion by rhizosphere bacteria. In: Waisel Y, Eshel A, Kafkazi U, editors. *Plant Roots: The Hidden Half*. New York: Marcel Dekker, pp. 769–781.

Karlson U, Frankenberger WT Jr. 1988. Effects of carbon and trace element addition of alkyl-selenide production by soil. *Soil Sci Soc Am J* **52**:1640–1644.

Kauffman J, Laughlin W, Baldwin R. 1986. Microbiological treatment of uranium mine waters. *Environ Sci Technol* **20**:243–248.

Kelley BC, Tuovinen OH. 1988. Microbial oxidation of minerals in mine tailings. In: Solomons W, Foerstner V, editors. *Chemistry and Biology of Solid Waste*. Berlin: Springer Verlag, pp. 33–53.

Kessi J. 2006. Enzymic systems proposed to be involved in the dissimilatory reduction of selenite in the purple non-sulfur bacteria *Rhodospirillum rubrum* and *Rhodobacter capsulatus*. *Microbiology* **152**:731–743.

Kessi J, Hanselmann KW. 2004. Similarities between the abiotic reduction of selenite with glutathione and the dissimilatory reaction mediated by *Rhodospirillum rubrum* and *Escherichia coli*. *J Biol Chem* **279**:50662–50669.

Kiremidjian-Schumacher L, Roy M, Wishe HI, Cohen MW, Stotzky G. 1994. Supplementation with selenium and human immune cell functions. II. Effect on cytotoxic lymphocytes and natural killer cells. *Biol Trace Elem Res* **41**:115–127.

Kotrebai M, Birringer M, Tyson JF, Block E, Uden PC. 2000. Selenium speciation in enriched and natural samples by HPLC-ICP-MS and HPLC-ESI-MS with perfluorinated carboxylic acid ionpairing agents. *Analyst* **125**:71–78.

Krafft T, Bowen A, Theis F, Macy JM. 2000. Cloning and sequencing of the genes encoding the periplasmic-cytochrome B-containing selenate reductase of *Thauera selenatis*. *DNA Seq* **10**:365–377.

Kramer U. 2005. Phytoremediation: Novel approaches to cleaning up polluted soils. *Curr Opin Biotechnol* **16**:133–141.

Kuepper H, Lombi E, Zhao FJ, McGrath SP. 2000. Cellular compartimentation of cadmium and zinc in relation to other elements in the hyperaccumulator *Arabidopsis halleri*. *Planta* **212**:75–84.

Kuepper H, Lombi E, Zhao FJ, Wieshammer G, McGrath SP. 2001. Cellular *compartmentation of nickel in the hyperaccumulators Alyssum lesbiacum*, Alyssum bertolonii and *Thlaspi goesingense*. *J Exp Bot* **52**:2291–2300.

Larsen PR, Davies TF, Hay ID. 1998. The thyroid gland. In: Wilson JD, Foster DW, Kronenberg HM, Larsen PR, editors. *Williams Textbook of Endocrinology*, ninth edition. Philadelphia: WB Saunders, pp. 389–515.

Lasat MM. 2002. Phytoextraction of toxic metals: A review of biological mechanisms. *J Environ Qual* **31**:109–120.

Lasat MM, Pence NS, Garvin DF, Ebbs SD, Kochian LV. 2000. Molecular physiology of zinc transport in the Zn hyperaccumulator *Thlaspi caerulescens*. *J Exp Bot* **51**:71–79.

Lawson S, Macy JM. 1995. Bioremediation of SeIV in oil refinery wastewater. *Appl Microbiol Biotechnol* **43**:762–765.

Lemly AD. 1997. Ecosystem recovery following selenium contamination in a freshwater reservoir. *Ecotoxicol Environ Safety* **36**:275–281.

Lesinger T, Margraff R. 1979. Secondary metabolites of the fluorescent pseudomonads. *Microbiol Rev* **43**:422–442.

Lin W, Okon Y, Hardy RWF. 1983. Enhanced mineral uptake by *Zea mays* and *Sorghum bicolor* roots inoculated with *Azospirillum brasilense*. *Appl Environ Microbiol* **45**:1775–1779.

Lin ZQ, Terry N. 2003. Selenium removal by constructed wetlands: Quantitative importance of biological volatilization in the treatment of selenium-laden agricultural drainage water. *Environ Sci Technol* **37**:606–615.

Lipman JG, Waksman SA. 1923. The oxidation of elemental Se by a new group of autotrophic microorganisms. *Science* **57**:60.

Lortie L, Gould WD, Rajan S, McCready RGL, Cheng KJ. 1992. Reduction of selenate and selenite to elemental selenium by a *Pseudomonas stutzeri* isolate. *Appl Environ Microbiol* **58**:4042–4044.

Losi ME, Frankenberger WT Jr. 1997. Reduction of selenium oxyanions by *Enterobacter cloacae* strain SLDa-1: Isolation and growth of the bacterium and its expulsion of selenium particles. *Appl Environ Microbiol* **63**:3079–3084.

Losi ME, Frankenberger WT Jr. 1998. Microbial oxidation and solubilization of precipitated elemental selenium in soil. *J Environ Qual* **27**:836–843.

Lovely DR. 1993. Dissimilatory metal reduction. *Annu Rev Microbiol* **47**:263–290.

Lu J, Pei H, Ip C, Lisk DJ, Ganther H, Thompson HJ. 1996. Effect of an aqueous extract of selenium-enriched garlic on in vitro markers and in vivo efficacy in cancer prevention. *Carcinogenesis* **17**:1903–1907.

Lugtenberg BJJ, de Weger LA. 1992. Plant root colonization by *Pseudomonas*. In: Galli E, Silver S, and Witholt B, editors. *Pseudomonas: Molecular Biology and Biotechnology*. Washington, DC: American Society for Microbiology, pp. 13–19.

Lynch JM, Whipps JM. 1990. Substrate flow in the rhizosphere. *Plant Soil* **129**:1–10.

Macy JM. 1994. Biochemistry of selenium metabolism by *Thauera selenatis* gen. nov. sp. nov. and use of the organism for bioremediation of selenium oxyanions in San Joaquin Valley drainage water. In: Frankenberger WT Jr, Benson S, editors. *Selenium in the Environment*. New York: Marcel Dekker, pp. 421–444.

Macy JM, Lawson S, DeMoll-Decker H. 1994. Bioremediation of selenium oxyanions in San Joaquin drainage water using *Thauera selenatis* in a biological reactor system. *Appl Microbiol Biotechnol* **40**:588–594.

Maher MJ, Santini J, Pickering IJ, Prince RC, Macy JM, George GN. 2004. X-ray absorption spectroscopy of selenate reductase. *Inorg Chem* **43**:402–404.

Martin JL, Shrift A, Gerlach ML. 1971. Use of ^{75}Se-selenite for the study of selenium metabolism in *Astragalus*. *Biochem* **10**:945–952.

Masscheleyn PH, Delaune R, Patrik WH Jr. 1990. Transformations of selenium as affected by sediment oxidation-reduction potential and pH. *Environ Sci Technol* **1**:91–96.

Mayland HF, James LJ, Panter KE, Sonderegger JL. 1989. Selenium in seleniferous environments. In: Jacobs LW, editor. *Selenium in Agriculture and the Environment*. SSSA Special Publication 23. Madison, WI: ASA and SSSA, pp. 15–50.

McGrath SP, Zhao FJ. 2003. Phytoextraction of metals and metalloids from contaminated soils. *Curr Opin Biotechnol* **14**:277–282.

McNeal JM, Balilestri LS. 1989. Geochemistry and occurrence of selenium: An overview. In: Jacobs LW, editor. *Selenium in Agriculture and the Environment*. SSSA Special Publication 23. Madison, WI: ASA and SSSA, pp. 1–13.

Medina D, Thompson H, Ganther H, Ip C. 2001. Se-Methylselenocysteine: A new compound for chemoprevention of breast cancer. *Nutr Cancer* **40**:12–17.

Mergeay M, Nies D, Schlegel HG, Gerits J, Charles P, Van Gijsegem F. 1985. *Alcaligenes eutrophus* CH34 is a facultative chemolithotroph with plasmid-bound resistance to heavy metals. *J Bacteriol* **162**:328–334.

Moore MD, Kaplan S. 1992. Identification of intrinsic high-level of resistance to rare earthoxides and oxyanions in members of the class Proteobacteria: Characterization of tellurite, selenite, and rhodium sesquioxide reduction in *Rhodobacter sphaeroides*. *J Bacteriol* **174**:1510–1514.

Moore MD, Kaplan S. 1994. Members of the family Rhodospirillaceae reduce heavy metal oxyanions to maintain redox poise during photosynthetic growth. *ASM News* **60**:17–23.

Mustacich D, Powis G. 2000. Thioredoxin reductase. *Biochem J* **346**(Pt 1):1–8.

Narasingarao P, Haeggblom MM. 2006. *Sedimenticola selenatireducens*, gen. nov., sp. nov., an anaerobic selenate-respiring bacterium isolated from estuarine sediment. *Systematic Appl Microbiol* **29**:382–388

Newton GL, Fahey RC. 1989. Glutathione in prokaryotes. In: Vinea J, editor. *Glutathione: Metabolism and Physiological Functions*. Boca Raton, FL: CRC Press, pp. 69–77.

Ng BH, Anderson JW. 1978. Synthesis of selenocysteine by cysteine synthase from selenium accumulator and non-accumulator plants. *Phytochemistry* **17**:2069–2074.

Nriagu JO. 1989. Global cycling of selenium. In: Milan Ihnat, editor. *Occurrence and Distribution of Selenium*. Boca Raton, FL: CRC Press, pp. 327–340.

Oremland R. 1991. Selenate removal from waste water. U.S. Patent 5,009,786.

Oremland RS, Hollibaugh JT, Maest AS, Presser TS, Miller LG, Culbertson CW. 1989. Selenate reduction to elemental selenium by anaerobic bacteria in sediments and culture: Biogeochemical significance of a novel, sulfate-independent respiration. *Appl Environ Microbiol* **55**:2333–2343.

Oremland RS, Steinberg NA, Maest AS, Miller LG, Hollibaugh JT. 1990. Measurement of in situ rates of selenate removal by dissimilatory bacterial reduction in sediments. *Environ Sci Technol* **24**:1157–1164.

Oremland RS, Switzer Blum J, Culbertson CW, Visscher PT, Miller LG, Dowdle P, Strohmaier FE. 1994. Isolation, growth and metabolism of an obligately anaerobic, selenate-respiring bacterium, strain SES-3. *Appl Environ Microbiol* **60**:3011–3019.

Orser CS, Salt DE, Pickering IJ, Prince R, Epstein A, Ensley BD. 1999. *Brassica* plants to provide enhanced human mineral nutrition: Selenium phytoenrichment and metabolic transformation. *J Med Food* **1**:253–261.

Parker DR, Page AL. 1994. Vegetation management strategies for remediation of selenium-contaminated soils. In: Frankenberger WT Jr, Benson S, editors. *Selenium in the Environment*. New York: Marcel Dekker, pp. 327–341.

Pickering IJ, Prince RC, Salt DE, George GN. 2001. Quantitative, chemically specific imaging of selenium transformation in plants. *Proc Nat Am Sci* **97**:10717–10722.

Pickering IJ, Wright C, Bubner B, Ellis D, Persans MW, Yu EY, George GN, Prince RC, Salt DE. 2003. Chemical form and distribution of selenium and sulfur in the selenium hyperaccumulator *Astragalus bisulcatus*. *Plant Physiol* **131**:1–8.

Pilon-Smits EAH, Hwang S, Lytle CM, Zhu Y, Tai JC, Bravo RC, Chen Y, Lustek T, Terry N. 1999. Overexpression of ATP sulfurylase in Indian mustard leads to increased selenate uptake, reduction and tolerance. *Plant Physiol* **119**:123–132.

Presser TS, Sylvester MA, Low WH. 1994. Bioaccumulation of selenium from natural geologic sources in western states and its potential consequences. *Environ Manag* **18**:423–436.

Price NM, Thompson PA, Harrison PJ. 1987. Selenium: An essential element for growth of the coastal marine diatom *Thalassiosira pseudonana* (Bacillariophyceae). *J Phycol* **23**:1–9.

Rabenstein DL, Tan KS. 1988. ^{77}Se NMR studies of bis(alkylthio)selenides of biological thiols. *Magn Reson Chem* **26**:1079–1085.

Rayman MP. 2000. The importance of selenium to human health. *Lancet* **2000**:233–241.

Rathgeber C, Yurkova N, Stackebrandt E, Beatty JT, Yurkov V. 2002. Isolation of tellurite- and selenite-resistant bacteria from hydrothermal vents of the Juan de Fuca Ridge in the Pacific Ocean. *Appl Environ Microbiol* **69**:4613–4622.

Roux M, Sarret G, Pignot-Paintrand I, Fontecave M, Coves J. 2001. Mobilization of selenite by *Ralstonia metallidurans* CH34. *Appl Environ Microbiol* **67**:769–773.

Roy M, Kiremidjian-Schumacher L, Wishe HI, Cohen MW, Stotzky G. 1994. Supplementation with selenium and human immune cell functions. I. Effect on lymphocyte proliferation and interleukin 2 receptor expression. *Biol Trace Elem Res* **41**:103–114.

Salt DE, Smith R, Raskin I. 1998. Phytoremediation. *Annu Rev Plant Physiol Plant Mol Biol* **49**:643–668.

Salt DE, Prince RC, Pickering IJ, Raskin I. 1995. Mechanisms of cadmium mobility and accumulation in Indian mustard. *Plant Physiol* **109**:1427–1433.

Sarathchandra SU, Watkinson JH. 1981. Oxidation of elemental selenium to selenite by *Bacillus megaterium*. *Science* **21**:600–601.

Sarret G, Avoscan L, Carriere M, Collins R, Geoffroy N, Carrot F, Coves J, Gouget B. 2005. Chemical forms of selenium in the metal-resistant bacterium *Ralstonia metallidurans* CH34 exposed to selenite and selenate. *Appl Environ Microbiol* **71**:2331–2337.

Schlekat CE, Dowdle PR, Lee BG, Luoma SN, Oremland RS. 2000. Bioavailability of particle-associated Se to the bivalve *Potamocorbula amurensis*. *Environ Sci Technol* **34**:4504–4510.

Seko Y, Imura N. 1997. Active oxygen generation as a possible mechanism of selenium toxicity. *Biomed Environ Sci* **10**:333–339.

Seko Y, Saito Y, Kitahara J, Imura N. 1989. Active oxygen generation by the reaction of selenite with reduced glutathione in vitro. In: Wendel A, editor. *Selenium in Biology and Medicine*. Berlin: Springer-Verlag, pp. 70-73.

Shrift A. 1969. Aspects of selenium metabolism in higher plants. *Annu Rev Plant Physiol* **20**:475–495.

Shrift A. 1973. Metabolism of selenium by plants and microorganisms. In: Klayman DK, Gunther WHH, editors. *Organic Selenium Compounds: Their Chemistry*. New York, NY: Wiley-Interscience, pp. 763–814.

Shrift A, Ulrich J. 1969. Transport of selenate and selenite into *Astragalus* roots. *Plant Physiol* **44**:893–896.

Siddique T, Zhang Y, Okeke BC, Frankenberger WT. 2006. Characterization of sediment bacteria involved in selenium reduction. *Bioresour Technol* **97**:1041–1049.

Siddique T, Arocena JM, Thring RW, Zhang Y. 2007. Bacterial reduction of selenium in coal mine tailings pond sediment. *J Environ Qual* **36**:621–627.

Spallholz JE. 1994. On the nature of selenium toxicity and carcinostatic activity. *Free Radic Biol Med* **17**:45–64.

Stadtman TC. 1990. Selenium biochemistry. *Annu Rev Biochem* **59**:111–127.

Stadtman TC. 1996. Selenocysteine. *Annu Rev Biochem* **65**:83–100.

Switzer-Blum J, Bindi AB, Buzzelli J, Stolz JF, Oremland RS. 1998. *Bacillus arsenicoselenatis* sp. nov., and *Bacillus selenitireducens*, sp. nov.: Two haloalkaliphiles from Mono Lake, California, which respire oxyanions of selenium and arsenic. *Arch Microbiol* **171**:19–30.

Terry N, Zayed AM, de Souza MP, Tarun AS. 2000. Selenium in higher plants. *Annu Rev Plant Physiol Plant Mol Biol* **51**:401–432.

Terry N, Karlson C, Raab TK, Zayed A. 1992. Rates of selenium volatilization among crop species. *J Environ Qual* **21**:341–344.

Thompson-Eagle ET, Frankenberger WT Jr. 1991. Selenium biomethylation in an alkaline, saline environment. *Water Res* **25**:231–240.

Thompson-Eagle ET, Frankenberger WT, Jr. Karlson U. 1989. Volatilization of selenium by *Alternaria alternata*. *Appl Environ Microbiol* **55**:1406–1413.

Tokunaga TK, Pickering IJ, Brown GEJ. 1996. Selenium transformations in ponded sediment. *Soil Sci Soc Am J* **60**:781–790.

Tomei FA, Barton LL, Lemanski CL, Zocco TG. 1992. Reduction of selenate and selenite to elemental selenium by *Wolinella succinogenes*. *Can J Microbiol* **38**:1328–1333.

Tomei FA, Barton LL, Lemanski CL, Zocco TG, Fink NH, Sillerud LO. 1995. Transformation of selenate and selenite to elemental selenium by *Desulfovibrio desulfuricans*. *J Ind Microbiol* **14**:329–336.

Torma AE, Habashi F. 1972. Oxidation of copper (II) selenite by *Thiobacillus ferrooxidans*. *Can J Microbiol* **18**:1780–1781.

Trelease SF, Trelease HM. 1939. Physiological differentiation in *Astragalus* with reference to selenium. *Am J Bot* **26**:530–535.

Turner RJ, Weiner JH, Taylor DE. 1998. Selenium metabolism in *Escherichia coli*. *Bio Metals*. **11**:223–227.

Ullrey DE. 1981. Selenium in the soil–plant–food chain. In Spallholz JE, Martin JL, Ganther HE, editors. *Selenium in Biology and Medicine* Westport: Avi Publishing Co., pp. 176–191.

Ursini F, Heim S, Kiess M, et al. 1999. Dual function of the selenoprotein PHGPx during sperm maturation. *Science* **285**:1393–1396.

Vallini G, Di Gregorio S, Lampis S. 2005. Rhizosphere-induced selenium precipitation for possible applications in phytoremediation of polluted Se effluents. *Zeitschrift Naturforschung C A J Biosci* **60**:349–356.

Vancura V, Hovadik A. 1965. Root exudates of plants II. Composition of root exudates of some vegetables. *Plant Soil* **22**:21–32.

Vinceti M, Wei ET, Malagoli C, Bergomi M, Vivoli G. 2001. Adverse health effects of selenium in humans. *Rev Environ Health* **16**:233–251.

Virupaksha TK, Shrift A. 1965. Biochemical differences between selenium accumulator and non-accumulator *Astragalus* species. *Biochem Biophys* **107**:69–80.

Wang Y, Bock A, Neuhierl B. 1999. Acquisition of selenium tolerance by a selenium non-accumulating *Astragalus* species via selection. *Biofactors* **9**:3–10.

Whanger PD. 2002. Selenocompound in plants and animals and their biological significance. *J Am College Nutr* **21**:223–232.

Whanger PD, Ip C, Polan CE, Uden PC, Welbaum G. 2000. Tumorigenesis, metabolism, speciation, bioavailability, and tissue deposition of selenium in selenium-enriched ramps (*Allium tricoccum*). *J Agric Food Chem* **48**:5723–5730.

Weres O, Jaoni AR, Tsao L.1989. The distribution, speciation, and geochemical cycling of selenium in a sedimentary environment, Kesterson Reservoir, California, U.S.A. *Appl Geochemistry* **4**:543–563.

Whelan BR. 1989. Testing a mechanistic model. VII. The effects of pH and of electrolyte on the reaction of selenite and selenate with a soil. *J Soil Sci* **40**:17–28.

White PJ, Bowen HC, Parmaguru P, Fritz M, Spracklen WP, Spiby RE, Meachan MC, Mead A, Harriman M, Trueman LJ, Smith BM, Thomas B, Broadley MR. 2004. Interactions between selenium and sulphur nutrition in *Arabidopsis thaliana*. *J Exp Bot* **55**:1927–1937.

Wu L, Huang ZZ, Burau RG. 1988. Selenium accumulation and selenium-salt co-tolerance in five grass species. *Crop Sci* **28**:517–522.

Ye Z, Huang L, Bell RW, Dell B. 2003. Low root zone temperature favours shoot partitioning into young leaves of oilseed rape (*Brassica napus*). *Physiol Plant* **118**:213–220.

Yokota A, Shigeoka S, Onishi T, Kitaoka S. 1988. Selenium as inducer of glutathione peroxidase in low-CO2-grown *Chlamydomonas reinhardtii*. *Plant Physiol* **86**:649–651.

Yurkov VV, Csotonyi JT. 2003. Aerobic anoxygenic phototrophs and heavy metalloid reducers from extreme environments. *Recent Res Dev Bacteriol* **1**:247–300.

Zahir ZA, Zhang Y, Frankenberger Jr, WT. 2003. Fate of selenate metabolized by *Enterobacter taylorae* isolated from rice straw. *J Agric Food Chem* **1**:3609–3613.

Zayed A, Lytle CM, Terry N.1998. Accumulation and volatilization of different chemical species of selenium by plants. *Planta* **206**:284–292.

Zayed A, Pilon-Smits E, de Souza M, Lin ZQ, Terry N. 2000. Remediation of selenium polluted soils and waters by phytovolatilization. In: Terry NN, Bañuelos G, editors. *Phytoremediation of Contaminated Soil and Water*. Boca Raton, FL: Lewis Publishers, pp. 61–83.

Zehr JP, Oremland RS. 1987. Reduction of selenate to selenide by sulfate-respiring bacteria: Experiments with cell suspensions and estuarine sediments. *Appl Environ Microbiol* **53**:1365–1369.

Zhang Y, Zahir ZA, Frankenberger WT Jr. 2003. Factors affecting reduction of selenate to elemental selenium in agricultural drainage water by *Enterobacter taylorae*. *J Agric Food Chem* **51**:7073–7078.

Zhu YL, Zayed AM, Qian JH, de Souza M, Terry N. 1999. Phytoremediation of trace elements by wetland plants: II. Water hyacinth. *J Environ Qual* **28**:339–344.

24 Environmental Contamination Control of Water Drainage from Uranium Mines by Aquatic Plants

CARLOS PAULO and JOÃO PRATAS

Earth Sciences Department, Faculty of Sciences and Technology of the University of Coimbra, 3000-272 Coimbra, Portugal

1 INTRODUCTION

Environmental pollution of soils and waters with trace metals is a global concern. Although trace metals are present in the earth's crust, and some natural contamination is related to them, over the years numerous anthropogenic activities have contributed for the anomalous load of metals to water and soils. The chemical contamination of freshwater environments (a result of the discharge of untreated waste, dumping of industrial effluent, or runoff from agricultural fields), besides promoting the ecological degradation, compromises the quality of the drinking water catchment areas.

The environmental responsibility of industrial activities is not the same as in the past, when the environmental policies were not a priority. Mining operations are an unavoidable example, and several thousand tons of untreated wastes in abandoned mines all over the world are potential threats to the environment. Water in tailing dams and ponds require special awareness, since they entail long-term treatment strategies, not only because of the extremely high concentrations of metals and low pH but also due to the permanent load of seepage waters with low flow rate [Lottermoser, 2003; Kunze et al., 2007]. In such case, the use of wastewater traditional methodologies may be unaffordable, and the remediation goals can never be accomplished.

Trace Elements as Contaminants and Nutrients: Consequences in Ecosystems and Human Health, Edited by M. N. V. Prasad
Copyright © 2008 John Wiley & Sons, Inc.

Today the emergent decontamination methods based on living organisms' ability to reduce, remove, and degrade the pollutants, such as phytoremediation, can be, in specific cases, a sustainable alternative because they have lower cost, a reduced environmental impact, and an increased public acceptance [Vanek and Schwitzguébel, 2003; Pilon-Smits, 2005]. Intensive research on constructed wetlands based on natural processes for mine effluent treatment is now proposed as an option for long-term treatment of mines. The recognition of accumulator plants as remediation tools and the successfull demonstration of some field applications have contributed to the growing acceptanee of ecological engineering approach [Kalin, 1998; Overall and Parry, 2004; Kunze et al., 2007]. Moreover, if properly designed, these systems are a suitable strategy for ecological restoration.

In this chapter, a review on the main environmental impacts and health hazards resulting specially from uranium element (U) and its mining activity will be made. Evidence of metal accumulation by aquatic plants and their use as biological polishing remediation systems will be reviewed, focusing in our investigation in Central Portugal. The plant selection methodology and a preliminary test of phytoremediation with aquatic plant species will be presented as illustration of the technique potential.

2   URANIUM MINING: ENVIRONMENTAL AND HEALTH

Uranium mining and milling operations are the main anthropogenic sources of U pollution worldwide. Besides the unaesthetic facet related with the mining activity, the major and most concerning environmental impact derives from the chemical reactivity and the radioactivity of the waste materials in tailings [Lottermoser, 2003; Sinclar et al., 2005]. This is even more significant in case of abandoned mines where there is no control and monitoring.

The tailings are the leftovers of rock ground resulting from U extraction process, including solid fragments of host rock and a large variety of chemicals used on ore concentration techniques. During U mining, several chemical procedures are applied and the release of harmful contaminants into soils and waters may occur. The use of acid lixiviants is sometimes required when low-grade ore is explored and milling is not economically viable. Methods as heap leaching (percolation of sulfuric acid through a pile of rock) or in situ leaching (sulfuric acid pumped into underground uranium deposits), if not correctly controlled, can lead to immediate or long-term discharge of metal pollutants [Lottermoser, 2003]. The disposal of the remaining of these liquids and of other chemicals used in milling allows the formation, in oxidant conditions, of highly soluble uranium chemical species, such as the uranyl cation UO_2^{2+}, in which U is in its highest (hexavalent) oxidation state. In this form, uranium can easily be dispersed into soils and waters, bioaccumutaled, and, consequently, transferred to the food chain. Other metals and radionuclides naturally associated with the exploited ore, such as thorium-230, radium-226, radon-222, arsenic (As), copper (Cu), lead (Pb), and zinc (Zn), can also be released increasing the risk of environmental impact [Lottermoser, 2003; Sinclar et al., 2005]. Health disorders, such as cancer and kidney diseases, are the common effects from trace metals direct poisoning by ingestion, inhalation, or dermal contact

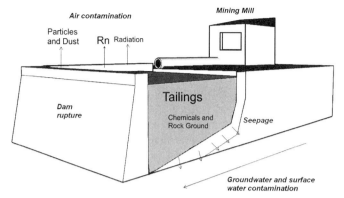

Figure 1. Hazards related with uranium mill tailings. (Based on Lottermoser, 2003 and Wise Uranium Project).

[ATSDR, 1999]. The sum of the major hazards and pathways of U mining is illustrated in Fig. 1 and Fig. 2.

The solid waste products in the tailings are particularly hazardous due to its radio-activity. Uranium has three natural isotopes, ^{238}U, ^{235}U, and the ^{234}U, with a relative

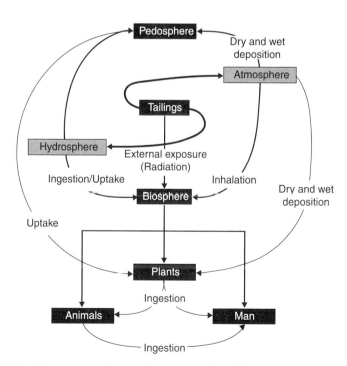

Figure 2. Potential pathways in environment and to human concentration from U mining wastes.

abundance of 99,278%, 0,717%, and 0,005%, respectively. None of them is stable and, therefore, U atoms undergo a spontaneous emission of radiation and/or particles (radioactivity decay). In this process, the initial U atoms (*parent nuclides*) are transformed into new atoms of a different element (*daughter nuclide*) by energy and mass loss in a series of 14 radioactive isotopes until a stable isotope of lead (Pb 210) is reached (see Table 1) [Bourdon et al., 2002; Malczewska et al., 2003].

The U decay products can also be found in the tailings, because they were a component of the original ore. Tailings are then enriched in radioactive isotopes such as thorium-230 and radium-226, the most abundant and also the more dangerous in long-term analysis. Nevertheless, this risk is minimized because their half-life is extremely long; for example, ^{238}U decay takes 4.5 billion years, ^{230}Th 75,690 years, and ^{226}Ra 1599 years [Bourdon et al., 2002; Vandenhove et al., 2006].

However, as ^{230}Th and ^{226}Ra decays, radon gas (^{222}Rn) is formed and emanated from rock and tailings to the surrounding environment. This gas is related to the major health hazard resulting from U mining (lung cancer) because of Rn short half-life (3.82 days). In this case, the ionizing radiation can be a source of toxicity to organisms because it can be strong enough to cause immediate damage to living cells [Bourdon et al., 2002; Lottermoser, 2003; Vandenhove et al., 2006].

TABLE 1. Decay Chain Series of 238U and 235U [Bourdon et al., 2002]

Decay Chain of ^{238}U			Decay Chain of ^{235}U		
Nuclide	Particles	Half-Life	Nuclide	Particles	Half-Life
^{238}U	α	44,683 billion years	^{235}U	α	0.70381 billion years
^{234}Th	β	24.1 days	^{231}Th	β	1.063 days
^{234}Pa	β	6.69 hrs	^{231}Pa	α	32760 years
^{234}U	α	245250 years	^{227}Ac	α, β	2177 years
^{230}Th	α	75690 years	^{227}Th	α	18.72 days
^{226}Ra	α	1599 years	^{223}Fr	α, β	22 min
^{222}Rn	α	3.823 days	^{223}Ra	α	11.435 days
^{218}Po	α, β	3.04 min	^{219}At	α, β	50 sec
^{218}At	α, β	1.6 sec	^{219}Rn	α	3.96 sec
^{218}Rn	α	35 msec	^{215}Bi	β	7.7 min
^{214}Pb	β	26.9 min	^{215}Po	α, β	1.78 msec
^{214}Bi	α, β	19.7 min	^{215}At	α	0.1 msec
^{214}Po	α	0.1637 msec	^{211}Pb	β	36.1 min
^{210}Tl	β	1.4 min	^{211}Bi	α, β	2.14 min
^{210}Pb	α, β	22.6 years	^{211}Po	α	0.516 sec
^{210}Bi	α, β	5.01 days	^{207}Tl	β	4.77 min
^{210}Po	α	138.4 days	^{207}Pb		Stable
^{206}Hg	β	8.2 min			
^{206}Tl	β	4.2 min			
^{206}Pb		Stable			

According to the Wise Uranium Project (http://www.wise-uranium.org/), there are at least 2352 million tons of U mill tailings in more than 20 different countries that need to be treated. More than 70% of these residues are distributed only by five countries: South Africa (700 Mt), Namibia (350 Mt), United States (235 Mt), Canada (202.13 Mt), and Germany (174.45 Mt).

2.1 Uranium Toxicity

The toxicological profile of uranium is, sometimes, merely related to radiation effects, but, in fact, it is the chemical toxicity of this element that can be responsible for major health effects in living organisms [Orlofff et al., 2004; Vandenhove et al., 2006]. As mentioned before, the radiation hazards are more likely to be related to the exposure to radon gas rather than U and other natural radionuclides [ATSDR, 1999; Orlofff et al., 2004; Bourdon et al., 2002; Brüske-Hohlfeld et al., 2006; Vandenhove et al., 2006]. The inhalation of radon gas has been associated to genotoxicity (for instance, in the lungs) and other chromosomal abnormalities in both human and animal studies [Jostes, 1996; Hussain et al., 1997].

However, the relationship between the increase of human cancer and genotoxicity in uranium mining areas, especially in miners, is frequently contradictory, because no significant direct cause–effect was found in most of the epidemiological studies [Jostes, 1996; ATSDR, 1999; Orloff et al., 2004; Weir, 2004; Wolf et al., 2004; Periyakaruppan et al., 2007; Schneider et al., 2007].

Recent studies in former uranium mining areas show that cancer rates did not increase among the local population after the beginning of uranium mining activity when compared with other areas [Boice et al., 2003, 2007]. However, when compared with the local population, former underground miners have an increase of lung cancer mortality, but it is most likely due to prior occupational exposure to both radon and to cigarette smoking [Boice et al., 2007]. It is probably the combination with other exposure factors, like cigarettes smoke and other carcinogenic elements as arsenic, that can affect and increase the radon-induced response among miners [Jostes, 1996; Kreuzer et al., 1999; Brüske-Hohlfeld et al., 2006; Grosche et al., 2006; Boice et al., 2007].

One the other hand, chemical toxicity of uranium is better studied, in particular in animal models. Uranium can enter the body three ways: by ingestion, by particules inhalation, and through the skin. The most efficient way of U absorption is by inhalation (40–66% of absorption of the total inhaled), whereas the amount of uranium absorbed by ingestion (contaminated water and food) only represents 0.3% to 6% of the total uranium ingested [Tansky, 2001; Malczewska et al., 2003; Periyakaruppan et al., 2007]. Almost negligible is the uranium absorption through the skin [Durakoviae, 1999; Malczewska et al., 2003].

The toxic effects of U in living organism are dependent on uranyl ions speciation and on their solubility. Natural uranium consists almost entirely of the ^{238}U in its hexavalent state (U^{6+}); in aqueous systems, this ion reacts with oxygen to form uranyl ion UO_2^{2+}, which is highly stable and soluble [Malczewska et al., 2003; Kalin et al., 2005]. Other soluble forms are UF_6, $UO_2(NO_3)_2$, UO_2Cl_2, UO_2F_2, and uranyl acetates, sulfates, and carbonates [Durakoviae, 1999].

Uranyl is effectively transported in the bloodstream through interactions with carbonate and protein ligands, forming stable complexes like uranyl hydrogen carbonate complex ($UO_2HCO_3^+$) and uranyl albumin complex [Weir, 2004; Van Horn and Huang, 2006]. In a minor extent, less soluble U forms (U^{4+}) can also be ingested or inhaled [Durakoviae, 1999]. Uranium is quickly delivered to several parts of the body, such as lungs, bones, and kidney, but the bioaccumulation rate is low because most of the uranium is excreted in urine within a few hours [ATSDR, 1999].

The UO_2^{2+} excretion rate is dependent of tubular urine pH, and changes in this factor may interfere with the main kidney function. The regulation of ions and molecules levels in blood, such as Na^+, Ca^{2+}, glucose, and amino acids, is controlled by a sodium–potassium exchanger ($Na^+/K^+/ATPase$) that can be affected by the biochemistry of UO_2^{2+} ions. At lower pH, the $UO_2HCO_3^+$ dissociates; as consequence, due to the similarities of UO_2^{2+} with Ca^{2+} and Mg^{2+}, uranium can mimic these cations in the transport process, reentering the bloodstream. The potential to increase K^+ permeability is enhanced because UO_2^{2+} binds more strongly than Ca^{2+}, which may inflict structural changes in cell membranes [Weir, 2004; Goldman et al., 2006; Vandenhove et al., 2006; Thiébault et al., 2007].

In addition to the uranyl direct effect in membrane transporters, as the dose increases, loss of cell polarity and integrity of the renal tubular epithelium cells can also take place, causing cellular necrosis and atrophy in the tubular wall and thereby leading to decreased reabsorption of calcium, phosphate, amino acids, and small proteins by the renal tubules [WHO, 2001; Kurttio et al., 2002; Kurttio et al., 2005; Goldman et al., 2006; Thiébault et al., 2007]. New findings in laboratory experiments in mice kidneys relate uranium toxicity with DNA cleavage and apoptsis by intrinsic pathway at the tubular ephithelium cells. Thiébault proved that the DNA cleavage caused by U is reversible at low concentration (200–400 μM), making it possible for cells to develop an adaptive response, but above 500 μM the destruction becomes irreversible and cells underwent necrosis [Thiébault et al., 2007].

Signs of acute uranium toxicity in humans, although very rare, include significant weight loss and hemorrhages in the eyes, legs, and nose [Weir, 2004].

Nephrotoxicity is most likely due to chronic exposure to uranium. Kidney injuries are reported in several studies as being directly associated with a continuous intake of U contaminated water, even if at low concentrations [Weir, 2004].

The effects of chronic renal insufficiency can be reflected to bone metabolism, thus inducing osteoporosis because of the increasing leakage of calcium, as verified in other heavy metals such as Cd [Kurttio et al., 2005]. Uranium is immobilized in the skeleton by calcium replacement in the hydroxyapatite complex of the bone crystal, as Ca and phosphate excretion increase [Weir, 2004; Kurttio et al., 2005].

In spite of the recognized uranium nephrotoxicity, studies on the effects of chronic exposure through drinking water in humans are very limited. Zamora and co-workers studied the effects of chronic ingestion of U-contaminated water comparing two Canadian communities: One group drank water from private wells with 2–781 μg/L of U ($n = 30$), whereas the other ($n = 20$) drank only tap water with low concentration of U ($<1 \mu g/L$ of U). The exposure time ranged from 1 to 33 years in the low-concentration group and from 3 to 59 years in the high-exposure group. The results showed that glucose excretion increased with uranium intake in the highly exposed community when compared with the control population, as well as the alkaline phosphatase and $\beta 2$-microglobulin. The same was not observed for other biomarkers of kidney function, such as creatinine and protein. The chronic ingestion of drinking water contaminated with U was strictly related to the damage of proximal tubule, where reabsortion of glucose takes place [Zamora et al., 1998].

Similar results were presented by other works where no effects on glomerular tubule (excretion of creatinine and albumin) were observed [Kurttio et al., 2002, 2006]. Kurttio reported in addition a possible relationship between greater diastolic and systolic blood pressures and cumulative uranium intake [Kurttio et al., 2006]. Recent studies on mice seem to corroborate these findings: Goldman and colleges were the first to show a direct inhibitory dose- and pH-dependent effect of uranyl on the glucose transport system in isolated apical membrane from kidney cortex [Goldman et al., 2006].

Although all laboratory test and epidemiological studies on human communities quoted in this text agree on the chemical toxicity of U, there is still some lack of definition on the severity of this element. Nevertheless, due to the evidence of possible health effects of U in humans, the definition of drinking water quality guidelines is important, particularly in areas with relatively high background levels and industrial contamination of uranium [Weir, 2004]. Some countries have already adopted specific guidelines based on the potential of chemical toxicity of uranium supported by the available epidemiological data, as well as practical considerations for water treatment [Weir, 2004; WHO, 2004]. In Canada, the uranium guidelines for U concentration on drinking water was defined as $20 \mu g/L$, based on total daily uranium intake by a 70-kg adult ($2.6 \mu g$) assuming that food is the main supplier ($2.0 \mu g$) and water represents only $0.6 \mu g$ [Weir, 2004]. A different value is used in the United States: the Environmental Protection Agency (EPA) defined $30 \mu g/L$ as the maximum acceptable U concentration in drinking water. The differences between these guidelines are mainly based on water treatment cost.

2.2 Uranium Mining History in Portugal

After 100 years of mining history in Portugal, the degradation of the abandoned mining areas is, nowadays, an important and concerning environmental issue. During those years, the economical input of mining lead to the establishment of important human settlements in their vicinities. Contaminated soils and water with uranium radionuclides and other heavy metals, as well as radiation increase in mining area, requires the development of long-term remediation strategies in order to minimize the risk of exposure.

The exploration of uranium ore started in 1917 after the discovery, in 1907, of the Urgeiriça deposit in Central Portugal. During the first activity years, radium salts were the main production, entirely exported to France [JEN, 1966].

With the beginning of the atomic era, the exploration strategy changed. Uranium became the main product of uranium mines, and Portugal was no exception. After 1945, the production of U_3O_8 concentrates became a profitable industry in central Portugal [JEN, 1966]. Until the end of uranium mining activity in Portugal (2001), more than 4370 tons of uranium was produced [Carvalho et al., 2005]. A total of 61 mines (see Fig. 3) of different sizes and characteristics were exploited during this period mostly by open pit mines, in mineralized Hercynian granites. Only a few were underground works [Carvalho et al., 2005]. U_3O_8 concentrates were made in the main ore milling plants (Urgeiriça, Quinta do Bispo, Senhora dos Remédios, and Bica), where the major environmental impacts are observed. As a consequence of milling, a large amount of tailings (about 13 Mton) were produced [Antunes et al., 2007].

A recent report on the effects of mining activities in Urgeiriça and their effects on populations and environment was recently published [MinUrar, 2004, 2005, 2007]. According to this study, the increase of contamination with U radionuclides and heavy metals in soils, waters, and plants is limited to the proximity of the mine, reinforcing the need for the development of remediation strategies. The use of private water wells and springs as drinking water supply and the use of stream water for agriculture use and, consequently, the contamination of horticulture plants were identified as main exposure pathways to the local population. The use of tailing material for construction building, the exposure to dust, and radon emanation in the tailing are other important pathways.

The epidemiological study of the exposed population in Urgeiriça revealed that in this area, there is a decrease of thyroid and male reproductive function, as well as

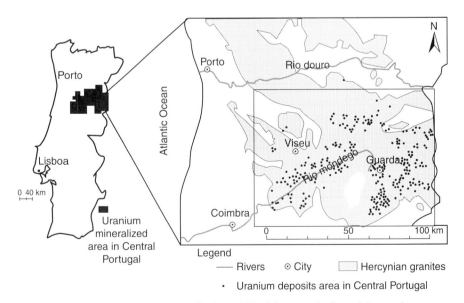

Figure 3. Geographical localization of U mining area in Central Portugal.

hematological and renal dysfunction. Basal chromosomal abnormalities and malignant neoplasia were also observed. However, when compared with a control population, a low statistical significance was found. Although the study did not conclude the cause of these health effects, it was suggested that the long-term exposure to the environmental contamination had an important role in the biological markers differences between the exposed and the control population.

3 PHYTOREMEDIATION OF METALS WITH AQUATIC PLANTS AS STRATEGIES FOR MINE WATER REMEDIATION

Phytoremediation has become an area of extreme research and development, and it stands now as a viable and promising option as a decontamination technique. The natural capacity that some plants have to tolerate, accumulate, or hyperaccumulate high metal concentration in their tissues is the basic concept of this metal decontamination process [Prasad and Freitas, 2003; Vanek and Schwitzguébel, 2003; Pilon-Smits, 2005].

Among the phytoremediation techniques, aquatic phytoremediation shows the highest potential. Studies with aquatic macrophytes communities are increasing considerably among the scientific community as a consequence of the growing importance that water resources management has to society. By definition, aquatic plants are those that complete their biological life cycle in water environments. This includes free-floating, submersed, and emergent life forms [Bracamonte and Domingo, 2002].

Their basic function in aquatic ecosystems is determinant, as they are not only responsible for oxygen production and for providing habitat and shelter for aquatic life, but they also actively manage the nutrient cycle, the sediment stabilization, and water quality through the mechanism of metal uptake [Brooks, 1998; Vardanyan and Ingole, 2006]. In fact, the ability of plants to accumulate metals from water, which may not be essential for their growth and development, has been observed in several studies performed in natural wetlands where the metal concentration in aquatic plants is several times higher than the concentrations levels on water.

This evidence has led to the generalized idea that metal hyperaccumulation in aquatic plants is not as rare as in terrestrial plants and that suitable and sustainable remediation strategies could be developed based on this characteristic [Rai et al., 1995; Brooks, 1998; Kamal et al., 2004; Fritioff and Greger, 2006; Nyquist and Greger, 2007]. Most of phytoremediation and bioindication studies use rooted aquatic plants, which are stationary and hence constantly exposed to metals in water; therefore, metals absorption is considered to be a continuous process [Samecka-Cymerman and Kempers, 2001; Vardanyan and Ingole, 2006].

Among the most common studied species are *Thypha sp.*, *Phragmites sp.*, *Lemna sp.*, and *Eichhornia sp.* but references to other metal accumulators macrophytes are also available. As result, the database on the remediation potential of aquatic plants is growing fast, and some examples of these metal accumulators are: *Eichhornia crassipes* (Mart.), *Solms* (U), *Lemna gibba* L. (As, U), *Azolla pinnata* R. Brown (Pb, Zn), *Myriophyllum aquaticum* (Vell.) Verdc. (Cu, Fe, Hg, Zn), *Myriophyllum spicatum* L. (Zn), *Ludwigina palustris* (L.) Ell. (Cu, Fe, Hg, Zn), *Ceratophyllum demersum* L. (Cu, Cr, Fe, Mn, Pb), and *Mentha aquatica* L. [Dushenkov et al., 1995; Rai et al., 1995; Zhu et al., 1999; Kamal et al., 2004; Mkandawire and Dudel, 2005; Mkandawire et al., 2005].

The development of sustainable strategies for water remediation based on the plant's ability to accumulate metals is focused on the study of plant physiology, environment, and also the biogeochemical cycle. All this knowledge can be combined and applied in man-made systems resembling the natural environment of the selected plants, the constructed wetland.

In the case of mine waste water, constructed wetlands systems have been effectively used since the 1980s in several countries [Gerth et al., 2005]. These systems combine the hydraulic, geochemical, and biological polishing processes based on the continuous metal removal by biomass growth and bacterial activity in the wetland sediments [Kalin, 1998; Kalin et al., 2005; Gerth et al., 2005]. The interest in this technique is still growing worldwide, and several constructed wetlands models are being developed. The use of algae is more usual than aquatic macrophytes because the tolerance to mine effluent characteristics (pH, contaminants) is higher in the first group [Kalin, 1998]. One of the latest success cases is the uranium Pölha mine constructed wetland (in Saxony, Germany) for mine drainage decontamination using algae (Characeae). This system is highly effective for reducing the metal concentration (namely, arsenic, iron, radium, and uranium) in water to acceptable legal limits with low impact and costs [Küchler et al., 2005; Kunze et al., 2007].

Some examples of macrophyte-based constructed wetland for mine water remediation are the one in the Schlema-Albedora uranium mine, in Germany, using *Phragmites australis* and *Carex gracillis* [Gerth et al., 2005], and the one in the Ranger Uranium Mine in Australia, using *Eleocharis dulcis* [Overall and Parry, 2004].

3.1 Uranium Accumulation in Aquatic Plants and Phytoremediation Studies

The available information about natural ability for U accumulation by aquatic plants is limited to a few species, and phytoremediation applications are even rarer in the literature. Despite all the advantages suggested for these remediation techniques (high kinetics rates, the hyperaccumulation of one or several metals, the possibility of in situ treatment of large volumes of water, and the significant reduction of secondary wastes), the information of field applications is rather sparse, revealing that there are still some limitations in transferring this technique from the laboratory studies to field applications [Brooks, 1998].

In mining areas, plants usually reflect the mineral composition of soil and waters. The identification of these plants is the most adopted procedure for mineral prospecting [Brooks, 1998; Pratas et al., 2005a]. The success of any phytoremediation technique depends also on the plant selection, and indigenous plants should be preferred since they are more tolerant to substrate and seasonal modifications [Brooks,

1998; Fritioff and Greger, 2006]. Sometimes, when this principle is not followed, some field applications are not functional.

Pettersson et al. [1993] identified the water lily (*Nymphaea violacea*) as an accumulator of several radionuclides when they observed high levels of ^{234,238}U, ^{228,230,232}Th, ^{226}Ra, ^{210}Pb, and ^{210}Po in the plant, waters, and sediments in the vicinity of the Ranger Uranium mine (Australia). Higher levels of these contaminants were detected in the roots and rhizomes rather than in the superior tissues. The correlation between sediment contamination and plant uptake was more significant than water/plant association. These findings suggest root uptake from sediment as the main uptake mechanism [Pettersson et al., 1993].

At the same mine, a recent attempt for phytoremediation of mine runoff water was tested using *Eleocharis dulcis*. Overall and Parry described the 6-ha constructed wetland filter as an effective process of U and Mg attenuation in water as the *E. dulcis* rapidly increase biomass and metal uptake [Overall and Parry, 2004]. The uranium concentration in water column and in sediment contributes differently for metal accumulation in the various tissues of *E. dulcis*. The aquatic plants mechanism for metal uptake depends on the type of plant, but it occurs essentially by direct absorption from the water column to the plant surface followed by passive or active transport across membranes and in a minor scale by root uptake [Rai et al., 1995; Brooks, 1998; Nyquist and Greger, 2007]. This is observed, especially in submersed species, because of their poorly developed root system, and it is also observed in free-floating plants [Brooks, 1998]. However, the root uptake in plants with highly developed root systems can also be effective, as proved by *E. dulcis* with higher uranium levels at roots when compared with stems. It has been suggested that the low accumulation in stems has an important advantage over this system because metal cycling and resuspension following the decay of stems is minimized [Overall and Parry, 2004]. Other studies state the opposite: Rhizospheric processes may affect the metals speciation in the sediment, enhancing their mobility by altering redox and pH conditions [Jacob and Otte, 2003].

Another example of a U accumulator plant is *Lemna gibba*, which has been identified by the work of Mkandawire [Mkandawire and Dudel, 2005]. This plant belongs to the Lemnacea family, which is one of the most studied in phytoremediation and intensively described in literature, namely, *Lemna minor* and *Spirodella polyrizha* [Miretzky et al., 2004; Pratas et al., 2005b; Rahman et al., 2007]. According to these studies, the fast growth rate, their widespread distribution in natural wetlands, the total independence from sediment, and the adaptation to stress conditions, like mine waters, make the Lemanacea family a good option for water treatment technologies, in spite of the constant need of biomass removal.

The potential use of *L. gibba* in uranium phytoremediation was first suggested after the observation of bioaccumulation of U and also As in the ponds of the Lengenfeld uranium mine (Germany) [Mkandawire and Dudel, 2005]. Concentrations of uranium ranged between 514 and 612 mg/kg DW (dry weight) in waters with 186–277 µg/L of U. The concentration of arsenic was also much higher than the surrounding water (plant, 61.7–1966 mg/kg DW; water, 3–265 µg/L of As). The tolerance to U contamination in water is reported by these authors as being possibly related with a biomineralisation process of U in *L. gibba* by exudation of uranyl oxalates as a response to oxidative stress [Mkandawire et al., 2005]. According to the authors, the U immobilization by these macrophytes would present a durable and clean technology for decontamination of polluted mine waters.

Other aquatic plants suggested as U accumulators are *Zostera japonica* and *Zostera marina* [Kondo et al., 2003]; *Phragmites australis* and *Carex gracillis*, [Gerth et al., 2005], and *Phragmites australis* and *Thypha latifolia* [Soudek et al., 2007].

A different use of aquatic plants for U phytoremediation was proposed by Bhainsa and D'Souza [2001] and by Shawky et al. [2005]. These authors propose the use of the dry roots biomass of water hyacinth (*Eichhornia crassipes*) for uranium sorption. In both studies, the high kinetics of U uptake is suggested as the principal advantage of their use. At least 54% of initial U concentration in water (200 mg/L) was removed by the dried roots of *E.crassipes* within 4 min of contact time; after 2 hr, 80% of U was removed [Bhainsa and D'Souza, 2001]. This remediation concept seems to be very efficient, and the use of a small reactor and the shorter residence time is a significant practical advantage [Bhainsa and D'Souza, 2001]. However, considering a continuous water remediation process, this could turn to be very expensive due to the saturation of the biomass, requiring frequent reactor substitution.

4 CASE STUDY: WATER DRAINAGE FROM URANIUM MINES CONTROL BY AQUATIC PLANTS IN CENTRAL PORTUGAL

As mentioned previously in this chapter, the uranium mines of central Portugal have some contribution to the contamination of soils and water, namely, in their vicinities, representing a cumulative impact on the area of contamination.

This evidence is particularly important because rivers, wells, and springs are very abundant in this region, integrating the hydrological system of the Mondego River, one of the most important river basins of Portugal. Because most of U mines are located in farming areas, the water is extensively used on agriculture and, in some cases, in animal and population drinking purposes [Antunes et al., 2007].

The actual state of degradation of most mines led us to study alternative and low-cost solutions based on phytoremediation techniques. The development of a biological polishing system for the final mine effluent is the main objective of this study.

A complete screening of the aquatic macrophytes occurring in the streams of the mining area was made, followed by laboratory experiments in order to determine the phytoremediation potential of these plants. With this procedure it was possible to gather information on abundance, geographical distribution, and tolerance to contamination of several species as well as the contamination of streams waters in U mining areas.

4.1 Selection of Aquatic Macrophytes: Field Studies

Water and plants were collected in running waters in the uranium mineralized area of Central Portugal (see Fig. 3) and, particularly, near the sources of effluent discharge from the uranium mines. The sampling was made in a month (May 2006), and more than 71 species from 41 different families were identified. A total of 185 field observations were made.

Water pH was determined with a pH meter in the field, and water was transported in clean and sterilized plastic bottles. The plants were also transported in clean plastic bags to the laboratory. All sample preparation and analysis was made in the geochemical laboratory of the Earth Science Department of the University of Coimbra as described in the following paragraphs. Plants were identified with the help of local herbarium and floras [Franco, 1971, 1984; Franco and Afonso, 1994, 1998; Sampaio, 1998; Bracamonte and Domingo, 2002].

For determination of U, As, and other metals, the water samples were filtered and the pH was adjusted to 2 with HNO_3, 65% (V/V).

The determination of the uranium mass concentration in water samples was made by standard fluorimetric analysis on "Fluorat-02-2M" analyzer. For quality control purpose, a certified reference water produced by the National Water Research Institute of Canada (reference TMDA-62) was tested.

To define the chemical characteristics of the stream water and also the occurrence of other heavy-metal contamination related to the mineralization, the elements As, Ca, Cu, Fe, K, Mg, Mn, Na, Ni, and Zn were measured by atomic absorption spectroscopy, with SOLAAR M Series equipment from Thermo–Unicam.

The preparation of plant material included, when appropriate, the separation in roots and aerial tissues, followed by rinse of the samples, first in tap water and then in distilled water, in order to withdraw all the residues. Afterwards they were dried in an oven at $60°C$. After drying, they were ground and crushed for later chemical analysis. Plant material was divided in two different samples for uranium and other metals analysis (As, Cu, Zn).

Fluorometry was the methodology adopted for the determination of uranium content in plants, as described in the work of Huffman and Riley [1970] and Van Loon and Barefoot [1989]. Certified Virginia tobacco leaves (reference CTA-VTL-2, Polish certified reference material) were also analyzed as a control.

For other elements analysis, plants were prepared by microwave digestion with an $HNO_3 - H_2O_2$ mixture in closed Teflon vases (Multiwave 3000, Anton Paar). The analysis was performed in the same way as stated for water.

4.1.1 Uranium and Other Metals Concentrations in Water.

A summary of the chemical analysis of water samples is presented in Tables 2 and 3. Uranium was the only metal to which it was possible to observe a significant anomalous water enrichment. The observations made within the 185 sampling sites showed that U was detected in the surficial waters in a wide range of concentrations (0.23–1220 $\mu g/L$). From geographical interpretation of the sampling locations (for example, see Fig. 4) it was clear that mine effluents have a significant contribution in the water contamination and therefore, the distinction between the sampling points affected by mine drainage was based on these criteria (15 observations). The natural background for U in superficial waters in the studied area was defined as being 1.8 $\mu g/L$ (determination made by PCA, not presented in this chapter). In most of the

TABLE 2. Water Chemical Analysis of Noncontaminated Streams ($n = 170$)

	pH	U (μg/L)	As (μg/L)	Ca (mg/L)	Cu (μg/L)	Fe (mg/L)	K (mg/L)	Mg (mg/L)	Mn (μg/L)	Na (mg/L)	Ni (μg/L)	Zn (mg/L)
Mean	6.34	1.76	4.60	8.68	6.22	0.27	3.43	4.44	100.66	18.24	12.25	0.02
SD	0.50	1.31	5.45	8.96	6.41	0.48	3.19	11.07	283.93	12.62	25.33	0.05
Max	8.40	9.39	40.18	48.03	23.77	2.41	13.66	80.71	1614.81	80.06	159.05	0.34
Min	5.65	0.23	0.15	1.17	0.42	0.00	0.34	0.85	0.54	8.06	0.89	0.01

TABLE 3. Water Chemical Analysis of Contaminated Stream Water by Mine Drainage ($n = 15$)

	pH	U (μg/L)	As (μg/L)	Ca (mg/L)	Cu (μg/L)	Fe (mg/L)	K (mg/L)	Mg (mg/L)	Mn (μg/L)	Na (mg/L)	Ni (μg/L)	Zn (mg/L)
Mean	5.84	139.40	3.21	40.32	26.88	1.06	2.36	11.09	510.67	162.20	37.35	3.04
SD	0.82	311.66	3.70	69.23	38.28	1.50	1.87	14.20	688.20	505.54	52.74	9.26
Max	7.30	1220.40	11.75	218.27	139.42	4.36	6.10	42.94	1697.41	1767.04	186.98	32.33
Min	4.48	11.32	0.21	0.21	2.66	0.04	0.03	0.92	3.09	7.57	2.29	0.01

sampled sites, the U concentration was below this value. Differences were only observed when comparing non-contaminated sites (Table 2) with streams directly fed by mine drainage (Table 3). The mean concentration of U in water was higher in these sampling sites (139.4 μg/L) due to two locations with high concentrations, near Cunha Baixa and Urgeiriça mine sites (see Fig. 4).

In Tables 2 and 3 the concentrations of the other analyzed cations are given for both the contaminated and non-contaminated streams, in order to demonstrate that

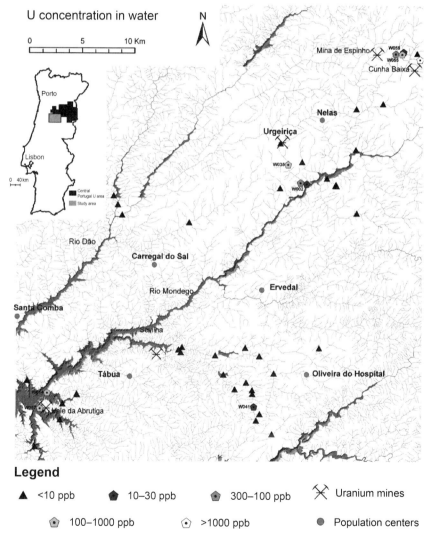

Legend

▲ <10 ppb   ⬟ 10–30 ppb   ⬢ 300–100 ppb   ✕ Uranium mines

⬠ 100–1000 ppb   ⊙ >1000 ppb   ● Population centers

Figure 4. Uranium concentration in water [ppb = μg/L] in the most contaminated are by mining influence in Central Portugal (number of sample sites in this area: 61).

the general water chemistry is basically the same. The differences, which may be apparent, are likely due to outliers and not significant given the differences between the number of samples.

4.1.2   U and Heavy Metal Accumulation in Aquatic Plants.

The metal analysis in plants from the uranium Region of Beiras (Central Portugal) revealed that U and As were highly accumulated by some species. The results of the analytical measurements on the most frequent aquatic plants in the studied area are presented in Table 4. High bioaccumulation levels were observed in several species and in a magnitude much higher than the concentration in the surrounding water. Prospects for future development of phytoremediation systems to U mine water decontamination is very high.

The highest concentrations of U were found in the submerged species *Callitriche stagnalis* Scop (1948 mg/kg DW), in *Potamogeton natans* L. (94.50 mg/kg DW), in *Potamogeton pectinatus* L. (364.84 mg/kg DW), and in the free-floating *Lemna minor* L. (42.46 mg/kg DW). In contrast, the measured concentrations in emergent plants such as *Apium nodiflorum* (L.) Lag., *Oenanthe crocata* L., *Typha latifolia* L., and *Juncus effusus* L were significantly lower when compared with the previously species, even in the roots.

Similar conclusions can be made for As bioaccumulation. Once more, in submerged and free-floating plants the concentrations were much higher than in emergent plants, with the exception of *O. crocata*. The highest concentrations of As were found in the Callitrachaeae family plants (*C. stagnalis*, 354.03 mg/kg DW; *C. brutia*; 436.92 mg/kg DW).

Equal behavior was reported by Robinson et al. [2006] in the Taupo Volcanic Zone, in New Zealand, where arsenic concentrations of 4215 mg/kg in *C. stagnalis* and 422 mg/kg (DW) in *C. petrie*, in dried biomass, was found in waters with high As concentration (mean of 90 ppb). Other metals, such as Cu and Zn, are also accumulated by plants from the Callitrichaceae family, namely, Cu by *C. verna* [Samekcka-Cymerman and Kempers, 2001] and Cu and Zn by *C. stagnalis* with maximum concentration values of 132 mg/kg and 1395 mg/kg in dry biomass, respectively [Paulo, 2006].

The abundance of *Callitriche stagnalis* and the ability to accumulate several heavy metals at the same time made this plant our first choice for biopolishing methodologies development.

Its selection was reinforced by the observation in Ribeira da Pantanha, a small stream that collects water for the Urgeiriça mine. The seepage and runoff from mine tailings is drained to this stream after treatment (acid neutralization, co-precipitation of radionuclides, and pH adjustment) [Carvalho et al., 2005]. However, high levels of U contamination are still detected downstream in Ribeira da Pantanha. Both concentrations in water and *C. stagnalis* are illustrated in Fig. 5.

As distance from tailings increase, the concentration in water and plants decrease. Although, the immobilization of this metal downstream is mainly related with the activity of sulphate and iron-reducing bacteria in anoxic sediments

TABLE 4. Uranium and Arsenic Concentrations (mg/kg Dw) on the Most Frequent Aquatic Plants in the Studied Area

Family/Plant	n	Uranium				Arsenic			
		Mean	SD	Max	Min	Mean	SD	Max	Min
Callitrichaceae									
Callitriche stagnalis Scop.	131	34.51	196.32	1948.41	0.01	20.38	45.67	354.03	0.12
Callitriche brutia Petagna	43	4.03	3.28	14.56	0.97	45.28	91.74	436.92	0.42
Callitriche hamulata Koch	26	2.67	1.43	6.35	0.28	21.55	37.62	160.37	0.16
Callitriche lusitanica Schotsman	21	4.56	11.28	52.23	0.23	21.78	19.77	62.44	3.73
Ranunculaceae									
Ranunculus trichophyllus Chaix	64	4.95	9.29	65.82	0.54	23.73	40.23	268.53	1.36
Ranunculus peltatus L.	5	5.11	6.03	15.11	0.76	33.90	40.91	103.98	3.80
Ranunculus penicilatus (Durmot.) Bab.	3	1.71	1.41	3.13	0.31	11.58	13.52	27.19	3.76
Ranunculus tripartitus DC.	6	1.45	1.42	3.73	0.16	6.40	7.84	20.27	1.00
Potamogetonaceae									
Potamogeton natans L.	11	15.33	33.60	94.50	0.17	2.97	4.41	13.04	0.12
Potamogeton pusillus L.	8	1.02	0.71	2.64	0.43	19.35	18.74	62.51	2.38
Potamogeton pectinatus L.	2	194.81	240.46	364.84	24.77	15.90	4.90	19.37	12.44
Lemnaceae									
Lemna minor L.	92	2.71	5.05	42.46	0.08	20.86	45.24	279.42	0.06
Lemna minuta Kunth	6	2.03	0.86	3.23	0.85	19.98	28.56	74.62	0.50
Spirodella polyrizha L.	5	4.10	3.20	9.65	1.81	10.63	20.09	46.48	0.42
Umbelliferae									
Apium nodiflorum (L.) Lag.	31	4.17	7.09	31.61	0.20	3.91	4.00	21.38	0.61
Oenanthe crocata L.	20	2.24	7.78	35.15	0.00	10.10	34.91	157.94	0.29
Typhaceae									
Typha latifolia L.	14								
Leafs		1.74	2.97	9.51	0.08	0.46	0.87	2.95	0.00
Roots		107.37	115.30	268.69	9.42	1.26	1.76	4.17	0.03
Juncaceae									
Juncus effusus L.	11								
Leafs		10.38	29.77	99.93	0.04	0.43	1.22	4.10	0.01
Roots		38.02	51.37	132.26	0.54	3.87	5.96	14.35	0.01

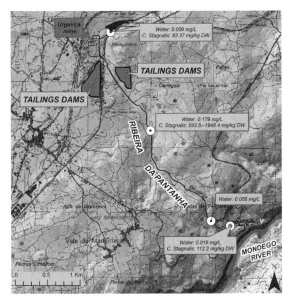

Figure 5. Water (mg/L) and *C. stagnalis* (mg/kg DW) concentration dowstream Ribeira da Pantanha.

[Schöner et al., 2005; Carvalho et al., 2005], the high levels of U observed in *C. stagnalis* suggest that these plants may have an important role in the biogeochemical uranium cycle in natural stream waters.

Other species with high metal uptake, such as *P. natans*, *P. pechinatus* and *L. minor*, may also be used as biological polishing applications either in monoculture systems or in a combined system with *C. stagnalis* resembling the natural systems.

4.2 Laboratory Experiments: Uranium Accumulation by *C. stagnalis*

To investigate the U accumulation by *C. stagnalis*, these plants were exposed during 7 days to different concentrations. By making use of this laboratory procedure, we intended to validate the field observations establishing a relationship between water contamination and plant accumulation.

A clean setup of plants cultures of *Callitriche stagnalis* was collected from a nonuraniferous area (Ribeira de Alveite, Poaires, Centre of Portugal; see Table 5).

They were kept in aquariums in the laboratory with water from that area, for a 48-hr acclimatization period. Plants at the same stage of growth, with similar size and nearly with the same fresh weight, were distributed in 300-ml glass containers.

For the uranium test, solutions of 40, 75, 150, 250, 500, 1000, 2000, and 4000 μg/L were prepared by contamination of water from Ribeira de Alveite with a solution of $UO_2(NO_3)_2$. Each container was filled with a fixed volume of 250 ml,

TABLE 5. Reference Stream—Ribeira de Alveite, Poaires: Water Chemical Characterization

U (μg/L)	Na (mg/L)	NH_4 (mg/L)	K (mg/L)	Mg (mg/L)	Ca (mg/L)	F (mg/L)	PO_4 (mg/L)	Cl (mg/L)	NO_2 (mg/L)	NO_3 (mg/L)	SO_4 (mg/L)
0.7	9.11	0.34	1.27	3.93	76.12	0.08	0.03	17.25	0.07	10.48	9.33

and the plants were added to this solution after being rinsed with a 10% sodium hypochlorite and washed in distilled water.

The tests were set up in a room with constant temperature ($20 \pm 3°C$), and plants were exposed only to natural light (12-hr photo-period). During the experiment, the solution pH was approximately 6.5, similar to the mean value of natural water from the studied area. Three replicates were used for each concentration set, including the controls.

Plants were harvested after 7 days, and they were immediately rinsed with distilled water. The water of each container was filtered and acidified. Analysis of U was performed as described in the previous section of this chapter.

The results are presented in Figs. 6 and 7. This test confirmed the ability of *C. stagnalis* to concentrate U and the bioaccumulation coefficient (BAC = concentration in plant/concentration in water) was 3.4×10^3 (average), confirming its high potential for uranium phytoremediation.

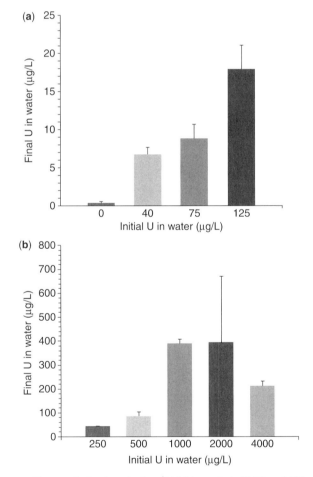

Figure 6. Average U concentration in water ($\mu g/L$) (**a**) in control, 40, 75, and 125 $\mu g/L$ container and (**b**) in 250, 500, 1000, 2000, and 4000 at $\mu g/L$ container. Error bars represent SD ($n = 3$).

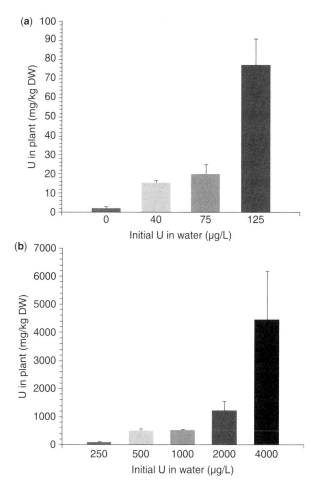

Figure 7. Average U concentration in *C. stagnalis* (mg/kg DW) (**a**) in control, 40, 75, and 125 μg/L container and (**b**) in 250, 500, 1000, 2000, and 4000 at μg/L container. Error bars represent SD ($n = 3$).

We observed a decrease in U concentration in water of 82% (on average). For U concentrations in water between 40 and 125-μg/L, it was possible with this simple experiment to decrease the U concentration to levels below the limit established by EPA (30 μg/L) (see Fig. 6a). Above 125 μg/L the same efficiency was not observed, probably due to metal saturation in plant tissue binding sites or to U toxic effects, which were not evaluated.

As mentioned previously, since uranium is not an essential element for plants, the main process of metal accumulation may be the passive uptake instead of a specific mechanism. The uptake is driven by physical–chemical interaction between metal cationic forms dissolved in water and negatively charged binding sites in plant cell wall surface because of the differentiate affinity with functional groups (ligands) like hydroxyl (—OH), phosphoryl (—PO$_3$O$_2$), amino (—NH$_2$), carboxyl (—COOH), and sulfhydryl (—SH) [Kalin et al., 2005; Nyquist and Greger, 2007]. Moreover, some of the dissolved cationic species (Ca^{2+}, Mg^{2+}) and uranyl ion may compete for these binding sites in the cell walls due to structural similarity [Kalin et al., 2005; Fritioff and Greger, 2006; Nyquist and Greger, 2007].

Although in these experiments it was not intended to define a toxicity threshold for *C. stagnalis* it was observed that for concentrations above 1 mg/L the plant decay was much faster than in the other concentrations. The exposure to uranium concentration in water higher than 2000 µg/L revealed toxic to *C. stagnalis* after 4 days when signals of intense chlorosis was observed. Similar oxidative stress signals were not observed for the concentrations of 40–1000 µg/L and in the control. However, the authors did not exclude the possibility that other factors, rather than metal concentration in water, produced these toxic effects in plants, namely, the possible lack of nutrients.

The rapid decay and decomposition of the plant in 2000 and 4000 µg/L solutions is certainly the cause for the high bioaccumulation verified in those solutions (see Fig. 7b), because of the high affinity between dissolved uranyl ion and organic matter [Bourdon et al., 2002]. This may also have implications on the technique success, that is, at these concentrations, U can be easily released from the plant and the water treatment goal is not achieved.

4.3 Phytoremediation Laboratory Prototype

After combining the information from both field studies and laboratory experiments, a laboratory-scale prototype of a multispecies biological polishing system was developed, as described in Paulo et al. [2006].

In this test, *C. stagnalis*, *P. natans*, and *P. pechinatus* plants were used to decontaminate 70 liters of a modified Murashige and Skoog hydroponic solution, with similar chemical composition of studied natural streams. This water was contaminated with 500 µg/L of U. The phytoremediation test lasted 15 days, and plant and water samples were collected after 1, 2, 7, and 15 days of continuous water circulation. The total biomass in this system was 90 g. Because of the short period of functioning, the biomass was the same at the end of the experiment.

The use of these plants in water treatment proved to be extremely efficient (see Figs. 8 and 9). All three species showed high U bioaccumulation and high uptake kinetics. The results described show a reduction of U concentration, in the water, from 500 to 72.32 µg/L of U, thus representing an efficiency of 85.54%. In *C. stagnalis* the concentration increased from 0.98 to 1567 mg/kg, in *P. natans* it increased from 3.46 to 270.9 mg/kg, and in *P. pectinatus* it increased from 2.63 to 1588 mg/kg.

This test suggested that these species have an important contribution in the uranium immobilization in natural streams systems by sequestering the metal in their tissues and, by this means, reducing the dispersion of the contaminant. However, to further improve our system, important issues need to be clarified for the validation of the phytoremediation potential of this plant assemblage. Uptake mechanism comprehension, metal retention time in plant, and saturation and decay processes are areas that need further investigation. In this test, we observed a decrease

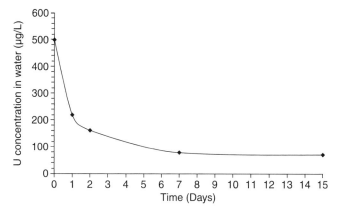

Figure 8. Uranium concentration in water during the phytoremediation tests [Paulo et al., 2006].

of U accumulation with time, particularly in *P. natans*, where U concentration remained stable after 2 days of exposure to the contaminated water. Assuming that U adsorption was the main accumulation mechanism, the observation may reflect the saturation of the metal-binding sites of *P. natans*, limiting its phytoremediation capacity in a long-term strategy. The same is expected to the other two species. The decay of plant material can also occur in a short time, and uranyl can be undesirably released from the biological material and be resuspended in solution.

The need of frequently harvesting and treatment of the contaminated biomass suggested by most of phytoremediation studies could become expensive and impracticable. This procedure is, in fact, the major disadvantage of the phytoremediation techniques.

This limitation can be surpassed through the full knowledge of the U biogeochemical cycle: We need to identify and control the interactions between plants, sediment, bacteria, and physical–chemical properties of the water in wetland systems

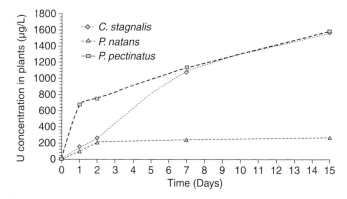

Figure 9. Uranium accumulation in plant biomass [Paulo et al., 2006].

[Kalin et al., 2005]. This ecological approach was suggested and effectively proved by Kalin and co-workers, advocating the importance of the biogenetic process of U ore body genesis in the immobilization of U in sediments. Plants actively filter U from water; and upon death, in combination with a low Eh environment, sulfates, and minerotrophic microbial systems, uranium can be immobilized as solid phases formed by chemical reactions with sulfates or other anions present in the sediments. The design of long-term ecological remediation approaches should be based on these natural models [Kalin et al., 2005].

The future of our investigation will follow this path. Moreover, the possible effects on the food chain should be the next step considered in future research.

5   FUTURE PROSPECTS OF WATER PHYTOREMEDIATION

Water resources are sensitive environmental systems. Their degradation can severely affect the ecosystems and, direct or indirectly, impact on the health and well-being of a community or area. The anomalous enrichment of water with trace metals, released from anthropogenic activities such as mining, is a determinant for the possible biomagnification of this contamination through the food chain.

The improvement of remediation techniques for mining operations must be a priority, as toxic effect information resulting from long-term exposure to trace metals becomes more available, along with the worldwide dependence of water.

The majority of the epidemiological studies presented in this chapter, despite the low statistical significance, agree that there are significant dissimilarities in biological functions when comparing the mining area population with a population of nonmining areas. Functional and histological kidney lesions caused by uranium toxicity are identified as the most common pathologies in laboratory assays. However, long-term data from epidemiological studies are still needed in order to better understand and validate the magnitude of uranium toxicity in human population. This information is a determinant to the establishment of water quality guidelines.

Despite the low efficiency of uranium absorption by ingestion, water is by far a much broader way of affecting people. The high mobility of uranyl cation in water enables a much higher dispersion than in the dust particles, for example. Inhalation is a much more dangerous contamination pathway in the case of mine workers and people who have direct contact with tailings. Water remediation and the information of populations' exposure risk to U must be adopted in contaminated areas where the population still uses private wells as a drinking water source.

The interest on phytoremediation techniques for mine water decontamination, such as constructed wetlands, creates a great expectation for future mine reclamation solutions. Some of the field applications mentioned in this chapter contributes significantly to this acceptance.

The study of plant communities, their interaction with pollutants, a better knowledge of plant accumulation mechanisms in the laboratory, and the means of metal immobilization in wetlands (biogeochemical cycles) are crucial steps of this research and a strong contribution for the improvement of constructed wetland efficiency.

Selection of indigeneous plants with high ability to concentrate metals is advantageous. Such selection must take into account features like biomass, growth rate, kinetics, and decay, among others. A possible integration in the food chain is a feature that shouldn't be neglected. For example, the use of the accumulator water-lily, identified in the vicinities of the Ranger Uranium mine, may be compromised because of its importance in the local aboriginal diet and, therefore, is a potential pathway for human contamination [Pettersson et al., 1993].

Constructed wetlands, based on natural wetlands dynamics, besides the attractive economical character, if correctly designed, will favor the natural capacity of ecosystems restoration, allied to the reasonable use of resources and to natural species and habitat conservation, based on sustainability concepts.

However, despite the advantages, phytoremediation techniques are still uncommon; and to state that water phytoremediation has a valid and competitive technology, the proven field applications must increase.

ACKNOWLEDGMENTS

This research was financially supported by the Portuguese Science Foundation, Project POCI/ECM/60750/2004. This work is part of the Master Thesis of the first author and he thanks Dr. João Pratas for the support, supervision and contribution to this work.

REFERENCES

Antunes SC, Pereira R, Gonçalves F. 2007. Acute and chronic toxicity of effluent water from an abandoned uranium mine. *Arch Environ Contam Toxicol* **53**(2):207–213.

ATSDR. 1999. Toxicological Profile for uranium. Atlanta, GA: U.S. Department of Health and Human Services, Public Health Service.

Bhainsa KC, D'Souza SF. 2001. Uranium (VI) biosorption by dried roots of *Eichhornia crassipes* water hyacinth. *J Environ Sci Health* **A369**:1621–163.

Boice JD, Mumma MT, Schweitzer S, Blot WJ. 2003. Cancer mortality in a Texas county with prior uranium mining and milling activities, 1950–2001. *Radiol Prot* **23**:247–262.

Boice JD, Mumma MT, Blot WJ. 2007. Cancer and noncancer mortality in populations living near uranium and vanadium mining and milling operations in Montrose County, Colorado, 1950–2000. *Radiat Res* **167**(6):711–726.

Bourdon B, Turner S, Henderson GM, Lunstron C. 2002. Introduction to U-series geochemistry. In: *Reviews in Mineralogy & Geochemistry* Vol. 52. The Geochemical Society, Mineralogical Society of America.

Bracamonte SC, Domingo LM. 2002. *Plantas aquáticas de las lagunas humedales de Castilla-la Mancha*. Madrid: Real Jardín Botànico.

Brooks RR. 1998. *Plants that Hyperaccumulate Heavy Metals: Their Role in Phytoremediation, Microbiology, Archeology, Mineral Prospecting and Phytomining*. Robert Brooks, editor Wallingford, UK: CAB international.

Brüske-Hohlfeld I, Rosario AS, Wolke G, Heinrich J, Kreuzer M, Kreienbrock L, Wichmann HE. 2006. Lung cancer risk among former uranium miners of the WISMUT Company in Germany. *Health Phys* **90**(3):208–216.

Carvalho F, Oliveira J, Madruga M, Lopes I, Libânio A, Machado L. 2005. Contamination of hydrographic basins. In: Merkel B, Hasche-Berge A, editors. *Uranium in the Environment: Mining Impact and Consequences.* Berlin: Springer-Verlag, pp. 691–702.

Durakoviae A. 1999. Medical effects of internal contamination with uranium. *Croatian Med J* **40**(1):49–66.

Dushenkov V, Kumar PBAN, Motto H, Raskin I. 1995. Rhizofiltration: The use of plants to remove heavy metals from aqueous streams. *Environ Sci Technol* **29**(5):1239–1245.

Franco JA. 1971. *Nova Flora de Portugal*, Vol. I. Lisboa: Escolar Editora.

Franco JA. 1984. *Nova Flora de Portugal*, Vol. II. Lisboa: Escolar Editora.

Franco JA, Afonso MR. 1994. *Nova Flora de Portugal*, Vol. III(I). Lisboa: Escolar Editora.

Franco JA, Afonso MR. 1998. *Nova Flora de Portugal*, Vol. III(II). Lisboa: Escolar Editora.

Fritioff Å, Greger M. 2006. Uptake and distribution of Zn, Cu, Cd, and Pb in an aquatic plant *Potamogeton natans*. *Chemosphere* **63**(2):220–227.

Gerth A, Hebner A, Kiessig G, Zellmer A. 2005. Passive treatment of minewater at the Schlema Alberoda site. In: Merkel B, Hasche-Berge A, editors. *Uranium in the Environment*: *Mining Impact and Consequences*. Berlin: Springer-Verlag, pp. 409–414.

Goldman M, Yaari A, Doshnitzki Z, Cohen-Luria R, Moran A. 2006. Nephrotoxicity of uranyl acetate: Effect on rat kidney brush border membrane vesicles. *Arch Toxicol* **80**(7):387–393.

Grosche B, Kreuzer M, Kreisheimer M, Schnelzer M, Tschense A. 2006. Lung cancer risk among German male uranium miners: a cohort study, 1946–1998. *Br J Cancer* **95**(9):1280–1287.

Huffman C, Riley LB. 1970. The fluorimetric method—Its use and precision for determination of uranium in the ash of plants. *U.S. Geol. Survey Prof. Paper* **700B**:181–183.

Hussain SP, Kennedy CH, Amstad P, Lui H, Lechner JF, Harris CC. 1997. Radon and lung carcinogenesis: Mutability of p53 codons 249 and 250 to 238Pu alphaparticles in human bronchial epithelial cells. *Carcinogenesis* **18**:121–125.

Jacob DL, Otte ML. 2003. Conflicting processes in the wetland plant rhizosphere: Metal retention or mobilization? *Water Air Soil Pollut: Focus* **3**(1):91–104.

JEN. 1966. *O urânio e outras matérias primas nucleares em Portugal*. Lisboa: Junta de Energia Nuclear.

Jostes RF. 1996. Genetic, cytogenetic, and carcinogenic effects of radon: A review. *Mutat Res* **340**(2–3):125–139.

Kalin M, Wheeler WN, Meinrath G. 2005. The removal of uranium from mining waste water using algal/microbial biomass. *J Environ Radioactivity* **78**:151–177.

Kalin M. 1998. Biological polishing of zinc in a mine waste management area. In: Geller W, Klapper H, Salomons W, editors. *Acid Mining Lakes: Acid Mine Drainage, Limnology and Reclamation*, Berlin, Springer-Verlag, pp. 321–334.

Kamal M, Ghaly AE, Mahmoud N, Cote R. 2004. Phytoaccumulation of heavy metals by aquatic plants. *Environment International*, **29**(11):1029–1039.

Kondo K, Kawabata H, Ueda S, Hasegawa H, Inaba J, Mitamura O, Seike Y, Ohmomo Y. 2003. Distribution of aquatic plants and absorption of radionuclides by plants through the leaf surface in brackish Lake Obuchi, Japan, bordered by nuclear fuel cycle facilities. *J Radioanal Nuclear Chem* **257**:305–312.

Kreuzer M, Grosche B, Brachner A, Martignoni K, Schnelzer M, Schopka HJ, Brüske-Hohlfeld I, Wichmann HE, Burkart W. 1999. The German uranium miners cohort study: Feasibility and first results. *Radiat Res* **152**:56–58.

Küchler A, Kiessig G, Kunze C. 2005. Passive biological treatment systems of mine waters at WISMUT sites. In: Merkel B, Hasche-Berge A, editors. *Uranium in the Environment: Mining Impact and Consequences*. Berlin: Springer-Verlag, pp. 329–340.

Kunze C, Kiebig G, Küchler A. 2007. Management of passive biological water systems for mine effluents. In: Marmiroli N, Samotokin B, Marmiroli M, editors. *Advanced Science and Technology for Biological Decontamination of Sites Affected by Chemical and Radiological Nuclear Agents*. NATO SCIENCE SERIES: IV Earth and Environmental Sciences, Vol. 75. Netherlands: Springer, pp. 177–196.

Kurttio P, Auvinen A, Salonen L, Saha H, Pekkanen J, Mäkeläinen I, Väisänen SB, Penttilä IM, Komulainen H. 2002. Renal effects of uranium in drinking water. *Environ Health Perspect* **110**(4):337–342.

Kurttio P, Komulainen H, Leino A, Salonen L, Auvinen A, Saha H. 2005. Bone as a possible target of chemical toxicity of natural uranium in drinking water. *Environ Health Perspect* **113**(1):68–72.

Kurttio P, Harmoinen A, Saha H, Salonen L, Karpas Z, Komulainen H, Auvinen A. 2006. Kidney toxicity of ingested uranium from drinking water. *Am J Kidney Dis* **47**(6):972–982.

Lottermoser BG. 2003. *Mine wastes—Characterization, Treatment and Environmental Impacts*. Berlin Heidelberg: Springer-Verlag.

Malczewska B, Myers O, Shuey C, Lewis J. 2003. *Recommendations for Uranium Health Based Groundwater Standard*. New Mexico: Environment Department, Ground Water Quality Bureau.

MinUrar. 2004, 2005, 2007. *Minas de Urânio e seus resíduos: Efeitos na saúde da população*. Relatório de progresso. Instituto Nacional DE Saúde Dr° Ricardo Jorge, Lisboa (Technical report).

Miretzky P, Saralegui A, Cirelli AF. 2004. Aquatic macrophytes potential for the simultaneous removal of heavy metals Buenos Aires, Argentina. *Chemosphere* **57**(8):997–1005.

Mkandawire M, Dudel EG. 2005. Accumulation of arsenic in Lemna gibba L. duckweed in tailing waters of two abandoned uranium mining sites in Saxony, Germany, *Sci Total Environ* **336**(13):81–89.

Mkandawire M, Dudel EG, Müller C. 2005. Possible mineralization of Uranium in Lemna gibba G3. n Uranium Mining Areas of Portugal. In: Merkel B, Hasche-Berge A, editors. *Uranium in the Environment: Mining Impact and Consequences*. Berlin: Springer-Verlag, pp. 496–505.

Nyquist J, Greger M. 2007. Uptake of Zn, Cu, and Cd in metal loaded *Elodea canadensis*. *Environ Exp Botany* **60**:219–226.

Orloff KG, Mistry K, Charp P, Metcalf S, Marino R, Shelly T, Melaro E, Donohoe AM, Jones RL. 2004. Human exposure to uranium in groundwater. *Environ Res* **94**(3):319–326.

Overall RA, Parry DL. 2004. The uptake of uranium by Eleocharis dulcis Chinese water chestnut in the Ranger Uranium Mine constructed wetland filter. *Environ Pollut* **132**:307–320.

Paulo C, Pratas J, Rodrigues N. 2006. Rhizofiltration of uranium from contaminated mine waters. In: Alpoim MC, Morais PV, Santos MA, Cristovao AJ, Centeno JA, Collery P, editors. *Metal Ions in Biology and Medicine* Vol. 9. Paris: John Libey, Eurotext, pp. 187–192.

Paulo C. 2006. Selecção de Plantas aquáticas e perspectivas na fitorremediação de escorrências uraníferas (Aquatic plant selection and perspective in uranium contaminated water Phytoremediation). Instituto Superior Técnico, Lisboa 119 pages (Master thesis).

Periyakaruppan A, Kumar F, Sarkar S, Sharma CS, Ramesh GT. 2007. Uranium induces oxidative stress in lung epithelial cells. *Arch Toxicol* **81**(6):389–395.

Pettersson HBL, Hancock G, Johnston A, Murray AS. 1993. Uptake of uranium and thorium series radionuclides by the waterlily, *Nymphaea violacea. J Environ Radioact* **19**:85–108.

Pilon-Smits E. 2005. Phytoremediation. *Annu Rev Plant Biol* **56**:1539.

Prasad MNV, Freitas H. 2003. Metal hyperaccumulation in plants—Biodiversity: Prospecting for phytoremediation technology. *Electronic J Biotechnol* **6**(3):285–321.

Pratas J, Prasad MNV, Freitas H, Conde L. 2005a. Plants growing in abandoned mines of Portugal are useful for biogeochemical exploration of arsenic, antimony, tungsten and mine reclamation. *J Geochem Exploration* **85**(3):99–107.

Pratas J, Rodrigues N, Paulo C. 2005b. Uranium accumulator plants from the centre of Portugal—their potential to phytoremediation. In: Merkel B, Hasche-Berge A, editors. *Uranium in the Environment: Mining Impact and Consequences.* Berlin: Springer-Verlag, pp. 477–482.

Rahman MA, Hasegawa H, Ueda K, Maki T, Okumura C, Rahman MM. 2007. Arsenic accumulation in duckweed *Spirodela polyrhiza* L.: A good option for phytoremediation. *Chemosphere* **69**(3):493–499.

Rai U, Sinha S, Tripathiv R, Chandra P. 1995. Wastewater treatability potential of some aquatic macrophytes: Removal of heavy metals. *Ecol Eng* **5**:512.

Robinson B, Kim N, Marchetti M, Moni C, Schroeter L, Dijssel C, Milne G, Clothier B. 2006. Arsenic hyperaccumulation by aquatic macrophytes in the Taupo Volcanic Zone, New Zealand. *Environ Exp Bot* **58**:206–215.

Samecka-Cymerman A, Kempers AJ. 2001. Bioindication of heavy metals with aquatic macrophytes: The case of a stream polluted with power plant sewages in Poland. *J Toxicol Environ Health* **62**(1):57–67.

Sampaio G. 1988. *Flora Portuguesa*, third edition. Lisbon: National Institute of Scientific Investigation.

Schneider J, Philipp M, Yamini P, Dörk T, Woitowitz HJ. 2007. ATM gene mutations in former uranium miners of SDAG Wismut: A pilot study. *Oncol Rep* **172**:477–482.

Schöner A, Sauter M, Büchel G. 2005. Uranium in natural wetlands: A hydrogeochemical approach to reveal immobilization processes In: Merkel B, Hasche-Berge A, editors. *Uranium in the Environment: Mining Impact and Consequences.* Berlin: Springer-Verlag, pp. 389–397.

Shawky S, Geleel MA, Aly A. 2005. Sorption of uranium by nonliving water hyacinth roots. *J Radioanal Nuclear Chem* **265**:81–74.

Sinclair G, Taylor G, Brown P. 2005. The effects of weathering and diagenetic processes on the geochemical stability of uranium mill tailings. In: Merkel B, Hasche-Berge A, editors. *Uranium in the Environment: Mining Impact and Consequences.* Berlin: Springer-Verlag, pp. 27–46.

Soudek P, Velenva S, Benesova D, Vanek T. 2007. From laboratory experiments to large scale aplication: An example of the phytoremediation of radionuclides. In: Marmiroli N, Samotokin B, Marmiroli M, editors. *Advanced Science and Technology for Biological Decontamination of Sites Affected by Chemical and Radiological Nuclear Agents. NATO Science Series IV Earth and Environmental Sciences*, Vol. 75. Netherlands. Springer, pp, 139–158.

Tansky RR. 2001. *Chemical and Radiological Toxicity of Uranium and Its Compounds.* Technical Report, Savannah River Company, US Department of Energy, EUA.

Thiébault C, Carrière M, Milgram S, Simon A, Avoscan L, Gouget B. 2007. Uranium Induces Apoptosis and Is Genotoxic to Normal Rat Kidney NRK52E Proximal Cells. *Toxicol Sci* **98**(2):479–487.

Van Horn J, Huang H. 2006. Uranium VI biocoordination chemistry from biochemical, solution and protein structural data. *Coord Chem Rev* **7–8**:765–775.

Van Loon JC, Barefoot RR. 1989. In: Van Loon JC, Barefoot RR, editors. *Analytical Methods for Geochemical Exploration*. San Diego: Academic Press.

Vandenhove H, Cuypers A, Van Hees M, Koppen G, Wannijn J. 2006. Oxidative stress reactions induced in beans *Phaseolus vulgaris* following exposure to uranium. *Plant Physiol Biochem* **44**(11–12):795–805.

Vanek T, Schwitzguébel J. 2003. Phytoremediation inventory: COST action 837. In: Vanek T, Schwitzguebel J, editors. *Phytoremediation Inventory COST Action 837*. Prague, Czech Republic: UOCHB AVCR.

Vardanyan LG, Ingole BS. 2006. Studies on heavy metal accumulation in aquatic macrophytes from Sevan Armenia and Carambolim India lake systems. *Environ Int* **32**(2):208–218.

Weir E. 2004. Uranium in drinking water, naturally. *CMAJ* **170**:6.

WHO. 2001. *Depleted Uranium: Sources, Exposure and Health Effects*. Full Report. Geneva: WHO (World Health Organization). Wise Uranium Project. Available at http://www.wise-uranium.org.

Wolf G, Arndt D, Kotschy Lang N, Obe G. 2004. Chromosomal aberrations in uranium and Coal miners. *Int J Radiat Biol* **80**(2):147–153.

Zamora ML, Tracy BL, Zielinski JM, Meyerhof DP. 1998. Chronic ingestion of uranium in drinking water: A study of kidney bioeffects in humans. *Toxicol Sci* **431**:68–77.

Zhu YL, Zayed AM, Quian JH, De Souza M, Terry N. 1999. Phytoaccumulation of trace metals by wetland plants: II. Water hyancinth. *J Environ Quality* **28**:339–344.

25 Copper as an Environmental Contaminant: Phytotoxicity and Human Health Implications

MYRIAM KANOUN-BOULÉ, MANOEL BANDEIRA DE ALBUQUERQUE, CRISTINA NABAIS and HELENA FREITAS

Centre for Functional Ecology, Department of Botany, University of Coimbra, 3001-455 Coimbra, Portugal

1 COPPER AND HUMANS: A RELATION OF 10,000 YEARS

Copper is one of the first metals used by humans along with gold, with a history of at least 10,000 years. Copper derives from the Latin *cuprum*, derived from the word *cyprium*, because Cyprus was the main source of Egyptian and Roman copper [Schroeder et al., 1966]. The earliest known objects date from about 6000 B.C. and were manufactured simply by hammering and heating native copper [Georgopoulos and Roy, 2001]. During the prehistoric Chalcolithic Period (derived from chalkos, the Greek word for copper), man discovered how to extract and use copper to produce ornaments and tools. As early as the fourth to third millennium B.C., copper was extracted from the Huelva region (Spain). Brass, an alloy of copper and zinc, was developed during Roman times. In the time of Pharaohs, Israel's Timna Valley provided copper used to treat infections and to sterilize water. Since then, man has made use of copper as a plumbing material.

Currently, copper is still widely used for electrical equipment; construction, such as roofing and plumbing; and industrial machinery, such as heat exchangers and alloys. Copper has also a wide range of other applications in agriculture (nutrients, pesticides, and fungicides), wood preservation, and medical applications [ATSDR, 1990].

Trace Elements as Contaminants and Nutrients: Consequences in Ecosystems and Human Health, Edited by M. N. V. Prasad
Copyright © 2008 John Wiley & Sons, Inc.

2   COPPER: IDENTITY CARD, MAIN SOURCES, AND ENVIRONMENTAL POLLUTION

Copper is a transition metal with an atomic mass of 63.54 Da and with two stable isotopes, ^{63}Cu and ^{65}Cu, with natural abundances of 69.2% and 30.8%, respectively. Copper can occur in one of four oxidation states—copper(0), copper(I), copper(II), and copper(III)—although trivalent copper is very rare. Copper(II) (the cupric ion) is the most important oxidation state of copper, generally the oxidation state found in water. Most cupric compounds and complexes are blue or green, and they are frequently soluble in water: Cu^{2+} (aq) is mildly hydrolyzed in near-neutral solution, forming the dimer $Cu_2(OH)_2^{2-}$.

Copper is present naturally in the environment in the elemental form, but most of the commercial production comes from sulfides and oxide minerals. The four main sources of copper are chalcocite (Cu_2S), covellite (CuS), malachite ($CuCO_3-Cu(OH)_2$), and chalcopyrite ($Cu_2S-Fe_2S_3$) [ATSDR, 1990; Georgopoulos and Roy, 2001].

Nriagu [1979c] estimated that the amount of copper entering annually in the biosphere was about 211,000 metric tons, whereas in a previous study [NAS, 1977] the value was estimated at 1.8 million metric tons. According to Nriagu [1979c], 80.7% of this copper was deposited in terrestrial compartments, 15.7% in the hydrosphere, and 3.6% in the atmosphere, whereas in 1980 the US Environmental Protection Agency gave the values 97%, 2.4%, and 0.04% for each compartment, respectively [Perwak et al., 1980].

2.1   Copper in the Atmosphere

Copper is released in the atmosphere in aerosol form. Depending on particle size, its removal is through bulk deposition, dry deposition, or rain by attachment to droplets within clouds [Schroeder et al., 1987]. The residence time for copper in the air is about 13 days, and it has been estimated that around 80% of the copper released in the atmosphere was deposited in terrestrial ecosystems [Nriagu, 1979c].

The US EPA conducted a detailed study of copper emissions into the atmosphere to estimate human exposure [Weant, 1985]. The main sources of emissions and the respective estimated copper emission rates, in metric tons per year, are given in Table 1.

2.2   Copper in the Hydrosphere

Copper is mainly transported into streams and waterways as runoff due either to natural weathering or to soil disturbances (68% of the total release of copper to water). Copper sulfate used in agriculture represented 13% of the copper released to water, and urban runoff contributed with 2% [Perwak et al., 1980]. Nolte [1988] established that in the absence of specific industrial sources, runoff can be the major factor contributing to copper pollution in rivers.

Input of copper into aquatic ecosystems mainly through waste discharges into saline waters, industrial discharges into freshwater, and leaching of antifouling marine paints and wood preservatives [Nriagu, 1979c] increased sharply during the past century and is 2–5 times higher than natural loadings [Nriagu, 1979d].

TABLE 1. Atmospheric Emission Sources and Emission
Rates of Copper [Georgopoulos and Roy, 2001;
Georgopoulos et al., 2001, 2001b; Nriagu and Pacyna, 1988]

Emission Sources of Copper	Emission Rates (metric tons per year)
Primary copper smelters	2100
Copper and iron ore processing	480–660
Iron and steel production	112–240
Combustion sources	45–360
Municipal incinerators	3.3–270
Secondary copper smelters	160
Copper sulfate production	45
Gray iron foundries	7.9
Primary lead smelting	5.5–65
Primary zinc smelting	24–340
Ferroalloy production	1.9–3.2
Brass and bronze production	1.8–36
Carbon black production	13–20

In the hydrosphere, copper can occur dissolved, in suspended particles or in the sediments. Only a small fraction of dissolved copper is present as free or hydrated cupric ion (Cu^{2+}). The majority will be found in a complex form with organic or inorganic ligands or associated with a colloidal phase. Copper bioavailability is affected by several natural processes like complexation/chelation and sorption/desorption as well as bacterial action [Harrison, 1985]. Copper forms stable complexes with bases, sulfur (or sulfur-containing ligands), organic anions, amino acids, amino sugar, alcohol, urea, and so on [Georgopoulos and Roy, 2001].

2.3 Copper in the Lithosphere and Pedosphere

Copper in soils may come from a variety of anthropogenic sources: mining and smelting activities; other industrial emissions and effluents; traffic; fly ash; dumped waste materials; contaminated dusts and rainfall; sewage and sludge from wastewater treatment plants; pig slurry; composted refuse; and agricultural fertilizers, pesticides, and fungicides [Nriagu, 1979c; Thornton, 1979; Ma, 1984; Nriagu and Pacyna, 1988; ATSDR, 1990; Roncero et al., 1992; Alva et al., 1995]. However, the main sources of copper inputs in soil are tailings and overburden from copper mines and mills. In tailings, copper is generally in the form of insoluble silicates and sulfides and represents the fraction that cannot be recovered from the ore [Perwak et al., 1980].

Nriagu [1979a] estimated that copper associated with soil organic matter was estimated as 2.4×10^{15} g which represents roughly 36% of the copper burden in soils. The residence time for copper in soils may be for as long as 1000 years and the flux of copper from land to ocean was estimated as 6.3×10^{12} g/yr [Nriagu, 1979c].

3   COPPER IN PLANTS

Copper is abundant in the environment and essential for the normal growth and metabolism of all living organisms [Schroeder et al., 1966]. Only trace amounts of copper are required for plant nutrition; however, at high concentrations, copper can be potentially toxic. Critical deficiency levels of copper are in the range of 1–5 mg/kg plant dry mass, and the threshold for toxicity is above 20–30 mg/kg dry mass [Marschner, 1995].

3.1   Metabolic Functions of Copper

Copper is an essential micronutrient for plants [Schroeder et al., 1966]. Due to its redox properties (electrochemical potential of -260 mV), copper is an essential element in many electron carriers involved in reactions occurring in the range from -420 to $+800$ mV. It is present in numerous proteins and enzymes like plastocyanin, cytochrome oxidase, polyphenol oxidase, monophenol oxidase, laccases, ascorbate oxidase, and superoxide dismutase (SOD) [Marschner, 1995]. More than 50% of the copper present in chloroplasts is bound to plastocyanin, an important component of the electron transport chain of photosystem I. The cytochrome oxidase is part of the mitochondrial electron transport chain and contains two copper atoms and two iron atoms. The enzymes of the group of polyphenol oxidases and laccases catalyze oxygenation reactions of plant phenols. Ascorbate oxidase catalyzes the oxidation of ascorbic acid to L-dehydroascorbic acid and may act as a terminal respiratory oxidase and is linked to redox systems together with glutathione [Marschner, 1995]. The several types of SOD isoenzymes are very important for the detoxification of superoxide radicals. The copper–zinc SOD, with an active site where both heavy metals are linked by a histidine residue, is localized in the stroma of chloroplasts. The copper atom is involved in the detoxification of O_2^- generated in photorespiration [Marschner, 1995]. Copper can also be involved as a cofactor in hormonal metabolism like ethylene production [Rodriguez et al., 1999].

In copper-deficient soils, copper is strongly held on inorganic and organic exchange sites and in complexes with organic matter [Thornton, 1979], causing reduced availability of copper to vegetation in these soils. The critical free concentration in nutrient media for copper deficiency ranges from 10^{-14} to 10^{-16} M. At a macroscopic level, typical symptoms of copper deficiency in terrestrial plants appear first at the top of young leaves and then extend downward the leaf margins. Twisted, malformed leaves exhibiting chlorosis or even necrosis can also be observed, along with reduced growth and abnormally dark coloration in rootlets [Gupta, 1979; Marschner, 1995].

Regarding the photosynthetic metabolism, copper deficiency was found to reduce photosystem I electron transport due to a decreased formation of plastocyanin [Baszynski et al., 1978; Shikanai et al., 2003]. A decrease in the activity of the photosystem II, in Cu-deficient plants, was also observed [Henriques, 1989]. As described by Baszynski et al. [1978], Henriques [1989], and Barón et al. [1992], Cu-deficient plants can exhibit a disintegration of the thylakoid membrane of chloroplasts, a decrease in pigment content, a reduced plastoquinone synthesis, and lower unsaturated C18 fatty acid contents. Plants with copper deficiency show a lower content of soluble carbohydrates, as expected from the role of copper in photosynthesis [Brown and Clark, 1977]. In legumes under low supply of copper, nodulation and N_2 fixation are reduced. This is probably due to an indirect effect involving a reduction of carbohydrate supply for nodulation and N_2 fixation.

3.2 Toxicity of Copper

When an oversupply of copper occurs, resulting from either its natural occurrence in soil or by anthropogenic release (mining, smelting, agriculture activities, waste disposal, etc.), a series of negative effects occur at cellular and whole plant levels. In agriculture, copper fungicide inputs have been reported as the main cause of high concentrations of copper in soils. The use of copper dates back to 1896, when the French farmers applied Bordeaux mixture (copper sulfate, lime, and water) to control fungal pests [Martin and Woodcock, 1983]. Nowadays, old vineyards and coffee fields are examples of crops where soils can still exhibit problems of copper toxicity for plants. Copper, like other trace metals, interacts with several cellular compounds and organelles. When present in excess, copper may cause injuries and even, in extreme cases, the death of the plant [Reichman, 2002; Shaw et al., 2004]. Plant response to copper toxicity may vary depending on various factors like the dose of exposure, the organic matter content in soil, the plant-specific copper tolerance, the tissue/organ exposed, the age of plant and the influence of other nutrients [Luo and Rimmer, 1995; Marschner, 1995; Reichman, 2002; Shaw et al., 2004].

3.2.1 Morphological Symptoms of Toxicity. Generally, the excess of copper induces a reduction of the total biomass of the plant and the development of chlorotic symptoms on the leaves. The reduction of growth and foliar chlorosis can be described as the most common visual symptom of copper toxicity in plants. They are the result of the negative effects of copper excess on cellular organelles affecting important physiologic processes like photosynthesis and respiration. Chlorosis is frequently reported as an initial symptom and, with increased exposure, necrosis can appear in the leaf tips and margins [Reichman, 2002; Yruela, 2005]. Plants tend to accumulate copper in their root tissues with small translocation to the shoots. Therefore alterations in root growth and morphology occur before any toxic symptoms appear on shoots [Quartacci et al., 2003; Sheldon and Menzies, 2005].

Roots are especially vulnerable to copper toxicity. Roots of *Origanum vulgare* subsp. *hirtum* exhibited severe reduction of length and root volume as well as generalized malformations after two months of growing in copper-enriched soil (0.3–25.5 mg/kg). At the microscopic level, malformed tissues and disintegration of the epidermal cells and more or less folded cortical cells were observed. Morphometric alterations occurred, such as decrease in relative volume of the epidermis and increase in the relative volumes of cortex, xylem, and phloem

[Panou-Filotheou and Bosabalidis, 2004]. Sheldon and Menzies [2005], analyzing the effects of copper excess in nutritive solution on growth and morphology of Rhode grass (*Chloris gayana* Knuth), also verified a severe inhibition of root growth. These authors reported that the roots were extremely stunted and the root cuticle was thickened, cracked, and brown. Arduini et al. [1995] assessed the effects of copper on root growth and morphology of stone pine (*Pinus pinea* L.) and maritime pine (*Pinus pinaster* Ait.) seedlings grown in culture solutions supplied with up to 5 μM CuSO$_4$, and observed that this concentration was enough to completely inhibit root growth of both species within 3 days. It also reported that cell membrane damage occurred after 10 days of exposure to 1 μM CuSO$_4$.

3.2.2 Copper-Induced Nutritional Imbalance. Copper enters into the cells in the form Cu^{2+} by diffusion or via membrane transporters. Evidence suggests that copper uses specific as well as nonspecific transporters and thus may compete with other nutrients and induce some physiologic deficiency [Reichman, 2002; Reid, 2001]. Copper-induced iron deficiency has been well-documented [Ouzounidou et al., 1995; Pätsikkä et al., 2002].

Copper-induced deficiency of nutrients, other than iron, can be observed in plants growing in media with high copper content. However, due to the conflicting results observed in the current literature, it is difficult to define with confidence which nutrients are affected by the excess of copper. Increased iron content was found in shoots and roots of maize (*Zea mays* L.) and reed (*Phragmites australis*) [Ali et al., 2002] under conditions of copper excess. Alaoui-Sossé et al. [2004] verified reductions in calcium, magnesium, and potassium contents in leaves of cucumber (*Cucumis sativus* L. cv. *Vert long maraicher*), but the root concentration of these nutrients were unaffected. On the other hand, no effects on iron and magnesium foliar contents were detected in chlorotic leaves of oregano, ensuring that the chlorosis was solely due to the copper itself [Panou-Filotheou et al., 2001]. The contents of zinc, calcium, and iron in leaves of *Lycopersicon esculentum* were reduced by increasing copper concentrations in a nutrient solution [Martins and Mourato, 2006]. Excess of copper may reduce or increase zinc accumulation in higher plants [Reichman, 2002]; however, there is no consensus among researchers regarding this nutrient. Copper can also induce phosphate deficiency due to the precipitation of insoluble metal–phosphate complexes [Reid, 2001].

3.2.3 Oxidative Stress. Reactive oxygen species (ROS) are formed as a natural by-product of the normal metabolism of oxygen and play an important role in cell signaling. However, under environmental stress conditions (including exposures to high levels of metal), ROS production can increase dramatically, resulting in significant damages to cell structures. ROS overproduction is commonly referred to as oxidative stress. Due to its redox properties, copper is considered as an efficient catalyst in the formation of ROS. The production of free radicals, such as hydroxyl and superoxide radicals will result in lipid peroxidation, depletion of sulfhydryls, altered calcium homeostasis and DNA damage [Drążkiewicz et al., 2004; Shaw et al., 2004; Yruela, 2005].

3.2.4 Lipid Peroxidation and Plasma Membrane Integrity. The oxidative degradation of lipids that are constituents of cell and organelle membranes occurs when free radicals "remove" electrons from these lipids, resulting in cell damage. It mainly affects the multiple double bonds of polyunsaturated fatty acids. The induced generation of ROS, like hydrogen peroxide and hydroxyl radical, has already been directly related to damages to plasma membrane and chloroplast membrane lipids [Devi and Prasad, 2004; Wang et al., 2004]. The excess of copper promoted the increase of lipid peroxidation in detached rice leaves [Chen and Kao, 1999], in intact leaves of barley [Vassilev et al., 2003], in intact roots of *Brassica juncea* L. [Wang et al., 2004], and in tomato [Mediouni et al., 2006].

The plasma membrane can be considered the first living structure to face the excessive incoming of trace metals and is particularly sensible to an excess of copper. Various studies have already demonstrated that the excess of copper can alter the plasma membrane permeability inducing to ionic leakage. Copper-induced K^+ leakage was detected in rice [Mahmood and Islam, 2006] and *Arabidopsis* [Murphy et al., 1999].

3.2.5 Photosynthesis. Once inside the cell, copper can interfere with the biosynthesis of the photosynthetic machinery changing the pigment and protein composition of photosynthetic membranes. A lower content of chlorophyll, inactivation of enzymes and proteins linked to photosynthetic process, and modifications of thylakoid membranes occur under copper toxicity.

Copper-induced chlorosis can result from the inhibition of pigment accumulation and retardation of chlorophyll integration into photosystems. Magnesium can be removed by copper in the chlorophyll, present in both antenna complexes and reaction centers, damaging the structure and function of the chlorophyll [Küpper et al., 2003]. Besides, low concentration of chlorophyll can predispose the plant to high photosensitivity in the photosystem II (photoinhibition) [Pätsikkä et al., 2002]. Decrease of chlorophyll content in plants after exposure to copper is frequently reported in the literature (tomato [Martins and Mourato, 2006], rice [Mahmood and Islam, 2006], radish [Chatterjee et al., 2006], etc.). However, any symptom induced by copper is dose-dependent. Xiong and Wang [2005], assessing the effects of copper on the cultivar Xiayangbai of chinese cabbage (*Brassica pekinensis* Rupr.), verified an increase of chlorophyll content [Chl a + Chl b] of 27.4% in plants growing at 1 mM/kg of soil in relation to control plants. No visual symptoms of toxicity were detected and the growth was unaffected. On the other hand, Xiong et al. [2006], evaluating the effects of 0.3 μM and 10.3 μM of copper in Hoagland solution in the same species, observed a reduction of chlorophyll content after six days of copper exposure.

Copper ions have a direct effect on the structure of the thylakoid membranes through peroxidation and oxidative stress, leading to the ineffectiveness of photosynthesis. Inhibition of the photosynthetic electron transport and O_2 evolution, as well as loss of photochemical activity of photosystem (PS) I and PS II, is commonly preceded by copper-induced changes in the chloroplast structure [Caspi et al., 1999; Mysliwa-Kurdziel et al., 2004]. In addition, several studies have pointed out that

PS II is more sensitive to copper than PS I [Pätsikkä et al., 2002; Vassilev et al., 2003; Yruela et al., 1996]. Enzymes related to the photosynthetic dark reactions are also affected by the excess of copper [Bertrand and Poirier, 2005]. Copper excess also caused a dramatic decline in the number and volume of chloroplasts from the mesophyll cells of *Origanum vulgare* subsp. *hirtum* [Panou-Filotheou et al., 2001].

Regarding carbohydrates allocation, the source–sink relationship seems to be affected by copper. Evidence has been reported in mature and expanding cucumber leaves, where the inhibition of growth led to both sucrose and starch accumulation and to a feedback inhibition of photosynthesis [Alaoui-Sossé et al., 2004; Vinit-Dunand et al., 2002].

3.2.6   Respiration. The excess of metallic ions can produce harmful effects on mitochondrial respiration, as observed by the changing of O_2 consumption and altered releasing of CO_2 in plants treated with copper. This can be the result of damages to mitochondria structure and function or of the increase in energy demand for ion uptake or export, production of specific metal-binding polypeptides, and so on. Enzymes related to dissimilation processes seem to be less affected by high levels of trace metals than enzymes related to photosynthetic processes; for this reason, most publications about metal physiology are focused on processes of energy fixation and carbon assimilation rather than on energy consumption [Lösch, 2004].

3.2.7   Mechanisms of Copper Tolerance. Trace metal tolerance in higher plants involves a set of mechanisms that aim the regulation of the intracellular metal concentration, avoiding its excessive accumulation at metabolic sensitive sites within the cells and tissues. Reduction of metal uptake seems to be a primary line defense against metal accumulation and its consequent effects. Subsequently, cellular mechanisms of metal detoxification work to guarantee the fine regulation of copper concentration inside different cells and tissues. At last, the enhancement of antioxidative defence plays an important role on the prevention of any damage caused by ROS occasionally produced when the homeostatic mechanisms are not enough to withstand the metal uptake [Clemens, 2001; Yruela, 2005].

3.2.8   Reduction of Copper Uptake. A way to prevent the effects of excess copper in the cells is the reduction of its influx to the cytosol. This can be accomplished by the release of extracellular exudates, retention of copper ions in the cell wall, and/or mycorrhiza action. Root exudates such as phenolic compounds, organic acids, amino acids, and so on, play a variety of functions, commonly related to metal uptake enhancement. However, some investigations have shown that some exudates can also reduce the metal uptake. Examples of releasing of root exudates for uptake control of nickel [Salt et al., 2000] and aluminum [Watanabe and Osaki, 2002] are known. The regulation of copper uptake has been related to the release of root exudates in wheat, rye, triticale, maize, and soybean [Nian et al., 2002]. The high affinity of copper to the carbonylic, carboxylic, phenolic, and sulfhydryl groups make copper ions to be strongly bound to cell walls, not only by means of ionic bonds with negatively charged sites but, above all, by coordination bonds with N, O, and S atoms [Quartacci et al., 2003]. This ability also prevents the excess of copper accumulation in physiologically active sites. Peng et al. [2005] report that the leaf cell wall of *Elsholtzia splendens*, a Chinese metallotolerant plant, retained between 29% and 38% of total copper, while in the root cell wall these values ranged between 45% and 61%. Ke et al. [2007], studying *Daucus carota* L. plants from a copper mine contaminated and noncontaminated field site, growing at increasing levels of copper (0–400 mg/kg of dry soil) confirmed that the leaf cell wall was able to retain between 41.5% and 54.6% of total leaf copper accumulated.

A mycorrhiza is a mutualistic symbiosis between fungi and plant roots. Different types of this symbiosis can be found in 90% of the terrestrial plants. Their symbiotic relationship is based on the transfer of carbon from plant to fungus and mineral nutrients (largely phosphorus and/or nitrogen) from the fungus to the host plant. Mycorrhizal fungi release molecules to solubilize nutrients, and therefore they can also solubilize metals and interfere with the toxic effect of heavy metals in plants [Cairney, 2000; Fomina et al., 2005; Meharg, 2003]. The mechanisms of mycorrhizal attenuation of copper toxicity are diverse and, given the diversity of mycorrhizal associations, probably species-dependent. An example of role of mycorrhizal in the reduction of copper toxicity is given by van Tichelen et al. [2001]. These authors showed that the fungi *Suillus bovines* and *Thelephora terrestris* were able to protect *Pinus sylvestris* against copper toxicity.

Plants and mycorrhizal associations can alleviate metal toxicity by two main strategies: *avoidance* and *compartmentalization mechanisms.* Avoidance mechanisms include (a) suppression of metal influx transporters, (b) enhancing of the metal efflux, (c) releasing of pollutant complex agents into the surrounding soil solution, (d) precipitation or binding of toxicants onto cell surfaces, and (e) releasing of extracellular enzymes that will alter the metal speciation so that it is not taken up or converted into a nontoxic species. The *compartmentalization strategy* consists in the translocation to parts of the plant or fungus with limited metabolic function where it is relatively nontoxic, or accumulated in the vacuole where they can be stored away from the cytoplasm [Khade and Adholeya, 2007; Meharg, 2003] (Fig. 1).

3.2.9 Copper Detoxification. Once inside the cytosol, free copper ions can be harmful to the cellular metabolism because of their reactivity and limited solubility. In order to guarantee the adequate copper concentration within the physiological limits, the cell may evolve a network of homeostatic mechanisms that contribute to metal detoxification by chelation, transport, and sequestration [Clemens, 2001; Dučić and Polle, 2005]. Thus, the surplus of copper can be sequestrated into the apoplast, stored into vacuoles or transferred to specialized cells, such as epidermal cells and trichomes [Yruela, 2005]. In this sense, chelators have contributed to metal detoxification by their capacity of buffering cytosolic metal, preventing its reactivity and allowing its quick intracellular transport. The main classes of copper/metal chelators known in plants are described below, along with copper-receptor proteins called chaperones.

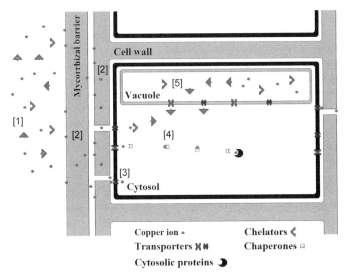

Figure 1. Cellular mechanisms of avoidance/tolerance of copper toxicity in higher plants. [1] Reduction of influx of copper ions to cytosol by root and mycorrhiza exudates; [2] binding to fungal sheath and cell walls; [3] active efflux of copper ions from the plasma membrane into the apoplast; [4] detoxification by chelation of various ligands and chaperones in the cytosol; [5] compartmentalization in the vacuole.

Phytochelatins [PC] are small metal-binding peptides of the general structure [γ-Glu-Cys]$_n$-Gly [$n = 2$–11] that are found in plants and certain fungi. They are synthesized from reduced glutathione (GSH) and are able to form complexes with several metals and metalloids (Cu, Cd, Zn, As, etc.). Glutathione acts as a precursor of chelators as well as an antioxidant. In fact, some authors have suggested that formation of PC can lead to a depletion of GSH and therefore oxidative stress. Despite being more commonly related to cadmium and arsenate detoxification, it has been discovered that copper is also a strong activator of PC biosynthesis, both in vivo and in vitro [Cobbett and Goldsbrough, 2002].

Metallothioneins [MT] are ubiquitous low-molecular-weight cysteine-rich polypeptides. MTs typically contain two metal-binding, cysteine-rich domains that give these metalloproteins a dumbbell shape, and their classification is based on the arrangement of cysteine [Cys] residues. Class I MTs contain 20 highly conserved Cys residues based on mammalian MTs and are widespread in vertebrates. Class II MTs are composed of those without the arrangement of Cys of Class I MTs and include all those from plants and fungi as well as nonvertebrate animals [Cobbett and Goldsbrough, 2002]. Metallothioneins play an important role in copper detoxification [Clemens, 2001; Hoof et al., 2001].

Carboxylic acids and amino acids represent potential ligands to metals due to the reactivity of metallic ions with S, N, and O. These compounds are related to transport through the xylem and vacuolar compartmentalization. Amino acids, such as histidine, nicotinamine, and γ-aminobutyric acid (GABA), have been described as effective ligands for copper transport in xylem sap [Liao et al., 2000; Pich and Scholz, 1996; Yang et al., 2005]. Special attention has also been given to free proline accumulation, which, in addition to chelation, can improve the metal tolerance acting as osmoregulator and antioxidant [Rai, 2002; Sharma and Dietz, 2006].

Copper chaperones are cytosolic copper soluble binding proteins specialized to deliver metal ions to organelles and metal-requiring proteins. Specific copper chaperones chelate "free" copper ions and deliver them to *copper pumps* for transportation into organelles (vacuole, mitochondria, chloroplasts, etc.) or directly to cytosolic copper-dependent proteins, preventing inappropriate copper interaction with other cellular components [Clemens, 2001; Hall, 2002; Yruela, 2005].

Several plant metal transporters have been identified. However, only two specific families of copper transporters are identified in plants to date. Mediated by post-Golgi or plasma membrane-located P-type ATPases (also known as CPx-ATPases), the efflux of copper ions has been related to copper homeostasis or copper resistance in a variety of cell types and organisms. Copper-transporting P-type ATPases, such as PAA1 and RAN1, have been identified in *Arabidopsis thaliana* [Tabata et al., 1997]. *Copper transporter* (COPT) proteins are another family of copper transporters, where five members of this family, COPT1-5, are found in *Arabidopsis thaliana*. Other possible copper transporters belong to the family involved in the transport of divalent ions, the *natural resistance-associated macrophage protein* [Nramp] family [Yruela, 2005].

3.2.10 Metallophytes. On sites with high concentrations of heavy metals (geogenic contamination, mine dumps, metal smelters), resistant plant species, named metallophytes, are found. The response of a resistant and sensitive population of *Minuartia hirsuta* (Caryophyllaceae) to increased concentrations of copper in nutrient solutions showed that the copper-resistant population had a root growth above the control level up to $80\,\mu M$ Cu [Ouzounidou et al., 1994]. The sensitive plants showed chlorosis and necrosis of young leaves in response to copper stress. Probably resistance is related to a more efficient sequestration of copper in the root cells.

Trace metal resistance involves a cost for the plants in terms of energy and other resources, implying that resistant individuals have a lower fitness in the absence of the heavy metal stress [Hagemeyer, 1999]. In fact, metal-resistant plants have lower growth rates in comparison with nonresistant plants [Wilson, 1988]. However, the "trade-off hypothesis," which states that plants under heavy metal stress invest energy in their resistance mechanisms, depressing other processes like growth or reproduction, is still under discussion. This is linked to the fact that the nature of the costs remains unknown [Harper et al., 1997].

3.3 Copper and Human Health

Copper is required by all living organisms for a number of metabolic functions such as respiration, free radical eradication, energy production, connective tissue formation, metabolism of iron, maturation of extracellular matrix, and neuropeptide and neuroendocrine signaling [Nriagu, 1979b, Pena et al., 1999; Patel et al., 2000; Failla et al., 2001]. Copper also plays an important catalytic function, primarily through copper-containing enzymes that reduce molecular oxygen [Institute of Medicine, 1990]. An adult healthy human body weighing 70 kg contains about 100 mg of copper located mainly in the skeleton and muscles [Turnlund, 1998; Ralph and McArdle, 2001]. The highest concentrations of copper are found in the liver, followed by the brain, kidney, and heart [Turnlund, 1998].

Copper is found in a variety of foods and is also available in the form of supplements. Copper-rich foods include shellfish, liver, nuts, legumes (e.g., lentils), mushrooms, barley, cooked tomato products, and chocolate. Drinking water and dietary supplements can also provide significant sources. The minimum requirement recommended by the World Health Organization (WHO) is 1.3 mg/day for an adult.

3.3.1   Copper Metabolism. The finding that copper was essential for hemoglobin in rats [Hart et al., 1928] marked the beginning of the research on copper metabolism in humans. Dietary copper absorption takes place across the mucosal membrane cells of the stomach and in the intestine [Institute of Medicine, 1990]. The absorbed copper is then transported via the portal blood to the liver [Turnlund, 1998] by albumin and transcuprein [Ralph and McArdle, 2001]. Excess copper is sequestered within the intestinal mucosal cells via metallothioneins [Danks, 1988].

In the liver, copper is incorporated in copper-requiring enzymes and proteins that are afterwards released in the bloodstream [Ralph and McArdle, 2001]. Much of the copper taken up by the liver is incorporated into ceruloplasmin [Turnlund, 1998]. The primary route of excretion of endogenous copper is through the bile, which is produced and excreted by the liver [Ralph and McArdle, 2001]. Copper of biliary origin and nonabsorbed oral copper are eliminated from the body in feces, although daily losses of copper are minimal in healthy individuals [Failla et al., 2001]. Biliary copper excretion is adjusted to maintain copper homeostasis [Institute of Medicine, 1990]. Copper is lost in small amounts in sweat and urine and by desquamation (flaking) of the skin [Turnlund, 1998; Danks, 1988].

Copper absorption is strongly influenced by the amount of dietary copper [Turnlund et al., 1989]. Although the amount absorbed increases with the increase of the amount in the diet, absorption is much more efficient and a higher percentage is absorbed when intake is low. As shown by Turnlund [1991], a 10-fold increase in dietary copper results in only twice as much copper absorbed. Likewise, the excretion of endogenous copper is markedly influenced by dietary copper intake [Turnlund et al., 1995a, b].

Copper is part of several proteins (Table 2), including many vital enzymes. Dietary copper is readily absorbed and distributed to copper-containing proteins, with little storage of excess copper in tissues compared with other trace elements, such as iron and zinc. The activity of copper containing enzymes decreases on account of copper depletion, because copper is not stored in the body.

3.3.2   Copper Homeostasis. Copper binds to histidine, cysteine, and methionine residues of proteins with high affinity, resulting in their deactivation [Camakaris

TABLE 2. Metabolic Function of Several Proteins with Copper as a Component

Protein/Enzyme	Function
Clotting factor V and VIII	Nonenzymatic components involved in the blood clotting process
Copper/zinc superoxide dismutase (Cu/Zn SOD)	Protects intracellular components from oxidative damage
Cytochrome c oxidase	Present in mitochondria, catalyzes the reduction of oxygen to water, used in the synthesis of adenosine triphosphate (ATP)
Diamine oxidase	Inactivates histamine released during allergic reaction and polyamines involved in cell proliferation
Dopamine β-hydroxylase	Catalyzes the conversion of dopamine to norepinephrine in the brain
Ferroxidase I (ceruloplasmin)	Mediate the efflux of stored iron from tissues to sites of hemoglobin synthesis
Ferroxidase II	Catalyzes the oxidation of ferrous iron
Lysyl oxidase	Uses lysine and hydroxylysine found in collagen and elastin to produce cross-linkages needed for the development of connective tissues including bone, teeth, skin, lung, and vascular network
Monoamine oxidase	Involved in serotonin degradation and the metabolism of catecholamines such as epinephrine, norepinephrine, and dopamine
Peptidylglycine-alpha-amidating monooxygenase (PAM)	Necessary for the bioactivation of peptides
Tyrosinase	Catalyzes the conversion of tyrosine to dopamine and the oxidation of dopamine to dopaquinone in melanin biosynthesis

et al., 1999]. Maintaining copper homeostasis requires a balance between copper uptake and distribution within cells and subsequent detoxification and removal.

Copper homeostasis is affected by several factors, including dietary copper intake and by interactions with other nutrients, such as zinc and iron [Institute of Medicine, 1990]. Pregnancy, inflammation, and chronic diseases, such as diabetes, also influence whole-body copper metabolism [Turnlund, 1998]. Genetic defects, such as Menkes disease, a hereditary disorder of copper metabolism characterized by the accumulation of copper in the gut epithelial cells, results in the misdistribution of copper to tissues activity of cuproenzymes [Pena et al., 1999]. Typically, individuals with this disorder die early in childhood. Wilson's disease, also a genetic disorder of copper metabolism, causes copper to accumulate in the liver, brain, and cornea [Turnlund, 1999]. Individuals with Wilson's disease typically suffer from severe neurological disorders, unless given copper chelating therapy or very low copper diets [Turnlund et al., 1999] (Fig. 2).

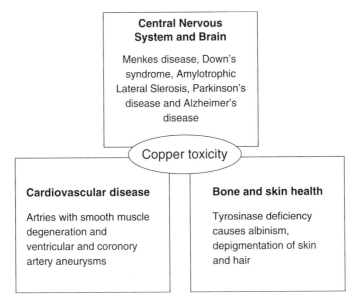

Figure 2. Copper homeostasis and its effect on the central nervous system, cardiovascular system, bones, and skin.

3.3.3 Copper Deficiency. In humans, copper deficiency in the absence of genetic disorders of copper metabolism is rare, but when it occurs, it is found more often in children than in adults. Patients nourished for prolonged periods with total parenteral nutrition [TPN] without added trace minerals are also at risk of copper deficiency [Institute of Medicine, 1990; Uauy et al., 1998]. Copper deficiency disorders may be acquired or inherited, but are more often induced by an imbalance between copper needs and dietary supply [Uauy et al., 1998]. It is estimated that in the United States, approximately 25% of the adult population may be copper-deficient [Klevay, 2000]. Symptoms of deficiency are decreased cupro-enzyme activity, eventually leading to the onset of abnormalities in neurological processes, connective tissue, and the skeletal and vascular systems [Failla et al., 2001]. Anemia, neutropenia, and osteoporosis can also be observed as a result of copper deficiency [Danks, 1988; Turnlund, 1999].

3.3.3.1 Copper Deficiency Induces Oxidative Stress. Under conditions of Cu deficiency, several components of the oxidant defence system can be compromised. Several studies showed that under an induced Cu deficiency the activity of CuZn-SOD decreased [Chen et al., 1994; Olin et al., 1994; Turnlund et al., 1997; Uriu-Adams and Keen, 2005]. In the same way, Cu deficiency decreases ceruloplasmin activity since failure to incorporate Cu renders it less stable [Hellman and Gitlin, 2002; Prohaska and Brokate, 2002]. Due to the functions of the enzyme, when ceruloplasmin is low, the incorporation of Fe to transferrin can be impaired, and Fe accumulates in tissues [Hellman and Gitlin, 2002].

A deficiency of Cu can also decrease the activities of certain non-Cu-containing enzymes of the oxidant defense system including catalase in heart and liver [Chen et al., 1994] and selenium-dependent glutathione peroxidase [Se-GPx] in liver and plasma tissues [Chen et al., 1994; Olin et al., 1994; Prohaska et al., 1992].

Copper deficiency can also interfere with other reactive oxygen species (ROS) scavengers. For instance, it was shown that metallothionein activity (which contains Cu and Zn) in kidney and liver tissues are sensitive to copper deficiency [Jankowski et al., 1993; Uriu-Adams and Keen, 2005]. On the contrary, the nonprotein glutathione is often increased in the liver and plasma of Cu-deficient animals [Chao and Allen, 1992; Olin et al., 1994; Zidenberg-Cherr et al., 1989], a change that is thought to represent an adaptive response to increased oxidative stress.

3.3.3.2 Copper Deficiency Induces Chronic Diseases. During the last decade, there has been increasing interest in the concept that deficiencies of copper can be an important factor in the development and evolution of chronic diseases such as cardiovascular dysfunction and diabetes (see review of Uriu-Adams and Keen [2005]).

Copper deficiency can induce several alterations in cardiac morphology, such as enlarged myocytes, derangement of myofibrils, fragmented basal laminae at capillary myocyte interfaces, and mitochondrial proliferation, swelling, and fragmentation [Heller et al., 2000; Kang et al., 2000; Klevay, 2000; Saari, 2000]. Additionally, electrocardiographic abnormalities and impaired contractile and mitochondrial respiratory function are observed in Cu-deficient hearts [Kang et al., 2000; Medeiros and Jennings, 2002; Saari, 2000]. In Cu-deficient animals and humans, aortas exhibit fissures and rupture due to altered architecture of elastic fibers, while coronary artery thrombosis and myocardial infarction increase mortality [Klevay, 2000; Shim and Harris, 2003]. Cu depletion experiments also show abnormalities in blood pressure and electrocardiograms in some human studies, although this finding is not universal [Institute of Medicine, 2002].

Studies from the Institute of Medicine [2000] showed that dietary copper deficiency in humans could alter glucose metabolism. Reciprocally, diabetes can disturb copper metabolism. It is reported that diabetes induces an increase of copper concentration in rodent liver and kidney [Chen and Failla, 1988; Uriu-Adams and Keen, 2005].

3.3.4 Copper Toxicity. Although copper is an essential micronutrient, normally subject to effective homeostatic control, excess dietary intakes can in some circumstances be toxic. Similarly to copper deficiency, many of the pathologic effects of copper overload are consistent with oxidative damage to the membrane or to macromolecules. Due to the redox properties of the metal, copper ions can induce the formation of reactive oxygen species (ROS) via the Haber–Weiss reaction. These oxidants will initiate a variety of auto-oxidative chain reactions that will be directly responsible, at membrane level, for oxidative alterations of proteins [Stadtman, 1990] and lipid peroxidation [Matoo et al., 1986]. In addition, Cu can bind directly to free thiols of cysteines, which can result in oxidation and subsequent cross-links between proteins, leading to impaired activity [Cecconi et al., 2002].

3.3.4.1   Acute Toxicity. Although acute toxicity cases can occur through the respiratory function or through the skin [World Health Organization, 1998], they are generally due to ingestion of beverages contaminated with high levels of copper or result from ingestion of high quantities of copper salts. In the last case, copper poisonings are often the result of single oral doses taken in suicide attempts and exceeding 20 g. Acute copper toxicity symptoms include headache, abdominal pain, vomiting and/or diarrhea, lethargy, tachycardia, respiratory difficulties, hemolytic anemia, and considerable kidney and liver damages ending with death [Davanzo et al., 2004; Srivastava et al., 2005; World Health Organization, 1998]. Although the mechanisms underlying copper poisoning are not all well understood yet, one of the most probable causes of copper-induced damages is the oxidant nature of the metal that will initiate important oxidative stress at multiple points of the body [Bremner, 1998].

3.3.4.2   Chronic Toxicity. Chronic dietary toxicity generally does not refer to contaminated water. Araya et al. [2003], [2004] demonstrated that the consumption of high quantities of water contaminated with low levels of copper could induce poisoning, but the mechanisms underlying this acute toxicity are still unidentified.

Chronic dietary copper toxicity is not generally considered as a significant human health concern. The upper level (UL) for Cu has been defined at a maximum of 10 mg/day by the Institute of Medicine [2000]. In accordance with this set value, several studies reported that chronic copper intakes until 10 mg Cu/day did not seem to induce significant toxicity [Mendez et al., 2004; Turnlund et al., 2004]. In the case of contaminated water consumption, it is important to note that a chronic consumption could represent and important risk for health of young children or individuals who are heterozygotic for Wilson's disease [Brewer, 2000; Eife et al., 1999].

Since the beginning of the 21st century, there has been an increasing interest in studying the relationships between an excessive amount of copper and the development of neurodegenerative diseases like Alzheimer's disease, Creutzfeldt–Jacob disease, and amyotrophic lateral sclerosis. Numerous studies already demonstrated that copper and the free radicals from copper-induced oxidative stress were involved in the initiation and/or development of neurodegeneration diseases, including Alzheimer's disease [Atwood et al., 2000; Opazo et al., 2002; Doraiswamy and Finefrock, 2004]. However, the role that copper plays in the progression of Alzheimer's disease is far from being elucidated; and, as reviewed by Maynard et al. [2005] and Doraiswamy and Finefrock [2004], numerous findings strongly suggest that copper deficiency may be of greater concern than copper excess.

3.3.5   Copper Interactions with other Nutrients and Vitamins

3.3.5.1   Zinc. High dietary zinc intake antagonizes copper availability by increasing its sequestration in the mucosal cell, and it increases fecal excretion of copper

[Hoffman et al., 1998]. Excessive zinc induces intestinal metallothioneins, which sequester both zinc and copper, but have a higher affinity for copper. As a result, copper is trapped in the mucosal cell, with a reduced transfer to the plasma pool [Ralph and McArdle, 2001]. This is the reason why high doses of zinc (0.05 mg before each meal) are used as therapy for the long-term treatment of patients with Wilson's disease [Lonnerdal, 1998].

3.3.5.2 Iron. Research findings suggest that copper is essential for efficient iron uptake and mobilization in animals. Copper-deficient swine developed microcyctic, hypochromic anaemia correctable by supplemental copper, but not by adding iron to the diet. Copper supplementation reduced the abnormal iron accumulation in these pigs and increased iron absorption into tissues [Pena et al., 1999]. Excessive iron can decrease copper status in several species. Anemia, often accompanied by accumulation of iron in the liver, has been reported in all species studied, including humans [Turnlund, 1999].

3.3.5.3 Ascorbic Acid. Ascorbic acid plays an important role in copper metabolism, intracellular transport of copper to the liver, and excretion [Lonnerdal, 1998]. In laboratory animals, ascorbic acid supplements have produced copper deficiency, and they may have similar effects in humans. Daily ascorbic acid supplements given to young men caused decreased ceruloplasmin activity [Turnlund, 1999; Danks, 1988; Lonnerdal, 1998]. However, at moderate and lower intakes of ascorbic acid, copper absorption is unaffected [Ralph and McArdle, 2001; Lonnerdal, 1998].

3.3.6 Risks of Human Exposure to Copper. Copper bioavailability is a key factor in the determination of dose exposure. To assess bioavailability, it is necessary to understand the interactions of the metal with other dietary components and the metabolic processes involved. Additionally, copper bioavailability might be affected by several drugs. However, little is still known about the nature and importance of these interactions.

Regarding copper interactions with nutrients, as previously described, high quantities of iron, zinc, and ascorbic acid are well known to affect copper bioavailability. Interactions between copper and other components have been extensively reviewed [Turnlund, 1999; Bodgen and Klevay, 2000].

To determine exposure risk, it is essential to consider all the potential sources of exposure. A conceptual model framework for the movement of copper from the environment to humans is given in Georgopoulos et al. [2001]. Exposure to copper occurs via three main routes: ingestion, inhalation and dermal. *Ingestion* exposure is generally due to contamination of food or drinking water with copper. Average copper concentrations in drinking water are estimated to be around 0.075 mg/liter, but copper concentrations are sometime above this value and, in that case, the 2 mg/day of copper, which is the maximum value estimated for daily copper input from food, are exceeded.

Inhalation exposure occurs through atmospheric dust contaminated with copper. As reported in Gerorgopoulos and Roy [2001], the annual atmospheric average copper concentrations generally range from 5 to $200 \, \text{ng/m}^3$ but values of $100 \, \text{mg/m}^3$ for a 24-hr period have already been observed. In that situation, and assuming that all the copper is attached to particles of inhalable size (which is unlikely to be the case), the average daily intake by inhalation, which generally ranges from 0.1 to 4 mg/day, would rise to 2000 mg.

Dermal exposure is not very frequent. In natural swimming waters, the complexation of copper by organic ligands will rapidly reduce the level of free copper. However, dermal exposures can be a risk for individuals working with agriculture or water treatment chemical products that are generally concentrated in soluble cupric salts.

4   FURTHER RESEARCH TOPICS

The environment is enriched with heavy metals mainly as a result of human activities. Once released from their geological or mineral frame, heavy metals potentially increase their toxic effects and may become a major environmental and public health issue. Unlike some organic pollutants that can be chemically and/or biologically transformed into other organic compounds less toxic (although not always), heavy metals cannot be "changed" only "masked" with chelators to reduce its toxic effects. The reasoning behind phytoremediation, which aims to use resistant plants to remove or stabilize heavy metals from the environment, has the drawback that metals are simply moving from one matrix (for example, the soil, mainly composed of silicate) to another matrix (plants which are mainly carbon structures). In any case, phytoremediation can be important to stabilize heavy metals, thus reducing their further spreading into the environment.

Although the symptoms of metal toxicity and deficiency in plants are fully described, the knowledge of the biochemical basis of metal toxicity symptoms in plants is still patchy. It is therefore relevant to understand the mechanisms of metal homeostasis, tolerance, and the evolution of metal resistance leading to tolerant/ resistant plant species (metallophytes). From the evolutionary point of view, the occurrence of hyperaccumulator plants is still an open question to be understood. Why do some plants accumulate more than 1000 ppm of heavy metals in the shoots? What is the evolutionary reasoning behind this accumulation? Is it simply a question of occupying a free place to live? In any case, hyperaccumulator plants constitute an important target to understand metal tolerance from a biochemical perspective.

High concentrations of heavy metals, namely copper, have a negative impact on human health. Symptoms of copper toxicity in humans can vary considerably, depending on the route of exposure, the concentration of copper, and the dose effectively received. Currently, the existing databases regarding copper exposure suggest that diet and drinking water are the main routes of exposure. This is directly connected with the research area of food quality, an important pathway of several contaminants. It is necessary to improve databases including populations large enough, so as to comprise individuals at higher risk of diseases caused by low or high exposure to heavy metals. It is also important to understand further the biochemical mechanisms behind heavy-metal deficiency and toxicity in humans, for prevention or curative purposes. Furthermore, additional efforts should also be made to determine biomarkers of early copper stress. As far as we know, humans were not able to develop resistant ecotypes to heavy metals, contrary to plants.

REFERENCES

Agency for Toxic Substances and Disease Registry (ATSDR). 1990. *Toxicological Profile for Copper.* Atlanta, GA: U.S. Public Health Service, TP-90–08.

Alaoui-Sossé B, Genet P, Vinit-Dunand F, Toussaint ML, Epron D, Badot PM. 2004. Effect of copper on growth in cucumber plants (*Cucumis sativus*) and its relationships with carbohydrate accumulation and changes in ion contents. *Plant Sci* **166**:1213–1218.

Ali NA, Bernal MP, Ater M. 2002. Tolerance and bioaccumulation of copper in *Phragmites australis* and *Zea mays. Plant Soil* **239**:103–111.

Alva AK, Graham JH, Anderson, CA. 1995. Soil pH and copper effects on young "Hamlin" orange trees. *Soil Sci Soc Am J* **59**:481–487.

Araya M, Chen B, Klevay LM, Strain JJ, Johnson L, Robson P, Shi W, Nielsen F, Zhu H, Olivares M, Pizarro F, Haber LT. 2003. Confirmation of an acute no-observed-adverse-effect and low-observed-adverse-effect level for copper in bottled drinking water in a multi-site international study. *Regul Toxicol Pharmacol* **38**:389–399.

Araya M, Olivares M, Pizarro F, Llanos A, Figueroa G, Uauy R. 2004. Community-based randomized double-blind study of gastrointestinal effects and copper exposure in drinking water. *Environ Health Perspect* **112**:1068–1073.

Arduini I, Godbold DL, Onnis A. 1995. Influence of copper on root growth and morphology of *Pinus pinea* L. and *Pinus pinaster* Ait. seedlings. *Tree Physiol* **15**:411–415.

Atwood CS, Scarpa RC, Huang X, Moir RD, Jones WD, Fairlie DP, Tanzi RE, Bush AI. 2000. Characterization of copper interactions with Alzheimer amyloid beta peptides: Identification of an attomolar-afinity copper binding site on amyloid beta 1–42. *J Neurochem* **75**:1219–1233.

Barón M, López-Jorgé J, Lachica M, Sadmann G. 1992. Changes in carotenoids and fatty acids in photosystem II of Cu-deficient pea plants. *Physiol Plant* **84**:1–5.

Baszynski T, Ruszkowska M, Król M, Tukendorf A, Wolinska D. 1978. The effect of copper deficiency on the photosynthetic apparatus of higher plants. *Z Pflanzenphysiol* **89**:207–216.

Bertrand M, Poirier I. 2005. Photosynthetic organisms and excess of metals. *Photosynthetica* **43**:345–353.

Bodgen JD, Klevay LM, eds. 2000. *Clinical Nutrition of the Essential Trace Elements and Minerals: The Guide for Health Prefessionals.* Totawa, NJ: Humana Press. Inc.

Bremner I. 1998. Manifestation of copper excess. *Am J Clin Nutr* **67**(suppl):p1069S–1073S.

Brewer GJ, 2000. Editorial: Is heterozygosity for a Wilson's disease gene defect and important underlying cause of infantile and childhood copper toxicosis syndromes? *J Trace Elem Exp Med* **13**:249–254.

Brown JC, Clark RB. 1977. Copper as essential to wheat reproduction. *Plant Soil* **48**:509–523.

Cairney JWG. 2000. Evolution of mycorrhiza systems. *Naturwissenschaften* **87**:467–475.

Camakaris J, Voskoboinik I, Mercer JF. 1999. Molecular mechanisms of copper homeostasis. *Biochem Biophys Res Commun* **261**:225–232.

Caspi V, Droppa M, Horváth G, Malkin S, Marder JB, Raskin VI. 1999. The effect of copper on chlorophyll organization during greening of barley leaves. *Photosynth Res* **62**:165–174.

Cecconi I, Scaloni A, Rastelli G, Moroni M, Vilardo PG, Costantino L, Cappiello M, Garland D, Carper D, Petrash JM, Del Corso A, Mura U. 2002. Oxidative modification of aldose reductase induced by copper ion. Definition of the metal–protein interaction mechanism. *J Biol Chem* **277**:42017–42027.

Chao PY, Allen KG. 1992. Glutathione production in copper-deficient isolated rat hepatocytes. *Free Radic Biol Med* **12**:145–150.

Chatterjee C, Sinha P, Dube BK, Gopal R. 2006. Excess copper-induced oxidative damages and changes in radish physiology. *Commun Soil Sci Plant Anal* **37**:2069–2076.

Chen LM, Kao CH. 1999. Effect of excess of copper on rice leaves: Evidence for involvement of lipid peroxidation. *Bot Bull Acad Sin* **40**:283–287.

Chen ML, Failla ML. 1988. Metallothionein metabolism in the liver and kidney of the streptozotocindiabetic rat. *Comp Biochem Physiol B* **90**:439–445.

Chen Y, Saari JT, Kang YJ. 1994. Weak antioxidant defenses make the heart a target for damage in copper-deficient rats. *Free Radic Biol Med* **17**:529–536.

Clemens S. 2001. Molecular mechanisms of plant metal tolerance and homeostasis. *Planta* **212**:475–486.

Cobbett C, Goldsbrough P. 2002. Phytochelatins and metallothioneins: Roles in heavy metal detoxification and homeostasis. *Annu Rev Plant Biol* **53**:159–182.

Danks DM. 1988. Copper deficiency in humans. *Annu Rev Nutr* **8**:235–257.

Davanzo F, Settimi L, Faraoni L, Maiozzi P, Travaglia A, Marcello I. 2004. Agricultural pesticide-related poisonings in Italy: Cases reported to the Poison Control Centre of Milan in 2000–2001. *Epidemiol Prev* **28**:330–337.

Devi SR, Prasad MNV. 2004. Membrane lipid alterations in heavy metals exposed plants. In: Prasad MNV, editor. *Heavy Metal Stress in Plants. From Biomolecules to Ecosystems*. India: Springer, pp. 127–145.

Doraiswamy PM, Finefrock AE. 2004. Metals in our minds: Therapeutic implications for neurodegenerative disorders. *Lancet Neurol* **3**:431–434.

Drążkiewicz M, Skórzyńska-Polit E, Krupa Z. 2004. Copper-induced oxidative stress and antioxidant defence in *Arabidopsis thaliana*. *Biometals* **17**:379–387.

Dučić T, Polle A. 2005. Transport and detoxification of manganese and copper in plants. *Braz J Plant Physiol* **17**:103–112.

Eife R, Weiss M, Barros V, Sigmund B, Goriup U, Komb D, Wolf W, Kittel J, Schramel P, Reiter K. 1999. Chronic poisoning by copper in tap water: I. Copper intoxications with predominantly gastrointestinal symptoms. *Eur J Med Res* **4**:219–223.

Failla ML, Johnson MA, Prohaska JR. 2001. Copper. In: *Present Knowledge in Nutrition*. Washington, DC: Life Sciences Institute Press, pp. 373–383.

Fomina MA, Alexander IJ, Colpaertc JV, Gadd GM. 2005. Solubilization of toxic metal minerals and metal tolerance of mycorrhizal fungi. *Soil Biol Biochem* **37**:851–866.

Georgopoulos PG, Roy A. 2001. Environmental copper: Its dynamics and human exposure issues. *J Toxic Environ Health Part B* **4**:341–394.

Georgopoulos PG, Roy A, Lioy MJ, Opiekun RE, Lioy PJ. 2001. *Copper: Environmental Dynamics and Requirements for Human Exposure Assessment.* New York: International Copper Association.

Gupta UC. 1979. Copper in agricultural crops. In: Nriagu JO, editor. *Copper in the Environment. Part 1: Ecological Cycling.* New York: John Wiley & Sons, pp. 255–288.

Hagemeyer J. 1999. Ecophysiology of plant growth under heavy metal stress. In: Prasad MNV, Hagemeyer J, editors. *Heavy Metal Stress in Plants: From Molecules to Ecosystems.* Berlin: Springer, pp. 157–181.

Hall JL. 2002. Cellular mechanisms for heavy metal detoxification and tolerance. *J Exp Bot* **53**:1–11.

Harper FA, Smith SE, Macnair MR. 1997. Where is the cost in copper tolerance in Mimulus guttatus? Testing the trade-off hypothesis. *Funct Ecol* **11**:764–774.

Harrison FL. 1985. Effect of physicochemical form on copper availability to aquatic organisms (special technical publication). *Aquatic Toxicology and Hazard Assessment*: Seventh Symposium ASTM, Philadelphia.

Hart EB, Steenbock J, Waddell J, Elvehjem CA. 1928. Iron in nutrition. VII. Copper as a supplement to iron for hemoglobin building in the rat. *J Biol Chem* **77**:797–812.

Heller LJ, Mohrman DE, Prohaska JR. 2000. Decreased passive stiffness of cardiac myocytes and cardiac tissue from copper-deficient rat hearts. *Am J Physiol Heart Circ Physiol* **278**:H1840–H1847.

Hellman NE, Gitlin JD. 2002. Ceruloplasmin metabolism and function. *Annu Rev Nutr* **22**:439–458.

Henriques FS. 1989. Effects of copper deficiency on the photosynthetic apparatus of sugar beet (*Beta vulgaris* L.). *J Plant Physiol* **135**:453–458.

Hoffman HN, Phyliky RL, Fleming CR. 1998. Zinc-induced copper deficiency. *Gastroenterology* **94**:508–512.

Hoof NALM, Hassinen VH, Hakvoort HWJ, Ballintijn KF, Schat H, Verkleij JAC, Ernst WHO, Karenlampi SO, Tervahauta AI. 2001. Enhanced copper tolerance in *Silene vulgaris* (Moench) Garcke populations from copper mines is associated with increased transcript levels of a 2b-type metallothionein gene. *Plant Physiol* **126**:1519–1526.

Institute of Medicine. 1990. *Nutrition During Pregnancy, Part 2: Nutrient Supplements.* Washington, DC: National Academy Press.

Institute of Medicine. 2000. Food and Nutrition Board. Micronodular cirrhosis and acute liver failure due to chronic copper self-intoxication. In: *Dietary Reference Intakes for Vitamin A, Vitamin K, Arsenic, Boron, Chromium, Copper, Iodine, Iron, Manganese, Molybdenum, Nickel, Silicon, Vanadium, and Zinc.* Washington, DC: National Academy Press.

Institute of Medicine. 2002. Food and Nutrition Board. Copper. In: *Dietary reference intakes: Vitamin A, Vitamin K, Arsenic, Boron, Chromium, Copper, Iodine, Iron, Manganese, Molybdenum, Nickel, Silicon, Vanadium, and Zinc.* Washington, DC: National Academy Press, pp. 224–257.

Jankowski MA, Uriu-Hare JY, Rucker RB, Keen CL, 1993. Effect of maternal diabetes and dietary copper on fetal development in rats. *Reprod Toxicol* **7**:589–598.

Kang YJ, Zhou ZX, Wu H, Wang GW, Saari JT, Klein JB. 2000. Metallothionein inhibits myocardial apoptosis in copper-deficient mice: Role of atrial natriuretic peptide. *Lab Invest* **80**:745–757.

Ke W, Xiong Z, Xie M, Luo Q. 2007. Accumulation, subcellular localization and ecophysiological responses to copper stress in two *Daucus carota* L. populations. *Plant Soil* **292**:291–304.

Khade SW, Adholeya A. 2007. Feasible bioremediation through arbuscular mycorrhizal fungi imparting heavy metal tolerance: A retrospective. *Biorem J* **11**:33–43.

Klevay LM. 2000. Cardiovascular disease from copper deficiency—A history. *J Nutr* **130**:489S–492S.

Küpper H, Šetlik I, Šetliková E, Ferimazova N, Spiller M, Küpper FC. 2003. Copper-induced inhibition of photosynthesis: Limiting steps of in vivo copper chlorophyll formation in Scenedesmus quadricauda. *Funct Plant Biol* **30**:1187–1196.

Liao MT, Hedley MJ, Woolley DJ, Brooks RR, Nichols MA. 2000. Copper uptake and translocation in chicory (*Cichorium intybus* L. cv *Grasslands Puna*) and tomato (*Lycopersicon esculentum* Mill. cv *Rondy*) plants grown in NFT system. II. The role of nicotianamine and histidine in xylem sap copper transport. *Plant Soil* **223**:243–252.

Lonnerdal B. 1998. Copper nutrition during infancy and childhood. *Am J Clin Nutr I* **67**:1046S–1053S.

Lösch R. 2004. Plant mitochondrial respiration under the influence of heavy metals. In: Prasad MNV, editor. *Heavy Metal Stress in Plants—From Biomolecules to Ecosystems.* India: Springer, pp. 182–200.

Luo Y, Rimmer DL. 1995. Zinc–copper interaction affecting plant growth on a metal-contamined soil. *Environ Pollut* **88**:79–83.

Ma WC. 1984. Sublethal toxic effects of copper on growth, reproduction and litter breakdown activity in the earthworm *Lumbricus rubellus*, with observations on the influence of temperature, and soil pH. *Environ Pollut* **33A**:207–219.

Mahmood T, Islam KR. 2006. Response of rice seedlings to copper toxicity and acidity. *J Plant Nutr* **29**:943–957.

Marschner H. 1995. *Mineral Nutrition of Higher Plants*, second edition. London: Academic Press.

Martin H, Woodcock D. 1983. *The Scientific Principles of Crop Protection*, seventh edition. Arnold: London.

Martins LL, Mourato MP. 2006. Effect of excess copper on tomato plants: Growth parameters, enzyme activities, chlorophyll, and mineral content. *J Plant Nutr* **29**:2179–2198.

Mattoo AK, Baker JE, Moline HE. 1986. Induction by copper ions of ethylene production in *Spirodela oligorrhiza*: Evidence for a pathway independent of 1-amminocyclopropane-1-carboxylic acid. *J Plant Physiol* **123**:193–202.

Maynard CJ, Bush AI, Masters CL, Cappai R, Li QX. 2005. Metals and amyloid-beta in Alzheimer's disease. *Int J Exp Pathol* **86**:147–159.

Medeiros DM, Jennings DB. 2002. Role of copper in mitochondrial biogenesis via interaction with ATP synthase and cytochrome c oxidase. *J Bioenerg Biomembr* **34**:389–395.

Mediouni C, Benzarti O, Tray B, Ghorbel MH, Jemal F. 2006. Cadmium and copper toxicity for tomato seedlings. *Agron Sustain Dev* **26**:227–232.

Meharg AA. 2003. The mechanistic basis of interactions between mycorrhizal associations and toxic metal cations. *Mycol Res* **107**:1253–1265.

Mendez MA, Araya M, Olivares M, Pizarro F, Gonzalez M. 2004. Sex and ceruloplasmin modulate the response to copper exposure in healthy individuals. *Environ Health Perspect* **112**:1654–1657.

Murphy AS, Eisinger WR, Shaff JE, Kochian LV, Taiz L. 1999. Early copper-induced leakage of K$^+$ from *Arabidopsis* seedlings is mediated by ion channels and coupled to citrate efflux. *Plant Physiol* **121**:1375–1382.

Mysliwa-Kurdziel B, Prasad MNV, Strzalka K. 2004. Photosynthesis in heavy metal stressed plants. In: Prasad MNV, editor. *Heavy Metal Stress in Plants. From Biomolecules to Ecosystems.* India: Springer, pp. 146–181.

National Academy of Sciences (NAS). 1977. *Copper. Committee on Medical and Biologic Effects of Environmental Pollutants.* Washington, DC: National Research Council, National Academy of Sciences.

Nian H, Yang ZM, Ahn SJ, Cheng ZJ, Matsumoto H. 2002. A comparative study on the aluminium- and copper-induced organic acid exudation from wheat roots. *Physiol Plant* **116**:328–335.

Nolte J. 1988. Pollution source analysis of river water and sewage sludge. *Environ Technol Lett* **9**:857–868.

Nriagu JO, editor. 1979a. *Copper in the Environment, Part 1: Ecological Cycling.* New York: John Wiley & Sons.

Nriagu JO, editor. 1979b. *Copper in the Environment, Part 2: Health Effects.* New York: John Wiley & Sons.

Nriagu JO. 1979c. The global copper cycle. In: Nriagu JO, editor. *Copper in the Environment, Part 1: Ecological Cycling.* New York: John Wiley & Sons, pp. 1–17.

Nriagu JO. 1979d. Copper in the atmosphere and precipitation. In: Nriagu JO, editor. *Copper in the Environment, Part 1: Ecological Cycling.* New York: John Wiley & Sons, pp. 45–75.

Nriagu JO, Pacyna JM. 1988. Quantitative assessment of worldwide contamination of air, water and soils by trace metals. *Nature (Lond)* **333**:134–139.

Olin KL, Walter RM, Keen CL. 1994. Copper deficiency affects selenoglutathione peroxidase and selenodeiodinase activities and antioxidant defense in weanling rats. *Am J Clin Nutr* **59**:654–658.

Opazo C, Huang X, Cherny RA, Moir RD, Roher AE, White AR, Cappai R, Masters CL, Tanzi RE, Inestrosa NC, Bush AI. 2002. Metalloenzyme-like activity of Alzheimer's disease betaamyloid. Cu-dependent catalytic conversion of dopamine, cholesterol, and biological reducing agents to neurotoxic H$_2$O$_2$. *J Biol Chem* **277**:40302–40308.

Ouzounidou G, Symeonidis L, Babalonas D, Karataglis S. 1994. Comparative responses of a copper-tolerant and a copper-sensitive population of *Minuartia hirsuta* to copper toxicity. *J Plant Physiol* **144**:109–115.

Ouzounidou G, Ciamporova M, Moutakas M, Karataglis S. 1995. Responses of maize (*Zea mays* L.) plants to copper stress. I. Growth, mineral content, and ultrastructure of roots. *Environ Exp Bot* **35**:167–176.

Panou-Filotheou H, Bosabalidis AM. 2004. Root structural aspects associated with copper toxicity in oregano (*Origanum vulgare* subsp. *hirtum*). *Plant Sci* **166**:1497–1504.

Panou-Filotheou H, Bosabalidis AM, Karataglis S. 2001. Effects of copper toxicity on leaves of oregano (*Origanum vulgare* subsp. *hirtum*). *Ann Bot* **88**:207–214.

Patel BN, Dunn RJ, David S. 2000. Alternative RNA splicing generates a glycosylphosphati-dylinositol-anchored form of ceruloplasmin in mammalian brain. *J Biol Chem* **275**:4305–4310.

Pätsikkä E, Kairavuo M, Šeršen F, Aro EM, Tyystjärvi E. 2002. Excess copper predisposes Photosystem II to photoinhibition in vivo by outcompeting iron and causing decrease in leaf chlorophyll. *Plant Physiol* **129**:1359–1367.

Pena MO, Lee J, Thiele DJ. 1999. A delicate balance: Homeostatic control of copper uptake and distribution. *J Nutr* **129**:1251–1260.

Peng HY, Yang XE, Tian SK. 2005. Accumulation and ultrastructural distribution of copper in *Elsholtzia splendens*. *J Zhejiang Univ Sci* **6**:311–318.

Perwak J, Bysshe S, Goyer M, Nelken L, Scow K. 1980. *Exposure and Risk Assessment for Copper*. Washington, DC: Office of Water Regulation Standards, US Environmental Protection Agency.

Pich A, Scholz G. 1996. Translocation of copper and other micronutrients in tomato plants (*Lycopersicon esculentum* Mill.): Nicotianamine-stimulated copper transport in the xylem. *J Exp Bot* **47**:41–47.

Prohaska JR, Sunde RA, Zinn KR. 1992. Livers from copper-deficient rats have lower gluta-thione peroxidase activity and mRNA levels but normal liver selenium levels. *J Nutr Biochem* **3**:429–436.

Prohaska JR, Brokate B. 2002. The timing of perinatal copper deficiency in mice influences offspring survival. *J Nutr* **132**:3142–3145.

Quartacci MF, Cosi E, Meneguzzo S, Sgherri C, Navari-Izzo F. 2003. Uptake and translocation of copper in Brassicaceae. *J Plant Nutr* **26**:1065–1083.

Rai VK. 2002. Role of amino acids in plant responses to stresses. *Biol Plant* **45**:481–487.

Ralph A, McArdle H. 2001. *Copper Metabolism and Copper Requirements in the Pregnant Mother, Her Fetus, and Children*. New York: International Copper Association.

Reichman SM. 2002. *The Responses of Plants to Metal Toxicity: A Review Focusing on Copper, Manganese and Zinc*. Melbourne: Australian Minerals & Energy Environment Foundation.

Reid RJ. 2001. Mechanisms of micronutrient uptake in plants. *Aust J Plant Physiol* **28**:659–666.

Rodriguez FI, Esch JJ, Hall AE, Binder BM, Schaller GE, Bleecker AB. 1999. A copper cofactor for the ethylene receptor ETR1 from Arabidopsis. *Science* **283**:996–998.

Roncero V, Duran E, Soler F, Masot J, Gomez L. 1992. Morphometric, structural, and ultra-structural studies of tench (*Tinca tinca* L.) hepatocytes after copper sulfate administration. *Environ Res* **57**:45–58.

Saari JT. 2000. Copper deficiency and cardiovascular disease: Role of peroxidation, glycation, and nitration. *Can J Physiol Pharmacol* **78**:848–855.

Salt DE, Kato N, Krämer U, Smith RD, Raskin I. 2000. The role of root exudates in nickel hiperaccumulation and tolerance in accumulator and non acummulator species of *Thlaspi*. In: Terry N, Banuelos GS, editors. *Phytoremediation of Contaminated Soil and Water*. Boca Raton, FL: CRC Press, pp. 189–200.

Schroeder HA, Nason AP, Tipton IH, Balassa JJ. 1966. Essential trace metals in man: Copper. *J Chronic Dis* **19**:1007–1034.

Schroeder WH, Dobson M, Kane DM, Johnson ND. 1987. Toxic trace elements associated with airborne particulate matter: A review. *J Air Pollut Control Assoc* **37**:1267–1285.

Sharma SS, Dietz KJ. 2006. The significance of amino acids and amino acid-derived molecules in plant responses and adaptation to heavy metal stress. *J Exp Bot* **57**:711–726.

Shaw BP, Sahu SK, Mishra RK. 2004. Heavy metal induced oxidative damage in terrestrial plants. In: Prasad MNV, editor. *Heavy Metal Stress in Plants: From Biomolecules to Ecosystems.* India: Springer, pp. 84–126.

Sheldon AR, Menzies NW. 2005. The effect of copper toxicity on the growth and root morphology of Rhodes grass (*Chloris gayana* Knuth.) in resin buffered solution culture. *Plant Soil* **278**:341–349.

Shikanai T, Müller-Moulé P, Munekage Y, Nyogi KK, Pilon M. 2003. PPA1, a P-type ATPase of *Arabidopsis*, functions in copper transport in chloroplasts. *Plant Cell* **15**:1333–1346.

Shim H, Harris ZL. 2003. Genetic defects in copper metabolism. *J Nutr* **133**:1527S–1531S.

Srivastava A, Peshin SS, Kaleekal T, Gupta SK. 2005. An epidemiological study of poisoning cases reported to the National Poisons Information Centre, All India Institute of Medical Sciences, New Delhi. *Hum Exp Toxicol* **24**:279–285.

Stadtman ER. 1990. Metal ion-catalyzed oxidation of proteins: Biochemical mechanisms and biological consequences. *Free Radical Bio Med* **9**:315–325.

Tabata K, Kashiwagi S, Mori H, Ueguchi C, Mizuno T. 1997. Cloning of a cDNA encoding a putative metal-transporting P-type ATPase from *Arabidopsis thaliana. Biochim Biophys Acta* **1326**:1–6.

Thornton I. 1979. Copper in soils and sediment. In: Nriagu JO, editor. *Copper in the Environment, Part 1: Ecological Cycling.* New York: John Wiley & Sons, pp. 171–216.

Turnlund JR. 1991. Bioavailability of dietary minerals to humans: The stable isotope approach. *Crit Rev Food Sci Nutr* **30**:387–396.

Turnlund J. 1998. Human whole-body copper metabolism. *Am J Clin Nutr* **67**:960–964.

Turnlund J. 1999. Copper. In: Shils ME, Olson JA, Shike M, Ross AC, editors. *Modern Nutrition in Health and Disease.* Baltimore, MD: Williams & Wilkins, pp. 241–252.

Turnlund JR, Keyes WR, Anderson HL, Acord LL. 1989. Copper absorption and retention in young men at three levels of dietary copper by use of the stable isotope 65Cu. *Am J Clin Nutr* **49:**870–878.

Turnlund JR, Keyes WR, Peiffer GL. 1995a. Copper excretion into the gastrointestinal tract at three levels of dietary copper studied with the stable isotope 65Cu. *FASEB J* **9**:A725 (abstract).

Turnlund JR, Keyes WR, Peiffer GL. 1995b. Copper absorption and retention from diets low and adequate in copper. *Am J Clin Nutr* **61**:908 (abstract 102).

Turnlund JR, Scott KC, Peifer GL, Jang AM, Keyes WR, Keen CL, Sakanashi TM. 1997. Copper status of young men consuming a low-copper diet. *Am J Clin Nutr* **65**:72–78.

Turnlund JR, Jacob RA, Keen CL, Strain JJ, Kelley DS, Domek JM, Keyes WR, Ensunsa JL, Lykkesfeldt J, Coulter J. 2004. Long-term high copper intake: Effects on indexes of copper status, antioxidant status, and immune function in young men. *Am J Clin Nutr* **79**:1037–1044.

Uauy R, Olivares M, Gonzalez M. 1998. Essentiality of copper in humans. *Am J Clin Nutr* **67**:952–959.

Uriu-Adams JY, Keen CL. 2005. Copper, oxidative stress, and human health. *Mol Aspects Med* **26**:268–298.

Van Tichelen KK, Colpaert JV, Vangronsveld J. 2001. Ectomycorrhizal protection of *Pinus sylvestris* against copper toxicity. *New Phytol* **150**:203–213.

Vassilev A, Lidon F, Campos PS, Ramalho JC, Barreiro MG, Yordanov I. 2003. Cu-induced changes in chloroplast lipids and photosystem 2 activity in barley plants. *Bulg J Plant Physiol* **29**:33–43.

Vinit-Dunand F, Epron D, Alaoui-Sossé B, Badot PM. 2002. Effects of copper on growth and on photosynthesis of mature and expanding leaves in cucumber plants. *Plant Sci* **163**:53–58.

Wang SH, Yang ZM, Yang H, Lu B, Li SQ, Lu YP. 2004. Copper-induced stress and antioxidative responses in roots of *Brassica juncea* L. *Bot Bull Acad Sinica* **45**:203–212

Watanabe T, Osaki M. 2002. Mechanisms of adaptation to high aluminium condition in native plant species growing in acid soils: A review. *Commun Soil Sci Plant Anal* **33**:1247–1260.

Weant GE. 1985. *Sources of Copper Air Emissions.* Research Triangle Park, NC: Air Energy Engineering Research Laboratory, US Environmental Protection Agency.

Wilson JB. 1988. The cost of heavy-metal tolerance: An example. *Evolution* **42**:408–413.

World Health Organization. 1998. *Guidelines for Drinking Water Quality.* Geneva: World Health Organization.

Xiong ZT, Wang H. 2005. Copper toxicity and bioaccumulation in Chinese cabbage (*Brassica pekinensis* Rupr.). *Environ Toxicol* **20**:188–194.

Xiong ZT, Liu C, Geng B. 2006. Phytotoxic effects of copper on nitrogen metabolism and plant growth in *Brassica pekinensis* Rupr. *Ecotox Environ Safety* **64**:273–280.

Yang XE, Peng HY, Tian SK. 2005. Gama-aminobutyric acid accumulation in *Elsholtzia splendens* in response to copper toxicity. *J Zhejiang Uni Sci* **6**:96–99.

Yruela I. 2005. Copper in plants. *Braz J Plant Physiol* **17**:145–156.

Yruela I, Pueyo JJ, Alonso PJ, Picorel R. 1996. Photoinhibition of Photosystem II from higher plants: Effect of copper inhibition. *J Biol Chem* **271**:27408–27415.

Zidenberg-Cherr S, Dreith D, Keen CL. 1989. Copper status and adriamycin treatment: Effects on antioxidant status in mice. *Toxicol Lett* **48**:201–212.

26 Forms of Copper, Manganese, Zinc, and Iron in Soils of Slovakia: System of Fertilizer Recommendation and Soil Monitoring

BOHDAN JURANI and PAVEL DLAPA

Department of Soil Science, Faculty of Natural Sciences, Comenius University, 842 15 Bratislava, Slovak Republic

1 FORMS OF TRACE ELEMENTS IN HETEROGENEOUS SOIL MATERIALS

The soil is a physically and chemically heterogeneous material. Its complexity embraces particles of different sizes where skeleton (>2 mm) and fine earth (<2 mm) particles are distinguished basically. The fine earth particles are commonly used for subsequent analyses (including trace element analysis) in soil science. An additional three classes with particle sizes <2 mm can further be distinguished in the fine earth: clay (<0.002 mm), silt (0.002–0.05 mm), and sand (0.05–2 mm). In mechanically heterogeneous material, individual soil particles can have very different chemical composition. Adjacent particles can be of organic nature or inorganic nature. Organic matter is generally composed of nonspecific organic compounds synthesized in the bodies of living organisms and of humified specific organic compounds forming humic and fulvic acids. Inorganic particles can also be very variable in soils and include chemically very different compounds such as silicates, oxi-hydroxides, carbonates, or highly soluble salts. The above-mentioned heterogeneity of soil materials can be observed in soil thin sections using a polarizing microscope where the variability in mineralogical composition (and thereby also

Trace Elements as Contaminants and Nutrients: Consequences in Ecosystems and Human Health, Edited by M. N. V. Prasad

variability in chemical composition) of mineral particles can easily be detected. Soil thin section study may also show that soil-forming processes often cause redistribution of individual chemical forms and lead to local accumulations of organic or mineral compounds in variable forms of coatings, hypocatings, quasicoatings, infillings, and nodules [Stoops and Vepraskas, 2003]. Such accumulations of organic compounds, iron and manganese hydrous oxides, clay minerals, and so on, which commonly act as sorbents for trace elements, may play a significant role in trace elements binding at specific positions inside heterogeneous soil material.

Heterogeneity of individual soil particles has led to the concept that bulk soil can be thought to contain different chemical fractions and that trace elements are distributed nonuniformly among these fractions [Davies, 1994]. Various extracting agents were used in order to dissolve selectively individual fractions of trace elements and thus to evaluate quantitatively their chemical forms of occurrence in a soil. Therefore several methods based on sequential extraction procedure were proposed until now. The most widespread procedure is that of Bascomb [1968], which was proposed for determination of iron forms in soils. He distinguished iron bound to silicates, crystallized oxides, and aged amorphous and gel amorphous hydroxides. Later McLaren and Crawford [1973] developed sequential extraction of copper forms present in soil solution, at exchange sites, at specific sorption sites, occluded in soil oxides, bound to soil organic matter, and in lattices of minerals. The sequential extraction procedure, which is most commonly used at present, was proposed by Tessier et al. [1979]. The modified Tessier procedure may consist of nine fractions as reported by Navarcik et al. [1997]: (1) redistilled water, pH 5.5, shaken for 1 hr, water-soluble; (2) 1 M $MgCl_2$, pH 7, shaken for 1 hr, exchangeable; (3) 0.025 M $Na_4P_2O_7$, shaken for 1 hr, bound to humic acids; (4) 1 M NaOAc + HOAc, shaken for 24 hr, bound to carbonates; (5) 0.04 M $NH_2OH \cdot HCl$, shaken for 24 hr, bound to Fe/Mn; (6) 30% H_2O_2 + HNO_3, pH 2, shaken for 24 hr, organically bound and bound to sulfates; (7) 2 M HNO_3, shaken for 24 hr; (8) 1 M NaOH, shaken for 24 hr; (9) residual fraction.

An example of the results obtained by sequential extraction procedure according to McLaren and Crawford [1973] are shown in Fig. 1. Within the soil profile of uncontaminated Ochrept, Cu was found to be concentrated in the surface horizons as a result of their sorption by the soil organic matter. In the topsoil, Cu was distributed mainly between organically bound (32–35%) and residual (56–59%) fractions, Fe-bound fraction contained only 9% of Cu. In the subsoil horizons, Cu occurred predominantly in the residual fraction (78–91%) and much less as organically bound (4–18%) and Fe-bound (2–5%). Different distribution of Cu was observed within the profile of contaminated Fluvent. Total Cu was distributed mainly as Fe-bound (38–60%), the organically bound fraction contained 16–42% of Cu, and 20–26% of Cu was present in residual fraction [Dlapa et al., 2000].

In soil tests, however, single extraction procedures are the most commonly used in order to evaluate mobile and mobilizable forms or (pseudo) total content of trace elements. These forms are related to actual availability of elements from various sources rather than being related to pools of elements extracted from chemically different soil fractions. The single extraction procedures are widely used for determination of limit values for contaminated soils with regard to various receptors. Extraction agents commonly used in single extracting agents are listed in Table 1.

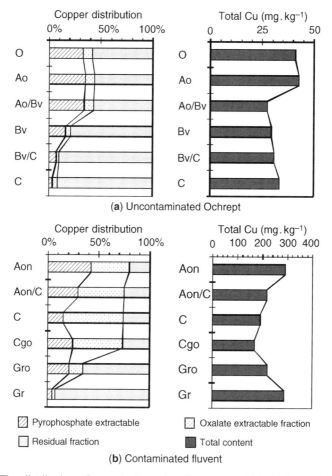

Figure 1. The distribution of organic (pyrophosphate extractable), hydrous oxide (oxalate extractable), and residual Cu fractions and their vertical distributions within soil profiles.

TABLE 1. Trace Element Fractions in Soils and Corresponding Extraction Agents

Fractions in Soil	Characteristic of Fraction	Commonly Used Extracting Media
Pseudo total (often referred to as total content)	=Nonactive + potentially active (mobilizable)	Strong acid solutions like *aqua regia*, HNO_3, HCl
Mobilizable	=Potentially active + active = potentially bioavailable or potentially leachable	Buffered and unbuffered complexing and chelating reagents like EDTA, DTPA
Mobile	Active (bioavailable and easily leachable)	Neutral unbuffered salt solution like $NaNO_3$, $CaCl_2$, NH_4NO_3

Source: Adapted from Gupta and Vollmer [1994].

Estimation of plant available forms of trace elements calls for a single extraction approach because plant roots uptake bioavailable pools of micronutrients from various chemical fractions of soils rather than uptake selectively from a single chemical pool of trace elements in the soil.

2 CONCEPT OF MICRONUTRIENTS USED IN AGRICULTURE OF FORMER CZECHOSLOVAKIA

High and stabile yields, with good nutritional and technological values, also need a proper level of available forms of micronutrients in soil. The main goal was to develop research regarding micronutrients in soil in late 1960s (and later on in Czechoslovakia). Many individual scientists have performed experimental studies concerning testing the micronutrient needs of plants; some of them used plant analysis, whereas the others preferred soil testing.

Development of methods based on plant analyses was derived mainly from principles of Bergmann's school to cure acute problems of micronutrients storage [Bergmann and Cumakov, 1977]. Due to better predictability, another group of scientists has preferred soil analyses. In the 1960s and 1970s, single (so-called "specific") extracting agents for single microelements were mainly, successfully used (Cu [Westerhoff, 1955], Mn [Schachtschabel, 1956], Zn [Wear and Sommer, 1948]). Therefore soil scientists, dealing with the problem of availability of micronutrients, have concentrated their efforts on the development of group extracting agents to have the chance to realize their investigation on an available form of micronutrients in agricultural soils of the whole country.

Recommendation of micronutrients application was based on specific limits for single elements and later also on specification only to certain "sensitive" crops.

Calculation of specific doses of micronutrient fertilizers is based on experiments, both with foliar application (spraying) and with soil application. Amount of soil-applied micronutrients depends also on some soil properties (e.g., soil texture).

For foliar application the time of application is also very important (specific stage of plant growth) to receive good results.

3 DETERMINATION OF AVAILABLE FORMS OF SOME MICRONUTRIENTS IN SOIL BASED ON THE RINKIS METHOD

In the 1970s in the former Czechoslovakia, the method of soil testing according to Rinkis [1972] was adopted, to receive the answer regarding the real need for plant micronutrients. This method was based on the following assumptions:

1. To receive basic information about optimal concentration of macro- and micronutrients in the soil for optimal plant growth, all basic studies were performed in inert substratum (quartz sand).

2. To compensate influence of soil properties (e.g., soil texture, soil organic matter content, soil pH, carbonate content, sesquioxide content) on nutrient concentrations in soil solution, Rinkis devised so-called correction coefficients for mentioned soil factors.

3. To receive basic information concerning available soil nutrient contents, Rinkis has used 1 M HCl as a group extracting agent with soil/solution ratio (by volume) of 1/5 for all nutrients.

The formal Rinkis calculation of needed micronutrients was based on the equation

$$A(\text{need of nutrient}) = B(\text{optimal level}) - C(\text{amount in 1M HCl in mg/liter})$$

where B value is individually corrected on soil factors. For calculation of the amount of element in (kg/ha), information about the depth of plow layer is need.

Table 2 shows ranges of data regarding soil Cu, Mn, and Zn contents extracted with 1 M HCl as determined in Latvia and Czechoslovakia; it also shows data calculated using correction coefficients, mentioned above [Cumakov and Jurani, 1980]. As clearly visible, there are rather large differences is measured soil Cu, Mn, and Zn contents between Latvia and Czechoslovakia, with the ranges being much higher and wider in Czechoslovakia. This is partly a result of completely different natural conditions in both countries, but also a result of modification of the Rinkis method used in Czechoslovakia. Soil/solution ratio in the case of Czechoslovakia's survey was 50 mg of soil and 250 ml 1 M HCl (mass/volume ratio), in the original Rinkis method the volume/volume ratio was used. Also an addition of aliquot amount of 2 M HCl to the soils containing carbonates makes a difference. Due to these differences and completely different soil conditions, and also thanks to a ready-made detail soil survey of all agricultural soils in Czechoslovakia (done in the 1960s), it was possible to prepare correction coefficients for main soil types and subtypes and, by subsequent calculation, to produce information about optimal concentration of micronutrients (Cu, Mn, and Zn) for main soil subtypes of Czechoslovakia. Using minimal and maximal values for a soil and single factors, border limits for each soil with "very low" and "very high" contents were obtained. The category of "medium" content was calculated on the basis of the middle 80% of all calculated values. The remaining 20% of measured values were used for a decision about "low" and "high" categories of

TABLE 2. Ranges of Measured and Calculated Contents (in mg/kg) of Cu, Mn, and Zn Extractable by 1 M HCl from Soils of Latvia and Former Czechoslovakia

Element	Latvia, Measured	Czechoslovakia Measured	Czechoslovakia Calculated
Cu	0.8–4.7	3.2–23.2	1.6–10.0
Mn	38.0–175.0	127.0–415.0	9.0–45.0
Zn	2.0–8.0	8.6–28.4	4.5–13.0

measured micronutrients. The values for "light" soil were obtained by multiplication of values by the coefficient 0.7, whereas the values for heavy soils were obtained by multiplication of values by coefficient 1.3. Calculated border limits for evaluation of the results done by Rinkis method are listed in Table 3.

Parallel to this approach, the so-called agronomical approach was prepared in order to establish limits for micronutrients in soil. They were based on three main presumptions:

1. After 3 years' application of micronutrients to the soil, the amount and dynamics of investigated micronutrients extracted by 1 M HCl were possible to evaluate.
2. The positive, but also negative, correlations between contents of micronutrients in plants and soils (extracting agent 1 M HCl) were observed.
3. There is a relationship between micronutrient contents in plants and reached yields.

Calculation using the relationship between soil micronutrient (1 M HCl) content and reached yield can solve the basic problem of limits for evaluation of micronutrients contents in soil. On yield curves, four categories of micronutrient contents in soil were established, in relation to the quantity of reached yield. Single categories were established on the basis of increase of the yield on yield curves, taking as the border limit amount of cereals 0.1 ton of grains, 1 ton of green matter for alfalfa and green fodder crops, 0.5 ton of bulbs for potatoes, and 1 ton of bulbs for sugar beet. Criteria were derived from 16 yield curves calculated for Cu, 8 for Mn, and 7 for Zn. Using such an approach, the category of "low content" of micronutrients was established. An additional category of "very low content" was established on the basis of occurrence of visual deficiency symptoms in plants. In such a category, the highest effect of micronutrient fertilizers application can be expected. "Medium content" was defined as that part of the curve where, under the influence of added soil micronutrients, an

TABLE 3. Calculated Limit Values for Cu, Mn, and Zn Contents Extractable by 1 M HCl

Element	Soil Texture Class	Content in Soil (mg/kg)				
		Very Low	Low	Medium	High	Very High
Cu	Light	<1.0	1.0–2.4	2.5–4.9	5.0–8.4	>8.4
	Medium	<1.5	1.5–3.4	3.5–7.0	7.1–12.0	>12.0
	Heavy	<2.0	2.0–4.4	4.5–9.1	9.2–15.6	>15.6
Mn	Light	<6.0	6.0–10.0	10.1–20.0	20.1–30.0	>30.0
	Medium	<9.0	9.0–15.0	15.1–30.0	30.1–45.0	>45.0
	Heavy	<12.0	12.0–20.0	20.1–40.0	40.1–60.0	>60.0
Zn	Light	<2.5	2.5–5.0	5.1–8.0	8.1–10.0	>10.0
	Medium	<4.0	4.0–7.0	7.1–10.0	10.1–15.0	>15.0
	Heavy	<5.5	5.5–9.0	9.1–13.0	13.1–20.0	>20.0

TABLE 4. Agronomical Criteria for Evaluation of Soil Micronutrient Contents Extractable by 1 M HCl

Element	Soil pH	Soil Texture Class	Content in Soil (mg/kg)				
			Very Low	Low	Medium	High	Very High
Cu		Light	<1.8	1.8–2.8	2.81–6.00	6.01–9.80	>9.80
		Medium	<2.0	2.0–3.5	3.51–9.00	9.01–14.00	>14.00
		Heavy	<3.0	3.0–5.0	5.01–12.00	12.01–18.20	>18.20
Mn	<5.0		<30	30–60	61–140	141–240	<240
	5.0–6.0	Light	<50	50–90	91–170	171–300	>300
	>6.0		<80	80–120	121–220	221–350	>350
	<5.0		<70	70–100	101–180	181–290	>290
	5.0–6.0	Medium	<100	100–130	131–230	231–350	>350
	>6.0		<120	120–170	171–300	301–420	>420
	<5.0		<100	100–160	161–240	241–350	>350
	5.0–6.0	Heavy	<130	130–190	191–280	281–420	>420
	>6.0		<150	150–230	231–350	351–500	>500
Zn		Light	<3.50	3.50–5.00	5.01–10.00	10.01–17.00	>17.00
		Medium	<4.50	4.50–7.00	7.01–14.00	14.01–20.00	>20.00
		Heavy	<6.00	6.00–9.00	9.01–16.00	16.01–25.00	>25.00

observed increase of yield does not exceed mentioned limits. Similarly, the regions of "high content" and "very high content" of micronutrients were defined. Influence of soil texture on availability of micronutrients and also on the level of soil limits was calculated on the basis of highly significant correlation between specific extraction agents (references) and 1 M HCl and proper regression equations. Soil pH was taken into account in the case of soil limits for Mn. Such constructed agronomical criteria of available soil micronutrient contents are listed in Table 4.

If we compare Tables 3 and 4, it is clearly visible that limits for Cu and Zn correspond rather well, but criteria for Mn derived from results of field experiments are nearly 10-times higher than those calculated from soil factors. Finally, it was decided to use agronomical criteria for evaluation of results obtained by soil extraction with 1 M HCl.

4   RESULTS OF MODIFIED RINKIS METHOD OF AVAILABLE COPPER, MANGANESE, AND ZINC IN SOILS OF SLOVAKIA

Within the years 1971–1975 the Central Agricultural Control and Testing Institute (CACTI) analyzed 13,044 soil samples from arable land of Slovakia and 1840 soil samples from meadows and pastures [Repka, 1977]. Altogether, a survey of 449,793 ha of arable land and 81,503 ha of meadows and pastures was accomplished. Air-dried and 2-mm sieved soil samples were used for modified

Rinkis extraction: 50 g of soil and 250 ml of 1 M HCl (mass/volume ratio). If the soil contained more than 1% carbonates, an aliquot amount of 2 M HCl on the dissolution of carbonates was added and subsequently 1 M HCl was added up to final volume of 250 ml. The suspension was shaken for 40 min on a rotation shaker (40 rotations per minute) and then filtered, and micronutrients Cu, Mn, and Zn were determined by AAS.

On average, in a sampled area of arable land, the researchers found 9.9 mg/kg of Cu, 293.0 mg/kg of Mn, and 12.9 mg/kg of Zn (done by the Rinkis method).

If, on the basis of "concept of micronutrients use," we are taking categories "low" + "very low" as those where an application of micronutrient fertilizers is necessary, then Cu fertilization needs 12.7% of arable land and 24.9% of meadows and pastures, Mn fertilization needs 8.4% of arable land and 6.3% of meadows and pastures, and Zn fertilization needs 34.3% of arable land and 23.4% of meadows and pastures.

The lowest amount of Cu occurs mainly in Alfisols (4.2 ± 1.8 mg/kg), while the highest amount occurs in Fluvents (13.9 ± 29.7 mg/kg); Fluvents and Ochrepts show the highest variability. The lowest amount of Mn occurs in Hapludalfs (254.2 ± 77.3 mg/kg), while the highest amount occurs in Rendolls (321.0 ± 223.3 mg/kg). The lowest amount of Zn occurs in Alfisols (6.8 ± 2.7 mg/kg), while the highest amount occurs in Fluvents (15.7 ± 16.9 mg/kg).

The main advantage of the determination of available forms of micronutrients, based on the Rinkis method, is the fact that in one extraction agent you can determine five micronutrients (B, Cu, Mn, Mo, Zn).

After a longer time of practical use of this method, also some serious failures and problems were detected:

1. This method is not suitable for determination of available forms of all micronutrients in soils containing carbonates.
2. This method is not suitable for determination of available forms of Mn and Fe, because no significant correlation was found between soil and plant micronutrient contents.

5 MORE SUITABLE METHOD FOR DETERMINATION OF PLANT AVAILABLE FORMS OF COPPER, MANGANESE, ZINC, AND IRON IN SOILS

As described in the previous section, the reasons concerning some imperfections of the modified Rinkis method, especially the unsuitability for soils containing carbonates and also the need to detect Fe deficiency, caused us to look at a new method without such problems but also less labor-intensive (like single, so-called specific extracting agents). Since the end of the 1960s to the mid-1970s, the first experiments appeared in which new specific substances—chelating agents—were used for detection of available forms of some micronutrients. With free metal ions in solution these organic substances create soluble complexes, thereby they reducing metal activity in solution. As a consequence, metal ions, having a tendency to supplement free metal ions in solution, are desorbing from soil surfaces or are dissolving from unstable solid phases. The amount of complexed metal ions in soil solution during extraction is a function of activity metal ion in soil solution (so-called intensity factor) and ability of the soil to compensate this ion, as long as its concentration in soil solution has decreased, due to any reason (capacity factor). The presence and role of natural chelates in soils have been reported by a number of authors. In practice, however, only synthetic chelating agents were adopted in trace element extraction tests.

The most commonly used chelating agents are ethylenediaminetetraacetic acid (EDTA) and diethylenetriaminepentaacetic acid (DTPA). These agents were also adopted in standard procedures. Norvell and Lindsay [1972] confirmed that the reactions of EDTA and DTPA in soils can be predicted from theoretical relationships, and Lindsay [1979] should be consulted for stability constants for metal–ligand reactions and related equilibrium calculations. Interaction of the chelating agents with soil sample generally proceeds through many dissolution, exchange, and complex-forming reactions. The concentration of any chelated element depends on all competing equilibrium reactions.

Out from the whole amount of suitable chelates, DTPA was considered as most suitable, offering most the favorable combination of stability constants for simultaneous complexing of Cu, Mn, Zn, and Fe. The first proposal on DTPA soil test for Zn and Fe was published in 1967 by Lindsay and Norvell [1967], and later it was completely evaluated also for Mn and Cu [Lindsay and Norvell, 1978]. Because micronutrient deficiencies are often observed in calcareous soils, properties of the extracting solution were modified in order to avoid excessive dissolution of $CaCO_3$ with occluded micronutrients not available to plants. Thus the DTPA solution was supplemented with triethanolamine (TEA) to provide a pH buffer and with $CaCl_2$ to convert chelate to $CaDTPA^{3-}$ form and to provide additional free Ca^{2+}.

The method was tested on 77 soil samples, including a vegetation experiment [Lindsay and Norvell, 1978]. The problems of the sampling and extraction procedure of DTPA was analyzed in detail by Soltanpour et al. [1976] and by Legget and Argyle [1983], who pointed out the importance of these problems. Silanpää [1982] also evaluated results of analysis of 3000 soil samples with this method, and he demonstrated significant correlation in vegetation pot experiments.

This method was evaluated also under Slovakian conditions [Jurani and Fiala, 1989]. The set of 17 soil samples represented a wide range of our soil conditions, from acid to slightly alkaline and from light to heavy textured, and it also included samples with low to rather high soil organic matter content.

6 LIMITS TO LINDSAY—NORVELL METHOD

At the beginning, some methodical aspects, like intensity of shaking, soil/solution ratio, volume, and shape of extraction vessel were tested. On the basis of literature sources [Legget and Argyle, 1983; Lindsay and Norvell, 1978; Soltanpur et al.,

TABLE 5. Contents (mg/kg) of Available Forms of Cu, Mn, Zn, and Fe Extracted by DTPA, in 17 Samples from Slovakia and 77 Samples from California

| Element | Slovakia | | California | |
	Average	Range	Average	Range
Cu	3.57	0.31–11.1	0.87	0.26–2.60
Mn	41.6	7.0–135.0	10.9	1.2–60.0
Zn	1.39	0.40–3.64	0.89	0.17–11.5
Fe	32.5	5.1–93.7	6.4	1.2–20.20

1976] and also on our own experiences [Jurani and Fiala, 1989], we recommended and used in practice the following procedure:

In a 100-ml standard plastic bottle, 10.0 g air-dried, 2-mm sieved soil is poured with 20 ml of 0.005 M DTPA + 0.1 M TEA + 0.01 M $CaCl_2$ prepared according to Lindsay and Norvell [1978]. Extraction was performed on a horizontal end-over-end shaker for 2 hr, 144 revolutions per minute, at the temperature 25°C. Filtration was performed using paper of medium density. Analytical determination of micronutrients was done by AAS.

The extracting solution is buffered at pH 7.3, in which excessive dissolution of $CaCO_3$ is avoided but the procedure is very sensitive to temperature during extraction, to soil/solution ratio, and to time of extraction. On the other hand, time of filtration within scale 0–30 min does not influence results significantly.

Table 5 shows average results and ranges for contents of Cu, Mn, Zn, and Fe in DTPA extracting agent in 17 samples from Slovakia and 77 soil samples from California [Lindsay and Norvell, 1978]. The results clearly show higher average values and also ranges, with the exception of Zn in soil samples from Slovakia.

For evaluation of convenience of this new method, correlation analysis of DTPA results, with results of specific extracting agents (extraction according to Westerhoff [1955] for Cu, extraction according to Schachtschabel [1956] for Mn, extraction

TABLE 6. Correlation Coefficients Between Contents Extracted by DTPA and Other Extracting Agents as Calculated for Cu, Mn, Zn, and Fe in 17 Samples from Slovakia

Element	Correlation Between	Correlation Coefficient	Statistical Significance $(1 - \alpha)$
Cu	DTPA—Westerhoff and	0.936	>0.999
	DTPA—Rinkis	0.945	>0.999
Mn	DTPA—Schachtschabel	0.979	>0.999
	and DTPA—Rinkis	−0.239	<0.950
Zn	DTPA—Wear–Sommer	0.727	>0.999
	and DTPA—Rinkis	0.906	>0.999
Fe	DTPA—Wear–Sommer	−0.223	<0.950
	and DTPA—Rinkis	0.569	>0.950

according to Wear and Sommer [1948] for Zn and Fe) and also with results of 1 M HCl extractions according to Rinkis [1972], was made (see Table 6). The obtained results showed very tight relationships with statistically significant correlation in most cases. Statistically insignificant relationships in the case of Mn and Fe confirmed the already known unsuitability of extraction agents other than DTPA.

For creation of limits for assessment of DTPA results, regression equations between specific extracting agents and the Lindsay–Norvell method as well as between 1 M HCl and the Lindsay-Norvell method were used (see Fig. 2). Finally, proposed limits are listed in Table 7.

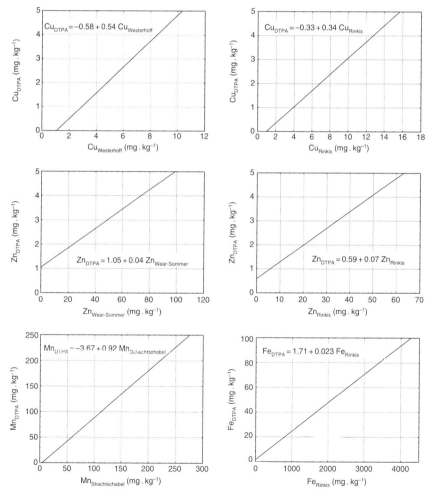

Figure 2. Regression relationships between DTPA extractable forms and other extractable forms of micronutrients in soils. The relationships between DTPA and Rinkis extracts for Mn as well as between DTPA and Wear–Sommer extracts for Fe are not shown in Fig. 1 because the relationships were not statistically significant.

TABLE 7. Limits for Soil Cu, Mn, Zn, and Fe Extracted by DTPA

Element	Criteria for Extracts in DTPA (mg/kg)		
	Low	Medium	High
Cu	<0.80	0.80–2.70	>2.70
Mn	<10.0	10.0–100.0	>100.0
Zn	<1.0	1.0–2.50	>2.50
Fe	<8.0	1.0–75.0	>75.0

Much later, in 1998, after international evaluation, this DTPA method was accepted as the International Standard method for the extraction of trace elements in soil samples (ISO 14870). This method is recommended especially for estimation of the availability of Cu, Fe, Mn, and Zn in soils to growing plants. It is preferably applicable to soils, having a pH greater than 6. In addition, toxic elements such as Cd, Cr, Ni, and Pb can also be determined in these extracts. Present-day soil monitoring also uses this method for evaluation of heavy metals in soil.

7   SOME RESULTS CONCERNING USING LINDSAY—NORVELL METHOD

Contents of available micronutrients in agricultural soils of Slovakia, using the DTPA method according to Lindsay and Norvell 1978, are reported by Kobza et al. [2002].

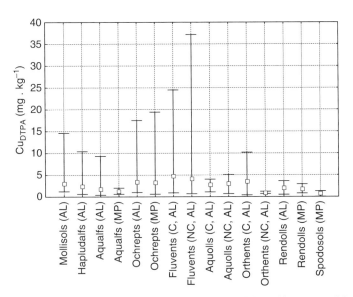

Figure 3. Contents of available Cu in agricultural soils of Slovakia extracted by DTPA. Average values and ranges of available contents are plotted for individual soil types (abbreviations: C, calcareous; NC, noncalcareous; AL, arable land; MP, meadows and pastures).

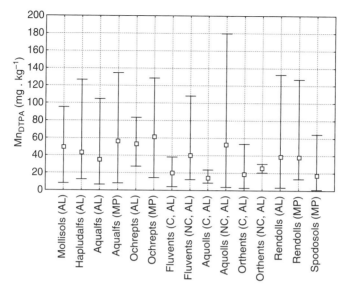

Figure 4. Contents of available Mn in agricultural soils of Slovakia extracted by DTPA. Average values and ranges of available contents are plotted for individual soil types (abbreviations: C, calcareous; NC, non-calcareous; AL, arable land; MP, meadows and pastures).

An average content of Cu occurs in rather narrow interval 0.7–4.8 mg/kg, but the whole range is much larger [from 0.02 mg/kg (low for Spodosols) to 37.2 mg/kg (high for noncalcareous Fluvents)]. An average content of Mn occurs in the interval 14.5–61.4 mg/kg, but the whole range is really very large [from 1.0 mg/kg (low for

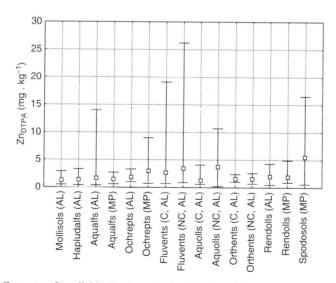

Figure 5. Contents of available Zn in agricultural soils of Slovakia extracted by DTPA. Average values and ranges of available contents are plotted for individual soil types (abbreviations: C, calcareous; NC, non-calcareous; AL, arable land; MP, meadows and pastures).

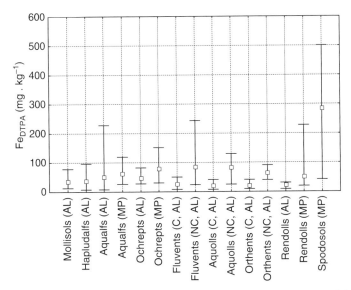

Figure 6. Contents of available Fe in agricultural soils of Slovakia extracted by DTPA. Average values and ranges of available contents are plotted for individual soil types (abbreviations: C, calcareous; NC, non-calcareous; AL, arable land; MP, meadows and pastures).

Spodosols) to 179.8 mg/kg (high for noncalcareous Aquolls)]. An average content of Zn occurs in rather narrow interval 1.3–5.6 mg/kg, but the whole range is much larger [from 0.2 mg/kg (low for noncalcareous Aquolls) to 26.2 mg/kg (high for noncalcareous Fluvents)]. An average content of Fe occurs in the interval 19.9–285.9 mg/kg, but the whole range shows much higher variability [from 7.8 (low for calcareous Aquolls) to 243.5 mg/kg (high for noncalcareous Fluvents)]. Detail results are shown in Figs. 3–6.

Results of available micronutrients survey in the Czech Republic, using the same method, were published by Nerad [1992]. According to this source, there is a low content of Cu (5.0%), Mn (4.4%), and Zn (5.9%) in agricultural soils in the Czech Republic. A survey was not performed for available Fe.

8   SYSTEM OF MICRONUTRIENTS APPLICATION: COPPER, MANGANESE, ZINC, AND IRON FOR AGRICULTURAL CROPS, RECOMMENDED IN SLOVAKIA

A soil sample represents an area where the correction of micronutrient contents has to be realized. Soil samples taken from a plow layer, in the case of specific crops such as vine, hop, and orchard (also from subsoil), are air-dried and passed through a 2-mm sieve prior to subsequent analysis. The extraction procedure was already described in the Section 6.

Obtained results have to be evaluated according to limits, which are listed in Table 6. Basic soil application of micronutrients has to be realized only to crops with suitable response to added microelement and to crops having also good tolerance (Table 8). The interval for repeated application, after fulfilling the previous two conditions, is 2 years in the case of Fe and 5–6 years for Cu, Mn, and Zn. These doses can be adjusted with regard to using organic fertilizers, macronutrients fertilization, and also agrochemical properties of the soil. In the case of soil conditions, causing unavailability of soil micronutrients, soil application is not recommended. Table 9 shows recommended doses used for soil micronutrient fertilization.

The recommended dose of Cu has to be decreased by 1/3 if over 40 ton/ha of manure or over 80 ton/ha of liquid manure will be applied to the soil. Also, the dose of Cu is decreased 20%, if over 130 kg/ha N in mineral fertilizers will be applied to the crop.

The recommended dose of Zn and Fe has to be decreased by 1/3 if over 40 ton/ha manure, or over 80 ton/ha of liquid manure will be applied to the soil. The

TABLE 8. Crop Reactions to Micronutrients Fertilization

Crops	Cu	Mn	Zn	Fe
Winter wheat	xx	xx	x	0
Winter barley	xx	xx	0	xx
Winter rye	x	x	0	0
Spring wheat	xx	xx	0	0
Spring barley	xx	x	0	xx
Oat	xx	xx	0	x
Maize	xx	x	xx	x
Sugar beet	xx	xx	x	xx
Potatoes	x	x	x	x
Alfalfa	xx	x	x	x
Clover	xx	x	x	x
Winter raps	xx	x	x	0
Sunflower	xx	x	0	0
Grean peas	x	xx	x	0
Bean	x	0	xx	xx
Lentil	x	0	0	0
Hop	x	x	xx	0
Apple tree	xx	xx	xx	—
Pear tree	xx	—	x	x
Plum tree	xx	x	xx	x
Peach tree	xx	xx	xx	xx
Apricot tree	x	xx	x	x
Cherry tree	x	xx	x	x
Vine	x	xx	xx	xx

xx good reaction.
x medium reaction.
0 low reaction.
— no data.

TABLE 9. Recommended Doses of Micronutrients for Soil Application (in kg/ha)

Element	Soil Texture Category		
	Light	Medium	Heavy
Cu	2.0–4.0	4.0–6.5	6.5–10.0
Mn	7.0	20.0	60.0
Zn	3.0–5.0	5.0–9.0	9.0–13.0
Fe	5.0–10.0	10.0–15.0	15.0–20.0

TABLE 10. Recommended Doses and Concentrations of Micronutrients for Foliar Applications

Micronutrient	Foliar Dose (kg/ha)	Concentration of Solution (%)
Cu	0.3–0.5	<1.0
Mn	0.5–1.0	<1.0
Zn	0.2–0.8	<1.0
Fe	0.5–3.0	<1.0

recommended dose of Zn and Fe has to be increased by 1/3 if the crop was limed with a dose of over 4 ton/ha. The dose has to be also increased by 1/3 if ameliorative fertilization of P is applied to the soil or if the content of available forms of P in soil exceeds 120 mg/kg.

For curing of low content of a certain micronutrient in soil, it is recommended to use single micronutrient fertilizers (i.e., fertilizer containing only a certain micronutrient that is lacking in soil) or technical salt. The proper dose has to be calculated on the basis of micronutrient content. To receive equable application, fertilizer (usually liquid) or technical salt is applied after dilution in water (in specific cases also in liquid industrial fertilizers) by spraying on soil. Recommended time of application is 2–4 weeks before sowing. For alfalfa, clover, and meadow application from fall to early spring is recommended.

Sometimes, due to lack of time or due to specific soil condition, soil application is not possible or not recommended. In such cases we recommend foliar application. In such cases it is necessary to hold on doses and recommended concentrations according to Table 10.

9   REMARKS TO THE SYSTEM USED FOR COPPER, MANGANESE, ZINC, AND IRON AVAILABLE FORMS DETERMINATION AND FERTILIZERS RECOMMENDATION

At the beginning of the described story of scientific interest regarding soil Cu, Mn, Zn, and Fe forms and content, in 1960s in former Czechoslovakia, the main target of specialists dealing with this topic was effort to increase yields of agricultural crops, their quantity, and, if possible, also quality. The developing industry of the country was able to satisfy the needs of agriculture in case of economical prosperity. All this was oriented to achieve food self-sufficiency in terms of agricultural products in order to attain food security for the country. The economic aspect of this strategy pressed scientists to find the cheapest way to solve this problem—such as one extracting agent for determinating many, if not all, available nutrients in soil—and also to find the way to get results, applicable for several years. When 1M HCl was used, due to unsatisfactory results obtained with this method, the search for a new method based on a soft extracting agent has started. As a result of success with the DTPA Lindsay Norvell method at the end of the 1980s, the use of micronutrient fertilization had an auspicious beginning, and also a market for single-element microfertilizers has developed.

Serious economic changes in the 1990s in connection with economic transformation, including a full market of food in the country, has changed previous targets. The priorities became health, healthy (uncontaminated) soil, and uncontaminated food with high biological quality. This new tendency also gave rise to a new task—that is, monitoring of the soil.

Unfortunately, we have not used all good experiences from the previous stage, and this is also a reason why our system of soil monitoring is still experiencing some difficulties. On the other hand, we are happy to have one, unified system, monitoring all (agricultural and forests) soils in the country.

10   NEW PRIORITIES IN RESEARCH OF TRACE ELEMENTS IN SOILS OF SLOVAKIA—SOIL MONITORING

Political and economical changes after 1989 in the former Czechoslovakia accelerated the realization of environmental monitoring including soil. Previously accumulated experience with the assessment of the contents of micronutrients could serve also for determination of "high" contents of metals and limit values, which are indicating normal values of element contents in soils. Discussions were organized in order to establish relatively large monitoring system. At present, two periods of the soil monitoring have been completed. The first period of the monitoring of soil properties in Slovakia was realized in 1992–1996, and the results were published by Linkes et al. [1997]. The results of the second period in 1997–2001 were summarized by Kobza et al. [2002]. At present, results of the third period (2002–2007) of the soil monitoring in Slovakia are under evaluation.

The main aim of the partial monitoring system for soil is to monitor soil contamination as well as soil properties, which are important from the standpoint of soil fertility and other environmental functions [Linkes et al., 1997]. The emphasis is on the monitoring of irreversible soil properties, while reversible soil properties belonging to soil regimes are not considered in such detail. Soil monitoring in Slovakia includes both agricultural and forest soils. The basic monitoring embraces 650 sites, which are sampled in 5-year intervals. The monitoring system has

TABLE 11. Contents of Cu in Monitored Soils of Slovakia Extractable by Different Extraction Agents [Linkes et al., 1997]

| Sampling Depth (cm) | Cu Contents in Soils | | | | | | | | |
| | Total Content | | | Extractable by 2 M HNO$_3$ | | | Extractable by 0.05 M EDTA | | |
	Mean	Min	Max	Mean	Min	Max	Mean	Min	Max
0–10	22.6	5.0	155.5	7.5	1.0	109.0	3.3	0.3	80.5
20–30	19.8	3.1	144.0	6.4	0.3	128.0			
35–45	17.3	1.0	157.1	5.2	0.6	112.8			

standardized parts: network of sampling sites, description of soil profiles, analytical procedures, and database structure and sample storage.

Slovakia is the country with occurrence of three types of increased contents of xenobiotics in the environment. The first type is represented by geochemical anomalies, which are of natural origin. The second type consists regions affected by loads from mining activities (so-called old ecological loads). These are, to some extent, eliminated by natural processes. The full extent and intensity of this type of contamination (deterioration) is difficult to determine because they are often a part of recent forest ecosystems. The third type is caused by recent industry. But generally it can be noticed that most of Slovakian soils do not have the character of contaminated soils.

Determination of threshold values looked very important from the beginning of the 1990s, and various approaches were tested in soil contaminant research. In the frame of soil monitoring, the attention is shifted also to other trace metals, many of which do not act as essential micronutrients. From the previous research experience, information on trace element contents in 2 M HNO$_3$ extracts was considered in 1994, when the Dutch ABC list was accepted and supplemented with the A' limit value for the trace element extracts in 2 M HNO$_3$. From the micronutrients discussed here, only Cu and Zn are considered to be potentially toxic and are monitored.

TABLE 12. Contents of Zn in Monitored Soils of Slovakia Extractable by Different Extraction Agents [Linkes et al., 1997]

| Sampling Depth (cm) | Zn Contents in Soils | | | | | | | | |
| | Total Content | | | Extractable by 2 M HNO$_3$ | | | Extractable by 0.05 M EDTA | | |
	Mean	Min	Max	Mean	Min	Max	Mean	Min	Max
0–10	64.3	11.0	1070.0	12.3	2.0	565.0	2.35	0.05	126.0
20–30	57.7	9.8	226.0	8.0	0.4	195.7			
35–45	47.6	5.0	175.0	6.2	0.8	124.0			

**TABLE 13. Average Contents of Cu and Zn Calculated for
Extract in 2 M HNO$_3$ in the Frame of the Monitoring of Soils
of Slovakia [Kobza et al., 2002]**

Soil Type	Cu (mg/kg)	Zn (mg/kg)
Orthents	10.1	8.2
Hapludalfs	7.3	9.3
Aqualfs	5.2	7.7
Mollisols	11.3	10.5
Aquolls	12.2	14.2
Fluvents	10.0	23.7
Ochrepts	6.7	10.9
Rendolls	8.0	19.6
Spodosols	4.1	22.3

For copper the A' limit value is 20 mg/kg for extract in 2 M HNO$_3$. Very high values (100–200 mg/kg) were found near mining and ore processing plants. But increased contents were found also in areas influenced by natural geochemical anomalies. Average content of Cu in soils of Slovakia is very close to Clark concentration (25 mg/kg) and is similar or lower than in most countries of Europe and North America.

Zinc has a limit value of 40 mg/kg for contents extractable with 2 M HNO$_3$. Very high contents were observed in areas of natural geochemical anomalies as well as in locations with metallurgic industry. Agricultural soils have average content slightly below the Clark concentration (71 mg/kg).

Results obtained by different methods from monitored soils are summarized in Tables 11 and 12. The efficiency of different extracting agents is clearly shown, but contamination limits for metal contents extracted by chelating agents have not been developed until recently in Slovakia. Average contents of Cu and Zn determined by extraction with 2 M HNO$_3$ are listed in Table 13 for individual soil types. In addition, contents of DTPA available micronutrients (Cu, Mn, Zn, and Fe) were analyzed in the second monitoring period, which were already discussed above (Figs. 3–6).

REFERENCES

Bascomb CL. 1968. Distribution of pyrophosphate-extractable iron and organic carbon in soils of various groups. *J Soil Sci* **19**:251–268.

Bergmann W, Cumakov A. 1977. *Key for Detection of Deficiencies in Plant Nutrition* (in Slovak). Bratislava: Priroda.

Cumakov A, Jurani B. 1980. Determination of the micronutrients fertilization requirements (The possibilities of applying the method of Rinkis in Czechoslovakia) (in Slovak). *Scientific Works of the Research Institute of Soil Science and Plant Nutrition Bratislava* **10**:175–184.

Davies BE. 1994. Soil chemistry and bioavailability with special reference to trace elements. In: Farago ME, editor. *Plants and the Chemical Elements*. New York: VCH Publishers, pp. 1–30.

Dlapa P, Kubova J, Medved J, Jurani B, Stresko V. 2000. Heavy metal fractionation in soils of different genesis. *Slovak Geol Mag* **6**:27–32.

Gupta SK, Vollmer MK. 1994. A new approach to assess and manage the hazards for contaminated soil. In: *Proceedings on Environmental Contamination*. Edinburgh: CEP Consultants Ltd., pp. 323–326.

ISO 14870. 1998. *Soil Quality—Extraction of Trace Elements by Buffered DTPA Solution*. International Organization for Standardization.

Jurani B, Fiala K. 1989. Verifying of suitability of DTPA use as group extractant for some micronutrients (in Slovak). *Scientific Works of the Research Institute of Soil Science and Plant Nutrition Bratislava* **15**:227–239.

Kobza J, et al. 2002. *Soil Monitoring of Slovak Republic. Present State and Development of Monitored Soil Properties* (in Slovak). Bratislava: Soil Science and Conservation Research Institute.

Legget GR, Argyle DP. 1983. The DTPA extractable Fe, Mn, Cu, and Zn from neutral and calcareous soils dried under different condition. *Soil Sci Amer J* **43**:518–522.

Lindsay W. 1979. *Chemical Equilibria in Soils*. New York: John Wiley & Sons.

Lindsay WL, Norvell WA. 1967. Equilibrium relationships of Zn^{2+}, Fe^{3+}, Ca^{2+}, and H^+ with EDTA and DTPA in soils. *Soil Sci Soc Am Proc* **33**:62–68.

Lindsay WL, Norvell WA. 1978. Development of a DTPA soil test for Zn, Fe, Mn, and Cu. *Soil Sci Am J* **42**:421–428.

Linkes V, et al. 1997. *Soil Monitoring of Slovak Republic. Present State of Monitored Soil Properties* (in Slovak). Bratislava: Soil Fertility Research Institute.

McLaren RG, Crawford DV. 1973. Studies on soil copper. I. The fractionation of copper in soils. *J Soil Sci* **24**:172–181.

Navarcik I, Cipakova A, Palagyi S. 1997. Behaviour of radionuclides and heavy metals in soils and possibility of elimination of their negative impact. In: Ronneau C, Bitchaeva O, editors. *Biotechnology for Waste Management and Site Restoration*. Dordrecht: Kluwer Academic Publishers, pp. 221–224.

Nerad J. 1992. *Trace Elements Content in Soils. Results of Trace Elements Content Survey on Soils of Arable Land* (in Czech). Prague: Central Control and Testing Institute for Agriculture.

Norvell WA, Lindsay WL. 1972. Reactions of DTPA chelates of iron, zinc, copper, and manganese with soils. *Soil Sci Soc Am Proc* **36**:778–788.

Repka F. 1977. *Results of Agrochemical Soil Testing on Micronutrients Content in Arable Land. Informative Report* (in Slovak). Bratislava: Central Agricultural Control and Testing Institute.

Repka F. 1978. *Micronutrients Content in Soils of Meadow and Pastures. Informative Report* (in Slovak). Bratislava: Central Agricultural Control and Testing Institute.

Rinkis GJ. 1972. *Optimalisation of Mineral Nutrition of Plants* (in Russian). Riga: Zinatne.

Schachtschabel P. 1956. Die Bestimmung des Mangansversorgsgrades der Boden. In: *Sixth International Congress on Soil Science, Paris*, p. 113.

Silanpää M. 1982. *Micronutrients and the Nutrient Status of Soil. A global study*. Rome: FAO.

Soltanpour PN, Khan A, Lindsay WL. 1976. Factors affecting DTPA—extractable Zn, Fe, Mn, and Cu from soils. *Commun Soil Sci Plant Anal* **7**:482–487.

Stoops G, Vepraskas MJ. 2003. *Guidelines for Analysis and Description of Soil and Regolith Thin Sections*. Madison, WI: Soil Science Society of America.

Tessier A, Campbell PGC, Bisson M. 1979. Sequential extraction procedure for the speciation of particulate trace metals. *Anal Chem* **51**:840–855.

Wear JI, Sommer AL. 1948. Acid-extractable zinc of soils in relation to the occurrence of zinc deficiency symptoms of corn: A method of analysis. *Soil Sci Soc Am Proc* **12**:143–144.

Westerhoff H. 1955. Beitrage zur Kupfer bestimmung im Boden. *Landwirtsch Forsch* **7**:190–192.

27 Role of Minerals in Halophyte Feeding to Ruminants

SALAH A. ATTIA-ISMAIL

Desert Research Center, Matareya, 11753 Cairo, Egypt

1 INTRODUCTION

Halophyte plants are those that grow naturally in saline environment. Halophytes grow in many arid and semi-arid regions around the world and are distributed from coastal areas to mountains and deserts. Saline conditions may include either salt-affected soil or saline water irrigation.

Economically speaking, the most productive halophytic plant species yield from 10 to 20 ton/ha of biomass when irrigated with seawater [Glenn et al., 1999]. Also, the productivity of cultivated halophytes, as compared with traditional crops, is high [Jaradat, 2003]. The average area of salt-affected soils might be calculated on the basis of FAO averages to be 450 million hectares. That means we are talking in about 4–5 billion tons of biomass. Halophytic plants have many uses: They can be used as animal feeds, vegetables, and drugs, as well as for sand dune fixation, wind shelter, soil cover, cultivation of swampy saline lands, laundry detergents, paper production, and so on [El Shaer and Attia-Ismail, 2002].

Feed shortage especially in coastal areas has led to the exploration and exploitation of marginal resources such as halophytic plants. Halophytes can, then, play an important role in the welfare of people in such regions.

As an animal feed component, halophytes are promising because they have the potentiality of being good feed resource. The potentialities of halophytes as animal feed components were recognized as early as the 1880s [Hutchings, 1965]. Yet, this potentiality does not go far because several constraints are limiting and have to be worked out. One of these constraints is the high contents of salts and consequently the high mineral composition.

Trace Elements as Contaminants and Nutrients: Consequences in Ecosystems and Human Health, Edited by M. N. V. Prasad
Copyright © 2008 John Wiley & Sons, Inc.

2   ASH AND MINERAL CONTENTS OF HALOPHYTES

The high content of ash is a typical characteristic of halophytes. Mineral contents differ in halophytes from those of regular forages. This has led to scientific concerns of with regard to what extent this may affect their feeding value to animals. Also, the mineral compositions of the high ash contents of halophytes have been controversial: Do they support animals' requirements of certain minerals, do they might exceed the requirements to the extent that they may represent a poisoning threat to animals? Mineral contents of halophytes vary considerably. The variations in mineral contents of halophytes depend on several factors. Among the many factors that cause the variations are the plant species, stage of growth, and season.

Table 1 shows the ash and mineral contents of different parts of some halophytes around the world. Halophytes may represent a good source of minerals to ruminant animals. High ash contents of halophytes, unless associated with nondigestible components such as silica, may compensate for mineral deficiency usually seen in animals in areas depending mainly on rangeland grazing as well as desert and coastal areas. Table 1 shows that ash contents of halophytes vary for whole plants, maturity stage, and plant part. It runs up to more than 40% of plant materials. Plant parts differ also in their ash contents. Leaves and twigs of *Atriplex repanda*, for instance, have 23.1%, whereas the fruits contain 32.65% (Table 1); this represents 41% increment in ash content. *Salicornia bigelovii* spikes have 66% more ash than the stem. The stem of *Sueda foliosa* has 117% more ash than leaves.

Stage of plant maturity affects ash contents as well. *Pappophorum caespitosum* is an example. Its ash contents decreased after fructification by 98.5%. *Trichloris crinita* is a contrast case where ash contents increased after fruitification by 152%. The variability in ash contents might be due to the physiological distribution and pools of minerals in different plant parts. Physiological and biochemical processes that take place in plants vary according to the plant species, and this may explain the contrast in the percentage of ash content increment or decrement due to the process of fruitification between both of *Pappophorum caespitosum* and *Trichloris crinita.*

3   FACTORS AFFECTING MINERAL CONTENTS OF HALOPHYTES

Halophyte plants either grow naturally in saline environment (salt-affected soils or under the circumstances of the irrigation with saline water) or need be cultivated. Either way has its impacts on ash contents and mineral composition of the plants. Minerals in halophytes differ widely according to the particularities of the environment they grow within. Type of soil (saline or sodic), level of salinity (in either soil or irrigation water), mineral profile of the soil, plant species, season, stage of growth, part of the plant, and so on, are among factors affecting ash and mineral contents of halophytes. Mineral contents are, therefore, affected accordingly.

TABLE 1. Ash and Mineral Contents of Some Halophytic Plants from Different Parts of the World

Halophytic Plants	Plant Part and/or Maturity Stage	Ash (%)	Ca (%)	P (%)	Na (%)	K (%)	Mg (%)	N (%)	S (%)	Zn (ppm)	Cu (ppm)	Fe (mg/kg)	Mn (ppm)
Acacia aneura	Leaves	5.15	1.39	0.07	0.01		0.26		0.17	76			863
Acacia bidwilli	Leaves		2.43	0.09	0.06		0.24		0.21	37			49
Acacia concurrens	Leaves		0.97	0.05	0.25		0.26		0.18	47			917
Acacia deanei	Leaves		0.70	0.09	0.09		0.27		0.15	37			64
Acacia fimbriata	Leaves		0.30	0.06	0.13		0.25		0.12	33			83
Acacia glaucocarpa	Leaves		1.01	0.05	0.08		0.23		0.14	45			53
Acacia leptocarpa	Leaves		0.97	0.06	0.25		0.37		0.18	75			622
Acacia melanoxylon	Leaves		0.81	0.09	0.16		0.28		0.16	64			457
Acacia salicina	Leaves		3.52	0.11	0.02		0.29		1.13	55			30
Acacia saligna	Whole	8.83	3.75		1.15	1.05	6.14	2.21		140.5			
Acacia stenophylla	Leaves		2.08	0.10	0.16		0.35		0.58	123			42
Anabasis brevifolia	Whole/flowering	17.5	3.19	0.08				1.52					
Aristida mendocina	Early fruitification	22.04	0.33	0.12				1.10					
Atriplex argentina	Leaves and twigs	6.95	0.67	0.08				2.71					
Atriplex atacamensis	Leaves and twigs	27	0.68	0.11				2.55					
Atriplex coquimbensis	Leaves and twigs	30.56	0.87	0.10				2.69					
Atriplex halimus	Whole plant	29.20	1.69	0.32	3.91	0.57	0.32	2.11	0.17	64	10	503	51
Atriplex lampa	Twigs	30	0.15	0.01				1.80					
Atriplex nummularia	Whole plant	18.91	2.08	1.17	4.99	2.99	15.63	2.03		133.5	60.52		
Atriplex repanda	Leaves and twigs	23.1	1.9	0.12				3.31					
	Fruits	32.65	0.48	0.14				1.65					

(Continued)

TABLE 1. *Continued*

Halophytic Plants	Plant Part and/or Maturity Stage	Ash (%)	Ca (%)	P (%)	Na (%)	K (%)	Mg (%)	N (%)	S (%)	Zn (ppm)	Cu (ppm)	Fe (mg/kg)	Mn (ppm)
Atriplex semibaccata	Leaves and twigs	26.6	0.57	0.09				3.00					
Cottea pappophoroides	Flowering	18–27	0.58	0.14				1.49					
Ephedra ochreata	Vegetative/twigs	25.8	1.94	0.20				0.99					
Glycyrrhiza inflata	Whole/flowering	22.63	0.22	0.14				1.76					
Haloxylon ammodendron	Whole/flowering	10.35	0.25	0.17				2.28					
Halostachys caspica	Whole/flowering	17.9	0.52	0.18				1.83					
Haloxylom salicornicum	Whole plant	16.16	4.00	0.15	5.01	1.74	0.33	2.8	0.07	73	18	603	80
Halocnemum strobilaceum	Whole plant	18–36	2.45	0.14	7.00	1.65	0.39	1.11	0.10	99.7	17	621	87
Kalidium foliatum	Whole/fruiting	30	0.48	0.08				3.03					
Karelinia caspica	Whole/flowering	20.9	0.85	0.11				1.85					
Larrea nitida	Vegetative/twigs	8.82	2.83	0.28				3.23					
Lyceum chilense	Frutification/shoots	9.93	1.37	0.13				2.21					
Lyceum tenuispinosum	Budding	8.11	4.41	0.13				1.76					
Nitraria retusa	Whole plant	9.99	1.96	0.22	5.35	0.66	0.36		0.14	32	11	567	62
Pappophorum caespitosum	Early fruitification	35.93	0.28	0.14				1.70					
	Past fruitification	18.1	0.44	0.14				1.45					
Pappophorum philippianum	Past fruitification	40.1	0.38	0.11				1.37					
Phragmites communis	Whole/flowering	43.41	0.24	0.06				1.92					
Reaumuria soongorica	Whole/fruiting	10.89	0.65	0.12				1.64					

Species	Part												
Salicornia bigelovii	Stem	25.1	1.01	0.08	6.44	0.75	0.40						
	Spikes	41.7	0.63	0.14	12.0	1.47	0.92						
Salsola tetranda	Whole plant	12.90	3.98	0.16	5.65	1.45	0.59	1.08	0.12	44	8.88	6.64	79
Sueda foliosa	Leaves	9.32	1.73	0.38				2.67					
	Stem	20.24	1.82	0.37				2.69					
Sueda fruticosa	Whole plant	30.2	2.11	0.41	4.06	1.29	0.30	1.94	0.20	55	13	674	88
Sueda physophora	Whole/fruiting	8.79	0.41	0.16				2.88					
Tamarix aphylla	Whole plant	9.71	3.73	0.16	2.75	0.78	0.43	1.94	0.12	38	13	274	60
Tamarix mannifera	Whole plant	8.06	1.44	0.01	2.60	0.8	0.46	1.22	0.09	38.5	16	291	52
Tamarex ramosissima	Whole/flowering	9.82	1.76	0.05				1.74					
Trichloris crinita	Early fruitification	14.32	0.27	0.13				1.54					
	Past fruitification	36.1	0.36	0.21				1.51					
Zygophyllum album	Whole plant	34.84	3.74	0.14	2.84	0.9	0.64	1.05	0.09	43.3	7.78	393	52

Source: Adapted from Abdelhameed et al. [2006], Attia-Ismail et al. [2003], Gabr [2002], Shawkat et al. [2001], Warren et al. [1990], Vercoe [1986], El-Bassosy [1983], and Kraidees et al. [1998].

4   SALT-AFFECTED SOILS

Saline soils are the soils that have high percentages of salts. Saline soils have an electrical conductivity at the saturation of soil extract of more than 4 dS/m [FAO, 1988]. Salt-affected soils are distributed all over the world. Salt-affected soils are classified into two main classes according to the type of salts present in soils: saline soils (solonchak) and sodic soils (solonetz) [Szabolcs, 1974]. FAO [2007] estimated that the solonchak soils area varies from 260 to 340 million hectares and are mostly found in Saharan Africa, East Africa, Central Asia, Australia, and South America. The solonetz soils are associated with solonchak soils and are estimated to be around 135 million hectares (http://www.fao.org/ag/agl/agll/wrb/wrbmaps/htm/). Saline soils were classified into three main groups according to the dominant cations: the sodium-dominated soils, the magnesium-dominated soils, and the calcium-dominated soils (http://www.fao.org/DOCREP/003/Y1899E/y1899e00. htm#toc). In each class of these soils the ratios of cations differ.

   The most common salts present in saline soils are sodium, potassium, calcium and magnesium in the form of chlorides, sulfates or bicarbonates. The sodium and chloride are present in the highest percentage particularly in highly saline soils. The most common sodic salts present in sodic soils are the carbonates [FAO, 1988] particularly sodium.

5   IRRIGATION WITH SALINE WATER

Saline water contains large amounts of dissolved salts. For water to be saline, it should include more than 1000 ppm dissolved salts; this value can be as high as 35,000 ppm for typical seawater. Saline water usually goes along with saline soils or in some cases independently of saline soils. Saline water varies in its salinity contents according to the source (seawater, underground water, or drainage water). Saline drainage waters may contain toxic levels of trace elements as well as the macro-minerals [Retana et al., 1993]. The minerals in saline water differ according to its source. Brackish water may contain toxic levels of certain minerals according to the effluent. Major elements of seawater are Ca, 419 ppm; Mg, 1304 ppm; Na, 10,710; K, 390; Cl, 19,350 (http://www.palomar.edu/oceanography/salty_ocean.htm).

6   SALINITY LEVEL

Salinity level either in irrigation water or contained in soils is a major determinant factor in plant growth and, therefore, in ash contents and mineral profile in plants. Brown et al. [2006] found that salinity level affects nutrient uptake in *Spartina alterniflora*. Daoud et al. [2001] mentioned that includer halophytes may accumulate salt and, therefore, increase mineral concentrations at increasing salinity level, while excluder halophytes may also accumulate salt but at lower rates. Table 2 shows that effect clearly. However, it appears that different parts of the plant respond differently.

TABLE 2. Effect of Salinity Level on Ash Contents of Different Parts of Some Halophytes [Daoud et al., 2001]

Plant Species	0%		25%		50%		75%		100%	
	Shoot	Root	Shoot	Root	Shoot	Root	Shoot	Root	Shoot	Root
Aster tripolium	24.04	13.48	28.13	14.4	31.11	12.9	32.0	16.88	34.59	18.02
Avicennia germinans	9.40	14.38	9.74	17.72	11.20	24.18	11.78	24.30	11.54	23.22
Batis maritima	25.88	21.18	45.22	32.28	54.68	44.88	46.20	46.44	56.34	31.2
Limoniastrum multiflorum	14.82	9.83	16.29	13.68	13.24	13.19	11.17	10.88	13.52	12.08
Rottboellia faciculata	6.48	32.52	8.06	21.36	9.22	33.28	10.72	25.76	8.08	25.6
Sesuvium verrucosum	31.46	34.56	34.24	36.54	37.02	35.78	42.48	36.08	38.56	34.38
Sesuvium portulacastrum	29.34	22.52	38.88	21.36	39.56	20.86	40.0	14.58	34.48	13.98
Spartina alterniflora	9.82	15.04	13.52	17.31	13.94	21.64	16.9	32.66	18.12	38.16

At higher salinity levels that reached seawater salinity, sodium concentrations increased in shoots while potassium concentrations decreased in roots (Table 2). Noticeably, the salinity level did not affect the concentrations of magnesium or calcium in their experiments. The distribution of ash and, therefore, mineral contents in different plant pools might be a mechanism by which the plants adapt to a saline environment [Gorham, 1996].

Parida et al. [2004] used *Bruguiera parviflora* to test the effect of salinity level tolerance, by adding increasing levels of sodium chloride, on some mineral contents of different parts of this mangrove (Table 3). The increase in sodium chloride concentrations led to increases in the concentrations of both of Na and Cl ions in all tested parts of the plant. Khan [2001] reported that the increase in sodium chloride uptake counteracts the uptake of potassium, calcium, and manganese in some plants.

Certain mineral ratios are affected by salinity level as a consequence of increasing salinity level like the K/Na, Ca/Na, and many other ratios.

7   PLANT SPECIES

Halophyte plant species differ in their ability to accumulate minerals [Reboreda and Cacador, 2007]. The ability of halophytes to accumulate minerals depends on the mechanism by which halophytes distribute minerals in different plant compartments. Minerals in halophytes are mainly accumulated in the roots with small quantities translocated to the stems and leaves [Windham et al., 2003]. Halophytes have higher potentiality to phytoextract minerals from the soil than do glycophytes [Jordan et al., 2002].

Legumes, generally, have different mineral profile (being high in calcium, potassium, magnesium) than grasses which tend to be higher in manganese and molybdenum when grown under the same conditions.

8   MINERAL ROLE IN RUMINANT NUTRITION

Minerals, in general, either major or minor, play a crucial role in the lives of animals and affect their production to a greater extent. Minerals have four vital functions in the bodies of animals. They have a structural function in bones and other structural tissues of the body, physiological function in body fluids as electrolytes, regulatory function as they regulate several metabolic processes in the body, and catalytic function when they enhance the enzyme activities [Underwood and Suttle, 1999].

9   RECOMMENDED MINERAL ALLOWANCES

Minerals required by animals for proper functioning of the body and for the optimal production are either macro- or micro-minerals. Macro-minerals are known to be Ca,

TABLE 3. Effect of Salinity Level on Mineral Contents of Different Parts of *Bruguiera parviflora* [Parida et al., 2004]

Plant Part	NaCl Level (nM)	Na (mg/g dry wt)	K (mg/g dry wt)	Ca (mg/g dry wt)	Mg (mg/g dry wt)	Fe (mg/g dry wt)	Cu (mg/g dry wt)	Mn (mg/g dry wt)	Cl (mg/g dry wt)
Leaf	0	14.47 ± 0.7	11.02 ± 0.1	12.31 ± 0.3	155.94 ± 1.9	325.21 ± 1.4	31.50 ± 0.3	103.10 ± 1.6	13.11 ± 0.1
	100	33.54 ± 0.3	12.26 ± 0.7	11.41 ± 0.2	162.49 ± 1.6	309.07 ± 1.8	28.50 ± 0.2	97.41 ± 0.8	38.53 ± 0.6
	200	38.63 ± 0.4	10.33 ± 0.6	9.99 ± 0.5	145.46 ± 1.2	254.11 ± 1.5	20.41 ± 0.5	53.06 ± 0.6	40.21 ± 0.4
	400	42.13 + 0.2	11.31 ± 0.2	7.81 ± .06	117.14 ± 1.4	295.75 ± 1.3	18.37 ± 0.64	7.10 ± 0.6	43.49 ± 0.4
Stem	0	9.48 ± 0.5	6.00 ± 0.5	2.49 ± 0.9	65.41 ± 0.7	137.1 ± 1.8	4.35 ± 0.7	17.05 ± 0.9	7.28 ± 0.1
	100	15.81 ± 0.7	7.86 ± 0.7	2.91 ± 0.6	61.54 ± 0.3	187.0 ± 0.8	5.69 ± 0.3	21.07 ± 0.4	21.91 ± 0.3
	200	16.70 ± 0.1	3.81 ± 0.3	2.26 ± 0.7	88.34 ± 0.5	190.5 ± 0.5	2.39 ± 0.5	14.01 ± 0.7	26.70 ± 0.6
	400	21.67 ± 0.2	2.71 ± 0.2	3.63 ± 0.5	298.41 ± 0.3	220.4 ± 1.5	1.48 ± 0.9	12.63 ± 0.5	38.54 ± 0.8
Root	0	7.21 ± 0.5	25.45 ± 0.4	2.74 ± 0.6	205.68 ± 0.8	632.1 ± 1.8	65.90 ± 0.9	28.40 ± 0.5	6.85 ± 0.7
	100	14.47 ± 0.3	17.78 ± 0.1	2.89 ± 0.3	159.55 ± 0.4	970.9 ± 1.1	51.10 ± 0.5	37.00 ± 0.5	18.37 ± 0.3
	200	17.25 ± 0.5	15.03 ± 0.9	2.45 ± 0.8	142.87 ± 0.5	383.9 ± 0.9	28.10 ± 1.2	33.07 ± 0.3	24.95 ± 0.5
	400	26.96 ± 0.8	11.20 ± 0.5	2.38 ± 0.3	116.78 ± 0.6	277.8 ± 0.6	14.52 ± 0.7	11.96 ± 0.7	31.47 ± 0.3

TABLE 4. Recommended Mineral Requirements of Sheep and Goats

Mineral	NRC	
	Lactating Goats, 1981	Sheep, 1985
Na (% of DM)	0.09–0.18	0.09–0.18
Cl (% of DM)	—	—
Ca (% of DM)	0.20–0.80	0.20–0.82
P (% of DM)	0.20–0.40	0.16–0.38
Mg (% of DM)	0.12–0.18	0.12–0.18
K (% of DM)	0.50–0.80	0.50–0.80
S (% of DM)	0.14–0.26	0.14–0.26
I (ppm)	0.10–0.80[a]	0.10–0.80
Fe (ppm)	30–50[a]	30–50
Cu (ppm)	10–20[a]	7–11
Mo (ppm)	0.50–1.0[a]	0.5
Co (ppm)	0.10–0.20[a]	0.1–0.2
Mn (ppm)	20–40[a]	20–40
Zn (ppm)	20–33[a]	20–33
Se (ppm)	0.10–0.20[a]	0.1–0.2

[a]Rick Machen, Texas Agriculture Extension Service.

P, Na, Cl, Mg, K, and S, while the micro-minerals are Fe, I, Cu, Zn, Mo, Mn, Co, and Se.

Recent studies have revealed some of the requirements of goats and sheep in dry areas. NRC [1981] recommended mineral requirements for goats and for sheep [NRC, 1985]. Table 4 shows some recommended mineral requirements for goats and sheep. Clear variations are present when matching mineral supply of halophytes to the requirements of these animal species. Depending on halophytes as the only mineral source for the animals may result in either mineral deficiency of some elements, toxicity of some others, or malabsorption of some other elements because of the presence of different proportions of them (e.g., Ca/P, K/Na, Ca/Na ratios). However, mineral requirements differ according to several conditions such as the season (summer versus winter, especially for the desert animal) and the physiological state (pregnant, lactating, growing, etc.).

10   MINERALS DEFICIENCY IN HALOPHYTE INCLUDED DIETS

Ash contents of halophytes may limit animal feed intake and digestibility coefficients of ruminant diets. Therefore, some conservation is placed on the use of these plant species in livestock ratios. Adequate trace mineral intake and absorption is required for a variety of metabolic functions including immunity, reproduction, and so on.

Animals, especially those inhabiting desert areas, suffer from high environmental heat stress. A typical characteristic of animals in desert areas is to dissipate heat through several adaptive mechanisms. One of them is to sweat. Sweating in such animals [Farid, 1989] is accompanied by the excretion of salts from the body. This author cited (from MacFarlane et al. [1963]) that camels' sweat is rich in bicarbonates (pH 8.2–8.5) and is particularly high in potassium, four times as much as sodium. When animals are exposed to high environmental temperatures, the sweating rate increases as a result (about 0.5 kg/p day in sheep and goats). Thus, higher concentrations of urea, sodium, potassium, and chloride might be found in the sweat of animals [Farid, 1989]. Sweat was found to contain different amounts of Mg, Na, K, Ca, and Cl in humans [Verde et al., 1982].

Animals in these areas suffer from mineral deficiency or are exposed to high concentrations of various minerals in either feedstuffs or drinking water (if brackish water is used) or both. For example, Walker [1980] found that the concentration of calcium and potassium is usually higher than that of other minerals, with the average being 1–1.5% in Southern African browsers. The deficiency results from either excessive sweating or through the feeding on halophyte plants that are either deficient in certain minerals or that may have imbalanced mineral ratios. The opposite applies for mineral toxicity. Figure 1 shows the symptoms of mineral deficiency in animals fed on *Atriplex halimus*. The clear calcium deficiency is manifested in these pictures [Khattab, 2007]. Figure 2 also shows lambs in a state of convulsion, exhibiting retracted head on chest with extension of limbs, while Fig. 3 shows lambs in a state of muscular deformities abduction at the foreleg. Figure 4 also shows lambs in a state of paralysis of the hind legs; these lambs exhibit incoordination of movement, tremors, convulsion, reluctance to move, aimless movement, and wandering.

Goats browsing naturally growing halophytes in Mexico [Haenlein and Ramirez, 2007] were deficient in minerals intake. They found a significant low supply of Mg, Cu, Mn, and Zn from the range plants browsed by goats compared to the requirements of these minerals. On the other hand, sheep selected different range plants and had deficiency in Ca, Mg, K, Cu, and Mn. Similarly, phosphorus deficiencies in acacias were reported by Vercoe [1986] and Craig et al. [1991] which led to an imbalance in the Ca/P ratio in foliage. *Panicum turgidum* was fed to camels in the southeastern region of Egypt [Badawy, 2005]. Mineral composition of this halophyte is

Figure 1. Manifestation of calcium deficiency in lambs fed *Atriplex halimus*. Courtesy of Dr. Khattab, Desert Research Center, Cairo, Egypt.

Figure 2. Lambs in a state of convulsion, retracted head on chest with extension of limbs. Courtesy of Dr. Khattab, Desert Research Center, Cairo, Egypt.

Figure 3. Lambs in a state of a muscular detormities adduction at the foreleg and abduction at the foreleg. Courtesy of Dr. Khattab, Desert Research Center, Cairo, Egypt.

Figure 4. Lambs in a state of paralysis of the hind legs. These lambs exhibit incoordination of movement, tremors, convulsion, reluctance to move, aimless movement, and wandering. Courtesy of Dr. Khattab, Desert Research Center, Cairo, Egypt.

presented in Table 5. The major problem encountered was the extreme low Ca/P ratio which reached 1:19 in wet season and 1:32 in dry season. This ratio is far below that generally recommended (2:1) for domestic animals. They found a similar trend in Mg concentrations. It tended to be higher in range during dry season than in wet season.

TABLE 5. Mineral Content in *Panicum turgidum* as Basal Diet for Camels

Mineral	Wet Season	Dry Season	DS + M
Ca%	1.69	1.62	1.60
P%	0.09	0.05	0.04
Mg%	0.23	0.27	0.26
Na%	0.35	0.42	0.40
K%	0.23	0.31	0.30

Source: Badawy [2005].

The levels of Mg and K in *Panicum turgidum* were inadequate compared with the dietary requirements of ruminants recommended by NRC [1981].

11 EXCESSIVE MINERALS IN LIVESTOCK RATIONS IN DRY AREAS

Some halophytes attain higher levels of some trace or major elements than do other traditional feedstuffs [Attia-Ismail, 2005]. It seems that some of their mineral concentrations are within normal ranges to the extent that they may support normal animal physiological functions. On the other hand, excessive concentrations may reach toxic levels to the animals. Certain mineral ratios are skewed. For instance, the calcium–phosphorus ratio present in most halophytes does not match that required by animals to support normal functions. In general, excessive sodium, chloride, and potassium contents of halophytes may have several drawbacks to the animal health. The capacity of the animal body to store excess electrolytes or to excrete in the feces is very little [Masters et al., 2007]. Marai et al. [1995] found that animals have low rectal temperature and high water retention when exposed to higher levels of sodium.

12 EFFECT OF HALOPHYTES FEEDING ON MINERAL UTILIZATION

Excess or deficient mineral intake may adversely affect diet and mineral utilization. However, the degree of absorption (bioavailability) of minerals may affect the state of mineral utilization by animals. The bioavailability of minerals is an important factor that has to be taken into consideration. It varies, however, among different minerals because of different conditions. Haenlein and Ramirez [2007] found that the degree of absorption of minerals differ and that it was low for Cu, Mn, Mg, Zn, Ca, and Fe.

Attia-Ismail et al. [2003] fed sheep a mixture of salt plants and found that Ca intake (mg/kg $BW^{0.75}$) of salt plant-fed animals was higher (Table 6) ($P < 0.05$) than that of a control group. In a previous study [Fahmy et al., 2001], animals consumed more salt plants (SP) than control (hay). Therefore, the amount of Ca intake

TABLE 6. Mineral Utilization of Sheep Fed a Mixture of Halophytic Plants [Attia-Ismail et al., 2003]

	Halophyte-Fed Sheep	Control Group
Calcium Utilization		
Ca intake (mg/BW$^{0.75}$)	914^a	484^b
Ca balance (mg/BW$^{0.75}$)	914^a	341^b
Serum Ca (mg%)	8.2^a	10.9^a
Sodium Utilization		
Na intake (mg/BW$^{0.75}$)	1546^a	1068^b
Na balance (mg/BW$^{0.75}$)	252^a	316^a
Serum Na (mg%)	421^a	431^a
Potassium Utilization		
K intake (mg/BW$^{0.75}$)	1138	1049
K balance (mg/BW$^{0.75}$)	8^b	61^a
Serum K (mg%)	25.5^a	61^b
Zinc Utilization		
Zn intake (mg/BW$^{0.75}$)	45.2^a	31.2^b
Zn balance (mg/BW$^{0.75}$)	40.4^a	28.6^b
Serum Zn (mg%)	1.22^a	0.85^b

[a,b]Values bearing different superscripts in the same row differ significantly ($p < 0.05$).

of the SP group and, hence, the balance were higher ($P < 0.05$) than control. Yet, serum Ca concentrations (mg%) of the SP group was almost similar to that of the control group. Sodium (Na) intake of salt plant-fed groups was, logically, higher than control group (Table 6).

Fahmy [1998] found Na intake to vary from 805 to 1507 mg/kg BW$^{0.75}$ for silages made of a mixture of salt plants; a range close to that was obtained in the work carried out by Attia-Ismail et al. [2003]. The Na balance in the SP-fed animals was, however, lower than that of the control group. The excreted Na in this experiment might be higher for the SP-fed animals than for the control group. Serum Na did not differ (Table 6). Potassium utilization followed the same trend as Na did except that serum K of control was higher than the SP-fed animals.

13   EFFECT OF MINERALS ON RUMEN FUNCTION

The role of mineral elements in rumen microbial activities has not generally received enough attention in work dealing with mineral metabolism in ruminants. Minerals, VFA, lactate, and glucose are the primary solutes in ruminal fluid. Short-chain fatty acids contribute to the osmotic pressure in the rumen. Major minerals contribute

TABLE 7. Blood Minerals Content of Barki Ewes Fed the Experimental Diets During Pregnancy and Lactation Periods (mean $\pm$ SE)

Items	Control	Mixture of *Atriplex halimus* and *Acacia saligna*	
		Fresh	Silage
Na (mmol/liter)	148.68 ± 3.5^c	173.15 ± 5.03^a	167.85 ± 2.38^b
K (mmol/liter)	4.68 ± 0.79^b	5.30 ± 0.43^a	167.85 ± 2.38^b
Ca (mg/dl)	11.57 ± 0.31^a	9.70 ± 0.57^b	9.01 ± 0.39^b
Zn (μg/dl)	96.33 ± 0.04^a	78.65 ± 0.05^b	77.03 ± 0.05^b
Mg (mg/dl)	2.53 ± 0.36^a	1.53 ± 0.02^b	1.26 ± 0.04^b
Cu (μg/dl)	117.5 ± 6.20^a	53.3 ± 3.05^b	47.5 ± 5.05^b
Se (μg/dl)	11.58 ± 4.01^b	22.60 ± 6.50^a	19.87 ± 5.93^a

[a-c]Values bearing different superscripts in the same column differ significantly ($P < 0.05$).

more than short-chain fatty acids to rumen osmolalrity [Bennink et al., 1978]. Usually, the osmotic pressure in the rumen is rather constant (250–280 mOsmol/kg), but it can increase up to 350–450 mOsmol/kg after feeding certain diets (e.g., lucerne pellets [Bennink et al., 1978]). It was concluded that the increment in potassium concentrations caused a rise in rumen osmolalrity to about 400 mOsmol/kg [Warner and Stacy, 1965]. Barrio et al. [1991] observed a closer relationship between osmolarity of rumen fluid and potassium concentration than between osmolalrity and sodium concentration in sheep. However, the increase in rumen osmotic pressure is followed by an increase in blood osmotic pressure [Dobson et al., 1976]. Camels fed on rations containing *Atriplex nummularia* showed higher blood Na values (448.3 and 454.3 mg/100 ml) than those of other tested groups [Kewan, 2003]. Khattab [2007] had the same results when *Atriplex halimus* was fed to sheep (Table 7).

Rumen volume increased when osmotic pressure was increased through sodium chloride increasing doses [Lopez et al., 1994]. Yet, it was concluded that the osmotic pressure effect is mainly exerted on the rate and extent at which water enters or leaves the rumen pool rather than on the absolute size of the rumen pool itself. It is, therefore, that the washout of the digesta from the rumen increases. Harrison et al. [1975] suggested that such an increase in the flow of digesta to the duodenum would increase the efficiency of microbial synthesis in the rumen, thereby increasing the flow of protein and starch to the duodenum.

Remond et al. [1993] found that the increase in osmolality after NaCl injection slightly decreased NH_3 absorption. The increase in osmolality from 260 to 420 mOsmol/liter was accompanied by a drop in pH [Hogan, 1961].

14 EFFECT OF MINERALS ON FEED INTAKE

Some researchers suggest that the Na and K chlorides play a controlling effect on feed intake. Bergen [1972] studied the possible role of osmotic pressure in altering voluntary feed intake when rumen osmolalrity was artificially elevated to more than 400 mOsm/kg with NaCl or NaAc just before feeding. He suggested that the short duration of elevated osmotic pressure when feeding roughages is not an important factor in controlling feed intake. Kato et al. [1979] reached the same conclusion when they infused sheep with potassium chloride or sodium chloride during meals. On the other hand, when Ternouth and Beattie [1971] infused NaCl, KCl, or sodium salts of organic acids into fistulated sheep, they found that the reduction in food intake is related to osmolalrity of the added fluid and that the reduction was more significant when potassium chloride was infused than with sodium chloride. They explained that on the basis of the probable relation of the slower absorption of potassium compared with sodium. An elevation in osmotic pressure in the rumen is sensed by the wall of the reticulo-rumen to inhibit feed intake [Carter and Grovum, 1990]. Forbes [1995] concluded that the relative importance of osmolalrity as a satiety factor in ruminants is still uncertain. However, the results of the effect of increased salt intake are inconsistent.

Fahmy et al. [2001] fed a mixture of sun-dried salt plants [22.5% as *Zygophyllum album*, 22.5% as *Halocnemum strobilaceum*, 45% as *Tamarix mannifera*, and 10% molasses (w/w)] to growing sheep. The mixture represented 40% of the total ration. The salt-plant-fed group had a higher feed intake than that of the Berseem hay-fed group. Total dry matter intake was comparable and did not differ significantly among animal groups. Khattab [2007] fed a mixture of fresh and silage of both *Atriplex halimus* and *Acacia saligna* to sheep lambs. Feed intakes were 1.57 for control, 1.45 for fresh atriplex and acacia, and 1.62 kg/hr/day for ensiled atriplex and acacia. Sodium intakes were 10.81, 24.13, and 21.54 while potassium intakes were 22.62, 15.74, and 11.69 g/hr/day, respectively.

15   EFFECT OF MINERALS ON WATER INTAKE AND NUTRIENT UTILIZATION

The salt content in halophytes can influence the animal's water requirement because additional water is required to excrete their high salt content. The high salt levels of the foliage increase the water requirements of grazing animals [Wilson, 1974]. Gihad [1993] reported that drinking water increased by 61.4% when sheep were fed *Atriplex halimus* instead of clover hay. Puri and Garg [2001] observed that supplementation of sodium chloride increased voluntary water intake, rumen volume, and outflow rate from the rumen. Rapid influx of water to neutralize osmotic pressure swells the ruminal papillae and can pull patches of the ruminal epithelium into the rumen by stripping the internal surface layers of the rumen wall from the underlying layers, as illustrated vividly in histological studies by Eadie and Mann [1970]. Swingle et al. [1996] observed that lambs fed halophyte diets consumed up to 110% more water per day and 50% more water per kilogram of dry matter intake.

16   EFFECT OF MINERALS ON MICROBIAL COMMUNITY IN THE RUMEN

It is proposed that the increased salt intake and, hence, the osmotic pressure of the rumen may affect the microorganism population and metabolism. Increasing osmotic pressure in the rumen is believed to be unfavorable to the growth of protozoa. Also, the increased rates of outflow are believed to contribute to decreased protozoal population [Warner and Stacy, 1977]. The artificial rise of rumen osmotic pressure [Bergen, 1972] up to 400 mOsmol/kg caused cellulose digestion to be inhibited. Shawket et al. [2001] reported that increased salt intake increased dilution rate of the rumen fluid phase and negatively affected the protozoal population. Kewan [2003] reported that the total protozoal count ($\times 10^3$) per milliliter was significantly higher ($P \leq 0.05$) for camels fed on rations that contained *Atriplex nummularia* than for those fed on *Acacia saligna* and treated rice straw rations. Total rumen protozoal count in camels fed on berseem hay ration were significantly ($P \leq 0.05$) higher than those fed on the other experimental rations.

In conclusion, halophytic plants may be used as a feed component in ruminant rations in spite of the precautions placed on their use in animal feeds. However, if these plants are treated the right way according to the limitations of each plant species, they would provide a plausible feed resource for ruminants in areas like arid, semi-arid, and coastal areas. This way the feed shortage in these areas will be relieved.

REFERENCES

Abdelhameed Afaf AE, Shawkat SM, Hafez IM. 2006. Physiological studies on the effect of feeding salt plants in ewes under semi arid conditions. 4th Sci. Conf. Physiol. Applic. for Anim. Wealth Dev., Cairo, Egypt, July 29–30, pp. 113–132.

Attia-Ismail SA. 2005. Factors limiting and methods of improving nutritive and feeding values of halophytes in arid, semi-arid and coastal areas. Conference on Biosaline Agriculture & High Salinity Tolerance, January 9–14, 2005, Mugla, Turkey.

Attia-Ismail SA, Fayed AM, Fahmy AA. 2003. Some mineral and nitrogen utilization of sheep fed salt plant and monensin. *Egyptian J Nutr Feeds* 6:151–161.

Badawy HS. 2005. Nutritional studies on camels fed on natural ranges in Shalatin–Halaib Triangle region. Ph.D. Cairo University.

Barrio JP, Bapat ST, Forbes JM. 1991. The effect of drinking water on food intake responses to manipulations of rumen osmolality in sheep. *Proc Nutr Soc* **50**:98A.

Bennink MR, Tyler TR, Ward GM, Johnson DE. 1978. Ionic milieu of bovine and ovine rumen as affected by diet. *J Dairy Sci* **61**:315–323.

Bergen WG. 1972. Rumen osmolality as a factor in feed intake control of sheep. *J Anim Sci* **34**:1054–1060.

Brown CE, Pezeshkia SR, DeLaune RD. 2006. The effects of salinity and soil drying on nutrient uptake and growth of Spartina alterniflora in a simulated tidal system. *Environ Exp Bot* **58**:140–148.

Carter RR, Grovum WL. 1990. A review of the physiological significance of hypertonic body fluids on feed intake and ruminal function: Salivation, motility and microbes. *J Anim Sci* **68**:2811–2832.

Craig GF, Bell DT, Atkins CA. 1991. Nutritional characteristics of selected species of Acacia growing in naturally saline areas of Western Australia. *Aust J Exp Agric* **31**:341–345.

Daoud S, Harrouni MC, Bengueddour R. 2001. Biomass production and ion composition of some halophytes irrigated with different seawater dilutions. First International Conference on Saltwater Intrusion and Coastal Aquifers—Monitoring, Modeling, and Management. Essaouira, Morocco, April 23–25, 2001.

Dobson A, Sellers AF, Gatewood VH. 1976. Absorption and exchange of water across rumen epithelium. *Am J Physio* **231**:1588–1594.

Eadie JM, Hyldgaard-Jensen SO, Mann RS. 1970. Observations on the microbiology and biochemistry of the rumen in cattle. *Br J Nutr* **24**:157.

El-Bassosy AA. 1983. A study of the nutritive value of some range plants from El-Saloom and Mersa Mattroh. Ph.D. Thesis, Faculty of Agriculture, Ain Shams University, Egypt.

El Shaer MH, Attia-Ismail SA. 2002. Halophytes as animal feeds: Potentiality, constraints, and prospects. International Symposium on Optimum Utilization in Salt Affected Ecosystems in Arid and Semi-arid Regions, Cairo, Egypt, April 8–11, 2002.

Fahmy AA. 1998. Nutritional studies on halophytes and agricultural wastes as feed supplements for small ruminants in Sinai. Ph.D. Thesis, Cairo University, Egypt.

Fahmy AA, Attia-Ismail SA, Fayed AM. 2001. Effect of Monensin on salt plant utilization and sheep performance. *Egyptian J Nutr Feeds* **4**:581–590.

FAO. 1988. Salt-affected soils and their management. By Abrol IP, Yadav JSP, and Massoud FI. *FAO Soils Bull.* **39**.

FAO. 2007. http://www.fao.org/ag/agl/agll/wrb/wrbmaps/htm/

Farid MFA. 1989. Water and minerals problems of the dromedary camel (an overview). *Options Méditerranéenn—Sér Sémin* **2**:111–124.

Forbes JM. 1995. Voluntary food intake and diet selection in farm animals. CAB International, Wallingford, Oxon OX 108DE, UK.

Gabr MG. 2002. First experience of Matrouh Resource Management Project in salt bush utilization for animal feeding. International Symposium on Optimum Utilization in Salt Affected Ecosystems in Arid and Semi Arid Regions, Cairo, Egypt, April 8–11, 2002, pp. 419–425.

Gihad EA. 1993. Utilization of high salinity tolerant plants and saline water by desert animals. In: Lieth H, Al-Masoom A, editors. Towards the rational use of high salinity tolerant plant, Klumer Academic Publishers. Dordrecht. T. Vs Vol. **1**:445–447.

Glenn EP, Brown JJ, Blumwald E. 1999. Salt tolerance and crop potential of halophytes. *Crit Rev Plant Sci* **18**:227–255.

Gorham J. 1996. Mechanisms of salt tolerance of halophytes. In: Choukr-Allah R, Malcolm CV, Hamdy A, editors. *Halophytes and Biosaline Agriculture*. New York: Marcel Dekker, pp. 31–53.

Haenlein GFW, Ramirez RG. 2007. Potential mineral deficiencies on arid rangelands for small ruminants with special reference to Mexico. *Small Ruminant Res* **68**:35–41.

Harrison DJ, Beever DE, Thomson DJ, Osbourn DF. 1975. Manipulation of rumen ermentation in sheep by increasing the rate of flow of water from the rumen. *J Agric Sci, Cambridge* **85**:93–101.

Hogan JP. 1961. The absorption of ammonia through the rumen of the sheep. *Aust J Biol Sci* **14**:448.

Hutchings SS. 1965. Grazing management of salt-desert shrub ranges in the Western United States. *Proc 9th Int Grass Cong* **2**:1619–1625.

Jaradat AA. 2003. Halophytes for sustainable biosaline. In: Al Sharhan AA, et al., editors. *Farming Systems in the Middle East.* Rotterdam, The Netherlands: A.A. Baikema/Swets & Zeitlinger. Desertification for the third millennium, pp. 187–203.

Jordan FL, Robin-Abbott MM, Raina M, Glenn P. 2002. A comparison of chelator-facilitated metal uptake by a halophyte and a glycophyte. *Environ Toxicol Chem* **21**:2698–2704.

Kato S, Sasaki Y, Tsuda T. 1979. Food intake and rumen osmolality in the sheep. *Ann Recherches Veterinaire* **10**:229–230.

Kewan KZ. 2003. Studies on camel nutrition. Ph.D. Thesis, Faculty of Agriculture, Alexandria University, Egypt.

Khan MA. 2001. Experimental assessment of salinity tolerance of *Ceriops tagal* seedlings and saplings from the Indus delta. *Pakistan Aquat Bot* **70**:259–268.

Khattab IMA. 2007. Studies on halophytic forages as sheep fodder under arid and semi arid conditions in Egypt. Ph.D. Thesis. Alexandria University.

Kraidees MS, Abouheif MA, Al-Saiady MY, Tag-Eldin A, Metwally H. 1998. The effect of dietary inclusion of halophyte *Salicornia bigelovii* Torr. on growth performance and carcass characteristics of lambs. *Anim Feed Sci Technol* **76**:149–159.

Lopez S, Hovell B, Macleod NA. 1994. Osmotic pressure, water kinetics and volatile fatty acid absorption in the rumen of sheep sustained by intragastric infusion. *British J Nutr* **71**:153–168.

MacFarlane WV, Morris RJH, Howard B. 1963. Turnover and distribution of water in desert camels, sheep, cattle and kangaroo. *Nature London* **197**:270–271.

Marai IFM, Habeeb AA, Kamal TH. 1995. Response of livestock to excess sodium intake. In: Phillips CJC, Chiy PC, editors. *Sodium in Agriculture.* Canterbury: Chalcombe Publications, pp. 173–180.

Masters DG, Benes SE, Norman HC. 2007. Biosaline agriculture for forage and livestock production. *Agric Ecosystems Environ* **119**:234–248.

NRC. 1981. *Nutrient Requirements of Goats.* Washington, DC: National Research Council, National Academic Press.

NRC. 1985. *Nutrient Requirements of Sheep.* Washington, DC: National Research Council, National Academic Press.

Parida AK, Das AB, Maritta B. 2004. Effect of salt on growth, ion accumulation, photosynthesis and leaf anatomy of the mangrove, *Bruguiera parviflora. Trees—Struct Funct* **18**:167–174.

Puri JP, Garg SK. 2001. Effect of osmotic agent supplementation in the diet of buffalo on some rumen functions and blood electrolytes. *Indian J Anim Sci* **71**(10):927–931.

Reboreda R, Cacador I. 2007. Halophyte vegetation influences in salt marsh retention capacity for heavy metals. *Environ Pollut* **146**:147–154.

Remond D, Chaise JP, Delval E, Poncet C. 1993. Net transfer of urea and ammonia across the ruminal wall of sheep. *J Amin Sci* **71**:2785–2792.

Retana J, Parker DR, Amrhein C, Page AL. 1993. Growth and trace element concentrations of five plant species grown in a highly saline soil. *J Environ Sci* **22**:805–811.

Shawkat SM, Khatab IM, Borhami BE, El-Shazly KA. 2001. Performance of growing goats fed halophytic pasture with different energy sources. *Egyptian J Nutr Feeds* **4**:215–264.

Swingle RS, Glenn EP, Squires V. 1996. Growth performance of lambs fed mixed diets containing halophyte ingredients. *Anim Feed Sci Technol* **63**:137–148.

Szabolcs I. 1974. Salt Affected Soils in Europe. The Hague: Martinus Nijhoff.

Ternouth JH, Beattie A. 1971. Studies of the food intake of sheep at a single meal. *Br J Nutr* **25**:153.

Underwood EJ, Suttle NF. 1999. *The Mineral Nutrition of Livestock*. New York: CABI Publishing.

Vercoe TK. 1986. Fodder potential of selected Australian tree species. In: Australian Acacias in Developing Countries, Proc. Of an international workshop held at the Forestry Training Center Gympie, Qld, Australia, 4–7 August, 1986 (J.W. Turnbull, Ed.), pp. 95–100. ACIAR Proceedings no. 16, Australian Center for International Research, Canberra.

Verde T, Shephard RJ, Corey P, Moore R. 1982. Sweat composition in exercise and in heat. *J Appl Physiol* **53**:1540–1545.

Walker BH. 1980. A review of browse and role in livestock production in Southern Africa. In: Le Houerou HN, editor. Browse in Africa, the current state of knowledge. ILCA Addis Ababa, Ehtiopia.

Warner ACI, Stacy BD. 1965. Solutes in the rumen of the sheep. *Quart J Exp Physiol* **50**:169–184.

Warner ACI, Stacy BD. 1977. Influence of ruminal and plasma osmotic pressure on salivary secretion in sheep. *Q J Expe Physiol* **62**:133–142.

Warren BE, Bunny CI, Bryant ER. 1990. A preliminary examination of the nutritive value of four saltbush (Atriplex) species. Proc. Austr.

Warren BE, Casson T, Ryall DH. 1994. Production from grazing sheep on revegetated saltland in Western Australia. In: Squires VR, Ayoub.

Wilson AD. 1974. Water requirements and water turnover of sheep grazing semi-arid pasture communities in New South Wales. *Aust J Agric Res* **25**:339.

Windham L, Weis JS, Weis P. 2003. Uptake and distribution of metals in two dominant salt marsh macrophytes, *Spartina alterniflora* (cordgrass) and *Phragmites australis* (common reed). *Estuar Coast Shelf Sci* **56**:63–72.

28 Plants as Biomonitors of Trace Elements Pollution in Soil

MUNIR OZTURK

Botany Department, Science Faculty, Ege University, 35100 Bornova, Izmir, Turkey

ERSIN YUCEL

Biology Department, Science Faculty, Anadolu University, 26470 Eskisehir, Turkey

SALIH GUCEL

Centre for Environmental Studies, Near East University, Nicosia, 33010 North Cyprus

SERDAL SAKÇALI

Biology Department, Faculty of Science & Arts, Fatih University, 34500 Hadimkoy, Istanbul, Turkey

AHMET AKSOY

Biology Department, Faculty of Science & Arts, Erciyes University, 38039 Kayseri, Turkey

1 INTRODUCTION

Studies on the application of trace elements to agricultural soils for improving plant growth have been undertaken for decades. However, heavy industrialization followed by waste disposal has reached a stage when these elements are accepted now as important environmental contaminants affecting our ecosystems adversely [Rao and Dubey, 1992; Sawidis et al, 1995a,b; Ross, 1994; Morselli et al., 2004]. The trace elements such as copper, zinc, manganese, and iron can serve as essential micronutrients, whereas cadmium, lead, chromium, and aluminum could prove dangerous [Roychowdhury et al., 2003; Shanker et al., 2005]. Chronic exposure to the trace elements can result in human health effects including kidney dysfunction from cadmium, skin and internal organ cancers from arsenic, and impaired development from lead [Ozturk and Turkan, 1992; Kartal et al., 1992, 1993; Bucher and Schenk,

Trace Elements as Contaminants and Nutrients: Consequences in Ecosystems and Human Health, Edited by M. N. V. Prasad

2000; Çinar & Elik, 2002; Del Rio et al., 2002; Alam et al., 2003, Banerjee, 2003]. This interaction is a part of the biogeochemical cycling of chemical elements involving the flow of elements from nonliving to the living in the ecosphere [Woolhouse, 1983; Friedland, 1989; Davidson et al.,1998]. If these elements occur in toxic levels in soils, their entry into the food chain through plants may lead to several health and environmental problems [Salt et al., 1995; Das et al., 1997]. However, the movement of trace elements depends on their amount and input sources, reaction with soil components, soil properties, and uptake rate by plants, as well as their solubility equilibrium [Cataldo and Wildung, 1978; Pendias and Pendias, 2000; Li et al., 2001; Lee et al., 2005; Remon et al., 2005; Çubukçu and Tüysüz, 2007].

The trace-element-tolerant plants are very well known indicators of their occurrence in soils and are even used for geobotanical prospecting for their deposits, but their degree of purity varies with different sites [Wittig and Baumen, 1992; Wittig, 1993]. The prospecting for trace elements in soils, particularly for natural resources, is performed either with accumulative indicators or with sensitive indicators [Brooks, 1993]. A specific plant cover develops at places accommodating some endemic species and metallophytes [Jonnalagadda and Nenzou, 1997]. The morphological, physiological, and other reactions of different types of terrestrial plants are now used for biomonitoring of trace elements as well as monitoring the environmental quality [Ozturk and Turkan, 1993; Aksoy and Ozturk, 1996, 1997; Aksoy, 1997; Singh, 2002].

For an understanding of the functioning of plants as monitors, we need to know the interactions of trace elements in the environment, their absorption by soils, and accumulation by plants [Baker and Walker, 1989; Schafer et al., 1998], together with the factors and mechanisms controlling availability and uptake [Streit and Stumm, 1993]. These reactions depend not only on the factor to be monitored but also on nutrients, water, age, sex, heritage, and species interactions. Although a lot of work is being done on the basic mechanisms and regulation of plant-based processes involved in the entry of toxic levels of trace elements into the food chain (through plants), the phytoremediation of contaminated soils, and mechanisms of tolerance by plants, there is still a big gap to be filled up about the role of trace elements in soils and their interactions with plants. The aim of this chapter is to give an overall evaluation of the trace element pollution in the soil–plant systems in the light of investigations undertaken by us as well as other workers around the world during the last few decades.

2   SOILS AND TRACE ELEMENTS

Soils are heterogeneous mixtures of different kinds of minerals, which include trace elements as well [Kabala and Singh, 2001; Iskandar and Kirkham, 2001; Pueyo et al., 2003; Mico et al., 2006]. However, human activities are leading to significant increases in the concentration of these elements in the environment, and soils are acting as a sink for them [EEA, 1995; Kelly et al., 1996; Yağdı et al., 2006]. The global dispersion of trace elements occurs through different sources such as traffic, emissions from different industrial complexes, remediation technologies of contaminated soils, smelting of ores, pesticides, coal combustion, and Pb-based paint atmospheric deposition, all adding to this contamination process in the urban, suburban, rural, and agricultural soils [Berrow and Ure, 1981; Verkleij, 1993; Al-Khashman, 2004]. Very high contamination of the soils is caused by mine activities resulting in a significant buildup in the levels of lead, cadmium, zinc, copper, chromium, aluminum, fluorine, mercury, selenium, arsenic, and so on [Verkleij, 1993; Bolan et al., 2005]. The reports published on the trace element accumulation in plants, soil, and street dust clearly bring to light the dangerous situation created for human health by their direct impact in the city centers and indirectly through a consumption of edible plants and animals coming from the fields situated near the roads with a high traffic load [Ozturk and Turkan, 1993]. A summary of the data published on the levels of concentrations of trace elements in the soils together with the plant cover supported by these sites alongside the highways with a heavy traffic load and alongside the sideways with a low traffic load is given in Table 1.

TABLE 1. Concentration of Trace Elements in the Soils

Formation Species	Metal and Concentrations (μg/g)					
	Highways		Sideways		Reference	
Trees						
Pinus halepensis	Zn	150	Zn	27	Fuentes et al. [2007]	
	Ni	6	Ni	0.1		
Populus alba (Surface)	Pb	305	Pb	21	Madejon et al. [2004]	
	Zn	583	Pb	58		
	Cd	4.29	Cd	1.56		
	Ni	18	Ni	17.3		
	Fe	35,508	Fe	25,960		
Populus alba (lowest part)	Pb	140	Pb	16	Madejon et al. [2004]	
	Zn	488	Zn	13		
	Cd	3.64	Cd	1.57		
	Ni	19	Ni	15		
	Fe	32,872	Fe	26,320		
Robinio pseudo-acacia	Pb	468	Pb	39	Celik et al. [2005]	
	Zn	506.43	Zn	10.67		
	Cd	7.36	Cd	0.48		
Populus canadensis	Pb	4.1	Pb	1	Celik et al. [2006]	
	Zn	1.9	Zn	0.1		
	Cd	0.2				
	Fe	1.8	Fe	0.6		
Elaeagnus angustifolia	Pb	485.26	Pb	40.21	Aksoy and Sahin [1999]	
	Zn	1,215.25	Zn	66.12		
Robinia pseudo-acacia	Pb	336.55	Pb	34.26	Aksoy et al. [2000b]	
	Zn	1,189	Zn	63		
	Fe	3,939.3	Fe	2,892.7		
Populus × *euramericana*	Zn	10,300			Sebastiani et al. [2004]	
	Fe	54,000				

(Continued)

TABLE 1. *Continued*

Formation Species	Metal and Concentrations ($\mu g/g$)				Reference
	Highways		Sideways		
Shrubs					
Ilex aquifolium	Pb	115	Pb	27	Samecka-Cymerman and Kempers [1999]
	Zn	862	Zn	470	
	Cd	2.25	Cd	0.71	
Rhododendron catawbiense	Pb	131	Pb	36	Samecka-Cymerman and Kempers [1999]
	Zn	1640	Zn	315	
	Cd	2.60	Cd	0.71	
	Ni	27.1	Ni	3.2	
Mahonia aquifolium	Pb	113	Pb	23	Samecka-Cymerman and Kempers [1999]
	Zn	743	Zn	213	
	Cd	2.15	Cd	0.69	
	Ni	16.5	Ni	3	
Lupinus albus	Pb	19,663			Castaldi et al. [2005]
	Zn	14667			
Primula vulgaris	Pb	24.01	Pb	1.83	Steinbörn and Breen [1999]
	Zn	525	Zn	101	
Teucrium scorodonia	Pb	15.18	Pb	3.65	Steinbörn and Breen [1999]
	Zn	637	Zn	125	
Succisa pratensis	Pb	25.53	Pb	1.86	Steinbörn and Breen [1999]
	Zn	359.77	Zn	75.1	
Ferns					
Pteris vittata	Pb	9.52			Fayigar et al. [2005]
	Zn	10.5			
Pteris vittata	Pb	8.1			Fayigar et al. [2005]
	Zn	0.81			
	Ni	7.4			
Mosses					
Hylocomium splendens	Pb	19.60	Pb	2.18	Steinbörn and Breen [1999]
	Zn	866.9	Zn	95	
Rhytidiadelphus loreus	Pb	4.17	Pb	1.14	
	Zn	342.8	Zn	141	
Soil					
	Pb	12	Pb	0	Yucel et al. [1995]
	Zn	38.5	Zn	7.5	
	Cd	1.880	Cd	0.344	
	Pb	23.03	Pb	14.43	Chronopoulos et al. [1997]
	Cd	74.23	Cd	29.70	
	Pb	1,198	Pb	25	Akbar et al. [2006]
	Zn	480	Zn	56.7	
	Cd	3.8	Cd	0.3	

(Continued)

TABLE 1. *Continued*

Formation Species	Metal and Concentrations ($\mu g/g$)			
	Highways		Sideways	Reference
	Pb 17.63		Pb 3.43	Steinbörn and Breen [1999]
	Zn 182		Zn 7.81	
	Pb 942		Pb 69	Aksoy et al. [1999]
	Zn 570		Zn 173	
	Cd 3.15		Cd 1.02	
	Pb 237		Pb 54	Tomasevic et al. [2004]
	Zn 215.6		Zn 122.7	
	Cd 1.73		Cd 1.34	
	Ni 73.6		Ni 62.4	
	Fe 284.9		Fe 215.4	
	Pb 401		Pb 15	Cicek and Koparal [2004]
	Zn 784		Zn 18	
	Cd 21.7		Cd 1.4	
	Ni 372		Ni 20.1	
	Fe 1,650		Fe 564	

3 PLANTS AS BIOMONITORS OF TRACE ELEMENTS

The flowering plants are frequently used for biomonitoring of trace elements [Seaward and Mashhour, 1991; Markert, 1993, 1994; Al-Shayeb et al., 1995; Dmuchowski and Bytnerowicz, 1995; Reeves and Baker, 2000; Pyatt, 1999, 2001; Lau and Luk, 2001; Yılmaz and Zengin, 2004; Palmieri et al., 2005; Rossini and Mingorance, 2006; Aksoy and Demirezen, 2006; Demirezen and Aksoy, 2004, 2006; Demirezen et al., 2007; Yılmaz et al., 2007], because they show a clear division into roots, shoots, and leaves, are much larger, and show no difficulty in separation of different organs and even tissues. In addition, the physiology, ecology, and morphology of higher plants is better known than those of lower plants. However, which group or species has to be chosen depends on the purpose of monitoring and on the type of ecosystem. The levels of trace elements in different ecosystems are affected by the following; the season; the vegetation type; the distance of plants from the source; the amount of precipitation; and the wind direction [Kovacs et al., 1993; Ouzounidou, 1994; Oncel et al., 2004; Onder and Şükrü, 2006]. Transpiration is an important motor for the transport of these elements from roots to leaves, because such plants in shade accumulate few trace elements compared to those under the sun; even leaves from the top of a tree may show a higher accumulation rate than those from the bottom; similarly, gall tissues contain much lower levels of trace elements than do normal plant tissues [Wittig, 1993]. In general, areas with a vegetation cover accumulate trace elements four times more than bare ones. In fact, 90% of the trace elements are retained by plants and only 10% go to the lower horizons of soil. Among the plants, accumulation rates in the herbaceous cover under trees are twice as high as in herbs growing in a treeless zone.

Different plant species growing in the same habitat can show different trace element levels. A great difference in trace element content exists between different organs [Wittig and Neite, 1989; Wittig and Baumen, 1992; Madejon et al., 2006]. Roots and rhizomes are often suitable indicators of soil-borne trace elements. Although this can be attributed to the differences in the uptake by roots, the leaf retention characteristics seem to play an effective role in this respect. Leaf anatomy has to be considered when comparing trace elemental concentrations in plants. The trace element levels in the leaves show marked seasonal variations, and hairy surfaces appear to be the most active accumulators as compared to smooth ones [Bereket and Yucel, 1990; Kutbay and Kilinç, 1991; Deu and Kreeb, 1993]. The levels of lead and cadmium are 10 times higher in the plants with hairy leaves compared to the smooth ones. *Thymus capitatus* with its hairy leaves accumulates lead three times more than the smooth leaved *Quercus coccifera* in the same habitat [Turkan and Ozturk, 1989]. The results show that hairy leaves of *Inula viscosa* and *Cistus creticus* accumulate very high levels of lead and cadmium. The same is found for the densely branching *Pinus brutia*. In particular, *Inula viscosa* appears to be a strong accumulator due to its rough and hairy leaves, with values of 80 µg/g of lead and 1.55 µg/g of cadmium [Turkan and Ozturk, 1989]. Kutbay and Kilinç [1991] have observed higher trace element levels in *Centaurea iberica* and *Plantago lanceolata*, both having hairy leaves. The same thing occurs in the case with the hairy-leaved *Lonicera xylosteum* compared to *L. tatarica* with smooth leaves.

The levels of trace element accumulation vary even among different parts of the plant species (Tables 2 and 3). While leaves of *Pyracantha coccinea*, for example, show values of 22 µg/g of lead, fruits contain only 8 µg/g [Turkan and Ozturk, 1989]. The lead content of leaves in tomato plants is around 48 µg/g, and fruits contain only 18 µg/g [Ozturk and Turkan, 1993]. Toker et al. [1990] found higher levels of lead, nickel, and cadmium in the leaves of *Pinus nigra* compared to its branches. In some findings, 147 µg/g of lead have been reported in the leaves, but only 48 µg/g in the tillers and 20 µg/g in the stems of cereals. Higher levels of lead have been observed in the leaf blades of *Acer pseudoplatanus* than in its petioles and fruits. The higher accumulation in the leaves can be attributed to their large surface area which is in regular contact with the surrounding atmosphere for a diffusion of gases (Table 3). The concentration of trace elements in the roots, stems, and leaves of different higher plants are given in Tables 2 and 3.

Recently, many reports have appeared stressing the fact that the bark of trees is a better indicator of trace element pollution than lichens. In many studies, good results have been obtained by the use of tree bark, because trees are important interceptors in forest ecosystems; hence their use as accumulative monitors is of great ecological relevance [Turkan and Henden, 1991; Walkenhorst et al., 1993]. Trees are also used for monitoring with the help of their rings [Hagemeyer, 1993]. The barks can show high accumulation power because of their perforations [Harju et al., 2002]. Türkan and Henden [1990] have obtained very high values in the samples taken from the sites very near to the pollution source: 14.4–55.6 µg/g lead, 0.38–1.33 µg/g cadmium, 2.55–11.69 µg/g nickel, 1.61–1.91 µg/g cobalt, 14.5–65 µg/g zinc, and 2.54–19.21 µg/g copper. Similar results have been reported for *Pinus*

TABLE 2. Concentrations of Some Trace Elements in the Roots and Stem of plants

Tree Species	Trace Element Concentrations ($\mu g/g$)			
	Highways		Sideroads	Reference
Pinus sylvestris	Pb	43	Pb	8.5	Samecka-Cymerman et al. [2006]
	Zn	94	Zn	14	
	Cd	2.1	Cd	0.33	
	Ni	4.5	Ni	1.6	
	Fe	1505	Fe	212	
	Fe	18	Fe	2.8	
Pinus nigra	Pb	82.2	Pb	0.6	Coskun [2006]
	Zn	57.7	Zn	6.36	
	Cd	0.68	Cd	0.04	
Populus alba (Stem)	Pb	2.06	Pb	1.54	Madejon et al. [2004]
	Zn	139	Zn	59.2	
	Cd	3.18	Cd	0.31	
	Ni	0.78	Ni	0.72	
	Fe	88	Fe	42.2	
Populus canadensis (bark)	Pb	36.5	Pb	15.5	Celik et al. [2006]
	Zn	1468	Zn	40	
	Cd	2	Cd	1.5	
	Fe	578	Fe	39	
Shrubs					
Empetrum nigrum	Ni	164	Ni	63	Monni et al. [2001]
Herbs					
Lupinus albus (root)	Pb	1032	Pb	595	Castaldi et al. [2005]
	Zn	5780	Zn	3720	
	Cd	36.84	Cd	10.23	
Lupinus albus (stem)	Pb	3.44	Pb	0.82	Castaldi et al. [2005]
	Zn	850	Zn	530	
	Cd	9.82	Cd	5.08	
Teucrium scorodonia	Pb	71	Pb	5.52	Steinbörn and Breen [1999]
	Zn	120	Zn	1.90	
Primula vulgaris,	Pb	373	Pb	35.6	
	Zn	130	Zn	10.69	
Succisa pratensis	Pb	830.28	Pb	95.65	
	Zn	22.67	Zn	75.17	

sylvestris, Robinia pseudoacacia, Tilia cordata, and *Acer platanoides*; however, values of lead in the bark samples of *Cryptomeria japonica* were very high, varying between 16.6 $\mu g/g$ and 159 $\mu g/g$ [Ozturk and Turkan, 1993].

Most of the higher plants are toxitolerant, successfully flourishing in the densely populated urban centers as well as around the industrial sectors or along roadsides [Kovacs et al., 1993; Ozturk and Turkan, 1993; Aksoy et al., 2000]. The species such as *Populus nigra italica, Pinus sylvestris, Picea abies, Fagus sylvatica,*

TABLE 3. Concentrations of Some Trace Elements in the Leaves

Formation/Species	Metal and Concentrations ($\mu g/g$)				
	Highways		Sideways		Reference
Pinus brutia	Pb	45			Turkan and Ozturk [1989]
	Zn	56			
	Cd	0.52			
Pinus nigra	Pb	39	Pb	8	Toker et al. [1990]
	Ni	11	Ni	6	
	Cd	0.28	Cd	0.40	
Pinus nigra	Pb	16	Pb	10	Bereket and Yucel [1990]
	Zn	122	Zn	86	
	Cd	0.73	Cd	0.68	
	Fe	513	Fe	101	
Pinus nigra ssp. *pallasiana*	Pb	55	Pb	0.1	Cicek and Koparal [2004]
	Zn	224.4	Zn	1.7	
	Cd	7.23	Cd	0.1	
	Ni	81.7	Ni	1.9	
	Fe	845.8	Fe	86.3	
Pinus pinaster	Pb	0.121	Pb	0.016	Aboal et al. [2004]
	Zn	15.04	Zn	0.04	
	Cd	0.187	Cd	0.033	
	Ni	81.7	Ni	1.9	
	Fe	115.8	Fe	11.8	
Pinus sylvestris	Pb	69	Pb	nd	McEnroe and Helmisaari
	Zn	140	Zn	5	[2001]
	Cd	88	Cd	nd	
	Fe	346	Fe	30	
	Ni	365	Ni	2	
Pinus canariensis	Pb	574	Pb	284	Tausz et al. [2005]
	Zn	39.2	Zn	20.4	
	Cd	39	Cd	15	
	Fe	719.4	Fe	442.6	
Cupressus sempervirens	Pb	445	Pb	237.7	El-Hasan et al. [2002]
	Zn	442.1	Zn	16.3	
	Cd	0.83	Cd	0.11	
	Fe	963	Fe	62	
	Ni	4.59	Ni	0.27	
Salix alba	Pb	55	Pb	0.1	Cicek and Koparal [2004]
	Zn	224.4	Zn	1.7	
	Cd	7.23	Cd	0.1	
	Ni	81.7	Ni	1.9	
	Fe	845.8	Fe	86.3	
Populus tremula	Pb	55	Pb	0.1	Cicek and Koparal [2004]
	Zn	224.4	Zn	1.7	
	Cd	7.23	Cd	0.1	
	Ni	81.7	Ni	1.9	
	Fe	845.8	Fe	86.3	

(Continued)

TABLE 3. *Continued*

Formation/Species	Metal and Concentrations (µg/g)				Reference
	Highways		Sideways		
Populus alba	Pb	5	Pb	3.93	Madejon et al. [2004]
	Zn	542.1	Zn	81.6	
	Cd	3.82	Cd	0.21	
	Fe	336.6	Fe	251	
	Ni	1.33	Ni	1.05	
Populus deltoides × maximowiczii)	Zn	3900	Zn	70	Sebastiani et al. [2004]
Populus × euramericana	Zn	3550	Zn	50	
Populus usbekistanica subsp. usbekistanica cv.	Pb	32	Pb	0	Yucel [1996]
	Zn	572.8	Zn	34	
	Cd	4.148	Cd	0.722	
Populus nigra ssp. nigra	Pb	26.83	Pb	0	Bereket and Yucel [1990]
	Zn	255.33	Zn	48.46	
	Cd	2.29	Cd	0.65	
Populus canadensis	Pb	40	Pb	14.5	Celik et al. [2006]
	Zn	246	Zn	43	
	Cd	1.5	Cd	0.5	
	Fe	486	Fe	135	
Robinia pseudoacacia	Pb	55	Pb	0.1	Cicek and Koparal [2004]
	Zn	224.4	Zn	1.7	
	Cd	7.23	Cd	0.1	
	Ni	81.7	Ni	1.9	
	Fe	845.8	Fe	86.3	
Corylus maxima	Pb	69.9	Pb	29.9	Kutbay and Kilinç [1991]
	Zn	129.9	Zn	10.5	
Robinia pseudo-acacia	Pb	206.2	Pb	15.11	Celik et al. [2005]
	Zn	139	Zn	11.53	
	Cd	3.70	Cd	0.32	
	Fe	3087	Fe	13.02	
Acer sp.	Pb	20.84	Pb	1.72	Baycu et al. [2006]
	Zn	36.99	Zn	7.80	
	Cd	0.83	Cd	0.05	
	Ni	3.39	Ni	0.48	
Aesculus sp.	Pb	8.91	Pb	0	
	Zn	130	Zn	12.62	
	Cd	0.75	Cd	0	
	Ni	2.72	Ni	0	
Ailanthus sp.	Pb	9.25	Pb	2.53	
	Zn	9.30	Zn	8.93	
	Cd	0.71	Cd	0	
	Ni	2.92	Ni	0.45	
Fraxinus sp.	Pb	1.65	Pb	1.65	
	Zn	91.03	Zn	10.14	
	Cd	0.52	Cd	0	
	Ni	7.47	Ni	0	

(Continued)

TABLE 3. *Continued*

Formation/Species	Metal and Concentrations (µg/g)				Reference
	Highways		Sideways		
Platanus sp.	Pb	16.81	Pb	0	
	Zn	57.14	Zn	9.88	
	Cd	0.69	Cd	0	
	Ni	5.24	Ni	0.85	
Populus sp.	Pb	2.57	Pb	1.42	
	Zn	92.60	Zn	37.38	
	Cd	1.35	Cd	0.09	
	Ni	3.90	Ni	0.61	
Robinia sp.	Pb	34.40	Pb	0	
	Zn	47.15	Zn	7.38	
	Cd	0.83	Cd	0	
	Ni	5.34	Ni	0	
Aesculus hippocastanum	Pb	20.3	Pb	5.35	Tomasevic et al. [2004]
	Zn	47.1	Zn	17.2	
	Cd	4.9	Cd	0.2	
	Fe	439.6	Fe	183.8	
Tilia sp.	Pb	11.4	Pb	1.88	
	Zn	28.6	Zn	15.2	
	Cd	1.4			
	Fe	324.5	Fe	105.9	
Malus domestica	Pb	3.37	Pb	2.53	Pinamonti et al. [1997]
	Zn	23.4	Zn	21	
	Cd	0.07	Cd	0.05	
	Ni	1.89	Ni	1.71	
Prunus amygdalus	Zn	62.3	Zn	11.4	Sawidis et al. [2001]
	Cd	1.56	Cd	0.80	
	Fe	233	Fe	111	
Juglans regia	Zn	30.7	Cd	9.1	
	Cd	1.30	Cd	0.25	
	Fe	452	Fe	104	
Pinus nigra	Zn	49.8	Zn	8.9	
	Cd	1.72	Cd	0.23	
	Fe	398	Fe	128	
Juniperus arizona	Zn	28.6	Zn	8.8	
	Cd	1.70	Cd	0.64	
	Fe	429	Fe	80	
Salix babylonica	Zn	94.7	Zn	20.7	
	Cd	1.82	Cd	0.92	
	Fe	265	Fe	92	
Populus nigra	Zn	139.1	Zn	26.1	
	Cd	1.75	Cd	0.46	
	Fe	171	Fe	52	
Robinia pseudoacacia	Zn	28.4	Zn	9.5	
	Cd	1.59	Cd	0.56	
	Fe	357	Fe	116	

(Continued)

TABLE 3. *Continued*

Formation/Species	Highways		Sideways		Reference
		Metal and Concentrations (µg/g)			
Ulmus minor	Zn	40.2	Zn	9.8	
	Cd	1.98	Cd	0.86	
	Fe	440	Fe	90	
Quercus infectoria	Pb	55	Pb	0.1	Cicek and Koparal [2004]
	Zn	224.4	Zn	1.7	
	Cd	7.23	Cd	0.1	
	Ni	81.7	Ni	1.9	
	Fe	845.8	Fe	86.3	
Quercus robur	P	0.456	Pb	0.056	Aboal et al. [2004]
	Zn	17.97	Zn	4.69	
	Cd	0.045	Cd	0.001	
	Fe	199.1	Fe	50.9	
	Ni	11.24	Ni	0.37	
Nerium oleander	Pb	27			Türkan and Ozturk [1989]
	Zn	23			
	Cd	0.5			
Pittosporum sinensis	Pb	50.57	Pb	15.51	Chronopoulos et al. [1997]
	Cd	29.96	Cd	33.33	
Nerium oleander	Pb	29.21	Pb	20.34	
	Cd	38.49	Cd	35.45	
Ilex aquifolium	Pb	23.9	Pb	7.8	Samecka-Cymerman, and
	Zn	840	Zn	191	Kempers [1999]
	Cd	1.35	Cd	0.51	
	Ni	6.75	Ni	0.91	
Mahonia aquifolium	Pb	15.2	Pb	3.8	
	Zn	764	Zn	402	
	Cd	1.39	Cd	0.55	
	Ni	2.36	Ni	0.55	
Rhododendron catawbiense.	Pb	22.1	Pb	4	
	Zn	1360	Zn	321	
	Cd	3.90	Cd	0.70	
	Ni	9.30	Ni	1.24	
Ligustrum japonicum	Zn	26.6	Zn	10.0	Sawidis et al. [2001]
	Cd	1.30	Cd	0.25	
	Fe	221	Fe	101	
Rosa canina	Zn	26.8	Zn	8.9	
	Cd	1.82	Cd	0.48	
	Fe	242	Fe	74	
Pyrocantha coccinea	Zn	37.5	Zn	13.9	
	Cd	1.15	Cd	0.46	
	Fe	262	Fe	113	
Empetrum nigrum	Ni	115	Ni	32 .	Monni et al. [2001]
Hedera helix	Pb	33			Kilinc and Kutbay [1991]
	Zn	40			

(Continued)

TABLE 3. *Continued*

Formation/Species	Metal and Concentrations ($\mu g/g$)				
	Highways		Sideways		Reference
Herbs					
Solanum nigrum ssp.	Pb	37			Kilinc and Kutbay [1991]
schultesii	Zn	41			
Mirabilis jalapa	Pb	33			
	Zn	12			
Xanthium strumarium	Pb	30			
	Zn	47			
Graminae	Pb	2.1			Viard et al. [2004]
	Zn	62			
	Cd	0.06			
Ballota acetabulosa	Fe	3950	Fe	1000	Henden et al. [1993]
Centaurea iberica	Pb	69.9	Pb	29.9	Kutbay and Kilinc [1991]
Malva neglecta	Zn	129.9	Zn	10.5	
Cistus creticus	Pb	56			Turkan and Ozturk [1989]
	Zn	82			
	Cd	1.36			
Plantago lanceolata	Pb	63			Kutbay and Kilinc [1991]
	Zn	119			
Centaurea iberica	Pb	70			
	Zn	109			
Xanthium spinosum	Pb	33			
	Zn	80			
Capsella bursa-pastoris	Pb	57	Pb	8	Aksoy et al. [1999]
	Zn	200	Zn	53	
	Cd	1.07	Cd	0.45	
Potamogeton pectinatus	Pb	237	Pb	151	Samecka-Cymerman, and
	Zn	272	Zn	246	Kempers [1996]
	Cd	1.5	Cd	1.1	
	Fe	15,200	Fe	14,200	
Myriophyllum spicatum	Pb	850	Pb	469	
	Zn	515	Zn	313	
	Cd	8.8	Cd	7.1	
	Fe	25,200	Fe	18,700	
Phragmites australis	Pb	9	Pb	0	Yucel et al. [1995]
	Zn	28.5	Zn	6.5	
	Cd	0.768	Cd	0.146	

Malus domestica, Ailanthus altissima, Sambucus nigra, and *Taraxacum officinale* have been used for passive monitoring by several workers [Porter, 1986; Wagner, 1993; Bargagli, 1993; Deu and Kreeb, 1993; Djingova and Kuleff, 1993]. Usually trace element concentrations decrease considerably within a distance of 50 m from

TABLE 4. Effects of Washing on the Removal of Trace Elements from the Plant Surfaces

Authors	Species	Material	Cd	Pb	Zn
Aksoy and Şahin [1999]	*Elaeagnus angustifolia*	Washed leaves	0.48–1.25 μg/g	15.4–65.20 μg/g	20.14–102.10 μg/g
		Unwashed leaves	0.50–3.45 μg/g	16.81–80.21 μg/g	22.08–231.26 μg/g
Aksoy et al. [2000b]	*Robinia pseudo-acacia*	Washed leaves	0.44–1.22 μg/g	14.89–62.42 μg/g	19–98 μg/g
		Unwashed leaves	0.47–3.39 μg/g	15.98–176.88 μg/g	21–242 μg/g
	Ballota acetabulosa	Unwashed leaves	1.50–15.3 mg/g	100–856 mg/g	94–1700 mg/g
	Nerium oleander	Unwashed leaves	0.4–5.3 ppm	7.2–23.4 ppm	11.5–27 ppm

733

the source due to the vertical diffusion and dispersion of aerosols in the atmosphere. In the leaves of tomato and tobacco plants, trace element concentrations decrease by about 50% in the area 50 m away from the source, but ~70–85% reductions have also been reported [Ozturk and Turkan, 1993]. In the area lying 50–100 m away from the source, only 10–30% of the trace elements can be found. However, no such tendency has been observed in carrot, strawberry, and mustard [Ozturk and Turkan, 1993]. However, the results on the concentration of trace element in the fruits in relation to distance are contradictory. Their accumulation by tomatoes shows a decrease with an increase of distance from the source being as high as 18 μg/g in the 10 m zone but only 3 μg/g just 50 m away [Ozturk and Turkan, 1993]. Similar findings have been reported in the products of 27 agricultural crop species growing alongside the highways. This decrease is observed in the vegetative parts of the plants, but no such correlation is observed in the fruits [Ozturk and Turkan, 1993]. If there is no uptake of trace elements by plants from the soil near a source, their level hardly goes beyond 10 ng/g. Trace element levels in *Olea europaea* plants growing around a polluting source are three times higher than in the trees away from the source, with accumulation varying between 22% and 64% alongside the source than away from it. In the case of *Pinus nigra* trees, values are higher in the heavily urbanized as compared to the rural sites [Ozturk and Turkan, 1993].

Many studies have been undertaken on the use of lower plants in particular lichens, mosses, and fungi. Very high concentrations of some trace elements in the *Hylocomium splendens*, *Rhytidiadelphus loreus*, *Pleurozium schreberi*, and other mosses collected from a highway have been reported, but their values on a sideway were comparatively lower [Garty, 1993; Steinbörn and Bren, 1999; Fernandez et al., 2002; Halıcı et al., 2003; Otvfs et al., 2003; Popescu et al., 2005; Samecka-Cymerman et al., 2006].

It is possible to remove a portion of trace elements from the plants by rinsing in particular the particles adhering outwardly to them (Table 4). More than 80% of lead can be removed from the surface of herbaceous plants by washing with water [Aksoy and Ozturk, 1997]. Some workers have succeeded in removing 86% of lead and 89% of zinc from the leaves of blackberry. In general, it is accepted that 30–95% of the trace elements, particularly lead, can be removed from the plants by washing. The unremovable part either remains attached to the cuticle in ion form or accumulates in the cracks and crevices of plants and does not completely enter the leaf [Ozturk and Turkan, 1993].

A specific plant cover develops on some ore outcrops and trace-element-rich spoil heaps, which embodies metal-resistant populations such as *Viola calaminaria*, *Thlaspi coerulescens*, *Armeria maritime*, *Agrostis capillaris*, *Silene vulgaris*, *Festuca ovina*, and *F. rubra*. It has been observed that in some mining areas only metal-tolerant genotypes of *Agrostis capillaris and Agrostis canina* grow, whereas *Molinia coerulea* and *Agrostis capillaries* usually cover the area around the zinc–cadmium smelter. Even vesicular–arbuscular mycorrhiza living in symbiosis with *Agrostis capillaries* also develop a resistance to the trace elements [Verkeleji, 1993].

4   CONCLUSIONS

The levels of trace elements play an important role in the environmental pollution status. The best way to fight this menace of pollution is to undertake periodic control of the levels of trace elements in our surroundings. This can easily be achieved by analyzing soils and different parts of evergreen plant species at regular intervals, which can serve as cheap biomonitors [Ernst et al., 2000]. Monitoring of trace elements in an ecosystem with the help of soils and plants gives realistic results because they are immobile, low in cost, and always available and need no servicing. They are better suited for monitoring ecosystem input and element fluxes than animals because they act as a quantitatively important sink and source [Wittig, 1993; Hagemeyer, 1993]. The ideal accumulative biomonitor is a species, in which internal concentrations accurately reflect external concentrations. Some species are accumulators or even hyperaccumulators, while others are excluders. To detect low concentrations of trace elements in the environment, hyperaccumulators are best suited. In highly polluted areas, species with a lower accumulation rate will give better differentiation. For some plant species or groups, great experience in active monitoring exists and exposure methods are standardized. The disadvantage with the plant systems is that their reaction not only depends on the quantity of the substance to be monitored but also on their age, state of health, soil type, soil moisture, nutrient status of the soil, precipitation, relative humidity, topography, and temperature.

Phytoremediation is a new technology employed for removing excessive toxic elements from the soil [Glass, 2000; Blaylock and Huang, 2000; Babaoğlu et al., 2004]. Hyperaccumulators are a special class of plants which have acquired the ability to accumulate trace elements with higher than 1% concentration in the foliar dry matter. The *Thlaspi* species is known as a hyperaccumulator, because it can extract and accumulate very high levels of toxic trace elements from the soil. Similarly, *Pteris vittata* is reported as a hyperaccumulator for arsenic [Abioye et al., 2004]. Transfer of bacterial mercury resistance genes to plants provide them the ability to grow on normally toxic Hg-containing substrates and has great potential as a method to remove hazardous bioavailable mercurials from contaminated environments. Marigold is able to take chromium in quite large quantities and could be suitable for phytoremediation of Cr-affected soil.

REFERENCES

Abioye O, Fayiga L, Ma Q, Cao X, Rathinasabapathi B. 2004. Effects of heavy metals on growth and arsenic accumulation in the arsenic hyperaccumulator *Pteris vittata* L. *Environ Pollut* **132**:289–296.

Aboal JR, Fernandez JA, Carballeria A. 2004. Oak leaves and pine needles as biomonitors of airborne trace elements pollution. *Environ Exp Bot* **51**:215–225.

Akbar KF, Hale WHG, Headley AD, Athar, M. 2006. Heavy metal contamination of roadside soils of northern England. *Soil Water Res* **1**(4):158–163.

Aksoy A, Demirezen D. 2006. *Fraxinus excelsior* as a biomonitor of heavy metal pollution. *Polish J Environ Studies* **15**(1):27–33.

Aksoy A. 1997. Kayseri-Kırşehir Karayolunda Yetişen Bitkilerde Ağır Metal Kirlenmesi. II. Ulusal Ekoloji ve Çevre Kongresi, Kırşehir.

Aksoy A, Ozturk M. 1996. *Phoenix dactylifera* L. as a biomonitor of heavy metal pollution in Turkey. *J Trace Microprobe Techniques* **14**:605–614.

Aksoy A, Ozturk M. 1997. *Nerium oleander* L. as a bomonitor of lead and other heavy metal pollution in Mediterranean environments. *The Sci Total Environ* **205**:145–150.

Aksoy A, Şahin, U. 1999. *Elaeagnus angustifolia* L. as a biomonitor of heavy metal pollution. *Tukish J Bot* **23**:83–87.

Aksoy A, Hale WHG, Dixon JM. 1999. *Capsella bursa-pastoris* (L.) Medic. as a biomonitor of heavy metal pollution. *Sci Total Environ* **226**:177–186.

Aksoy A, Celik A, Ozturk M, Tulu, M. 2000. Roadside plants as possible indicators of heavy metal pollution in Turkey. *Chemia I Inzynieria Ekol* **11**:1152–1155.

Aksoy A, Şahin U, Duman F, 2000b. *Robinia pseudo-acacia* L. as a posssible biomonitor of heavy metal pollution in Kayseri. *Turk J Bot* **24**:279–284.

Alam MGM, Snow ET, Tanaka A. 2003. Arsenic and heavy metal contamination of vegetables grown in Samta Village, Bangladesh. *Sci Total Environ* **308**:83–96.

Al-Khashman OA. 2004. Heavy metal distribution in dust, street dust and soils from the work place in Karak Industrial Estate, Jordan. *Atmos Environ* **38**:6803–6812.

Al-Shayeb SM, Al-Rajhi MA, Seaward MRD. 1995. The date palm (*Phoenix dactylifera* L.) as a biomonitor of lead and other elements in arid environments. *Sci Total Environ* **168**:1–10.

Babaoğlu M, Gezgin S, Topal A, Sade B, Dural H. 2004. *Gypsophila sphaerocephala* Fenzi ex Tchihat.: A boron hyperaccumulator plant species that may phytoremediate soils with toxic B levels. *Turk J Bot* **28**:273–278.

Baker AJM, Walker PL. 1989. Ecophysiology of metal uptake by tolerant plants. In: Shaw, AJ, editor. *Heavy Metal Tolerance in Plants: Evolutionary Aspects.* Boca Raton, FL: CRC Press, pp. 155–177.

Banerjee ADK, 2003. Heavy metal levels and solid phase speciation in street dusts of Delhi, India. *Environ Pollut* **123**:95–105.

Bargagli R. 1993. Plant leaves and lichens as biomonitors of natural or anthropogenic emissions of mercury. In: Markert B, editor. *Plants as Biomonitors/Indicators for Heavy Metals in the Terrestrial Environment.* Weinheim: VCH Press, pp. 461–481.

Baycu G, Tolunay D, Ozden H, Gunebakan S. 2006. Ecophysiological and seasonal variations in Cd, Pb, Zn, and Ni concentrations in the leaves of urban deciduous trees in Istanbul. *Environ Pollut* **143**:545–554.

Bereket G, Yucel E. 1990. Monitoring of heavy metal pollution of traffic origin in Eskisehir. Doğa-Tr. *J Chem* **14**:266–271.

Berrow ML, Ure AM. 1981 The determination of metals and metalloids in soil. *Environ Technol Lett* **2**:485–502.

Blaylock JM, Huang JW. 2000. Phytoextraction of metals. In: Raskin L, Ensley BD, editors. *Phytoremediation of Toxic Metals: Using Plants to Clean Up the Environment.* New York: John Wiley & Sons, pp. 53–70.

Bolan NS, Adriano DC, Naidu R, Mora M, Santiagio M. 2005. Phosphorus–trace element interactions in soil–plant systems. In: Sims JT, Sharpley AN, editors *Phosphorus: Agriculture and the Environment.* American Society of Agronomy, Crop Science Society of American, and Soil Science Society of America Agronomy Monograph no. 46, pp. 317–352.

Brooks RR. 1993. Geobotanical and biogeochemical methods for detecting mineralization and pollution from heavy metals in Oceania, Asia, and the Americas. In: Markert B, editor. *Plants as Biomonitors/Indicators for Heavy Metals in the Terrestrial Environment.* Weinheim: VCH Press, pp. 127–154.

Bucher AS, Schenk MK. 2000. Toxicity level for phytoavailable zinc in compost peat substrates. *Science Horticulture (Amsterdam)* **83**:339–352.

Castaldi P, Santona L, Melis P. 2005. Heavy metal immobilization by chemical amendments in a polluted soil and influence on white lupin growth. *Chemosphere* **60**:365–371.

Cataldo DA, Wildung RE. 1978. Soil and Plant Factors Influencing the Accumulation of Heavy Metals by Plants. *Environ Health Perspect* **27**:149–159.

Celik A, Kartal AA, Akdogan A, Kaska Y. 2005. Determining the heavy metal pollution in Denizli (Turkey) by using Robinio pseudo-acacia L. *Environ Int* **31**:105–112.

Celik S, Yucel E, Celik S, Ozturk M. 2006. *Populus × canadensis Moench as bioindicator of heavy metal pollution in the West Black Sea Region of Turkey.* IVth Balkan Botanical Congress, 20–26 June, Sofia, Bulgaria.

Chronopoulos J, Haidouti C, Chronopoulou-Sereli A, Massas, I. 1997. Variations in plant and soil lead and cadmium content in urban parks in Athens, Greece. *Sci Total Environ* **196**:91–98.

Cicek A, Koparal AS. 2004. Accumulation of sulphur and heavy metals in soil and tree leaves sampled from the surroundings of Tunc-bilek thermal power plant. *Chemosphere* **57**:1031–1036.

Cinar T, Elik A. 2002. Determination of heavy metals in biocollectors as indicator of environmental pollution. *Int J Environ Anal Chem* **82**:321–329.

Coskun M. 2006. Toxic Metals in the Austrian Pine Bark (*Pinus nigra* Arnold.) in Thrace Region, Turkey. *Environ Monit Assess* **121**:173–179.

Çubukçu A, Tüysüz, N. 2007. Trace element concentrations of soils, plants and waters caused by a copper smelting plant and other industries, Northeast Turkey. *Environ Geol* **52**(1): 93–108.

Das P, Samantaray S, Rout GR. 1997. Studies on cadmium toxicity in plants. A review. *Environ Pollut* **98**:29–36.

Davidson CM, Duncan AL, Littlejohn D, Ure AM, Garden LM. 1998. A critical evaluation of the three-stage BCR sequential extraction procedure to assess the potential mobility and toxicity of heavy metals in industrially contaminated land. *Anal Chim Acta* **363**:45–55.

Del Rio M, Font R, Almela C, Velez D, Montoro R, De Haro A. 2002. Heavy metals and arsenic uptake by wild vegetation in the Guadiamar river area after the toxic spill of the Aznalcollar mine. *J Biotechnol* **98**:125–137.

Demirezen D, Aksoy A. 2004. Acumulation of metal in *Typha angustifolia* (L.) and *Potomageton pectinatus* (L.) living in Sultan Marsh (Kayseri, Turkey). *Chemosphere* **56**:685–696.

Demirezen D, Aksoy A. 2006. Common hydrophytes as bioindicators of iron and manganese pollutions. *Ecol Indicators* **6**:388–393.

Demirezen D, Aksoy A, Uruc K. 2007. Effect of pollution density on growth, biomass and nickel accumulation capacity of *Lemma gibba* (Lemnaceae). *Chemosphere* **66**:553–557.

Deu M, Kreeb KH. 1993. Seasonal variations of foliar metal content in three fruit tree species. In: Markert B, editor. *Plants as Biomonitors/Indicators for Heavy Metals in the Terrestrial Environment.* Weinheim: VCH Press, pp. 577–592.

Djingova R, Kuleff I. 1993. Monitoring of heavy metal pollution by *Taraxacum officinale*. In: Markert B, editor. *Plants as Biomonitors/Indicators for Heavy Metals in the Terrestrial Environment.* Weinheim: VCH Press, pp. 435–460.

Dmuchowski W, Bytnerowicz A. 1995. Monitoring environmental pollution in Poland by chemical analysis of Scots Pine (*Pinus sylvestris* L.) needles. *Environ Pollut* **87**:84–104.

El-Hasan T, Al-Omari H, Jiries A, Al-Nasir F. 2002. Cypress tree (*Cupressus semervirens* L.) bark as an indicator for heavy metal pollution in the atmosphere of Amman City, Jordan. *Environ Int* **28**:513– 519.

Ernst WHO, Nilisse HJM, Ten Brookum WM. 2000. Combination toxicology of metal-enriched soils: physiological responses of-and Cd resistant ecotypes of *Silene vulgaris* on polymetallic soils. *Environ Exp Bot* **43**:55–71.

EEA (European Environmental Agency). 1995. Soil pollution by heavy metals. In: *Europe's Environment, the Dobris Assessment.* Luxembourg: Office des Publications, p. 676.

Fayiga, AO, Ma LQ. 2005. Using phosphate rock to immobilize metals in soil and increase arsenic uptake by hyperaccumulator *Pteris vittata. Sci Total Environ* **XX**:1–9.

Fernandez JA, Ederra A, Nunez E, Martinez-Abaigar J, Infante M, Heras P. 2002. Biomonitoring of metal deposition in northern Spain by moss analysis. *Sci Total Environ* **300**:115–27.

Friedland AJ. 1989. The movement of metals through soils and ecosystems. In: Shaw AJ, editor. *Heavy Metal Tolerance in Plants: Evolutionary Aspects.* Boca Raton, Fl; CRC Press, pp: 7–19.

Fuentes D, Disante KB, Valdecantos A, Cortina J, Vallejo VR. 2007. Response of *Pinus halepensis* Mill. seedlings to biosolids enriched with Cu, Ni and Zn in three Mediterranean forest soils. *Environ Pollut* **145**:316–323.

Garty, J. 1993. Lichens as biomonitors of heavy metal pollution. In: Markert B, editor. *Plants as Biomonitors/Indicators for Heavy Metals in the Terrestrial Environment.* Weinheim: VCH Press, pp. 191–264.

Glass DJ. 2000. Economical potential of phytoremediation. In: Raskin I, Ensley BD, editors. *Phytoremediation of Toxic Metals: Using Plants to Clean Up the Environment.* New York: John Wiley & Sons, pp. 15–31.

Hagemeyer J. 1993. Monitoring trace metal pollution with tree rings: A critical reassessment. In: Markert B. editor. *Plants as Biomonitors/Indicators for Heavy Metals in the Terrestrial Environment.* Weinheim: VCH Press, pp. 541–564.

Halıcı MG, Aksoy A, ve Demirezen D. 2003. *Protoparmeliopsis muralis* (Schreb.) M. Choisy Liken türü kullanarak Erciyes Dağı ve çevresinde ağır metal kirliliğinin ölçülmesi, I. Ulusal Erciyes Sempozyumu Kayseri.

Harju L, Saarela KE, Rajander J, Lill OJ, Lindroos A, Heselius SJ. 2002. Environmental monitoring of trace elements in bark of Scots pine by thick-target PIXE. *Nucl Instrum Methods Phys Res B* **189**:163–167.

Henden E, Turkan I, Celik U, Kıvılcım S. 1993. Ağır Metal Kirlenmesinin Bir Monitörü Olarak *Ballota acetobulosa* (L.) Bentham, I. Ulusal Ekoloji ve Çevre Kongresi Ankara.

Iskandar K, Kirkham MB. 2001. *Trace Elements in Soil: Bioavailability, Flux, and Transfer.* Lewis Publications, January, p. 304.

Jonnalagadda SB, Nenzou G. 1997. Studies on arsenic rich mine tips: II The heavy element uptake by vegetation. *J Environ Sci Health* **A32**:455–464.

Kabala C, Singh, BR. 2001. Fractionation and mobility of copper, lead and zinc in soil profiles in the vicinity of a copper smelter. *J Environ Quality* **30**:485–492.

Kartal S, Elci L, Dogan M. 1992. Investigation of lead, nickel, cadmium and zinc pollution of traffic in Kayseri. *Fresenius Environ Bull* **1**:28–35.

Kartal S, Elci L, Kilicel F. 1993. Investigation of soil pollution levels of zinc, copper, lead, nickel, cadmium and manganese around Cinkur Plant in Kayseri. *Fresenius Environ Bull* **2**:614–619.

Kelly J, Thornton I, Simpson PR. 1996. Urban geochemistry: A study of influence of anthropogenic activity on heavy metal content of soils in traditionally industrial and non-industrial areas of Britain. *Appl Geochem* **11**:363–370.

Kovacs M, Turcsanyi G, Penksza K, Kaszab L, Szoke P. 1993. Heavy metal accumulation by ruderal and cultivated plants in a heavily polluted district of Budapest. In: Markert B, editor. *Plants as Biomonitors/Indicators for Heavy Metals in the Terrestrial Environment.* Weinheim: VCH Press, pp. 495–506.

Kutbay HG, Kilinç M. 1991. Heavy metal pollution in plants growing along motor roads. In: Ozturk M, Erdem U, Gork G, editors. *Urban Ecology.* Izmir, Turkey: Ege University Press, pp. 62–66.

Lau OW, Luk SF. 2001. Leaves of *Bauhinia blakeana* as indicators of atmospheric pollution in Hong Kong. *Atmos Environ* **35**:3113–3120.

Lee P-K, Yu Y-H, Yun S-T, Mayer B. 2005. Metal contamination and solid phase partitioning of metals in urban roadside sediments. *Chemosphere* **60**:672–689.

Li X, Poon C-S, Liu PS. 2001. Heavy metal contamination of urban soils and street dusts in Hong Kong. *Appl Geochem* **16**:1361–1368.

Madejon P, Moranon T, Murillo JM, Robinson B. 2004. White poplar (*Populus alba*) as a biomonitor of trace elements in contaminated riparian forests. *Environ Pollut* **132**:145–155.

Madejon P, Moranon T, Murillo JM. 2006. Biomonitoring of trace elements in the leaves and fruits of wild olive and holm oak trees. *Sci Total Envirom* **355**:187–203.

Markert B. 1993. *Plants as Biomonitors/Indicators for Heavy Metals in the Terrestrial Environment.* Weinheim: VCH Press.

Markert B. 1994. Plants of biomonitors—potential advantages and problems. In: Adriano DC, Chen ZS, Yang SS, editors. *Biochemistry of Trace Elements.* Norwood, NY: Science of Technology Letters, pp. 601–613.

McEnroe NA, Helmisaari HS. 2001. Decomposition of Coniferous forest litter along a heavy metal pollution gradient, South-West Finland. *Environ Pollut* **113**:11–18.

Mico C, Recatala L, Peris M, Sanchez J. 2006. Assessing heavy metal sources in agricultural soils of a European Mediterranean area by multivariate analysis. *Chemosphere* **65**:863–872.

Monni S, Uhlig C, Hansen E, Magel E. 2001. Ecophysiological responses of *Empetrum nigrum* to heavy metal pollution. *Environ Pollut* **112**:121–129.

Morselli L, Brusori B, Passarini F, Bernardi E, Francaviglia R, Gatelata L. 2004. Heavy metal monitoring at a Mediterranean natural ecosystem of Central Italy trends in different environmental matrixes. *Environ Int* **30**(2):173–181.

Oncel MS, Zedef V, Mert S. 2004. Lead contamination of roadside soils and plants in the highways between Istanbul and Sakarya, NW Turkey. *Fresenius Environ Bull* **13**:1525–1529.

Onder S, Şükrü, D. 2006. Air borne heavy metal pollution of *Cedrus libani* (A. Rich.) in the city centre of Konya (Turkey). *Atmos Environ* **40**:1122–1133.

Otvfs E, Pazmandi T, Tuba Z. 2003. First national survey of atmospheric heavy metal deposition in Hungary by the analysis of mosses. *Sci Total Environ* **309**:151–160.

Ouzounidou G. 1994. Copper-induced changes on growth metal content and photosynthetic function of *Alyssum montanum* L. plants. *Environ Exp Bot* **34**:165–72.

Ozturk MA, Turkan I. 1992. Ağır Metaller Canlılar İçin Bir Yük mü?, In Kiziroğlu, İ editor. II. *Uluslararası Ekoloji ve Çevre Sorunları Sempozyumu.* Ankara: Desen Ofset A.Ş, pp. 134–140.

Ozturk M, Turkan I. 1993. Heavy metal accumulation by plants growing alongside the motorroads: A.case study from Turkey. In: Markert B, editor. *Plants as Biomonitors: Indicators for Heavy Metals in the Terrestrial Environment.* Germany: VCH Publishers, pp. 515–522.

Palmieri RM, La Pera L, Di Bela G, Dugo G, 2005. Simultaneous determination of Cd(II), Cu(II), Pb(II) and Zn(II) by derivative stripping chronopotentiometry in *Pittosporum tobira* leaves: A measurement of local atmospheric pollution in Messina (Sicily, Italy). *Chemosphere* **59**:1161–1168.

Pendias AK, Pendias H. 2000. *Trace Elements in Soils and Plants.* Boca Raton: CRC Press, 413 pp.

Pinamonti F, Stringari G, Gasperi F, Zorzi G. 1997. The use of compost: Its effects on heavy metal levels in soil and plants. *Resour, Conserv Recyling* **21**:129–143.

Popescu CS, Gheboıanu A, Badica T, Gugiu MM, Constantinescu O, Vargolici M, Bancuta I. 2005. Air quality study by the Pıxe method and mosses as bıoındıcators. IV. *Romanian Reports Phys* **58**(4):409–414.

Porter JR. 1986. Evaluation of washing procedures for pollution of *Ailanthus altissima* leaves. *Environ Pollut* **B12**:195–202.

Pueyo M, Sastre J, Hernandez E, Vidal M, Lopez-Sanchez JF, Rauret G. 2003. Prediction of trace element mobility in contaminated soils by sequential extraction. *J Environ Qual* **32**:2054–2066.

Pyatt FB. 1999. Comparison of foliar and stem bioaccumulation of heavy metals by Corsican pines in the Mount Olympus Area of Cyprus. *Ecotoxicol Environ Saf* **42**:57–61.

Pyatt FB. 2001. Copper and lead bioaccumulation by *Acacia retinoides* and *Eucalyptus torquatain* sites contaminated as a consequence of extensive ancient mining activities in Cyprus. *Ecotoxicol Environ Saf* **50**:60–464.

Rao MV, Dubey PS. 1992. Occurrence of heavy metals in air and their accumulation by tropical plants growing around an industrial area. *Sci Total Environ* **126**:1–16.

Reeves RD, Baker AJM. 2000. Metal accumulating plants. In: Raskin I, Ensley BD, editors. *Phytoremediation of Toxic Metals: Using Plants to Clean Up the Environment.* New York: John Wiley & Sons, pp. 193–229.

Remon E, Bouchardon J-L, Cornier B, Guy B, Leclerc J-C, Faure O. 2005. Soil characteristics, heavy metal availability and vegetation recovery at a former metallurgical landfill: Implications in risk assessment and site restoration. *Environ Pollut* **137**:316–323.

Ross MS. 1994. Sources and forms of potentially toxic metals in soil-plant systems. In: Ross MS, editor. *Toxic Metals in soil–Plant Systems*. Chichester: John Wiley & sons, pp. 3–25.

Rossini OS, Mingorance MD. 2006. Assessment of airborne heavy metal pollution by above-ground plant parts. *Chemosphere* **65**:177–182.

Roychowdhury T, Tokunage H, Ando M. 2003. Survey of arsenic and other heavy metals in food composites and drinking water and estimation of dietary intake by the villages from an arsenic affected area of the west Bengal, India. *The Sci Total Environ* **308**:115–135.

Salt DE, Prince RC, Pickering IJ, Raskin I. 1995. Mechanisms of cadmium mobility and accumulation in Indian mustard. *Plant Physiol* **109**:1427–1433.

Samecka-Cymerman A, Kempers J. 1996. Bioaccumulation of heavy metals by aquatic macro-phytes around Wroclaw, Poland. *Ecotoxicol Environ Saf* **35**:242–247.

Samecka-Cymerman A, Kempers AJ. 1999. Bioindication of heavy metals in the town Wroclaw (Poland) with evergreen plants. *Atmos Environ* **33**:419–430.

Samecka-Cymerman A, Kosior G, Kempers AJ. 2006. Comparison of the moss *Pleurozium schreberi* with needles and bark of *Pinus sylvestris* as biomonitors of pollution by industry in Stalowa Wola (southeast Poland). *Ecotoxicol Environ Saf* **65**:108–117.

Sawidis T, Chettri MK, Zachariadis GA, Stratis JA. 1995a. Heavy metals in aquatic plants and sediments from water systems in Macedonia, Greece. *Ecotoxicol Environ Saf* **32**:73–80.

Sawidis T, Marnasidis A, Zachariadis G, Stratis J. 1995b. A study of air pollution with heavy-metals in Thessaloniki city (Greece) using trees as biological indicators. *Arch Environ Contam Toxicol* **28**:118–124.

Sawidis T, Chettri MK, Papaioannou A, Zachariadis G, Stratis J. 2001. A study of metal distribution from lignite fuels using trees as biological monitors. *Ecotoxicol Environ Saf* **48**:27–35.

Schafer J, Hannker D, Eckhardt JD. 1998. Uptake of traffic-related heavy metals and platinum group elements (PGE) by plants. *Sci Total Environ* **215**:59– 67.

Seaward MRD, Mashhour MA. 1991. Oleander (*Nerium oleander* L.) as a monitor of heavy metal pollution. In: Ozturk AM, Erdem U, Gork G, editors. *Urban Ecology*. Izmir, (Turkey): Ege University Press, pp. 48–61.

Sebastiani L, Scebba F, Tognetti R. 2004. Heavy metal accumulation and growth responses in poplar clones Eridano (*Populus deltoides* × *maximowiczii*) and I-214 (*P.* × *euramericana*) exposed to industrial waste. *Environ Exp Bot* **52**:79–88.

Shanker AK, Cervantes C, Loza-Tavera H, Avudainayagam S. 2005. Chromium toxicity in plants. *Environ Int* **31**:739–753.

Singh K. 2002. Biogeochemistry of trace elements: A report of the International Conference held at the Technical University, Vienna, Austria in 1999. *Crop Sci* **42**:297–298.

Steinborn M, Bren J. 1999. Heavy metals In soil and vegetation at Shallee Mine, Silvermines, Co. Tipperary. *Biology and Environment Proceedings of the Royal Irish Academy*, 99b, No. 1: 37–42.

Streit B, Stumm W. 1993. Chemical properties of metals and the process of bioaccumulation in terrestrial plants. In: Markert B, editors. *Plants as Biomonitors/Indicators for Heavy Metals in the Terrestrial Environment*. Weinheim: VCH Press, pp. 29–62.

Tausz M, Trummer W, Goessler W, Wonisch A, Grill D, Naumann S, Jimenez MS, Morales D. 2005. Accumulating pollutants in conifer needles on an Atlantic island—A case study with *Pinus canariensis* on Tenerife, Canary Islands. *Environ Pollut* **136**:397–407.

Toker MC, Temizer A, Yalçın I. 1990. Ankarada bazı Yol Ortası Refüjlerde Yetiştirlen Çamlarda Ağır metal (Pb. Ni, Cd) Birikimi. *C.Ü. Fen Bilimleri Derg* **13**:25–40.

Tomasevic M, Rajsic S, Dordevic D, Tasic M, Krstic J, Novakovic V. 2004. Heavy metals accumulation in tree leaves from urban areas. *Environ Chem Lett* **2**:151–154.

Türkan I, Henden E. 1990. Monitoring of heavy metals using bark samples of pine (*Pinus sp.*). In: Ozturk M, Erdem U, Gork G, editors. *Urban Ecology.* İzmir, Turkey: p. 72.

Turkan I, Ozturk M. 1989. Lead contamination in the plants growing near motor roads. *J Fac Sci Ege Univ* **2**:25–33.

Verkleij JAC. 1993. The effects of heavy metal stress on higher plants and their use as biomonitors. In: Markert B. editor. *Plants as Biomonitors/Indicators for Heavy Metals in the Terrestrial Environment.* Weinheim: VCH Press, pp. 413–424.

Viard B, Pihan F, Promeyrat S, Pihan, JC. 2004. Integrated assessment of heavy metal (Pb, Zn, Cd) highway pollution: Bioaccumulation in soil, Gramineae and land snails. *Chemosphere* **55**:1349–1359.

Wagner G. 1993. Large-scale screening of heavy metal burdens in higher plants. In: Markert B. editor. *Plants as Biomonitors/Indicators for Heavy Metals in the Terrestrial Environment* Weinheim: VCH Press, pp. 425–434.

Walkenhorst AI, Hagemeyer J, Breckle SW. 1993. Passive monitoring of air borne pollutants, particularly trace metals, with tree bark. In: Markert B, editor. *Plants as Biomonitors/Indicators for Heavy Metals in the Terrestrial Environment.* Weinheim: VCH Press, pp. 523–540.

Wittig R. 1993. General aspects of biomonitoring heavy metals by plants. In: Markert B, editor. *Plants as Biomonitors/Indicators for Heavy Metals in the Terrestrial Environment.* Weinheim: VCH Press, pp. 3–28.

Wittig R, Neite H. 1989. Distribution of lead in the soils of *Fagus sylvatica* forests in Europe. In: Ozturk MA, editor. *Plants and Pollutants in Developed and Developing Countries.* Izmir, Turkey: Ege University Press, pp. 199–206.

Wittig R, Baumen T. 1992. Schwermetallrasen (*Violetum calaminariae rhenanicum* Ernst 64) im engeren Stadtgebiet von Stolberg/Rheinland. *Acta Biol Benrodis* **4**:67–81.

Woolhouse HW. 1983. Toxicity and tolerance in the responses of plants to metals. In: Lange O, Nobel PS, Osmond CB, Ziegler H, editor. *Encyclopedia of Plant Physiology*, Vol. 12: *Physiological Plant Ecology III* Berlin: Springer-Verlag, pp. 245–289.

Yagdı K, Kacar O, Azkan N. 2000. Heavy metal contamination in soils and its effects in agriculture. *J Agri Faculty OMU* **15**(2):109–115.

Yılmaz R, Sakcali S, Yarci C, Aksoy A, Ozturk M. 2007. *Aesculus hippocastanum* L. as a biomonitor of heavy metal pollution. *Pakistan J Bot*, (in press).

Yılmaz S, Zengin, M. 2004. Monitoring environmental pollution in Erzurum by chemical analysis of Scots pine (*Pinus sylvestris* L.) needles. *Environ Int* **29**:1041–1047.

Yucel E, Dogan F, Ozturk, M. 1995. Heavy metal status of Porsuk stream in relation to public health. *Ekoloji* **17**:29–32.

Yucel E. 1996. Investigation on Pb, Cd ve Zn pollution from traffic using Asian Populus (*Populus usbekistanica* Kom. subsp. *usbekistanica* cv. "Afghanica" in Kutahya City (Turkey). *Turk J Bot* **20**:2: 113–116.

29 Bioindication and Biomonitoring as Innovative Biotechniques for Controlling Trace Metal Influence to the Environment

BERND MARKERT

Department of Environmental High Technology, International Graduate School (IHI) Zittau, D-02763 Zittau, Germany[1]

1 INTRODUCTION

Environmental chemicals affect biological systems at different levels of organization, from individual enzyme systems through cells, organs, single organisms, and populations to entire ecosystems [Baker and Brooks, 1989; Bargagli, 1998; Markert et al., 2003a,b, 2004; Prasad, 2004; Fränzle et al., 2007]. As a rule, the latter do not just react to single substances or parameters [Clemens et al., 2002]; they show species-specific and situation specific sensitivity to the whole constellation of factors and parameters acting on them at their location [Baker and Brooks, 1989; Broadley et al., 2007; Callahan et al., 2006; Greger, 2004; Reimann et al., 2003; Zechmeister et al., 2006]. Information on the sensitivity and specificity of such reactions provide a basis for planning the use and evaluating the result of effect-related biological measuring techniques [Schröder et al., 1996, 2007].

An objective of prophylactic environmental protection must be to obtain and evaluate reliable information on the past, present, and future situation of the environment. Besides the classic global observation systems such as satellites and instrumental measurement techniques such as trace gas and on-line water monitoring, increasing use should be made of bioindicative systems that provide integrated information permitting prophylactic care of the environment and human health. In the last

---

[1]Present address: Fliederweg 17, D-49733 Haren-Erika, Germany.

Trace Elements as Contaminants and Nutrients: Consequences in Ecosystems and Human Health, Edited by M. N. V. Prasad

25 years, bioindicators have shown themselves to be particularly interesting and intelligent measuring systems. As long ago as 1980, Müller considered the "bioindicative source of information" one of the pillars of modern environmental monitoring, since "bioindication is the breakdown of the information content of biosystems, making it possible to evaluate whole areas" [Fränzle et al., 2007].

The application of organism-based surveillance methods (biomonitoring/bioindication) can extend our knowledge of mechanisms that are involved in (matter-based) interactions between the organism and its environment [Wünschmann, 2007]. Embracing and combining topics and methods from environmental chemistry, toxicology, and ecology, ecotoxicology succeeded in characterizing and quantifying effects of chemicals during the last decades by means of novel methods. The mere multitude and diversity of chemicals that make their way into the environment (by 2003, some 5 million compounds were known [Fent, 2003], with an additional 500–1000 newly introduced compounds per year which then are broadly used) brings about an immense task to ecotoxicology, an inter- and transdisciplinary task that nowadays surpasses the challenges to both human and veterinary toxicology [Fränzle, 1999]. Being a part of environmental sciences, ecotoxicology includes many specific methodological innovations which only in their combination permit to meet societal demands for protection of environment [Oehlmann and Markert, 1999]. Here, the main tasks to be done before novel chemical compounds can be accepted even for laboratory amounts and purposes (biotests; e.g., Fomin et al. [2003]) include prospective risk evaluations, but also include measurements of effects damaging organisms and entire populations in open, "free" environments (the latter achieved by means of bioindication; e.g., Markert et al. [2003a]). Using organisms of different trophic levels (plants, animals, microorganisms) to determine detrimental effects of chemicals does not represent the only peculiarity typical of both lines of work, but these also differ with respect to statements on the situation of terrestrial and aquatic environmental compartments (Fig. 1).

To give an insight into this highly interesting research area, a short introduction in definitions on related terms is given, followed by a comparison of measurements

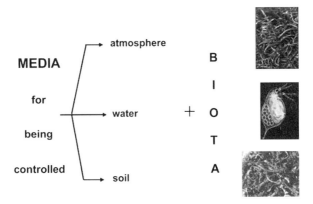

Figure 1. Environmental media and their bioindication using various living organisms (e.g., mosses, daphnia, earthworms).

performed by spectrometers and bioindicators, some examples of bioindication results in the field, and some critical remarks about what bioaccumulation data really tell us.

2 DEFINITIONS

In the following we will give some definitions summarized in Markert et al. [1997, 2003b]:

A **bioindicator** is an organism (or part of an organism or a community of organisms) that contains information on the **quality** of the environment (or a part of the environment). A **biomonitor**, on the other hand, is an organism (or part of an organism or a community of organisms) that contains information on the **quantitative** aspects of the quality of the environment. The clear differentiation between bioindication and biomonitoring using the qualitative/quantitative approach makes it comparable to instrumental measuring systems (see Section 3). Such effects (information bits) of bioindicators (biomonitors) may include changes in their morphological, histological, or cellular structure, their metabolic–biochemical processes (including accumulation rates), their behavior, or their population structure. **Accumulation indicators/monitors** are organisms that accumulate one or more elements and/or compounds from their environment. **Effect** or **impact indicators/monitors** are organisms that demonstrate specific or unspecific effects in response to exposure to a certain element or compound or a number of substances. According to the paths by which organisms take up elements or compounds, various mechanisms contribute to overall accumulation (**bioaccumulation**), depending on the species-related interactions between the indicators/monitors and their biotic and abiotic environment. **Biomagnification** is the term used for absorption of the substances from nutrients via the epithelia of the intestines. It is therefore limited to heterotrophic organisms and is the most significant contamination pathway for many land animals except in the case of metals that form highly volatile compounds (e.g., Hg, As) and are taken up through the respiratory organs (e.g., trachea, lungs). **Bioconcentration** means the direct uptake of the substances concerned from the surrounding media—for example, the physical environment, through tissues or organs (including the respiratory organs). Besides plants, which can only take up substances in this way (mainly through roots or leaves), bioconcentration plays a major role in aquatic animals. The same may also apply to soil invertebrates with a low degree of solarisation when they come into contact with the water in the soil.

Active bioindication (biomonitoring) means when bioindicators (biomonitors) bred in laboratories are exposed in a standardised form in the field for a defined period of time. At the end of this exposure time the reactions provoked are recorded or the xenobiotics taken up by the organism are analyzed. In the case of **passive** bioindication (biomonitoring), organisms already occurring naturally in the ecosystem are examined for their actions. This classification of organisms (or communities of these) has been devised according to their "origin".

Various newer methods (biomarkers, biosensors, biotests in general) have been introduced into the application field of bioindication, besides the classical floristic, faunal, and biocoenotic investigations that primarily record unspecific reactions to pollutant exposure at higher organismical levels of bioindication.

Biomarkers are measurable biological parameters at the suborganismic (genetic, enzymatic, physiological, morphological) level in which structural or functional changes indicate environmental influences in general and also indicate the action of particular enzymes in qualitative and sometimes also in quantitative terms. Examples are: enzyme or substrate induction of cytochrome P-450 and other Phase I enzymes by various halogenated hydrocarbons; the incidence of forms of industrial melanism as markers for air pollution; tanning of the human skin caused by UV radiation; and changes in the morphological, histological, or ultrastructure of organisms or monitor organs (e.g., liver, thymus, testicles) following exposure to pollutants. A **biosensor** is a measuring device that produces a signal in proportion to the concentration to a defined group of substances through a suitable combination of a selective biological system, (e.g., enzyme, antibody, membrane, organelle, cell, or tissue) and a physical transmission device (e.g., potentiometric or amperometric electrode, optical or optoelectronic receiver).

Biomarkers and biosensors can be used as a **biotest** (bioassay) that describes a routine toxicological–pharmacological procedure for testing the effects of agents (environmental chemicals, pharmaceuticals) on organisms, usually in the laboratory but occasionally in the field under standardized conditions (with respect to biotic and abiotic factors). In the broader sense, the definition covers cell and tissue cultures when used for testing purposes, enzyme tests or tests using microorganisms, and plants and animals in the form of single-species or multispecies procedures in model ecological systems (e.g., microcosms and mesocosms). In the narrower sense, the term only covers single-species and model system tests, while the other procedures may be called suborganismic tests. Bioassays use certain biomarkers or—less often—specific biosensors and can be used in bioindication or biomonitoring.

The term **tolerance** can be described as desired resistance of an organism or community by unfavorable abiotic (climate, radiation, pollution) or biotic factors (parasites, pathogens), where adaptive physiological changes (e.g., enzyme induction, immune response) can be observed [Oehlmann and Markert, 1997]. Unlike tolerance, **resistance** is a genetically derived ability to withstand stress [Oehlmann and Markert, 1997]. This means that all tolerant organisms are resistant, but not all resistant organisms are tolerant. **Sensitivity** of an organism or community means its suseptibility to biotic or abiotic changes. Sensitivity is low if the tolerance or resistance to an environmental stressor is high, and sensitivity is high if the tolerance or resistance is low.

3   COMPARISION OF INSTRUMENTAL MEASUREMENTS AND THE USE OF BIOINDICATORS WITH RESPECT TO HARMONISATION AND QUALITY CONTROL

In Fig. 2 a rough comparison is made in between instrumental methods used in the lab and biotechniques by application of bioindicators or biomonitors. Common problems are related to a representative sampling procedure that can introduce an error up to

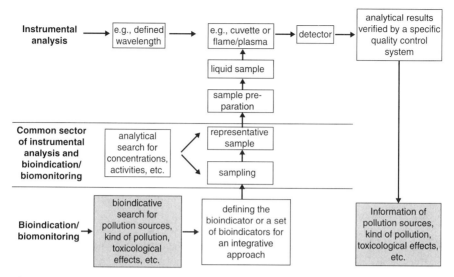

Figure 2. Comparison of measurements performed by spectrometers and bioindicators/biomonitors. In practice, instrumental measurements are often an integral part of bioindication and biomonitoring methods (from Markert et al. [2003a,b]). A full instrumental flow chart for instrumental chemical analysis of environmental samples can be found in Markert [1996].

1000% if general rules of representative collection of the measuring samples will not be taken under strong consideration [Markert, 1996]. It must be clearly stated that instrumental measuring techniques are often an integrated part of bioindication, especially when the parameter under investigation (e.g., the concentration of a trace metal) is considered. In both (instrumental and biotechniques) the sampling procedure is the essential part of the overall procedure. The sampling procedure has to be evaluated in between all (international) participants—at least of the actual investigation—to obtain comparability.

In the same way as the sampling procedure for instrumental measurements and biotechniques have to be under control, terms of precision (reproducibility) and accuracy (the "true" value) have to be checked with string standards during the overall monitoring procedure. We will not go in detail in this chapter, but methods to observe precision and accuracy can be obtained in all textbooks of analytical sciences (see, e.g., Markert et al. [2003a] and Markert [1996]). Basically, two methods are now used to check the accuracy of analytical results:

1. The use of certified standard reference materials [commercially available samples with a certified content of the compound (e.g., a trace metal) to be measured and a matrix similar to the original samples to be measured in the laboratory].
2. The use of independent analytical procedures.

4   EXAMPLES FOR BIOMONITORING

In this overview chapter, only a few examples of bioindication methods are presented (mainly the use of mosses for trace metal deposition from the atmosphere). Considerable information and examples of monitoring chemical elements by plants and animals are given in Markert et al. [2003a].

4.1   Mosses for Atmospheric Pollution Measurements

In addition to instrumental measuring techniques, the use of mosses have long been an accepted method of monitoring human-induced atmospheric inputs of heavy metals and other elements into the environment [Rühling and Tyler, 1968; Smodis, 2003; Herpin et al., 2004].

Mosses represent one of the simplest forms of terrestrial plants. This is directly connected with the peculiarities of their water metabolism. The dominant phase of the bryophyte life cycle is represented by gametophyte generation. Consequently, evolution of structural complexity has been partly limited by the dependence of sexual reproduction on free water for the dispersal of the motile antherozoids.

For some 10 years now, epiphytic or epigeic plants—mostly mosses and lichens, but also Bromeliaceae (Tillandsia) in Latin America—are being employed as bioindicators, their use being accepted not only in the scientific community but also among political or economic authorities in both Europe and Southern America [Bargagli, 1998; Freitas et al., 2000; Pignata et al., 2002; Schröder et al., 2003; Zechmeister et al., 2003; Markert et al., 2004]. This state of matters differs from those in other countries and continents (China, Australia, North America) where plant-based biomonitoring was developed for scientific purposes or on regional rather than national scales (only). Thus no comprehensive description may be given, but one has to consider unlike developments of these techniques [Zechmeister, 1995; Djingova and Kuleff, 2000; Garty, 2000a,b; Herpin et al., 2001; Wolterbeek, 2002; Djingova et al., 2003; Fränzle, 2003].

This high diversity in developments of bioindication and biomonitoring was not produced arbitrarily but owes to fundamental (and thus acceptable!) reasons [Markert et al., 2007]:

- Historical ones, with corresponding methods spreading from Scandinavia via Europe to the rest of the world
- Geographical, climatic, and geogenic distinctions that influence each living organism and population (hard skills)
- Different mental and psychological attitudes depending on intercultural attitudes among scientists of different origins who encounter each other, for example, in international meetings for harmonization and procedures (soft skills)

During the period 1997–2002, IAEA, by establishing the IAEA CRP project "On Validation and Application of Plants as Biomonitors of Trace Element

Atmospheric Pollution," analyzed by nuclear and related techniques, gave an example of what can be done [Smodis, 2003]. This project united some larger number of well-experienced bioindication experts from various countries with beginners and users of this technique all over the world [Markert et al., 2007]. All the participants had a large intellectual bargain in terms of

- Harmonization among workgroups spread all over the globe
- Assuring and controlling data quality
- Accepting and tolerating cultural features typical of the investigated region

During recent years, meetings rendered application of these biotechniques sustainable—for example, producing a Latin American bioindication program of its own (ARCAL) also funded by IAEA. It is to develop bioindication techniques for uses on the entire continent.

The heavy metals in moss survey provides data on accumulation of at least As, Cd, Cu, Fe, Hg, Ni, Pb, V, and Zn in naturally growing mosses throughout Europe [Schröder and Pesch, 2004]. This method of determining regional deposition pattern in relation to time is based on the typical physiological and morphological characteristics of the mosses such as the uptake of elements from wet and dry deposition via the plant surface and direct incorporation of these in cellular structures by passive exchange processes. Additionally, moss monitoring had financial and practical advantages, so that it can be theoretically applied worldwide. This has led to a systematic extension of the use of mosses in environmental monitoring [Herpin et al., 2001, 2004; Schröder and Pesch, 2004].

In Germany the first large-scale survey using mosses to determine heavy-metal inputs was carried out in 1990/1991 and repeated in 1995/1996 [Herpin et al., 2004]. Samples of *Pleurozium schreberi, Scleropodium purum, Hypnum cupressiforme*, and *Hylocomium splendens* were taken at a total of 1026 sites in Germany. In the 1995/1996 monitoring campaign, 95% of the original sites of the 1990/1991 study were resampled. The results from 1995/96 display local elevated values and many cases affected by known sources of heavy-metal emissions. The industrialized and urban regions of Germany show up clearly in the 1995/96 moss-monitoring results: the Ruhr area, parts of Saarland and Baden Württemberg, as well as areas in eastern Germany. Relatively low values for many elements were found in large areas of Lower Saxony and Bavaria [Siewers and Herpin, 1998]. Comparing the results of the 1990/91 and 1995/96 moss-monitoring programmes, most elements (except cadmium, copper, and zinc) show a decrease in concentration over the relevant period. Especially in the former GDR, chromium, iron, titanium, and vanadium decrease significantly. This is a reflection of the closure of and/or technological improvements in large power plants and due to the fact that lignite has given way to other fuels. Vanadium and nickel, typical constituents of crude oil, also show a decrease in the western part and thus document the type of fuel consumed [Siewers and Herpin, 1998]. The UNECE moss survey gives proof that an elevated atmospheric input in Eastern Europe still occured in 2000/2001 [Buse et al., 2003].

Levels of Cd, Fe, Zn, and Pb in Southern Poland were four to six times those given as basic or background concentrations. Among the 27 states participating in this study, Bulgaria and Poland still had considerably higher environmental burdens than the Western European states [Buse et al., 2003]. A more detailed comparison of European countries is given by Schröder and Pesch [2004].

The results given can be converted into quantitative deposition rates using the formula

$$D = c \cdot \frac{A}{E_x}$$

where D is the estimated deposition, c is the measured concentration in moss, A is the biomass increase/year, and E is the efficiency factor of uptake by the element.

Figure 3 gives deposition amounts of different elements for the years 1995 and 1996 as compared to data from "classical" deposition samplers [Zechmeister, 1997]. For example, for Cd there is a decrease of deposition burdens and a relatively good agreement between data from deposition samplers and from biomonitoring covering the same period of time.

4.2   Is There a Relation Between Moss Data and Human Health?

The field of environmental medicine (in relation to human toxicology) is a branch of medicine that is concerned with identifying, investigating, diagnosing, and preventing impairment of health and well-being and with identifying, investigating, and minimizing risks caused by definable spheres of interaction between man and the environment [Markert et al., 2007]. Can we estimate, calculate, or extrapolate the health effects of pollutants from results of environmental biomonitoring as, for example, given by Wappelhorst et al. [2000] in Fig. 4? The difficulties arise

Element	Mechanical sampler average values (upper Elbe valley)	Biomonitor *Pleurozium schreberi* ERN	
	1994–1995	1995	1996
Cd	180	130	80
Cu	9,500	4,200	3,400
Pb	8,800	2,100	1,700
Tl	37	12	11
Zn	47,000	15,000	10,000

$[\mu g \cdot m^{-2} \cdot a^{-1}]$

Figure 3. Estimates of spatial deposition for 1995–1996, calculated from element contents of *Pleurozium schreberi* compared to data obtained by a mechanical sampler in 1994–1995 (data from Wappelhorst [1999]). ERN, Euroregion Neisse.

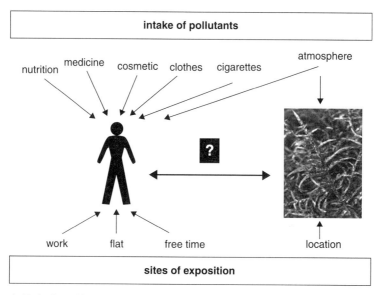

Figure 4. Paths by which pollutants are taken in by human beings and moss. Unlike mosses, human beings are exposed to pollutants in numerous places and ingest substances by several routes (after Wappelhorst et al. [2000] and Heim et al. [2002]).

mainly by the different ways of human input of substances (food, cosmetic, tobacco smoke, etc.) and the different places where these substances will be taken up day by day [Wappelhorst et al., 2000].

Additionally, Wappelhorst et al. [2000] tried to compare atmospheric trace element concentrations by moss monitoring with the incidence of various types of diseases. The authors correlated figures for various diseases with appropriate

Pleurozium schreberi

ICD 401-405: Hypertony and blood high pressure

- by higher Tl concentrations?

age	55–60	60–65	65–70	70–75	75–80	80–85	>85
Be	−0,18	−0,12	−0,03	−0,12	−0,05	−0,10	−0,05
Bi	0,20	**0,59**	**0,76**	0,69	**0,81**	**0,82**	**0,86**
Cs	0,42	**0,76**	**0,85**	**0,84**	**0,90**	**0,94**	**0,92**
Mn	−0,09	−0,36	−0,57	−0,48	−0,60	−0,63	−0,71
Na	0,36	0,06	−0,25	−0,12	−0,31	−0,36	−0,49
Tl	0,44	**0,81**	**0,96**	**0,92**	**0,98**	**0,99**	**0,99**

Figure 5. Coefficients of correlation between essential and secondary hypertension (ICD 401-405) and the element concentrations in the moss *Pleurozium schreberi* depending of the age [from Wappelhorst et al., 2000]. Bold: significant correlations.

element concentrations in the mosses. They found that there are indications for an existing relation between, for example, the thallium content of mosses and the occurrence of cardiovascular disease (Fig. 5).

The significant correlations found between the mosses and the incidence of a disease (Fig. 5) can only provide indications as to the possible causes of the disease. Causality is not taken into account when the correlation coefficients are calculated. This means that correlations can never prove that a connection exists [Wappelhorst et al., 2000].

5   WHAT DO BIOACCUMULATION DATA REALLY TELL US?

As telling as environmental data derived from bioindication may be, Wünschmann [2007] still states that results from single-species tests will give only limited information on effects of chemical substances in higher biological integration levels (populations, biocoenoses, or ecosystems). Accordingly, problems even arise when just extrapolating data obtained in one species of plant or animal to another one (even if it belongs to the same genus); the same holds for transfer of (laboratory) test results to the freeland situation that is distinguished by more complex structures and corresponding timescales [Fränzle, 1999].

Thus ecotoxicology, focused on identifying and evaluating effects of hazardous compounds on ecosystems, can only partly meet the precautionary ends of toxicology (concerned with human health). This necessitates working on human-based samples in the framework of a broad-viewing investigation focused on such effects which possibly affect humans [Markert et al., 2007]. There is a chance to meet the comprehensive precautionary expectations of toxicology only if investigations are combined into a biointegrative approach in a systematical manner. Thus both temporal trends of environmental burdening and newly developing centers of pollution can be identified. For this purpose, Markert et al. [2003b] designed the multimarkered-bioindication concept (MMBC; Fig. 6); its approach depends on some combination of ecotoxicological data sets with those from human medicine (especially toxicology). This method, which is based on "tool boxes" (cf. the explanations for Fig. 6), thus implies an approach integrating different instrumental and bioindicative methods.

As presented already by Markert et al. [2007], Fig. 6 represents one proposal of a complete dynamics environmental monitoring system supported by bioindication to integrate human toxicological and ecotoxicological approaches. It can recombine its measurement parameters according to the particular system to be monitored or the scientific frame of reference. The two main subjects of investigation—man and the environment—and the disciplines human toxicology and ecotoxicology derived from them are associated with various "toolboxes" and sets of tests ("tools," e.g., bioassays) for integrated environmental monitoring [Markert et al., 2003b]. The system shown in Fig. 5 consists of six toolboxes. The first two are derived mainly from environmental research: DAT (for data) and TRE (for trend). DAT contains, as a set, all the data available from the (eco-)system under investigation—that is, including data acquired by purely instrumental means (for example, from meteorological devices). DAT also contains maximum permissible concentrations of substances

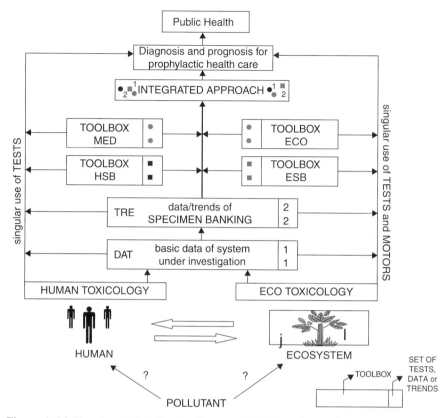

Figure 6. Multimarkered Bioindication Concept (MMBC), an integrative approach on human health care which draws upon a multidisciplinary input organized along several integrated and functional "windows." MMBC is one way that bioindicative toolboxes may be organized in a hierarchical framework for purposes of human and ecotoxicology. In toolboxes MED and ECO, respectively, there are single sets of tests for functional combination in order to obtain an integrated approach toward some specified scientific problem. Toolboxes HSB (human specimen banking) and ESB (environmental specimen banking) derive from years of research using sample banks for both human and environmental toxicology; thus they can supplement MED and ECO by important information on toxicological and ecotoxicological features of environmental chemicals. This integrated approach is not only to link all the results but to corroborate them by data already available from (eco-)systems research, toxicology, environmental monitoring, and specimen banks. Toolboxes TRE and DAT provide parameters required to accomplish this [Markert et al., 2003b].

in drinking water, food, or air at the workplace, along with the data for the relevant ADI ("acceptable daily intake") and NO(A)EL ("no observed (adverse) effect level"). The toolbox TRE contains data on trends; these have been compiled mainly from years of investigations by national environmental sample banks, or from information available from long-term national and international studies (e.g., Ellenberg et al. [1986]). Specific conclusions and trend forecasts can then be prepared using the subsequent toolboxes HSB (human specimen banking) and ESB (environmental specimen banking). The toolbox MED (medicine) contains all methods usually employed in hematological and chemical clinical investigations of subchronic and chronic toxicity, whereas ECO is largely made up of all the bioindicative testing systems and monitors relevant to ecosystems which may be combined to suit the particular situation to be monitored [Markert et al., 2003b, 2007].

By relating data from all the toolboxes with some network, it must be achieved to assess average health risks to certain parts of the population or at least upper limits of future risks posed by pollutants [Markert et al., 2007]. For this kind of risk assessment, all the information on kinds of effects, dose–effect relationships, and toxicological limits derived therefrom by present level of scientific knowledge are combined and used [WHO, 1996]. Although toxicological experiments on humans would be unethical, corresponding data pertinent to toxic risks can be obtained from workplace experiences and cases of accidental, homicidal, or suicidal poisoning. For both statistical reasons and evaluation of subacute-dose effects that might yet bring about diseases, results of epidemiological surveys that compare exposed groups to control groups must be added. Recent information technology allows for development and use of simulation models that integrate all these data, integrating a large number of parameters that are not directly linked to each other.

Because the method by which the MMBC combines functional and integrated windows for prophylactic health care was outlined in more detail above, we refer to the corresponding literature rather than repeating ourselves here [Markert et al., 2003]. An integration of data on ecological quality and human health takes knowledge of the sites, methods, and locations where and how the data were obtained, producing metadata upon superposition of these pieces of information. By combination of such metadata with geostatistical information including and using GIS techniques, some integration of environmental monitoring data concerning exposure to and effects brought about by contaminants was achieved in Germany. This metadata system contains some 800 items producing, beyond a detailed consideration of "classical" topics of environmental monitoring (water, air, soils, plants, animals, landscapes), some picture of environmental quality and human health [Schröder et al., 2003].

Analysis of biogenic samples in either biomonitoring or bioindication will produce data; these, however, must not be taken as pieces of information on the "state of the environment" directly, except for measurements of atmospheric deposition (by means of mosses, Tillandsiae, and the like). Even then, no organism might enrich all the elements from the environment by some identical bioconcentration factor BCF, but there will always be selectivity with drawbacks in biomonitoring [Fränzle and Markert, 2007a,b].

6   FUTURE OUTLOOK: BREAKING "MENTAL" BARRIERS BETWEEN ECOTOXICOLOGISTS AND MEDICAL SCIENTISTS

For purposes of prophylactic health care, work and approaches of ecotoxicologists and of medical doctors should be integrated rather than simply joined, let alone working separatedly along different paths as we used to do for the last 20 years [Markert et al., 2007]. In German traditions of landscape ecology, there is some precedent for combining geosciences, ecology, and medical sciences, dating back many years [Jusatz and Flohn, 1937; Jusatz, 1958; Mueller, 1974; Schweinfurth, 1974; cf. Meade, 1977]. Culture, human health, and environmental quality are also related to each other as we should acknowledge [Warren and Harrison, 1984]. There are still some obstacles in changing from recent eco- or humantoxicological methods of bioindication toward the above-postulated integrative approach. Part of the dilemma is that ecologists and medical professionals usually neither communicate nor cooperate with each other. Let us give one example: There is one (most modern and otherwise fine) textbook on toxicology (*Lehrbuch der Toxikologie*, Marquard and Schaefer [2004]) in which 100 contributors present excellent chapters on most relevant topics within our common scientific areas of interest yet almost completely ignore or miss results obtained by ecotoxicologists (who are organized in SETAC). There are two imminent steps to overcome this problem:

1. Toxicologists of the various branches of study should be educated at academia using the same integrative textbooks (e.g., that by Fomin et al. [2003] on practical doing and uses of biotests).
2. Join people from different scientific branches by common scientific projects, like those outlined in Fig. 7.

Especially concerning the latter topic, a common terminology along with common goals and methods are required which get over the "barriers" among scientific disciplines in order to eventually initiate successful common scientific research.

In addition, an improved understanding of the implications and "meaning" of biomonitoring data is warranted [Fränzle and Markert, 2007a,b]. For exposition to metal ions, but presumably also applicable to metal-free electrophilic species such as aldehydes, ketones, or boron compounds, a model of bioaccumulation was designed which allows for prediction of effects (although, once again, only for taxonomical species already under investigation [mainly green plants]) from bioaccumulation or, conversely, for an estimation of environmental levels of the corresponding ions/molecules from amounts detected in biomass [Fränzle, 2007]. Thus, both exposition risks and the state of the environment (both aquatic and soil media) may be inferred. Our most recent research addresses effects beyond a simple correlation with stabilities of complexes of the corresponding metal ions, giving some measure for modification along the uptake pathways. This meets several purposes: It quantitatively investigates matter transport in living organisms comprehensively, beyond biochemical detail and without requiring the latter refining of bioindication and corresponding statements (exposition) and risk assessment, and eventually provides taxonomic and ecotoxicological information.

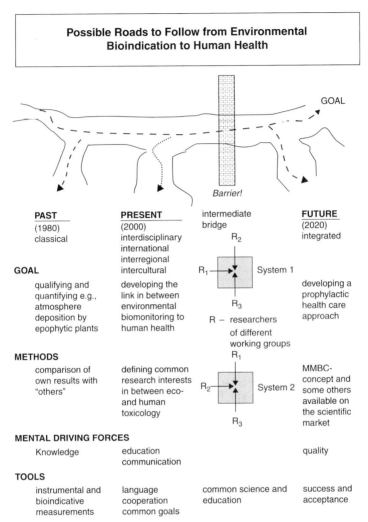

Figure 7. The link between environmental monitoring and human-health research including obstacles that might be encountered on various fields and levels when going this way (from Markert et al., 2007). In the past, analytical scientists, ecotoxicologists, and environmental physicians were unlikely to cooperate, also producing a gap of knowledge (technical and published) as well as of methodology. Now we are in a position to overcome this gap by transdisciplinary communication, producing a basis for more intensive collaboration along common ends for research and thereafter education. Such common scientific interest can be increased if (though different and specific) research methods are applied to similar problems, promoting common learning and deeper insight into the same interdisciplinary problems (for which the middle column joining present and future situations of scientific work stands as a symbol).

REFERENCES

Baker AJM, Brooks RR. 1989. Terrestrial higher plants which hyperaccumulate metallic elements—A review of their distribution, ecology and phytochemistry. *Biorecovery* **1**:86–126.

Bargagli R, editor. 1998. *Trace Elements in Terrestrial Plants—An Ecophysiological Approach to Biomonitoring and Biorecovery.* Heidelberg: Springer.

Broadley MR, White PJ, Hammond JP, Zelko I, Lux A. 2007. Zinc in plants, Tansley review. *New Phytol* **173**:677–702.

Buse A, Norris D, Harmens H, Büker P, Ashenden T, Mills G. 2003. Heavy metals in European mosses: 2000/2001 survey. UNECE ICP Vegetation. Centre for Ecology & Hydrology.

Callahan DL, Baker AJM, Kolev SP, Wedd AG. 2006. Metal ion ligands in hyperaccumulating plants. *J Biol Inorg Chem* **1**:2–12.

Clemens S, Palmgren G, Kraemer U. 2002. A long way ahead: Understanding and engineering plant metal accumulation. *Trends Plant Sci* **7**:309–315.

Djingova R, Kuleff I. 2000. Instrumental techniques for trace analysis. In: Markert B, Friese K, editors. *Trace Elements: Their Distribution and Effects in the Environment.* Amsterdam: Elsevier, pp. 137–185.

Djingova R, Kovacheva P, Wagner G, Markert B. 2003. Distribution of platinum group elements among different plants along some highways in Germany. *Sci Total Environ* **308**:235–246.

Ellenberg H, Mayer R, Schauermann J. 1986. *Ökosystemforschung. Ergebnisse des Solling Projektes.* Stuttgart: Ulmer Verlag.

Fent K. 2003. *Ökotoxikologie. Umweltchemie, Toxikologie, Ökologie.* Stuttgart: Thieme-Publisher.

Fomin A, Oehlmann J, Markert B. 2003. *Praktikum zur Ökotoxikologie—Grundlagen und Anwendungen biologischer Testverfahren.* Landsberg: Ecomed.

Fränzle O. 1999. Ökotoxikologie im Spannungsfeld von Ökologie und Toxikologie. In: Oehlmann J, Markert B, editors. 1999. *Ökotoxikologie. Ökosystemare Ansätze und Methoden.* Landsberg: Ecomed-Verlagsgesellschaft, pp. 23–48.

Fränzle O. 2003. Bioindicators and environmental stress assessment. In: Markert BA, Breure AM, Zechmeister HG, editors. *Bioindicators and Biomonitors.* Amsterdam: Elsevier, pp. 41–84.

Fränzle S. 2007. Prinzipien und Mechanismen der Verteilung und Essentialität von chemischen Elementen in pflanzlicher Biomasse—Ableitungen aus dem Biologischen System der Elemente. Habilitationsschrift. Deutschland: Hochschule Vechta.

Fränzle S, Markert B. 2007a. Metals in Biomass: From the biological system of elements to reasons of fractionation and element use, *Environ Sci Pollut Res,* online first, DOI: http://dx.dol.org/10.1065/espr2006.12.372.

Fränzle S, Markert B. 2007b. What does bioindication really tell us? Analytical data in their natural environment. *Ecol Chem Eng* **14**(1):7–23.

Fränzle S, Markert B, Wünschmann S. 2007. Dynamics of trace metals in organisms and ecosystems: prediction of metal bioconcentration in different organisms and estimation of exposure risks. *Environ Pollut* (in press).

Freitas MC, Reis MA, Alves LC, Wolterbeek HT. 2000. Nuclear analytical techniques in atmospheric trace element studies in Portugal. In: Markert B, Friese K, editors. *Trace Elements/Their Distribution and Effects in the Environment.* Amsterdam: Elsevier, pp. 187–213.

Garty J. 2000a. Environment and elemental content of lichens. In: Markert B, Friese K, editors. *Trace Elements/Their Distribution and Effects in the Environment.* Amsterdam: Elsevier, pp. 245–276.

Garty J. 2000b. Trace metals, other chemical elements and lichen physiology: research in the nineties. In: Markert B, Friese K, editors. *Trace Elements/Their Distribution and Effects in the Environment.* Amsterdam: Elsevier, pp. 277–322.

Greger M. 2004. Metal availability, uptake, transport and accumulation in plants. In: Prasad MNV, editor. *Heavy Metal Stress in Plants: From Biomolecules to Ecosystems*, Second edition, pp. 1–27. Heidelberg: Springer.

Heim M, Wappelhorst O, Markert B. 2002. Thallium in terrestrial environments—Occurrence and effects. *Ecotoxicology* **11**:369–377.

Herpin U, Siewers U, Kreimes K, Markert B. 2001. Biomonitoring—Evaluation and assessment of heavy metal concentrations from two German moss surveys. In: Burga CA, Kratochvil A, editors. *General and Applied Aspects on Regional and Global Scales.* Dordrecht: Kluwer Academic Publishers, pp. 73–95.

Herpin U, Siewers U, Markert B, Rosolen V, Breulmann G, Bernoux M. 2004. Second German heavy-metal survey by means of mosses, and comparision of the first and second approach in Germany and other European countries. *Environ Sci Pollut Res* **11**(1):57–66.

Jusatz HJ. 1958. Die Bedeutung der landschaftsökologischen Analyse für geographisch-medizinische Forschung. *Erdkunde* **XII**:284–289.

Jusatz HJ, Flohn H. 1937. Geomedizin und Geographie. *Petermanns Geogr Mitt* **83**:1–5.

Markert B. 1996. *Instrumental Element and Multi-element Analysis of Plant Samples—Methods and Applications.* New York: John Wiley & Sons.

Markert B, Oehlmann J, Roth M. 1997. Biomonitoring von Schwermetallen–eine kritische Bestandsaufnahme. *Z. Ökologie Naturschutz* **6**:1–8.

Markert B, Breure T, Zechmeister H. editors. 2003a. *Bioindicators and Biomonitors: Principles, Concepts and Applications.* Amsterdam: Elsevier.

Markert B, Breure T, Zechmeister H. 2003b. Definitions, strategies and principles for bioindication/biomonitoring of the environment. In: Markert B, Breure T, Zechmeister H, editors. 2003. Bioindicators & Biomonitors. Principles, Concepts and Applications. Amsterdam: Elsevier, pp. 3–39.

Markert B, Fraenzle S, Fomin A. 2004. From the biological system of elements to biomonitoring. In: Merian E, Anke M, Ihnat M, Stoeppler M, editors. *Elements and Their Compounds in the Environment*, Vol. 1: *General Aspects.* Weinheim: Wiley/VCH, pp. 235–254.

Markert B, Wünschmann S, Fraenzle S, Wappelhorst O, Weckert V, Breulmann G, Djingova R, Herpin U, Lieth H, Schröder W, Siewers U, Steiness E, Wolterbeek B, Zechmeister H. 2007. On the road from environmental biomonitoring to human health aspects: Monitoring atmospheric heavy metal deposition by epiphytic/epigeic plants: Present status and future needs. *Int J Environ Pollut* (in press).

Marquard H, Schaefer S, editors. 2004. *Lehrbuch der Toxikologie.* Stuttgart: Wissenschaftliche Verlagsgesellschaft mbH, p. 1348.

Meade MS. 1977. Medical geography as human ecology. *Geogr Rev* **67**:379–393.

Mueller P. 1974. Beiträge der Biogeographie zur Geomedizin und Ökologie des Menschen. In: *Fortschritte der geomedizinischen Forschung*. Beihefte: Geographische Zeitschrift, pp. 88–109.

Mueller P. 1980. *Biogeography*. Stuttgart: Ulmer.

Oehlmann J, Markert B. 1997. Humantoxikologie. In: *Eine Einführung für Apotheker, Ärzte, Natur- und Ingenieurwissenschaftler*. Stuttgart: Wissenschaftliche Verlagsgesellschaft.

Oehlmann J, Markert B. 1999. Ökotoxikologie. In: *Ökosystemare Ansätze und Methoden*. Landsberg: Ecomed.

Pignata ML, Gudino GL, Wannaz ED, Pla RR, Gonzales CM, Carreras HA, Orellana L. 2002. Atmospheric quality and distribution of heavy metals in Argentina employing *Tillandsia capillaris* as a biomonitor. *Environ Pollut* **120**:59–68.

Prasad MNV, editor. 2004. *Heavy Metal Stress in Plants: From Biomolecules to Ecosystems*. Heidelberg: Springer.

Reimann C, Siewers U, Tarvainen T, Bityokova L, Erikson J, Gilucis A, Gregorauskiene V, Lukashev K, Martinian N, Pasieczna A. 2003. *Agricultural Soils in Northern Europe: A Geochemical Atlas*. Stuttgart: E. Schweitzebart'sche Verlagsbuchhandlung.

Ruehling A, Tyler G. 1968. An ecological approach to the lead problem. *Environ Pollut* **121**:321–342.

Schröder W, Fraenzle O, Keune H, Mandy P, editors. 1996. *Global Monitoring of Terrestrial Ecosystems*. Berlin: Ernst und Sohn Verlag.

Schröder W, Schmidt G, Pesch R. 2003. Harmonization of environmental monitoring. Tools for examination of methodical comparability and spatial representativity, Gate to Environmental and Health Sciences, July 2003, pp. 1–13 [DOI: http://dx.doi.org/10.1065/ehs2003.07.010].

Schröder W, Pesch R. 2004. The 1990, 1995 and 2000 heavy metals in mosses survey in Germany and other European Countries. Trends and statistical aggregation of metal accumulation indicators, DOI: http://dx.doi.org/10.1065/ehs2004.06.011.

Schröder W, Hornsmann I, Pesch R, Schmidt G, Fränzle S, Wünschmann S, Heidenreich H, Markert B. 2007. Moosmonitoring als Spiegel der Landnutzung? Stickstofffund Metallakkumulation zweier Regionen Mitteleuropas, Zeitschrift für Umweltwissenschaften und Schadstofforschung, online first, DOI: http://dx.doi.org/10.1065/ehs2006.11.155.

Schweinfurth U. 1974. Geoökologische Überlegungen zur geomedizinischen Forschung. *Fortschritte der geomedizinischen Forschung*. Beihefte: Geographische Zeitschrift, pp. 30–43.

Siewers U, Herpin U. 1998. Schwermetalleinträge in Deutschland, Moos-Monitoring 1995/96: *Geologisches Jahrbuch*, Reihe D, Heft SD 2. Stuttgart: E. Schweizerbarth'sche Verlagsbuchhandlung.

Smodis B. 2003. IAEA approaches to assessment of chemical elements in atmosphere. In: Markert B, Breure AM, Zechmeister HG, editors. *Bioindicators and Biomonitors: Principles, Concepts and Applications*. Amsterdam: Elsevier, pp. 875–902.

Warren A, Harrison CM. 1984. People and the ecosystem. Biogeography as a study of ecology and culture. *Geoforum* **15**:365–381.

Wappelhorst O. 1999. Charakterisierung atmosphärischer Deposition in der Euroregion Neiße durch ein terrestrisches Biomonitoring. Dissertation, Internationales Hochschulinstitut Zittau.

Wappelhorst O, Kühn I, Oehlmann J, Markert B. 2000. Deposition and disease: A moss monitoring project as an approach to ascertaining potential connections. *Sci Total Environ* **249**:243–256.

Wolterbeek HT. 2002. Biomonitoring of trace element air pollution: Principles, possibilities and perspectives. *Environ Pollut* **120**:11–21.

World Health Organization (WHO). 1996. *Biological Monitoring of Chemical Exposure in the Workplace, Guidelines*, Vol. 1 and 2: Geneva.

Wünschmann S. 2007. Bestimmung chemischer Elemente im System Nahrung/Muttermilch— ein Beitrag zur Bioindikation? Dissertation, University of Vechta.

Zechmeister HG. 1995. Correlation between altitude and heavy metal deposition in the Alps. *Environ Pollut* **89**:73–80.

Zechmeister HG. 1997. *Schwermetalldeposition in Österreich erfaßt durch Biomonitoring mit Moosen*. Umweltbundesamt, Wien, Monographien Band 94.

Zechmeister HG, Grodzinska K, Szarek-Lukaszewska. 2003. Bryophytes. In: Markert B, Breure AM, Zechmeister HG, editors. *Bioindicators and Biomonitors: Principles, Concepts and Applications*. Amsterdam: Elsevier, pp. 329–375.

Zechmeister HG, Dullinger S, Hohenwallner D, Riss A, Hanus-Illnar A, Scharf S. 2006. Pilot study on road traffic emissions (PAHs, heavy metals) measured by using mosses in a tunnel experiment in Vienna, Austria, *Environ Sci Pollut Res* **13**:398–405.

BIODIVERSITY INDEX

Abutilon theophrasti, 71
Acacia aneura, 703
A. auriculaeformis, 198
A. bidwilli, 703
A. concurrens, 703
A. deanei, 703
A. fimbriata, 703
A. glaucocarpa, 703
A. leptocarpa, 703
A. melanoxylon, 703
A. saligna, 716, 717
A. stenophylla, 703
Acalypha australis, 71
Acer platanoides, 727
A. pseudoplatanus, 726
A. rubrum, 282
Acer sp., 729
Achillea nobilis, 287
Acroptilon repens, 286, 287
Aegilops taushii, 171
Aesculus hippocastanum, 730
Aesculus sp., 729
Agaricus arvensis, 130
A. bisporus, 131
A. silvaticus, 131
Agaricus sp., 130
Agrobacterium tumefaciens, 174
Agropyron cristatum, 284
A. repens, 293
Agrostis, 282
Agrostis canina, 734
A. capillaries, 734
A. vinealis, 287
Ailanthus altissima, 732
Ailanthus sp., 729
Allium, 607

Allium bisulcatus, 607
A. sepa, 142
Alpinia zerumbet, 105
Alternanthera philoxeroides, 142, 147
A. pungense, 142
Alyssum lesbiacum, 384
Amanita muscaria, 17
Amaranthus lividus, 70
A. retroflexus, 70, 282, 283, 286
A. spinosus, 142
Ambrosia trifida, 69
Anabasis brevifolia, 703
Ancathia igniaria, 286
Anethum graveolens, 111
Angelica sylvestris, 288
Anthyllis cytisoides, 201
Apium nodiflorum, 638, 639
Arabidopsis, 191, 201, 248, 251, 450, 659
Arabidopsis halleri, 204, 418, 427, 428,
    431, 438
A. lyrata, 431
A. thaliana, 2, 86, 87, 200, 202, 204, 243,
    245, 249, 388, 414, 416–418, 420,
    421–423, 426–432, 451–463
Ariplex nummularia, 703
Aristida mendocina, 703
Armeria maritime, 437, 734
Armillariella mellea, 131
Artemisia argyi, 69
A. annua, 147
A. dracunculus, 286, 287
A. frigida, 286, 290
A. gracilescens, 286
A. lavandulaefolia, 69
A. scoparia, 69, 286, 287, 290
A. sieversiana, 286

Trace Elements as Contaminants and Nutrients: Consequences in Ecosystems and Human Health, Edited by M. N. V. Prasad
Copyright © 2008 John Wiley & Sons, Inc.

A. vulgaris, 142

Arthraxon hispidus, 70

Aspergillus, 48, 252, 253

Aspergillus nidulans, 237

Aster, 605

Aster tripolium, 707

Astragalus, 110, 605

Astragalus bisculatus, 612

Astragalus membranaceus, 110

Atriplex, 605

Atriplex argentina, 703

A. atacamensis, 703

A. cana, 287, 292

A. coquimbensis, 703

A. halimus, 703, 711, 715, 716

A. lampa, 703

A. nummularia, 715, 717

A. repanda, 702, 703

A. semibaccata, 704

Atropa acuminata, 103

Avena sativa, 196

Avicennia germinans, 707

A. marina, 192

Azadirachta indica, 149

Azolla filiculoides, 13

A. pinnata, 632

Bacillus sp. strain SF-1, 598

B. megaterium, 602

B. selenitireducens strain MLS-10, 598

B. arsenicoselenatis strain E-1H, 598

Bacopa monnieri, 142, 147

Ballota acetabulosa, 732, 733

Barbus graellsii, 359

Batis maritima, 707

Berberis sibirica, 287

Berteroa incana, 286, 287

Beta vulgaris, 105

Betula, 197

Betula pendula, 6, 13, 15, 287

Bidens pilosa, 69

B tripartita, 69

Biovar trifolii, 198

Biscutella laevigata, 197

Boletus edulis, 131

Brachypodium pinnata, 288

Brassica, 607

B. napus, 103, 195, 192, 245, 423, 605

B. napus var. Hyola, 418, 611

B. oleracea, 87, 105

B. pekinensis, 659

B. rapa, 103, 104

Brassica juncea, 124, 142, 174, 195, 196, 198, 384, 387, 605, 606, 609, 612, 659

Brassica species, 198

Brassica spp, 193

Brugmansia candida, 103

Bruguiera parviflora, 708

Caenorhabtitis elegans, 432, 453

Calamagrostis epigeios, 287

Callitriche brutia, 639

C. hamulata, 639

C. lusitanica, 639

C. stagnalis, 638, 639, 640, 641, 644, 645

C. petrie, 640

C. verna, 640

Calocybe gambosa, 131

Calotropis gigantean, 142

Calystegia hederacea, 71

Camellia sinensis, 109, 142

Campanula sibirica, 288

Cannabis ruderalis, 286, 287

C. sativa, 70

Cantharellus cibarius, 131

C. lutescens, 131

C. tubaeformis, 131

Capsella bursa-pastoris, 732

Caragana pumila, 286, 292

Carex gracillis, 632, 634

C. rigescens, 70

Cassia angustifolia, 109, 147

C. siamea, 142

Castilleja, 605

Catharanthus roseus, 103, 105

Caulerpa racemosa, 129, 130

Cellulomonas sp. 32, 325

Cenanthe brutia, 640

Centaurea iberica, 726, 732

Centella asiatica, 142

Cephalanoplos setosum, 69

Cerastium fontanum, 282

Ceratocarpus arenarius, 286

Ceratophyllum demersum, 13, 194, 632

Ceriodaphnia, 357

Chamaenerion angustifolium, 290

Chamomilla recutita, 103

Chenopodium album, 70, 143
C. botrys, 286
C. hybridum, 70
Chlamydomonas reinhardtii, 423, 456, 458, 460
Chlorella protothecoides, 247
C. vulgaris, 197, 354
Chloris gayana, 658
Chrysanthemum cineraefolium, 147
Cirsium pendulum, 69
Cistus creticus, 726, 732
Cladosporium cladosporioides, 149
Cleistogenes squarrosa, 286
Clostridium sticklandii, 597
Cochlearia pyrenaica, 204
Cochlearia, 191
Cocos nucifera, 282
Coffea arabica, 105
Colchicum autumnale, 103
Comandra, 605
Comelina sinensis, 108
Commelina communis, 70
Conyza canadensis, 246
Corylus maxima, 729
Corynebacterium, 603
Cotoneaster melanocarpus, 287
Cottea pappophoroides, 704
Crataegus clorocarpa, 287
Cryptomeria japonica, 727
Cucumis sativus, 658
Cupressus sempervirens, 728
Cyanidioschizon merole, 423
Cymbopogon flexuosus, 105, 147
C. martinii, 147
C. winterianus, 147
Cynodon dactylon, 192

Daphnia, 350
Datura innoxia, 103, 143, 199
D. stramonium, 103, 104
Daucus carota, 661
Delphinium dictyocarpum, 287
Deschampsia caespitosa, 194
D. flexuosa, 12, 15, 16, 18
Desulfovibrio baculatus, 597
D. desulfuricans, 598
Dianthus rigidus, 286, 290
D. versicolor, 288
Digitalis obscura, 103, 105

Dioscorea bulbifera, 105
Distichlis spicata, 610
Dreissena polymorpha, 353
Dunaliella tertiolecta, 199

Echinacea purpurea, 108
Echinochloa crusgalli, 70
Eichhornia crassipes, 201, 632, 634
Eichhornia sp., 632
Elaeagnus angustifolia, 723, 733
Elaeis guineensis, 196
Eleocharis dulcis, 632, 633
Elsholtzia splendens, 661
Empetrum nigrum, 727, 731
*Enterobacter cloacae SLD*1a-1, 598
E. taylorae, 598
Ephedra distachya, 286, 287, 290
Ephedra ochreata, 704
Erigeron canadensis, 286
Erwinia chrysanthemi, 243
Erysimum cheiranthoides, 103
Escherichia coli, 236, 597, 601
Eucalyptus citridora, 143
E. globulus, 109
E. tereticornis, 282
Euphorbia cyparissias, 196

Fagopyrum esculentum, 143
Fagus sylvatica, 732
Festuca arundinacea, 282, 611
F. ovina, 734
F. pretense, 293
F. rubra, 282, 734
F. valesiaca, 284, 286, 290
Foeniculum vulgare, 109
Fontinalis antipyretica, 201
Fragaria viridis, 287
Fraxinus sp., 729
Fritillaria meleagroides, 288
Fumaria vaillantii, 286, 287
Funaria hygrometrica, 192

Galium ruthenicum, 286
Glomus mosseae, 196
Gloriosa superba, 105
Glycine max, 193, 235, 236, 244, 245, 247, 248, 255–257
Glycine soja, 70
Glycyrrhiza inflata, 704

G. uralensis, 70
Gracilaria lemaneiformis, 129
Grayia, 605
Grindelia, 605
Gutierrezia, 605
Gynostemma pentaphyllum, 109

Halocnemum strobilaceum, 704, 716
Halostachys caspica, 704
Haloxylon ammodendron, 704
H. salicornicum, 704
Haloxylon ammodendron, 704
Hedera helix, 731
Helianthus annuus, 143
H. tuberosus, 69
Helictotrichon desertorum, 286
Hemidesmus indicus, 143, 147
Hieracium virosum, 287
Hizikia fusiforme, 129
Holcus lanatus, 194, 199
Hordeum vulgare, 196, 423
Humulus scandens, 70
Hydrocotyle umbellate, 147
Hylocomium splendens, 724, 734, 749
Hymenoscyphus ericae, 237
Hyosciamus albus, 103
H. niger, 103
Hypericum perforatum, 143
Hypnum cupressiforme, 749
Hyptis suaveolens, 103

Ilex aquifolium, 724, 731
I. vomitoria, 103
Inula viscosa, 726
I. britannica, 69
Ipomea aquatica, 129, 143
Iris scariosa, 290

Juglans regia, 730
Juncus balticus, 610
J. effusus, 638, 639
Juniperus, 282
Juniperus arizona, 730
J. sabina, 287
J. virginiana, 192

Kalidium foliatum, 704
Kalimeris integrifolia, 69
Karelinia caspica, 704

Kjellmaniella crassifolia, 129, 130
Kochia scoparia, 286, 290
K. sieversiana, 290

Lactarius deliciosus, 130
Lactuca indica, 69
L. sativa, 252
Laminaria japonica, 129, 130
L. setchellii, 129, 130
Larrea nitida, 704
Lawsonia inermis, 143
Leccinum aurantiacum, 131
L. scabrum, 131
Lemna gibba, 632, 633, 634
L. minuta, 639
L. trisulca, 13
L. minor, 198, 633, 638, 639, 641
Lemna sp., 632
Lens esculenta, 247
Leonurus heterophyllus, 70
Lepidium apetalum, 71
L. densiflorum, 286
Lepidium latifolium, 286, 287, 290
Lepista sp., 130
Ligustrum japonicum, 731
Limoniastrum multiflorum, 707
Limonium chrysocomum, 287
L. suffrutescens, 290
Linum usitatissimum, 193
Liquidambar stryaciflua, 282
Liriodendron tulipifera, 282
Lolium perenne, 12, 15, 282
Lonicera pallasii, 287
L. tatarica, 726
Lonicera xylosteum, 726
Lotus angustissimus, 290
L. japonicus, 87
Ludwigina palustris, 632
Lupinus albus, 724, 727
L. angustifolius, 103
L. luteus, 243, 246, 248
Lyceum chilense, 704
L. tenuispinosum, 704
Lycoperdon perlatum, 130, 131
Lycopersicun esculentum, 197, 201, 236, 658
Lygeum spartum, 201

Machaeran hera, 605
Macrolepiota procera, 131

M. rhacodes, 130
Macrolepiota sp., 130
Mahonia aquifolium, 724, 731
Mallotus philipensis, 108
Malus domestica, 730, 732
M. xiaojinensis, 246
Malva neglecta, 732
Matricaria chamomilla, 105, 109
M. recutita, 287
Medicago, 293
Medicago sativa, 245, 247
M. truncatula, 247
Melica transsilvanica, 287, 288
Menha piperita, 109
Mentha aquatica, 632
M. arvensis, 105, 147
M. cardiaca, 147
M. citrata, 147
M. haplocalyx, 70
M. piperata, 143, 147
M. spicata, 147
Metaplexis japonica, 71
Methanococcus vannielii, 597
Michelia champak, 108
Minuartia hirsute, 663
Mirabilis jalapa, 732
Molinia coerulea, 12, 15, 18, 734
Morinda, 605
Morus alba, 109
Myriophyllum aquaticum, 632
M. spicatum, 632, 732
Myrtilli folium, 110

Nepturia, 605
Nerium odoratum, 143
N. oleander, 731, 733
Neurospora crassa, 237
Nicotiana rustica, 102, 103
N. tabacum, 105, 194, 246, 252, 253, 387, 388
Nigella sativa, 199
Nitraria retusa, 704
Nordostachys jatamansi, 149
Nymphaea violacea, 633

Ocimum sanctum, 144, 147
O. tenuiflorum, 105
Oenanthe crocata, 638–640
Oenothera biennis, 70

Olea europaea, 734
Onobrychis, 293
Oonopsis, 605
Origanum syriacum, 103
O. vulgare subsp. hirtum, 657, 660
Oryza sativa, 129, 236, 237, 252, 253, 422, 423, 451

Padus avium, 287, 288
Panax ginseng, 105
P. quinquefolium, 105
Panicum turgidum, 711, 713
Papaver bracteatum, 105
Pappophorum caespitosum, 702, 704
P. philippianum, 704
Parnassia palustris, 288
Paspalum distichum, 192
P. notatum, 199
Pelargonium graveolens, 144, 147
Pelargonium sp., 145
Perilla frutescens, 69
Peucedamum terebinthaceum, 70
Phaseolus, 197
Phaseolus limensis, 256
P. polyanthus, 173
P. vulgaris, 173, 191, 193, 194, 195, 197, 244, 247, 252, 255, 257
Phragmites australis, 193, 287, 290, 611, 632, 634, 658, 732
P. communis, 704
Phragmites sp., 632
Phyllanthus amarus, 105, 147
P. niruri, 144
Phytophthora infestans, 242
Picea abies, 105, 732
P. mariana, 282
Pinus brutia, 726, 728
P. canariensis, 728
P. halepensis, 723
P. nigra, 726, 727, 728, 730, 734
P. nigra ssp. pallasiana, 728
P. pinaster, 658, 728
P. pinea, 658
P. ponderosa, 282
P. radiata, 282
Pinus sylvestris, 13, 15, 192, 195, 287, 288, 661, 727, 728, 732
Pisum sativum, 244, 247
Pittosporum sinensis, 731

Plantago asiatica, 70
P. depressa, 70
P. lanceolata, 70, 109, 236, 726, 732
Platanus sp., 730
Platichthys flesus, 363
Pleurotus ostreatus, 131
Pleurozium schreberi, 734, 749
Polygonum aviculare, 69, 286, 287
Polypogon monspeliensis, 610
P. bungeanum, 69
P. hydropiper, 69
P. lapathifolium, 69
P. orientale, 71
P. roseoviride, 69
P. viscosum, 70
Populus, 430, 730
Populus alba, 723, 727, 729
P. balsamifera subsp. trichocarpa, 87
P. canadensis, 723, 727, 729
P. deltoides x maximowiczii, 729
P. deltoids, 431
P. nigra, 287, 288, 730
P. euramericana, 723, 729
P. nigra italica, 732
P. nigra ssp. nigra, 729
P. tremula, 287, 728
P. trichocarpa, 431
*P. usbekistanica subsp. usbekistanica
     cv.*, 729
Porphyra sp., 129
Portulaca oleracea, 71
Potamogeton natans, 638, 639, 641, 645
P. pechinatus, 641, 645
P. pectinatus, 200, 638, 639, 645, 732
P. pusillus, 639
Potentilla acaulis, 286
P. nudicaulis, 288
P. paradoxa, 70
Primula vulgaris, 724, 727
Prunus amygdalus, 730
Psathyrostachys juncea, 290
Pseudomonas, 602
Pseudomonas stutzeri, 599
Psoralea degelenica, 288
Pteris vittata, 724, 735
Punica granatum, 217, 219
Pyracantha coccinea, 726

Pyrobaculum arsenaticum, 598
P. coccinea, 731

Quercus, 282
Quercus coccifera, 726
Q. infectoria, 731
Q. robur, 192, 731

Ralstonia metallidurans, 599, 600
Ranunculus chinensis, 71
R. peltatus, 639
R. penicilatus, 639
R. trichophyllus, 639
R. tripartitus, 639
Reaumuria soongorica, 288, 704
Rhizobium leguminosarum, 198
Rhizoctonia, 223
Rhodiola sachalinensis, 102
Rhodobacter sphaeroides, 601
Rhododendron catawbiense, 724, 731
Rhynchostegium riparioides, 198
Rhytidiadelphus loreus, 724, 734
Ribes saxatile, 287, 288
Robinia pseudo-acacia, 723, 727, 729,
     730, 733
Robinia sp., 730
Rosa canina, 109, 731
Rottboellia faciculata, 707
Rozites caperata, 131
Rumex acetosa, 105
R. confertus, 290
Ruppia maritima, 610
Russula cyanoxantha, 131

Saccharomyces cerevisiae, 418, 419,
     432, 459
Sagittaria sagittifolia, 288
Salicornia bigelovii, 702, 705
Salix alba, 728
S. babylonica, 730
S. cinerea, 292
S. tetrasperma, 144
Salsola, 293
Salsola tetranda, 705
Salvelinus alpinus, 363
Salvia folium, 110
S. miltiorrhiza, 105

S. officinalis, 109
S. plebeia, 70
Sambucus nigra, 732
Sanguisorba officinalis, 287, 292
Scapania, 192
Scenedesmus subspicatus, 354
Schizosaccharomyces pombe, 452, 453
Schoenoplectus acutus, 610
S. robustus, 610
Scleropodium purum, 749
Sebertia acuminate, 384
Secale cereale, 303
*Sedimenticola selenatireducens strain
    AK4OH*1, 599
Selenihalanaerobacter shriftii, 598
Sesbania cannabina, 199
S. rostrata, 199
Sesbania species, 199
Sesuvium portulacastrum, 707
S. verrucosum, 707
Setaria italica, 198
S. viridis, 286
Silene suffrutescens, 290
S. vulgaris, 193, 199, 200,
    201, 734
Silphium perfoliatum, 69
Solanum dulcamara, 286
S. nigrum, 144
S. nigrum ssp. schultesii, 732
S. tuberosum, 242, 246
Sonchus brachyotus, 69
Spartina alterniflora, 610, 706, 707
S. patens var. Flageo, 611
Spinach, 236
Spiraea hypericifolia, 286, 292
Spirodella polyrizha, 633, 639
Sporobulus airoides, 611
Spruce, 13
Stanleya, 605
Stenotrophomonas maltophilia, 599
*Stenotrophomonas maltophilia
    SeITE*02, 599
Stipa capillata, 284, 286, 292
S. kirghisorum, 287
S. sareptana, 286, 290
Stylosanthes hamata, 428, 432
Succisa pratensis, 724, 727

Sueda foliosa, 702, 705
S. fruticosa, 705
S. physophora, 705
Suillus bovines, 661
S. variegatus, 131
*Sulfurospirillum barnesii strain
    SES*-3, 597

Tamarix aphylla, 705
T. mannifera, 705, 716
T. ramosissima, 705
Taraxacum officinale, 12, 14, 109, 144, 732
Taxus chinensis, 105
Teucrium scorodonia, 724, 727
Thalassiosira weissflogii, 384, 449
Thalia dealbata, 610
Thauera selenatis, 598, 608
Thelephora terrestris, 661
Thiobacillus ferrooxidans, 194, 602
Thlaspi arvense, 202
T. bulbosum, 204
T. caerulescens, 192, 195, 196, 197, 200,
    201, 202, 204, 418, 420, 423,
    426–428, 431, 436, 438,
    456, 734
T. goesingense, 204, 428, 431
T. idahoense, 204
T. liaceum, 204
T. magellanicum, 204
T. montanum, 204
T. ochroleucum, 202, 204
T. parvifolium, 204
T. praecox, 204
T. rotundifolium, 204
T. rotundifolium subsp.cepaeifolium, 204
T. stenocarpum, 204
T. violascens, 204
Thlaspi, 191, 199, 735
Thymus capitatus, 726
Typha latifolia, 634
Typha sp., 632
Tilia sp., 730
Tilia cordata, 727
T. vulgaris, 109
Tradescantia bracteata, 282
Trichloris crinita, 702, 705
Tricholoma terreum, 130

Trifolium, 293
Trifolium repens, 282
T. subterraneum, 200, 220
Triticum aestivum, 199, 303
T. aestivum, 252
T. dicoccon, 171
T. spelta, 128
T. vulgare, 328
Typha angustifolia, 290
T. latifolia, 144, 610, 611, 638, 639
T. laxmannii, 288

Ulmus minor, 731
Ulva lactuca, 129
Undaria pinnatifida, 129
Urtica dioica, 109, 286, 287
U. urens, 292

Vaccinium, 195
Vaccinium myrtillus, 11, 12, 15,
    18, 126
V vitis-idaea, 11, 12, 15, 18, 126
Valeriana altaica, 288
Vetiveria zizanioides, 144,
    145, 147

Viburnum opulus, 287, 288
Vigna mungo, 243
V. ungiculata, 247, 245, 248
Viola, 144
Viola calaminaria, 734
V. montana, 288
Vitex, 293
Vitis vinifera, 149

Withania somnifera, 147
Wolinella succinogenes, 598

Xanthium spinosum, 732
X. strumarium, 286, 287, 732
Xanthomonas, 602
Xenopus, 422
Xylorhiza, 605

Zea mays, 87, 144, 196, 199, 244, 248, 253,
    422, 658
Ziziphora clinopodioides, 287
Zostera japonica, 634
Z. marina, 196, 199, 634
Zygophyllum
    album, 705, 716

SUBJECT INDEX

ABC transporter, 393, 449, 451–453, 457–460, 462–463
Acceptable daily intake (ADI), 753
Accumulation capability, 72
Accumulation indicators/monitors, 745
Accumulators, 38, 42, 125, 129, 130, 131, 192, 605, 607, 632, 634, 726, 735
Acquired immune deficiency syndrome (AIDS), 219
Acrodermatitis enteropathica, 34, 35
Active bioindication (biomonitoring), 745
Adsoption, 282, 296, 329, 395, 396, 567, 595, 596, 645
Adsoption-desorption, 57, 62
Agnostic, 151
Agrochemicals, 37
Agro-ecosystems, 55, 56, 60, 65, 67, 177
Alellopathy, 101
Allergen, 151
Allergenicity, 90
Alzheimer disease, 99, 668
Amelioration, 47, 177, 241, 396, 483, 488
Amensalism, 101
Anemia, 26, 27, 30, 31, 32, 35, 88, 162, 164, 166, 233, 237, 239, 240, 241, 251, 471, 472, 473, 477, 479, 480, 481, 482, 484, 487, 501, 512, 517, 519, 545, 628, 666, 668, 669
Anoxia, 30
Antagonists, 34, 42
Anthropogenic, 37, 56, 57, 139, 185, 344, 345, 374, 376, 377, 396, 495, 498, 501, 504, 542, 556, 571, 595, 600, 623, 624, 646, 655, 657
Anthropogenic emissions, 185, 345, 556
Anthropogenic processes, 345

Antibacterial agent, 43
Antifouling agents, 34
Antimetabolite, 387
Antioxidant, 27, 33, 165, 168, 198, 205, 217, 219, 235, 243, 352, 353, 354, 356, 358, 382, 383, 384, 451, 517, 518, 528, 543, 544, 601, 603, 604, 605, 662, 663
Apathy, 88
Apoplasmic, 433, 434, 435
Apoplast, 43, 73, 122, 388, 420, 426, 433, 436, 437, 438, 454, 661
Aquatic
  macrophytes, 192, 631, 632, 635
  plants, 6, 129, 349, 350, 623, 624, 631, 632, 633, 634, 636
Arbuscular mycorrhizal, 193, 194, 197, 200, 201, 396, 734
Aricultural Pollution Resources, 59
Aromatherapy, 138
Arsenate compound, 500
Arsenic, 23, 55, 129, 167, 195, 389, 495, 500, 524, 624, 721
Arsenite, 129, 237, 481, 500–501, 528
Arsenium, 518
Arthritis, 33, 36, 162, 175, 512, 516, 519
Ascorbate-Glutathione cycle, 382
Ash and mineral content of halophytes, 702
Atmospheric deposition, 37, 177, 377, 561, 564, 586, 588, 723, 754
Atomic absorption spectrometry, 43, 154
Atomic emission spectrometry, 43
ATP sulfurylase, 450, 606
ATP-binding cassette (ABC), 48, 393, 451, 452
Autotrophy, 16
Available forms of some micronutrients, 682
Ayurveda, 138

Trace Elements as Contaminants and Nutrients: Consequences in Ecosystems and Human Health, Edited by M. N. V. Prasad
Copyright © 2008 John Wiley & Sons, Inc.

Bhasma, 151, 152, 153
Bhasmikarana, 151, 152, 153
Bioaccessibility, 126, 130
Bioaccumulation, 1, 130, 192, 195, 313,
    349, 548, 503, 562, 628, 634, 636,
    640, 641, 644, 645, 745, 752, 755
Bioavailability, 6, 38, 39, 42, 44, 46, 47, 48,
    62, 74, 85, 88, 90, 91, 92, 145, 152,
    161, 172, 193, 233, 234, 240, 241, 243,
    256, 282, 283, 328, 300, 310, 311, 325,
    349, 350, 351, 396, 548, 549, 596, 612,
    655, 669, 713
Biocoenosis, 16
Bioconcentration, 2, 4, 6, 17, 47, 192, 548,
    609, 745, 754
Biofortification, 47, 48, 49, 161, 169, 170,
    178, 234, 241, 416
Biofortified crops, 169
Biogeochemistry 18, 56, 296, 297
Bio-geo-diversity, 161
Bioindication and biomonitoring, 743, 745,
    748
Bioindicative aspects, 555
Bioindicator, 364, 556, 744, 745, 746, 748
Biological Concentration Factor
    (BCF), 2, 4
Biological reaction in selenium cycling, 596
Biological system of elements, 1, 6
Biomagnification, 548, 646, 745
Biomarkers, 353, 356, 363, 629, 671,
    745, 746
Biosensor, 745, 746
Biosphere, 2, 185, 205, 295, 654
Biotest, 744–746, 755
Biotransformation, 356, 548

Ca^{2+} channels, 450
Ca^{2+}/Cation antiporters, 432
Ca^{2+}/H^+ antiport, 450
Ca-dependant Cd/Pb-tolerance, 433
Cadmia fornacum
Cadmium, 24–25, 28–30, 37–43, 45–46, 55,
    128, 147, 149, 194, 196, 199–201, 234,
    237–238, 254, 296, 363, 373–378,
    380–382, 384–396, 424, 433, 449,
    452–453, 546, 458, 462, 495–497,
    511–516, 519, 524, 532–537, 541, 547,
    557, 560, 564, 569, 586, 593, 662, 721,
    723, 726–727, 734, 749

Cadmium
    Chloride, 513
    toxicity, 30, 234, 378, 382, 392, 515, 534,
        535
    transporters, 393
Calcareous soils, 186, 204, 218, 220,
    227, 687
Calcification, 31
Calculation of transfer factor, 570
Cancer, 27, 29, 30, 33, 44, 99, 110, 162, 163,
    164, 165, 166, 175, 219, 283, 335, 336,
    374, 389, 392, 457, 474, 495, 496, 549,
    500, 501, 503, 504, 505, 506, 507,
    508, 511, 512, 513, 514, 517, 518, 519,
    529, 532, 540, 541, 542, 550, 604, 607,
    627, 721
Carbonic anhydrase, 183, 384, 449
Carcinogen, 25, 26, 30, 31, 33, 336, 374, 389,
    396, 499, 508, 512, 513, 517, 518, 519,
    530, 532, 537, 543, 604, 605
Carcinogenesis 110, 506, 507, 532, 542, 543
Cardiomyopathy 26, 30, 33, 485, 486,
    532, 604
Cardiovascular diseases 30, 33, 140, 161,
    219, 548
Cation diffusion facilitator (CDF) 191, 393,
    428–432, 450
Cation exchange capacity (CEC) 43, 128,
    177, 389
Cation/Proton Antiporters, 393
CAX, 415, 432, 450
CDF family, 393, 428–429, 431–432
Ceruloplasmin, 477, 664
Chaperone-docking site, 430
Chaperons 9, 427, 450, 534, 661
Chelating agents 514, 515, 628, 686,
    687, 697
Chelator therapy, 514
Chemical element composition in mother's
    milk, 582
Chemopreventive agents, 111
Chromic acid, 509–510
Chromium, 24–26, 30–31, 38, 42–43, 55,
    163, 166, 176–177, 236, 495–496,
    505, 508–511, 519, 526, 537–542,
    721, 723, 735, 749
Chromium VI, 537
Chryptochromes 101
Circadian rhythms 101, 325

Class I human carcinogen, 374
Class III metallothioneins, 384
Cobalt transporter, 427
Codex committee on food additives and
    contaminants (CCFAC), 39
Combustion emissions, 37
Commensalism, 101
Comparison of instrumental measurements
    and the use of bioindicators, 746
Competitive exclusion principle (CEP), 11,
    12, 16
Complexation, 62
Composition nutrient diagnosis (CND), 219
Concentration ratio (CR), 301
Concept of micronutrients, 682
Copper chaperone, 437, 663
Copper, 663
    Acetoarsenate, 500
    Atmosphere, 654
    Deficiency, 666
    Detoxification, 661
    Homeostasis, 664
    Human health, 663
    Hydrophere, 654
    Induced nutritional imbalance, 658
    Interactions with other nutrients and
        vitamins, 668
    Lithosphere and pedosphere, 655
    Metabolism, 664
    Plants, 656
    Toxicity, 667
Copts, 420–421
Cosmaceutical, 139
Cretinism, 23, 31, 267, 268, 273,
    274, 275
Crop breeding technology, 74
Cytochrome C, 378, 528, 531, 536, 541

Deferasirox, 487
Deferiprone (DFP), 483, 487, 488
Deferiprone (L1), 483
Deferrioxamine (DFO), 487
Dermatosis, 29, 504
Desorption, 57, 62, 154, 300, 595, 655
Detoxification of Se oxyanions, 599
Diabetes, 36, 99, 110, 165, 166, 176, 177,
    333, 485, 501, 513, 519, 665, 667
Diagnosis and recommendation integrated
    system (DRIS), 219

Diarrhea 31, 34, 142, 143, 166, 287, 390,
    510, 668
Diethylenetriaminepentaacetic acid (DTPA),
    687
Diatary fibers, 46
Dissociation, 62
Dissolution, 57
DNA repair, 381, 382, 392, 517, 518, 531,
    536, 541, 542, 547
DNA-binding proteins, 204, 543
Dravyaguna, 150
Dyserythropoiesis, 474

Echocardiography, 486
Echological loads, 696
Ecological niche 16, 18
Ecosystem, 18, 281, 283, 357, 374, 396,
    517, 524, 595, 608, 610, 624, 631,
    646, 654, 696, 721, 725, 726, 735, 743,
    745, 752, 754
Ecotype, 42, 87, 194, 199, 671
Effects of
    Minerals in rumen function, 714
    Minerals on feed intake, 715
    Minerals on microbial community in the
        rumen, 716
    Minerals on water intake and nutrient
        utilization, 716
Electrochemical ligand parameter, 5
Element concentration detected in clostrum
    and mature milk, 570
Elicitors, 106
Enanthemas, 46
Encephalopathy 31
Endemic, 23
Endocrinopathy 485
Enrichment capability 72
Enterohepatic circulation, 47
Environmental
    Chemistry, 556
    Contaminants, 35, 373, 653, 721
    Medicine, 556
    Performance index (EPI), 562
    Specimen banking (ESB), 753
Enzymatic antioxidants, 382
Erosion, 44
Erythrocytic protoporphyria
    (EPP) 473, 480
Erythroid, 471–483

Erythropoisis, 471–474, 479–480,
    484–485, 488
Erythropoietic
    Cell, 480
    Regulation, 479
Essential and nonessential elements, 11
Essential elements 3, 4, 6, 8, 9, 11, 24, 25, 47,
    102, 238, 385, 386, 416, 557, 572, 574,
    575, 578, 585
Ethnopharmacology, 137
Ethylenediaminetetraacetic acid (EDTA), 687
Euroregion, 561
Examples of biomonitoring, 748
Excessive minerals 713
Excluders 141, 386, 735

Factors affecting mineral contents of
    halophytes, 702
Farm yard manure (FYM), 167, 179
Fe (III) reductase, 435
Femtomolar, 1
Ferritin 31, 36, 48, 172, 174, 175, 233,
    234, 239, 240, 241, 242, 243–249,
    251–257, 474, 475, 476, 478, 480,
    483, 484, 486–488
Ferroportin 1 (FPN1), 477
Ferrosochelatase, 480
Floristic composition, 281
Fluorosis 23, 26, 31, 167
Food and milk samples from
    mothers, 564, 566
Food Chain 23, 26, 28, 31, 32, 37, 38, 39, 42,
    43, 46, 48, 65, 175, 254, 283, 284, 295,
    298, 334, 344, 346, 349, 358, 389, 391,
    396, 438, 498, 499, 524, 595, 611, 624,
    646, 647, 722
Fortification, 48, 85, 89, 93, 167, 176, 179,
    204, 233, 257, 267, 271, 278
Free radicals (FR) 73, 175, 204, 234, 380,
    519, 528, 536, 549, 658, 659, 668
Fulminate, 497
Fungicides 34, 37, 60, 512, 545, 653, 655

Gastrointestinal problems 27, 28, 30, 31,
    34, 604
General control nonrepressible subfamily,
    460–461
Genetic regulation, 233
Genomics, 93

Genotoxic 30, 380
Genotypes 35, 86, 89, 92, 102, 107, 126, 129,
    131, 132, 170, 171, 191, 194, 195, 198,
    200, 201, 223, 241, 507, 734
Geological correlates, 44
Geomedicine, 167
Global environmental 177, 344
Glutathione, 393, 600
Glutathione metabolic
    enzymes, 382
Glutathioselenol, 600
Glutathione synthase, 460
Goiter 23, 26, 31, 162, 164, 166, 267, 268,
    270, 271–278
Golden rice 234
Green
    Revolution, 162
    Tongue, 34
Greenockite (Cds), 374
Gygrophytes 292

H$^+$ ATPase, 434
Halophyte included diets, 710
Heavy metals tolerance factor 1, 452
Heme carrier protein 1, 476
Heme oxygenase (HO), 475, 481
Hemochromatosis, 31, 36, 475, 476, 477,
    486, 514
Hemoglobin 31, 35, 45, 347, 471, 472, 473,
    474, 475, 478, 479, 480, 482, 483, 503,
    510, 517, 518, 574, 664
Hepatocytes, 363, 476, 478, 481,
    485, 514
Hepatosplenomegaly, 474
Hepcidin (HAMP), 477, 484
Herbal
    Drugs, 139
    Extracts, 149
    Medicine, 139
    Products, 99
Herbicides, 34, 60, 456, 463, 500
Heterogeneous soil materials, 679, 681
High-throughput screening for induced
    point mutations (TILLING), 87
Histopathology, 363
Homeopathy, 138
Homeostasis, 82, 84, 86, 204, 233, 235, 239,
    242, 248, 257, 524, 526, 528, 543, 544,
    659, 663–665, 670

Human health, 161
    Effects, 343
    Implications, 653
    Nutrition, 24
    Specimen banking (HSB), 753
Hydrogen peroxide radical, 526
Hydrogen selenide, 175, 600
Hydroponics, 102, 103, 105, 300, 461
Hyperaccumulation, 14, 15, 191, 195, 201,
        204, 392, 418, 427, 431, 438, 606,
        607, 632, 633
Hyperaccumulators, 42, 56, 71, 72, 73, 75,
        108, 110, 124, 145, 191, 195, 198, 199,
        200, 201, 202, 423, 427, 431, 436, 456,
        605, 606, 607, 609, 670, 735
Hypercalcemia, 141
Hypersensitive response (HR), 106
Hypersensitivity, 34, 35, 431
Hypersusceptibility, 35
Hypertension, 30, 31, 166, 287, 484, 486,
        516, 532
Hyperthermophilic creatures, 2
Hyperglycemic activity, 110
Hypogonadism, 485
Hypothyroidism, 31, 33, 268, 485

Idiopathic copper toxicosis, 30
Immunocompetence, 33
Immunoestimulant, 108
Impact indicators/monitors, 745
Indicators, 141, 270, 275, 278, 353, 562,
        722, 726, 745
Indigenous system of medicine, 138
Industry Pollution Resources, 57
Inflammation, 33, 477, 483, 484, 486, 508,
        510, 531, 564, 667
Insecticides, 60, 500, 512
Intelligence quotient (IQ), 268
Intentional Contamination, 45, 46, 48
International council for control of Iodine
        deficiency disorders (ICCIDD), 275
International institute for tropical agriculture
        (IITA), 172
Inter simple sequence repeat, 383
Iodine deficiency, 23, 31, 33, 42, 44, 164,
        166, 267, 268, 269, 270, 271, 272, 273,
        275, 276, 278
Iodine deficiency disorders (IDD), 23, 42,
        267, 268, 271, 272, 276

Iodine Fortification, 267
Iodine-induced hyperthyroidism (IIH), 267
Ionomics, 83, 86, 87, 93
IRE- binding protein (IRP), 478
Iron deficiency, 31, 45, 86, 88, 141, 164, 166,
        184, 198, 233, 237, 238, 239, 240, 241,
        251, 257, 389, 473, 477, 486, 658
Iron
    deficiency anemia (IDA), 88, 233, 239,
        240, 473, 477
    dependent regulatory sequence, (IDRS)
        251
    regulated transporters, 393, 418, 450, 477
    regulatory proteins, 475, 476, 478, 484
    regulatory RNA binding proteins (IRPs),
        248, 251
    response elements (IREs), 478
    toxicity, 235
    responsive elements (IRE), 248, 251
irrigation with saline water, 706
isaflavonoids, 219
Isotopes, 121, 126, 296, 298, 626, 654
Itai-itai, 30, 42, 65, 162, 390

Kashin–Beck disease, 33, 44, 604
Keshan disease, 23, 33, 44, 166, 604
Kupffer cells, 479

Lanthanides, 126
Leaching, 44
Lead, 515, 544
    Acetate, 517
    Phosphate, 517
Leguminous plants, 48, 219
Lethargy, 34, 668
Lindsay-Norvell method, 687, 690
Lipid hydroperroides (ROOH), 378
Lipid peroxidation (LPO), 352–354, 356,
        358, 363, 378, 379, 386, 528, 533, 548,
        549, 658, 659, 667
Lipid peroxidation and plasma membrane
        integrity, 659
Lipoxygenase, 379
Lithophilic element, 297
Low affinity cation transporter (LCT), 393,
        433, 450

Macroalgae, 129, 130
Malnutrition, 84, 85, 162, 168, 171, 234

Malondialdehyde, 353, 379, 380, 538
MATE family, 393
Maximum contamination level
    (MCL), 175
Maximum Permissible level (MPL), 175
Mean cell volume (MCV), 483
Mean corpuscular hemoglobin
    (MCH), 472, 483
Mean corpuscular volume (MCV), 472
Mechanism of copper tolerance, 660
Medical geochemistry, 23, 44
Medicinal and aromatic plants (MAP), 137,
    139, 141, 147
Medicinal plants, 101, 102, 106, 107, 108,
    110, 137, 138, 139, 140, 161
Megaloblastic, 473
Membrance transport protein, 416
Menkes' disease, 30, 35, 164, 665
Mercuric chloride, 497, 499
Mercuric sulfide, 497
Mercury, 497, 545
Mesophytes 286, 292
Metabolic function of copper, 656
Metal
    Accumulation, 47, 81, 85, 199, 254, 393,
        624, 633, 643, 660
    Bioindication, 584
    Sequestration, 17, 202, 254, 352, 394,
        430, 431
    Tolerance proteins (MTP1), 429
Metalloenzymes, 25, 154, 183, 191
Metallomics, 152–154
Metallophytes, 195, 196, 663, 670, 722
Metalloprotoporphyrins, 472
Metallothioneins, 196, 204, 252, 255, 352,
    354, 358, 363, 384, 393, 450, 514, 530,
    534, 535, 540, 543, 585, 662, 664, 667,
    669
Metals efflux proteins, 416, 423
Metals uptake protein families, 416
Methyl mercury, 499
Microbe-Induced Bioremediation, 608
Microbial assimilatory
    reduction, 597
Microbial dissimilatory
    reduction, 597
Microelements, 24, 85, 86, 217, 682
Micronutrient application, 692
Micronutrient initiative (MI), 275

Micronutrients, 24, 47, 55, 81, 82, 84, 86, 93,
    102, 104, 108, 162, 168, 219, 233, 239,
    416, 422, 426, 433, 435, 436, 513, 682,
    683, 684, 685, 686, 687, 688, 690, 692,
    693, 695, 696, 697, 721
Minamata disease, 357
Mineral
    Allowances, 708
    Deficiency, 23, 161, 702, 710, 711
Minerals in halophytes, 701
Mitogen-activated protein kinase (MAPK),
    449, 528
Mode of toxic action, 529
Molecular markers, 172
Monogastric animals 48, 204
Morphological symptoms of toxicity, 657
Mosses for atmospheric pollution
    measurements, 748
Mother milk, 555
Multidrugs resistance associated protein
    (MRP) subfamily, 48, 454–457
Mutation, 18, 36, 86, 87, 92, 204, 243, 283,
    381, 382, 424, 430, 471, 472, 474, 507,
    508, 536, 540, 541, 543
Mutualism, 101
Myocardial fibrosis, 23

Na^+/Ca^{2+} exchangers, 432
National Agricultural Research and
    Extension Systems 167, 172
Natural
    Attenuation, 396
    Phytocenosis, 286
Natural Pollution Resources, 57
Nauseaa, 26, 27, 34, 46, 151, 152, 153, 512
Necrosis, 23, 205, 358, 363, 514, 534, 628,
    656, 657, 663
Nephrotoxicity, 495, 629
Neurochemical transmission, 140, 161
Neutraceuticals, 139
Nickel 24, 25, 27, 32, 38, 39, 42, 43, 47, 55,
    81, 201, 374, 437, 505, 509, 510, 512,
    519, 532, 541–545, 547, 586, 588,
    594, 595, 660, 726, 727, 749
Nicotinamine synthase, 436
Ni-hyperaccumulator, 431
Nickel, 542
No observed (adverse) effect level [NO
    (A)EL], 753

Nonsymbiotic, 17
Non-transferrin – bound iron
    (NTBI), 479, 480
Nonenzymatic antioxidants, 383
Nramp family, 393, 416, 663
Nutriomics, 83
Nutritional Security, 161

Omnipresent, 286, 544
Ordovician, 18
Osteoporosis, 30, 163, 166, 389, 390, 512,
    629, 666
Overexpression, 48, 89, 92, 93, 174, 191,
    234, 253, 254, 393, 420, 427, 431, 451,
    477, 483, 534, 535
Oxidation, 2, 8, 17, 18, 33, 57, 62, 65, 151,
    234, 247, 249, 297, 325, 363, 379, 380,
    381, 383, 497, 501, 504, 505, 509, 517,
    523, 526, 529, 535, 539, 544, 545, 569,
    580, 594, 596, 597, 601, 602, 624, 654,
    656, 665, 668
Oxidation of reduced Se forms, 602
Oxidative damages, 33, 204, 234, 235, 242,
    252, 352, 353, 378, 379, 380, 383, 528,
    535, 544, 603, 665, 667
Oxidative stress, 31, 193, 194, 235, 239, 249,
    254, 352, 353 363, 378, 379, 383, 384,
    386, 455, 482, 483, 508, 519, 526, 528,
    529, 530, 531, 533, 534, 535, 540, 542,
    548, 580, 634, 644, 658, 659, 662, 666,
    667, 668

P-type ATPase, 35, 393 423–424, 663
P_{1B} ATPase, 423–427
Parenchymal liver cell, 479
Paris Green, 500
Passive bioindication Biomonitoring), 745
Pathyam, 151
PCBs or DDT, 577
Peroxidase, 386
Peroxidation, 352–354, 356, 358, 363,
    378–379, 386 397, 528, 533,
    548–549, 658–659, 667
Pharmaceuticals, 138, 139, 545, 746
Pharmacological activity, 108, 110
Pharmacopeia, 111, 137, 152, 153
Phloem, 434
Phosphate fertilizer, 374, 377–378, 389
Phospholipases, 379

Photooxidation, 9
Photoperiod, 327
Photoreceptors, 101
Photosynthesis, 2, 4, 5, 8, 14, 18, 66, 75, 191,
    205, 220, 234, 249, 255, 435, 460, 601,
    657, 659, 660
Photosynthetic organs, 6, 11
Phototropins, 101
Physiology of lactation, 559
Phytase, 46, 48, 92, 204, 252, 253
Phytoalexin, 106
Phytochelatins (PCs), 73, 122, 383, 384, 386,
    388, 393, 436, 450, 451, 530, 538, 662
Phytochemicals, 99, 102, 104, 106, 107, 151
Phytochromes, 101
Phytoclean, 68
Phytoconstituents, 138
Phytodegradation, 68
Phytoestrogen, 217, 219, 220, 221, 222, 225,
    226, 227
Phytoextraction, 56, 68, 72, 73, 74, 75, 131,
    147, 196, 311, 393, 394, 609, 611
Phytoferritin, 233, 239, 251
Phytomedicine, 137
Phytopharmaceutical formulations, 108
Phytopharmaceuticals, 111, 139
Phytoremediation, 47, 49, 56, 60, 64, 68, 73,
    197, 198, 284, 384, 394, 438, 457, 609,
    611, 612, 624, 631, 632, 633, 634, 635,
    636, 641, 645, 646, 647, 670, 722, 735
Phytoremediation Laboratory Prototype, 645
Phytoremediation studies, 632
Phytosiderophores, 45, 122, 132, 198, 238,
    392, 422, 435
Phytostabilization, 609, 611, 612
Phytotherapy, 138
Phytotoxicity, 197, 201, 314, 316, 438, 653
Phytovolatilization, 68
Plant secondary metabolism, 99, 100, 104
Plant's defense, 99
Plant-Microbe Interaction, 611
Plants as Biomonitors of Trace Elements,
    721, 748
Plasma-membrane transporters, 238
Plasmodesmata, 433
Pleiotropic drug resistance
    subfamily, 458–459
Polar sunrise mercury depletion
    incidence, 346

Polygenic, 45
Polymorphisms, 35, 86, 416
Polyvinylchloride, 376
Precipitation, 57, 62, 235, 474, 482, 547, 594, 612, 640, 658, 661, 735
Protein carriers, 15
Protein Oxidation, 379
Proteoliposomes, 431
Proteomics, 93
Protochlorophillide reductase, 379
Proton slippage, 417
Protoporphyrin IX, 473, 481, 517
Provisional weekly tolerable (PWTI), 108
Psychomotor development, 88
Pulmonary hypertension, 484, 486

Quantitative Trait Loci (QTL), 47, 87

Radioisotopes, 131, 295
Radionuclides, 121, 124, 281–284, 295, 297, 298, 300–303, 308, 311, 314–315, 318, 321, 324–326, 328, 334, 336, 587, 624, 627, 630–631, 633, 640
Radiophytoremediation, 284
Random amplified polymorphic DNA, 381
Rare earth elements (REE), 3, 10, 296, 297
Rare occurance, 287
RAS proteins, 534
Rasa, 150
Reactive nitrogen species (RNS), 525, 604
Reactive oxygen species (ROS), 106, 175, 242, 353, 378, 392, 397, 450, 507, 528, 530–531, 542, 548, 600, 603, 658, 667, 668
Recommendad daily allowance (RDA), 35, 270
Reducing equivalents, 506, 597, 601
Reduction of copper uptake, 660
Reference dose (RfD), 348
Renal calculi, 33
Resorption, 14, 558, 575, 586–587
Respiration, 66, 188, 234–235, 529, 508, 656–657, 660, 664
Retention, 14
Rhizosphere degradation, 68
Rhizosphere Microorganisms, 321, 324
Rhizospheres, 6, 14–15, 43, 65–66, 68, 72, 121, 192, 238, 253, 208, 309–310, 317, 321, 324–325, 327, 387, 606, 609–612, 649

Rinkis calculation, 683
Rinkis method, 683–686
Risks of human exposure to copper, 669
Ruminant nutrition, 708
Ruminants, 101, 701, 715, 716–717

Salinity level, 706, 708
Secondary toxic messengers, 380
Sedimentation, 56, 60, 147
Seed coat technology, 75
Selenium, 518, 593
    environmental concern, 607
    in humans and animals, 603
    in plants, 605
    in the environment, 594
    methylation and demethylation, 602
    phytoremediation processes, 611
    plant-assisted bioremediation, 609
    volatilization, 602
Selenocysteine methyltransferase (SMT), 609
Selenodiglutathione, 600, 607
Selenoethers, 600
Selenols, 600
Sensitivity, 746
Sequestrants, 15
Serum ferritin, 240, 484, 486
Sewage sludge, 13, 37–38, 60, 147, 192, 197–198, 377, 389–390, 395
Sickle cell disease (SCD) 484–485
Siddha, 138–140
Siderophore, 48, 122, 132, 197–198, 235, 476
Silurian, 18
Site-directed mutagenesis, 193, 430
Slat affected soils, 706
Soil Amendments, 300, 394
Soil organic matter (SOM), 4, 177, 300, 655, 680, 683, 687
Soils and trace elements, 722
Specific extracting agents, 682, 686, 688–689
Splalerite (Zinc sulfide, ZnS), 374
Sporadic occurance, 287
Super oxide anion radical, 526
Superoxide, 378
Symplasmic, 235–437
Symplast, 122, 239, 388, 433, 436–437
Synaptic transmission, 140, 161
Synergistic, 151, 387, 543, 578, 612
Systemic acquired resistance, 106

Terrestrial ecosystem, 18, 654
Thalassemia, 240, 471–474, 479, 480, 482–486, 488–489
Thalassemia intermedia, 471, 474, 479, 482, 483–484
Thalassemia major, 471, 479, 484, 488
The Mercury sunrise, 346
Thermotolerant, 48
Thyroglobulin, 270
Thyroid, 25, 27, 31, 110, 164, 166, 267, 268–270, 275, 485, 506, 603, 631
Thyroid stimulating hormone (TSH), 270–271, 275
Tolerance capability, 72
Tonoplast, 418, 420, 431, 433, 456
Total goiter prevalence (TGP), 270
Total goiter rate (TGR), 270, 273–275, 277
Toxic heavy metals, 56, 496, 497, 585
Toxicokinetics, 347
Traditional healing Plants, 137
Transfer factor, 127, 283–284, 557–558, 570, 574, 577, 583–584
Transfer of chemical element, 558, 560
Transfer of trace metals, 555
Transferrin, 30, 36, 178, 475–478, 487, 510, 580, 583, 585, 667
Transferrin receptor, 36, 475–478, 480
Transgenic plants, 48, 90, 191, 201, 234, 456, 459, 609
Transition metals, 215, 233, 242, 423, 428, 430–431
Translation, 55, 60, 66, 83, 248, 251, 384, 460, 461, 478, 481, 483, 597
Translocation capability, 72
Transportation, 55, 65–66, 72, 145, 501, 577, 578, 663
Trophic chains, 4, 18
Type I ATPase 423
Type II ATPase 423
Type III ATPase 423
Type V ATPase 423
Type VI ATPase 423

Unani, 110, 112, 138–139
Unit occurance group, 287
Universal salt iodization (USI), 271, 278

Uranium Accumulation in Aquatic Plants, 632
Uranium Mines, 623, 630, 632–635, 647,
Uranium Mining History, 630
Uranium Toxicity, 627, 628, 644, 647
Urinary iodine concentration or excretion (UIC/E), 270

Vascular diseases, 29, 30, 33, 140, 161, 219, 548
Veerya, 150
Vipaka, 150
Volatilization, 57, 396, 498, 582, 596, 602, 603, 605, 608, 609, 612

Water drainage, 623, 636
Water phytoremediation, 646
Wildlife Protection Act, 150
Wilson's disease, 30, 665, 668, 669
World Health Organization (WHO), 111, 137, 179, 348, 391, 501, 664, 668

Xenobiotic compounds, 384
Xenobiotics, 456, 523, 696, 745
Xerophytes, 286, 292, 293
Xerophytic grasses, 284
X- linked inherited disease, 30
X-ray fluorescence spectrometry, 43
Xylem, 43, 72, 122–123, 200, 202, 238, 384, 386, 388, 426–427, 434–437, 533, 606, 657, 662–663

YSL family, 414, 421, 422, 423

Zinc, 517
    Flowers, 373
    Fortification, 204
    Hyperaccumulation, 431
    Intervention, 188
    Metalloproteins,
    Regulated transporter (ZIP), 418, 4
    Transporting genes, 189
    Transporters, 201, 202
    Transporter, 429
ZIP family, 36, 393, 418–420, 450